OBJECTIVE BIOCHEMISTRY

Question Bank for Competitive Examinations

(Strictly According to the Latest Syllabus)

OBJECTIVE BIOCHEMISTRY

Question Bank : For Competitive Examinations

Strictly According to the Latest Syllabus

Useful for:

ICAR, ASRB, NET, JRF, SRF, ARS, NAARM, IARI, NDRI,
CSIR-JRF-NET, GATE, ICMR, DBI, UPCATET, JNU,
Entrance Examination for Admission to
M.Sc., Ph.D., Pharmacy Nursing, Physiotherapy, and Paramedical Programmes
UPSC Administrative Service (Pre and Mains)
State Public Service Commission (Pre and Mains)
Undergraduate and Postgraduate Courses of Various Indian Universities

R.S. Sengar
Reshu Chaudhary

NEW INDIA PUBLISHING AGENCY
New Delhi – 110 034

NEW INDIA PUBLISHING AGENCY

101, Vikas Surya Plaza, CU Block, LSC Market
Pitam Pura, New Delhi 110 034, India
Phone: + 91 (11)27 34 17 17 Fax: + 91(11) 27 34 16 16
Email: info@nipabooks.com
Web: www.nipabooks.com

Feedback at feedbacks@nipabooks.com

ISBN: 978-93-81450-37-6

Composed and Designed by NIPA

डा. आर.एस.कुरील
कुलपति

Dr. R.S. Kureel
Vice-Chancellor

नरेन्द्र देव कृषि एवं प्रौद्योगिक विश्वविद्यालय,
कुमारगंज, फैजाबाद-224 229
(उ.प्र.),भारत
Narendra Deva University of Agriculture &Technology
Kumarganj, Faizabad-224 229 (U.P.), India

Foreword

Competitive Examinations are the order of the day. All Colleges conducting professional courses at PG level are admitting students based on common entrance examination, which is of objective type. In Biochemistry, Biotechnology, Life Science, Pharmacy, M. Pharm. admissions are based on qualifying the GATE entrance examination conducted by Govt, of India. In this book, the authors have done good work in preparing several objective questions which help the students to face the subject in the examination with poise and confidence.

The book is well balanced and consists of multiple choice questions from all the important topics like carbohydrate, protein and lipid metabolism and other important Biochemical aspects. The typesetting and quality of printing is good. The author is also well experienced in taking up this type of work.

The main purpose of this book is to bring together objective questions of the subject matter to provide up to date information for better understanding of the readers. It would serve as a useful reference material for researchers, students and consultants changed in science.

I recommend this book to all the students preparing for competitive examination and admitting in Biochemistry, Biotechnology, Life Science and also for Medical and Pharmacy College/Universities students.

R.S. Kureel

Preface

The preparation of any competitive examination involves three phases, first the thorough study of all of the relevant topics, second revision phase and third the most important is the practice phase. I am honoured to present look, "Objective Biochemistry" in the form of Question Bank before our keen readers. Here my main aim is to provide, authentic and qualitative questions to the competitors to equip and judge themselves prior to the different National Level examinations in the field of Biochemistry. I have brought out this book basically for students who plan to appear for Biochemistry descipline in the entrance examinations, (held by ICAR, ASRB, -NET, SRF, ARS, GATE, ICMR, CSIR-SRF-NET, NAARM, DBT, JNU, UPCATET, IARI, NDRI, UPSC, PCS etc.) as well as other specific tests in the field of Medical, Pharmacy, Physiotherapy, Nursing, UPCATET and other Paramedical (PG level). There is a dearth of good entrance manual of Biochemistry for the above said examinations. Hence, I have prepared an exhaustive Question bank of around 30,000 MCQs with answers covering a wide spectrum of basic Biochemical topics of the subject.

Some of the important topics which are given a good coverage include Atoms and Molecule, Water and pH buffer, Biological Energy Generation and Thermodynamics, Carbohydrate and its metabolism, Amino Acid, Protein and its metabolism, Lipid and its metabolism, Nucleic acids, Enzymes and Enzyme Kinetics, Hormone and growth substances, Pigments, Tools and Techniques, Vitamins and Agriculture Biochemistry. The objective questions are prepared based on the background taken from previous question papers of Professional medical and Paramedical competitive entrance examinations. I opted systematic presentation of questions which leads more to conceptual understanding rather than physical memory. I am optimistic that it would help students and readers in self improvement.

Though I have tried my best to minimize the typographical errors in the manuscript but still some might have cropt in. I thankfully invite the readers to bring these in my notice, so that they may be rectified in next edition.

I convey my gratitude to the entire management of New India Publishing Agency, New Delhi for taking all pains in the publication of the book.

Contents

Chapter-1

Atoms, Molecules and Chemical Arithmetic

Fill in the blanks

1. Matter exist in................states.

Ans. three

2. Physical properties of water and ice are...........

Ans. different

3. Chemical properties of water and ice are...........

Ans. similar

4. Element is the.............form of matter.

Ans. simplest

5. Hydrogen is the...........element.

Ans. lightest

6. In a chemical compound, the constituent elements are present in a................

Ans. fixed ratio

7. The constituents of a mixture........... their identity.

Ans. retain

8. Camphor on heating undergoes.........

Ans. sublimation

9. Distillation is the process of.......... followed by......................

Ans. vaporization, condensation

10. Homogenous mixtures are called............

Ans. solutions

11. In chromatographic technique, the............is termed stationary phase and the.............is termed mobile phase.

Ans. adsorbent, liquid

12. The stationary phase in paper chromatography is................

Ans. water adsorbed on paper

13. The most common adsorbent in thin layer chromatography is..............

Ans. alumina

14. The molar volume of all gases under NTP conditions is..................

Ans. 22.4 L

15. The Avogadro's number represents...............

Ans. 6.02×10^{23} chemical units

16. For a monoatomic element, a mole is equivalent to..............

Ans. 6.02×10^{23} atoms

17. 18 g sample of water contains............. mole(s) of water, or............ moles of hydrogen and........moles of oxygen atoms.

Ans. one, two, one

18. 1 mole of any gas at NTP occupies ....L of volume.

Ans. 22.4

19. The mass of 1×10^{22} molecules of $CuSO_4.5H_2O$ is..............

Ans. 4.14 g

20. 3 g of a salt molecular mass 30 is dissolved in 250 ml of water. The molarity of the solution is..............

Ans. 0.4 mol/1

21. The mass equivalent to $1/12^{th}$ that of ${}_6C^{12}$ atom is called..............

Ans. atomic mass units

22. The molar mass of any substance contain.............chemical units of that substance.

Ans. 6.02×10^{23}

23. The substance which gets used up first in any reaction is called ....

Ans. limiting reagent

24. The ratio, no. of moles of a solute/volume of the solution in litres, is called..........

Ans. molarity

25. In a mixture containing A and B, the mole fraction of A is given by............

Ans. $n_A/ (n_A + n_B)$

26. The charge and mass of the electron are............................ coulombs and .... kg respectively.

Ans. 60×10^{19}, $.911 \times 10^{-31}$

27. The minimum energy required to eject electron from the surface of a metal is called ....

Ans. threshold energy, work function

28. The energy possed by one mole of photons is called ....

Ans. one Einstein

29. The expression for radial probability is ....

Ans. $4\pi r^2\psi^2$

30. Matter and raditions behave as particles and as ....

Ans. waves

31. The lowest energy state of an electron in an atom is known as ....

Ans. ground state

32. The region in space occupied by the electron is termed as an ....

Ans. orbital

33. The dual character of the electron was suggested by ....

Ans. De Broglie

34. The particles of light are called ....

Ans. photons

35. Bond length is .... proportional to bond order.

Ans. inversely

Multiple Choice Questions

1. The perfix mega is
   (a) 10^3 (b) 10^{-6}
   (c) 10^6 (d) 10^9 **Ans. c**
2. The SI unit of length is
   (a) Meter (b) Centimeter
   (c) Nanometer (d) Kilometer **Ans. a**
3. Impure table salt can be purified by
   (a) Crystallization (b) Filtration
   (c) Sublimation (d) Distillation **Ans. a**
4. A solid which is soluble in a solvent can be obtained from its solution by
   (a) Sublimation (b) Crystallization
   (c) Fractional crystallization (d) Filtration **Ans. b**
5. Chromatography may be used for the purification/separation of the constituents of a mixture consisting of
   (a) Solids (b) Gases
   (c) Liquids (d) All the above **Ans. c**
6. Essential oils can be extracted by
   (a) Simple distillation (b) Sublimation
   (c) Steam distillation (d) Fractional distillation **Ans. c**
7. A mixture of four liquids having their boiling points differing by only a few degrees, can be separated by
   (a) Fractional distillation
   (b) Separating funnel
   (c) Steam distillation
   (d) Distillation under reduced pressure **Ans. a**
8. The symbolic representation of a molecule of an element or compound is called
   (a) Structural formula (b) Empirical formula
   (c) Molecular formula (d) Geometric formula **Ans. c**
9. One mole of helium gas represents
   (a) 3.011×10^{23} He_2 (b) 6.023×10^{23} He
   (c) 6.023×10^{23} He_2 (d) 12.069×10^{23} He **Ans. b**

10. The statement that the elements always combine in certain fixed ratios belong to the law of

(a) Reciprocal proportions (b) Conservation of mass

(c) Definite proportions (d) Multiple proportions **Ans. c**

11. The modern basis of expressing atomic and molecular masses is based on

(a) Carbon-12 (b) Oxygen-16

(c) Hydrogen-1 (d) Chlorine-35.5 **Ans. a**

12. One mole of glucose contains.........moles of carbon

(a) Twelve (b) One

(c) Six (d) Two **Ans. c**

13. Four one litre flasks are separately filled with the gases H_2, He, O_2 and O_3 at the same temperature and pressure. The ratio of total number of atoms of these gases present in different flasks would be

(a) 1:2:2:1 (b) 1:1:1:1

(c) 2:1:2:3 (d) 3:2:2:1 **Ans. c**

14. A molal solution is the one that contains one mole of a solute in

(a) One litre of the solution (b) 1000 g of the solvent

(c) One litre of the solvent (d) 22.4 litres of the solution

Ans. b

15. In which mode of expression, the concentration of a solution remains independent of temperature?

(a) Molarity (b) Molality

(c) Formality (d) Normality **Ans. b**

16. The equivalent mass of $MnSO_4$ is half its molecular mass when it is converted to

(a) MnO_2 (b) Mn_2O_3

(c) MnO_4^- (d) MnO_4^{-2} **Ans. a**

17. M is the molecular weight of $KMnO_4$. The equivalent weight of $KMnO_4$ when it is converted into K_2MnO_4 is

(a) M/3 (b) M

(c) M/5 (d) M/7 **Ans. b**

18. If 0.50 mole of $BaCl_2$ is mixed with 0.20 moles of Na_3PO_4 the maximum number of moles of $Ba_3(PO_4)_2$ that can be formed is

(a) 0.10 (b) 0.70

(c) 0.20 (d) 0.50 **Ans. a**

19. 100 ml of 0.1 N solution of $AgNO_3$ are added to 100 ml of 1.1 N solution of KCl. The concentration of nitrate ion in the resultant mixture is

(a) 0.1 N (b) 0.05 N

(c) 0.2 N (d) Reduced to zero **Ans. b**

20. If 250 ml of 0.25 M NaCl solution is diluted with water to a volume of 500 ml, the new concentration of the solution is

(a) 0.08333 M (b) 0.167 M

(c) 0.125 M (d) 0.0167 M **Ans. c**

21. The number of molecules in 4.25 g of ammonia is approximately

(a) 2.0×10^{23} (b) 1×10^{23}

(c) 1.5×10^{23} (d) 2.5×10^{23} **Ans. c**

22. The number of molecules in 32 g of oxygen is

(a) 3.2×10^{23} (b) 3.2×10^{16}

(c) 6.02×10^{23} (d) 6.02×10^{10} **Ans. c**

23. The mole of CO_2 contains

(a) 18.1×10^{23} molecules of CO_2 (b) 6.02×10^{23} atoms of C

(c) 6.02×10^{23} atoms of O (d) 3g atoms of CO_2 **Ans. b**

24. Number of atoms in 1.4 gram negative is

(a) 6.02×10^{22} (b) 1.20×10^{22}

(c) 3.01×10^{23} (d) 6.02×10^{23} **Ans. a**

25. Neutrons are present in all atoms except

(a) He (b) C

(c) H (d) Ne **Ans. c**

❑❑❑

Chapter-2

Atomic Structure

Fill in the blanks

1. Cathode rays when subjected to a magnetic field, get attracted towards ....pole.

Ans. north

2. Millikan's method is used for determining the...........of an electron.

Ans. charge

3. The ratio e/m for cathode rays is found to be equal to that for............

Ans. electron

4. Alpha particles are...............

Ans. He^{2+}

5. The atomic model which assumed an atom to be a sphere of uniform positive charge into which negatively charged electrons were embedded is known as............atomic model.

Ans. Thomson

6. A radioactive isotope of hydrogen is............. . It contains........... protons and .... neutrons in its nucleus.

Ans. tritium, 1, 2

7. Neutron was discovered by............

Ans. J. Chedwick

8. The atomic number of the element having maximum number of unpaired 3p electrons is............

Ans. 15

9. The atoms of different element which have the same mass number are called..........

Ans. isobars

10. $_{17}Cl^{35}$ and $_{17}Cl^{37}$ are..............

Ans. isotopes

11. Emission spectra are obtained when electrons jump from.........energy levels to............level.

Ans. higher, lower

12. In the atomic absorption and atomic emission spectra, the corresponding special lines appear at.............. wavelength.

Ans. same

13. The spectral lines observed in the region 1800-2000 nm fall in the ....series.

Ans. Brackett

14. The............is a measure of probability of finding an electron at a given point of space in an atom.

Ans. ψ^2

15. The transition of electron in hydrogen atom from fourth to the first energy state emits a spectral line which falls in...........series.

Ans. Lyman

16. The uncertainty principle and the concept of wave nature of matter was proposed by............and...........respectively.

Ans. Heisenberg, de Broglie

17. The de Broglie wavelength of matter particle.......with an............in their mass.

Ans. decreases, increase

18. The Heisenberg uncertainty principle has given rise to the........concept for the existence of electrons in any atom.

Ans. orbital

19. Energy of an electron depends upon the...........quantum number of the shell in which the electron resides.

Ans. principal

20. Energy of an electron goes on..........as the value of n increases.

Ans. increasing

21. The................quantum number determines the shape of an orbital.

Ans. azimuthal

22. The............quantum number determines the orientation of an orbital in space.

Ans. magnetic

23. The 3p electrons will have n = ... and l = ...

Ans. 3, 1

24. The p-orbitals are not............ symmetrical.

Ans. spherically

25. The s-orbitals, the orbital angular momentum is...........

Ans. zero

26. The 1s and 2s orbitals are........in shape, and the three p orbitals are ....shaped.

Ans. spherical, dumb-bell

27. h/π is the angular momentum in the.........shell of helium.

Ans. 2nd

28. For the orbitals for which n+1 values are the same, the orbital having .... value of...........is filled first.

Ans. lower, n

29. The lowest energy state of an atom is called the............

Ans. ground state

30. Each orbital can accommodate at the maximum of two electrons with opposite spins. This statement is termed.........

Ans. Pauli's exclusion

31. According to..........no two electrons in an atom will have all the four quantum numbers.

Ans. Pauli's exclusion principle

32. There are.............outer electrons in a carbon atom, and the ground state electronic configuration of carbon is...............

Ans. four, 2, 4 ($1s^2$, $2s^2$, $2p^2$)

33. The electronic configuration of fluoride is......it has.........unpaired electrons.

Ans. $1s^2$, $2s^2$, $2p^5$, one

34. The half-filled and completely-filled orbitals are...............stable.

Ans. more

Multiple Choice Questions

1. The essential constituent of matter is
   (a) Neutron (b) Electron
   (c) Proton (d) None **Ans. b**

2. Cathode rays are produced when high voltage is applied across a gas under
   (a) Very low pressure (10^{-3} Torr)
   (b) High pressure
   (c) Low pressure (≈ 0.1 Torr)
   (d) Ordinary pressure **Ans. a**

3. The e/m ratio of an electron is.......lines that of an alpha particle
   (a) 920 (b) 1840
   (c) 3680 (d) 2 **Ans. c**

4. The average distance of an electron in an atom from the nucleus is of the order of
   (a) 10^{-13} cm (b) 1 cm
   (c) 10^{-8} cm (d) 1 mm **Ans. c**

5. Which of the following is not correct
   (a) $E/v = h$ (b) $\lambda = v/c$
   (c) $E = mc^2$ (d) $\lambda = hp$ **Ans. b**

6. Which of the following has largest de Broglie wavelength provided all have equal velocity
   (a) NH_3 molecule (b) CO_2 molecule
   (c) Electron (d) Proton **Ans. c**

7. According to the uncertainty principle
   (a) $\lambda = h/p$ (b) $E = mc^2$
   (c) $\Delta x.\Delta p = h/4\pi$ (d) $\Delta x.\Delta p = h/6\pi$ **Ans. c**

8. The uncertainty principle was first stated by
   (a) Planck (b) Pauli
   (c) Heisenberg (d) de Broglie **Ans. c**

9. It is impossible to determine the position as well as momentum of an electron in an atom. This principle is known as
   (a) Pauli's exclusion principle

(b) de Broglie's law
(c) Heisenberg's uncertainty principle
(d) Aufbau principle **Ans. c**

10. Bohr atomic model can explain
(a) The spectrum of hydrogen molecule
(b) The spectrum of hydrogen atom only
(c) Spectrum of an atom or ion containing one electron only
(d) The solar spectrum **Ans. c**

11. Which electronic level would allow a hydrogen atom to absorb a photon but not to emit a photon?
(a) 1s (b) 3s
(c) 2p (d) 3d **Ans. a**

12. The principal quantum number of an atom is related to the
(a) Spin angular momentum
(b) Orbital angular momentum
(c) Size of the shell
(d) Orientation of the orbital in space **Ans. c**

13. When the azimuthal quantum number l =1, the shape of the orbital will be
(a) Double dumb-shell (b) Spherical
(c) Dumb-shell (d) Circular **Ans. c**

14. Magnetic quantum number specifies
(a) Orbital orientation (b) Orbital size
(c) Orbital shape (d) Nuclear stability **Ans. a**

15. The correct set of quantum numbers for the unpaired electron of chlorine atom is (in the order as n, l, m)
(a) 2, 1, 1 (b) 2, 1, 0
(c) 3, 0, 0 (d) 3, 1, 1 **Ans. d**

16. Which of the following sets of quantum numbers is not allowable (in the order n, l, m and s)
(a) 2, 1, +1, -1/2 (b) 2, 1, 0, +1/2
(c) 2, 2, -1, +1/2 (d) 3, 2, 0, -1/2 **Ans. c**

17. The total number of electrons that can be accommodated in all the orbitals having principal quantum number 2 and azimuthal quantum number 1 is

(a) 6 (b) 2

(c) 4 (d) 8 **Ans. a**

18. The number of electrons which can be accommodated in an orbital is

(a) Three (b) One

(c) Two (d) Four **Ans. c**

19. The orbital diagram in which Aufbau principle is violated is

(a) 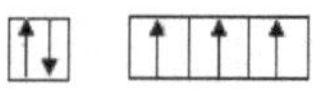

(b) 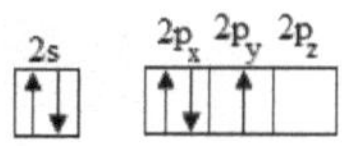

(c) 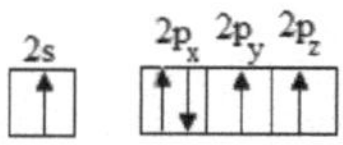

(d) 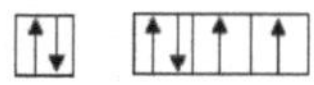 **Ans. c**

20. Nitrogen has the electronic configuration $1s^2\ 2s^2\ 2p^1_x\ 2p^1_y\ 2p^1_z$ and not $1s^2\ 2s^2\ 2p^2_x\ 2p^1_y\ 2p^0_z$. This is determined by

(a) Hund's rule (b) Aufbau Principle

(c) Pauli's exclusion principle (d) Uncertainty principle **Ans. a**

21. In a given atom no two electrons can have the same values for all the four quantum numbers. This is called

(a) Hund's rule (b) Pauli's exclusion principle

(c) Aufbau Principle (d) Uncertainty principle **Ans. b**

22. How many unpaired electrons are there in Ni^{2+}

(a) 4 (b) 0

(c) 2 (d) 6 **Ans. c**

23. Which of the elements having the following outermost electronic configuration has atomic number of 22? The inner orbitals may be assumed to be complete

(a) $2p^5$ (b) $3p^6$

(c) $5s^1$ (d) $3d^2$ **Ans. d**

24. The correct ground state electronic configuration of chromium atom is

(a) $[Ar]\ 3d^6\ 4s^0$ (b) $[Ar]\ 3d^5\ 4s^1$

(c) $[Ar]\ 3d^4\ 4s^2$ (d) $[Ar]\ 4d^5\ 4s^1$ **Ans. b**

25. The electronic configuration of silver atom in ground state is

(a) [Kr] $4d^{10}$ $5s^1$ (b) [Ar] $3d^{10}$ $4s^1$

(c) [Xe] $4f^{14}$ $5d^0$ $6s^1$ (d) [Kr] $4d^9$ $5s^2$ **Ans. a**

26. Correct set of four quantum numbers for the valence (outermost) electron of rubidium (Z = 37) is

(a) 5, 1, 1, +1/2 (b) 5, 0, 0, +1/2

(c) 6, 0, 0, +1/2 (d) 5, 1, 0, +1/2 **Ans. b**

27. The outermost electronic configuration of the most electronegative element is

(a) ns^2 np^5 (b) ns^2 np^2

(c) ns^2 np^6 (d) ns^2 np^4 **Ans. a**

28. Pick out the isoelectronic structures from the following : I. CH_3^+, II. H_3O^+, III. NH_3, IV. CH_3^-

(a) II, III and IV (b) I and II

(c) III and IV (d) I and III **Ans. a**

29. Na^+ ion is isoelectronic with

(a) Ca^{2+} (b) Li^+

(c) Ba^{2+} (d) Mg^{2+} **Ans. d**

30. No two electrons in an atom of an element have

(a) Identical set of all four quantum numbers

(b) Same principal quantum number

(c) Same magnetic quantum number

(d) Same azimuthal quantum number **Ans. a**

31. Isotopes are atoms showing same

(a) Mass number (b) Atomic Mass

(c) Number of neutrons (d) Atomic number **Ans. d**

32. The ratio of area covered by second orbit to the first orbit is

(a) 1:1 (b) 1:16

(c) 8:1 (d) 16:1 **Ans. d**

33. Find the magnetic moment of a divalent ion in aqueous solution, if its atomic number is 25

(a) 3.0 BM (b) 4.9 BM

(c) 5.9 BM (d) 6.9 BM **Ans. c**

34. When the azimuthal quantum number has the value of 2, the number of orbitals are

(a) 3 (b) 0

(c) 7 (d) 5 **Ans. d**

35. Which of the following is not paramagnetic.

(a) S^{2-} (b) NO

(c) S_2^- (d) N_2^- **Ans. a**

36. When two A o's combine, they form

(a) One M.O. (b) Two M.O.'s

(c) Three M.O.'s (d) Four M.O.'s **Ans. b**

37. The bond order in hydrogen molecule is

(a) One (b) Two

(c) Three (d) Four **Ans. b**

38. The bond order in oxygen molecule is

(a) One (b) Two

(c) Three (d) Four **Ans. b**

39. The maximum number of electrons that can be present in a molecular orbital is

(a) One (b) Two

(c) Three (d) Four **Ans. b**

40. As compared to bonding m.o., the antibonding m.o has

(a) Higher energy (b) Lower energy

(c) Equal energy (d) Unpredictable value of energy

Ans. a

❑❑❑

Chapter-3

Bonding and Molecular Structure

Fill in the blanks

1. All noble gas except helium have………………… electrons in their outermost shell.

Ans. eight

2. In electron-dot structures, the dots represent…………electrons.

Ans. valence

3. A substance which can easily lose electrons is said to be…….and a substance which can easily gain electrons is said to be……………..

Ans. electropositive, electronegative

4. The gain, loss or sharing of electrons during chemical bond formation affects only the…………

Ans. valence electrons

5. When a molecule is formed, the constituent atoms generally try to attain…………configuration.

Ans. noble gas

6. The number of electrons gained or lost by an atom to acquire noble gas configuration is called the …………of the atom.

Ans. electrovalency

7. The bonding of atoms is called ………………bonding and the forces of attraction between oppositely charged ions are known as……….forces.

Ans. ionic, coulombic

8. Ionic compounds are highly soluble in…………

Ans. polar liquids

9. ....ionization energy and......... electron affinity generally favour the formation of ionic compounds.

Ans. low, high

10. The bond formed by the mutual sharing of two electrons is called.........bond.

Ans. covalent

11. The bond formed by the mutual sharing of two electron pairs is called.........bond.

Ans. double bond

12. The numerical value of the number of electrons which an atom contributes for sharing purposes is called the........of the element in that compound.

Ans. covalency

13. The single-bond length is...........than double-bond length.

Ans. greater

14. Electronegativity of an element is a measure of its electron..........power in a chemical bond.

Ans. attracting

15. The nature of bonding in $NH_3.BF_3$ is...............

Ans. coordinate

16. The number of sigma bonds in HCN molecule is...............

Ans. two

17. There are..............π bonds in nitrogen molecule.

Ans. two

18. The............bonds are non-directional, while...........bonds are directional.

Ans. ionic, covalent

19. ....hybrid orbitals of nitrogen atom are involved in the formation of ammonium ion.

Ans. sp^3

20. The.............hybridization gives trigonal (planar) shape of the molecule.

Ans. sp^2

21. The compound having both sp and sp^2 hybridized carbon atom is............

Ans. propadiene

22. The shape of NH_3 molecule is.........

Ans. pyramidal

23. The shape of CH_3^+ is.............

Ans. trigonal planner

24. The angle between two covalent bonds is maximum in..............

Ans. CO_2

25. The net dipole moment of BF_3 is..............

Ans. zero

26. The strength of hydrogen bond depends upon the........... and........... of the atom bonded to hydrogen.

Ans. size, electronegativity

27. o- nitrophenol is......soluble in water as compound to m- and p- isomers.

Ans. less

28. Ortho nitro phenol has................hydrogen bond while p-nitro phenol has...........hydrogen bond.

Ans. intramolecular, intermolecular

29. In SF_4 molecule involves...............hybridization.

Ans. sp^3d

30. The shape of ClF_3 molecule is..................

Ans. T shaped

31. The shape of BrF_5 molecule is.....................

Ans. square pyramidal

32. If a molecule contains unpaired electrons it is .... in nature.

Ans. paramagnetic

33. Side wise overlap of two p-orbitals results in the formation of .... molecular orbital.

Ans. Pi

34. Bond length is ....proportional to bond order.

Ans. inversely

35. The molecule is stable if bond order is ....

Ans. positive.

Multiple Choice Questions

1. The number of valence electrons in an atom of element A is I and that in the atom of B is 7, answer the following

   (I) The compound formed between A and B is of the type

   (a) AB (b) AB_3

   (c) AB_2 (d) A_2B **Ans. a**

   (II) The compound between A and B is

   (a) Covalent

   (b) Highly soluble in non-polar solvents

   (c) Highly volatile

   (d) A good conductor in solution **Ans. d**

2. In hydrogen molecule, the two hydrogen atoms are held by

   (a) Electrovalent bond (b) Coordinate bond

   (c) Covalent bond (d) Hydrogen bond **Ans. c**

3. Which of the following is an ionic compound?

   (a) HCN (b) CCl_4

   (c) KCl (d) CO_2 **Ans. c**

4. Which of the following is a property of a covalent compound?

   (a) It exists as ions

   (b) It is soluble in water

   (c) It has low melting and boiling points

   (d) It has high heat of vaporization or fusion **Ans. c**

5. Which of the following has electrovalent linkage?

   (a) $SiCl_4$ (b) $MgCl_4$

   (c) CH_4 (d) BF_3 **Ans. b**

6. The types of bond present in $CuSO_4.5H_2O$ are only

   (a) Electrovalent, covalent and coordinate covalent

   (b) Electrovalent and covalent

   (c) Covalent and coordinate covalent

   (d) Electrovalent and coordinate covalent **Ans. d**

7. In which of the following, ionic, covalent, and coordinate bonds are present?

   (a) Sodium cyanide (b) Water

   (c) Ammonia (d) Potassium bromide **Ans. a**

8. If a molecule MX_3 has zero dipole moment, the sigma bonding orbitals used by M (atomic number less than 21) are

(a) sp^2 hybrid (b) Pure p

(c) sp hybrid (d) sp^3hybrid **Ans. a**

9. The bonds present in N_2O_5 are

(a) Only ionic (b) Only covalent

(c) Covalent and ionic (d) Covalent and coordinate

Ans. b

10. The total number of electrons that take part in bond formation in N_2 is

(a) 6 (b) 10

(c) 2 (d) 4 **Ans. a**

11. Which of the following statements is wrong?

(a) d^2p^2 hybrid orbitals are all at 90^0 to one another

(b) Hybridization is the mixing of atomic orbitals prior to their combining into molecular orbital

(c) d^2sp^3 hybrid orbitals are directed towards the concerns of a regular tetrahedron

(d) sp^2 hybrid orbitals are formed from two p-atomic, and one s-atomic orbitals **Ans. d**

12. The compound 1, 3-butadiene has

(a) Only one sp hybridized carbon atom

(b) sp, sp^2 and sp^3 hybridized carbon atoms

(c) Only sp^2 hybridized carbon atoms

(d) Both sp^3 and sp^2 hybridized carbon atoms **Ans. c**

13. The molecule having one unpaired electron is

(a) O_2 (b) NO

(c) CO (d) CN^- **Ans. b**

14. On hybridization of one s- and one p- orbital we get

(a) Two orbitals at 180^0

(b) Four orbitals directed tetrahedrally

(c) Two mutually perpendicular orbitals

(d) Three orbitals in a plane **Ans. a**

15. Which of the following bonds is non-polar?

(a) O-H (b) F-F

(c) N-H (d) C-H **Ans. b**

16. NH_3 has a higher boiling point than expected because

(a) It has intermolecular hydrogen bonds

(b) With water, it forms NH_4OH

(c) It has strong intermolecular covalent bonds

(d) Its density decreases on freezing **Ans. a**

17. The bond between carbon atom (1) and carbon atom (2) in the compound $N = {}^1C\text{-}{}^2CH = CH_2$ involves the hybrids

(a) sp and sp^2 (b) sp and sp

(c) sp^2 and sp^2 (d) sp^3 and sp **Ans. a**

18. Which of the following has a bond formed by sp-sp^3 hybrid orbitals?

(a) $HC \equiv CH$ (b) $CH_3 - C \equiv CH$

(c) $CH_3 - CH - CH = CH_2$ (d) $CH_3 - CH = CH - CH_3$ **Ans. b**

19. The type of hybrid orbitals used by Cl atom in ClO_2^- is

(a) sp (b) sp^3

(c) sp^2 (d) None of these **Ans. b**

20. The C – H bond distance is longest in

(a) C_2H_6 (b) $C_2H_4Br_2$

(c) C_2H_4 (d) C_2H_2 **Ans. d**

21. The species in which the central atom use sp^2 hybrid orbitals in its bonding is

(a) NH_3 (b) PH_3

(c) SbH_3 (d) CH_3^+ **Ans. d**

22. Amongst the following, the molecule that is linear is

(a) SO_2 (b) ClO_2

(c) CO_2 (d) NO_2 **Ans. c**

23. CO_2 is isostructural with

(a) C_2H_2 (b) $HgCl_2$

(c) $SnCl_2$ (d) NO_2 **Ans. a**

24. The linear structure is assumed by

(a) $SnCl_4$ (b) CS_2

(c) NCO^- (d) NO_2^+ **Ans. b**

25. The molecule with zero dipole moment is

(a) Methylene chloride (b) Methyl chloride

(c) Carbon tetrachloride (d) Chloroform **Ans. c**

26. Carbon tetrachloride has zero dipole because of

(a) Its regular tetrahedral structure

(b) Similar size of carbon and chloride atoms

(c) Its planar structure

(d) Similar electrons affinities of carbon and chlorine **Ans. a**

27. The maximum number of hydrogen bonds a water molecule can form is

(a) 3 (b) 2

(c) 4 (d) 1 **Ans. c**

28. The hydrogen bond is strongest in

(a) F – H ··· F (b) S – H ··· S

(c) O – H ··· O (d) F – H ··· O **Ans. a**

29. Which one of the following does not have hydrogen bond

(a) Phenol (C_6H_5OH) (b) Liquid HCl

(c) Liquid NH_3 (d) Water (H_2O) **Ans. b**

30. Which of the following molecule/ion does not contain unpaired electrons

(a) N_2^+ (b) O_2^{2-}

(c) O_2 (d) B_2 **Ans. b**

31. The angle between covalent bonds is maximum in

(a) CH_4 (b) BF_3

(c) PF_3 (d) NH_3 **Ans. b**

32. CO_2 is isostructural with

(a) $SnCl_2$ (b) $ZnCl_2$

(c) $HgCl_2$ (d) C_2H_4 **Ans. c**

33. Which of the following is paramagnetic

(a) O_2^- (b) CN^-

(c) CO (d) NO^+ **Ans. a**

34. Among the following paramagnetic compound is

(a) NO_2O_2 (b) O_3

(c) N_2O (d) KO_2 **Ans. d**

35. The species having bond order different from that in CO is

(a) NO^- (b) NO^+

(c) CN^- (d) N_2 **Ans. a**

❑❑❑

Chapter-4

Water and pH buffer

Fill in the blanks

1. Mineral element present in cytokinins is...............

Ans. iron

2. .... is the pH value of pure water.

Ans. 7

3. The total calcium of the human body is about...........kg.

Ans. 1–1.5

4. The predominant action of plasma is.......

Ans. Cl^-

5. The daily water allowance for normal adult (60 kg) is about.............. ml.

Ans. 1800–2500

6. Silica and calcium occur in............. and.................

Ans. protistan shells, human bones

7. ....is the pH of blood.

Ans. 7.4

8. The daily water allowance for normal infant is about.........ml.

Ans. 330-1000

9. The percentage of calcium present in extracellular fluid is...............

Ans. 1

10. Cardiac arrest may occur due to over doses of................

Ans. potassium

Multiple Choice Questions

1. pH of a solution is equal to

(a) $-[\log OH^-]$ (b) $\log [H^+]$
(c) $-\log [H^+]$ (d) $\log [OH^-]$ **Ans. c**

2. pH of NaCl solution is

(a) 7 (b) 8
(c) 6 (d) 0 **Ans. a**

3. pH of lemon juice is formed to be below 7, lemon juice is

(a) Acidic (b) Basic
(c) Neutral (d) Amphoteric **Ans. a**

4. pH of 0.1 M HCl solution is

(a) Equal to 0.2 M HCl solution
(b) More than 0.2 M HCl solution
(c) Less than 0.2 M HCl solution
(d) Nothing can be said **Ans. c**

5. pH of water is 7 because

(a) Water has no OH^- ions (b) In water $[H^+] = [OH^-]$
(c) In water $[OH^-] < [H^+]$ (d) In water $[OH^-] > [H^+]$ **Ans. b**

6. Dilute NaOH oil solution is

(a) Acidic (b) Basic
(c) Neutral (d) Amphoteric **Ans. b**

7. The two colours seen at the extreme ends of the pH chart are

(a) Red and blue (b) Red and green
(c) Green and blue (d) Orange and green **Ans. a**

8. A student observed that the colour of pH paper changes to green when the dipped it in water. She added a few drops of concentrated hydrochloric acid to the water. The colour of pH paper would turn to

(a) Light red (b) Dark blue
(c) Lemon yellow (d) Apple green **Ans. a**

9. A student tested the pH of distilled water and find that the colour of the pH paper changed to green. He checked the pH again after dissolving a pinch or common salt in it. The colour of the pH paper this time was

(a) Green (b) Blue
(c) Red (d) Yellow **Ans. a**

10. Bottle A contains oxalic acid and bottle B contains sodium carbonate solution. When pH paper is dipped in each of the solutions, the colour seen in A and B respectively be

(a) Blue, orange (b) Orange, green

(c) Green, blue (d) Orange, blue **Ans. d**

11. What should be used to measure pH of a solution?

(a) Dry blue litmus paper (b) Acidified $K_2Cr_2O_7$ paper

(c) Lead acetate paper (d) Moist blue litmus paper

Ans. d

12. What can be said about pH of 0.1 M HCl and 0.1 M H_2SO_4

(a) pH of HCl is lower than H_2SO_4

(b) pH of both is same

(c) pH of HCl is higher than H_2SO_4

(d) Nothing can be said **Ans. c**

13. A beaker contains KOH solution and blue litmus solution. Excess HCl is added to beaker. What will be the find colcur of solution in beaker?

(a) Colourless (b) Yellow

(c) Red (d) Blue **Ans. c**

14. The two colours at the extreme end of pH chart are

(a) Red and blue (b) Green and yellow

(c) Green and purple (d) Violet and red **Ans. a**

15. pH paper is used to

(a) Find hydronium ion concentration of sample

(b) Find reducing nature of sample

(c) Find oxidizing nature of sample

(d) Find calorific value of sample **Ans. a**

16. Acids dissolve in water to produce

(a) OH^- ion (b) H^+ ion

(c) CO_3^{-2} ion (d) Salt **Ans. b**

17. What is the pH of acetic acid

(a) Below 7 (b) 0

(c) Above 7 (d) 7 **Ans. a**

18. If pH of solution changes from 3 to 4 the solution becomes

(a) Less acidic (b) More acidic

(c) Strongly basic (d) Basic **Ans. a**

19. What is pH value of the M/1000 HCl?

(a) pH= 2 (b) pH = 4

(c) pH = 0 (d) pH = 3 **Ans. d**

20. What is pH value of the M/1000 NaOH?

(a) pH = 11 (b) pH = 10

(c) pH = 6 (d) pH = 7 **Ans. a**

21. What is the pH of Sea water?

(a) 8.5 (b) 8.0

(c) 9.0 (d) 6.5 **Ans. a**

22. What is the pH of the solution when hydrogen ion concentration is 1.0×10^{-7} Mole?

(a) 0 (b) 10

(c) 6 (d) 7 **Ans. a**

23. The solution turns red litmus to blue. The pH of the solution is

(a) Less than 7 (b) More than 7

(c) Equal to 7 (d) None **Ans. b**

24. Which acid is present in orange?

(a) Acetic acid (b) Sulphuric acid

(c) Citric acid (d) Nitric acid **Ans. c**

25. What is pH of blood?

(a) 0 (b) 2

(c) 4 (d) 7.4 **Ans. d**

26. What is the mathematical expression of pH?

(a) pH = - Log $[H^+]$ (b) pH = Log H^+

(c) pH = - Log $[H^-]$ (d) pH = Log $[H_3O^+]$ **Ans. a**

27. What is the ionic product of distilled water?

(a) 10^{-12} Moles/litre (b) 10^{-17} Moles/litre

(c) 10^{-14} Moles/litre (d) 0 **Ans. b**

28. The concentration of H^- ions in distilled water is

(a) 10^{-7} Moles/litre (b) 10^{-6} Moles/litre

(c) 10^{-4} Moles/litre (d) 10^{0} Moles/litre **Ans. a**

29. A solution has pH 4. The solution will be

(a) Neutral (b) Basic

(c) Acidic (d) Ionic **Ans. c**

30. What is the hydrogen ion concentration of a solution when its pH = 4?

(a) 10^{-2} Moles/litre (b) 10^{-3} Moles/litre

(c) 10^{-6} Moles/litre (d) 10^{-4} Moles/litre **Ans. d**

31. What will be the pH of a basic solution?

(a) pH > 7 (b) pH < 7

(c) pH = 7 (d) pH = 0 **Ans. a**

32. What will be the pH of a acidic solution?

(a) pH = 7 (b) pH > 7

(c) pH = 1 (d) pH < 7 **Ans. d**

33. What is relationship between pH and pOH?

(a) pH + pOH = 12 (b) pH + pOH = 14

(c) 0 (d) -1 **Ans. b**

34. What do you mean by pOH?

(a) pOH = - Log $[H^+]$ (b) pOH = - Log $[H^-]$

(c) pOH = - Log $[OH^-]$ (d) pOH = - Log $[10^{-1}]$ **Ans. c**

35. How the pH of pure water is affected by rise in temperature?

(a) Decreases (b) Increases

(c) No effect (d) Neither increase nor decrease

Ans. a

36. The total body water in various subjects is relatively constant when expressed as percentage of the lean body mass and is about

(a) 30% (b) 40%

(c) 50% (d) 70% **Ans. d**

37. The percentage of water contained in the body of an individual is less because of

(a) High fat content (b) Low fat content

(c) High protein content (d) Low protein content **Ans. a**

38. In intracellular compartment the fluid present in ml/kg body weight is about

(a) 100 (b) 200
(c) 400 (d) 330 **Ans. d**

39. In extra cellular compartment, the fluid present in ml/kg of body weight is about

(a) 120 (b) 220
(c) 270 (d) 330 **Ans. c**

40. Fluid present in dense connective tissue and cartilage in ml/kg body weight is about

(a) 10 (b) 20
(c) 45 (d) 55 **Ans. c**

41. The total body water in ml/kg body weight in average normal young adult male is about

(a) 200 (b) 400
(c) 600 (d) 1000 **Ans. c**

42. The fluid present in bones which can not be exchanged readily because of relative avascularity is about

(a) 20 ml/kg (b) 25 ml/kg
(c) 45 ml/kg (d) 60 ml/kg **Ans. c**

43. Water derived in gm from complete oxidation of each gm of carbohydrate is about

(a) 0.15 (b) 0.25
(c) 0.35 (d) 0.55 **Ans. d**

44. The oxidation of 100 gm of fat yields

(a) 50 gm water (b) 107 gm water
(c) 150 gm water (d) 200 gm water **Ans. b**

45. Each gm of protein on complete oxidation yields

(a) 0.21 gm water (b) 0.31 gm water
(c) 0.41 gm water (d) 0.51 gm water **Ans. c**

46. The daily total body water derived from oxidation of food stuffs is about

(a) 100 ml (b) 300 ml
(c) 600 ml (d) 1000 ml **Ans. b**

47. The daily water allowance for normal infant is about

(a) 100–200 ml (b) 250–300 ml

(c) 330–1000 ml (d) 1000–2000 ml **Ans. c**

48. Insensible loss of body water of normal adult is about

(a) 50–100 ml (b) 100–200 ml

(c) 300–500 ml (d) 600–1000 ml **Ans. d**

49. The predominant cation of plasma is

(a) Na^+ (b) K^+

(c) Ca^+ (d) Mg^{++} **Ans. a**

50. Vasopressin (ADH)

(a) Enhance facultative reabsorption of water

(b) Decreases reabsorption of water

(c) Increases excretion of calcium

(d) Decreases excretion of calcium **Ans. a**

51. Enhanced facultative reabsorption of water by Vasopressin is mediated by

(a) Cyclic AMP (b) Ca^{++}

(c) Cyclic GMP (d) Mg^{++} **Ans. a**

52. Action of kinins is to

(a) Increase salt excretion

(b) Decrease salt retention

(c) Decrease water retention

(d) Increase both salt and water excretion **Ans. d**

53. The activity of kinins is modulated by

(a) Prostaglandins (b) Ca^{++}

(c) Increased cAMP level (d) Increased cGMP level **Ans. a**

54. An important cause of water intoxication is

(a) Nephrogenic diabetes insipidus

(b) Renal failure

(c) Gastroenteritis

(d) Fanconi syndrome **Ans. b**

55. Minimum excretory urinary volume for waste products elimination during 24 hrs is

(a) 200–300 ml (b) 200–400 ml
(c) 500–600 ml (d) 800 ml **Ans. c**

56. In primary dehydration

(a) Intracellular fluid volume is reduced
(b) Intracellular fluid volume remains normal
(c) Extracellular fluid volume is much reduced
(d) Extracellular fluid volume is much increased **Ans. a**

57. An important cause of secondary dehydration is

(a) Dysphagia (b) Oesophageal varices
(c) Muscle cramps (d) Gastroenteritis **Ans. d**

58. Important finding of secondary dehydration is

(a) Intracellular oedema (b) Cellular dehydration
(c) Thirst (d) Muscle cramps **Ans. a**

59. Urine examination in secondary dehydration shows

(a) Ketonuria (b) Low specific gravity
(c) High specific gravity (d) Albuminuria **Ans. b**

60. Daily requirement of calcium for normal adult human is

(a) 100 mg (b) 800 mg
(c) 2 g (d) 4 g **Ans. b**

61. Normal total serum calcium level varies between

(a) 4–5 mg (b) 9–11 mg
(c) 15–20 mg (d) 50–100 mg **Ans. b**

62. The element needed in quantities greater than 100 mg for human beings is

(a) Calcium (b) Zinc
(c) Selenium (d) Cobalt **Ans. a**

63. The mineral present in the human body in larger amounts than any other cation is

(a) Sodium (b) Calcium
(c) Potassium (d) Iron **Ans. b**

64. The percentage of the total body calcium present in bones is

(a) 1 (b) 11

(c) 55 (d) 99 **Ans. d**

65. The percentage of calcium present in extracellular fluid is

(a) 1 (b) 5

(c) 10 (d) 50 **Ans. a**

66. The physiologically active form of calcium is

(a) Protein bond (b) Ionised

(c) Complexed with citrate (d) Complexed with carbonate

Ans. b

67. The normal concentration of calcium in C.S.F is

(a) 1.5–2.5 mg/100 ml (b) 2.5–4 mg/100 ml

(c) 4.5–5 mg/100 ml (d) 9–10 mg/100 ml **Ans. c**

68. Absorption of calcium is increased on a

(a) High protein diet (b) Low protein diet

(c) High fat diet (d) Low fat diet **Ans. a**

69. Calcium absorption is interfered by

(a) Protein in diet (b) Phytic acid in cereals

(c) Alkaline intestinal pH (d) Vitamin D **Ans. b**

70. Calcium absorption is increased by

(a) Vitamin D (b) Vitamin C

(c) Vitamin K (d) Vitamin E **Ans. a**

71. In serum product of Ca x P (in mg/100ml) in children is normally

(a) 20 (b) 30

(c) 50 (d) 60 **Ans. c**

72. In ricket, the product of Ca x P (in mg/100 ml) in serum is below

(a) 30 (b) 50

(c) 70 (d) 100 **Ans. a**

73. In man, the amount of calcium in gms filtered in 24 hrs period by the renal glomeruli is

(a) 5 (b) 10

(c) 15 (d) 20 **Ans. b**

74. The percentage of the calcium eliminated in feces is

(a) 10–20 (b) 30–40

(c) 50–60 (d) 70–90 **Ans. d**

75. The maximal renal tubular reabsorptive capacity for calcium (T_{mca}) in mg/min is about

(a) 1.5 ±0.1 (b) 4.99 ±0.21

(c) 5.5 ±1.2 (d) 10.2 ±2.2 **Ans. b**

76. Renal ricket is caused by renal tubular defect (usually inherited) which interferes with reabsorption of

(a) Calcium (b) Phosphorous

(c) Sodium (d) Chloride **Ans. b**

77. After operative removal of the parathyroid glands resulting into hypoparathyroidism the concentration of the serum calcium may drop below

(a) 11 mg (b) 10 mg

(c) 9 mg (d) 7 mg **Ans. d**

78. One of the principal cations of soft tissue and body fluids is

(a) Mg (b) S

(c) Mn (d) Co **Ans. a**

79. The normal concentration of magnesium in whole blood is

(a) 0–1 mg/100 ml (b) 1–2 mg/100 ml

(c) 2–4 mg/100 ml (d) 4–8 mg/100 ml **Ans. c**

80. The normal concentration of magnesium in C.S.F is about

(a) 1 mg/100 ml (b) 3 mg/100 ml

(c) 5 mg/100 ml (d) 8 mg/100 ml **Ans. b**

81. The magnesium content of muscle is about

(a) 5 mg/100 ml (b) 10 mg/100 ml

(c) 21 mg/100 ml (d) 50 mg/100 ml **Ans. c**

82. Intestinal absorption of magnesium is increased in

(a) Calcium deficient diet (b) High calcium diet

(c) High oxalate diet (d) High phytate diet **Ans. a**

83. Deficiency of magnesium may occur with

(a) Alcoholism (b) Diabetes mellitus

(c) Hypothyroidism (d) Advanced renal failure **Ans. a**

84. Hypermagnesemia may be observed in
(a) Hyperparathyroidism (b) Diabetes mellitus
(c) Kwashiorkar (d) Primary aldosteronism **Ans. b**

85. Na^+/K^+ ATPase along with ATP requires
(a) Ca (b) Mn
(c) Mg (d) Cl **Ans. c**

86. The principal cation in extracellular fluid is
(a) Sodium (b) Potassium
(c) Calcium (d) Magnesium **Ans. a**

87. The normal concentration of sodium (in mg/100 ml) of human plasma is
(a) 100 (b) 200
(c) 250 (d) 330 **Ans. d**

88. A decrease in serum sodium may occur in
(a) Adrenocortical insufficiency
(b) Hypoparathyroidism
(c) Hyperparathyroidism
(d) Thyrotoxicosis **Ans. a**

89. Hypernatremia may occur in
(a) Diabetes insipidus (b) Diuretic medication
(c) Heavy sweating (d) Kidney disease **Ans. a**

90. The metabolism of sodium is regulated by the hormone
(a) Insulin (b) Aldosterone
(c) PTH (d) Somatostatin **Ans. b**

91. The principal cation in intracellular fluid is
(a) Sodium (b) Potassium
(c) Calcium (d) Magnesium **Ans. b**

92. The normal concentration of potassium in whole blood is
(a) 50 mg/100 ml (b) 100 mg/100 ml
(c) 150 mg/100 ml (d) 200 mg/100 ml **Ans. d**

93. The normal concentration of potassium in human plasma in meq/I is about
(a) 1 (b) 2
(c) 3 (d) 5 **Ans. d**

94. The normal concentration of potassium in cells in mg/100 ml is about
(a) 100 (b) 200
(c) 350 (d) 440 **Ans. d**

95. Potassium content of nerve tissue in mg/100 ml is about
(a) 200 (b) 330
(c) 400 (d) 530 **Ans. d**

96. Potassium content of muscle tissue in mg/100 ml is about
(a) 50–100 (b) 100–150
(c) 250–400 (d) 150–200 **Ans. c**

97. One of the symptoms of low serum potassium concentration includes
(a) Muscle weakness (b) Confusion
(c) Numbness (d) Tingling of extremities **Ans. a**

98. Potassium metabolism is regulated by the hormone
(a) Aldosterone (b) PTH
(c) Somatostatin (d) Estrogen **Ans. a**

99. A high serum potassium, accompanied by a high intracellular potassium occurs in
(a) Adrenal insufficiency (b) Any illness
(c) Gastrointestinal losses (d) Cushing's syndrome **Ans. a**

100. Hypokalemia occurs in
(a) Cushing's syndrome (b) Addison's disease
(c) Renal failure (d) Advanced dehydration **Ans. a**

101. Cardiac arrest may occur due to over doses of
(a) Sodium (b) Potassium
(c) Zinc (d) Magnesium **Ans. b**

102. The normal concentration of chloride in mg/100 ml of whole blood is about
(a) 200 (b) 250
(c) 400 (d) 450 **Ans. b**

103. The normal concentration of chloride in mg/100 ml of plasma is about
(a) 100 (b) 200
(c) 365 (d) 450 **Ans. c**

104. The normal concentration of chlorine in mg/100 ml of C.S.F is about

(a) 200 (b) 250

(c) 300 (d) 440 **Ans. d**

105. Hypokalemia with an accompanying hypochloremic alkalosis may be observed in

(a) Cushing's syndrome (b) Addison's disease

(c) Hyptothyroidism (d) Malnutrition **Ans. a**

106. Hypercholremia is associated with

(a) Hyponatremia (b) Hypernatremia

(c) Metabolic alkalosis (d) Respiratory acidosis **Ans. b**

107. The exclusive function of iron in the body is confined to the process of

(a) Muscular contraction (b) Nerve excitation

(c) Cellular respiration (d) Blood coagulation **Ans. c**

108. The normal concentration of bicarbonate in blood is

(a) 21 meq/L (b) 24 meq/L

(c) 26 meq/L (d) 30 meq/L **Ans. c**

109. At the pH of blood 7.4, the ratio between the carbonic acid and bicarbonate fractions is

(a) 1 : 10 (b) 1 : 20

(c) 1 : 30 (D) 1 : 40 **Ans. b**

110. A 0.22 M solution of lactic acid (pKa 3.9) was found to contain 0.20 M in the dissociated form and 0.02 M undissociated form, the pH of the solution is

(a) 2.9 (b) 3.3

(c) 4.9 (d) 5.4 **Ans. c**

111. Important buffer system of extracellular fluid is

(a) Bicarbonate/carbonic acid

(b) Disodium hydrogen phosphate/sodium dihydrogen phosphate

(c) Plasma proteins

(d) Organic Phosphate **Ans. a**

112. The pH of body fluids is stabilized by buffer systems. The compound which will be the most effective buffer at physiologic pH is

(a) Na_2HPO_4 pKa = 12.32 (b) Na_2HPO_4 pKa=7.21

(c) NH_4OH pKa = 7.24 (d) Citric acid pKa = 3.09 **Ans. b**

113. The percentage of CO_2 carrying capacity of whole blood by hemoglobin and oxyhemoglobin is

(a) 20 (b) 40

(c) 60 (d) 80 **Ans. c**

114. The normal serum CO_2 content is

(a) 18–20 meq/L (b) 24–29 meq/L

(c) 30–34 meq/L (d) 35–38 meq/L **Ans. b**

115. The carbondioxide carrying power of the blood residing within the red cells is

(a) 50% (b) 60%

(c) 85% (d) 100% **Ans. c**

116. Within the red blood cells the buffering capacity contributed by the phosphates is

(a) 5% (b) 10%

(c) 20% (d) 25% **Ans. d**

117. The normal ratio between the alkaline phosphate and acid phosphate in plasma is

(a) 2 : 1 (b) 1 : 4

(c) 20 : 1 (d) 4 : 1 **Ans. d**

118. The oxygen dissociation curve for hemoglobin is shifted to the right by

(a) Decreased O_2 tension (b) Decreased CO_2 tension

(c) Increased CO_2 tension (d) Increased pH **Ans. c**

119. Bohr effect is

(a) Shifting of oxyhemoglobin dissociation curve to the right

(b) Shifting of oxyhemoglobin dissociation curve to the left

(c) Ability of hemoglobin to combine with O_2

(d) Exchange of chloride with carbonate **Ans. a**

120. Chloride shift is

(a) H ions leaving the RBC in exchange of Cl^-

(b) Cl^- leaving the RBC in exchange of bicarbonate

(c) Bicarbonate ion returns to plasma and exchanged with chloride which shifts into the cell

(d) Carbonic acid to the plasma **Ans. c**

121. Of the total body water, intracellular compartment contains about

(a) 50% (b) 60%

(c) 70% (d) 80% **Ans. c**

122. Osmotically active substances in plasma are

(a) Sodium (b) Chloride

(c) Proteins (d) All of these **Ans. d**

123. Osmotic pressure of plasma is

(a) 80–100 milliosmole/litre (b) 180–200 milliosmole/litre

(c) 280–300 milliosmole/litre (d) 380–400 milliosmole/litre

Ans. c

124. Contribution of albumin to colloid osmotic pressure of plasma is about

(a) 10% (b) 50%

(c) 80% (d) 90% **Ans. c**

125. The highest concentration of proteins is present in

(a) Plasma (b) Intracellular fluid

(c) Interstitial fluid (d) Transcellular fluid **Ans. c**

126. Oncotic pressure of plasma is due to

(a) Proteins (b) Chloride

(c) Sodium (d) All of these **Ans. a**

127. Oncotic pressure of plasma is about

(a) 10 mm of Hg (b) 15 mm of Hg

(c) 25 mm of Hg (d) 50 mm of Hg **Ans. c**

128. Oedema can occur when

(a) Plasma Na and Cl are decreased

(b) Plasma Na and Cl are increased

(c) Plasma proteins are decreased

(d) Plasma proteins are increased **Ans. c**

129. Colloid osmotic pressure of intracellular fluid is

(a) Equal to that of plasma (b) More than that of plasma

(c) Less than that of plasma (d) Nearly zero **Ans. b**

130. The water produced during metabolic reactions in an adult is about

(a) 100 ml/day (b) 300 ml/day

(c) 500 ml/day (d) 700 ml/day **Ans. b**

131. The daily water loss through gastrointestinal tract in an adult is about
(a) Less than 100 ml/day (b) 200 ml/day
(c) 300 ml/day (d) 400 ml/day **Ans. a**

132. Recurrent vomiting leads to loss of
(a) Potassium (b) Chloride
(c) Bicarbonate (d) All of these **Ans. b**

133. Obligatory reabsorption of water
(a) Is about 50% of the total tubular reabsorption of water
(b) Is increased by antidiuretic hormone
(c) Occurs in distal convoluted tubules
(d) Is secondary to reabsorption of solutes **Ans. d**

134. Antidiuretic hormone
(a) Is secreted by hypothalamus
(b) Secretion is increased when osmolality of plasma decreases
(c) Increases obligatory reabsorption of water
(d) Acts on distal convoluted tubules and collecting ducts **Ans. d**

135. Urinary water loss is increased in
(a) Diabetes mellitus (b) Diabetes insipidus
(c) Chronic glomerulonephritis (d) All of these **Ans. d**

136. Diabetes insipidus results from
(a) Decreased insulin secretion
(b) Decreased ADH secretion
(c) Decreased aldosterone secretion
(d) Unresponsiveness of osmoreceptors **Ans. b**

137. Thiazide diuretics inhibit
(a) Carbonic anhydrase
(b) Aldosterone secretion
(c) ADH secretion
(d) Sodium reabsorption in distal tubules **Ans. d**

138. Furosemide inhibits reabsorption of sodium and chloride in
(a) Proximal convoluted tubules (b) Loop of Henle
(c) Distal convoluted tubules (d) Collecting ducts **Ans. b**

139. A diuretic which is an aldosterone antagonist is
(a) Spironolactone (b) Ethacrynic acid
(c) Acetazolamide (d) Chlorothiazide **Ans. a**

140. In a solution having a pH of 7.4, the hydrogen ion concentration is
(a) 7.4 nmol/L (b) 40 nmol/L
(c) 56 nmol/L (d) 80 nmol/L **Ans. b**

141. At pH 7.4, the ratio of bicarbonate: dissolved CO_2 is
(a) 1 : 1 (b) 10 : 1
(c) 20 : 1 (d) 40 : 1 **Ans. c**

142. Quantitatively, the most significant buffer system in plasma is
(a) Phosphate buffer system
(b) Carbonic acid-bicarbonate buffer system
(c) Lactic acid-lactate buffer system
(d) Protein buffer system **Ans. b**

143. In a solution containing phosphate buffer, the pH will be 7.4, if the ratio of monohydrogen phosphate : dihydrogen phosphate is
(a) 4 : 1 (b) 5 : 1
(c) 10 : 1 (d) 20 : 1 **Ans. a**

144. pKa of dihydrogen phosphate is
(a) 5.8 (b) 6.1
(c) 6.8 (d) 7.1 **Ans. c**

145. Buffering action of haemoglobin is mainly due to its
(a) Glutamine residues (b) Arginine residues
(c) Histidine residues (d) Lysine residues **Ans. c**

146. Respiratory acidosis results from
(a) Retention of carbon dioxide
(b) Excessive elimination of carbon dioxide
(c) Retention of bicarbonate
(d) Excessive elimination of bicarbonate **Ans. a**

147. Respiratory acidosis can occur in all of the following except
(a) Pulmonary oedema (b) Hysterical hyperventilation
(c) Pneumothorax (d) Emphysema **Ans. b**

148. The initial event in respiratory acidosis is
(a) Decrease in pH
(b) Increase in pCO_2
(c) Increase in plasma bicarbonate
(d) Decrease in plasma bicarbonate **Ans. b**

149. Respiratory alkalosis can occur in
(a) Bronchial asthma (b) Collapse of lungs
(c) Hysterical hyperventilation (d) Bronchial obstruction **Ans. c**

150. The primary event in respiratory alkalosis is
(a) Rise in pH
(b) Decrease in pCO_2
(c) Increase in plasma bicarbonate
(d) Decrease in plasma chloride **Ans. b**

151. Anion gap is the difference in the plasma concentrations of
(a) (Chloride) – (Bicarbonate)
(b) (Sodium) – (Chloride)
(c) (Sodium + Potassium) – (Chloride +Bicarbonate)
(d) (Sum of cations) – (Sum of anions) **Ans. c**

152. Normal anion gap in plasma is about
(a) 5 meq/L (b) 15 meq/L
(c) 25 meq/L (d) 40 meq/L **Ans. b**

153. Anion gap is normal in
(a) Hyperchloraemic metabolic acidosis
(b) Diabetic ketoacidosis
(c) Lactic acidosis
(d) Uraemic acidosis **Ans. a**

154. Anion gap is increased in
(a) Renal tubular acidosis
(b) Metabolic acidosis resulting from diarrhoea
(c) Metabolic acidosis resulting from intestinal obstruction
(d) Diabetic ketoacidosis **Ans. b**

155. Anion gap in plasma is because
(a) Of differential distribution of ions across cell membranes
(b) Cations outnumber anions in plasma
(c) Anions outnumber cations in plasma
(d) Of unmeasured anions in plasma **Ans. b**

156. Salicylate poisoning can cause
(a) Respiratory acidosis
(b) Metabolic acidosis with normal anion gap
(c) Metabolic acidosis with increased anion gap
(d) Metabolic alkalosis **Ans. c**

157. Anion gap of plasma can be due to the presence of all the following except
(a) Bicarbonate (b) Lactate
(c) Pyruvate (d) Citrate **Ans. a**

158. All the following features are found in blood chemistry in uncompensated lactic acidosis except
(a) pH is decreased (b) Bicarbonate is decreased
(c) pCO_2 is normal (d) Anion gap is normal **Ans. d**

159. All the following statements about renal tubular acidosis are correct except
(a) Renal tubules may be unable to reabsorb bicarbonate
(b) Renal tubules may be unable to secrete hydrogen ions
(c) Plasma chloride is elevated
(d) Anion gap is decreased **Ans. d**

160. All the following changes in blood chemistry can occur in severe diarrhea except
(a) Decreased pH (b) Decreased bicarbonate
(c) Increased pCO_2 (d) Increased chloride **Ans. c**

161. During compensation of respiratory alkalosis, all the following changes occur except
(a) Decreased secretion of hydrogen ions by renal tubules
(b) Increased excretion of sodium in urine
(c) Increased excretion of bicarbonate in urine
(d) Increased excretion of ammonia in urine **Ans. d**

162. Blood chemistry shows the following changes in compensated respiratory acidosis

(a) Increased pCO_2 (b) Increased bicarbonate
(c) Decreased chloride (d) All of these **Ans. d**

163. Metabolic alkalosis can occur in

(a) Severe diarrhoea
(b) Renal failure
(c) Recurrent vomiting
(d) Excessive use of carbonic anhydrase inhibitors **Ans. c**

164. Which of the following features are present in blood chemistry in uncompensated metabolic alkalosis except?

(a) Increased pH (b) Increased bicarbonate
(c) Normal chloride (d) Normal pCO_2 **Ans. d**

165. One joule is the energy required to

(a) Raise the temperature of 1 gm of water by 1°C
(b) Raise the temperature of 1 kg of water by 1°C
(c) Move a mass of 1 gm by 1 cm distance by a force of 1 Newton
(d) Move a mass of 1 kg by 1 m distance by a force of 1 Newton

Ans. d

166. Organic compound of small molecular size is

(a) Urea (b) Uric acid
(c) Creatinine (d) Phosphates **Ans. a**

167. Organic substance of large molecular size is

(a) Starch (b) Insulin
(c) Lipids (d) Proteins **Ans. d**

168. Body water is regulated by the hormone

(a) Oxytocin (b) ACTH
(c) FSH (d) Epinephrine **Ans. a**

169. Calcium is required for the activation of the enzyme

(a) Isocitrate dehydrogenase (b) Fumarase
(c) Succinate thiokinase (d) ATPase **Ans. d**

170. Cobalt is a constituent of

(a) Folic acid (b) Vitamin B_{12}

(c) Niacin (d) Biotin **Ans. b**

171. Calcium absorption is inferred by

(a) Fatty acids (b) Amino acids

(c) Vitamin D (d) Vitamin B_{12} **Ans. a**

172. The average of pH of urine is

(a) 5.6 (b) 6.0

(c) 6.4 (d) 7.0 **Ans. b**

173. Human body has 70% water. Water content of human embryo is

(a) 70% (b) 75-80%

(c) 90-95% (d) 50-60% **Ans. c**

174. Which one has maximum water content

(a) Human (b) Horse

(c) Jelly Fish (d) Coral **Ans. c**

175. Percentage of water present in extracellular fluid is

(a) 35% (b) 45%

(c) 55% (d) 75% **Ans. b**

176. Total amount of water present in human body cells is

(a) 20-22 liters (b) 18-19 liters

(c) 14-15 liters (d) 10-12 liters **Ans. a**

177. What property of water keeps electrolytes in ionic state

(a) Cohesion (b) Lattice structure

(c) High density (d) High electric constant **Ans. d**

178. Water functions as hydroskeleton in

(a) Sponges (b) Earthworms

(c) Human body (d) Plants **Ans. b**

179. Water has a high cohesion force. This is required in

(a) Ascent of sap

(b) Excretion

(c) Ionization

(d) Rapid dispersal of dissolved substances **Ans. a**

180. In dehydration synthesis, water is a
(a) Reactant (b) Product
(c) Lubricant (d) Reagent **Ans. b**

181. Water has maximum density at
(a) Room temperature (b) 4°C
(c) 0°C (d) -4°C **Ans. b**

182. Which one is high in case of water
(a) Surface tension
(b) Specific heat
(c) Heat of vaporization and heat of fusion
(d) All the above **Ans. d**

183. Water has high specific heat due to
(a) Its dipole nature
(b) Smaller angle between hydrogen atoms and oxygen atom
(c) Hydrogen bonds amongst molecules
(d) All the above **Ans. c**

184. Water is most abundant component of organisms because
(a) It is liquid (b) It is incompressible
(c) It is universal solvent (d) It is colourless **Ans. c**

185. Water (H_2O) is liquid while all others of equivalent and even higher molecular weight molecules are gaseous because of the presence of
(a) Covalent bonding between hydrogen and oxygen
(b) Electrostatic attraction amongst water molecules
(c) Hydrogen bonds
(d) Ionic bonds **Ans. c**

186. Ice is lighter than water due to
(a) Lattice aggregates occupy more space
(b) Cold water is extremely dense
(c) Absence of hydrogen bond in ice
(d) Absence of aggregates in ice **Ans. a**

187. Property of adhesion of water molecules to cell walls is due to
(a) Hydrogen bond (b) Dipole nature
(c) Ionization of water (d) All the above **Ans. a**

188. Framework elements are
(a) 6 (b) 4
(c) 3 (d) 2 **Ans. c**

189. Framework elements take part in
(a) Synthesis of protoplasm
(b) Synthesis of cell wall
(c) Formation of storage products
(d) Both (b) and (c) **Ans. d**

190. Framework elements are
(a) Nonmetal minerals (b) Nonminerals
(c) Mixed (d) Metals **Ans. b**

191. Big four elements are
(a) Carbon, hydrogen, oxygen and sulphur
(b) Carbon, oxygen, sulphur and phosphorous
(c) Carbon, nitrogen, sulphur and phosphorous
(d) Carbon, hydrogen, oxygen and nitrogen **Ans. d**

192. Protoplasmic elements are
(a) Proteins and enzymes (b) Nucleic acids
(c) Lipids (d) All the above **Ans. d**

193. Besides the big four, protoplasmic elements include
(a) Sulphur and phosphorus (b) Sodium and potassium
(c) Calcium and magnesium (d) Chromium and selenium
Ans. a

194. Buffers against pH changes are
(a) Monobasic phosphate (b) Dibasic phosphate
(c) Carbonic acid (d) All the above **Ans. d**

195. Fluoride ion
(a) Prevents tooth decay (b) Nonessential
(c) Toxic even in small quantity (d) Both (b) and (c) **Ans. a**

196. Dibasic phosphate (HPO_4^{2-}) is a buffer against
(a) Toxic chemicals (b) Strong acid
(c) Strong base (d) Weak bases and acids **Ans. c**

197. Monobasic phosphate (HPO_4^{2-}) is a buffer against

(a) Weak acid (b) Weak base

(c) Strong base (d) Strong acid **Ans. d**

198. Iodine occurs in human body as

(a) Thyroxine (b) Inorganic iodide

(c) Protein bound iodine (d) All the above **Ans. d**

199. Maximum amount of manganese is found in

(a) Ribosomes (b) Lysosomes

(c) Mitochondria (d) Nucleus **Ans. c**

200. Ca^{2+} and Mg^{2+}

(a) Reduce excitability of nerves and muscles

(b) Increase excitability of nerves and muscles

(c) Conduct nerve impulses

(d) Both (b) and (c) **Ans. a**

❑❑❑

Chapter-5

Biological Energy Generation and Thermodynamics

Fill in the blanks

1. The components of Biogas are CO, CH_4 and..............

Ans. hydrogen

2. ....involves raising of forests by communities of people for obtaining fuel an non-agricultural land.

Ans. Social forestry

3. Dead scum of *Botryococcus braunii* contains about......................... hydrocarbons.

Ans. 70%

4. Hydrogen generated by biological means is called...............

Ans. biohydrogen

5. ....found that certain plants of family Euphorbiaceae convert a substantial amount of photosynthate into petroleum substitutes.

Ans. Melvin Calvin

6. Gobar gas regional station is located at...............(U.P.).

Ans. Etawah

7. Production of Biogas involves..........major steps.

Ans. three

8. *Synechococcus* is a promising cyanobacterium for production of...........fuel.

Ans. hydrogen

9. Hog fuel is a mixture of..........and bark waste.

Ans. wood

10. A major quantity of wood is used as...........

Ans. fire wood

11. A system having only one phase is called ....

Ans. homogenous

12. A system which can exchange heat but no matter with its surroundings is called...............................system.

Ans. closed

13. ....is an example of intensive property.

Ans. Temperature

14. Height of a building is a............................function.

Ans. state

15. The process in which a system after having undergone a change returns back to its initial state is called.......................

Ans. cyclic

16. The process under constant volume condition is known as................

Ans. isochoric

17. E denotes the...........................of the system.

Ans. internal energy

18. The relation, $\Delta E = q + w$ represents the.......................of thermodynamics.

Ans. first law

19. The heat absorbed during a process under constant pressure conditions is called......................and is denoted by ....

Ans. enthalpy, H

20. A stable compound has.........................enthalpy of formation.

Ans. negative

21. Combustion is an.....................reaction. Then, ΔH_{comb} is...................

Ans. exothermic, negative

22. If a reaction is written in the reversed manner, its enthalpy change remains unchanged in.......................but changes its....................

Ans. magnitude, sign

23. Bond breaking is an............................reaction.

Ans. endothermic

24. For a spontaneous process, ΔG is........................

Ans. negative

25. Entropy change at fusion point is....................

Ans. positive

26. In the expression, $\Delta G = \Delta H - T\Delta S$, the parameter..........................predicts the spontaneity of a reaction.

Ans. ΔG

27. The equilibrium constant of a reaction is related to the standard free energy change by the relation, ΔG^0 = ...

Ans. -2.303 RT log K

28. Fill in the appropriate terms

	Sign of			Nature of reaction
	ΔH	ΔS	ΔG	
a)	-	+	-	....
b)	+	+	+	Non-spontaneous
c)	....	....	-	Spontaneous
d)	-	-	- (low temperature)	....
e)	-	-	+ (high temperature)	....
f)	....	....	....	Non-spontaneous at all temperatures

Ans. a) spontaneous, c) -, +, d) spontaneous, e) non-spontaneous, f) +,-,+

29. Law of conservation of energy is known as ....

Ans. The First Law of thermodynamics

30. The heat change occurring at constant pressure is called ....

Ans. enthalpy change

Multiple Choice Questions

1. Respiratory quotient depends upon
   (a) The extent to which substrate is broken down
   (b) Nature of the respiratory substrate
   (c) Amount of oxygen present in the respiratory substrate
   (d) All the above **Ans. d**
2. Which of the following statement is wrong
   (a) $RQ = CO_2/O_2 = 1$ (Carbohydrates)
   (b) RQ = Volume of O_2 absorbed/volume of CO_2 evolved
   (c) $RQ = CO_2/O_2 < 1$ (Carbohydrates)
   (d) $RQ = CO_2/O_2 > 1$ (organic acids) **Ans. b**
3. Floating respiration utilizes – – –as substrate
   (a) Carbohydrates (b) Proteins
   (c) Fats (d) Starch **Ans. a**
4. Protoplasmic respiration utilizes – – – as substrate
   (a) Carbohydrates (b) Proteins
   (c) Fats (d) Organic acids **Ans. b**
5. Alcoholic fermentation differs other anaerobic respirate in being
   (a) Intracellular (b) Intercellular
   (c) Extracellular (d) Internucleolar **Ans. c**
6. One molecule of glucose contains about – –calories of energy in the form of bands
   (a) 69,000 (b) 6,900
   (c) 6,90,000 (d) 69,00,000 **Ans. a**
7. In anaerobic respiration kreb's cycle is replaced by
   (a) Floating respiration (b) Enter-dudoroff pathway
   (c) Photorespiration (d) Hexose-monophosphate shunt **Ans. d**
8. The inhibition of sugar breakdown due to the presence of oxygen under aerobic conditions is called
   (a) Horecker's effect (b) Pasteur's effect
   (c) Dixon's effect (d) Hill's effect **Ans. b**

9. An energy funding pathway which is resistant to cyanide is

(a) Kreb's (b) Hexose-monophosphate shunt

(c) Glycolysis (d) Entner-Duodroff **Ans. b**

10. The addition of a phosphate group to a compound is called

(a) Phosphorylation (b) Phosphorogenesis

(c) Phosphorylasis (d) Photophosphorylation **Ans. a**

11. The form of energy that is used universally by living organisms is

(a) Heat (b) Solar

(c) Chemical (d) ATP **Ans. c**

12. A molecule that loses a hydrogen atom is said to have been oxidized, because a hydrogen atom contains

(a) An ion (b) An electron

(c) A proton (d) A nucleus **Ans. b**

13. Which one of the following is an enzyme that contains a channel which allows protons expelled from a bacterial cell to flow back cross the cytoplasmic membrane

(a) ATPase (b) Dehydrogenase

(c) Proteinase (d) Fumarase **Ans. a**

14. What two compounds are required for CO_2 fixation

(a) Fumaric acid and ATP

(b) Hydrogen sulphide and oxygen

(c) Cytochrome and ATP

(d) $NADPH_2$ and ATP **Ans. d**

15. In glycolysis in addition to pyruvic acid and ATP two molecules of - - - are made

(a) $NADH_2$ (b) NADH

(c) ATP (d) ATP and NADH **Ans. b**

16. *Streptococcus lactis* uses the $NADH_2$ produced during glycolysis is to reduce

(a) Pyruvic acid to lactic acid (b) Lactic acid to pyruvic acid

(c) Acetaldehyde to ethanol (d) NADP to $NADPH_2$ **Ans. a**

17. In alcoholic fermentation by yeasts the $NADH_2$ produced during glycolysis is used to reduce

(a) Pyruvic acid to lactic acid (b) Lactic acid to pyruvic acid
(c) Acetaldehyde to ethanol (d) NADP to $NADPH_2$ **Ans. c**

18. Even through most intestinal bacteria are fermentation their growth is not limited by the low efficiency of fermentation because they have

(a) Unlimited supply of nutrients
(b) Unlimited supply of O_2
(c) Limited supply of CO_2
(d) None of the above **Ans. a**

19. Which of the following is an example of an energized nucleotide

(a) UMP (b) dCTPQ
(c) GMP (d) dCMP **Ans. b**

20. The biosynthesis of both RNA and proteins is dependent upon the nucleotide sequence of

(a) mRNA (b) tRNA
(c) DNA (d) rRNA **Ans. c**

21. CO_2 fixation by the calvin cycle requires energy source. It also requires $NADPH_2$ or $NADH_2$ as a donor

(a) GTP, oxygen (b) ATP, oxygen
(c) Light electron (d) ATP electron **Ans. d**

22. In the initial reaction of the calvin cycle, a 5-carbon compound and a 1-carbon compound react to form

(a) Three 2-carbon compounds
(b) Two 3-carbon compounds
(c) One 4-carbon compound, and one 2-carbon compound
(d) One 5-carbon compound, and one 1-carbon compound **Ans. a**

23. The energy source that directly causes bacterial flagella to rotate is called the

(a) GTP hydrolysis (b) Torsional strain
(c) Proton motive force (d) ATP hydrolysis **Ans. c**

24. In order for glycolysis to continue NAD must be generated through

(a) Respiration (b) Fermentation
(c) (a) or (b) (d) Both **Ans. c**

25. Which of the following statements is correct

(a) Synthesis of cell constituents is on energy liberating process

(b) Dissimilation is an energy requiring process

(c) The energy liberated by dissimilation is used directly for synthesis of cell constituents with no interlining energy trapping system

(d) Dissimilation of nutrients provides the budding block synthesis of cell constituents with no intervening energy trapping system

Ans. d

26. The rate of most enzyme catalyzed reactions increase for each 10^0C increase in temperature up to the temperature at which the enzyme begins to be damaged

(a) 0.5 (b) 2

(c) 4 (d) 10 **Ans. b**

27. A chemoheterotroph is an organism that

(a) Dissimilates organic substrates to obtain energy

(b) Is exemplified by Anabaena cylindrical

(c) Dissimilates inorganic nutrients to obtain energy

(d) Uses light as its source of energy **Ans. a**

28. In the thylakoid membranes of cyanobacteria, algae and green plants the thylakoid membranes two kinds of chlorophyll-centres called

(a) PSI, PSII (b) PSIII, PSO

(c) PSI, PSIII (d) None of the above **Ans. a**

29. The mitochondria of brown fat have a unique protein in their inner membrane called

(a) Thermogenin (b) Uncoupling protein (UCP)

(c) Both (a) and (b) (d) None of the above **Ans. c**

30. Brown fat dissipate heat energy because

(a) Oxidation process inside the Golgi complex

(b) Oxidation process is fast as compared with yellow fats

(c) The protons inside the membrane remain in the matrix

(d) The protons pass through F_0F_1 complex to yield ATP **Ans. c**

31. Phosphofructokinase I is inhibited only by

(a) ATP only (b) Hexokinase

(c) ATP & citrate both (d) Isocitrate dehydrogenase

Ans. c

32. Mutations in the mitochondrial DNA are known to cause the human disease called

(a) MERRF (myoclonicepilepsy and ragged-red fiber disease)

(b) PKU

(c) UCP

(d) LHON (Leber's hereditary optic neuropathy) **Ans. d**

33. Mitochondria probably evolved from

(a) Eubacteria (b) Archebacteria

(c) Endosymbiotic Bacteria (d) Parasymbiotic Bacteria **Ans. c**

34. Which of the following is called the Hill's reagent

(a) Oxidoreductase

(b) ATP-synthase

(c) 1,6 dichlorophenol

(d) 2,6-dichlorophenol indophenols **Ans. d**

35. A compound that does not release pi on hydrolysis is

(a) Phosphocreatine (b) Thioesters

(c) ATP (d) Glucose-6-phosphate **Ans. b**

36. If carbon dioxide contents of atmosphere is a or high as 300 parts per million

(a) Plants would grow for sometime and then die

(b) All plants would be killed

(c) The plants would not grow properly

(d) The plants would thrive well **Ans. a**

37. When a molecule of pyruvic acid is subjected to anaerobic oxidation there is

(a) Loss of 3 molecules of ATP (b) Loss of 6 molecules of ATP

(c) Gain of 2 molecules of ATP (d) Gain of 4 molecules of ATP

Ans. b

38. A characteristic feature of ripening of some fruits (like banana) is a sudden increase in respiration which is known as

(a) Climatic (b) Photorespiration

(c) Climacteric (d) Anthesis **Ans. c**

39. Which of the following respiration correctly explain the function of mitochondria

(a) CO_2 + ADP + Phosphate $\xrightarrow{(M)}$ H_2O + AMP + O_2

(b) CO_2 + ATP + Phosphate $\xrightarrow{(M)}$ H_2O + ATP + O_2

(c) O_2 + ADP + Phosphate $\xrightarrow{(M)}$ H_2O + ATP + CO_2

(d) O_2 + ADP + Phosphate $\xrightarrow{(M)}$ H_2O + ATP + CO_2 **Ans. c**

40. Substrate for photorespiration is

(a) Malic acid (b) Fumaric acid

(c) Glycolate (d) Serine **Ans. c**

41. Which of the following can respire in absence of oxygen is

(a) Potato (b) *Chlorella*

(c) *Yeast* (d) Fish **Ans. c**

42. The general name that its given for an enzyme that catalyze the transfer of a functional group from one position to another on the same molecule is

(a) Transmutase (b) Isomerase

(c) Mutase (d) Pyruvate kinase **Ans. c**

43. The cycle of reactions that includes glucose conversion to lactate in muscle and lactate conversion to glucose in liver called

(a) Allosteric regulation (b) Cori cycle

(c) Glucogenesis (d) Lactogenesis **Ans. b**

44. Zymase was discovered by - - - and for the first time obtained from

(a) Buchner, yeast (b) Pasteur, crow

(c) Hill, Sparrow (d) Altman, opuntia **Ans. a**

45. Which of the following statements is true

(a) Respiration is carried out by only leaf cells

(b) The end product of anaerobic respiration is CO_2 + pyruvic acid

(c) Substrate phosphorylation occurs when, α-keto gluteric acid changes to succinic acid

(d) Respiration in plants takes place only in night **Ans. c**

46. Energy of glucose is released by respiration was worked out by

(a) Sutton and Boveri (b) Hill

(c) Kreb's (d) Hatch and Slack **Ans. c**

47. Match the following

R. Altmann	(a)	DNA in mitochondria
Noss and Nass	(b)	Photoperiodism
C.D. Duve	(c)	Mitochondria
Garner and Allard	(d)	Lysosomes

Choose the correct answer

(a) a, c, b, d (b) a, b, d, c

(c) a, b, c, d (d) c, a, d, b **Ans. d**

48. Tick the correct statement

(a) Respiration is a catabolic process in which breakdown of carbohydrates only takes place

(b) Respiration is a catabolic process in which breakdown of carbohydrates, fats, proteins takes place

(c) Respiration is a catabolic process in which breakdown of carbohydrates, fats, proteins takes place in the presence of certain enzymes and energy is released

(d) Respiration in plants takes place through leaves only **Ans. c**

49. Calorific value of biogas is

(a) 100-120 Mj/m^3 (b) 23-28 Mj/m^3

(c) 11-15 Mj/m^3 (d) 1-2 Mj/m^3 **Ans. b**

50. Who of the following scientist showed that *Saccharomyces cerevisae* fermentation products such as beer and butter milk

(a) Schatz (b) S. Waksman

(c) A. Fleming (d) L. Pasteur **Ans. d**

51. In which of the following forms animal energy is available

(a) Biofuels (b) Fossil fuels

(c) Animal wastes (d) HMP and DAP **Ans. d**

52. One of the country where alcohol has been very successfully used as motor fuel is

(a) Germany (b) U.S.A.

(c) Brazil (d) Japan **Ans. c**

53. Biogas contains
(a) 50-70% methane (b) 80-95% ethane
(c) Traces of methane (d) 30-40% methane **Ans. a**

54. Bioenergy is the energy obtained from
(a) Coal (b) Biogas
(c) Petroleum (d) Natural gas **Ans. b**

55. Energy crops are
(a) Sugarcane or sugarbeet, potato and tomato
(b) Sugarcane, potato and tapioca
(c) Sugarcane, Tapioca and cauliflower
(d) Millets, banana and tomato **Ans. b**

56. What type of fuel are coal, petrol and natural gas
(a) Biofuels (b) Fossil
(c) Electric fuels (d) Liquid fuels **Ans. b**

57. What wood is most suited as fuelwood
(a) Monocot (b) Dicot
(c) Gymnospermous (d) All **Ans. b**

58. Which of these can be used in place of petrol
(a) Ethanol (b) Propanol
(c) Butanol (d) Methanol **Ans. a**

59. Which of these is poor source of fuel wood
(a) *Acacia nilotica* (b) *Azadirachta indica*
(c) *Qercus* (d) *Pinus roxburghii* **Ans. d**

60. Which of these is a petroleum plant
(a) Euphorbia (b) Maize
(c) Sugarcane (d) Potato **Ans. a**

61. A species selected for energy plantation must not be
(a) Hardy (b) Slow growing
(c) Local (d) Resistant to pests **Ans. b**

62. Maximum alcohol is obtained from
(a) Maize (b) Sugarcane
(c) Wood (d) Tapioca **Ans. b**

63. Fuel wood is
(a) Renewable resource of energy
(b) A source of petroleum products
(c) A source of biogas
(d) A source of alcohol **Ans. a**

64. Vitamin B_2 is obtained from
(a) *Acetobactor* (b) *Penicilium*
(c) *Ashbya gossypi* (d) *Aspergillus* **Ans. c**

65. Antibiotics are
(a) Syrups (b) Toxins
(c) Medicines (d) Plants **Ans. c**

66. Source of bioenergy are
(a) Solar energy (b) Animal energy and fossil fuels
(c) Animal energy and biofuels (d) Animal energy **Ans. c**

67. Raising of crops for the production of ethanol is known as
(a) Energy cropping (b) Agro-technology
(c) Biomass production (d) Energy plantation **Ans. a**

68. Petroleum yielding plants belong to
(a) Asclepiadaceae (b) Apocyanaceae
(c) Euphorbiaceae (d) All of above **Ans. d**

69. Source of liquid hydrocarbons is
(a) Asclepiadeceae (b) Apocyanaceae
(c) Euphorbiaceae (d) All of the above **Ans. d**

70. Pyrolysis of wood yields
(a) Charcoal (b) Charcoal and gas
(c) Alcohol (d) Charcoal, gas and oil **Ans. d**

71. Waksman and Woodruff were responsible for the isolation of
(a) Penicillin (b) Neomycin
(c) Streptomycin (d) Actinomycin **Ans. d**

72. Which of the following organic acid was produced by fermentation
(a) Oxalic acid (b) Propionic acid
(c) Lactic acid (d) Citric acid **Ans. c**

73. Vitamin B_{12} was first of all isolated by

(a) Funk (b) Mellanby

(c) Albert Valnszent (d) McCollum **Ans. a**

74. Rennet is used in

(a) Bread making (b) Cheese making

(c) Fermentation (d) Synthesis of antibiotics

Ans. b

75. Germinating barley is employed for preparation of

(a) Wine (b) Cheese

(c) Beer (d) Lactic acid **Ans. c**

76. Hybridomas are employed for

(a) Production of somatic hybrid

(b) Synthesis of antibiotics

(c) Killing cancer cells

(d) Synthesis of monoclonal antibodies **Ans. d**

77. Cells obtained from cancerous tumors are known as

(a) Monoclonal cells (b) Myelomas

(c) Hybridomas (d) Lymphocyte **Ans. b**

78. The components of biogas are

(a) CO_1, C_2H_5, H_2 (b) CO_1, CH_4, H_2

(c) CO_2, CH_4, H_2 (d) C_2, C_2H_5, O_2 **Ans. c**

79. Biogas production rate is limited by

(a) Conversion of monomer to organic acids

(b) Production of methane

(c) Digestion of cellulose

(d) None of these **Ans. c**

80. Gas produced as a result of anaerobic fermentation of agricultural wastes is

(a) Biogas (b) Methane

(c) Carbon monoxide (d) Gobargas **Ans. a**

81. Energy cropping does not involve raising of

(a) *Tapioca* (b) *Euphorbia*

(c) Sugarcane (d) Potato **Ans. b**

82. The first step in biogas production is
(a) Aerobic breakdown of polymers
(b) Generation of methane
(c) Anaerobic breakdown of polymers
(d) Conversion of monomers to organic acids **Ans. c**

83. Energy plantation differ from energy cropping in being
(a) Renewable (b) Source of wood
(c) Exportable (d) Substitute for petroleum
Ans. b

84. Dextran is used in the
(a) Blood clotting (b) Bleeding
(c) Blood transfusion (d) Blood preservation **Ans. c**

85. Rennet enzyme was purified by
(a) Payen and Persoz (b) C. Hansen
(c) A. Fleming (d) S.A. Waksman **Ans. b**

86. Enzymes can be immobilized by
(a) Cross-linking enzyme molecule
(b) Entrapping them in gel
(c) Covalently attaching to a solid support
(d) All the above methods **Ans. d**

87. The concepts of antibiotics was given by
(a) Babes (b) Waksman
(c) Vuilemin (d) Pasteur **Ans. c**

88. Which of these is used in biotechnology
(a) Both cattle and yeast (b) Yeast
(c) Cattle (d) Neither cattle nor yeast
Ans. a

89. The conversion of cortexolone to prenisolone involves
(a) Dehydrogenation hydroxylation
(b) Hydroxylation
(c) Dehydrogenation
(d) Hydroxylation, dehydrogenation **Ans. d**

90. Which enzymes are used for manufacturing detergents

(a) Proteases (b) Lactases

(c) Glucoamylases (d) Amylases **Ans. a**

91. Kohler and Milstein developed biotechnology for the production of

(a) Myelomas (b) Monoclonal antibodies

(c) Immobilized enzymes (d) Steroid conversion **Ans. b**

92. Utility of fungi for steroid conversion was demonstrated by

(a) Marry and Peterson (b) Pasteur and Joubert

(c) Waksman and Woodruff (d) Kohler and Milstein **Ans. a**

93. For rapid production of alcohol, immobilized yeast cells are kept in

(a) Calcium alginate beads (b) Silica gel

(c) Porcelain alginate beads (d) Wire netting **Ans. a**

94. Hybridomas are result of fusion of

(a) Male reproductive cell with myeloma

(b) Female reproductive cell with myeloma

(c) Normal antibody producing cell with myeloma

(d) Abnormal antibody producing cell with myeloma **Ans. c**

95. The antibiotics obtained from *Streptomyces* are

(a) Novobiocin, Spiramycin

(b) Chloramphenicol, erythromycin

(c) Leucomycin, Neomycin

(d) All of these **Ans. d**

96. The enzyme diastase was identified by

(a) S.A. Waskman (b) Payen and Persoz

(c) A. Fleming (d) C. Hansen **Ans. b**

97. Gasohol mixture is useful as

(a) Manure (b) Automobile fuel

(c) Fermentor (d) All of these **Ans. b**

98. Good firewood must not be

(a) Highly combustible (b) Easy to dry

(c) Resinous (d) Non smoky **Ans. c**

99. Which one of these is not correctly matched

(a) Animal energy-Used most efficiently
(b) Biogas-Produced from dung
(c) Latex-Source of liquid hydrocarbon
(d) Ethanol-Used as gasoline **Ans. a**

100. Which enzymes are used for converting corn starch into high fructose syrup

(a) Gluco isomerases (b) Glucoamylases
(c) Amylases (d) All of these **Ans. d**

101. The first acid to be produced using fermentation was

(a) Lactic acid (b) Malic acid
(c) Acetic acid (d) Citric acid **Ans. a**

102. The method or growing micro-organisms as a thin layer on nutrient medium is known as

(a) Suspended growth (b) Thin layer growth system
(c) Support growth system (d) None of these **Ans. c**

103. Most of Eubacteriales antibiotics come from

(a) *Rhizobium* (b) *Streptococcus*
(c) *Bacillus* (d) *Pseudomonas* **Ans. c**

104. Woodruff in 1942 isolated

(a) Bacitracin (b) Streptothricin
(c) Penicillin (d) Novobiocin **Ans. b**

105. Hepatitis B vaccine is

(a) First generation vaccine (b) An interferon
(c) Third generation vaccine (d) Second generation vaccine **Ans. d**

106. Shikonin is the name of

(a) A red dye used in the manufacture of cosmetics
(b) Vitamin
(c) Enzyme
(d) Steroids **Ans. a**

107. Who recognized the petroleum plants

(a) De Vries (b) Darwin
(c) Mendel (d) Melvin Calvin **Ans. d**

108. Which organism is useful in the preparation of Roquefort cheese
(a) *Rhizopus* (b) *Yeast*
(c) *Aspergillus* (d) *Peniclillum* **Ans. d**

109. Which of these participates in the manufacturing of dextrans
(a) *Lactobacillus* (b) *Mucor*
(c) *Leuconostoc* (d) *Pseudomonas* **Ans. c**

110. Which one of these is not correctly matched
(a) Methane-Biogas production
(b) Ethanol-Energy cropping
(c) Fuelwood-DAP
(d) Producer gas-Wood gasification **Ans. c**

111. Small pox vaccine was developed by
(a) Funk (b) Jenner
(c) Pasteur (d) Fleming **Ans. b**

112. Restriction enzyme was discovered by
(a) Waksman (b) Smith and Nathans
(c) Berg (d) A. Fleming **Ans. b**

113. Yeasts are economically important because
(a) These are used in tobacco and tea factories
(b) These spread diseased in plants
(c) These spread disease in animals
(d) These are used in bakeries and breweries **Ans. d**

114. Tissue culture technique has been successfully used for producing
(a) Alcoholic bevarages (b) Shikonin
(c) Cheese (d) Insulin **Ans. b**

115. The enzyme TPA is used to
(a) Strengthen tissues (b) Dissolve blood clots
(c) Maintain turgor pressure (d) Increase plasma **Ans. b**

116. The number of antibiotics obtainable from *Streptomyces girseus* are
(a) 20 (b) 40
(c) 30 (d) 10 **Ans. b**

117. The bread is soft and porous when the yeast cells are mixed in the lump of wheat flour, because

(a) Yeast is soft and flour also become soft

(b) Yeast produces benzoic acid

(c) Yeast produces lactic acid

(d) Evolution of CO_2 makes the bread spongy **Ans. d**

118. Humulin is

(a) Human insulin (b) A new digestive enzyme

(c) A form of chitin (d) A powerful antibiotic **Ans. a**

119. Tissue plasminogen activator (TPA) is

(a) An enzyme

(b) An electric device

(c) A vitamin

(d) A chemical that stimulates tissue differentiation **Ans. a**

120. The fruit juice turn bitter in taste if they are kept in open place for some time because of

(a) Fermentation of the juice by yeast

(b) Some internal factors

(c) Bacteria of the atmosphere react with the juices

(d) All the above three statements are correct **Ans. a**

121. *Torulopis utilis* is

(a) Microorganism that yields third generation vaccines

(b) An important intestinal commensal

(c) A food yeast

(d) Employed for synthesis of citric acid **Ans. c**

122. Chloramphenicol and erythromycin (broad spectrum antibiotics) are produced by

(a) *Penicillium* (b) *Rhizobium*

(c) *Streptomyces* (d) *Nitrobacter* **Ans. c**

123. Streptomycin was first isolated in 1944-45 by

(a) A. Fleming (b) Leuwenhoek

(c) Waksman (d) Burkholder **Ans. c**

124. Growing crops for production of alcohol I called

(a) Energy cropping (b) Energy gardening

(c) Energy plantation (d) Energy cultivation **Ans. a**

125. Steroids are used in

(a) Energy cropping

(b) Treatment of autoimmune disease

(c) Birth control

(d) Energy cultivation **Ans. c**

126. Chloromycetin is produced by

(a) *Streptomyces venezuelae* (b) *Streptomyces ramosus*

(c) *Streptomyces erythrias* (d) *Bordetella pertusis* **Ans. a**

127. Heating of milk or any other liquid at 65°C and then sudden cooling is known as

(a) Pasteurization (b) Preservation

(c) Sterilization (d) Fermentation **Ans. a**

128. Most of the microorganisms which produce antibiotics live in the soil because

(a) They cannot get nutrition outside the soil

(b) Darkness favours synthesis of antibiotics

(c) By the phenomenon of antibiotics their growth, nutrition and survival value are enhanced in competitive world of microflora of the soil

(d) No one easily misuse their antibiotics **Ans. c**

129. Broad spectrum antibiotic is that which

(a) Acts on all bacteria and virus

(b) Acts on both pathogens and hosts

(c) It effective in very small amounts

(d) Acts on a variety of pathogenic microorganisms **Ans. d**

130. Terramycin is obtained from

(a) *Streptomyces ramosues* (b) *Streptomyces venzuelac*

(c) *Streptomyces giseus* (d) *Streptomyces aureofacilens*

Ans. a

131. Triple antigen of DPT is meant for

(a) A vaccine against tetanus whooping cough and deptheria

(b) Mixture of viruses that causes tetanus diphtheria and whooping cough

(c) A vaccine against malaria, typhoid and cancer

(d) Immobilized enzymes **Ans. a**

132. A compound produced by an organism that inhibits the growth of another microorganism is called

(a) Antiseptic (b) Antibodies

(c) Antibiotic (d) Penicillin **Ans. c**

133. Third generation vaccines are

(a) Attenuated pathogen (b) Pathogen relative

(c) Synthetic antigen (d) Monoclonal antigen **Ans. c**

134. Rennin used in cheese industry is

(a) Antibiotic (b) Alkaloid

(c) Enzyme (d) Inhibitor **Ans. c**

135. A compound which is produced by an organism and inhibits the growth of other organisms is called

(a) Antibiotic (b) Antigen

(c) Antibody (d) Antiallergic **Ans. a**

136. Antibiotics inhibit the growth of or destroy

(a) Bacteria and fungi (b) Bacteria and viruses

(c) Bacteria, algae and viruses (d) Bacteria, fungi and viruses

Ans. a

137. The biological catalysts which initiate and speed up the rate of reaction without being used or changed are called

(a) Enzymes (b) Proteins

(c) Hormones (d) Antibiotics **Ans. a**

138. Which one of the following is used in the production of alcohol

(a) *Leuconoste citrovorum* (b) *Saccharomyces cerevisiae*

(c) *Torulopis utilis* (d) *Clostridium botulinum*

Ans. b

139. Payen and Persoz discovered
(a) Diastase (b) Lactose
(c) Protease (d) Rennet **Ans. a**

140. For biogas production, besides, dung which of the following weed is recommended in our country
(a) *Hydrilla* (b) *Solanum nigum*
(c) *Eichornia crassipes* (d) *Mangifera indica* **Ans. c**

141. Which of the following is the nonconventional source of energy
(a) Solar radiations
(b) Coal
(c) Petroleum
(d) Electricity from nuclear power plants **Ans. a**

142. The value of a resource depends upon
(a) Quantity of resource (b) Type of society and its needs
(c) Quality of resource (d) None of the above **Ans. b**

143. Antibiotic fumagalin is used against
(a) Tonsillitis
(b) Typhoid
(c) Some bacteria like *Salmonella* and *Shigella*
(d) None of the above **Ans. c**

144. Fruits, meats milk etc. are dried for preservation at room temperature by the process of
(a) Freezing (b) Dehydration
(c) Pasteurization (d) Vernalisation **Ans. b**

145. Cheese and yoghurt are products of the process
(a) Distillation (b) Dehydration
(c) Pasteurization (d) Vernalization **Ans. c**

146. The pioneer country in the production of fuel alcohol is
(a) Saudi Arabia (b) Iran, Iraq
(c) Brazil (d) Japan **Ans. c**

147. The main reason why antibiotics could not solve all the problems of bacteria mediated disease is
(a) Inactivation of antibiotics by bacterial enzymes
(b) Decreases efficiency of immune system

(c) The development of mutant strains resistant

(d) Insensitivity of the individual following prolonged exposure to antibiotics **Ans. c**

148. With the exception of water which one is possibly the most important accessory chemical substance in industrial processes

(a) Liquid nitrogen (b) Ethanol

(c) Rubber (d) Petroleum **Ans. a**

149. The term antibiotic was coined by

(a) Salman Waksman (b) Louis Pastiur

(c) Edward Jenner (d) Alexender Fleming **Ans. a**

150. The organism used for alcoholic fermentation is

(a) *Penicillium* (b) *Pseudomonas*

(c) *Aspergillus* (d) *Sacchromyces* **Ans. d**

151. Citric acid is produced by

(a) *Aspergillus niger* (b) *Streptococcus lactic*

(c) *Acetobacter suboxydans* (d) *Candida utilis* **Ans. a**

152. Which enzymes should be added to the detergents in order to remove proteinaceous stains

(a) Amylases (b) Proteases

(c) Zymases (d) Pectolytic enzymes **Ans. b**

153. An enzyme useful in dissolving blood clots in heart attack patients is

(a) Progesterone (b) Cholesterol

(c) Tissue plasminogen activator (d) Thrombin **Ans. c**

154. The microorganism grown on molasses and sold as a food flavouring substance is

(a) *Acetobactor* (b) *Rhizopus*

(c) *Saccharomyces* (d) *Lactobacillus* **Ans. c**

155. During steroid conversion *Rhizopus stolonifer* brings about

(a) Dehydrogenation (b) Deamination

(c) Hydroxylation (d) Carboxylation **Ans. c**

156. One of the following vaccines is produced through genetic engineering

(a) Small pox (b) Tetanus

(c) Hepatitis B (d) Typhoid **Ans. c**

157. A substance obtained from calf stomach is used in the manufacture of

(a) Cheese (b) Curd

(c) Fructose (d) Insulin **Ans. a**

158. The Noble prize for hybridoma technology was awarded to

(a) Kohler and Milstein

(b) Temin and Baltimore

(c) Huber, Michel and Dissenhofer

(d) Pasteur **Ans. a**

159. Coenzyme FMN and FAD are the derivatives of

(a) ETS (b) Respiration

(c) Vitamin B_{12} (d) Vitamin B_2 **Ans. d**

160. Vaccines produced by recombinant DNA technology are known as

(a) V generation (b) III generation

(c) II generation (d) I generation **Ans. c**

161. Substance with core structure made up of one-carbon and three-carbon ring are

(a) Insulin (b) Steroids

(c) Vitamins (d) Antibiotics **Ans. b**

162. *Pseudomonas* had been particularly useful in the production of

(a) Cobalamine (b) Acetic acid

(c) Cholesterol (d) Hormones **Ans. a**

163. Both power and manure are provided by

(a) Energy crops (b) Acetic acid

(c) Petro crops (d) Biogas **Ans. d**

164. Major source of liquid hydrocarbon is

(a) *Cocos nucifera* (b) *Calotropis gigantean*

(c) *Solanum tubersum* (d) *Euphorbia anisyphilitica*

Ans. b

165. City refuse is mainly composed of

(a) Lipids (b) Chitin

(c) Cellulose (d) Lignin **Ans. c**

166. Antibiotics inhibit the growth of - - - or destroy
(a) Bacteria (b) Bacteria and viruses
(c) Bacteria, fungi and viruses (d) Bacteria, algae and viruses
Ans. a

167. Who of the following established the scientific basis of vaccination
(a) Edward Jenner (b) Louis Pasteur
(c) George Kohlar (d) Cesar Milstein **Ans. a**

168. Small pox vaccine was developed by
(a) Louis Pasteur (b) Salman Waksman
(c) Cesar Milstein (d) Jenner **Ans. d**

169. Steroids are used in
(a) Treatment of hormonal imbalance
(b) Birth control
(c) Treatment of autoimmune disease
(d) All the above **Ans. d**

170. Antibiotic cephalosporin is got from
(a) Fungus (b) Bacterium
(c) Alga (d) Mycoplasma **Ans. a**

171. Who demonstrated for the first time that microorganisms could produce substances capable of inhibiting growth of other organisms
(a) Louis Pasteur (b) Babes
(c) Kohlar (d) Cesar Milstein **Ans. b**

172. Insulin
(a) Growth inhibitor
(b) A vitamin
(c) A hormone made by α-cells of islets of langerhans of pancreas in human body
(d) It is a steroid **Ans. c**

173. Which of the following are biofuels
(a) Petroleum (b) Coal
(c) Natural gas (d) Wood **Ans. b**

174. The sap of which is good substitute for diesel oil
(a) *Cobaifera langsdorfii* (b) *Ficus elastic*
(c) *Nerium* (d) *Vinca* **Ans. a**

175. The gas obtained by gasification of wood is called
(a) Natural gas (b) Fuel gas
(c) Producer gas (d) Biogas **Ans. c**

176. Biogas can be good substitute for
(a) Fuel wood (b) Petroleum and oil
(c) Coal (d) Charcoal **Ans. b**

177. Maximum alcohol is obtained from which of the following energy crop
(a) Maize (b) Wood
(c) Tapioca (d) Sugarcane **Ans. d**

178. The residue left after biogas generation can be
(a) Used in boilers (b) Used as manure
(c) Used to generate energy (d) Used as insulator **Ans. b**

179. Biogas is produced by anaerobic breakdown of biomass or agricultural waste by methanogenic bacteria it is a
(a) Multiple step process (b) Three step process
(c) Two step process (d) One step process **Ans. b**

180. The first step in biogas production is
(a) Conversion of organic acids present in biomass into methane
(b) Anaerobic breakdown of complex compound present in biomass into monomers
(c) Aerobic breakdown of complex compounds into monomers
(d) Conversion of polymers into organic acids **Ans. b**

181. The substitute of petroleum could be
(a) Acetone (b) Methanol
(c) Ethanol (d) Butanol **Ans. c**

182. Monoclonal antibodies are produced by
(a) Somatic cells
(b) Sarcoma cells
(c) Germ cells
(d) Hybrid of carcinogen cell and bone cell **Ans. d**

183. The lignin present in biomass is not digested, then which control the rate of biogas production

(a) Digestion of cellulose

(b) Digestion of hemicelluloses

(c) Conversion of monomers into organic acids

(d) Production of CO_2 **Ans. a**

184. A byproduct of brewing industry is food yeast or vitaminised food. It contains

(a) Fats (b) 40-50% proteins

(c) Vitamins of B complex (d) All of these **Ans. c**

185. More than 70% eubacteriales antibiotics are obtained from

(a) *Polymyxa* (b) *Streptomyces*

(c) *Bacillus* (d) *Pseudomonas* **Ans. b**

186. Processed cheese was formerly prepared with the help of

(a) Rennet from calf's stomach

(b) Ficin obtained from sap of fig trees

(c) Rennet obtained from stomach of sheep and goats

(d) Both (b) and (c) **Ans. d**

187. Streptokinase (TPA) help in

(a) Increasing O.P. of cell (b) Clearing blood clots

(c) Increasing plasma (d) Dissolving tissues **Ans. b**

188. Who isolated rennet, a cheese producing enzyme from calf's stomach

(a) Waksman (b) Smith and Norths

(c) Hensen (d) Fleming **Ans. c**

189. Pulse polio immunization is being carried out with the aim to

(a) Contaion polio

(b) Treat polio

(c) Eradicate polio

(d) Prevent deformities due to polio **Ans. c**

190. Conversion of cortexolone steroid into prednisolone drug involves a microbe mediated process called

(a) Deoxygenation (b) Hydroxylation

(c) Dehydrogenation (d) Hydrogenation **Ans. b**

191. Genetically engineered human insulin is synthesized with the help of

(a) *Pseudomonas* (b) *Rabbit*

(c) *Rhizopus* (d) *E. coli* **Ans. d**

192. Shikonin is a

(a) Steroid produced by biotechnology

(b) Enzyme produced in tissue culture

(c) Vitamin produced by genetic engineering

(d) A red dye produced in tissue culture for using in cosmetics

Ans. d

193. It is beneficial to have insulin by biotechnology, because

(a) It can be produced in mass (b) It is non allergic

(c) It is less expensive (d) All of the above **Ans. d**

194. Human insulin has 51 amino acids in two chains, mark the correct statement

(a) A chain has 30 and B has 21 amino acids

(b) A chain has 21 and B has 20 amino acids

(c) A chain has 21 and B has 30 amino acids

(d) A chain has 20 and B has 31 amino acids **Ans. c**

195. Tissue culture technique has been biotechnologically successful in producting

(a) Beverages (b) Insulin

(c) Cheese (d) Shikonin **Ans. d**

196. Dextrins are plasma substitutes and used during blood transfusion. These are formed by polymerization of simple sugars through the agency of

(a) *Leuconostoc* (b) *Streptococcus*

(c) *Rhizopus* (d) *Aspergillus* **Ans. a**

197. The phenomenon of antibiotics (secretion of one microbe inhibits the growth of other microbe) was discovered by

(a) Babes (b) Fleming

(c) Vuillenim (d) Pasteur **Ans. a**

198. Vitamin B_{12} is cobalt containing vitamin produced through fermentation by

(a) *Pseudomonas* (b) *Propionibacterium*

(c) Both (a) and (b) (d) None of the above **Ans. c**

199. Concept of antibiosis between two organisms was proposed by
(a) Waksman (b) Vuillenin
(c) Babes (d) Fleming **Ans. b**

200. An Indian born molecular biologist developed a super strain of *Pseudomonas* bacteria which is capable of eating up oil spills. The scientist is
(a) Gobind ji (b) H.G. Khorana
(c) A.M. Chakrborty (d) N. Karmaker **Ans. c**

201. Which vitamin was first to be produced by using biotechnology
(a) Vitamin B_2 (b) Vitamin B_1
(c) Vitamin C (d) Vitamin A **Ans. c**

202. First vaccine for small pox was developed by
(a) Jenner (b) R. Koch
(c) Waksman (d) Pasteur **Ans. a**

203. Modified antibiotics are obtained through
(a) Genetic engineering (b) Vaccination
(c) Aminocentosis (d) Ultrafilteration **Ans. a**

204. Who among the following established the scientific basis of vaccination
(a) L. Pasteur (b) E. Jenner
(c) Von Behring (d) G. Kohlar **Ans. a**

205. Which one of the following pairs is correctly matched
(a) *Chrysanthemum* - Natural pyrethroid
(b) *Leucaena* - Dye
(c) *Acacia* - Essential oil
(d) *Cassuarina* - Fodder **Ans. a**

206. Superbug a biotechnological feat is useful in the
(a) Espionage (b) Biodegradation of plastics
(c) Treatment of cancer (d) Cleaning oil spillage **Ans. d**

207. Which one of the following pairs serve as biofertilizer
(a) *Aspergillus* and *Actinomycetes*
(b) *Frankia* and *Nostoc*
(c) *Albugo* and *Nostoc*
(d) *Pseudomonas* and *E. coli* **Ans. b**

208. *Mycorrhizae* fungi mobilize nutrients from soil to roots of plants. The nutrients in question is

(a) Organic acid (b) Phosphorus

(c) Potassium (d) Manganese **Ans. b**

209. A recently discovered anticancer drug is obtained from

(a) Taxus (b) Tagetes

(c) Tamarix (d) Thea **Ans. a**

210. Which one of the following sets of plant have shown promise as potential hydrocarbon (petroleum) sources

(a) *Simmondsia chinesis, Euphorbia abyssinica, E. terucali*

(b) *Hardwickia pinnata, Capaifera longsofii, Euphorbia resinfera*

(c) *Capaifera multiyuga, Euphorbia lathyrus, Calotropis procera*

(d) *Sesamum indicum, Linum usitatissimum, Arachis hypogeal*

Ans. c

211. For energy plantation, selection of suitable species is the most crucial aspect. Consider the following statements in this regard. The species must have (a) Good cropping ability (b) High calorific value of wood (c) Economic importance (d) Capacity for early flowering. Of these statements

(a) (a) and (b) are correct (b) (a) and (c) are correct

(c) (c) and (d) are correct (d) (a), (b), (c) and (d) are correct

Ans. a

212. Which one of the following is a correct statement about biogas

(a) Biogas contains 10-90% of CO_2 and traces of H_2, N_2 and H_2S

(b) Biogas is formed by facultative anaerobic microorganisms

(c) Methane is produced by methanogenic bacterial in 3 steps

(d) Generally cellulose digestion is the rate limiting process in biogas production **Ans. d**

213. Gobar gas contains mainly

(a) CH_4+ CO_2 (b) CH_4+ O_2

(c) CO_2+ H_2 (d) CO_2+ SO_2 **Ans. a**

214. Margarine is prepared from

(a) Ground nut (b) Coconut

(c) Almond (d) Cashewnut **Ans. b**

215. Youghurt is produced by

(a) *Lactobacillus bulgaricus* only

(b) *Lactobacillus acidophilus*

(c) *Streptococcus themophilus* only

(d) Both *Lactobacillus bulgaricus* and *Streptococcus themophilus*

Ans. d

216. Cheese are usually classified on the basis of

(a) Texture (b) Flavor

(c) Colour (d) All of the above **Ans. d**

217. The prerequisites for biotechnological production of antibiotics is

(a) To search an antibiotic producing microorganism

(b) To isolate the antibiotic gene

(c) To join antibiotic with *E. coli* plasmid

(d) All of the above **Ans. d**

218. The main technique involved in agricultural biotechnology is called

(a) Tissue culture (b) Transformation

(c) Plant breeding (d) DNA replication **Ans. a**

219. The new strain of bacteria produced by biotechnology in alcohol industry is

(a) *Escherichia coli* (b) *Saccharomyces cerevisiae*

(c) *Bacillus subtilis* (d) *Pseudomonas putida* **Ans. d**

220. Important objective of biotechnology in agriculture section is

(a) To produce pest resistant varieties of plant

(b) To increase the nitrogen content

(c) To decrease the seed number

(d) To increase the plant weight **Ans. a**

221. Yeast *Saccharomyces cervisiae* is used in the industrial production of

(a) Butanol (b) Citric acid

(c) Tetracycline (d) Ethanol **Ans. d**

222. A bioenergy source obtained by fermentation to supplement fossil fuel petrol is

(a) Methane (b) Diesel

(c) Ethanol (d) Kerosene **Ans. c**

223. Most of the petrocrops belong to family
(a) Leguminocae (b) Euphorbiaceae
(c) Rutaceae (d) Malvaceae **Ans. b**

224. As a result of pyrolysis, which one of the following is produced
(a) Alcohol (b) Charcoal
(c) Charcoal and gas (d) Charcoal, gas and oil **Ans. d**

225. Who is credited with identifying petrocrops
(a) Swaminathan (b) Calvin
(c) Krebs (d) Borlaug **Ans. b**

226. Credit for bringing green revolution to India goes to
(a) B.P. Pal (b) Norman Borlaug
(c) M.S. Swaminathan (d) K.C. Mehta **Ans. c**

227. Vaccines are
(a) Curative medicines
(b) Monoclonal antibodies
(c) Treated bacteria or viruses or one of the proteins
(d) MHC (major histocompatibility complex proteins) **Ans. c**

228. Penicillin was discovered by
(a) Walkman (b) Dubois
(c) Robert Koch (d) A. Fleming **Ans. d**

229. Yeast is used in preparation of
(a) Acetic acid (b) Cheese
(c) Curd (d) Ethyl alcohol **Ans. d**

230. Enzymes are basically
(a) Vitamins (b) Fat
(c) Protein (d) Carbon **Ans. c**

231. One bacterium which has found extensive use in genetic engineering work in plants is
(a) *Agrobacterium tumefaciens* (b) *Bacillus coagulens*
(c) *Clostridium septicum* (d) *Xanthomonas citri* **Ans. a**

232. Industrial production of ethanol from starch is brought about by a certain species of
(a) *Sacchromyces* (b) *Penicillium*
(c) *Azotobacter* (d) *Lactobacillus* **Ans. a**

233. Respiration is

(a) Breaking down of complex organic substances into simple substances

(b) Transformation of potential energy into kinetic energy

(c) Libration of energy

(d) All of these **Ans. d**

234. Respiration is regarded as

(a) Anabolic process (b) Catabolic process

(c) Reduction process (d) Synthetic process **Ans. b**

235. Kreb's cycle takes place in

(a) Mitochondria (b) Chloroplast

(c) Ribosome (d) Endoplasmic reticulum

Ans. a

236. Respiration takes place

(a) In green parts of the plant only

(b) In all the living cells of the plants

(c) In living and dead cells of the plants

(d) In those parts of the plant which are above the soil **Ans. b**

237. Respiration and photosynthesis both require

(a) Glucose (b) Cytochromes

(c) Chlorophyll (d) Sun light **Ans. b**

238. Different steps in respiration are controlled by

(a) Enzymes (b) Sugars

(c) Auxins (d) Kinins **Ans. a**

239. The universal hydrogen acceptor is

(a) COA (b) ATP

(c) FMN (d) NAD (coenzyme) **Ans. d**

240. During complete metabolism of glucose the number of ATP formed from corresponding number of ADP molecules, OR the net gain of energy from one gram of glucose during aerobic respiration is

(a) 2 ATP (b) 36 ATP

(c) 38 ATP (d) 16 ATP **Ans. b**

241. Synthesis of ATP in mitochondria requires
(a) NADP (b) O_2
(c) FMN (d) Pyruvic acid **Ans. b**

242. Largest amount of phosphate bond energy is produced in the process of respiration during
(a) Anaerobic respiration (b) Tricarboxylic acid cycle
(c) Glycolysis (d) None of these **Ans. b**

243. The cycle in which the pyruvic acid is broken down in presence of oxygen is known as
(a) Anaerobic respiration
(b) Glycolysis
(c) Kreb's cycle or tricarboxylic acid cycle
(d) None of these **Ans. c**

244. In hexose monophosphate shunt, the number of CO_2 molecule evolved is
(a) Much lesser than glycolysis (b) Same as in glycolysis
(c) Less than glycolysis (d) More than glycolysis **Ans. d**

245. Evolution of CO_2 is more than in take of oxygen when
(a) Fats are respired (b) Glucose is respired
(c) Sucrose is respired (d) Organic acids are respired
Ans. d

246. Alternate name for Kreb's cycle is
(a) EMP (b) Glyoxalate cycle
(c) Citric acid cycle (d) Glycolate cycle **Ans. c**

247. Pentose phosphate pathway an alternative pathway of respiration was elucidated by
(a) Blackman (b) Horckar
(c) Dickens (d) Warburg **Ans. d**

248. In an organism utilizing carbohydrates as its source of energy anaerobically the R.Q. is likely to be
(a) 1.0 (b) 0.9
(c) 0.7 (d) Infinity **Ans. d**

249. To a living organism which of the following has the greater amount of available energy per molecule

(a) CO_2 (b) H_2O

(c) ATP (d) ADP **Ans. c**

250. Alcohol is produced during the process of

(a) Photosynthesis (b) Aerobic respiration

(c) Anaerobic respiration (d) None of these **Ans. c**

251. Energy currency or energy reservoir of the cells is

(a) AMP (b) RNA

(c) DNA (d) ATP **Ans. d**

252. ATP synthesis occurs on the

(a) Outer membrane of mitochondrion

(b) Inner membrane of mitochondrion

(c) Matrix

(d) None of these **Ans. b**

253. The reaction of Kreb's cycle takes place

(a) In ER

(b) In the cytoplasm

(c) On the surface of mitochondrion

(d) In the matrix of mitochondrion **Ans. d**

254. Mitochondria are the site of

(a) Phosphorylation (b) Photolysis

(c) Oxidative phosphorylation (d) Starch synthesis **Ans. c**

255. The site of Kreb's cycle is or Kreb's cycle takes place in

(a) Lysosomes (b) Vasicles of ER

(c) Mitochondrial matrix (d) Dictyosomes **Ans. c**

256. Instantaneous source of energy is

(a) Protein (b) Fats

(c) Amino acid (d) Glucose **Ans. d**

257. Kreb's cycle with the formation of a six carbon compound by reaction between

(a) Fumaric acid and pyruvic acid

(b) Oxaloacetic acid and acetyl CoA
(c) Malic acid and acetyl CoA
(d) Succinic acid and pyruvic acid **Ans. b**

258. The number of molecules of pyruvic acid formed from one molecule of glucose at the end of glycolysis is
(a) 2 (b) 3
(c) 4 (d) 1 **Ans. a**

259. In germinating castor seeds the RQ is
(a) Zero (b) Less than one
(c) One (d) More than one **Ans. b**

260. The other name of glycolysis is
(a) TCA pathway (b) HMS pathway
(c) Carbon pathway (d) EMP pathway **Ans. d**

261. Glucose is converted into pyruvic acid in
(a) HMS pathway (b) Kreb cycle
(c) Glycolysis (d) TCA pathway **Ans. c**

262. The enzyme which converts glucose to glucose 6-phosphate is
(a) Hexokinase (b) Glucose synthetase
(c) Glucose 6-phosphate (d) Phosphorylase **Ans. a**

263. Which is also formed alongwith ATP in glycolysis?
(a) $NADPH_2$ (b) $NADH_2$
(c) FAD (d) $FADH_2$ **Ans. b**

264. RQ for glucose is
(a) 1 (b) 2
(c) 0.05 (d) 0.5 **Ans. a**

265. Pyruvate dehydrogenase is used in converting
(a) Glucose to pyruvic acid (pyruvate)
(b) Pyruvate to glucose
(c) Pyruvate to acetyl CoA
(d) Pyruvic acid to lactic acid **Ans. c**

266. During hydrolysis from one ATP molecules the energy given by one molecule of ATP is

(a) 721 k.cal (b) 7600 k.cal

(c) 1000 k.cal (d) 7.3 k.cal **Ans. d**

267. An indispensable role in energy metabolism is played by

(a) Phosphorus (b) Calcium

(c) Potassium (d) Sodium **Ans. a**

268. The respiration of young maturing seeds is quite high but as water content decreases during further maturation respiration

(a) Decrease steadily (b) Increases steadily

(c) Stops completely (d) Remains high **Ans. a**

269. Respiration is an

(a) Exothermic process (b) Endothermic process

(c) Anabolic process (d) Endergonic process **Ans. a**

270. Fermentation is an

(a) Incomplete oxidation of carbohydrate

(b) Complete oxidation of carbohydrate

(c) Anaerobic respiration

(d) None of these **Ans. a**

271. Conversion of pyruvic acid into ethyl alcohol is facilitated by the enzyme

(a) Phosphotase

(b) Carboxylase

(c) Dehydrogenase

(d) Carboxylase and dehydrogenase **Ans. d**

272. In fermentation

(a) Energy is liberated

(b) Energy is consumed

(c) Energy is stored

(d) Energy is neither consumed nor librated **Ans. a**

273. The end product of fermentation is

(a) Pyruvic acid and H_2O (b) H_2O and CO_2

(c) C_2H_5OH and CO_2 (d) CO_2 **Ans. c**

274. Number of ATP molecules formed during aerobic respiration due to break down of glucose molecule is

(a) 18 (b) 28

(c) 38 (d) 4 **Ans. c**

275. In Opuntia in night respiratory quotient (RQ) will be

(a) Zero (b) One

(c) Less than one (d) More than one **Ans. a**

276. During respiration pyruvic acid is formed by

(a) TCA cycle (b) Kreb's cycle

(c) Glycolysis (d) None of these **Ans. c**

277. Which of the following processes is used in the conversion of pyruvate to acetyl CoA

(a) Oxidative phosphorylation (b) Oxidative dehydration

(c) Oxidative decarboxylation (d) Oxidative dehydrogenation

Ans. c

278. Pyruvic acid before entering the Kreb's cycle is changed into

(a) Citric acid (b) Succinic acid

(c) Acetyl CoA (d) Malic acid **Ans. c**

279. If the respirable material in respiroscope is some organic acid, what will happen to the level of sodium chloride ?

(a) It will rise (b) It will fall

(c) It will remain as such (d) It will show variation **Ans. b**

280. If CO_2 is given off in respiration why does the amount of CO_2 in the atmosphere remains relatively constant?

(a) CO_2 is converted in photosynthesis to carbohydrates

(b) CO_2 is split up during photosynthesis

(c) CO_2 is a buffer

(d) CO_2 forms carbonate rocks **Ans. a**

281. Hexose monophosphate shunt is

(a) A process by which new products are synthesized

(b) Pentose phosphate pathway

(c) Conversion of glucose into pyruvic acid

(d) The sum of all chemical transformations **Ans. b**

282. High energy compounds

(a) Are produced when ATP loses two of its phosphate groups

(b) Refer to sugar the oxidation of which release large amounts of energy during respiration

(c) Are produced in respiration only

(d) Are compounds which link with exergonic and endergonic process

Ans. d

283. The principal molecules that stores energy is

(a) DPM (b) DNA

(c) ATP (d) Protein molecules **Ans. c**

284. When ATP is converted into ADP it releases

(a) Enzymes (b) Energy

(c) Hormones (d) Secretion **Ans. b**

285. Kreb's cycle involves or is the formation of

(a) ATP from ADP (b) Energy from pyruvic acid

(c) Pyruvic acid from glucose (d) Lactic acid from glucose

Ans. b

286. ATP formation occurs when there is change from or ATP is produced during the reaction

(a) Pyruvic acid → Acetyl CoA

(b) Glucose → Glucose 6-phosphate

(c) Fructose 6-phosphate → Fructose 1-6-diphosphate

(d) Glucose 6-phosphate → Fructose 6-phosphate **Ans. b**

287. R.Q. is less than 1, when

(a) Sucrose is oxidized (b) Fats are oxidized

(c) Organic acids are oxidized (d) Glucose is oxidized **Ans. b**

288. Oxidative phosphorylation occurs in

(a) Chloroplast (b) Feroxysome

(c) Mitochondria (d) Centrioles **Ans. c**

289. The formation of Acetyl-CoA from pyruvic acid is the result of its

(a) Dehydration (b) Dephosphorylation

(c) Reduction (d) Oxidative decarboxylation

Ans. d

290. Besides the net gain of 2 ATP molecules in glycolysis which other molecules are simultaneously formed?

(a) $NADH_2$ (b) $FADH_2$
(c) $NADP_2H_2$ (d) $FAMH_2$ **Ans. a**

291. Complete oxidation of 1 gram mole of glucose gives rise to

(a) 6,860 cals (b) 68,600 cals
(c) 6,86,000 cals (d) 68,60,000 cals **Ans. c**

292. In presence of cyanides, azides and carbon mono-oxides rate of respiration

(a) Increase (b) Decrease
(c) Remain of same (d) None of these **Ans. b**

293. Which of the following is the product of phosphorylation?

(a) DPGA (b) PGA
(c) Pyruvic acid (d) Fructose 1-6 diphosphate
Ans. d

294. In how many steps is CO_2 released in aerobic respiration?

(a) One (b) Three
(c) Six (d) Twenty **Ans. b**

295. In respiration pyruvic acid is

(a) A result of protein break down
(b) Formed only when oxygen is available
(c) One of the product of Kreb's cycle
(d) Broken down into a two carbon fragment and carbon dioxide
Ans. d

296. Which among the following is the most appropriate reason for storing green coloured apples at low temperature?

(a) The rate of respiration is reduced
(b) The rate of photosynthesis and respiration are reduced
(c) Rate of photosynthesis is reduced
(d) Respiration and photosynthesis are completely inhibited **Ans. a**

297. Organelles which are regarded as power house of the cell and in which oxidative reactions of respiratory process takes place are

(a) Mitochondria (b) Endoplasmic reticulum
(c) Ribosome (d) Chloroplast **Ans. a**

298. In succulents respiratory quotient is always less than one because of
(a) Incomplete reduction (b) Incomplete oxidation
(c) Complete oxidation (d) Complete reduction **Ans. b**

299. The end product of anaerobic catabolism of carbohydrates or end product of glycolysis or EMP pathway
(a) Starch is converted into glucose
(b) Glucose is converted into pyruvic acid
(c) Proteins is converted into glucose
(d) Glucose is converted into fructose **Ans. b**

300. In the process of respiration in plants 180 gms of sugar plus 192 gms of oxygen produces
(a) 528 gms of CO_2, 216 gms of water and 1,348 cals of energy
(b) 264 gms of CO_2, 198 gms of water and 674 cals of energy
(c) 132 gms of CO_2, 54 gms of water and 337 cals of energy
(d) Large amount of CO_2, not water and not energy **Ans. b**

301. In aerobic respiration
(a) Glucose can not be used directly
(b) Glucose is a preferred substrate
(c) Glucose is not be used directly
(d) Any material which burns in air may be used as a substrate
Ans. b

302. The end product of fermentation when sugars are used as raw materials are
(a) CO_2 and H_2O (b) C_2H_5OH and CO_2
(c) H_2O and Acetic acid (d) Alcohol **Ans. b**

303. The end product of aerobic respiration in plants are
(a) H_2O and energy (b) CO_2, H_2O and energy
(c) Starch and O_2 (d) Sugar and O_2 **Ans. b**

304. The incomplete break down of sugars in anaerobic respiration results in the formation of
(a) Fructose and water (b) Glucose and carbon dioxide
(c) Alcohol and carbon dioxide (d) Water and carbon dioxide
Ans. c

305. Which statement is wrong

(a) All angiosperms are with seed development but all seed plants are not angiosperms

(b) Stomatal opening is influenced by many factors and potassium

(c) The factors which effect the respiration influence the photosynthesis but reverse is not possible

(d) All enzymes are protein but all proteins are not enzymes

Ans. c

306. After respiration the conversion of energy is mainly in the form of

(a) ADP (b) AMP

(c) Glucose (d) ATP **Ans. d**

307. Link between glycolysis and Kreb's cycle is, OR which of the following forms is the connecting link between glycolysis and citric acid and α-oxidation of fatty acids

(a) Oxaloacetic acid (b) Succinic acid

(c) Acetyl CoA (d) Citric acid **Ans. c**

308. Common immediate source of energy in cellular activity is

(a) ATP (b) NAD

(c) RNA (d) DNA **Ans. a**

309. ATP is or stand for a/an

(a) Hormone

(b) Protein

(c) Molecule which contain high energy phosphate bonds

(d) Enzyme which brings about oxidation **Ans. c**

310. Oxidation of succinate of formulate in the Kreb's cycle is due to

(a) Addition of oxygen to it (b) Removal of hydrogen from it

(c) Loss of electron from it (d) None of these **Ans. d**

311. R.Q. (Respiratory Quotient) is represented by

(a) N/C (b) C/N

(c) O_2/CO_2 (d) CO_2/O_2 **Ans. d**

312. The most energy in the cell is liberated by oxidation of carbohydrates when

(a) Pyruvic acid is converted into acetyl CoA

(b) Pyruvic acid is converted into CO_2 and H_2O

(c) Sugar is converted into pyruvic acid
(d) Glucose is converted into pyruvic acid **Ans. d**

313. Total ATP production during EMP pathway is
(a) 24 ATP molecules (b) 38 ATP molecules
(c) 4 ATP molecules (d) 8 ATP molecules **Ans. c**

314. Food is converted into energy in
(a) Nucleolus (b) Chloroplast
(c) Mitochondria (d) Nucleus **Ans. c**

315. Enzymes taking part in glycolysis are present in or glycolysis takes place in
(a) Cytoplasm (b) Mitochondria
(c) Vacuole (d) None of these **Ans. a**

316. Oxidative phosphorylation and photophosphorylation both require the electron carriers
(a) Oxygen (b) Carbon dioxides
(c) Water (d) Cytochromes **Ans. d**

317. Importance of Kreb's cycle is
(a) Production of vitamins
(b) Production of amino acids
(c) To encourage glycolysis
(d) Production of ATP molecules through oxidative phosphorylation **Ans. d**

318. Respiratory quotients refers to
(a) The number of glucose molecules breaking up and produces energy in one hour
(b) The ratio between the amount CO_2 produced to the amount of O_2 absorbed
(c) Amount of ATP produced to the amount lost during respiration
(d) None of these **Ans. b**

319. The correct sequence in Kreb's cycle is
(a) Isocitric acid $\rightarrow$ α - keto gluteric acid $\rightarrow$ oxalo succinic acid
(b) Isocitric acid $\rightarrow$ oxalosuccinic acid $\rightarrow$ α - keto gluteric acid

(c) Oxalo succinic acid → citric acid → α - keto gluteric acid
(d) α - keto gluteric acid → isocitric acid → oxalo succinic acid
Ans. b

320. The equation $\Delta G = \Delta H - T\Delta S$ depends on
(a) Constant temperature and pressure
(b) Constant temperature and volume
(c) Constant pressure and volume
(d) None of these **Ans. a**

321. Some autotropic bacteria utilize which of the following reaction?
(a) $Fe^{+++} \rightarrow Fe^{++}$ (b) $Fe^{-+} \rightarrow Fe^{+++}$
(c) $Cu \rightarrow Cu^{++}$ (d) $Mg^{++} \rightarrow Mg$ **Ans. b**

322. Lipmann introduced the symbol Ⓟ indicating
(a) Phosphate bond (b) Low energy phosphate bond
(c) High energy phosphate bond (d) Energy phosphate bond
Ans. c

323. The number of high energy phosphate contained by ATP is
(a) 1 (b) 2
(c) 3 (d) None of the above **Ans. b**

324. AMP contains
(a) High energy phosphate bond
(b) Phosphate bond
(c) Energy phosphate bond
(d) Low energy phosphate bond **Ans. d**

325. The free energy is derived from respiratory chain oxidation within
(a) Cytosol (b) Microsomes
(c) Mitochondria (d) Lysosomes **Ans. c**

326. In TCA Cycle at the succinyl thiokinase step there is the direct generation of
(a) One - Ⓟ (b) Two - Ⓟ
(c) Three - Ⓟ (d) All of these **Ans. a**

327. Phosphagens act as storage forms of
(a) Low - energy phosphate (b) High - energy phosphate
(c) Energy phosphate (d) No energy phosphate **Ans. b**

328. Phosphagens maintain ATP concentration in
(a) Muscle (b) Liver
(c) Kidney (d) Heart **Ans. a**

329. When ATP becomes abundant, its concentration can cause
(a) Direct reaction (b) Irreversible reaction
(c) Reverse reaction (d) All of these **Ans. c**

330. The enzyme adenylate kinase is said to be
(a) Glucokinase (b) Pyruvate kinase
(c) Phosphoglycerokinase (d) Myokinase **Ans. d**

331. The activation of long chain fatty acids produces
(a) Inorganic phosphate (b) Organic phosphate
(c) Adenosine diphosphate (d) No phosphate **Ans. a**

332. The activation via the pyrophosphate pathway results in the loss of
(a) One - Ⓟ (b) Two - Ⓟ
(c) Three - Ⓟ (d) None of these **Ans. b**

333. Which of the following take part in phosphorylations in the cell?
(a) UTP (b) GTP
(c) CTP (d) All of these **Ans. d**

334. Adenylate kinase is a specialized
(a) Monophosphate kinase (b) Diphosphate kinase
(c) Triphosphate kinase (d) All of these **Ans. a**

335. Oxidases are conjugated proteins having the prosthetic group
(a) Magnesium (b) Manganese
(c) Copper (d) Iron **Ans. c**

336. Cytochrome oxidase is
(a) a_3 (b) aa_3
(c) a (d) None of these **Ans. b**

337. Cytochrome oxidase is poisoned by
(a) Cyanide (b) Sulphide
(c) Sulphite (d) Sulphate **Ans. a**

338. Phenolase is an enzyme containing
(a) Cobalt (b) Iron
(c) Zinc (d) Copper **Ans. d**

339. Monoamine oxidase oxidizes
(a) Epinephrine (b) Norepinephrine
(c) Glucagon (d) Glutathione **Ans. a**

340. Aerobic dehydrogenases have the prosthetic group
(a) ATP (b) NAD^+
(c) FAD^+ (d) $NADP^+$ **Ans. c**

341. Uricase catalyzes the oxidation of uric acid to
(a) Carbon dioxide (b) Ammonia
(c) Glyoxal (d) Allantoin **Ans. d**

342. Xanthine dehydrogenase converts purine bases to
(a) Uric acid (b) Hypoxanthine
(c) Xanthine (d) Urea **Ans. a**

343. Alcohol dehydrogenase from liver contains
(a) Copper (b) Zinc
(c) Sodium (d) Potassium **Ans. b**

344. NADP – linked dehydrogenases in the extra-mitochondria are found to synthesize
(a) Urea (b) Steroid
(c) Ascorbic acid (d) Niacin **Ans. b**

345. The endoplasmic reticulum cytochrome
(a) b (b) c
(c) a (d) P-450 **Ans. d**

346. Mono-oxygenases are found in the
(a) Mitochondria (b) Microsomes
(c) Nuclei (d) Cytosol **Ans. b**

347. The mitochondrial superoxide dismutase contains
(a) Mg^{++} (b) Zn^{++}
(c) Mn^{++} (d) Co^{++} **Ans. c**

348. When substrates are oxidized through an NAD – linked dehydrogenase, the P : O ratio is
(a) 1 (b) 2
(c) 3 (d) 4 **Ans. c**

349. The antibiotic piericidin A inhibits the site of respiratory chain
(a) I
(b) II
(c) III
(d) All of the above **Ans. a**

350. The uncoupling agent of oxidative phosphorylation
(a) Antimycin
(b) Dicoumarol
(c) Barbiturates
(d) Penicillin **Ans. b**

351. The oxidative phosphorylation in intact mitochondria is completely blocked by
(a) Streptomycin
(b) Gentamycin
(c) Puromycin
(d) Oligomycin **Ans. d**

352. The high-energy compound is
(a) UDPG
(b) ATP
(c) ADP
(d) Arginine phosphate **Ans. a**

353. The R.Q. of mixed diets is about
(a) 0.72
(b) 0.75
(c) 0.80
(d) 0.85 **Ans. d**

354. The BMR is influenced by the factor
(a) Sex
(b) Climate
(c) Habit
(d) All of these **Ans. d**

355. The BMR is subnormal in
(a) Addison's disease
(b) Hyperthyroidism
(c) Hypothyroidism
(d) All of these **Ans. a**

356. The BMR is below normal in
(a) Starvation
(b) Lipoid nephrosis
(c) Both of (a) and (b)
(d) None of these **Ans. c**

357. The BMR is above normal in
(a) Diabetes insipidus
(b) Leukemia
(c) Polycythemia
(d) All of the above **Ans. d**

358. When lard, glucose and protein are combined, the percentage of SDA is less than the sum of their individual SDA by
(a) 20
(b) 22
(c) 24
(d) 25 **Ans. b**

359. When protein and lard are combined, the percentage of SDA is less than the sum of their individual SDA by

(a) 45 (b) 50

(c) 54 (d) 58 **Ans. c**

360. The SDA is decreased more by

(a) Carbohydrate (b) Fat

(c) Protein (d) All of these **Ans. b**

361. When different amino acids are fed, high SDA is found to be produced by

(a) Glycine (b) Alanine

(c) Phenylalanine (d) All of these **Ans. d**

362. The glands of internal secretion have no direct influence upon the SDA of

(a) Protein (b) Fat

(c) Carbohydrate (d) All of these **Ans. a**

363. Thyroidectomy in animals reduces the SDA of

(a) Carbohydrate (b) Fat

(c) Both (a) and (b) (d) None of these **Ans. c**

364. It is unfavourable to heavy muscular work by a diet rich in

(a) Carbohydrate (b) Protein

(c) Fat (d) All of these **Ans. b**

365. The extra heat is accelerated in the performance of work with

(a) Carbohydrate (b) Fat

(c) Both (a) and (b) (d) None of these **Ans. c**

366. When new tissue is formed, the usual SDA is not exerted by

(a) Carbohydrate (b) Fat

(c) Vitamins (d) Protein **Ans. d**

367. A man of 70 kg weight and 180 cm height with a surface area of 1.8 square meter, aged 30 and a BMR of 40 will have a basal metabolism of

(a) 1750 calories (b) 1800 calories

(c) 1875 calories (d) 1925 calories **Ans. b**

368. Lumberman entails the expenditure of

(a) 8000 calories (b) 9000 calories
(c) 10000 calories (d) 12000 calories **Ans. a**

369. In long bicycle races in 24 hours, there is the expenditure of

(a) 8000 calories (b) 9000 calories
(c) 10000 calories (d) 12000 calories **Ans. c**

370. Above the average for the six days, the percentage of Sunday's consumption is

(a) 24 (b) 28
(c) 32 (d) 36 **Ans. d**

371. Women have a basal metabolic rate lower than that of men by

(a) 7% - 10% (b) 8% - 12%
(c) 10% - 15% (d) 12% - 20% **Ans. a**

372. The average supplement for muscular work in calories is

(a) 400 (b) 600
(c) 800 (d) 1000 **Ans. b**

373. The daily average caloric requirement of the adult male is

(a) 2000 (b) 2200
(c) 2500 (d) 3000 **Ans. d**

374. If the NPU (Net Protein Utilization) is low, the requirements for proteins are

(a) Low (b) High
(c) Moderate (d) None of these **Ans. b**

375. The relatively essential amino acid

(a) Glycine (b) Serine
(c) Alanine (d) Histidine **Ans. d**

376. Vegetable proteins are almost associated with a large amount of

(a) Minerals (b) Vitamins
(c) Carbohydrate (d) Fat **Ans. c**

377. Animal proteins are associated with fat but very little carbohydrate except

(a) Milk (b) Peas
(c) Beans (d) Potatoes **Ans. c**

378. The foodstuffs having the "protein-sparing" effect

(a) Carbohydrate (b) Fat

(c) Both (a) and (b) (d) None of these **Ans. a**

379. Protein has a more catalytic function in the form of

(a) Hormones (b) Enzymes

(c) Antigens (d) Antinogens **Ans. b**

380. Proteins differ in "biologic value" depending on their contents of

(a) Non-essential amino acids (b) Semi-essential amino acids

(c) Essential amino acids (d) All of these **Ans. c**

381. Good quality proteins include

(a) Soyabeans (b) Peanuts

(c) Potatoes (d) All of these **Ans. d**

382. High quality proteins include

(a) Eggs (b) Dairy products

(c) Kidney (d) All of these **Ans. d**

383. Proteins are also important sources of

(a) Nitrogen (b) Sulphur

(c) Phosphorus (d) All of these **Ans. d**

384. The protective substances in a diet are

(a) First class proteins (b) Mineral elements

(c) Vitamins (d) All of these **Ans. d**

385. The bulk of the food is reduced by

(a) Carbohydrate (b) Fat

(c) Protein (d) Minerals **Ans. b**

386. The daily consumption of fat in grams is usually over

(a) 50 (b) 60

(c) 80 (d) 90 **Ans. c**

387. Animal fats have a better biologic value because they contain the vitamins

(a) A (b) D

(c) Both (a) and (b) (d) None of these **Ans. c**

388. The major symptoms of essential fatty acid deficiency are

(a) A scaly dermatitis (b) Hair loss

(c) Poor wound healing (d) All of these **Ans. d**

389. In societies where fat is the principal food for energy intake, the population tends to develop

(a) Coronary heart disease (b) Obesity

(c) Cancer of the bowel (d) All of these **Ans. d**

390. Carbohydrate is the cheapest source of food and for producing energy it can be readily

(a) Digested (b) Absorbed

(c) Utilized (d) All of these **Ans. d**

391. A process in which no heat is transferred to and from the system is called

(a) Isothermal (b) Adiabatic

(c) Isochoric (d) Isolated **Ans. b**

392. Internal energy of an ideal gas depends on

(a) Pressure (b) Temperature

(c) Volume (d) None of these **Ans. b**

393. Which of the following describes the first law of thermodynamics?

(a) $q = \Delta E - w$ (b) $\Delta H = q + w$

(c) $\Delta E = \Delta H + P\Delta V$ (d) $\Delta H = q - w$ **Ans. d**

394. The factor which does not influence the heat of reaction is

(a) The physical state of the reactants and products

(b) The temperature of the reaction

(c) The method by which the change is brought about

(d) Conditions such as constant volume or constant pressure

Ans. c

395. For an endothermic reaction

(a) $H_r < H_p$ (b) $H_r < H_p$

(c) $H_r = H_p$ (d) $H_r \geq HP$

Where r is for reactant and p for products. **Ans. b**

396. The enthalpy change for the reaction, $N_2(g)+3H_2(g) \rightarrow 2NH_3(g)$ is –93.7 kJ. Then, the above is reaction is

(a) Endothermic (b) Exothermic

(c) Isothermal (d) Reversible **Ans. b**

397. The relationship between enthalpy (H), internal energy (E), pressure (P) and volume (V) of a system is

(a) E = H+PV (b) H = E+PV

(c) PV = H-E (d) H = E+P+V **Ans. b**

398. Given the following thermochemical equations

$M + \frac{1}{2} O_2 \rightarrow MO + 351.4$ kJ

$X + \frac{1}{2} O \rightarrow XO + 90.8$ kJ

The heat of reaction (Q) for the following reaction, $M + XO \rightarrow MO + X + Q$ is

(a) 442.2 kJ (b) 260.7 kJ

(c) -442.2 kJ (d) 260.6kJ **Ans. d**

399. The enthalpy of formation of $SO_2(g)$ is -296.7 kJ. The energy required for the decomposition of 1 mole of $SO_2(g)$ is

(a) 148.35 kJ (b) 296.7 kJ

(c) 593.4 kJ (d) -148.35 kJ **Ans. b**

400. In which of the following neutralization reactions the enthalpy of neutralization is the highest

(a) HCl and NaOH (b) NH_4OH and HCl

(c) CH_3COOH and NaOH (d) CH_3COOH and NH_4OH

Ans. a

401. At 298 K, the enthalpy of formation of liquid water is -285.9 kJmol^{-1}, and that of water vapours is - 242.8 kJmol^{-1}. The enthalpy of vaporisation of water at 268 K is

(a) 43.1 kJmol^{-1} (b) -43.1 kJmol^{-1}

(c) 242.8 kJmol^{-1} (d) -242.8 kJmol^{-1} **Ans. a**

402. The heat of combustion of carbon is -394 kJ/mol. The heat evolved in the combustion of 6.02×10^{22} atom of carbon is

(a) 3940 kJ (b) 394 kJ

(c) 39.4 kJ (d) 0.394 kJ **Ans. c**

403. The difference between the enthalpies of reaction at constant pressure and constant volume for the reaction

$2C_6H_6(I) + 15\ O_2(g) \rightarrow 12\ CO_2(g) + 6H_2O(I)$ at 25°C in kJ is

(a) -7.43 (b) +3.72

(c) -3.72 (d) +7.43 **Ans. a**

404. The difference in ΔH and ΔE for the combustion of ethane at 27°C is

(a) 8.314 x 300 (-3) $Jmol^{-1}$ (b) 8.314x27 (-3) $Jmol^{-1}$

(c) 8.314 x 300 (-5) $Jmol^{-1}$ (d) Zero **Ans. d**

405. When a liquid boils, there is an increase in

(a) Entropy (b) Heat of vaporization

(c) Kinetic energy (d) Potential energy **Ans. a**

406. The entropy change for a spontaneous process always

(a) Increases (b) Decreases

(c) Keeps changing (d) Remains unchanged **Ans. a**

407. The direction of a chemical reaction is the direction of decreases of

(a) Gibb's free energy (b) Entropy

(c) Enthalpy (d) None **Ans. a**

408. For the process, $H_2O(s)$ (ice) ↔ $H_2O(l)$ (water) at 273 K, which of the following is correct?

(a) $G_{ice} = G_{water} = 0$ (b) $G_{ice} = G_{water} \neq 0$

(c) $G_{ice} > G_{water}$ (d) $G_{ice} < G_{water}$ **Ans. b**

409. The free energy change for a reversible reaction at equilibrium is

(a) Zero (b) Small positive

(c) Small negative (d) Large positive **Ans. a**

410. For a reaction, A→B+C, ΔG = - 30.4 kJ mol^{-1}. The reaction is

(a) Spontaneous (b) Non-spontaneous

(c) Explosive (d) At equilibrium **Ans. a**

411. Choose the correct answer. A thermodynamic state function is a quantity

(a) Used to determine heat changes

(b) Whose value is independent of path

(c) Used to determine pressure volume work

(d) Whose value depends on temperature only **Ans. b**

412. For the process to occur under adiabatic conditions, the correct condition is

(a) $\Delta T = 0$ (b) $\Delta p = 0$

(c) $q = 0$ (d) $w = 0$ **Ans. c**

413. The enthalpies of all elements in their standard states are

(a) Unity (b) Zero

(c) <0 (d) Different for each element

Ans. b

414. $\Delta U^{\theta}$of combustion of methane is -X kJ mol^{-1}. The value of ΔH^{θ} is

(a) $=\Delta U^{\theta}$ (b) $>\Delta U^{\theta}$

(c) $<\Delta U^{\theta}$ (d) =0 **Ans. c**

415. The enthalpy of combustion of methane, graphite and dihydrogen at 298 K are, -890.3 kJ mol^{-1} -393.5 kJ mol^{-1}, and -285.8 kJ mol^{-1} respectively. Enthalpy of formation of CH_4(g) will be

(a) -74.8 kJ mol^{-1} (b) -52.27 kJ mol^{-1}

(c) +74.8 kJ mol^{-1} (d) +52.27 kJ mol^{-1} **Ans. a**

416. A reaction, $A+B \rightarrow C+D + q$ is found to have a positive entropy change. The reaction will be

(a) Possible at high temperature

(b) Possible only at low temperature

(c) Not possible at any temperature

(d) Possible at any temperature **Ans. d**

417. In thermodynamics, a process is called reversible when

(a) Surroundings and system change into each other

(b) There is no boundary between system and surroundings

(c) The surroundings are always in equilibrium with the system

(d) The system changes into surroundings spontaneously **Ans. c**

418. For which one of the following equations is $\Delta_r H°$ is equal to $\Delta_r N°$ for the product?

(a) $N_2(g) + O_3(g) \rightarrow N_2O_3$

(b) $CH_4(g) + 2Cl_2(g) \rightarrow CH_2Cl_2(g) + 2HCl(g)$

(c) $Xe(g) + 2F_2(g) \rightarrow XeF_4(g)$

(d) $2CO(g) + O_2(g) \rightarrow 2CO_2(g)$ **Ans. c**

419. Enthalpy of $CH_4 + ½ O_2 \rightarrow CH_3OH$ is negative. If enthalpy of combustion of CH_4 and CH_3OH are x and y respectively, then which relation is correct?

(a) x > y (b) x < y

(c) x = y (d) x ≠ y **Ans. b**

420. For the reaction $C_3H_8(g) + 5O_2(g) \rightarrow 3CO_2(g) + 4H_2O(I)$ at constant temperature, ΔH - ΔU is

(a) +RT (b) –3RT

(c) +3RT (d) –RT **Ans. b**

421. The enthalpy changes for a reaction does not depend upon

(a) Use of different reactants for the same product

(b) The nature of intermediate reaction steps

(c) The differences in the initial and final temperature of the involved substances

(d) The physical state of the reactants and products **Ans. b**

422. Which of the following reaction defines $\Delta_f H°$?

(a) C (diamond) + $O_2(g) \rightarrow CO_2(g)$

(b) $½H_2(g) + ½ F_2(g) \rightarrow HF(g)$

(c) $N_2(g) + 3H_2(g) \rightarrow 2NH_3(g)$

(d) $CO(g) + ½ O_2(g) \rightarrow CO_2(g)$ **Ans. b**

423. The bond energies of C-C, C=C, H-H and C-H linkages are 350, 600, 400 and 410 kJ per mol respectively. The heat of hydrogenation of ethylene is

(a) -170 kJ mol^{-1} (b) -260 kJ mol^{-1}

(c) -400 kJ mol^{-1} (d) -40 kJ mol^{-1} **Ans. a**

424. One mole of a non-ideal gas undergoes a change of state (2.0 atm. 3.0 L, 95 K) → 4.0 atm, 5.0 L, 245K) with a change in internal energy, ΔU = 30.0L atm. The change in enthalpy (ΔH) of the process in L atm is

(a) 40.0

(b) 42.3

(c) 44.0

(d) Not defined because pressure is not constant **Ans. c**

425. In which of the following entropy is increasing?

(a) Crystallisation of sugar from sugar solution

(b) $N_2(g) + 3H_2(g) - 2NH_3(g)$

(c) Adding naphthalene to benzene

(d) $H_2 + I_2 \rightarrow 2HI$ **Ans. c**

426. For the equilibrium $H_2O(l)$ $H_2O(g)$ at 1 atm and 298 K

(a) Standard free energy change is equal to zero ($\Delta G^0 = 0$)

(b) Free energy change is less than zero ($G° < 0$)

(c) Standard free energy change is less than zero ($\Delta G^0 < 0$)

(d) Standard free energy change is greater than zero ($\Delta G^0 > 0$)

Ans. a

427. If the bond energies of H - H, Br - Br and H - Br are 433, 192 and 364 kJ mol^{-1} respectively, the ΔH^0 for the reaction $H_2(g) + Br_2(g) \rightarrow 2HBr(g)$ is

(a) +103 kJ (b) 261 kJ

(c) -103 kJ (d) -261 kJ **Ans. c**

428. To neutralize completely 20 ml of 0.1 M aqueous solution of phosphorous acid (H_3PO_3), the volume of 0.1 1M aqueous KOH required is

(a) 10 ml (b) 20 ml

(c) 40 ml (d) 60 ml **Ans. c**

429. The enthalpies of combustion of carbon and carbon monoxide are -393.5 and -283 kJ mol^{-1} respectively. The enthalpy of formation of carbon monoxide is

(a) 110.5 kJ mol^{-1} (b) 676.5 kJ mol^{-1}

(c) –676.5 kJ mol^{-1} (d) –110.5 kJ mol^{-1} **Ans. d**

430. Considering entropy (S) as a thermodynamic parameter, the criterion for the spontaneity of any process is

(a) $\Delta S_{system} - \Delta S_{surrounding} > 0$ (b) $\Delta S_{system} > 0$

(c) $\Delta S_{surrounding} > 0$ (d) $\Delta S_{system} + \Delta S_{surrounding} > 0$

Ans. d

431. The enthalpy of vaporization of a liquid is 30 kj mol^{-1} and entropy of vaporization is 75 $Jmol^{-1}$. The boiling point of the liquid at 1 atm is

(a) 250 K (b) 400 K

(c) 450 K (d) 600 K **Ans. b**

432. Two moles of an ideal gas is expanded isothermally and reversibly from 1 L to 10 L at 300 K. The enthalpy change (in kJ) for the process is

(a) 11.4 kJ (b) -11.4 kJ

(c) 0 kJ (d) 4.8 kJ **Ans. b**

433. An ideal gas expands in volume from $1 \times 10^{-3} m^3$ to $1 \times 10^{-2} m^3$ at 300 K against a constant pressure of 1×10^5 Nm^{-2}. The work done is

(a) -900 J (b) -900 kJ

(c) 270 kJ (d) 900 kJ **Ans. a**

434. A cylinder of gas supplied by Bharat Petroleum is assumed to contain 14 kg of butane. If a normal cooking, butane gas in the cylinder last for .... days (ΔH_c of C_4H_{10} = –2658 kJ per mole)

(a) 15 days (b) 32 days

(c) 40 days (d) 50 days **Ans. b**

435. Values of ΔH and ΔS for different reactions are given below:

Reaction	ΔH (kJ mol^{-1})	ΔS (JK^{-1} mol^{-1})
I	+98.0	+14.8
II	-55.5	-84.8
III	+28.3	-17.0
IV	-40.5	+24.6
V	+34.7	0.0

On the basis of these values which one of these will be spontaneous at all temperatures?

(a) Reaction I and V (b) Reaction II

(c) Reaction III (d) Reaction IV **Ans. d**

436. What would be the heat released when an aqueous solution containing 0.5 mol of HNO_3 is mixed with 0.3 mol of OH^- (enthalpy of neutralization is -57.1 kJ)

(a) 28.5 kJ (b) 17.1 kJ

(c) 2.85 kJ (d) 1.7 kJ **Ans. b**

437. ΔH and ΔS for a reaction are +30.558 kJ mol^{-1} and 0.066 kJ mol^{-1} at 1 atm pressure. The temperature at which free energy is equal to zero and the nature of reaction below this temperature are

(a) 483 K, spontaneous (b) 443 K, non-spontaneous

(c) 443 K, spontaneous (d) 463 K, non-spontaneous

Ans. d

438. Five moles of a gas is put through a series of changes as shown graphically in a cyclic process. The processes $A \rightarrow B$, $B \rightarrow C$ and $C \rightarrow A$ respectively are:

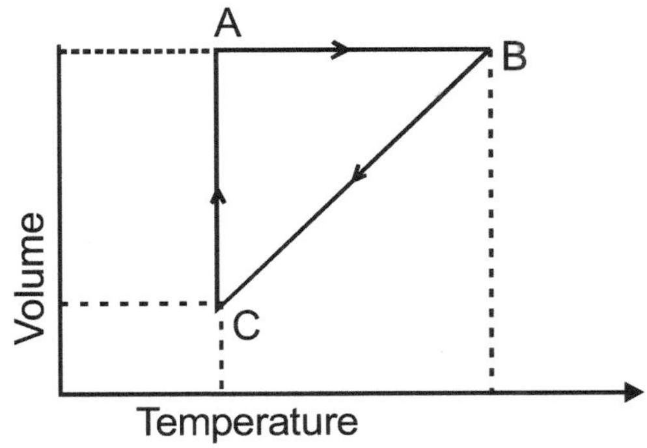

(a) Isochoric, isobaric, isothermal
(b) Isobaric, isochoric, isothermal
(c) Isothermal, isobaric, isochoric
(d) Isochoric, isothermal, isobaric **Ans. a**

439. For a spontaneous reaction, the ΔG, equilibrium constant (K) and E°_{cell} will be respectively

(a) - ve, >1, +ve (b) +ve, >1, -ve
(c) -ve, <1, -ve (d) +ve, <1, +ve **Ans. a**

440. If the bond dissociation energies of XY, X_2 and Y_2 (all diatomic molecules) are in the ratio of 1:1:0.5 and $\Delta_f H$ for the formation of XY is -200 kJ mol^{-1}, the bond dissociation energy of X_2 will be

(a) 100 kJ mol^{-1} (b) 200 kJ mol^{-1}
(c) 800 kJ mol^{-1} (d) 400 kJ mol^{-1} **Ans. c**

441. The absolute enthalpy of neutralization of the reaction $MgO(s) + 2HCl(aq) \rightarrow MgCl_2(aq) + H_2O(l)$ will be

(a) -57.3 kJ mol^{-1} (b) Greater than -57.3 kJ mol^{-1}
(c) Less than -57.3 kJ mol^{-1} (d) 57.3 kJ mol^{-1} **Ans. c**

442. A reaction occurs spontaneously if

(a) $T\Delta S < \Delta H$ and both ΔH and ΔS are +ve
(b) $T\Delta S > \Delta H$ and both ΔH is +ve and ΔS are -ve
(c) $T\Delta S > \Delta H$ and both ΔH and ΔS are −ve
(d) $T\Delta S = \Delta H$ and both ΔH and ΔS are −ve **Ans. c**

443. Which of the following pairs of chemical reaction is certain to result in spontaneous reaction

(a) Exothermic and increasing disorder

(b) Exothermic and decreasing disorder

(c) Endothermic and increasing disorder

(d) Endothermic and decreasing disorder **Ans. a**

444. The enthalpy of hydrogenation of cyclohexane is -119.5 kJ mol^{-1}. If resonance energy of benzene is -150.4 kJ mol-1, its enthalpy of hydrogenation would be

(a) -208.1 kJ mol^{-1} (b) -269.9 kJ mol^{-1}

(c) -358.5 kJ mol^{-1} (d) -508.9 kJ mol^{-1} **Ans. a**

445. The enthalpy and entropy change for the reaction $Br(l) + Cl_2(g) \rightarrow 2BrCl(g)$ 30 $kJmol^{-1}$ respectively will be in equilibrium is

(a) 273 K (b) 450 K

(c) 300 K (d) 285.7 K **Ans. c**

446. Assume each reaction is carried out in an open container. For which reaction will $\Delta H = \Delta U$?

(a) $C(s) + 2H_2O(g) \rightarrow 2H_2(g) + CO_2(g)$

(b) $PCl_5(g) \rightarrow PCl_3(g) + Cl_2(g)$

(c) $2CO(g) + O_2(g) \rightarrow 2CO_2(g)$

(d) $H_2(g) + Br_2(g) \rightarrow 2HBr(g)$ **Ans. d**

447. The reaction A to B is not feasible but on charging entropy through a series of steps

$A \rightarrow C \rightarrow D \rightarrow B$

$\Delta S\ (A \rightarrow C) = 50$ ev,

$\Delta S\ (C \rightarrow D) = 30$ ev,

$\Delta S\ (B \rightarrow D) = 20$ ev.

The entropy change for $A \rightarrow B$ would be

(a) 100 ev (b) 60 ev

(c) - 60 ev (d) - 100 ev **Ans. b**

448. An ideal gas is allowed to expand both reversibly and irreversibly in an isolated system. If T_i is the initial temperature and T_f is the final temperature which of the following is correct?

(a) $T_f > T_i$ for reversible process but $T_f = T_i$ for irreversible process

(b) $(T_f)_{rev} = (T_f)_{irrev}$

(c) $T_f = T_i$ for both reversible and irreversible processes

(d) $(T_f)_{irrev} = (T_f)_{rev}$ **Ans. d**

449. The standard enthalpy of formation, $\Delta_f H°$ at 298 K for methane, $CH_4(g)$ is -74.8 kJ mol^{-1}. The additional information required to determine the average energy for C-H bond formation would be

(a) Latent heat of vaporisation of methane

(b) The first four ionisation energies of carbon and electron gain enthalpy of hydrogen

(c) The dissociation energy of hydrogen molecules, H_2

(d) The dissociation energy of H_2 and enthalpy of sublimation of carbon **Ans. d**

450. The enthalpy change for the following processes are listed below

$Cl_2(g) = 2Cl(g)$: 242.3 kJ mol^{-1}

$I_2(s) = 2I(g)$: 151.0 kJ mol^{-1}

$ICl(g) = I(g) + Cl(g)$: 211.3 kJ mol^{-1}

$I_2(s) = I_2(g)$: 62.76 kJ mol^{-1}

Given that the standard states for iodine and chlorine are $I_2(s)$ and $Cl_2(g)$, the standard enthalpy of formation for ICl(g) is

(a) -16.8 kJ mol^{-1} (b) +16.8 kJ mol^{-1}

(c) +244.8 kJ mol^{-1} (d) -14.6 kJ mol^{-1} **Ans. b**

451. $(\Delta H - \Delta U)$ for the formation of carbon monoxide (CO) from its elements at 298 K is (R = 8.314 JK^{-1} mol^{-1})

(a) 1238.78 J mol^{-1} (b) -2477.57 J mol^{-1}

(c) 2477.57 J mol^{-1} (d) -1238.78 J mol^{-1} **Ans. a**

452. The decomposition of limestone $CaCO_3(g) \leftrightarrow CaO(s) + CO_2(g)$ is non-spontaneous at 298 K. The $\Delta H°$ and $\Delta S°$ values for the reaction are 176.0 kJ and 160 JK^{-1} respectively. At what temperature the decomposition becomes spontaneous?

(a) At 1000 K (b) Below 500°C

(c) Above 827°C (d) At 500°C **Ans. c**

453. Heat of combustion of carbon monoxide at constant volume and at 17°C is -67, 710 calories. The heat of combustion at constant pressure is

(a) -68,000 cal (b) -67,800 cal

(c) -67,050 cal (d) +68,500 cal **Ans. a**

454. For the process: $H_2O(l)$ (1 bar, 373 K) → $H_2O(g)$ (1 bar, 373 K), the correct set of thermodynamic parameters is

(a) $\Delta G = 0, \Delta S = +ve$ (b) $\Delta G = 0, \Delta S = -ve$

(c) $\Delta G = +ve, \Delta S = 0$ (d) $\Delta G = -ve, \Delta S = +ve$ **Ans. a**

455. Assuming the water vapour is an ideal gas, the internal energy change (ΔU) when 1 mole of water is vapourised at 1 bar pressure and 100°C (given molar enthalpy of vaporisation of water at 1 bar and 373 K = 41 kJ mol^{-1} and R = 8.3 $Jmol^{-1}$ K^{-1}) will be

(a) 3.7904 kJ mol^{-1} (b) 37.904 kJ mol^{-1}

(c) 41.00 kJ mol^{-1} (d) 4.100 kJ mol^{-1} **Ans. b**

456. Identify the correct statement regarding a spontaneous process

(a) Endothermic processes are never spontaneous

(b) Exothermic processes are always spontaneous

(c) Lowering of energy in the reaction process is the only criterion for spontaneity

(d) For a spontaneous process in an isolated system, the change in entropy is positive **Ans. d**

457. In conversion of limestone to lime, $CaCO_3(s) \rightarrow CaO(s) + CO_2(g)$ the values of ΔH^0 and ΔS^0 are + 179.1 kJ mol^{-1} and 160.2 J/K respectively at 298 K and 1 bar. Assuming that ΔH^0 and ΔS^0 do not change with temperature, the temperature above which conversion of limestone to lime will be spontaneous is

(a) 1200 K (b) 845 K

(c) 1118 K (d) 1008 K **Ans. c**

458. Consider the following reactions

(i) $H^+(aq) + OH^-(aq) \rightarrow H_2O(l)$ $\Delta H = -x_1$ kJ mol^{-1}

(ii) $H_2(g) + \frac{1}{2} O_2(g) \rightarrow H_2O(l)$ $\Delta H = -x_2$ kJ mol^{-1}

(iii) $CO_2(g) + H_2(g) \rightarrow CO(g) + H_2O(l)$ $\Delta H = -x_3$ kJ mol^{-1}

(iv) $C_2H_2(g) + 5/2\, O_2(g) \rightarrow 2CO_2(g) + H_2O(l)$ $\Delta H = -x_4$ kJ mol^{-1}

Enthalpy of formation of $H_2O(l)$ is

(a) $+x_1$ kJ mol^{-1} (b) $-x_2$ kJ mol^{-1}

(c) $+x_3$ kJ mol^{-1} (d) $-x_4$ kJ mol^{-1} **Ans. b**

459. Given that the bond energies of H - H and Cl - Cl are 430 kJ mol^{-1} and 240 kJ mol^{-1} respectively and $\Delta_f H$ for HCl is - 90 kJ mol^{-1}, bond enthalpy of HCl is

(a) 245 kJ mol^{-1} (b) 290 kJ mol^{-1}

(c) 380 kJ mol^{-1} (d) 425 kJ mol^{-1} **Ans. d**

460. Using the following thermochemical equations

(i) S(Rh) + 3/2 O_2(g) → SO_3(g); ΔH = -2x kJ/mol

(ii) SO_2(g) + ½ O_2(g) → SO_3(g); ΔH = -y kJ/mol

Find out the heat of formation of SO_2(g) in kJ/mol

(a) (2 x+y) (b) (x−y)

(c) (2x/y) (d) (y-2x) **Ans. d**

461. The lattice enthalpy and hydration enthalpy of four compounds are given below

Compound	**Lattice Enthalpy (kJ/mol)**	**Hydration Enthalpy (kJ/mol)**
P	+780	-920
Q	+1012	-812
R	+828	-878
S	+632	-600

The pair of compounds which is soluble in water is

(a) P and R (b) Q and R

(c) P and Q (d) Q and S **Ans. a**

462. 4.48 L of an ideal gas at STP requires 12 calories to raise it temperature by 15°C at constant volume. The C_p of the gas is

(a) 3 cal (b) 4 cal

(c) 7 cal (d) 6 cal **Ans. d**

463. Change in internal energy, when 4 kJ of work is done on the system and 1 kJ of heat is given out by the system is

(a) +1 kJ (b) -5 kJ

(c) +5 kJ (d) +3 kJ **Ans. d**

464. The average molar heat capacities of ice and water are respectively 37.8 J/mol and 75.6 J/mol and the enthalpy of fusion of ice is 6.012 kJ/mol. The amount of heat required to change 10 g of ice at -10°C to water at 10°C would be

(a) 2376 J (b) 4752 J

(c) 3970 J (d) 1128 J **Ans. c**

465. For the gas phase reaction $PCl_5(g) \rightarrow PCl_3(g) + Cl2(g)$. Which of the following conditions are correct?

(a) $\Delta H = 0$ and $\Delta S < 0$ (b) $\Delta H > 0$ and $\Delta S > 0$

(c) $\Delta H < 0$ and $\Delta S < 0$ (d) $\Delta H > 0$ and $\Delta S < 0$ **Ans. b**

466. Which of the following are not state functions

(i) q+w (ii) q

(iii) w (iv) H - TS

(a) (i) and (iv) (b) (ii), (iii) and (iv)

(c) (i), (ii) and (iii) (d) (ii) and (iii) **Ans. d**

467. Bond dissociation enthalpy of H_2, Cl_2 and HCl are 434, 242 and 431 kJ/mol respectively. Enthalpy of formation of HCl is

(a) 93 kJ/mol (b) -245 kJ/mol

(c) -93 kJ/mol (d) 245 kJ/mol **Ans. c**

468. Standard entropy of X_2, Y_2 and XY_3 are 60, 40 and 50 JK^{-1} mol^{-1} respectively. For the reaction $\frac{1}{2} X_2 + 3/2\, Y_2 \rightarrow XY_3$, ΔH = - 30 kJ, to be at equilibrium, the temperature will be

(a) 1000 K (b) 1250 K

(c) 500 K (d) 750 K **Ans. d**

469. Oxidising power of chlorine in aqueous solution can be determined by parameters indicated below

$$\frac{1}{2} Cl_2(g) \xrightarrow{\frac{1}{2}\Delta_{dies}H} Cl(g) \xrightarrow{\Delta_{eg}H} Cl^-(g) \xrightarrow{\Delta_{hyd}H} Cl^-(aq)$$

The energy involved in the conversion of $\frac{1}{2} Cl_2(g)$ to Cl-(aq) (using data, $\Delta_{dies}HCl_2$ = 240 kJ mol^{-1}, $\Delta_{eg}HCl$ = - 349 kJ mol^{-1}, $\Delta_{hyd}HCl^-$ = -381 kJ mol^{-1}) will be

(a) + 120 kJ mol^{-1} (b) + 152 kJ mol^{-1}

(c) - 610 kJ mol^{-1} (d) - 850 kJ mol^{-1} **Ans. c**

470. Sodium chloride is soluble in water but not in benzene because

(a) $\Delta H_{hydration} < \Delta H_{Lattice\ energy\ in\ water}$ and
$\Delta H_{solvation} > \Delta H_{Lattice\ energy\ in\ benzene}$

(b) $\Delta H_{hydration} > \Delta H_{Lattice\ energy\ in\ water}$ and
$\Delta H_{solvation} < \Delta H_{Lattice\ energy\ in\ benzene}$

(c) $\Delta H_{hydration} = \Delta H_{Lattice\ energy\ in\ water}$ and
$\Delta H_{solvation} < \Delta H_{Lattice\ energy\ in\ benzene}$

(d) $\Delta H_{hydration} < \Delta H_{Lattice\ energy\ in\ water}$ and
$\Delta H_{solvation} = \Delta H_{Lattice\ energy\ in\ benzene}$ **Ans. b**

471. The enthalpy of neutralization of oxalic acid by a strong base is - 25 k cal/mol. The enthalpy of neutralization of strong acid and strong base is - 13.7 k cal/equiv. The enthalpy of dissociation of

$H_2C_2O_4 \rightarrow 2H^+ + C_2O_4^{2-}$ is

(a) 1.0 kcal/mol (b) 2.4 kcal/mol

(c) 18.55 kcal/mol (d) 117 kcal/mol **Ans. b**

472. Entropy changes for the process $H_2O(l) \rightarrow H_2O(g)$ at normal pressure and 274 K are given below

$\Delta S_{system} = -22.13$, $\Delta S_{surroundings} = +22.05$

The process is non-spontaneous because

(a) ΔS_{system} is -ve (b) $\Delta S_{surroundings}$ is +ve

(c) $\Delta S_{universe}$ is -ve (d) $\Delta S_{system} \neq \Delta S_{surroundings}$ **Ans. c**

473. A reaction cannot take place spontaneously at any temperature when

(a) Both ΔH and ΔS are positive

(b) Both ΔH and ΔS are negative

(c) ΔH is negative and ΔS is positive

(d) ΔH is positive and ΔS is negative **Ans. d**

474. The heats of atomisation of $PH_3(g)$ and $P_2H_4(g)$ are 954 kJ mol^{-1} and 1485 kJ mol^{-1} respectively. The P - P bond energy in kJ mol^{-1} is

(a) 213 (b) 426

(c) 318 (d) 107 **Ans. a**

475. Pick out the wrong statement

(a) Standard free energy of formation of all elements is zero

(b) A process accompanied by decrease in entropy is spontaneous under certain conditions

(c) The entropy of a perfectly crystalline substance at absolute zero is zero

(d) A process that leads to increase in free energy will be spontaneous

Ans. d

476. Given the $dU = TdS - pdV$ and $H = U + pV$. Which one of the following relations is true?

(a) $dH = dS + Vdp$ (b) $dH = SdT + Vdp$

(c) $dH = -SdT + Vdp$ (d) $dH = dU + pdV$ **Ans. a**

477. The values of ΔH and ΔS for the reaction, C(graphite) + $CO_2(g) \rightarrow 2CO(g)$ are 170 kJ and 170 J/K respectively. This reaction will be spontaneous at

(a) 910 K (b) 1110 K

(c) 510 K (d) 710 K **Ans. b**

478. From the following bond energies:

H - H bond energy : 431.37 kJ mol^{-1}

H - H bond energy : 431.37 kJ mol^{-1}

H - H bond energy : 431.37 kJ mol^{-1}

H - H bond energy : 431.37 kJ mol^{-1}

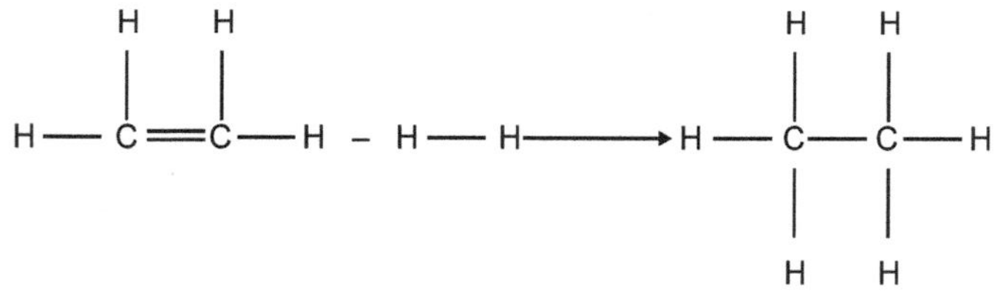

Calculate the bond enthalpy of the following reaction:

(a) -243.6 kJ/mol (b) -120.0 kJ/mol

(c) 553.0 kJ/mol (d) 1523.6 kJ/mol **Ans. b**

479. In a fuel cell methanol is used as fuel and oxygen is used as an oxidiser. The reaction is:

$CH_3OH(l) + 3/2\, O_2(g) \rightarrow CO_2(g) + 2H_2O(l)$

At 298 K, standard Gibb's energies of formation for $CH_3OH(l)$, $H_2O(l)$ and $CO_2(g)$ are - 166.2, -237.2 and -394.4 kJ/mol respectively. If standard enthalpy of combustion of methanol is - 726 kJ/mol, efficiency of the fuel cell will be

(a) 80% (b) 87%

(c) 90% (d) 97% **Ans. d**

480. On the basis of the following thermochemical data ($\Delta_f G^o H+(aq) = 0$)

$H_2O(l) \rightarrow H^+(aq) + OH^-(aq)$, $\Delta H = 57.32$ kJ

$H_2(g) + ½ O_2(g) \rightarrow H_2O(l)$, $\Delta H = -286.2$ kJ

The value of enthalpy of formation of OH^- at 25°C is

(a) -22.88 kJ (b) -228.88 kJ

(c) +228.88 kJ (d) -245.52 kJ **Ans. b**

481. Thermodynamically the most stable form of carbon is

(a) Diamond (b) Graphite

(c) Fullerenes (d) Coal **Ans. b**

482. How many grams of ice at 0°C can be melted by the addition of 500 J of heat? (The molar heat of fusion for ice is 6.02 kJ mol-^{1})

(a) 0.0831 g (b) 1.50 g

(c) 3.01 g (d) 12.0 g **Ans. b**

483. Which of these species has a standard enthalpy of formation equal to zero

(a) $F_2(g)$ (b) F(g)

(c) HF(aq) (d) $F^-(aq)$ **Ans. a**

484. A 1.0 g sample of substance A at 100°C is added to 100 mL of H_2O at 25°C. using separate 100 mL portions of H_2O, the procedure is repeated with substance B and then with substance C. How will the final temperature of the water compare

Substance	Specific Heat
A	0.60 $Jg^{-1}\,{}^0C^{-1}$
B	0.40 $Jg^{-1}\,{}^0C^{-1}$
C	0.20 $Jg^{-1}\,{}^0C^{-1}$

(a) $T_C > T_B > T_A$ (b) $T_B > T_A > T_C$

(c) $T_A > T_B > T_C$ (d) $T_A = T_B = T_C$ **Ans. c**

485. As $O_2(l)$ is cooled at 1 atm pressure, it freezes to form solid I at 54.5 K. At a lower temperature, solid I rearranges to solid II, which has a different crystal structure. Thermal measurements show that for the phase transition of solid I to solid II, $\Delta H = -743.1$ J mol^{-1} and $\Delta S = -17.0$ JK^{-1} mol^{-1}. At what temperature are solids I and II in equilibrium

(a) 2.06 K (b) 31.5 K

(c) 43.7 K (d) 53.4 K **Ans. c**

486. The standard enthalpy of formation of $H_2(g)$ and $Cl_2(g)$, HCl(g) are 218 kJ mol^{-1}, 121.68 kJ mol^{-1} and - 92.31 kJ mol^{-1} respectively. Calculate standard enthalpy change in kJ for ½ $H_2(g)$ + ½ $Cl_2(g) \rightarrow HCl(g)$

(a) +431.99 (b) -262.14

(c) -431.99 (d) +247.37 **Ans. b**

487. Which of the following thermodynamic relation is correct

(a) dG = VdP - SdT (b) dU = PdV + TdS

(c) dH = - VdP + TdS (d) dG = VdP + SdT **Ans. a**

488. Enthalpy of vaporization of benzene is +35.3 kJ mol^{-1} at its boiling point, 80°C. The entropy change in the transition of the vapour to liquid at its boiling point (in $JK^{-1}mol^{-1}$) is

(a) -441 (b) -100

(c) +441 (d) +100 **Ans. b**

489. Based on the first law of thermodynamics, which one of the following is correct

(a) For an isochoric process $\Delta U = -q$

(b) For an adiabatic process $\Delta U = -w$

(c) For an isothermal process $q = +w$

(d) For a cyclic process $q = -w$ **Ans. d**

490. Calculate the standard enthalpy change (in kJ mol^{-1}) for the reaction, $H_2(g) + O_2(g) \rightarrow H_2O_2(g)$, given, that bond enthalpies of H - H, O = O, O - H and O - O (in kJ mol^{-1}) are respectively 438, 498, 464 and 138

(a) -130 (b) -65

(c) +130 (d) -334 **Ans. a**

491. A certain reaction is at equilibrium at 82°C and the enthalpy change for this reaction is 21.3 kJ. The value of ΔS (in JK^{-1} mol^{-1}) for the reaction is

(a) 55.0 (b) 60.0

(c) 68.5 (d) 120.0 **Ans. b**

492. The species which by definition has zero standard molar enthalpy of formation at 298 K is

(a) $Br_2(g)$ (b) $Cl_2(g)$

(c) $H_2O(g)$ (d) $CH_4(g)$ **Ans. b**

493. The standard enthalpy of formation of NH_3 is - 46.0 kJ mol^{-1}. If the enthalpy of formation of H_2 from its atoms is -436 kJ mol^{-1} and that of N_2 is -712 kJ mol^{-1}, the average bond enthalpy of N-H bond in NH_3 is

(a) +1056 kJ mol^{-1} (b) -1102 kJ mol^{-1}

(c) -964 kJ mol^{-1} (d) +352 kJ mol^{-1} **Ans. d**

494. For a particular reversible reaction at temperature T, ΔH and ΔS were found to be both +ve. If T_e is the temperature at equilibrium, the reaction would be spontaneous when

(a) T_e is 5 times T (b) $T = T_e$

(c) $T_e > T$ (d) $T > T_e$ **Ans. d**

495. Which one among the following pairs does not represent example for intensixe property

(a) Temperature and density

(b) Pressure and molar volume

(c) Molar heat capacity and density

(d) Heat capacity and enthalpy **Ans. d**

496. Pick out the wrong statement

(a) The standard free energy of formation of elements is zero

(b) A process that leads to increase will be spontaneous

(c) A process accompanied by decrease in entropy will be non-spontaneous under normal conditions

(d) Enthalpy of combustion is always negative **Ans. b**

497. The values of ΔH and ΔS for a reaction are respectively 30 kJ mol^{-1} and 100JK^{-1} mol^{-1}. Then the temperature above which the reaction will become spontaneous is

(a) 300 K (b) 30 K

(c) 100 K (d) 300°C **Ans. a**

498. The standard enthalpy of formation of $C_2H_4(g)$, $CO_2(g)$ and $H_2O(l)$ are 52, - 394 and -286 kJ mol^{-1} respectively. Then the amount of heat evolved by burning 7 g of $C_2H_4(g)$ is

(a) 1412 kJ (b) 9834 kJ

(c) 353 kJ (d) 706 kJ **Ans. c**

499. The standard enthalpy of formation of which of the following is zero

(a) CH_3OH (b) HF

(c) H_2O (d) F_2 **Ans. d**

500. For an ideal gas, which of the following is true

(a) $\left(\frac{\partial U}{\partial V}\right)_T > 0$ (b) $\left(\frac{\partial P}{\partial V}\right)_T = 0$

(c) $\left(\frac{\partial U}{\partial V}\right)_T = 0$ (d) $\left(\frac{\partial (PV)}{\partial V}\right)_T > 0$ **Ans. c**

501. Which is a state function

(a) ΔS (b) ΔU

(c) ΔG (d) All of these **Ans. d**

502. Under equilibrium conditions, which of the following is true

(a) $\Delta S = 0$ (b) $\Delta H > 0$

(c) $\Delta S > 0$ (d) $\Delta H < 0$ **Ans. a**

503. Given:

$C_2H_6(g) \rightarrow 2C(g) + 6H(g); \Delta H = 712$ kcal

The C - C bond energy is 112 kcal, what is the C - H bond energy

(a) 88 kcal (b) 12 kcal

(c) 50 kcal (d) 100 kcal **Ans. d**

504. For the equation $C_p - C_v = R$, the significance of R is

(a) Work done per mole per Kelvin

(b) Heat absorbed per mole per Kelvin

(c) Heat released per mole per Kelvin

(d) Work done per mole per degree Celsius **Ans. a**

505. The amount of heat released when 20 ml 0.5 M NaOH is mixed with 100 ml 0.1 M HCl is x kJ. The heat of neutralization is

(a) -110 x kJ/mol (b) -50 x kJ/mol

(c) +100 x kJ/mol (d) +50 x kJ/mol **Ans. a**

506. One mole of an ideal gas expands at a constant temperature of 300 k from an initial volume of 10 litre to a final volume of 20 litre. The work done in expanding the gas is

(a) 750 Joule (b) 1728 Joule

(c) 1500 Joule (d) 3456 Joule **Ans. b**

507. Given:

$NH_3(g) + 3Cl_2(g) \leftrightarrow NCl_3(g) + 3HCl(g) - \Delta H_1$

$N_2(g) + 3H_2(g) \leftrightarrow 2NH_3(g)$ $\quad - \Delta H_2$

$H_2(g) + Cl_2(g) \rightarrow 2HCl(g)$ $\quad + \Delta H_3$

The heat of formation of NCl_3 in terms of ΔH_1, ΔH_2 and ΔH_3 is

(a) $\Delta H_f = - \Delta H_1 + \frac{1}{2} \Delta H_2 - 3/2\ \Delta H_3$

(b) $\Delta H_f = - \Delta H_1 + \frac{1}{2} \Delta H_2 + 3/2\ \Delta H_3$

(c) $\Delta H_f = - \Delta H_1 - \frac{1}{2} \Delta H_2 - 3/2\ \Delta H_3$

(d) None of these **Ans. d**

508. Molar heat capacity of aluminium is 25 $JK^{-1}\ mol^{-1}$. The heat necessary to raise the temperature of 54 g of aluminium (Atomic mass 27 g mol^{-1}) from 30°C to 50°C is

(a) 1.5 kJ (b) 0.5 kJ

(c) 1.0 kJ (d) 2.5 kJ **Ans. c**

509. A reaction is spontaneous at low temperature but non-spontaneous at high temperature. Which of the following is true for the reaction

(a) $\Delta H > 0, \Delta S > 0$ (b) $\Delta H < 0, \Delta S > 0$

(c) $\Delta H > 0, \Delta S = 0$ (d) $\Delta H < 0, \Delta S < 0$ **Ans. d**

510. Four grams of graphite is burnt in a bomb calorimeter of heat capacity of 30 kJ K^{-1} in excess of oxygen at 1 atmospheric pressure. The temperature rises from 300 to 304 K. what is the enthalpy of combustion of graphite (in kJ mol^{-1})

(a) 360 (b) 1440

(c) -360 (d) -1440 **Ans. c**

511. At the sublimation temperature, for the process $CO_2(s) \leftrightarrow CO_2(g)$

(a) ΔH, ΔS and ΔG are all positive

(b) $\Delta H > 0, \Delta S > 0$ and $\Delta G = 0$

(c) $\Delta H > 0, \Delta S > 0$ and $\Delta G > 0$

(d) $\Delta H < 0, \Delta S > 0$ and $\Delta G < 0$ **Ans. b**

512. The enthalpy of solution of sodium chloride is 4 kJ mol^{-1} and its enthalpy of hydration of ions is -784 kJ mol^{-1}. Then the lattice enthalpy of NaCl (in kJ mol^{-1}) is

(a) +788 (b) +4

(c) +398 (d) +780 **Ans. a**

513. The amount of heat evolved when 500 cm^3 of 0.1 M HCl is mixed with 200 cm^3 of 0.2 M NaOH is

(a) 1.292 kJ (b) 2.292 kJ
(c) 0.292 kJ (d) 22.9 kJ **Ans. b**

514. During the adsorption of Krypton on activated charcoal at low temperature

(a) $\Delta H < 0$ and $\Delta S < 0$ (b) $\Delta H > 0$ and $\Delta S < 0$
(c) $\Delta H > 0$ and $\Delta S > 0$ (d) $\Delta H < 0$ and $\Delta S > 0$ **Ans. a**

515. For the reversible reaction, A(s) + B(g) $\leftrightarrow$ C(g) + D(g), ΔG^0 = - 350 kJ which one of the following statements is true

(a) The reaction is thermodynamically non-feasible
(b) The entropy change is negative
(c) Equilibrium constant cannot be greater than one
(d) The reaction should be instantaneous **Ans. c**

516. Enthalpy change for the reaction, 4H(g) $\rightarrow$ $2H_2$(g) is -869.6 kJ. The dissociation energy of H-H bond is

(a) -434.8 kJ (b) -869.6 kJ
(c) +434.8 kJ (d) +217.4 kJ **Ans. c**

517. Which of the following is correct option for free expansion of an ideal gas under adiabatic condition

(a) $q = 0, \Delta T \neq 0, w = 0$ (b) $q \neq 0, \Delta T = 0, w = 0$
(c) $q = 0, \Delta T = 0, w = 0$ (d) $q = 0, \Delta T < 0, w \neq 0$ **Ans. c**

418. The entropy of change involved in the isothermal reversible expansion of 2 moles of an ideal gas from a volume of 10 dm^3 to a volume of 100 dm^3 at 27°C

(a) 38.3 J mol^{-1} K^{-1} (b) 35.8 J mol^{-1} K^{-1}
(c) 32.3 J mol^{-1} K^{-1} (d) 42.3 J mol^{-1} K^{-1} **Ans. a**

519. The value of enthalpy change (ΔH) for the reaction:

$C_2H_5OH(l) + 3O_2(g) \rightarrow 2CO_2(g) + 3H_2O(l)$

at 27°C is -1366.5 kJ mol^{-1}. The value of internal energy change for the above reaction at this temperature will be

(a) -1369.0 kJ (b) -1364.0 kJ
(c) -1361.5 kJ (d) -1371.5 kJ **Ans. b**

❑❑❑

Chapter-6

Carbohydrates

Fill in the blanks

1. A certain amount of......................is needed to initiate any.............. reaction.

Ans. energy, chemical

2. Carbon......................hydrogen and....................are the principle (major) elements.

Ans. oxygen, nitrogen

3. Common hexoses are glucose, ...and......................

Ans. fructose, galactose

4. Biologically important disaccharides are ...., ... and................

Ans. maltose, sucrose, lactose

5. Glycogen is the main reserve food material of............... cells. It is known as the............

Ans. animal, animal starch

6. Starch is the reserve food material of.......................cells. It has two components: amylose and..................

Ans. plant, amylopectin

7. Some important heteropolysaccharides include hyalouronic acid, heparin, chondroitin sulphate etc. which are together grouped as..............

Ans. mucopolysaccharide

8. ....of connective tissue actin and myosin of muscles and...............of scales, feathers, hairs, etc. are fibrous proteins.

Ans. collagen, keratin

9. Most common monosaccharides present in the nucleus are..................

Ans. pentoses

10. Cellulose is a unbranched chain with...............4 bonds.

Ans. β-1

11. Lactose molecule is composed of...............

Ans. glucose + galactose

12. Iodine test is used to detect..............

Ans. carbohydrates

13. On hydrolysis of cane sugar, we can obtain...................

Ans. glucose + fructose

14. A non-reducing commercial sugar is............

Ans. sucrose

15. In fruits/vegetables such as "Bhindi" the mucilage is made up of...............and.............

Ans. galactose, mannose

16. The "repeating unit" of glycogen is......................

Ans. glucose

17. Ribose sugar is present in...................and................

Ans. RNA, ATP

18. Table sugar is...............

Ans. sucrose

19. Glycogen is a homopolymer made of.......................

Ans. glucose units

20. The number of "ends" in a glycogen molecule would be...............

Ans. equal to the number of branches plus one

21. Chitin is a....................

Ans. nitrogenous polysaccharide

22. Carbohydrates contain................, ... and ....

Ans. carbon, hydrogen, oxygen

23. Carbohydrates are the major source of..................in our food.

Ans. energy

24. Carbohydrates contain hydrogen and oxygen exactly in the same ratio as in water but still they are not considered as....................

Ans. hydrates

25. Human beings cannot digest.............while the animals can do so easily.

Ans. cellulose

26. A monosaccharide is represented by a general formula.................

Ans. $C_nH_{2n}O_n$

27. A typical ketohexose is...................

Ans. fructose

28. Glucose and galactose are...................of each other.

Ans. isomeric

29. Deoxyribose is a.................

Ans. aldopentose

30. Starch is a...............and is described by the formula................

Ans. polysaccharide, $(C_6H_{10}O_5)n$

31. Glucose reduces the................solution to.................and................ therefore indicates the presence of...............group.

Ans. Fehling, red cuprous oxide, aldehyde

32. The nutrient that is used in the body as a source of energy and is a raw material for growth is....................

Ans. carbohydrate

33. Lactose has the same molecular formula as..................

Ans. sucrose

34. Animal starch is known as....................

Ans. glycogen

35. Sucrose is a.........................and is broken down into...................... and..........during hydrolysis.

Ans. disaccharide, glucose, fructose

36. A very high molecular mass carbohydrate is..............

Ans. cellulose

37. On fermentation, sugars get converted to.................... and ....

Ans. ethanol, CO_2

38. The disaccharide present in milk is.........

Ans. lactose

39. Disaccharides are formed by the condensation of two monosaccharides linked to each other by a...................

Ans. glycoside linkage

40. Sucrose is converted into glucose and fructose by the enzyme.............

Ans. invertase

41. ....carbohydrate compound is found abundantly in nature.

Ans. Cellulose

42. The change in the optical rotation (with time) of freshly prepared solution of sugar is known as..................

Ans. mutarotation

43. Starch is hydrolysed to maltose by an enzyme named........................

Ans. diastase

44. The polysaccharides which are used as food components are................

Ans. gums

45. ....is known as grape sugar.

Ans. Glucose

46. ....sugar is found in fruit juices.

Ans. Fructose

47. Upon hydrolysis, the.................. sugar found in milk yields glucose and galactose.

Ans. lactose

48. ....is the polysaccharide occurring in the tuberous roots of Dahlia.

Ans. Inulin

49. Water soluble carbohydrates are known as.....................

Ans. sugars

50. Ribose sugar differs from deoxyribose sugar in having one.............. atom extra.

Ans. oxygen

51. ....contain the largest proportion of cellulose among natural materials.

Ans. Cotton fibres

52. Cellulose is an amorphous polymer of...............

Ans. glucose

53. Slimy substance of bhindi is a................

Ans. mucopolysaccharide

54. Carbohydrates containing an aldehyde group are called................. and those containing a keto group are called...............

Ans. Aldoses, ketoses

55. Anomers differ in configuration at the................carbon.

Ans. glycosidic

56. Milk sugar is.....................

Ans. lactose

57. The polysaccharides which are acidic in nature are called..............

Ans. gums

58. Most of the naturally occurring sugars have...............configuration while most of the naturally occurring............have L-configuration.

Ans. D-, α-amino acids

59. Cellulose is a linear polymer of.............

Ans. β-D(+) Glucose

60. Ecentric starch grains are found in................

Ans. Potato tubers

61. The polysaccharides which are acidic in nature are called ....

Ans. Gums

Multiple Choice Questions

1. Glucose and galactose are two isomeric monosaccharide known as
   (a) Anomers (b) Epimers
   (c) Enantiomers (d) Conformers **Ans. b**
2. Glucose and mannose are epimers. This mean that
   (a) They are mirror images of each other
   (b) One is an aldose and other a ketose
   (c) They rotate plane polarized light in opposite directions
   (d) They differ only in the configuration of one carbon atom **Ans. d**
3. The sugar residues of amylose are
   (a) In α-1, 4-linkages (b) In β-1, 4-linkages
   (c) Galactose units only (d) Fructose units only **Ans. a**
4. The chitin in the fungal call wall is
   (a) A protein (b) A glycoprotein
   (c) A polysaccharide (d) A lipopolysaccharide **Ans. c**
5. Which of the following statement about sucrose is false?
   (a) No free aldehyde or ketone group sucrose
   (b) Contain an α-1, 2-glycosidic bond
   (c) Oxygen bridge between glucose and fructose is between the anomeric carbons
   (d) Sucrose can reduce Fehling's solution **Ans. d**
6. Which of the following statement is correct?
   (a) Anomers differ in configuration at the glycosidic carbon
   (b) Epimers differ in configuration at any carbon other than glycosidic carbon
   (c) Both
   (d) None **Ans. c**
7. Which of the following statement is incorrect?
   (a) Amylase is a water soluble fraction of starch
   (b) Amylopectin is a water insoluble fraction of starch
   (c) Amylopectin fraction has both α-1, 4 and α-1, 6-glycosidic linkage
   (d) None **Ans. d**

8. Which is not true about monosaccharides?

(a) Colourless (b) Soluble in water

(c) Sweet in taste (d) Soluble in non-polar solvent

Ans. d

9. Glucose gives the silver mirror test with ammonical $AgNO_3$ because it contains group

(a) Ketonic (b) Amide

(c) Ester (d) Aldehydic **Ans. d**

10. Starting from a glucose residue in glycogen, how many net ATP molecules will be formed in the glycolysis of the residues to pyruvate

(a) One (b) Two

(c) Three (d) Four **Ans. c**

11. The greatest amount of body glycogen can be found in which of the following human tissues?

(a) Liver (b) Kidney

(c) Skeletal muscle (d) Cardiac muscle **Ans. c**

12. Glycogen synthase is characterized by all of the following statements except

(a) The enzyme exists in active and inactive forms

(b) Uridine diphosphate glucose is a substrate

(c) It is activated by phosphorylation

(d) It requires a primer strand of glycogen **Ans. c**

13. Which of the following prosthetic group is required by pyruvate carboxylase, an enzyme involved in gluconeogenetic pathway

(a) Pyridoxal phosphate (b) Biotin

(c) Tetrahydrofolate (d) TPP **Ans. b**

14. Which carbohydrate is used in silvering of mirrors?

(a) Sucrose (b) Cellulose

(c) Fructose (d) Glucose **Ans. d**

15. Sugars have the suffix

(a) -ol (b) -al

(c) -ose (d) -one **Ans. c**

16. Carbohydrates have

(a) A bitter taste

(b) A sour taste

(c) A sweet taste

(d) Some have a sweet taste while some are tasteless **Ans. d**

17. The carbohydrate which cannot be hydrolysed by the human digestive system is

(a) Starch (b) Glycogen

(c) Cellulose (d) All of the above **Ans. c**

18. Cane sugar on hydrolysis gives

(a) Glucose + Fructose (b) Fructose + Fructose

(c) Glucose + Lactose (d) Glucose + Glucose **Ans. a**

19. A carbohydrate which cannot be hydrolysed to simpler compounds is called

(a) Monosaccharide (b) Polysaccharide

(c) Disaccharide (d) Trisaccharide **Ans. a**

20. Glucose reacts with acetic anhydride to form

(a) Monoacetate (b) Tetra-acetate

(c) Penta-acetate (d) Hexa-acetate **Ans. c**

21. Glucose molecule reacts with X number of molecules of phenylhydrazine to yield osazone. The value of X is

(a) Three (b) Two

(c) One (d) Four **Ans. a**

22. Which carbohydrate is an essential constituent of plant cells?

(a) Starch (b) Cellulose

(c) Sucrose (d) Vitamins **Ans. b**

23. The commonest disaccharide has the molecular formula

(a) $C_{12}H_{22}O_{11}$ (b) $C_{10}H_{18}O_9$

(c) $C_{10}H_{20}O_{10}$ (d) $C_{10}H_{32}O_{11}$ **Ans. a**

24. Which of the following is the sweetest sugar?

(a) Sucrose (b) Glucose

(c) Fructose (d) Maltose **Ans. c**

25. In polysaccharides, the linkage connecting monosaccharide units is called

(a) Glycoside linkage (b) Nucleoside linkage

(c) Glycogen linkage (d) Peptide linkage **Ans. a**

26. α-D(+)-glucose and β-D(+)-glucose are

(a) Enantiomers (b) Geometrical isomers

(c) Epimers (d) Anomers **Ans. d**

27. Mutarotation does not occur in

(a) Sucrose (b) D-Glucose

(c) L-Glucose (d) None of these **Ans. a**

28. Sanger's reagent is used for the identification of

(a) N-Terminal of a peptide chain

(b) C-Terminal of a peptide chain

(c) Side chain of amino acids

(d) Molecular weight of the peptide chain **Ans. a**

29. Which is correct statement?

(a) Starch is polymer of α-glucose

(b) Amylase is a component of cellulose

(c) Proteins are composed of only one type of amino acids

(d) In cyclic structure of furanose, there are four carbons and one oxygen atom **Ans. a**

30. Ribose is a

(a) Monosaccharide (b) Polysaccharide

(c) Polypeptide (d) Disaccharide **Ans. a**

31. Which of the following is a reducing agent?

(a) Sucrose (b) Glucose

(c) Fructose (d) Glycogen **Ans. b**

32. An example of a ketohexose is

(a) Fructose (b) Glucose

(c) Galactose (d) Sucrose **Ans. a**

33. Which of the following monosaccharides is a pentose?

(a) Glucose (b) Fructose

(c) Arabinose (d) Galactose **Ans. c**

34. Glucose cannot be classified as

(a) A hexose (b) A carbohydrate

(c) A oligosaccharide (d) An aldose **Ans. c**

35. A sugar which reduces the Fehling's solution to red cuprous oxide is

(a) Sucrose (b) Glucose

(c) Fructose (d) Lactose **Ans. c**

36. Disaccharides are formed by the condensation of two monosaccharides linked to each other by a

(a) Peptide bond (b) Ester linkage

(c) Glycoside linkage (d) Hydrogen bond **Ans. c**

37. Sucrose is converted into glucose and fructose by the enzyme

(a) Zymase (b) Invertase

(c) Diastase (d) Maltase **Ans. b**

38. On fermentation, sugars get converted to

(a) Ethanol+CO_2 (b) Carbon dioxide

(c) Glucose (d) Fructose **Ans. a**

39. The disaccharide present in milk is

(a) Sucrose (b) Maltose

(c) Lactose (d) Cellobiose **Ans. c**

40. Which one of the following compounds is found abundantly in nature?

(a) Fructose (b) Starch

(c) Glucose (d) Cellulose **Ans. d**

41. The change in the optical rotation (with time) of freshly prepared solution of sugar is known as

(a) Specific rotation (b) Inversion

(c) Rotatory motion (d) Mutarotation **Ans. d**

42. Starch is hydrolysed to maltose by an enzyme named

(a) Lipase (b) Diastase

(c) Zymase (d) Invertase **Ans. b**

43. Cellulose is a polymer containing the monomer

(a) Ribose (b) Glucose

(c) Fructose (d) Starch **Ans. b**

44. The polysaccharides which are used as food components are
(a) Cellulose (b) Gums
(c) Ribose (d) Pectins **Ans. b**

45. Mineral functioning as chemical messenger is
(a) Calcium (b) Magnesium
(c) Manganese (d) Molybdenum **Ans. a**

46. The minimum carbon number in a carbohydrate is
(a) 6 (b) 4
(c) 3 (d) 1 **Ans. c**

47. The ratio between hydrogen and oxygen in a carbohydrate is
(a) 5:1 (b) 4:3
(c) 3:1 (d) 2:1 **Ans. d**

48. $C_nH_{2n}O_n$ is the formula of
(a) Fatty acid (b) Fat
(c) Glycerol (d) Carbohydrate **Ans. d**

49. Deoxyribose is
(a) $C_5H_{10}O_5$ (b) $C_5H_{10}O_4$
(c) $C_6H_{12}O_6$ (d) $C_6H_{12}O_5$ **Ans. b**

50. Simplest form of carbohydrate is
(a) Carbon (b) Starch
(c) Monosaccharide (d) Cane sugar **Ans. c**

51. A monosaccharide is
(a) Lactose (b) Sucrose
(c) Ribose (d) Maltose **Ans. c**

52. Monosaccharides have carbons
(a) 20 (b) 10-15
(c) 2-10 (d) 3-7 **Ans. d**

53. The most common carbohydrate monomer is
(a) Maltose (b) Lactose
(c) Glucose (d) Galactose **Ans. c**

54. A nonreducing sugar is
(a) Glucose (b) Sucrose
(c) Fructose (d) Maltose **Ans. b**

55. The sugar present in milk is
(a) Casein (b) Maltose
(c) Lactose (d) Heparin **Ans. c**

56. Xylitol is
(a) Artificial sweetener (b) Nutritive sweetener
(c) Nonnutritive sweetner (d) Both (a) and (b) **Ans. c**

57. Triose carbohydrate is
(a) Glycerol (b) Glyceraldehydes
(c) Dihydroxy acetone (d) Both (b) and (c) **Ans. d**

58. Reducing sugars have
(a) Free aldehyde (b) Bound aldehyde
(c) Free aldehyde or ketone (d) Bound ketone **Ans. c**

59. Galactose is
(a) Pentose sugar (b) Hexose sugar
(c) Heptose sugar (d) Triose carbohydrate **Ans. b**

60. Deoxyribose is a
(a) Pentose sugar (b) Tetrose carbohydrate
(c) Hexose sugar (d) Triose carbohydrate **Ans. a**

61. A ketose sugar is
(a) Glucose (b) Fructose
(c) Glyceraldehyde (d) Both (a) and (b) **Ans. b**

62. Laevulose present in honey is
(a) Disaccharide (b) Glucose
(c) Fructose (d) Pentose **Ans. c**

63. L-Fructose is found in
(a) Grape (b) Honey
(c) Most fruits (d) Lemon **Ans. b**

64. In Fehling's or Bendict's solution, a reducing sugar causes conversion of
(a) Ferric to ferrous state (b) Ferrous to ferric state
(c) Cuprous to cupric state (d) Cupric to cuprous state
Ans. d

65. Use of artificial sweetener saccharin has been discontinued because it is
(a) Carcinogenic
(b) Liable to decompose in hot weather of tropics
(c) Bitter in the beginning
(d) All the above **Ans. a**

66. The commonly used artificial sweetener is
(a) Acesulfame K (b) Cyclamate
(c) Aspartame (d) Saccharin **Ans. c**

67. Mannose is
(a) Disaccharide (b) Hexosan monosaccharide
(c) Pentosan monosaccharide (d) A tetrose **Ans. b**

68. Which one is sugar acid
(a) Vitamin C (b) Vitamin K
(c) Vitamin D (d) Vitamin E **Ans. a**

69. Cellobiose, the hydrolytic breakdown product of cellulose is
(a) A monosaccharide (b) A disaccharide
(c) A tetrasaccharide (d) A trisaccharide **Ans. b**

70. Inulin is a polymer of
(a) Glucose (b) Galactose
(c) Fructose (d) Arabinose **Ans. c**

71. Raffinose is a
(a) Monosaccharide (b) Disaccharide
(c) Trisaccharide (d) Tetrasaccharide **Ans. c**

72. Most sugars are dextrorotatory which means they
(a) Tilt light to right (b) Rotate polarized light to right
(c) Tilt light to left (d) Rotate polarized light to left
Ans. b

73. $C_nH_{2n}O_2$ is the formula of
(a) Carbohydrate (b) Fatty acid
(c) Fat (d) Nucleic acid **Ans. b**

74. Which one yields twice as many calories as carbohydrate per unit weight
(a) Protein (b) Fat
(c) Amino acids (d) Vitamins **Ans. b**

75. Cellulose is a

(a) Polypeptide (b) Polysaccharide

(c) Polynucleotide (d) Disaccharide **Ans. b**

76. Cellulose is formed by union of repeated residues of

(a) Amino acids (b) Lipids

(c) Glucose (d) Fructose **Ans. c**

77. A molecule of cellulose is made of about

(a) 50,000-100,000 glucose molecules

(b) 150,000-200,000 glucose molecules

(c) 300,000-400,000 glucose molecules

(d) 6,000-10,000 glucose molecules **Ans. d**

78. Which one is a polysaccharide is

(a) Starch (b) Cellulose

(c) Glycogen (d) All the above **Ans. d**

79. A fibrous polysaccharide is

(a) Glycogen (b) Starch

(c) Cellulose (d) Collagen **Ans. c**

80. A heteropolysaccharide is

(a) Glycogen (b) Starch

(c) Cellulose (d) Chitin **Ans. d**

81. Which is an unbranched glucan

(a) Cellulose (b) Starch

(c) Glycogen (d) All the above **Ans. a**

82. An anticoagulant mucopolysaccharide commonly present in animal body is

(a) Chondroitin sulphate (b) Keratin sulphate

(c) Heparin (d) Hyaluronic acid **Ans. c**

83. Which one is a homopolysaccharide?

(a) Starch (b) Cellulose

(c) Glycogen and inulin (d) All the above **Ans. d**

84. A fructosan storage polysaccharide is

(a) Amylose (b) Amylopectin

(c) Inulin (d) Callose **Ans. c**

85. The smallest storage polysaccharide is

(a) Inulin (b) Amylose

(c) Amylopectin (d) Dextrose **Ans. a**

86. Dahlia starch is used for

(a) Study of digestive enzymes (b) Study of kidney function

(c) Preparation of noodles (d) Manufacture of chocolates

Ans. b

87. A structural polysaccharide is

(a) Chitin (b) Hyaluronic acid

(c) Heparin (d) Keratan sulphate **Ans. a**

88. A mucopolysaccharide is

(a) Slime, phycocolloid and pectin

(b) Mucin, callose and heparin

(c) Hemicelluloses, pectin and mucin

(d) Hyaluronic acid, chondroitin sulphate and mucin **Ans. a**

89. Where are starch gains located

(a) Lysosomes (b) Mitochondria

(c) Amyloplasts (d) Protoplastids **Ans. c**

90. Iodine gives colour

(a) Purple with amylase

(b) Bluish black with amylopectin

(c) Bluish black with amylase and purple with amylopectin

(d) Red with amylase and blue with amylopectin **Ans. c**

91. Glycogen is stored inside cells of

(a) Kidneys and liver

(b) Liver and muscles

(c) Spleen and liver

(d) Pancreas and storage connective tissue **Ans. b**

92. Glycogen occurs in the form of

(a) Rounded grains (b) Ellipsoid grains

(c) Elongated granules (d) Flat ellipsoid granules **Ans. d**

93. Glycogen granules are located inside

(a) Amyloplasts (b) Mitochondria

(c) Cytoplasm (d) Lysosomes **Ans. c**

94. Glycosidic linkage at place of branching in starch and glycogen is

(a) $\alpha\ 1\to 6$ (b) $\alpha\ 1\to 4$

(c) $\beta\ 1\to 4$ (d) $\beta\ 1\to 6$ **Ans. a**

95. Main chain of glycogen and starch is helically coiled with each turn of helix having

(a) 10-14 glucose residues (b) 8-10 glucose residues

(c) 6 glucose residues (d) 4 glucose residues **Ans. c**

96. Glycosidic linkage present between adjacent glucose units of starch or glycogen is generally

(a) $\beta\ 1 \to 4$ (b) $\alpha\ 1 \to 4$

(c) $\alpha\ 2 \to 1$ (d) $\beta\ 2 \to 1$ **Ans. b**

97. Linkage present in between glucose residues of cellulose is

(a) $\beta\ 1 \to 4$ (b) $\beta\ 1 \to 6$

(c) $\alpha\ 1 \to 4$ (d) $\alpha\ 1 \to 6$ **Ans. a**

98. The most abundant organic molecule is

(a) RUBISCO (b) Starch

(c) Cellulose (d) Chitin **Ans. c**

99. The monomer units of chitin and fungus cellulose are

(a) Mannitol (b) Acetyl glucosamine

(c) Ascorbic acid (d) Glucouronic acid **Ans. b**

100. A mucopolysaccharide that functions as cell cement and lubricant is

(a) Heparin (b) Hyaluronic acid

(c) Keratin sulphate (d) Chondroitin sulphate **Ans. b**

101. A polysaccharide employed in tissue culture is

(a) Cellulose (b) Starch

(c) Glycogen (d) Agar-agar **Ans. d**

102. Compound starch grains are present in

(a) Potato (b) Rice

(c) Oat (d) All the above **Ans. d**

103. Hilum of a starch grain is formed of
(a) A special micro-grain of starch
(b) Proteinaceous centre
(c) Lipid centre
(d) Nucleic acid centre **Ans. b**

104. What is wrong
(a) Cellulose is most abundant organic molecule
(b) Chitin is the second most abundant organic molecule
(c) Cellulose is the most abundant heteropolysaccharide
(d) Chitin is the most abundant heteropolysaccharide **Ans. c**

105. Rayon and cellophane are formed of
(a) Cellulose xanthate (b) Cellulose nitrate
(c) Cellulose acetate (d) Carboxymethyl cellulose
Ans. a

106. A salt of cellulose used in propellent explosives is
(a) Cellulose acetate (b) Cellulose nitrate
(c) Cellulose superphosphate (d) Cellulose hypoxanthate **Ans. b**

107. Cellulose acetate is raw material for
(a) Photographic films (b) Tericot
(c) Shatter proof glass (d) All the above **Ans. d**

108. Smoothening agent in ice-creams and brightening agent in detergents is
(a) Cellulose hypoxanthate (b) Cellulose xanthate
(c) Carboxymethyl cellulose (d) All the above **Ans. c**

109. The general formula of monosaccharides is
(a) $C_nH_{2n}O_n$ (b) $C_{2n}H_2O_n$
(c) $C_nH_2O_{2n}$ (d) $C_nH_{2n}O_{2n}$ **Ans. a**

110. The general formula of polysaccharides is
(a) $(C_6H_{10}O_5)_n$ (b) $(C_6H_{12}O_5)_n$
(c) $(C_6H_{10}O_6)n$ (d) $(C_6H_{10}O_6)_n$ **Ans. a**

111. The aldose sugar is
(a) Glycerose (b) Ribulose
(c) Erythrulose (d) Dihydoxyacetone **Ans. a**

112. A triose sugar is
(a) Glycerose (b) Ribose
(c) Erythrose (d) Fructose **Ans. a**

113. A pentose sugar is
(a) Dihydroxyacetone (b) Ribulose
(c) Erythrose (d) Glucose **Ans. b**

114. The pentose sugar present mainly in the heart muscle is
(a) Lyxose (b) Ribose
(c) Arabinose (d) Xylose **Ans. a**

115. Polysaccharides are
(a) Polymers (b) Acids
(c) Proteins (d) Oils **Ans. a**

116. The number of isomers of glucose is
(a) 2 (b) 4
(c) 8 (d) 16 **Ans. d**

117. Two sugars which differ from one another only in configuration around a single carbon atom are termed
(a) Epimers (b) Anomers
(c) Optical isomers (d) Stereoisomers **Ans. a**

118. Isomers differing as a result of variations in configuration of the –OH and –H on carbon atoms 2, 3 and 4 of glucose are known as
(a) Epimers (b) Anomers
(c) Optical isomers (d) Steroisomers **Ans. a**

119. The most important epimer of glucose is
(a) Galactose (b) Fructose
(c) Arabinose (d) Xylose **Ans. a**

120. α-D-glucose and β-D-glucose are
(a) Stereoisomers (b) Epimers
(c) Anomers (d) Keto-aldo pairs **Ans. c**

121. α-D-glucose $+ 112^0 \rightarrow + 52.5^0 \leftarrow + 19^0$ β- D-glucose for glucose above represents
(a) Optical isomerism (b) Mutarotation
(c) Epimerisation (d) D and L isomerism **Ans. b**

122. Compounds having the same structural formula but differing in spatial configuration are known as

(a) Stereoisomers (b) Anomers

(c) Optical isomers (d) Epimers **Ans. a**

123. In glucose the orientation of the —H and —OH groups around the carbon atom 5 adjacent to the terminal primary alcohol carbon determines

(a) D or L series (b) Dextro or levorotatory

(c) α and β anomers (d) Epimers **Ans. a**

124. The carbohydrate of the blood group substances is

(a) Sucrose (b) Fucose

(c) Arabinose (d) Maltose **Ans. b**

125. Erythromycin contains

(a) Dimethyl amino sugar (b) Trimethyl amino sugar

(c) Sterol and sugar (d) Glycerol and sugar **Ans. a**

126. A sugar alcohol is

(a) Mannitol (b) Trehalose

(c) Xylulose (d) Arabinose **Ans. a**

127. The major sugar of insect hemolymph is

(a) Glycogen (b) Pectin

(c) Trehalose (d) Sucrose **Ans. c**

128. The sugar found in DNA is

(a) Xylose (b) Ribose

(c) Deoxyribose (d) Ribulose **Ans. c**

129. The sugar found in RNA is

(a) Ribose (b) Deoxyribose

(c) Ribulose (d) Erythrose **Ans. a**

130. The sugar found in milk is

(a) Galactose (b) Glucose

(c) Fructose (d) Lactose **Ans. d**

131. Invert sugar is

(a) Lactose

(b) Sucrose

(c) Hydrolytic products of sucrose
(d) Fructose **Ans. c**

132. Sucrose consists of
(a) Glucose + glucose (b) Glucose + fructose
(c) Glucose + galactose (d) Glucose + mannose **Ans. b**

133. The monosaccharide units are linked by $1 \rightarrow 4$ glycosidic linkage in
(a) Maltose (b) Sucrose
(c) Cellulose (d) Cellobiose **Ans. a**

134. Which of the following is a non-reducing sugar?
(a) Isomaltose (b) Maltose
(c) Lactose (d) Trehalose **Ans. d**

135. Which of the following is a reducing sugar?
(a) Sucrose (b) Trehalose
(c) Isomaltose (d) Agar **Ans. c**

136. A dissaccharide formed by 1,1-glycosidic linkage between their monosaccharide units is
(a) Lactose (b) Maltose
(c) Trehalose (d) Sucrose **Ans. c**

137. A dissaccharide formed by α–1, 4-glycosidic linkage between their monosaccharide units is
(a) Lactose (b) Maltose
(c) Trehalose (d) Sucrose **Ans. b**

138. Mutarotation refers to change in
(a) pH (b) Optical rotation
(c) Conductance (d) Chemical properties **Ans. b**

139. A polysaccharide which is often called animal starch is
(a) Glycogen (b) Starch
(c) Inulin (d) Dextrin **Ans. d**

140. The homopolysaccharide used for intravenous infusion as plasma substitute is
(a) Agar (b) Inulin
(c) Pectin (d) Starch **Ans. a**

141. The polysaccharide used in assessing the glomerular filteration rate (GFR) is

(a) Glycogen (b) Agar
(c) Inulin (d) Hyaluronic acid **Ans. c**

142. The constituent unit of inulin is

(a) Glucose (b) Fructose
(c) Mannose (d) Galactose **Ans. b**

143. The polysaccharide found in the exoskeleton of invertebrates is

(a) Pectin (b) Chitin
(c) Cellulose (d) Chondroitin sulphate **Ans. b**

144. Which of the following is a heteroglycan?

(a) Dextrins (b) Agar
(c) Inulin (d) Chitin **Ans. b**

145. The glycosaminoglycan which does not contain uronic acid is

(a) Dermatan sulphate (b) Chondroitin sulphate
(c) Keratan sulphate (d) Heparan sulphate **Ans. c**

146. The glycosaminoglycan which replace it contain uronic acid is

(a) Hyaluronic acid (b) Heparin
(c) Chondroitin sulphate (d) Dermatan sulphate **Ans. b**

147. Keratan sulphate is found in abundance in

(a) Heart muscle (b) Liver
(c) Adrenal cortex (d) Cornea **Ans. d**

148. Repeating units of hyaluronic acid are

(a) N-acetyl glucosamine and D-glucuronic acid
(b) N-acetyl galactosamine and D-glucuronic acid
(c) N-acetyl glucosamine and galactose
(d) N-acetyl galactosamine and L- iduronic acid **Ans. a**

149. The approximate number of branches in amylopectin is

(a) 10 (b) 20
(c) 40 (d) 80 **Ans. d**

150. In amylopectin the intervals of glucose units of each branch is

(a) 10–20 (b) 24–30
(c) 30–40 (d) 40–50 **Ans. b**

151. A polymer of glucose synthesized by the action of *Leuconostoc mesenteroids* in a sucrose medium is

(a) Dextrans (b) Dextrin
(c) Limit dextrin (d) Inulin **Ans. a**

152. Glucose on reduction with sodium amalgam forms

(a) Dulcitol (b) Sorbitol
(c) Mannitol (d) Mannitol and sorbitol **Ans. b**

153. Glucose on oxidation does not give

(a) Glycoside (b) Glucosaccharic acid
(c) Gluconic acid (d) Glucuronic acid **Ans. a**

154. Oxidation of galactose with conc. HNO_3 yields

(a) Mucic acid (b) Glucuronic acid
(c) Saccharic acid (d) Gluconic acid **Ans. a**

155. A positive Benedict's test is not given by

(a) Sucrose (b) Lactose
(c) Maltose (d) Glucose **Ans. a**

156. Starch is a

(a) Polysaccharide (b) Monosaccharide
(c) Disaccharide (d) None of these **Ans. a**

157. A positive Seliwanoff's test is obtained with

(a) Glucose (b) Fructose
(c) Lactose (d) Maltose **Ans. b**

158. Osazones are not formed with the

(a) Glucose (b) Fructose
(c) Sucrose (d) Lactose **Ans. c**

159. The most abundant carbohydrate found in nature is

(a) Starch (b) Glycogen
(c) Cellulose (d) Chitin **Ans. c**

160. Impaired renal function is indicated when the amount of PSP excreted in the first 15 minutes is

(a) 20% (b) 35%
(c) 40% (d) 45% **Ans. a**

161. An early feature of renal disease is
(a) Impairment of the capacity of the tubule to perform osmotic work
(b) Decrease in maximal tubular excretory capacity
(c) Decrease in filtration factor
(d) Decrease in renal plasma flow **Ans. a**

162. ADH test is based on the measurement of
(a) Specific gravity of urine
(b) Concentration of urea in urine
(c) Concentration of urea in blood
(d) Volume of urine in ml/minute **Ans. a**

163. The specific gravity of urine normally ranges from
(a) 0.900–0.999 (b) 1.003–1.030
(c) 1.000–1.001 (d) 1.101–1.120 **Ans. b**

164. Specific gravity of urine increases in
(a) Diabetes mellitus (b) Chronic glomerulonephritis
(c) Compulsive polydypsia (d) Hypercalcemia **Ans. a**

165. Fixation of specific gravity of urine to 1.010 is found in
(a) Diabetes insipidus (b) Compulsive polydypsia
(c) Cystinosis (d) Chronic glomerulonephritis
Ans. d

166. Addis test is the measure of
(a) Impairment of the capacity of the tubule to perform osmotic work
(b) Secretory function of liver
(c) Excretory function of liver
(d) Activity of parenchymal cells of liver **Ans. a**

167. Number of stereoisomers of glucose is
(a) 4 (b) 8
(c) 16 (d) None of these **Ans. c**

168. Maltose can be formed by hydrolysis of
(a) Starch (b) Dextrin
(c) Glycogen (d) All of these **Ans. d**

169. α –D–Glucuronic acid is present in

(a) Hyaluronic acid (B) Chondroitin sulphate

(c) Heparin (d) All of these **Ans. c**

170. Fructose is present in hydrolysate of

(a) Sucrose (b) Inulin

(c) Both of the above (d) None of these **Ans. c**

171. Ribulose is a

(a) Ketotetrose (b) Aldotetrose

(c) Ketopentose (d) Aldopentose **Ans. c**

172. A carbohydrate, commonly known as dextrose is

(a) Dextrin (b) D-Fructose

(c) D-Glucose (d) Glycogen **Ans. c**

173. A carbohydrate found only in milk is

(a) Glucose (b) Galactose

(c) Lactose (d) Maltose **Ans. c**

174. A carbohydrate, known commonly as invert sugar, is

(a) Fructose (b) Sucrose

(c) Glucose (d) Lactose Ans. b

175. A heteropolysaccharide among the following is

(a) Inulin (b) Cellulose

(c) Heparin (d) Dextrin **Ans. c**

176. The predominant form of glucose in solution is

(a) Acyclic form (b) Hydrated acyclic form

(c) Glucofuranose (d) Glucopyranose **Ans. d**

177. An L-isomer of monosaccharide formed in human body is

(a) L-fructose (b) L-Erythrose

(c) L-Xylose (d) L-Xylulose **Ans. d**

178. Hyaluronic acid is found in

(a) Joints (b) Brain

(c) Abdomen (d) Mouth **Ans. a**

179. The carbon atom which becomes asymmetric when the straight chain form of monosaccharide changes into ring form is known as

(a) Anomeric carbon atom (b) Epimeric carbon atom
(c) Isomeric carbon atom (d) None of these **Ans. a**

180. The smallest monosaccharide having furanose ring structure is

(a) Erythrose (b) Ribose
(c) Glucose (d) Fructose **Ans. b**

181. Which of the following is an epimeric pair?

(a) Glucose and fructose (b) Glucose and galactose
(c) Galactose and mannose (d) Lactose and maltose **Ans. b**

182. α-Glycosidic bond is present in

(a) Lactose (b) Maltose
(c) Sucrose (d) All of these **Ans. b**

183. Branching occurs in glycogen approximately after every

(a) Five glucose units (b) Ten glucose units
(c) Fifteen glucose units (d) Twenty glucose units **Ans. b**

184. N-Acetylglucosamnine is present in

(a) Hyaluronic acid (b) Chondroitin sulphate
(c) Heparin (d) All of these **Ans. a**

185. Iodine gives a red colour with

(a) Starch (b) Dextrin
(c) Glycogen (d) Inulin **Ans. c**

186. Amylose is a constituent of

(a) Starch (b) Cellulose
(c) Glycogen (d) None of these **Ans. a**

187. Synovial fluid contains

(a) Heparin (b) Hyaluronic acid
(c) Chondroitin sulphate (d) Keratin sulphate **Ans. b**

188. Gluconeogenesis is decreased by

(a) Glucagon (b) Epinephrine
(c) Glucocorticoids (d) Insulin **Ans. d**

189. Lactate formed in muscles can be utilized through
(a) Rapoport-Luebeling cycle (d) Glucose-alanine cycle
(c) Cori's cycle (d) Citric acid cycle **Ans. c**

190. Glucose-6-phosphatase is not present in
(a) Liver and kidneys (b) Kidneys and muscles
(c) Kidneys and adipose tissue (d) Muscles and adipose tissue **Ans. d**

191. Pyruvate carboxylase is regulated by
(a) Induction (b) Repression
(c) Allosteric regulation (d) All of these **Ans. d**

192. Fructose-2, 6-bisphosphate is formed by the action of
(a) Phosphofructokinase-1
(b) Phosphofructokinase-2
(c) Fructose bisphosphate isomerase
(d) Fructose-1, 6-bisphosphatase **Ans. b**

193. The highest concentrations of fructose are found in
(a) Aqueous humor (b) Vitreous humor
(c) Synovial fluid (d) Seminal fluid **Ans. d**

194. Glucose uptake by liver cells is
(a) Energy-consuming (b) A saturable process
(c) Insulin-dependent (d) Insulin-independent **Ans. d**

195. Renal threshold for glucose is decreased in
(a) Diabetes mellitus (b) Insulinoma
(c) Renal glycosuria (d) Alimentary glycosuria **Ans. c**

196. Active uptake of glucose is inhibited by
(a) Ouabain (b) Phlorrizin
(c) Digoxin (d) Alloxan **Ans. b**

197. Glucose-6-phosphatase is absent or deficient in
(a) Von Gierke's disease (b) Pompe's disease
(c) Cori's disease (d) McArdle's disease **Ans. a**

198. Debranching enzyme is absent in
(a) Cori's disease (b) Andersen's disease
(c) Von Gierke's disease (d) Her's disease **Ans. a**

199. McArdle's disease is due to the deficiency of
(a) Glucose-6-phosphatase (d) Phosphofructokinase
(c) Liver phosphorylase (d) Muscle phosphorylase **Ans. d**

200. Tautomerisation is
(a) Shift of hydrogen (b) Shift of carbon
(c) Shift of both (d) None of these **Ans. a**

201. In essential pentosuria, urine contains
(a) D-Ribose (b) D-Xylulose
(c) L-Xylulose (d) D-Xylose **Ans. c**

202. Action of salivary amylase on starch leads to the formation of
(a) Maltose (b) Maltotriose
(c) Both of the above (d) Neither of these **Ans. c**

203. Congenital galactosaemia can lead to
(a) Mental retardation (b) Premature cataract
(c) Death (d) All of the above **Ans. d**

204. Uridine diphosphate glucose (UDPG) is
(a) Required for metabolism of galactose
(b) Required for synthesis of glucuronic acid
(c) A substrate for glycogen synthetase
(d) All of the above **Ans. d**

205. Catalytic activity of salivary amylase requires the presence of
(a) Chloride ions (b) Bromide ions
(c) Iodide ions (d) All of these **Ans. a**

206. The following is actively absorbed in the intestine
(a) Fructose (b) Mannose
(c) Galactose (d) None of these **Ans. c**

207. An amphibolic pathway among the following is
(a) HMP shunt (b) Glycolysis
(c) Citirc acid cycle (d) Gluconeogenesis **Ans. c**

208. Cori's cycle transfers
(a) Glucose from muscles to liver (b) Lactate from muscles to liver
(c) Lactate from liver to muscles (d) Pyruvate from liver to muscles
Ans. b

209. Excessive intake of ethanol increases the ratio

(a) NADH : NAD^+ (b) NAD^+ : NADH

(c) $FADH_2$: FAD (d) FAD : $FADH_2$ **Ans. a**

210. Ethanol decreases gluconeogenesis by

(a) Inhibiting glucose-6-phosphatase

(b) Inhibiting PEP carboxykinase

(c) Converting NAD^+ into NADH and decreasing the availability of pyruvate

(d) Converting NAD^+ into NADH and decreasing the availability of lactate **Ans. c**

211. Glycogenin is

(a) Uncoupler of oxidative phosphorylation

(b) Polymer of glycogen molecules

(d) Protein primer for glycogen synthesis

(d) Intermediate in glycogen breakdown **Ans. c**

212. During starvation, ketone bodies are used as a fuel by

(a) Erythrocytes (b) Brain

(c) Liver (d) All of these **Ans. b**

213. Animal fat is in general

(a) Poor in saturated and rich in polyunsaturated fatty acids

(b) Rich in saturated and poor in polyunsaturated fatty acids

(c) Rich in saturated and polyunsaturated fatty acids

(d) Poor in saturated and polyunsaturated fatty acids **Ans. b**

214. In the diet of a diabetic patient, the recommended carbohydrate intake should preferably be in the form of

(a) Monosaccharides (b) Disaccharides

(c) Polysaccharides (d) All of these **Ans. c**

215. Obesity increases the risk of

(a) Hypertension (b) Diabetes mellitus

(c) Cardiovascular disease (d) All of these **Ans. d**

216. Worldwide, the most common vitamin deficiency is that of

(a) Ascorbic acid (b) Folic acid

(c) Vitamin A (d) Vitamin D **Ans. b**

217. Consumption of iodised salt is recommended for prevention of
(a) Hypertension (b) Hyperthyroidism
(c) Endemic goitre (d) None of these **Ans. c**

218. Restriction of salt intake is generally recommended in
(a) Diabetes mellitus (b) Hypertension
(c) Cirrhosis of liver (d) Peptic ulcer **Ans. b**

219. Polyuria can occur in
(a) Diabetes mellitus (b) Diarrhoea
(c) Acute glomerulonephritis (d) High fever **Ans. b**

220. Specific gravity of urine is raised in all of the following except
(a) Diabetes mellitus (b) Diabetes insipidus
(c) Dehydration (d) Acute glomerulonephritis
Ans. d

221. Specific gravity of urine is decreased in
(a) Diabetes mellitus (b) Acute glomerulonephritis
(c) Diarrhoea (d) Chronic glomerulonephritis
Ans. b

222. Heavy proteinuria occurs in
(a) Acute glomerulonephritis (b) Acute pyelonephritis
(c) Nephrosclerosis (d) Nephrotic syndrome **Ans. b**

223. Mucopolysaccharides are
(a) Homopolysaccharides (b) Hetropolysaccharides
(c) Proteins (d) Amino acids **Ans. a**

224. Bence-Jones protein precipitates at
(a) 20°–40° C (b) 40–60° C
(c) 60°–80° C (d) 80°–100° C **Ans. b**

225. Serum cholesterol is decreased in
(a) Endemic goitre (b) Thyrotoxicosis
(c) Myxoedema (d) Cretinism **Ans. b**

226. The heptose ketose sugar formed as a result of chemical reaction in HMP shunt
(a) Sedoheptulose (b) Galactoheptose
(c) Glucoheptose (d) Mannoheptose **Ans. a**

227. The number of isomers of glucose is

(a) 4 (b) 8

(c) 12 (d) 16 **Ans. d**

228. The epimers of glucose is

(a) Fructose (b) Galactose

(c) Ribose (d) Deoxyribose **Ans. b**

229. The intermediate in hexose monophosphate shunt is

(a) D-Ribulose (b) D-Arobinose

(c) D-xylose (d) D-lyxose Ans. a

230. Honey contains the hydrolytic product of

(a) Lactose (b) Maltose

(c) Inulin (d) Starch **Ans. c**

231. On boiling Benedict's solution is not reduced by

(a) Sucrose (b) Lactose

(c) Maltose (d) Fructose **Ans. a**

232. Glycosides are found in many

(a) Vitamins (b) Drugs

(c) Minerals (d) Nucleoproteins **Ans. b**

233. Galactose on oxidation with conc. HNO_3 produces

(a) Gluconic acid (b) Saccharic acid

(c) Saccharo Lactone (d) Mucic acid **Ans. d**

234. The distinguishing test between monosaccharides and disaccharides is

(a) Bial's test (b) Selwanoff's test

(c) Barfoed's test (d) Hydrolysis test **Ans. c**

235. Cellulose is made up of the molecules of

(a) α-glucose (b) β-glucose

(c) Both of the above (d) None of these **Ans. a**

236. Iodine solution produces no color with

(a) Cellulose (b) Starch

(c) Dextrin (d) Glycogen **Ans. a**

237. Glycogen structure includes a branch in between – – – glucose units

(a) 6–12 (b) 8–14

(c) 6–10 (d) 12–18 **Ans. d**

238. Amylose contains glucose units

(a) 100–200 (b) 200–300

(c) 300–400 (d) 500–600 **Ans. c**

239. N-acetylneuraminic acid is an example of

(a) Sialic acid (b) Mucic acid

(c) Glucuronic acid (d) Hippuric acid **Ans. c**

240. In place of glucuronic acid chondroitin sulphate B contains

(a) Gluconic acid (b) Gulonic acid

(c) Induronic acid (d) Sulphonic acid **Ans. c**

241. Blood group substances consist of

(a) Lactose (b) Maltose

(c) Fructose (d) Mucose **Ans. c**

242. The component of cartilage and cornea is

(a) Keratosulphate (b) Chondroitin sulphate

(c) Cadmium sulphate (d) Antimony sulphate **Ans. a**

243. Benedict's test is less likely to give weakly positive results with concentrated urine due to the action of

(a) Urea (b) Uric acid

(c) Ammonium salts (d) Phosphates **Ans. b**

244. Active transport of sugar is depressed by the agent

(a) Oxaloacetate (b) Fumarate

(c) Malonate (d) Succinate **Ans. c**

245. The general test for detection of carbohydrates is

(a) Iodine test (b) Molisch test

(c) Barfoed test (d) Osazone test **Ans. b**

246. Glucose absorption may be decreased in

(a) Oedema (b) Nephritis

(c) Rickets (d) Osteomalitis **Ans. a**

247. Glycogen synthetase activity is depressed by

(a) Glucose (b) Insulin

(c) Cyclic AMP (d) Fructokinase **Ans. c**

248. The branching enzyme acts on the glycogen when the glycogen chain has been lengthened to between glucose units

(a) 1 and 6 (b) 2 and 7
(c) 3 and 9 (d) 6 and 11 **Ans. d**

249. Cyclic AMP is formed from ATP by the enzyme adenylate cyclase which is activated by the hormone

(a) Insulin (b) Epinephrine
(c) Testosterone (d) Progesterone **Ans. b**

250. Hexokinase has a high affinity for glucose than

(a) Fructokinase (b) Galactokinase
(c) Glucokinase (d) All of the above **Ans. c**

251. Dihydroxyacetone phosphate and glyceraldehyde-3-phosphate are inter-converted by

(a) Triose isomerase
(b) Phosphotriose isomerase
(c) Diphosphotriose isomerase
(d) Dihydroxyacetone phosphorylase **Ans. b**

252. Citrate is converted to isocitrate by aconitase which contains

(a) Ca^{++} (b) Fe^{++}
(c) Zn^{++} (d) Mg^{++} **Ans. b**

253. The reaction succinyl CoA to succinate requires

(a) CDP (b) ADP
(c) GDP (d) $NADP^+$ **Ans. b**

254. The carrier of the citric acid cycle is

(a) Succinate (b) Fumarate
(c) Malate (d) Oxaloacetate **Ans. d**

255. UDPG is oxidized to UDP glucuronic acid by UDP dehydrogenase in presence of

(a) FAD^+ (b) NAD^+
(c) $NADP^+$ (d) ADP^+ **Ans. b**

256. Galactose is phosphorylated by galactokinase to form

(a) Galactose-6-phosphate (b) Galactose-1, 6 diphosphate
(c) Galactose-1-phosphate (d) All of these **Ans. c**

257. The conversion of alanine to glucose is termed

(a) Glycolysis (b) Oxidative decarboxylation

(c) Specific dynamic action (d) Gluconeogenesis **Ans. d**

258. The blood sugar raising action of the hormones of suprarenal cortex is due to

(a) Gluconeogenesis

(b) Glycogenolysis

(c) Glucagon-like activity

(d) Due to inhibition of glomerular filtration **Ans. a**

259. Under anaerobic conditions the glycolysis one mole of glucose yields ___________ moles of ATP

(a) One (b) Two

(c) Eight (d) Thirty **Ans. b**

260. Which of the following metabolite integrates glucose and fatty acid metabolism?

(a) Acetyl CoA (b) Pyruvate

(c) Citrate (d) Lactate **Ans. a**

261. Cerebrosides consist of mostly of this sugar

(a) Glucose (b) Fructose

(c) Galactose (d) Arabinose **Ans. c**

262. Glucose will be converted into fatty acids if the diet has excess of

(a) Carbohydrates (b) Proteins

(c) Fat (d) Vitamins **Ans. a**

263. The purple ring of Molisch reaction is due to

(a) Furfural (b) Furfural + α Naphthol

(c) °C Naphthol (d) Furfurol + H_2SO_4 + α-Naphthol **Ans. b**

264. One of the following enzymes does not change glycogen synthase a to b

(a) Glycogen synthase kinases 3, 4, 5

(b) Ca^{2+} calmodulin phosphorylase kinase

(c) Ca^{2+} calmodulin dependent protein kinase

(d) Glycogen phosphorylase a **Ans. d**

265. In EMP pathway -2-phosphoglycerate is converted to
(a) Phospho enol pyruvate
(b) Enol pyruvate
(c) Di hydroxy acetone phosphate (DHAP)
(d) 1,3 bisphosphoglycerate **Ans. a**

266. An aneplerotic reaction which sustains the availability of oxaloacetate is the carboxylation of
(a) Glutamate (b) Pyruvate
(c) Citrate (d) Succinate **Ans. b**

267. Specific test for ketohexoses
(a) Seliwanoff's test (b) Osazone test
(c) Molisch test (d) None of these **Ans. a**

268. Two important byproducts of HMP shunt are
(a) NADH and pentose sugars
(b) NADPH and pentose sugars
(c) Pentose sugars and 4 membered sugars
(d) Pentose sugars and sedoheptulose **Ans. b**

269. Pyruvate dehydrogenase complex and α-ketoglutarate dehydrogenase complex require the following for their oxidative decarboxylation
(a) CoASH and Lipoic acid
(b) NAD^+ and FAD
(c) CoASH and TPP
(d) CoASH, TPP, NAD^+, FAD, Lipoate **Ans. d**

270. The four membered aldose sugar phosphate formed in HMP shunt pathway is
(a) Xylulose P (b) Erythrulose P
(c) Erythrose P (d) Ribulose P **Ans. c**

271. Cane sugar (Sucrose) injected into blood is
(a) changed to fructose
(b) changed to glucose
(c) undergoes no significant change
(d) changed to glucose and fructose **Ans. c**

272. Pentose production is increased in
(a) HMP shunt (b) Uromic acid pathway
(c) EMP pathway (d) TCA cycle **Ans. a**

273. Conversion of Alanine to carbohydrate is termed
(a) Glycogenesis (b) Gluconeogenesis
(c) Glycogenolysis (d) Photosynthesis **Ans. b**

274. The following is an enzyme required for glycolysis
(a) Pyruvate kinase (b) Pyruvate carboxylase
(c) Glucose-6-phosphatase (d) Glycerokinase **Ans. a**

275. Our body can get pentoses from
(a) Glycolytic pathway (b) Uromic acid pathway
(c) TCA cycle (d) HMP shunt **Ans. d**

276. Conversion of glucose to glucose-6- phosphate in human liver is by
(a) Hexokinase only
(b) Glucokinase only
(c) Hexokinase and glucokinase
(d) Glucose-6-phosphate dehydrogenase **Ans. c**

277. The normal glucose tolerance curve reaches peak is
(a) 15 min (b) 1 hr
(c) 2 hrs (d) 2½ hrs **Ans. b**

278. Oxidative decarboxylation of pyruvate requires
(a) $NADP^+$ (b) Cytochromes
(c) Pyridoxal phosphate (d) CoASH **Ans. d**

279. Glucose tolerance is increased in
(a) Diabetes mellitus (b) Adrenalectomy
(c) Acromegaly (d) Thyrotoxicosis **Ans. b**

280. Glucose tolerance is decreased in
(a) Diabetes mellitus (b) Hypopituitarisme
(c) Addison's disease (d) Hypothyroidism **Ans. a**

281. During glycolysis, Fructose 1, 6 diphosphate is decomposed by the enzyme
(a) Enolase a (b) Fructokinase
(c) Aldolase (d) Diphosphofructophosphatose
Ans. c

282. The following enzyme is required for the hexose monophosphate shunt pathway

(a) Glucose-6-phosphatase

(b) Phosphorylase

(c) Aldolase

(d) Glucose-6-phosphate dehydrogenase **Ans. d**

283. Dehydrogenase enzymes of the hexose monophosphate shunt are

(a) NAD^+ specific (b) $NADP^+$ specific

(c) FAD specific (d) FMN specific **Ans. b**

284. Glycogen is converted to glucose-1-phosphate by

(a) UDPG transferase (b) Branching enzyme

(c) Phosphorylase (d) Phosphatase **Ans. c**

285. Which of the following is not an enzyme involved in glycolysis?

(a) Euolase (b) Aldolose

(c) Hexokinase (d) Glucose oxidase **Ans. d**

286. Tricarboxylic acid cycle to be continuous requires the regeneration of

(a) Pyruvic acid (b) oxaloacetic acid

(c) α-oxoglutaric acid (d) Malic acid **Ans. b**

287. Dehydrogenation of succinic acid to fumaric acid requires the following hydrogen carrier

(a) NAD^+ (b) $NADP^+$

(c) flavoprotein (d) Glutathione **Ans. c**

288. The tissues with the highest total glycogen content are

(a) Muscle and kidneys (b) Kidneys and liver

(c) Liver and muscle (d) Brain and Liver **Ans. c**

289. Rothera test is not given by

(a) β-hydroxy butyrate (b) Bile salts

(c) Glucose (d) None of these **Ans. a**

290. Gluconeogenesis is increased in the following condition

(a) Diabetes insipidus (b) Diabetes Mellitus

(c) Hypothyroidism (d) Liver diseases **Ans. b**

291. The oxidation of lactic acid to pyruvic acid requires the following vitamin derivative as the hydrogen carrier

(a) Lithium pyrophosphate (b) Coenyzme A

(c) NAD^+ (d) FMN **Ans. c**

292. Physiological glycosuria is met with in

(a) Renal glycosuria (b) Alimentary glycosuria

(c) Diabetes Mellitus (d) Alloxan diabetes **Ans. b**

293. Two examples of substrate level phosphorylation in EM pathway of glucose metabolism are in the reactions of

(a) 1,3-bisphosphoglycerate and phosphoenol pyruvate

(b) Glucose-6-phosphate and Fructo-6-phosphate

(c) 3-phosphoglyceraldehyde and phosphoenolpyruvate

(d) 1,3-diphosphoglycerate and 2-phosphoglycerate **Ans. a**

294. The number of molecules of ATP produced by the total oxidation of acetyl CoA in TCA cycle is

(a) 6 (b) 8

(c) 10 (d) 12 **Ans. d**

295. Substrate level phosphorylation in TCA cycle is in step

(a) Isocitrate dehydrogenase (b) Malate dehydrogenase

(c) Aconitase (d) Succinate thiokinase **Ans. d**

296. Fatty acids cannot be converted into carbohydrates in the body as the following reaction is not possible

(a) Conversion of glucose-6-phosphate into glucose

(b) Fructose 1, 6-bisphosphate to fructose-6- phosphate

(c) Transformation of acetyl CoA to pyruvate

(d) Formation of acetyl CoA from fatty acids **Ans. c**

297. Tissues form lactic acid from glucose. This phenomenon is termed as

(a) Aerobic glycolysis (b) Oxidation

(c) Oxidative phosphorylation (d) Anaerobic glycolysis **Ans. d**

298. One molecule of glucose gives ______ molecules of CO_2 in EMP-TCA cycle

(a) 6 (b) 3

(c) 1 (d) 2 **Ans. a**

299. One molecule of glucose gives ______ molecules of CO_2 in one round of HMP shunt.

(a) 6 (b) 1

(c) 2 (d) 3 **Ans. b**

300. The 4 rate limiting enzymes of gluconeogenesis are

(a) Glucokinase, Pyruvate carboxylae phosphoenol pyruvate carboxykinase and glucose-6-phosphatase

(b) Pyruvate carboxylase, phosphoenol pyruvate carboxykinase, fructose1,6-diphosphatase and glucose-6-phosphatase

(c) Pyruvate kinase, pyruvate carboxylase, phosphoenol pyruvate carboxykinase and glucose-6-phosphatase

(d) Phospho fructokinase, pyruvate carboxylase, phosphoenol pyruvate carboxykinase and fructose 1, 6-diphosphatase **Ans. b**

301. For glycogenesis, Glucose should be converted to

(a) Glucuronic acid (b) Pyruvic acid

(c) UDP glucose (d) Sorbitol **Ans. c**

302. Fluoride inhibits ______ and arrests glycolysis

(a) Glyceraldehyde- 3- phosphate dehydrogenase

(b) Aconitase

(c) Enolase

(d) Succinate dehydrogenase **Ans. c**

303. One of the following statement is correct

(a) Glycogen synthase 'a' is the phosphorylated

(b) cAMP converts glycogen synthase b to 'a'

(c) Insulin converts glycogen synthase b to a

(d) UDP glucose molecules interact and grow into a Glycogen tree **Ans. c**

304. Amylo 1, 6-glucosidase is called

(a) Branching enzyme (b) Debranching enzyme

(c) Glucantransferase (d) Phosphorylase **Ans. b**

305. Glucose enters the cells by

(a) Insulin independent transport (b) Insulin dependent transport

(c) Enzyme mediated transport (d) Both (a) and (b) **Ans. d**

306. Glycogen while being acted upon by active phosphorylase is converted first to

(a) Glucose

(b) Glucose 1-phosphate and Glycogen with 1 carbon less

(c) Glucose-6-phosphate and Glycogen with 1 carbon less

(d) 6-Phosphogluconic acid **Ans. c**

307. When O_2 supply is inadequate, pyruvate is converted to

(a) Phosphopyruvate (b) Acetyl CoA

(c) Lactate (d) Alanine **Ans. c**

308. Reactivation of inactive liver phosphorylase is normally favoured by

(a) Insulin (b) Epinephrine

(c) ACTH (d) Glucagon **Ans. d**

309. Before pyruvic acid enters the TCA cycle it must be converted to

(a) Acetyl CoA (b) Lactate

(c) α-ketoglutarate (d) Citrate **Ans. a**

310. The hydrolysis of Glucose-6-phosphate is catalysed by a specific phosphatase which is found only in

(a) Liver, intestines and kidneys (b) Brain, spleen and adrenals

(c) Striated muscle (d) Plasma **Ans. a**

311. The formation of citrate from oxalo acetate and acetyl CoA is

(a) Oxidation (b) Reduction

(c) Condensation (d) Hydrolysis **Ans. c**

312. Which one of the following is a rate limiting enzyme of gluconeogenesis?

(a) Hexokinase (b) Phsophofructokinase

(c) Pyruvate carboxylase (d) Pyruvate kinase **Ans. c**

313. The number of ATP produced in the succinate dehydrogenase step is

(a) 1 (b) 2

(c) 3 (d) 4 **Ans. b**

314. Which of the following reaction gives lactose?

(a) UDP galactose and glucose (b) UDP glucose and galactose

(c) Glucose and Galactose (d) Glucose, Galactose and UTP

Ans. a

315. UDP Glucuronic acid is required for the biosynthesis of
(a) Chondroitin sulphates (b) Glycogen
(c) Lactose (d) Starch **Ans. a**

316. Which one of the following can covert glucose to vitamin C?
(a) Albino rats (b) Humans
(c) Monkeys (d) Guinea pigs **Ans. a**

317. Which one of the following cannot convert glucose to Vitamin C?
(a) Albino rats (b) Dogs
(c) Monkeys (d) Cows **Ans. c**

318. Transketolase has the coenzyme
(a) NAD^+ (b) FP
(c) TPP (d) Pyridoxol phosphate **Ans. c**

319. Two conditions in which gluconeogenesis is increased are
(a) Diabetes mellitus and atherosclerosis
(b) Fed condition and thyrotoxicosis
(c) Diabetes mellitus and Starvation
(d) Alcohol intake and cigarette smoking **Ans. c**

320. Acetyl CoA is not used for the synthesis of
(a) Fatty acid (b) Cholesterol
(c) Pyruvic acid (d) Citric acid **Ans. c**

321. The total glycogen content of the body is about ______ gms.
(a) 100 (b) 200
(c) 300 (d) 500 **Ans. c**

322. The total Glucose in the body is ________ gms.
(a) 10–15 (b) 20–30
(c) 40–50 (d) 60–80 **Ans. b**

323. Pyruvate kinase requires ______ ions for maximum activity
(a) Na^+ (b) K^+
(c) Ca^{2+} (d) Mg^{2+} **Ans. b**

324. ATP is 'wasted' in Rapoport-Lueberring cycle in RBCs as otherwise it will inhibit
(a) Phosphoglucomutase
(b) Phosphohexo isomerase

(c) Phosphofructo kinase
(d) Phosphoenol pyruvate carboxy kinase **Ans. c**

325. The following co-enzyme is needed for the oxidative decarboxylation of ketoacids
(a) $NADP^+$ (b) TPP
(c) Folate coenzyme (d) Biotin coenzyme **Ans. b**

326. Synthesis of Glucose from amino acids is termed as
(a) Glycolysis (b) Gluconeogenesis
(c) Glycogenesis (d) Lipogenesis **Ans. b**

327. The following examples are important heteropolysaccharides except
(a) Amylopectin (b) Heparin
(c) Peptidoglycan (d) Hyaluronic acid **Ans. a**

328. Which of the following features are common to monosaccharides?
(a) Contain asymmetric centres
(b) Are of 2 types – aldoses and ketoses
(c) Tend to exist as ring structures in solution
(d) Include glucose, galactose and raffinose **Ans. c**

329. Polysaccharides
(a) Contain many monosaccharide units which may or may not be of the same kind
(b) Function mainly a storage or structural compounds
(c) Are present in large amounts in connective tissue
(d) All of these **Ans. d**

330. The absorption of glucose in the digestive tract
(a) Occurs in the small intestine
(b) Is stimulated by the hormone Glucagon
(c) Occurs more rapidly than the absorption of any other sugar
(d) Is impaired in cases of diabetes mellitus **Ans. a**

331. UDP-Glucose is converted to UDPGlucuronic acid by
(a) ATP (b) GTP
(c) $NADP^+$ (d) NAD^+ **Ans. b**

332. The enzymes involved in Phosphorylation of glucose to glucose 6-phosphate are

(a) Hexokinase (b) Glucokinase

(c) Phosphofructokinase (d) Both (a) and (b) **Ans. d**

333. In conversion of Lactic acid to Glucose, three reactions of Glycolytic pathway are circumvented, which of the following enzymes do not participate?

(a) Pyruvate Carboxylase

(b) Phosphoenol pyruvate carboxy kinase

(c) Pyruvate kinase

(d) Glucose-6-phosphatase **Ans. c**

334. The normal resting state of humans, most of the blood glucose burnt as "fuel" is consumed by

(a) Liver (b) Brain

(c) Kidneys (d) Adipose tissue **Ans. b**

335. A regulator of the enzyme Glycogen synthase is

(a) Citric acid (b) 2, 3-bisphosphoglycerate

(c) Pyruvate (d) GTP **Ans. c**

336. Which of the following compound is a positive allosteric modifier of the enzyme pyruvate carboxylase?

(a) Biotin (b) Acetyl CoA

(c) Oxaloacetate (d) ATP **Ans. a**

337. A specific inhibitor for succinate dehydrogenase is

(a) Arsinate (b) Malonate

(c) Citrate (d) Cyanide **Ans. b**

338. Most of the metabolic pathways are either anabolic or catabolic. Which of the following pathways is considered as "amphibolic" in nature?

(a) Glycogenesis (b) Glycolytic pathway

(c) Lipolysis (d) TCA cycle **Ans. d**

339. Transketolase activity is affected in

(a) Biotin deficiency (b) Pyridoxine deficiency

(c) PABA deficiency (d) Thiamine deficiency **Ans. b**

340. The following metabolic abnormalities occur in Diabetes mellitus except

(a) Increased plasma FFA

(b) Increased pyruvate carboxylase Activity

(c) Decreased lipogenesis

(d) Decreased gluconeogenesis **Ans. b**

341. A substance that is not an intermediate in the formation of D-glucuronic acid from glucose is

(a) Glucoss-1-P (b) 6-Phosphogluconate

(c) Glucose-6-P (d) UDP-Glucose **Ans. b**

342. The hydrolysis of Glucose-6-P is catalysed by a phosphatase that is not formed in which of the following?

(a) Liver (b) Kidney

(c) Muscle (d Small intestine **Ans. c**

343. An essential for converting Glucose to Glycogen in Liver is

(a) Lactic acid (b) GTP

(c) CTP (d) UTP **Ans. d**

344. Which of the following is a substrate for aldolase activity in Glycolytic pathway?

(a) Glyceraldehyde-3-P (b) Glucose-6-P

(c) Fructose-6-P (d) Fructose1, 6-bisphosphate **Ans. d**

345. The ratio that approximates the number of net molecule of ATP formed per mole of Glucose oxidized in presence of O_2 to the net number formed in abscence of O_2 is

(a) 4 : 1 (b) 10 : 2

(c) 12 : 1 (d) 18 : 1 **Ans. b**

346. The Primaquin sensitivity types of haemolytic anaemia has been found to relate to reduced R.B.C activity of which enzyme?

(a) Pyruvate kinase deficiency

(b) Glucose-6-phosphatase deficiency

(c) Glucose-6-P dehydrogenase deficiency

(d) Hexokinase deficiency **Ans. c**

347. Which of the following hormones is not involved in carbohydrate metabolism?

(a) Cortisol (b) ACTH

(c) Glucogon (d) Vasopressin **Ans. d**

348. Which of the following enzymes in Glycolytic pathway is inhibited by fluoride?

(a) Glyceraldehyde-3-P dehydrogenase

(b) Phosphoglycerate kinase

(c) Pyruvate kinase

(c) Enolase **Ans. d**

349. Out of 24 mols of ATP formed in TCA cycle, 2 molecules of ATP can be formed at "substrate level" by which of the following reaction?

(a) Citric acid → Isocitric acid (b) Isocitrate → Oxaloacetate

(c) Succinic acid → Fumarate (d) Succinylate → Succinic acid

Ans. d

350. Which of the following statements regarding T.C.A cycle is true?

(a) It is an anaerobic process

(b) It occurs in cytosol

(c) It contains no intermediates for Gluconeogenesis

(d) It is amphibolic in nature **Ans. d**

351. An allosteric enzyme responsible for controlling the rate of T.C.A cycle is

(a) Malate dehydrogenase (b) Isocitrate dehydrogenase

(c) Fumarase (d) Aconitase **Ans. b**

352. The glycolysis is regulated by

(a) Hexokinase (b) Phosphofructokinase

(c) Pyruvate kinase (d) All of these **Ans. d**

353. How many ATP molecules will be required for conversion of 2-molecules of Lactic acid to Glucose?

(a) 2 (b) 4

(c) 8 (d) 6 **Ans. d**

354. Which of the following enzyme is not involved in HMP shunt?

(a) Glyceraldehyde-3-P dehydrogenase

(c) Glucose-6-P-dehydrogenase

(c) Transketolase
(d) Phosphogluconate dehydrogenase **Ans. a**

355. In presence of the following cofactor, pyruvate carboxylase converts pyruvate to oxaloacetate

(a) ATP, Protein and CO_2 (b) CO_2 and ATP
(c) CO_2 (d) Protein **Ans. a**

356. For conversion of oxaloacetate to phosphoenol pyruvate, high energy molecule is required in the form of

(a) GTP only (b) ITP only
(c) GTP (or) ITP (d) None of these **Ans. c**

357. If the more negative standard reduction potential of a redox pair, the greater the tendency to

(a) To lose electrons (b) To gain electrons
(c) To lose/gain electrons (d) To lose and gain electrons
Ans. a

358. Electron transport and phosphorylation can be uncoupled by compounds that increase the permeability of the inner mitochondrial membrane to

(a) Electrons (b) Protons
(c) Uncouplers (d) All of these **Ans. b**

359. The more positive the $E_{0'}$, the greater the tendency of the oxidant member of that pair to

(a) Lose electrons (b) Gain electrons
(c) Lose (or) gain electrons (d) Lose and gain electrons **Ans. b**

360. The standard free energy of hydrolysis of terminal phosphate group of ATP is

(a) –7,300 cal/mol (b) –8,300 cal/mol
(c) 10,000 cal/mol (d) +7,300 cal/mol **Ans. a**

361. The transport of a pair of electrons from NADH to O_2 via the electron transport chain produces

(a) –52,580 cal (b) –50,580 cal
(c) 21,900 cal (d) +52,580 cal **Ans. d**

362. Sufficient energy required to produce 3 ATP from 3 ADP and 3 Pi is

(a) –21,900 cal (b) 29,900 cal
(c) 31,900 cal (d) 39,900 cal **Ans. a**

363. The free energy change, ΔG

(a) Is directly proportional to the standard free energy change, ΔG

(b) Is equal to zero at equilibrium

(c) Can only be calculated when the reactants and products are present at 1mol/1 concentrations

(d) Is equal to –RT in K_{eq} **Ans. b**

364. Under standard conditions

(a) The free energy change ΔG°, is equal to 0

(b) The standard free energy change ΔG, is equal to 0

(c) The free energy change, ΔG°, is equal to the standard free energy change,ΔG°

(d) K_{eq} is equal to 1 **Ans. c**

365. An uncoupler of oxidative phosphorylation such as dinitrophenol

(a) Inhibits electron transport and ATP synthesis

(b) Allow electron transport to proceed without ATP synthesis

(c) Inhibits electron transport without impairment of ATP synthesis

(d) Specially inhibits cytochrome b **Ans. b**

366. All of the following statements about the enzymic complex that carries out the synthesis of ATP during oxidative phosphorylation are correct except

(a) It is located on the matrix side of the inner mitochondrial membrane

(b) It is inhibited by oligomycin

(c) It can exhibit ATPase activity

(d) It can bind molecular O_2 **Ans. d**

367. Glucokinase

(a) Is widely distributed and occurs in most mammalian tissues

(b) Has a high K_m for glucose and hence is important in the phosphorylation of glucose primarily after ingestion of a carbohydrate rich meal

(c) Is widely distributed in Prokaryotes

(d) None of these **Ans. b**

368. The reaction catalysed by phosphofructokinase

(a) Is activated by high concentrations of ATP and citrate

(b) Uses fruitose-1-phosphate as substrate

(c) Is the rate-limiting reaction of the glycolytic pathway

(d) Is inhibited by fructose 2, 6-bisphosphate **Ans. c**

369. Compared to the resting state, vigorously contracting muscle shows

(a) An increased conversion of pyruvate to lactate

(b) Decreased oxidation of pyruvate of CO_2 and water

(c) A decreased $NADH/NAD^+$ ratio

(d) Decreased concentration of AMP **Ans. a**

370. Which one of the following would be expected in pyruvate kinase deficiency?

(a) Increased levels of lactate in the R.B.C

(b) Hemolytic anemia

(d) Decreased ratio of ADP to ATP in R.B.C

(d) Increased phosphorylation of Glucose to Glucose-6-phosphate **Ans. b**

371. Which one of the following statements concerning glucose metabolism is correct?

(a) The conversion of Glucose to lactate occurs only in the R.B.C

(b) Glucose enters most cells by a mechanism in which Na^+ and glucose are co-transported

(c) Pyruvate kinase catalyses an irreversible reaction

(d) An elevated level of insulin leads to a decreased level of fructose 2, 6-bisphosphate in hepatocyte **Ans. c**

372. Which one of the following compounds cannot give rise to the net synthesis of Glucose?

(a) Lactate (b) Glycerol

(c) α-ketoglutarate (d) Acetyl CoA **Ans. b**

373. Which of the following reactions is unique to gluconeogenesis?

(a) Lactate Pyruvate

(b) Phosphoenol pyruvate

(c) Oxaloacetate phosphoenol pyruvate

(d) Glucose-6-phosphate Fructose-6-phosphate **Ans. c**

374. The synthesis of glucose from pyruvate by gluconeogenesis

(a) Requires the participation of biotin

(b) Occurs exclusively in the cytosol

(c) Is inhibited by elevated level of insulin
(d) Requires oxidation/reduction of FAD **Ans. a**

375. The conversion of pyruvate to acetyl CoA and CO_2
(a) Is reversible
(b) Involves the participation of lipoic acid
(c) Depends on the coenzyme biotin
(d) Occurs in the cytosol **Ans. b**

376. Pasteur effect is
(a) Inhibition of glycolysis
(b) Oxygen is involved
(c) Inhibition of enzyme phosphofructokinase
(d) All of these **Ans. d**

377. How many ATPs are produced in the conversion of phosphoenol pyruvate to citrate?
(a) 1 (b) 2
(c) 4 (d) 6 **Ans. c**

378. Reduced glutathione functions in R.B.Cs to
(a) Produce NADPH
(b) Reduce methemoglobin to hemoglobin
(c) Produce NADH
(d) Reduce oxidizing agents such as H_2O_2 **Ans. d**

379. Phenylalanine is the precursor of
(a) L-DOPA (b) Histamine
(c) Tyrosine (d) Thyroxine **Ans. c**

380. D-Mannose is present in some plant products like
(a) Resins (b) Pectins
(c) Mucilage (d) Gums **Ans. d**

381. Galactose is a main constituent of
(a) Milk sugar (b) Honey
(c) Cane sugar (d) Chitin **Ans. a**

382. Glucosamine is an important constituent of
(a) Homopolysaccharide (b) Heteropolysaccharide
(c) Mucopolysaccharide (d) Dextran **Ans. c**

383. Glycogen is present in all body tissues except
(a) Liver (b) Brain
(c) Kidney (d) Stomach **Ans. b**

384. Iodine test is positive for starch, dextrin and
(a) Mucoproteins (b) Agar
(c) Glycogen (d) Cellulose **Ans. c**

385. Human heart muscle contains
(a) D-Arabinose (b) D-Ribose
(c) D-Xylose (d) L-Xylose **Ans. c**

386. Barfoed's solution is not reduced by
(a) Glucose (b) Mannose
(c) Sucrose (d) Ribose **Ans. c**

387. Cori cycle is
(a) Synthesis of glucose (b) Reuse of glucose
(c) Uptake of glycose (d) Both (a) & (b) **Ans. d**

388. Cane sugar is known as
(a) Galactose (b) Sucrose
(c) Fructose (d) Maltose **Ans. b**

389. Which of the following is not reducing sugar?
(a) Lactose (b) Maltose
(c) Sucrose (d) Fructose **Ans. c**

390. α–D-Glucose and β–D-glucose are related by
(a) Epimers (b) Anomers
(c) Multirotation (d) Ketoenol pair **Ans. b**

391. The stable ring formation in D-Glucose involves
(a) C-1 and C-4 (b) C-1 and C-2
(c) C-1 and C-5 (d) C-2 and C-5 **Ans. c**

392. Reduction of Glucose with Ca^{++} in water produces
(a) Sorbitol (b) Dulcitol
(c) Mannitol (d) Glucuronic acid **Ans. a**

393. Starch and glycogen are polymers of
(a) Fructose (b) Mannose
(c) α–D-Glucose (d) Galactose **Ans. c**

394. Reducing ability of carbohydrates is due to
(a) Carboxyl group (b) Hydroxyl group
(c) Enediol formation (d) Ring structure **Ans. a**

395. Which of the following is not a polymer of glucose?
(a) Amylose (b) Inulin
(c) Cellulose (d) Dextrin **Ans. b**

396. The carbohydrate reserved in human body is
(a) Starch (b) Glucose
(c) Glycogen (d) Inulin **Ans. c**

397. A disaccharide linked by α-1-4 Glycosidic linkages is
(a) Lactose (b) Sucrose
(c) Cellulose (d) Maltose **Ans. d**

398. Simplest from of carbohydrate is
(a) Carbon (b) Starch
(c) Monosaccharide (d) Cane sugar **Ans. c**

399. The sweetest chemical is
(a) Fructose (b) Saccharin
(c) Monellin (d) Thaumatin **Ans. d**

400. Reducing sugars have
(a) Free aldehyde (b) Bound aldehyde
(c) Free aldehyde or ketone (d) Bound ketone **Ans. c**

401. Mannitol is
(a) Amino acid (b) Amino alcohol
(c) Sugar alcohol (d) Sugar acid **Ans. c**

402. Glycosidic bond
(a) C – O – C (b) CONH
(c) >C = O (d) CHO **Ans. a**

403. Glycogen is similar to
(a) Amylase (b) Amylopectin
(c) Cellulose (d) Chitin **Ans. b**

404. Gums are degradation product of
(a) Carbohydrates in cell wall (b) Proteins in cell wall
(c) Amino acids in cell (d) Lipids in cell membrane
Ans. a

405. A mucopolysaccharide of high commercial value is
(a) Agar (b) Carrageenin
(c) Alginic acid (d) All of these **Ans. d**

406. Mucilages are polysaccharides formed from
(a) Galactose + glucose (b) Galactose + galactose
(c) Galactose + lactose (d) Galactose + mannose **Ans. d**

407. Which of the following heteropolysaccharides act as an anticoagulant?
(a) Heparin (b) Gamma globulin
(c) Chondroitin 4-sulphate (d) All of these **Ans. a**

408. Which one of the following is a homopolysaccharide?
(a) Glycogen (b) Dextran
(c) Chitin (d) All of the above **Ans. d**

409. A carbohydrate unique to arthropods is
(a) Chitin (b) Hyaluronic acid
(c) Chondriotin sulphate (d) Waxes **Ans. a**

410. Which one of the following is a 5-C pentose?
(a) Fructose (b) Glyceraldehydes
(c) Dihydroxyacetone (d) Ribose **Ans. d**

411. Which of the followings is not a macromolecule?
(a) Amino acid (b) Protein
(c) Nucleic acid (d) Starch **Ans. a**

412. The major part of dry weight of plant is made up of
(a) Fat (b) Protein
(c) Carbohydrate (d) Nucleic acid **Ans. c**

413. Which is not a pentose sugar?
(a) Arabinose (b) Xylose
(c) Xylulose (d) Mannose **Ans. d**

414. Mannitol is a

(a) Sugar alcohol (b) Protein

(c) Lipid (d) None of these **Ans. a**

415. Which one of the following has no free aldehyde or ketone group?

(a) Fructose (b) Maltose

(c) Sucrose (d) Galactose **Ans. c**

416. Which one is not a homopolysaccharide

(a) Glycogen (b) Dextran

(c) Hyaluronic acid (d) Chitin **Ans. c**

417. The carbohydrates absent in plants are

(a) Galactose (b) Lactose

(c) Glycogen (d) All of these **Ans. c**

418. Pectin is

(a) Waste product (b) Excretory product

(c) Phytolectin (d) Secretory product **Ans. d**

419. Cellulose in plant cell wall is made up of

(a) Unbranched chain glucose, molecules linked by $\alpha, 1 \rightarrow 6$ glycosidic bond

(b) Unbranched chain glucose, molecules linked by $\beta, 1 \rightarrow 4$ glycosidic bond

(c) Branched chain glucose, molecules linked by $\beta, 1 \rightarrow 6$ glycosidic bond in straight chain and $\alpha, 1 \rightarrow 4$ at the site of branching

(d) Branched chains have $\alpha, 1 \rightarrow 4$ bond and $\beta, 1 \rightarrow 6$ glycosidic bond both **Ans. b**

420. Cellulose is

(a) Homoglycan pentosan

(b) Homoglycan hexosan (glucosan)

(c) Heteroglycan mucopolysaccharide

(d) Disaccharide **Ans. b**

421. The stored food found in muscles is

(a) Protein (b) Glycogen

(c) Lipid (d) Phosphorus **Ans. b**

422. Energy rich compound that is produced during biological oxidation of glucose is

(a) Adenosine triphosphate (b) Pyruvic acid
(c) Acetoacetate (d) Adenosine monophosphate

Ans. a

423. In submerged leaves, respiration takes place through

(a) Stomata present on ventral surface of leaf
(b) General surface of the leaf
(c) Stomata present on dorsal surface of the leaf
(d) Stomata present on both surfaces of the leaf **Ans. b**

424. The respiratory substrates used during respiration are carbohydrates, proteins, organic acids and fats which are used in the order of

(a) Fats, carbohydrates, protein, organic acid
(b) Carbohydrates, fats, proteins, organic acids
(c) Carbohydrates, proteins, organic acids, fats
(d) Protein, organic acid, carbohydrates **Ans. b**

425. What would you suggest an athlete to take before going to participate in an event?

(a) Butter (b) Sugar
(c) Bread (d) Cheese **Ans. b**

426. One day old cane sugar juice was found by the customer to have an alcoholic smell. It may be due to

(a) Aerobic respiration (b) Fermentation
(c) Putrefaction (d) Fungal growth **Ans. b**

427. Which of the following is important in carbohydrate and fat metabolism?

(a) Pyruvic acid (b) Citric acid
(c) Acetyl coenzyme A (d) Carbon dioxide **Ans. c**

428. Organelles which are regarded as "Power house" of the cell and in which the oxidative reactions of the respiratory process takes place are

(a) Mitochondria (b) Endoplasmic reticulum
(c) Ribosomes (d) Chloroplasts **Ans. a**

429. Conversion of 2-phosphoglyceric acid into 2-phosphoenol pyruvic acid requires

(a) Na^+ (b) K^+
(c) Ca^{++} (d) Mg^{++} **Ans. d**

430. During conversion of one pyruvic acid molecule into one molecule of acetyl CoA, how many molecules of CO_2 are produced?

(a) Four (b) Three

(c) Two (d) One **Ans. d**

431. Conversion of oxaloacetic acid to citric acid requires one molecules of

(a) CO_3 (b) CO_2

(c) CO (d) H_2O **Ans. d**

432. Which of the following compounds mediates the flow of electrons from reduced flavoproteins to respiratory cytochromes?

(a) Cytochrome a_6 (b) Coenzyme

(c) Cytochrome b_6 (d) Cytochrome a **Ans. b**

433. A fresh red berry kept in a cupboard can perform

(a) Photosynthesis (b) Transpiration

(c) Respiration (d) Both (b) and (c) **Ans. d**

434. Pyruvic acid has...carbon atoms

(a) 5 (b) 4

(c) 3 (d) 2 **Ans. c**

435. Life without air is

(a) Anaerobic (b) Impossible

(c) Reduction (d) Free from oxidative damage

Ans. a

436. Fishes is an aquarium thrive better if green plants are growing there because they

(a) Can feed on then

(b) Like green surroundings

(c) Inhale oxygen released by green plants

(d) Inhale CO_2 releases by green plants **Ans. c**

437. If soaked pea seeds are kept in four respiratory flasks, they would germinate best, if the flask contains

(a) Hydrogen (b) Oxygen

(c) Nitrogen (d) Carbon dioxide **Ans. b**

438. The synthesis of ATP in both photosynthesis and respiration is essentially an oxidation process involving the removal of energy from

(a) Cytochromes (b) Phytochromes

(c) Electrons (d) Oxygen **Ans. c**

439. In hexose monophosphate shunt, the net formation of ATP molecules is
(a) 38 (b) 36
(c) 35 (d) 34 **Ans. c**

440. Energy released per gram would be
(a) Highest when rice starch is respiratory substrate
(b) Highest when wheat starch is respiratory substrate
(c) Highest when potato starch is respiratory substrate
(d) Same in all the cases **Ans. d**

441. Respiration is accompanied by
(a) Decrease in size (b) Decrease in weight
(c) Increase in weight (d) No change in weight **Ans. b**

442. Zymase is
(a) Enzyme complex (b) Acetaldehyde dehydrogenase
(c) Pyruvate dehydrogenase (d) Pyruvate decarboxylase **Ans. a**

443. Phosphorylation of glucose with the help of ATP and hexokinase produces
(a) Glucose 1-6-diphosphate (b) Glucose 1-phosphate
(c) Glucose 6-phosphate (d) Fructose 1-6-bisphosphate
Ans. c

444. Number of ATP molecules formed from complete oxidation of fructose 1, 6- diphosphate is
(a) It produces ATP
(b) It uses ATP
(c) End products are CO_2 and H_2O
(d) NADH is produced **Ans. c**

445. Glycolysis
(a) Produces ATP (b) Uses ATP
(c) NADH is produced (d) All of these **Ans. d**

446. ATP molecules produced respectively by $NADH_2$ and $FADH_2$ during electron transport are
(a) 1 and 1 (b) 2 and 3
(c) 3 and 2 (d) 3 and 3 **Ans. c**

447. Tricarboxylic acids of Kreb's cycle are
(a) Citric acid, isocitric acid and malic acid
(b) Citric acid, isocitric acid and oxalosuccinic acid

(c) Succinic acid, fumaric acid and citric acid

(d) Oxalosuccinic acid, citric acid and α- ketoglutaric acid **Ans. b**

448. In aerobic respiration of a glucose molecule, citric acid cycle synthesize how many ATP molecules?

(a) 38 (b) 36

(c) 24 (d) 12 **Ans. c**

449. In muscles, anaerobic conditions change pyruvic acid to

(a) $C_3H_6O_3$ (b) C_2H_4O

(c) $C_3H_4O_3$ (d) C_2H_5OH **Ans. a**

450. R.Q. of germinating castor seed is

(a) 1 (b) >1

(c) <1 (d) 0 **Ans. c**

451. Older stem of woody plants have structures which allow air to diffuse into the respiring cells. The structures are called

(a) Stomata (b) Pores

(c) Lenticels (d) Leaf scars **Ans. c**

452. If a starved plant is provided with hexoses the rate of respiration will

(a) Increase (b) Decrease

(c) Decrease or increase (d) No change **Ans. a**

453. Mineral activator needed for the enzymes carboxylase of TCA cycle is

(a) Mo (b) Mg

(c) Mn (d) Fe **Ans. c**

454. Which of the scientific paper would you assign to plant physiology?

(a) Cell and cell division (b) Evergreen forest of India

(c) Embryo culture of plants (d) Respiratory activities in plants **Ans. d**

455. Link between carbohydrate and fat metabolism or between glycolysis, Kreb's cycle and α- oxidation of fatty acid is

(a) Pyruvic acid (b) Oxaloacetate

(c) Acetyl CoA (d) Citric acid **Ans. c**

456. Which 5 carbon organic acid of the Kreb's cycle is a key compound in the nitrogen metabolism of a cell?

(a) Fumaric acid (b) Citric acid

(c) α-ketoglutaric acid (d) Oxalosuccinic acid **Ans. c**

457. Fermentation

(a) Begins with glucose (b) Net gain of only 2 ATP

(c) Occurs in cytosol (d) All of these **Ans. d**

458. The greatest contributor of electrons to the electron transport system is

(a) Glycolysis (b) Oxygen

(c) Kreb's cycle (d) Transition reaction **Ans. c**

459. Substrate level phosphorylation takes place in the

(a) Kreb's cycle and transition reaction

(b) Glycolysis and Kreb's cycle

(c) Glycolysis and the electron transport system

(d) Electron transport system and the transition reaction **Ans. b**

460. Fatty acids are broken down to

(a) Pyruvate molecules, which take electron to the transport system

(b) Acetyl groups, which enter Kreb's cycle

(c) Amino acids which excrete ammonia

(d) All of these **Ans. b**

461. Of the 36 ATP molecules that are produced during the complete break down of glucose most are formed by

(a) Substrate level phosphorylation

(b) Chemiosmotic phosphorylation

(c) The electron transport system

(d) Both (b) and (c) **Ans. d**

462. In eukaryotes, net gain of ATP in complete oxidation of one molecule of glucose is

(a) 20 ATP (b) 56 ATP

(c) 38 ATP (d) 46 ATP **Ans. c**

463. Glyoxysomes are concerned with the metabolism of

(a) Proteins (b) Fats

(c) Carbohydrate (d) All of these **Ans. b**

464. The value of R.Q. decrease in germinating seeds

(a) When there is a shift from fat to carbohydrate respiration

(b) When there is a shift from protein to carbohydrate respiration

(c) When there is a shift from aerobic respiration to anaerobic, the respiration becomes totally anaerobic

(d) When there is a shift from carbohydrate to fat respiration **Ans. d**

465. Energy released during aerobic respiration is more than the energy released during anaerobic respiration, how many times more energy is released in former than the latter

(a) 19 times (b) 18 times

(c) 36 times (d) 8 times **Ans. b**

466. Glycolysis occurs in

(a) Cytosol (b) Mitochondria

(c) Golgi complex (d) Ribosomes **Ans. a**

467. Which is the final electron acceptor in respiration?

(a) Hydrogen (b) Oxygen

(c) Cytochrome (d) Dehydrogenase **Ans. b**

468. Glycolysis and Kreb's cycle are linked by

(a) Acetyl CoA (b) Citric acid

(c) Pyruvic acid (d) Oxaloacetic acid **Ans. a**

469. R.Q. is less than one in the case of

(a) Glucose (b) Fats

(c) Organic acid (d) Sucrose **Ans. b**

470. The site of glycolysis in a cell is

(a) Chromosome (b) Nucleus

(c) Cytoplasm (d) Chloroplast **Ans. c**

471. Respiration can be defined as a/an

(a) Anabolic process by which animal cell utilize oxygen and carbon dioxide to form ATP

(b) Anabolic process by which animal cell utilize oxygen, produce carbon dioxide and convert the released energy to ATP

(c) Catabolic process by which animal cells utilize carbon dioxide produce oxygen and convert the released energy to ATP

(d) Catabolic process by which animal cell utilize oxygen produce carbon dioxide and convert the released energy to ATP **Ans. d**

472. The formula for the process of glycolysis is

(a) $2H_2O + 6CO_2 \leftrightarrow C_6H_{12}O_6 + 6O_2$

(b) $C_6H_{12}O_6 \rightarrow 2C_3H_4O_3 + 4H$
(c) $C_6H_{12}O_6 + 6CO_2 \rightarrow 6CO_2 + 6H_2O$
(d) None of these **Ans. b**

473. Upon the oxidation of the one molecule of pyruvate by mitochondrial respiration, the molecules of ATP generated are
(a) 38 (b) 30
(c) 8 (d) 15 **Ans. d**

474. Plants, but not animals, can converts fatty acids to sugars by a series of reactions known as the
(a) Glyoxylate cycle (b) Photosynthesis
(c) Kreb's cycle (d) Glycolysis **Ans. a**

475. When yeast ferments glucose, the products are
(a) Water and CO_2 (b) Methanol and water
(c) Methanol and CO_2 (d) Ethanol and CO_2 **Ans. d**

476. The respiratory substrate yielding maximum number of ATP molecules among the following is
(a) Amylose (b) Betogenic amino acid
(c) Glucose (d) Glycogen **Ans. d**

477. The major role of phosphorus in plant metabolism is
(a) To generate metabolic energy
(b) To evolve oxygen during photosynthesis
(c) To evolve carbon dioxide during respiration
(d) To create anaerobic conditions **Ans. a**

478. Biological oxidation in Kreb's cycle involves
(a) N_2 (b) CO_2
(c) O_2 (d) SO_2 **Ans. c**

479. Oxidative phosphorylation occurs during the process of
(a) Protein synthesis (b) N_2 - fixation
(c) Respiration (d) Transpiration **Ans. c**

480. Mitochondria supply most of the necessary biological energy by
(a) Breaking down of sugar
(b) Oxidizing substrates of TCA cycle
(c) Reducing NADP
(d) Breaking down of protein **Ans. b**

481. EMP pathway and pentose phosphate pathway

(a) May occur simultaneously (b) Occur together
(c) Do not occur simultaneously (d) Are mutually exclusively

Ans. a

482. For each molecules of glucose respired, TCA cycle must rotate

(a) Five times (b) Four times
(c) Thrice (d) Twice **Ans. d**

483. Which component of ATPase is embedded in mitochondrial membrane?

(a) F_0 (b) F_1
(c) Both a and b (d) None of these **Ans. b**

484. R.Q. of malic acid is

(a) 0.7 (b) 1
(c) 1.33 (d) 4 **Ans. c**

485. Enzymes of oxidative phosphorylation are found in

(a) Endoplasmic reticulum (b) Chloroplast
(c) Mitochondria (d) Golgi bodies **Ans. c**

486. Vinegar is synthesized from alcohol by

(a) *Lactobacillus bacteria* (b) *Bacilli aceti* bacteria
(c) *Azotobactor* (d) *Rhizobium* **Ans. b**

487. What is common between photosynthesis and respiration?

(a) Cytochromes (b) Light
(c) H_2O (d) Temperature **Ans. a**

488. In which of the following process, the end product is pyruvic acid?

(a) Kreb's cycle (b) Glycolysis
(c) Fermentation (d) Oxidative phosphorylation

Ans. b

489. One of products of aerobic respiration is

(a) Malic acid (b) Lactic acid
(c) Pyruvic acid (d) Ethyl alcohol **Ans. c**

490. The end product anaerobic respiration in plants is

(a) Pyruvic acid (b) Malic acid
(c) Ethyl alcohol (d) Methyl alcohol **Ans. c**

491. Which one of the following products is formed during glycolysis of glucose?

(a) Pyruvic acid (b) Carbon dioxide

(c) Citric acid (d) Ethanol **Ans. a**

492. Which group of the following scientists discovered the EMP pathway of glycolysis?

(a) Embden, Mayerhoff and Paranas

(b) Emerson, Hoffman and Peterison

(c) Embden, Morrisonal and Picher

(d) Avery, Maleofd and mc Carthy **Ans. a**

493. Hexose monophosphate pathway (HMP) is shunt of glycolysis. It is common in animal cells of liver and adrenal cortex though found in prokaryotes also. This pathway occurs in

(a) Cytoplasm and produced 36 ATP only

(b) Cytoplasm and mitochondria both and produces 38 ATP

(c) Cytoplasm and produces 35 ATP directly

(d) Mitochondria and produces 35 ATP directly **Ans. a**

494. Pyruvic acid is formed in which phase of respiration?

(a) EMP pathway (b) Kreb's cycle

(c) Anaerobic respiration (d) ETS **Ans. a**

495. Respiration of starved leaves (consuming proteins) is called

(a) Pinotorespiration (b) Oxidative phosphorylation

(c) Flowting respiration (d) Protoplasmic respiration **Ans. d**

496. If germination seeds are kept in a bottle which is connected with a tube containing calcium water, then latter will turn

(a) Red (b) Yellowish

(c) Unchanged (d) Milky **Ans. d**

497. Maximum amount of energy in a cell is released by oxidation of carbohydrates when

(a) CO_2 and H_2O are formed from pyruvic acid

(b) Pyruvic acid is formed

(c) C_3H_3OH and CO_2 are formed

(d) Acetyl CoA is formed from pyruvic acid **Ans. a**

498. How many phosphorylation (ATP) formation, occur as electrons move across the ETC from $FADH_2$ to O_2?

(a) 0 (b) 2

(c) 3 (d) 1 **Ans. b**

499. Enzymes involved in oxidative decarboxylation of pyruvic acid found in

(a) Oxysomes (b) F_0-F_1 particles

(c) Cytoplasm (d) Matrix of mitochondria **Ans. d**

500. Oxidative phosphorylation (synthesis of ATP during respiration in mitochondrion) requires O_2 and occurs in

(a) Stroma

(b) Inner membrane of mitochondria

(c) Outer membrane of mitochondria

(d) Matrix of mitochondria **Ans. b**

501. Which cycle operates in photorespiration?

(a) Calvin cycle (b) HSK cycle

(c) Glycolate cycle (d) Kreb's cycle **Ans. c**

502. If value of RQ is >1, then respiratory substrate is

(a) Castor seeds (b) Fat/fatty acid

(c) Malic acid/organic acid (d) Gram seeds **Ans. c**

503. Seeds can be preserved if they are stored

(a) After drying in steam (b) Absolutely fresh

(c) After boiling in water (d) With minimum moisture

Ans. d

504. Action of ATPase needs the presence of

(a) Mg^{++} and K^+ (b) Cu^{++} and Fe^+

(c) Na^+ and K^+ (d) Ca^{++} and Mg^{++} **Ans. c**

505. Flow of electrons in ETS in $NADH_2$ → EMN → FeS protein → CoQ → Cyto b → Cyto C_1→ Cyto C → Cyto a → Cyto a_3 → O_2. At three steps ATP is formed (oxidative phosphorylation) where does II ATP is formed?

(a) Cyto C_1 and C (b) Cyto a and Cyto a_3

(c) Cyto b and Cyto C_1 (d) Between $NADH_2$ and FMN

Ans. c

506. Kreb's cycle was discovered by Kreb in pigeon muscles in 1940. Which step is called Gateway step or link reaction in respiration?

(a) ETS terminal oxidation (b) Formation of Acetyl CoA

(c) Glycolysis (d) Citric acid formation

Ans. b

507. According to Mitchell's chemiosmotic coupling theory of ATP formation, passage of 2e$^-$ from $NADH_2$ over ETC releases energy to push out 3 pairs of protons (proton pump). These protons then pass in wardly through F_1 head (ATPase) to form ATP. How many ATP are formed when one pair of protons return to inner chamber through F_1 of $F_0 - F_1$ particle?

(a) 2 ATP (b) 36 ATP

(c) 1 ATP (d) 3 ATP **Ans. c**

508. In Kreb's cycle

(a) OAA undergoes 4 oxidation and 2 decarboxylations

(b) Acetyl coenzyme A undergoes 4 oxidation and 2 carboxylations

(c) Pyruvic acid undergoes 4 oxidation and 2 decarboxylations

(d) TCA undergoes 4 oxidations and 4 decarboxylations **Ans. b**

509. Which is formed through phosphorylation in glycolysis?

(a) 1,3-bisphosphoglyceric acid (b) Fructose 1, 6-bisphosphate

(c) Both are correct (d) GAP **Ans. c**

510. In Kreb's cycle the first product is citric acid which is a 6 carbon compound. It is formed by a condensing irreversible reaction between

(a) OAA and Acetyl CoA (b) OAA and pyruvic acid

(c) Pyruvic acid and acetyl CoA (d) OAA and citrate synthetase

Ans. a

511. Amphibolic cycle that occurs only in aerobic condition is

(a) Kreb's cycle (b) E-D pathway

(c) Glycolysis (d) EMP pathway **Ans. a**

512. Which step of glycolysis requires pyruvic kinase, Mg^{++}, K^+ and ADP?

(a) Cleavage of Fructose 1,6-biphosphate

(b) Conversion of PEP into pyruvic acid

(c) Conversion of 3 PGA into 2 PGA

(d) None of these **Ans. b**

513. One molecule of glucose requires 2 ATP to get phosphorylated to form Fructose 1,6-biphosphate in glycolysis. How many ATP are used in the same process if the substrate is fructose?

(a) 0 (b) 1

(c) 2 (d) 4 **Ans. c**

514. Which of the following steps in glycolysis is irreversible?

(a) PEP → Pyruvic acid

(b) Fructose 6-PO_4 → Fructose 1,6-biPO_4

(c) Glucose → Glycose -6PO_4

(d) All of these **Ans. d**

515. Which is wrong for glycolysis?

(a) CO_2 is produced

(b) ATP is formed

(c) ATP is used

(d) Substrate level phosphorylation occurs **Ans. a**

516. The atom within each cytochrome molecule that actually accepts and release electron is

(a) Mg (b) Zn

(c) C (d) Fe **Ans. d**

517. Oxaloacetic acid (OAA) is a…carbon compound

(a) 5 (b) 6

(c) 4 (d) 2 **Ans. c**

518. In the formation of Acetyl CoA from pyruvic acid in mitochondria, Pyruvic acid gets

(a) Decarboxylated (b) Oxidised

(c) Both (a) and (b) (d) Reduced and isomerised

Ans. c

519. The importance of glycolysis is that

(a) It provides building material for the synthesis of cellular components

(b) It degrades glucose to generate ATP

(c) Both (a) and (b)

(d) None of these **Ans. c**

520. Chemosynthetic bacteria resemble the photosynthetic bacteria in
(a) Synthesizing assimilatory power
(b) Synthesizing carbohydrate through calvin cycle
(c) Oxidizing inorganic compounds for synthesizing ATP
(d) None of these **Ans. b**

521. Ratio of ATP in aerobic and anaerobic respiration is
(a) 4:1 (b) 18:1
(c) 18:2 (d) 36:1 **Ans. b**

522. More energy is stored in per unit weight of
(a) Protein (b) Glucose
(c) Fat (d) Carbohydrate **Ans. c**

523. Importance of Kreb's cycle is
(a) To accelerate glycolysis
(b) Production of amino acid
(c) Production of ATP
(d) Production of vitamins and amino acids **Ans. c**

524. Ratio of CO_2 produced in aerobic and anaerobic respiration is
(a) 4:1 (b) 1:1
(c) 2:1 (d) 3:1 **Ans. d**

525. In ETS, electron move from
(a) Higher ψ w to ψ w
(b) Higher (–) to higher (+) potential
(c) Higher (+) to lower (-) redox potential
(d) None of these **Ans. b**

526. Conversion of α-ketoglutaric acid into succinic acid is a step of
(a) ED pathway (b) TCA cycle
(c) HMP pathway (d) EMP pathway **Ans. b**

527. KCN (potassium cyanide) stops cell activity by
(a) Coagulating protein carriers
(b) Blocking trachea
(c) Preventing transfer of e^- from cytochrome a_3 to O_2
(d) Preventing diffusion of O_2 **Ans. c**

528. In the first step of glycolysis, glucose fructose is phosphorylated. This phosphate group comes from

(a) ATP (b) Nucleotide

(c) ADP (d) Inorganic PO_4 **Ans. a**

529. How many stages an aerobic breakdown of glucose requires?

(a) 4 (b) 3

(c) 2 (d) 1 **Ans. a**

530. Glyceraldehydes 3-phosphate (PG Ald or GAP) is oxidised during glycolysis happens to the hydrogenation and electron that are removed during its oxidation?

(a) They oxidize NAD^+

(b) They reduce NAD^+

(c) They are removed

(d) They are transferred to pyruvic acid **Ans. b**

531. The cut surface of an apple fruit turn brown when exposed. If cut surface is dipped on ascorbic acid (Vitamin C) browning does not occur because

(a) Ascorbic acid prevents release of polyphenols

(b) Ascorbic acid prevents oxidation of cut surface

(c) Ascorbic acid inhibits oxidase enzyme

(d) Ascorbic acid over comes injury **Ans. c**

532. Before entering into Kreb's cycle, pyruvic acid undergoes oxidative decarboxylation to form

(a) Acetyl CoA (b) CoA

(c) Citric acid (d) OAA **Ans. a**

533. Where does glycolysis and TCA (Kreb's cycle) occur in cell?

(a) Glycolysis in cytoplasm and TCA cycle in oxisomes of mitochondrion

(b) Glycolysis in cytosol and TCA cycle in stroma of chloroplast

(c) Glycolysis in cytosol and TCA cycle in matrix of mitochondria

(d) Glycolysis in matrix of mitochondrion and TCA cycle on membrane of mitochondrion **Ans. c**

534. The production of ATP by oxidative phosphorylation is driven by energy from

(a) Cytochromes

(b) $NADH_2$

(c) CoA

(d) Diffusion of protons in mitochondrion **Ans. d**

535. At the end of Kreb's cycle, most of energy removed from glucose molecule is transferred to

(a) Citric acid (b) ATP

(c) OAA (d) $NADH_2/FADH_2$ **Ans. d**

536. Mitochondria are power house (miniature batteries) of the cell. They are seat of

(a) Cellular respiration

(b) Internal respiration

(c) Biological respiration

(d) All of the above but never with external respiration **Ans. d**

537. The main and intermediate source of bioenergy which is obtained by a cell from catabolic reaction and is important means of transfer of energy is

(a) Glycogen (b) Pyruvic acid

(c) ATP (d) Glucose **Ans. c**

538. In ETC, ATP is not formed in which of the following steps?

(a) Cyto a-a_3 (b) Cyto b - Cyt c

(c) Cyto c - Cyto a (d) $NADH_2 \rightarrow FAD$ **Ans. c**

539. The first 5-C dicarboxylic acid in Kreb's Cycle is

(a) Acetyl CoA (b) Citric acid

(c) α - ketoglutaric acid (d) OAA **Ans. c**

540. Climacteric rise is a phenomenon encountered in respiration of

(a) Aquatics (b) Fleshy fruits

(c) All fruits (d) All plants **Ans. b**

541. In Kreb's cycle, the H^+ removed at succinate level is accepted by

(a) FAD^+ (b) FMN

(c) NAD^+ (d) ADP **Ans. a**

542. Phosphorylation of hexose sugar takes place with the help of

(a) TPN (b) NADP

(c) NAD (d) ATP and Mg^{++} **Ans. d**

543. Number of ATP molecules synthesized in photorespiration is

(a) 0 (b) 2
(c) 16 (d) 34 **Ans. a**

544. Pasteur effect is

(a) Reduction in oxidation rate even in the presence of plenty of O_2
(b) Shifting of aerobic conditions to anaerobic or vice versa
(c) Conversion of ADP to ATP
(d) Sterilization at 63^{0}C for 30 minutes **Ans. b**

545. Glyoxylate cycle (conversion of fats in sugars) is a modified

(a) HMP pathway (b) EMP pathway
(c) TCA cycle (d) None of these **Ans. c**

546. Proteins enter into Kreb's cycle through

(a) OAA (b) α - ketoglutarate
(c) Both a and b (d) None of these **Ans. c**

547. If fructose 1,6-biphosphate is oxidised in aerobic respiration the ATP production will be

(a) 38 (b) 36
(c) 32 (d) 20 **Ans. a**

548. Correct sequence of events in Kreb's cycle is

(a) Acetyl CoA → citric acid → α -ketoglutaric acid → succinic acid → fumaric acid → malic acid → OAA
(b) Acetyl CoA → citric acid → malic acid → α-ketoglutaric acid → succinic acid → OAA
(c) Acetyl CoA → citrate → pyruvate→ α-ketoglutarate → succinate → malate → fumarate → OAA
(d) None of these **Ans. a**

549. RQ is measured with Ganong's respiroscope. If in it the level remains constant during experiment, the RQ value comes to unity. When it shows rise in Hg level, then

(a) RQ value will be infinity (b) Value of RQ will be 0
(c) Value of RQ will be < 1 (d) RQ value will be unity **Ans. c**

550. When production of CO_2 is more than in take of O_2 the respiratory substrate is

(a) Fat/protein (b) Glucose
(c) Sucrose (d) Organic acid **Ans. d**

551. ATP synthesis occurs in mitochondria during cellular respiration. In the energy transfer process from ADP to ATP donor of phosphate is

(a) Glucose 6-phosphate (b) Creatine phosphate

(c) Inorganic phosphate (d) GTP **Ans. c**

552. O_2 during terminal oxidation in ETS gets reduced to H_2O. Here cytochrome oxidase (cyto a_3) is present which has Fe^{++} and Cu^{++} both. This oxygen here acts as

(a) Terminal acceptor of both electron and proton

(b) Terminal acceptor of proton H^+

(c) Terminal acceptor of e^-

(d) O_2 is molecular and accepts only two hydrogen atoms **Ans. a**

553. Match list I with II and selects the correct answer using the codes given below the lists

List I	List II
A. Conversion of glucose to pyruvate	(1) Glycolysis
B. Conversion of pyruvate to acetyl CoA	(2) Oxidative phosphorylation
C. Break down of acetyl CoA to yield reduced coenzymes	(3) Kreb's cycle
D. Synthesis of ATP	(4) Oxidative decarboxylation

Codes

	A	B	C	D
(a)	1	4	3	2
(b)	3	4	2	2
(c)	4	3	2	1
(d)	3	1	4	2

Ans. a

554. The very first step in the formation of acetyl CoA is the

(a) Formation of a complex between thiamine pyrophosphate and pyruvate

(b) Release of the acetyl group from lipoic acid CoA

(c) Decarboxylation of pyruvate

(d) Transfer of electron from reduced lipoic acid to NAD^+ **Ans. c**

555. The proton pumps energy to cell through
(a) pH gradient and electropotential gradient
(b) ATP and ADP
(c) Auxins and hormones
(d) Carbohydrates and lipids **Ans. b**

556. During the glyoxylate cycle, four carbon oxaloacetate is generated from two carbon acetate. The correct sequence in which the intermediate compounds appear from citrate is
(a) Citrate, succinate, fumarate, glyoxylate
(b) Citrate, fumarate, glyoxylate, malate
(c) Citrate, isocitrate, glyoxylate, malate
(d) Citrate, isocitrate, malate, glyoxylate **Ans. c**

557. Kreb's cycle also called TCA or citric acid cycle. It is also called metabolic sink as it is
(a) Common pathway for carbohydrates and organic acids only
(b) Common pathway for carbohydrates, fats and proteins (amino acids)
(c) Common pathway for carbohydrates and fats only
(d) None of these **Ans. b**

558. Breakdown of pyruvic acid in Kreb's cycle occurs
(a) Anaerobic respiration (b) Fermentation
(c) Aerobic respiration (d) Photosynthesis **Ans. c**

559. Kreb's cycle occurs in
(a) Chloroplast (b) Ribosomes
(c) Cytoplasm (d) Mitochondria **Ans. d**

560. Mitochondria are the site for
(a) Carboxylation (b) Transpiration
(c) Oxidative phosphorylation (d) Photophosphorylation **Ans. c**

561. How many ATP will be produced during the production of one molecule of Acetyl CoA from one molecule of pyruvic acid
(a) 3 ATP (b) 5 ATP
(c) 8 ATP (d) 38 ATP **Ans. a**

562. At the end of glycolysis, which one of the following change takes place?

(a) Starch into glucose (b) Proteins into glucose

(c) Glucose into fructose (d) Glucose into pyruvic acid

Ans. d

563. Cytochrome functions in

(a) Oxidation of glucose (b) Production of energy

(c) Transfer of electrons (d) Growth **Ans. c**

564. The importance of Kreb's cycle is in the production of

(a) Acetyl CoA (b) Water

(c) ATP (d) ADP **Ans. c**

565. Vant Hoff's law states that

(a) Respiration rate is constant respective of change in temperature

(b) Respiration stops with increase of temperature beyond 10^0C

(c) Respiration rate increase two or three times for every rise of 10^0C

(d) Respiration rate decrease two or three times for every rise of 10^0C

Ans. c

566. EMP (Embden, Mayerhof and Parnas) is the biochemical pathway in

(a) Glycolysis (b) Kreb's cycle

(c) Photosynthesis (d) None of these **Ans. a**

567. Anaerobic process after glycolysis is called

(a) Kreb's cycle (b) TCA cycle

(c) Calvin cycle (d) None of these **Ans. d**

568. The high boiling point of water is advantageous to living organisms because

(a) The environment seldom reaches the boiling point of water

(b) Organisms can easily boil off enough water to keep themselves cool

(c) It allows organisms to spread heat evenly through out their bodies

(d) Organisms can absorb a great deal of heat before they reach the boiling point **Ans. d**

569. The mechanism of ATP formation both in chloroplast and mitochondria is explained by

(a) Relay pump theory of Goldewski

(b) Munch's pressure/mass flow model

(c) Chemiosmotic theory of Mitchell
(d) Cholodny went's model **Ans. c**

570. Respiratory enzymes are located in
(a) Mitochondrial matrix (b) Pre-mitochondrial space
(c) Cristae (d) Outer membrane **Ans. a**

571. Fermentation is represented by the equation
(a) $C_6H_{12}O_6 \rightarrow 6H_2O + 6O_2 + 673$ Kcal
(b) $6CO_2 + 12H_2O \xrightarrow[\text{chlorophyll}]{\text{light}} 6H_2O + O_2 + C_6H_{12}O_6$
(c) $6CO_2 + 6H_2O \rightarrow C_6H_{12}O_6 + 6O_2$
(d) $C_6H_{12}O_6 \rightarrow 2C_2H_5OH + 2CO_2 + 18$ Kcal **Ans. d**

572. Heat energy of plant is measured in
(a) Grams (b) Decibels
(c) Calories (d) Pounds **Ans. c**

573. The high energy phosphate produced in Kreb's cycle when succinyl CoA is converted into succinate is
(a) ATP (b) GTP
(c) CTP (d) UTP **Ans. b**

574. Anaerobic respiration takes place in
(a) Vacuole (b) Cytoplasm
(c) Nucleus (d) Ribosome **Ans. b**

575. Enzymes α-ketoglutarate dehydrogenase bring about
(a) Reduction only (b) Oxidation only
(c) Decarboxylation only (d) Oxidation and decarboxylation
Ans. d

576. Maximum calories are produced by
(a) Vitamins (b) Proteins
(c) Carbohydrates (d) Fats **Ans. d**

577. The process of release of energy by chemical break down of glucose in the absence of oxygen is termed as
(a) Anaerobiosis (b) Aerobiosis
(c) Ornithine cycle (d) Deamination **Ans. a**

578. RQ value of 4 may be expected for the complete oxidation of
(a) Glucose (b) Oxalic acid
(c) Malic acid (d) Tartaric acid **Ans. b**

579. Mitochondrial matrix contains the enzymes for
(a) Respiration
(b) TCA cycle
(c) Glycolysis and TCA cycle
(d) TCA cycle and electron transport **Ans. b**

580. What would happen if $NADH_2$ is oxidised to form water in a single step?
(a) 3 ATP produced
(b) Most of the energy would be wasted in heat form
(c) 5 ATP produced
(d) Cell would burn **Ans. a**

581. In eukaryotes, the electron transport chain and enzymes of Kreb's cycle or TCA cycle are found in
(a) Nucleolus (b) Endoplasmic reticulum
(c) Mitochondria (d) Ribosomes **Ans. c**

582. Lactic acid fermentation is characterized by lack of production of
(a) ATP (b) $NADH_2$
(c) CO_2 and $NADH_2$ (d) CO_2 **Ans. c**

583. The number of glucose molecule required to produce 38 ATP molecules under anaerobic conditions by an yeast cell is
(a) 38 (b) 19
(c) 1 (d) 2 **Ans. b**

584. The food products are kept in cold storage or under refrigeration because
(a) It become very tasty
(b) Can be used in off season
(c) Food products maintain their freshness, longevity, taste etc due to least respiration
(d) They remain very cool **Ans. c**

585. The number of oxygen atoms utilized during aerobic oxidation of one molecule of pyruvate is
(a) 12 (b) 6
(c) 5 (d) 3 **Ans. c**

586. Fermentation is anaerobic production of
(a) Protein and acetic acid
(b) Alcohol, lactic acid or similar compounds
(c) Ethers and acetones
(d) Alcohol and lipoproteins **Ans. b**

587. Protein is mostly the respiratory substrate in
(a) Floating respiration (b) Protoplasmic respiration
(c) Anaerobic respiration (d) Respiration of oily seeds
Ans. b

588. Oxidative phosphorylation involves simultaneously oxidation and phosphorylation to finally form
(a) Pyruvate (b) ATP
(c) DPN (d) NADP **Ans. b**

589. Which of the following is the Key intermediate compound linking glycolysis to the Kreb's cycle?
(a) ATP (b) Acetyl CoA
(c) NADH (d) Malic acid **Ans. b**

590. In Kreb's cycle, FAD participates as electron acceptor during conversion of
(a) Succinyl CoA to succinic acid
(b) α-ketoglutarate to succinyl CoA
(c) Succinic acid to fumaric acid
(d) Fumaric acid to malic acid **Ans. c**

591. The major link in transfer of chemical energy is ATP - ADP system was first described by Lipman in
(a) 1940 (b) 1950
(c) 1960 (d) 1970 **Ans. a**

592. The end products of fermentation are
(a) O_2 and acetaldehyde (b) O_2 and ethyl alcohol
(c) CO_2 and ethyl alcohol (d) CO_2 and acetaldehyde **Ans. c**

593. The inner wall of mitochondria contains
(a) Enzymes of Kreb's cycle (b) Enzymes of respiratory chain
(c) DNA (d) Ribosomes **Ans. b**

594. The energy released by one molecule of glucose during aerobic respiration is

(a) 521 kg cal (b) 573 kg cal
(c) 621 kg cal (d) 673 kg cal **Ans. d**

595. Oxidative phosphorylation occurs during

(a) Photosynthesis (b) Respiration
(c) Transpiration (d) Protein synthesis **Ans. b**

596. The chemical linking of glycolysis and the Kreb's cycle is

(a) Pyruvic acid (b) Acetyl CoA
(c) Citric acid (d) Succinic acid **Ans. b**

597. Plants die during winter due to frost because

(a) Photosynthesis does not occur at low temperature
(b) Respiration stops
(c) Dehydration and mechanical injury occur in tissue
(d) Rate of respiration increase **Ans. c**

598. Cytochromes present in cells are

(a) Carbon acceptors (b) Hydrogen acceptors
(c) Nitrogen acceptors (d) Oxygen acceptors **Ans. b**

599. The enzymes which takes part in glycolysis are found in

(a) Mitochondria (b) Cytoplasm
(c) Mitochondria and cytoplasm (d) Vacuole **Ans. b**

600. Chitin is a ....

(a) Polysiccharides (b) Nitrogenous polysaccharides
(c) Lipoprotein (d) Protein **Ans. b**

601. The 'repeating unit' of glycogen in

(a) Fructose (b) Mannose
(c) Glucose (d) Galactose **Ans. c**

❑❑❑

Chapter-7

Amino Acids and Proteins

Fill in the blanks

1. Glycerides are esters of.................

Ans. glycerol

2. Imino acids differ from amino acids in having..............group instead of $-NH_2$.

Ans. -NH

3. ....is the world's most common protein.

Ans. Ribulose biphosphate carboxylase

4. Denaturation involves conversion of.........into............proteins.

Ans. Globular, fibrous

5. Most of the amino acids exist as...............

Ans. zwitter ions

6. The acidic character of glycine is due to.....................

Ans. $-NH_3^+$ group

7. Carbonic..................is an example of metallo-protein.

Ans. anhydrase

8. Casein, the milk protein is........................

Ans. Phosphoprotein

9. In one molecule of albumin the number of amino acid is.............

Ans. 610

10. Plasma proteins which contain more than 4% hexosamine are...........

Ans. mucoproteins

11. The hydrogen bonds between peptide linkages of protein molecules are interfered by.........................

Ans. guanidine

12. The number of helices present in a collagen molecule is............

Ans. 3

13. In a protein molecule the disulphide bond is not broken by..............

Ans. denaturation

14. The lipoprotein associated with activation of LCAT is.....................

Ans. IDL

15. Normal range of plasma total phospholipids is.....................

Ans. 1.8-5.8 mmol/l

16. In................tissue prostaglandins decrease lipogenesis.

Ans. adipose

17. The main site of urea synthesis in mammals is...................

Ans. liver

18. The end product of protein digestion in GIT is.........

Ans. amino acid

19. A compound serving a link between citric acid cycle and..............is fumarate.

Ans. urea cycle

20. The limiting amino acid of rice is..................

Ans. lysine

21. The limiting amino acid of fish protein is...............

Ans. tryptophan

22. A conjugated protein present in the egg yolk is......................

Ans. vitellin

23. The apolipoprotein which from the integral component of VLDL is....................

Ans. B-100

24. IgE has a tendency to attach to.................. and ....

Ans. Basophils, mast cells

25. Bence-Jones protein is a......... of .... chains.

Ans. dimer, light

26. After accounting for SDA the net gain of energy from 25 gm of proteins is about..................

Ans. 70 kcal

27. After accounting for SDA the net gain of energy from 25 gm of carbohydrates is about............

Ans. 95 kcal

28. After accounting for SDA the net gain of energy from 100 gm of fat is about............

Ans. 780 kcal

29. Protein content of meat is about................

Ans. 20%

30. Protein content of rice is about.............

Ans. 7%

31. ....is a basic amino acid.

Ans. Arginine

32. The amino acid................is contains a guanidine of group.

Ans. arginine

33. The amino acid not containing the usual -COOH group is...........

Ans. taurine

34. Branched chain amino acids are.............., ... and ....

Ans. valine, leucine and isoleucine

35. Hopkins-cole test is for identification of...................

Ans. tryptophan

36. The neutral amino acid is................

Ans. leucine

37. The amino acid.............containing hydroxyl group.

Ans. threonine

38. Transmethylation of guanido acetic acid gives...............

Ans. creatine phosphate

39. Haemoglobin has a high content of...............amino acid.

Ans. histidine

40. The example of scleroproteins.................

Ans. elastin

41. Isoelectric pH of an amino acid is that pH at which it has a................ charge.

Ans. nil

42. Abnormal chain of amino acid in sickle cell anaemia is..............

Ans. Beta chain

43. Systematic name of lysine is................

Ans. 2,6 diaminohexanoic acid

44. Protein is a polymer of...................

Ans. amino acids

45. Insulin is a protein which control.............

Ans. metabolic pathways

46. Isoelectric point of amino acid is used for..................

Ans. precipitation

47. Xanthoproteic test is positive in proteins containing.................. amino acid.

Ans. aromatic

48. Insulin is a protein which controls..............

Ans. digestion

49. Biuret test is specific for...........

Ans. two peptide linkage

50. .... Amino acid has no symmetric carbon atoms.

Ans. Glycine

Multiple Choice Questions

1. An example of a structural protein which is also an enzyme is
   (a) Actin (b) Myosin
   (c) Trypsin (d) Troponin **Ans. b**
2. Difference between R- and T- states of haemoglobin is
   (a) Dimeric and tetrameric structure
   (b) Oxy and deoxy states
   (c) Binding to ferrous and ferric ions
   (d) Folded and unfolded states **Ans. b**
3. Non-polar amino acid residues are found mostly
   (a) In the core of proteins (b) On the surface of proteins
   (c) On alpha helices (d) In no specific region **Ans. a**
4. Proteins may be separated according to size by
   (a) Reverse phase chromatography
   (b) Ion exchange chromatography
   (c) Molecular exclusion chromatography
   (d) Isoelectric focusing **Ans. c**
5. Which of the following is an amino acid?
   (a) Histidine (b) Glycine
   (c) Cysteine (d) Proline **Ans. d**
6. Which one of the following is the least important factor as far as protein folding is concerned?
   (a) Hydrophobic effect (b) Hydrogen bonding
   (c) Electrostatic interactions (d) Quaternary association **Ans. d**
7. Which of the following has a quaternary structure?
   (a) α-chymotrypsin (b) Haemoglobin
   (c) Insulin (d) Myoglobin **Ans. b**
8. Transamination is a
   (a) Irreversible process (b) Reversible process
   (c) Both of the above (d) None of the above **Ans. b**
9. An alfa helix represents
   (a) Primary structure of a protein
   (b) Secondary structure of certain parts of a protein

(c) Tertiary structure of a protein
(d) Aggregation of proteins **Ans. b**

10. Most abundant protein in the human body is
(a) Haemoglobin (b) Keratin
(c) Collagen (d) Immunoglobulin **Ans. c**

11. At isoelectric point, net mobility of amino acid is
(a) Towards cathode (b) Towards anode
(c) Zero (d) None **Ans. c**

12. What does a protein lose when it denatures?
(a) Its primary structure (b) Its peptide bonds
(c) Its sequence of amino acids (d) Its three dimensional shape
Ans. d

13. For a protein to have a quaternary structure, it must
(a) Have four amino acids
(b) Consist of two or more polypeptide subunits
(c) Consist of four polypeptide subunits
(d) Have at least four disulfide bridges **Ans. b**

14. Enormous diversity of protein molecules is due to
(a) Sequence of amino acids (b) R-group of amino acids
(c) Peptide bonds (d) Amino group of amino acids
Ans. a

15. Disulfide bridge forms between two cystiene residues as a result of
(a) Oxidation of sulfhydral group
(b) Reduction of sulfhydral group
(c) Amide formation
(d) None **Ans. a**

16. Thermal denaturation of protein involves
(a) Conformational change in the protein
(b) Random cleavage of the peptide bond
(c) Covalent modification of certain amino acids
(d) Increase in its isoelectric point **Ans. a**

17. The shape of the polypeptide backbone making α-helix in proteins is

(a) Like the English letter "a" (b) Right handed spiral

(c) Left handed spiral (d) Like the Greek letter "a"

Ans. b

18. The mean molecular weight of an amino acid in a typical globular protein is

(a) 70 (b) 110

(c) 150 (d) 90 **Ans. b**

19. Which amino acid residue is most likely to be found in the interior of a water soluble globular protein?

(a) Ser (b) Arg

(c) Val (d) Asp **Ans. c**

20. Which one of the following is unlikely to be water-soluble protein?

(a) Ribonuclease (b) Myoglobin

(c) Collagen (d) Lysozyme **Ans. c**

21. What is the most common approach for the determination of precise 3-D structure of globular proteins?

(a) Circular dischroism (b) Mass spectrometry

(c) Infrared spectroscopy (d) X-ray diffraction **Ans. d**

22. What is untrue about histamine?

(a) It is produced on decarboxylation of histamine

(b) It cause allergy

(c) It lowers blood pressure

(d) Diphenyl hydramine hydrochloride activates histamine **Ans. d**

23. Which of the following electrophoretic analytic procedures does not depend on the charge of the protein?

(a) Moving boundry electrophoresis

(b) Zone electrophoresis

(c) Polyacrylamide gel electrophoresis in sodium dodecyl sulfate

(d) Isoelectric focusing **Ans. c**

24. The peptide bond has a backbone of atoms in which of the following sequences?

(a) C-N-N-C (b) C-C-C-N

(c) C-C-N-C (d) N-C-C-C **Ans. c**

25. In two – dimentional electrophoresis, proteins are separated
(a) First by their charge and then by their mass
(b) First by their mass and by their charge
(c) Only on the basis of their charge
(d) Only on the basis of their mass **Ans. a**

26. Proteins are polyamides of
(a) β-Amino acids (b) α-Amino acids
(c) α-Hydroxy acids (d) β-Hydroxy acids **Ans. b**

27. Which of the following is a protein?
(a) Lecithin (b) Glycogen
(c) Keratin (d) Nucleic acid **Ans. c**

28. Which of the following has a branched chain structure?
(a) Amylopectin (b) Amylase
(c) Cellulose (d) Nylon **Ans. a**

29. Antibodies are
(a) Carbohydrates (b) Proteins
(c) Lipids (d) Enzymes **Ans. b**

30. The main structural feature of proteins is
(a) Ether linkage (b) Ester linkage
(c) Peptide linkage (d) All the three above **Ans. c**

31. Enzymes are
(a) Fatty acids (b) Vitamins
(c) Proteins (d) None of these **Ans. c**

32. Which one of the following is not a protein?
(a) Wool (b) Nail
(c) Hair (d) DNA **Ans. d**

33. The helical structure of protein is stabilized by
(a) Peptide bonds (b) Dipeptide bonds
(c) Hydrogen bonds (d) Vander Waals forces **Ans. c**

34. The acid showing salt-like character in aqueous solution is
(a) Acetic acid (b) Benzoic acid
(c) Formic acid (d) α-Aminoacetic acid **Ans. d**

35. Which of the following is not an amino acid?
(a) Glycine (b) Alanine
(c) Histidine (d) Benzidine **Ans. d**

36. Amino acids are the building blocks of
(a) Carbohydrates (b) Vitamins
(c) Fats (d) Proteins **Ans. d**

37. The helical structure of protein is stabilized by
(a) H-Bonding (b) Van der Waals' forces
(c) Ionic bond (d) Covalent bonding **Ans. a**

38. Which amino acid is achiral?
(a) Alanine (b) Valine
(c) Proline (d) Histidine **Ans. b**

39. What structural feature distinguishes proline from other natural α-amino acids?
(a) It is optically inactive (b) It contains aromatic group
(c) It is a dicarboxylic acid (d) It has a secondary amine
Ans. d

40. Which statement is incorrect about peptide bond?
(a) C-N bond length in proteins is longer than usual bond length of C-N bond
(b) Spectroscopic analysis shows planar structure of –CO-NH-group
(c) C-N bond length in proteins is smaller than usual bond length of C-N bond
(d) None of the above **Ans. a**

41. The simplest amino acid is
(a) Cystine (b) Alanine
(c) Glycine (d) Valine **Ans. c**

42. Proteins are made up by varied arrangements of
(a) Amino acids (b) Fats
(c) Nucleosides (d) Phospholipids **Ans. c**

43. Which of the following food stuffs contains nitrogen?
(a) Carbohydrates (b) Fats
(c) Proteins (d) None of the above **Ans. c**

44. Insulin, a hormone, chemically is

(a) A fat (b) A steroid

(c) A protein (d) A carbohydrate **Ans. c**

45. The enzyme pepsin hydrolyses

(a) Proteins to amino acids

(b) Fats to fatty acids

(c) Glucose to ethyl alcohol

(d) Polysaccharides to monosaccharides **Ans. a**

46. Disruption of the native conformation of a protein is called

(a) Hydrolysis (b) Emulsification

(c) Denaturation (d) Condensation **Ans. c**

47. Glycine is

(a) Enzyme (b) Hormone

(c) Fatty acid (d) Amino acid **Ans. d**

48. An essential amino acid is

(a) Tryptophan (b) Phenylalanine

(c) Leucine (d) All the above **Ans. d**

49. A nonessential amino acid is

(a) Lysine (b) Methionine

(c) Serine (d) Isoleucine **Ans. c**

50. Number of protein amino acids is

(a) 20 (b) 16

(c) 32 (d) 64 **Ans. a**

51. Number of amino acids is

(a) 20 (b) 120

(c) 200 (d) 500 **Ans. c**

52. Rice is deficient in

(a) Lysine (b) Alanine

(c) Glycine (d) Leucine **Ans. a**

53. Essential amino acids were discovered by

(a) Evans and Burr (b) Hopkins

(c) Beevers (d) Sutherland **Ans. b**

54. Protein amino acids are
(a) Laevorotatory
(b) Dextrorotatory
(c) Laevorotatory except glycine which is nonrotatory
(d) Laevorotatory except glycine which is dextrorotatory **Ans. c**

55. Number of essential amino acids for human requirement is
(a) 4 (b) 6
(c) 7 (d) 8 **Ans. d**

56. Number of semi-indispensable amino acids is
(a) 4 (b) 3
(c) 2 (d) 1 **Ans. c**

57. An aromatic amino acid is
(a) Tryptophan (b) Tyrosine
(c) Phenylalanine (d) All the above **Ans. d**

58. A functional but nonprotein amino acid is
(a) Ornithine (b) Citrulline
(c) Diaminopimelic acid (d) All the above **Ans. d**

59. A bond between two amino acids is
(a) Amide bond (b) Ionic bond
(c) Hydrophobic linkage (d) All the above **Ans. a**

60. Antibiotics contain
(a) Protein amino acids (b) No amino acids
(c) Nonprotein amino acids (d) Both (a) and (b) **Ans. c**

61. Aspartame is
(a) Natural peptide sweetener
(b) Artificial peptide sweetener
(c) Natural carbohydrate sweetener
(d) Artificial carbohydrate sweetener **Ans. b**

62. Protein is a
(a) Macromolecule (b) Steroid
(c) Fat (d) Micromolecule **Ans. a**

63. A protein molecule is formed of

(a) Chain of amino acids (b) Chain of fatty acids

(c) Chain of monosaccharides (d) Chain of oligosaccharides

Ans. a

64. Number of amino acids present in hormone adrenocorticotrophin is

(a) 10 (b) 70

(c) 58 (d) 39 **Ans. d**

65. Primary structure of protein is due to

(a) Hydrogen bond (b) Peptide bond

(c) – S – S linkages (d) Ionic bonds **Ans. b**

66. How many amino acid molecules form a molecule of haemocyanin (hemocyanin), a blood pigment of snails

(a) 68 (b) 72

(c) 4500 (d) 8200 **Ans. d**

67. Secondary structure of protein is

(a) α-helix (b) β-pleated

(c) β-helix (d) Both (a) and (b) **Ans. d**

68. Protein can be denatured by

(a) Carbon dioxide (b) Carbon monoxide

(c) Heat (d) Oxygen **Ans. c**

69. Quaternary structure is present in

(a) Haemoglobin (b) Histone

(c) Globulin (d) Elastin **Ans. a**

70. The protein present in milk is

(a) Casein (b) Glutelin

(c) Myosin (d) Globulin **Ans. a**

71. A complete protein is one which possesses

(a) All nonessential amino acids

(b) All essentials amino acids

(c) All the 20 amino acids

(d) All the amino acids with complex structure **Ans. b**

72. The most abundant molecule of protoplast is

(a) Proteins (b) Lipids

(c) Carbohydrates (d) Nucleic acids **Ans. a**

73. Molecular weight of smallest protein molecule ACTH (adrenocorticotropic hormone) is

(a) 5600 (b) 4500

(c) 3100 (d) 6200 **Ans. b**

74. In α-helix secondary structure, hydrogen bonds lie between imide group of one amino acid and carbonyl group of

(a) 2nd amino acid (b) 3rd amino acid

(c) Fourth amino acid (d) Fifth amino acid **Ans. c**

75. Which is type of secondary protein structure

(a) α-helix (b) β-pleated

(c) Collagen helix (d) All the above **Ans. d**

76. In β-pleated secondary protein structure there are

(a) Two or more polypeptide chains

(b) Hydrogen bonds between adjacent polypeptide chains

(c) Parallel or antiparallel polypeptides

(d) All the above **Ans. d**

77. Haemoglobin is

(a) Monomeric protein (b) Oligomeric protein

(c) Chromoprotein (d) Both (b) and (c) **Ans. d**

78. Primary structure of polypeptide is estabilised or secondary structure of polypeptide is maintained by

(a) Hydrogen bonds (b) Disulphide bonds

(c) Chromoprotein (d) Both (b) and (c) **Ans. a**

79. Protomers are

(a) Primitive proteins (b) Protein subunits

(c) Protein aggregates (d) None of the above **Ans. b**

80. Final structure in fibrous proteins is

(a) Secondary structure (b) Tertiary structure

(c) Quaternary structure (d) Primary structure **Ans. a**

81. Quaternary structure is found in

(a) Simple monomeric proteins (b) Conjugate monomeric proteins

(c) Oligoproteins (d) Both (b) and (c) **Ans. c**

82. Which forms nail hair and feathers

(a) Keratin sulphate (b) Keratin

(c) Collagen (d) Elastin **Ans. b**

83. Oxygen transporting protein is haemoglobin. Which one is oxygen storing protein

(a) Myoglobin (b) Actin

(c) Myosin (d) Carcinogen **Ans. a**

84. Mucin present in saliva is a

(a) Mucoprotein (b) Mucopolysaccharide

(c) Derived protein (d) Small polysaccharide **Ans. a**

85. Proteins present in milk, egg and meat are

(a) Partially complete (b) Complete

(c) Incomplete (d) Both (b) and (c) **Ans. b**

86. A globular protein is

(a) Elastin (b) Keratin

(c) Albumin (d) Collagen **Ans. c**

87. Amino acids are

(a) Amphoteric (b) Amphiatic

(c) Amphipathic (d) None of these **Ans. a**

88. Tryptophan is involved in the formation of

(a) Heme (b) Vitamin niacin and IAA

(c) Hormones (d) Urea formation in liver **Ans. b**

89. Amino acids with longest side chain are

(a) Arginine and lysine (b) Lysine and leucine

(c) Arginine and glycine (d) Tryptophan and valine **Ans. a**

90. Which one of these is imino (not amino) acid?

(a) Glycine (b) Phenylalanine

(c) Proline (d) Lysine **Ans. c**

91. Excess of amino acids are stored in

(a) Kidney (b) Liver

(c) Spleen (d) None of these **Ans. d**

92. The difference between one amino acid and another is found in the

(a) Carboxyl group (b) Amino group

(c) R group (d) Peptide bond **Ans. c**

93. Which amino acid is denoted by symbol F?

(a) Phenylalanine (b) Proline

(c) Tryptophan (d) Methionine **Ans. a**

94. A semi indispensable amino acid for human nutrition is

(a) Arginine (b) Lysine

(c) Valine (d) Leucine **Ans. a**

95. Relationship between amino acid and protein is similar to one found between

(a) Glucose and fructose (b) Nucleotides and nucleic acid

(c) Nucleosides and nucleic acid (d) Purines and pyrimidines

Ans. b

96. Non essential amino acid is

(a) Non need in the diet

(b) Not essential for growth

(c) Not synthesized in body

(d) Not required for protein synthesis **Ans. a**

97. Amino acid with two asymmetric carbon atom is

(a) Valine (b) Histidine

(c) Isoleucine (d) Serine **Ans. c**

98. An amino acid metabolized by brain only is

(a) Alanine (b) Glutamic acid

(c) Glutamine (d) Histidine **Ans. b**

99. Which of the following amino acids is not utilized by brain?

(a) Valine (b) Glycine

(c) Methionine (d) Phenylalanine **Ans. b**

100. Chocolate and cheese are avoided while going bed as they cause sleeplessness due to

(a) Glycine (b) Tyrosine

(c) Tryptophan (d) Sugar **Ans. b**

101. Which of the following amino acid is unable to form a proper peptide bond?

(a) Glutamine (b) Asparagines

(c) Glycine (d) Proline **Ans. d**

102. If the molecular mass of an amino acid is 150 daltons, the molecular mass of a tripeptide will be

(a) 450 (b) 486

(c) 504 (d) 414 **Ans. d**

103. Substance forming 80% of cytoplasm in plant cells is

(a) Proteins (b) Water

(c) Fats (d) Minerals **Ans. b**

104. Most of the water found in the young cell occurs in

(a) Cell wall (b) Nucleus

(c) Cytoplasm (d) Vacuoles **Ans. c**

105. The most abundant protein is found in

(a) Chloroplasts (b) Mitochondria

(c) Liver (d) Brain **Ans. a**

106. Starving person will first use

(a) Fats (b) Blood protein

(c) Muscle protein (d) Glycogen **Ans. d**

107. Protein specificity is due to

(a) Types of amino acids (b) Sequence of amino acids

(c) Time of synthesis (d) Quantity **Ans. b**

108. Glycogen is a polymer of

(a) Galactose (b) Glucose

(c) Fructose (d) Sucrose **Ans. b**

109. A protein having both structural and enzymatic traits is

(a) Myosin (b) Collagen

(c) Trypsin (d) Actin **Ans. a**

110. A plant proteinase is

(a) Papain (b) Trypsin

(c) Pepsin (d) Urease **Ans. a**

111. Which one is not an essential amino acid?

(a) Leucine (b) Lysine

(c) Methionine (d) Alanine **Ans. d**

112. The most abundant element in living beings is

(a) Oxygen (b) Nitrogen

(c) Carbon (d) Hydrogen **Ans. a**

113. Blood clotting is helped by

(a) Na^+ (b) K^+

(c) Ca^{2+} (d) Mg^{2+} **Ans. c**

114. The most diverse chemicals are

(a) Polysaccharides (b) Lipids

(c) Proteins (d) Sugars **Ans. d**

115. An amino acid has three ionizable groups with pKa's of 2.0, 10.5 and 3.8. What is the pI of this amino acid?

(a) 6.25 (b) 9.05

(c) 5.43 (d) 2.9 **Ans. d**

116. Which of the following statement is not true?

(a) The pI is the pH value at which a protein has overall charge of +1

(b) At a pH value equal to pI, a protein will not move in the electric field

(c) An acidic protein will have a pI less than 7

(d) A basic protein will have a pI greater than 7 **Ans. a**

117. Collagen is a fibrous protein consisting of three peptide chains interwined in a triple helix. What repetitive motif is responsible for collagens ability to form the triple helix?

(a) Ala-X-Y (b) Cys-X-Y

(c) Gly-X-Y (d) Pro-X-Y **Ans. c**

118. If there are 20 kinds of amino acids and they can be assembled in any order, then how many different proteins of 100 amino acids long can theoretically be produced?

(a) 20x100 (b) 20^{100}

(c) 100^{20} (d) 20x20x100 **Ans. b**

119. The property of resonance as applied to protein structure is responsible for

(a) The prevention of rotation about the alpha carbon

(b) The partial double bond character of the side chain-alpha carbon bond

(c) The planar nature of the peptide bond

(d) Repulsion of electron pair **Ans. c**

120. Which of the following statements is false?

(a) Amino acids tend to be least soluble in water at their isoelectric point

(b) Sickle cell Hb and normal Hb have same value of negative charge

(c) Amino acids are made visible on the chromatogram by treatment with ninhydrin

(d) The net charge on an amino acid is a function of the pH of the solution **Ans. b**

121. An α-helix is made up of 78 amino acid residues. What will be the value of its axial length and also length when polypeptide chain is fully extended respectively?

(a) 117Å, 117Å (b) 281Å, 117Å

(c) 117Å, 281Å (d) None **Ans. c**

122. Which one of the following peptides lacks a chiral center at the first position, is devoid of a hydrogen in the backbone amide at the second position and the side-chain of the third residue has a pKa close to the physiological pH?

(a) Gly-Pro-Tyr (b) Pro-Gly-Tyr

(c) Pro-Gly-His (d) Gly-Pro-His **Ans. d**

123. When you boil an egg, you convert the albumin into a white solid mass. In chemical terms you would say that?

(a) The protein was dehydrated by heat

(b) The protein was denatured by heat

(c) The protein was cross-linked by heat

(d) The protein was degraded by heat **Ans. b**

124. Thermal denaturation of proteins involves

(a) Conformational change in the protein

(b) Random cleavage of the peptide bonds

(c) Covalent modification of certain amino acids

(d) Increase in its isoelectric point **Ans. a**

125. Which one of the following replacements of an amino acid by another amino acid actually represents substitution of an atom by another atom?

(a) Ser → Cys (b) Gly → Phe

(c) Gly → Ala (d) Asp → Asn **Ans. a**

126. Which pair of amino acids will have the highest absorbance at 280 nm? (Assume equimolar concentrations)

(a) Thr and His (b) Phe and Pro

(c) Trp and Tyr (d) Phe and His **Ans. c**

127. Which one of the following statements about protein secondary structure is correct?

(a) An α-helix is primarily stabilized by ionic interactions between the side chains of the amino acids

(b) β-sheets exist only in antiparallel form

(c) β-turns often contain proline

(d) An α-helix can be composed of more than one polypeptide chain **Ans. c**

128. Choose the mispair

	Fibrous Protein	Most abundant amino acids
(a)	Silk fibrion	Gly, Ala, Ser
(b)	Collagen	Phe, Ser, Pro
(c)	Elastin	Gly, Ala, Val
(d)	α-Keratin (wool)	Ser, Glu, Cys

Ans. b

129. Choose the incorrect statement with regard to collagen molecule

(a) It is formed from three α-chains

(b) Glycine accounts for one-third of amino acids found within collagens

(c) It contains less hydroxyproline than elastin

(d) Ascorbate is essential for hydroxylation of proline **Ans. c**

130. Histines have very high percentage of arginine and lysine residues. For this class of proteins, which of the following reagents would be a suitable choice for generating peptides in the determination of the amino acids sequence of the protein?

(a) Cynogen bromide (b) Thermolysin

(c) Trypsin (d) N-bromosuccinamide **Ans. c**

131. Which of the following is not a feature of α-keratin?

(a) Presence of –S-S- cross link between adjacent α-helices

(b) Presence of intrachain H-bonding

(c) Contains predominant amino acid residues like Phe, Cys, Met

(d) Presence of intrachain –S-S- cross link **Ans. d**

132. According to Beer-Lambert's law, what will be the absorbance when $I = I_0/10$?

(a) 0 (b) 1

(c) 0.1 (d) 100 **Ans. b**

133. Since the pK values for aspartic acid are 2.0, 3.9, and 10.0, it follows that the isoelectric point (pI) is which of the following?

(a) 3.0 (b) 3.9

(c) 5.9 (d) 6.0 **Ans. a**

134. Enzyme-catalyzed hydrolysis of proteins produces amino acids of the form

(a) D- (b) DL-

(c) L- (d) All the above **Ans. c**

135. The ionizable groups of amino acids are at least

(a) 1 (b) 2

(c) 3 (d) 4 **Ans. b**

136. Which of the following is not true about alanine

(a) At pH 1, the overall charge is +1

(b) At pH 1, it will move towards cathode

(c) An equimolar mixture of D and L alanine does not rotate the plane polarized light

(d) Contains branched side chain **Ans. d**

137. Electrophoresis of a purified protein named X in the presence of sodium dodecyl sulfate and 2-mercaptoethanol, shows a single band of 45 kDa. In gel filtration column chromatography, protein X elutes between alcohol dehydrogenase (160 kDa) and beta-amylase (190 kDa). How many identical subunits is protein X composed of

(a) One (b) Two
(c) Three (d) Four **Ans. d**

138. Although protein concentration can be estimated by measuring the absorbance of a protein solution at 280 nm, this technique does not yield an accurate measure of absolute protein concentration. Why?

(a) Protein solutions refract light optimally at 280 nm, thus interfering with the measurement
(b) Proteins scatter light optimally at this wavelength, thus interfering with the concentration measurement
(c) The aromatic amino-acid content varies from protein to protein
(d) Cysteine residues absorb light optimally at this wavelength **Ans. c**

139. Protein structures are confirmed by weak bonds, the example of which is

(a) Hydrophobic (b) Disulphide
(c) Peptide (d) All the above **Ans. a**

140. How many different polypeptides are there those contain one residue each of Gly, Asp, Tyr, Cys and Leu?

(a) 625 (b) 120
(c) 32 (d) None **Ans. b**

141. Presence of specific functional group in a amino acid gives a specific colour reaction and used as a test for the amino acid. Choose the incorrect combinations

	Name of test	Amino acid
(a)	Hopkins cole	Tryptophan
(b)	Pauly	Histidine
(c)	Nitroprusside	Cysteine
(d)	Millon	Arginine

Ans. d

142. α-helix is disrupted by certain amino acid like

(a) Proline (b) Arginine
(c) Histidine (d) Lysine **Ans. a**

143. Which of the following is not correct about the collagen triple-helix

(a) Is rich in glycine (b) Is rich in proline

(c) Is rich in hydroxyproline (d) Is an α helix **Ans. d**

144. In the small intestine trypsin hydrolyzes peptide linkage containing

(a) Arginine (b) Histidine

(c) Serine (d) Aspartate **Ans. a**

145. Chymotrypsin in the small intestine hydrolyzes peptide linkage containing

(a) Phenylalanine (b) Alanine

(c) Methionine (d) Valine **Ans. a**

146. Which of the following statement is false?

(a) Homozygotes for HbS results in an elevated BPG content in their erythrocytes

(b) Hill constant, n can never be larger than the number of ligand binding sites on the protein

(c) Peptide groups, with few exceptions, assume the trans conformation

(d) None of the above **Ans. d**

147. Transmination is a

(a) Irreversible process (b) Reversible process

(c) Both (a) and (b) (d) None **Ans. b**

148. A solution of tryptophan has an absorbance at 280 nm of 0.54 in a 0.5 cm path length cuvette. Given the absorbance coefficient (ε) for tryptophan is $5.4 \times 10^3 M^{-1} cm^{-1}$, the concentration of the solution is

(a) 0.2 mM (b) 20 mM

(c) 1×10^{-3} M (d) 0.1 mM **Ans. a**

149. Changes in protein conformation can be detected by

(a) Ultraviolet absorption spectroscopy

(b) Flurescence emission

(c) Circular dichroism

(d) All the above **Ans. c**

150. Histones have very high percentage of arginine and lysine residues (15-30%). For this class of proteins which of the following reagents would be suitable choice for generating peptides in the determination of the amino acid sequence of the protein

(a) Cyanogens bromide (b) Thermolysin

(c) Trypsin (d) N-bromosuccinamide **Ans. c**

151. The following type of interaction is mainly responsible for aggregation of proteins in dilute solutions

(a) Hydrogen bonds (b) Hydrophobic interactions

(c) Disulphide bonds (d) Peptide bonds **Ans. b**

152. Of the groups of amino acids indicated below, which one group of amino acids can be phosphorylated?

(a) Thr, Tyr, Asn (b) Ser, Thr, Tyr

(c) Ser, Phe, Asn (d) Ser, Thr, Phe **Ans. b**

153. Shown below are interactions that are found in proteins. Which one of them does not involve a covalent linkage?

(a) Cystenyl-amide interaction (b) Cys-Cys interaction

(c) Asp-Asp interaction (d) Peptide bond **Ans. a**

154. In peptide bond, resonance energy stabilization contributes to

(a) Coplanarity (b) Trans alignment

(c) Polarity (d) Repulsion of electron pair

Ans. a

155. The elution of a protein from size exclusion chromatography depends on

(a) The size of the protein alone

(b) The size and shape of the protein

(c) The size and surface charge distribution of the protein

(d) The shape and surface charge distribution of the protein **Ans. b**

156. The peptide, Ala-Arg-Gln-Met-Thr-Trp-Lys-Val, upon treatment with cyanogens bromide (CNBr) produces

(a) Ala-Arg-Gln-Met + Thr-Trp-Lys-Val

(b) Ala-Arg-Gln-Met - Thr-Trp + Lys-Val

(c) Ala-Arg + Gln-Met-Thr-Trp-Lys-Val

(d) Ala-Arg-Gln + Met-Thr-Trp-Lys-Val **Ans. a**

157. Binding of oxygen to hemoglobin occur when Fe is in the following oxidation state

(a) +1 (b) +2

(c) +3 (d) +4 **Ans. b**

158. In one arginine has a molecular weight of 174 Daltons, then what would be the molecular weight (in Daltons) of a circular polymer of 38 arginines?

(a) 6612 (b) 5928

(c) 5946 (d) 7278 **Ans. b**

159. Which of the following statements about oxygen binding and release by haemoglobin is correct?

(a) On binding oxygen, the iron of the heme prosthetic group is oxidized to the ferric state

(b) Lowering the pH accelerates the release of oxygen from oxyhemoglobin

(c) A high concentration of 2, 3-diphospho-glycerate in the erythryocyte enhances the binding of oxygen by hemoglobin

(d) Oxygen binding by any one of the four heme groups occurs independently of the other three **Ans. b**

160. Which of the amino acid pairs have two chiral centres?

(a) Proline and arginine (b) Leucine and isoleucine

(c) Isoleucine and threonine (d) Methionine and cystiene

Ans. c

161. Which of the following cofactors derived from a vitamin is involved in the formation of hydroxyproline during collagen synthesis?

(a) Pyridoxal phosphate (vitamin B_6)

(b) Biotin

(c) Thiamine pyrophosphate (vitamin B_1)

(d) Ascorbic acid (vitamin C) **Ans. d**

162. Collagen occurs in different forms, which are usually classified on the basis of the

(a) Type of carbohydrate present

(b) Cystein content

(c) Hydroxyproline and hydroxylysine content

(d) Types of peptide chains present **Ans. d**

163. The lipoproteins with the highest electrophoretic mobility and the lowest triglyceride content are

(a) Very low density lipoproteins

(b) Low density lipoproteins

(c) Intermediate density lipoprotein

(d) High density lipoproteins **Ans. d**

164. Which statement about amino acid is correct?

(a) They always have at least one amino group and at least one carboxyl group

(b) In the formation of proteins, it is a condensation reaction that links the amino group of one amino acid to the variable side chain of the adjacent amino acid

(c) The variable side chains of all of the amino acids are highly reactive and carry a charge at neutral pH

(d) The peptide bond that links amino acids together in a protein is a type of ionic bond, which explains why proteins are unstable at high temperatures **Ans. a**

165. Which statement about proteins is correct?

(a) Protein quaternary structure is determined solely by the primary amino acid sequence

(b) Protein sequences that span a lipid bilayer membrane usually contain a number of charged amino acids

(c) Examples of protein secondary structure include an alpha helix or a beta pleated sheet

(d) Protein tertiary structure is the result of the interaction of two or more independent polypeptide chains **Ans. c**

166. Difference between the R- and T-states of hemoglobin is

(a) Dimeric and tetrameric structure

(b) Oxy and deoxy states

(c) Binding to ferrous and ferric ions

(d) Folded and unfolded states **Ans. b**

167. An example of a trimeric protein is

(a) Lysozyme (b) Haemoglobin

(c) Keratin (d) Collagen **Ans. d**

168. Using circular dichroism one can obtain

(a) Relative proportions of α-helices, β-sheets and loops in proteins

(b) Three dimensional structure of a protein

(c) Amino acid sequence of a protein

(d) Quaternary structure of a protein **Ans. a**

169. Non polar amino acid residues are found mostly

(a) In the core of proteins (b) On the surface of proteins

(c) On alpha helices (d) In no specific region **Ans. a**

170. The secondary structure of a protein cannot be determined by one of the following techniques

(a) NMR spectroscopy (b) X-ray crystallography

(c) Fluorescence spectroscopy (d) Circular dichroism **Ans. c**

171. In cell-free lysates of *E. coli,* radioactive isotope, [γ-^{32}P] ATP can be used to label the following

(a) Closed circular duplex DNA

(b) Protein containing serine, threonine or tyrosine

(c) Protein devoid of serine, threonine and tyrosine

(d) Peptidoglycan **Ans. b**

172. Which one of the following is an example of a non-covalent interaction in proteins?

(a) Salt bridge (b) Disulfide bridge

(c) Peptide bond (d) Phosphodiester bond **Ans. a**

173. The following amino acid has a nonpolar and aliphatic R group

(a) Tyrosine (b) Histidine

(c) Glutamine (d) Leucine **Ans. d**

174. For determining the C-terminal amino acid of a polypeptide chain, the reagent that would be useful is

(a) Trypsin (b) Carboxypeptidase

(c) Phenyl isothiocyanate (d) 1 (N) HCl **Ans. b**

175. Which statement is incorrect?

(a) Solubility of polypeptides depends upon relative polarity of their R-group

(b) Higher ionizable group in polypeptides make it more soluble

(c) Isoelectric point of histones is very high (about 10.8)

(d) Selenocysteine is derived from cysteine **Ans. d**

176. The closest estimate for the number of amino acid residues in a protein with a molecular weight of 85 kDa is

(a) 710 (b) 7100

(c) 8500 (d) 85 **Ans. a**

177. Which one of the following bonds in proteins has a partial double bond character?

(a) C_α - C (b) C_α - N

(c) C – N (d) C – O **Ans. c**

178. Which of the following statement is incorrect?

(a) Mb consists of a single polypeptide chain in addition to a heme prosthetic group

(b) The red colour associated with the oxygenated forms of Mb and Hb is due to heme

(c) Hb has a higher affinity for oxygen compared to Mb as seen from the oxygen binding curves

(d) The α and β chains of the Hb tetramer are slightly shorter than the single chain of Mb **Ans. c**

179. Molecular weight of an unknown protein can be found out by

(a) Electrophoresis

(b) Ion-exchange chromatography

(c) Affinity chromatography

(d) None of the above techniques **Ans. a**

180. N-terminal amino acids are usually determined by Sanger's method using

(a) Ninhydrin reagent

(b) Fluro-2,4-dinitrobenzene/dansyl chloride

(c) Hydrazine

(d) Concentrated nitric acid **Ans. b**

181. Cleavage of the following peptide with chymotrypsin

^+H_3N-Gly-Arg-Ala-Ser-Phe-Gly-Asn-Lys-Trp-Glu-Val-COO^- results in

(a) 2 framgments (b) 3 fragments

(c) 4 fragments (d) No cleavage **Ans. b**

182. The average molecular weight of an amino acid in a protein is

(a) 125 (b) 120

(c) 110 (d) 137 **Ans. c**

183. Which pair of amino acids absorbs UV light strongly at 280 nm?

(a) Thr and His (b) Cys and Asp

(c) Gln and Pro (d) None of the above **Ans. d**

184. The catabolism of haemoglobin

(a) Involves the removal of phytol chain

(b) Involves the oxidative cleavage of the porphyrin ring

(c) Results in the liberation of CO_2

(d) None **Ans. b**

185. In sickle cell anemia, the basis of the malfunction of the haemoglobin molecule is

(a) Substitution of a single amino acid

(b) Incorrect secondary structure

(c) Faulty binding of the heme groups

(d) Reduced affinity for oxygen **Ans. a**

186. Disulfide bonds most often stabilize the native structure of

(a) Extracellular proteins (b) Dimeric proteins

(c) Intracellular proteins (d) Multisubunit proteins **Ans. a**

187. The helices in the amino acid supersecondary structure are held together primarily by

(a) Charge-charge interactions

(b) Covalent cross links

(c) Favourable R group interactions

(d) Main chain hydrogen bonding **Ans. c**

188. The quaternary structure of human haemoglobin is best described as a

(a) Dimer of two myoglobin subunits

(b) Tetramer of identical subunits

(c) Tetramer of four different subunits

(d) Tetramer of two different subunits **Ans. d**

189. Collagen is best described as

(a) An α-helical structural protein

(b) A coiled-coil found in hair

(c) A cross-linked globular protein

(d) A triple-helical fibrous protein **Ans. d**

190. The mirror image of right handed α-helix with all L-amino acids will appear as

(a) Left handed α-helix with L-amino acids

(b) Left handed α-helix with D-amino acids

(c) Right handed α-helix with L-amino acids

(d) Right handed α-helix with D-amino acids **Ans. b**

191. Aspartame, the synthetic sweetener is

(a) A dipeptide

(b) Also secreted by the posterior pituitary gland in response to thirst

(c) A glucose derivative

(d) Not digested (i.e. hydrolyzed) by humans **Ans. a**

192. Choose the incorrect pair

	Compound	Precursor amino acids
(a)	Porphyrin	Serine
(b)	Kreatine	Glycine, Arginine, Methionine
(c)	Glutathione	Glycine, Glutamate, Cysteine
(d)	Spermine	Methionine, Ornithine **Ans. a**

193. On a Ramachandran plot the entries for hemoglobin would be clustered around

(a) All four corners

(b) The extended chain conformation the extended chain conformation

(c) The left-handed α-helix conformation

(d) The right-handed α-helix conformation **Ans. d**

194. Choose the incorrect statement?

(a) Aspartate has the smallest pI value

(b) Arginine has the largest pI value

(c) pKa value of side chain is maximum for aspartate

(d) Side chain of isoleucine has maximum hydropathy value **Ans. c**

195. Hydrogen bonds in α-helices are
(a) More numerous than van der Waals interactions
(b) Not present at phenylalanine residues
(c) Analogous to the steps in a spiral staircase
(d) Roughly parallel to the helix axis **Ans. d**

196. The peptide bond in proteins is
(a) Planar, but rotates to three preferred dihedral angles
(b) Nonpolar, but rotates to three preferred dihedral angles
(c) Nonpolar, and fixed in a trans conformation
(d) Planar, and usually found in a trans conformation **Ans. d**

197. The molecular formula for glycine is $C_2H_5O_2N$. What would be the molecular formula for a linear oligomer made by linking ten glycine molecules together by condensation synthesis?
(a) $C_{20}H_{50}O_{20}N_{10}$ (b) $C_{20}H_{32}O_{11}N_{10}$
(c) $C_{20}H_{40}O_{10}N_{10}$ (d) $C_{20}H_{68}O_{29}N_{10}$ **Ans. b**

198. The forces that maintain the three-dimensional structure of a protein is mainly
(a) Non-covalent (b) Covalent
(c) Coordinate (d) Covalent and non-covalent
Ans. a

199. SDS is used in poly acrylamide gel electrophoresis of a mixture of proteins for their efficient separation on the gel. SDS in this experiment is used to
(a) Solubilize the protein
(b) Stabilize the protein
(c) Decrease the surface tension of buffer
(d) Have uniform charge density on the proteins **Ans. d**

200. If the molecular mass of an amino acid is 150 daltons, the molecular mass of a tripeptide will be
(a) 450 (b) 486
(c) 504 (d) 414 **Ans. d**

201. Sodium dodecyl sulphate upon addition into water can form spherical micelles. In this micelle the hydrocarbon tail would be
(a) Cleaved on micelle formation
(b) In the interior of the micelle

(c) Towards the surface of the micelle

(d) None of the above **Ans. b**

202. The peptide bond is not

(a) A partial double bond (b) A polar

(c) Planar (d) Trans in proteins **Ans. b**

203. The high solubility of amino acids in water is due to

(a) Presence of side chain

(b) Dipolar ion structure

(c) Unipolarity

(d) Hydrophilic nature of amino group **Ans. b**

204. Choose the incorrect statement

(a) At isoelectric point, protein has minimum solubility

(b) Among standard amino acids, tryptophan has highest molecular weight

(c) At low pH, O_2 affinity of Hb increases

(d) Histidine contains three ionizable groups **Ans.c**

205. The rate-limiting step in the biosynthetic pathway for catecholamines is

(a) The hydroxylation of tyrosine

(b) The hydroxylation of phenylalanine to tyrosine

(c) The formation of dopamine

(d) The reduction of biopterin **Ans. a**

206. Epinephrine is formed from nonepinephrine by

(a) Decarboxylation (b) Oxidative deamination

(c) O-methylation (d) N-methylation **Ans. d**

207. Which of the following amino acids is the precursor for the thyroid hormones T_3 and T_4?

(a) Tryptophan (b) Lysine

(c) Tyrosine (d) Histidine **Ans. c**

208. Which of the following reactions in the biosynthesis of porphyrins is the rate-limiting step?

(a) δ- Aminolevulinate synthetase

(b) Porphyrinogen synthetase

(c) Uroporphyrinogen synthetase

(d) Decarboxylation of uroporphyrinogen III **Ans. a**

209. All of the following statements regarding the degradation of porphyrins are correct except

(a) The formation of biliverdin is the first step in the degradation of heme

(b) Reduction of biliverdin yields bilirubin

(c) Bilirubin is conjugated to glucuronic acid in the liver

(d) Free (unconjugated) bilirubin may be reduced to urobilinogen in the liver **Ans. d**

210. Creatine is formed from which of the following amino acids?

(a) Glycine, lysine and methionine

(b) Glycine, arginine and S-adenosylmethionine

(c) Glycine, ornithine and S-adenosylmethionine

(d) Glycine and S-adenosylmethionine **Ans. b**

211. The relative viscosity of a protein solution is affected by

(a) Temperature, shape and molecular weight

(b) Temperature and molecular weight

(c) Molecular weight

(d) Temperature and shape **Ans. a**

212. Which one of the following pairs refers to paralogous proteins

(a) Rat myoglobin and rat hemoglobin

(b) Rat myoglobin and mouse myoglobin

(c) Rat myoglobin and mouse hemoglobin

(d) Rat myoglobin and rat cytochrome **Ans. a**

213. The rate of migration of a protein in an SDS-Polyacrylamide gel is not influenced by

(a) Size of the protein (b) Charge of the protein

(c) Pore size of the gel (d) Strength of the electric field

Ans. b

214. The concept of molecular clock implies that

(a) Many proteins show a constant rate of change with time

(b) Organisms evolve at a constant rate

(c) One can date evolutionary event with molecules alone

(d) All molecules change at the same rate in evolution **Ans. a**

215. Anion exchangers preferentially bind proteins with

(a) Acidic isoelectric points

(b) Basic isoelectric points

(c) No charge

(d) Abundant hydrophobic groups **Ans. b**

216. Which of the following amino acids can exist as diastereomers?

(a) Isoleucine and leucine (b) Isoleucine and valine

(c) Threonine and serine (d) Isoleucine and threonine

Ans. a

217. All proteins contain the

(a) Same 20 amino acids

(b) Different amino acids

(c) 300 Amino acids occurring in nature

(d) Only a few amino acids **Ans. a**

218. Proteins contain

(a) Only L- α- amino acids (b) Only D-amino acids

(c) DL-Amino acids (d) Both (a) and (b) **Ans. a**

219. The optically inactive amino acid is

(a) Glycine (b) Serine

(c) Threonine (d) Valine **Ans. a**

220. At neutral pH, a mixture of amino acids in solution would be predominantly

(a) Dipolar ions (b) Nonpolar molecules

(c) Positive and monovalent (d) Hydrophobic **Ans. a**

221. The true statement about solutions of amino acids at physiological pH is

(a) All amino acids contain both positive and negative charges

(b) All amino acids contain positively charged side chains

(c) Some amino acids contain only positive charge

(d) All amino acids contain negatively charged side chains **Ans. a**

222. pH (isoelectric pH) of alanine is

(a) 6.02 (b) 6.6

(c) 6.8 (d) 7.2 **Ans. a**

223. Sulphur containing amino acid is

(a) Methionine (b) Leucine

(c) Valine (d) Asparagine **Ans. a**

224. An example of sulphur containing amino acid is

(a) 2-Amino-3-mercaptopropanoic acid

(b) 2-Amino-3-methylbutanoic acid

(c) 2-Amino-3-hydroxypropanoic acid

(d) Amino acetic acid **Ans. a**

225. All the following are sulphur containing amino acids found in proteins except

(a) Cysteine (b) Cystine

(c) Methionine (d) Threonine **Ans. d**

226. An aromatic amino acid is

(a) Lysine (b) Tyrosine

(c) Taurine (d) Arginine **Ans. b**

227. The functions of plasma albumin are

(a) Osmosis (b) Transport

(c) Immunity (d) both (a) and (b) **Ans. a**

228. Amino acid with side chain containing basic groups is

(a) 2-Amino 5-guanidovaleric acid

(b) 2-Pyrrolidine carboxylic acid

(c) 2-Amino 3-mercaptopropanoic acid

(d) 2-Amino propanoic acid **Ans. a**

229. An example of α-amino acid not present in proteins but essential in mammalian metabolism is

(a) 3-Amino 3-hydroxypropanoic acid

(b) 2-Amino 3-hydroxybutanoic acid

(c) 2-Amino 4-mercaptobutanoic acid

(d) 2-Amino 3-mercaptopropanoic acid **Ans. c**

230. An essential amino acid in man is

(a) Aspartate (b) Tyrosine

(c) Methionine (d) Serine **Ans. c**

231. Non essential amino acids

(a) Are not components of tissue proteins

(b) May be synthesized in the body from essential amino acids

(c) Have no role in the metabolism

(d) May be synthesized in the body in diseased states **Ans. b**

232. Which one of the following is semiessential amino acid for humans?

(a) Valine (b) Arginine

(c) Lysine (d) Tyrosine **Ans. b**

233. An example of polar amino acid is

(a) Alanine (b) Leucine

(c) Arginine (d) Valine **Ans. c**

234. The amino acid with a nonpolar side chain is

(a) Serine (b) Valine

(c) Asparagine (d) Threonine **Ans. b**

235. A ketogenic amino acid is

(a) Valine (b) Cysteine

(c) Leucine (d) Threonine **Ans. c**

236. An amino acid that does not form an α-helix is

(a) Valine (b) Proline

(c) Tyrosine (d) Tryptophan **Ans. b**

237. An amino acid not found in proteins is

(a) β-Alanine (b) Proline

(c) Lysine (d) Histidine **Ans. a**

238. In mammalian tissues serine can be a biosynthetic precursor of

(a) Methionine (b) Glycine

(c) Tryptophan (d) Phenylalanine **Ans. b**

239. A vasodilating compound is produced by the decarboxylation of the amino acid

(a) Arginine (b) Aspartic acid

(c) Glutamine (d) Histidine **Ans. d**

240. Biuret reaction is specific for

(a) -CONH-linkages (b) -$CSNH_2$ group

(c) -(NH)NH_2 group (d) All of these **Ans. a**

241. Sakaguchi's reaction is specific for

(a) Tyrosine (b) Proline

(c) Arginine (d) Cysteine **Ans. c**

242. Million-Nasse's reaction is specific for the amino acid

(a) Tryptophan (b) Tyrosine

(c) Phenylalanine (d) Arginine **Ans. b**

243. Ninhydrin with evolution of CO_2 forms a blue complex with

(a) Peptide bond (b) α-Amino acids

(c) Serotonin (d) Histamine **Ans. b**

244. The most of the ultraviolet absorption of proteins above 240 nm is due to their content of

(a) Tryptophan (b) Aspartate

(c) Glutamate (d) Alanine **Ans. a**

245. Which of the following is a dipeptide?

(a) Anserine (b) Glutathione

(c) Glucagon (d) α-Lipoprotein **Ans. a**

246. Which of the following is a tripeptide?

(a) Anserine (b) Oxytocin

(c) Glutathione (d) Kallidin **Ans. c**

247. A peptide which acts as potent smooth muscle hypotensive agent is

(a) Glutathione (b) Bradykinin

(c) Tyrocidine (d) Gramicidin-s **Ans. b**

248. A tripeptide functioning as an important reducing agent in the tissues is

(a) Bradykinin (b) Kallidin

(c) Tyrocidine (d) Glutathione **Ans. d**

249. An example of metalloprotein is

(a) Casein (b) Ceruloplasmin

(c) Gelatin (d) Salmine **Ans. b**

250. Carbonic anhydrase is an example of
(a) Lipoprotein (b) Phosphoprotein
(c) Metalloprotein (d) Chromoprotein **Ans. c**

251. An example of chromoprotein is
(a) Haemoglobin (b) Sturine
(c) Nuclein (d) Gliadin **Ans. a**

252. An example of scleroprotein is
(a) Zein (b) Keratin
(c) Glutenin (d) Ovoglobulin **Ans. b**

253. Casein, the milk protein is
(a) Nucleoprotein (b) Chromoprotein
(c) Phosphoprotein (d) Glycoprotein **Ans. c**

254. An example of phosphoprotein present in egg yolk is
(a) Ovoalbumin (b) Ovoglobulin
(c) Ovovitellin (d) Avidin **Ans. c**

255. A simple protein found in the nucleoproteins of the sperm is
(a) Prolamine (b) Protamine
(c) Glutelin (d) Globulin **Ans. b**

256. Histones are
(a) Identical to protamine
(b) Proteins rich in lysine and arginine
(c) Proteins with high molecular weight
(d) Insoluble in water and very dilute acids **Ans. b**

257. The protein present in hair is
(a) Keratin (b) Elastin
(c) Myosin (d) Tropocollagen **Ans. a**

258. The amino acid from which synthesis of the protein of hair keratin takes place is
(a) Alanine (b) Methionine
(c) Proline (d) Hydroxyproline **Ans. b**

259. In one molecule of albumin the number of amino acids is
(a) 510 (b) 590
(c) 610 (d) 650 **Ans. c**

260. Plasma proteins which contain more than 4% hexosamine are
(a) Microglobulins (b) Glycoproteins
(c) Mucoproteins (d) Orosomucoids **Ans. c**

261. After releasing O_2 at the tissues, hemoglobin transports
(a) CO_2 and protons to the lungs (b) O_2 to the lungs
(c) CO_2 and protons to the tissue (d) Nutrients **Ans. a**

262. Ehlers-Danlos syndrome characterized by hypermobile joints and skin abnormalities is due to
(a) Abnormality in gene for procollagen
(b) Deficiency of lysyl oxidase
(c) Deficiency of prolyl hydroxylase
(d) Deficiency of lysyl hydroxylase **Ans. a**

263. Proteins are soluble in
(a) Anhydrous acetone (b) Aqueous alcohol
(c) Anhydrous alcohol (d) Benzene **Ans. b**

264. A cereal protein soluble in 70% alcohol but insoluble in water or salt solution is
(a) Glutelin (b) Protamine
(c) Albumin (d) Gliadin **Ans. d**

265. Many globular proteins are stable in solution inspite they lack in
(a) Disulphide bonds (b) Hydrogen bonds
(c) Salt bonds (d) Non polar bonds **Ans. a**

266. The hydrogen bonds between peptide linkages of protein molecules are interfered by
(a) Guanidine (b) Uric acid
(c) Oxalic acid (d) Salicylic acid **Ans. a**

267. Globular proteins have completely folded, coiled polypeptide chain and the axial ratio (ratio of length to breadth) is
(a) Less than 10 and generally not greater than 3–4
(b) Generally 10
(c) Greater than 10 and generally 20
(c) Greater than 10 **Ans. a**

268. Fibrous proteins have axial ratio
(a) Less than 10
(b) Less than 10 and generally not greater than 3–4
(c) Generally 10
(d) Greater than 10 **Ans. d**

269. Each turn of α-helix contains the amino acid residues (number)
(a) 3.6 (b) 3.0
(c) 4.2 (d) 4.5 **Ans. a**

270. Distance traveled per turn of α-helix in nm is
(a) 0.53 (b) 0.54
(c) 0.44 (d) 0.48 **Ans. b**

271. Along the α-helix each amino acid residue advances in nm by
(a) 0.15 (b) 0.10
(c) 0.12 (d) 0.20 **Ans. a**

272. The number of helices present in a collagen molecule is
(a) 1 (b) 2
(c) 3 (d) 4 **Ans. c**

273. In proteins the α-helix and β-pleated sheet are examples of
(a) Primary structure (b) Secondary structure
(c) Tertiary structure (d) Quaternary structure **Ans. b**

274. The α-helix of proteins is
(a) A pleated structure
(b) Made periodic by disulphide bridges
(c) A non-periodic structure
(d) Stabilized by hydrogen bonds between NH and CO groups of the main chain **Ans. c**

275. At the lowest energy level α-helix of polypeptide chain is stabilized
(a) By hydrogen bonds formed between the H of peptide N and the carbonyl O of the residue
(b) Disulphide bonds
(c) Non polar bonds
(d) Ester bonds **Ans. a**

276. Both α-helix and β-pleated sheet conformation of proteins were proposed by

(a) Watson and Crick (b) Pauling and Corey

(c) Waugh and King (d) Y.S. Rao **Ans. b**

277. The primary structure of fibroin, the principal protein of silk worm fibres consists almost entirely of

(a) Glycine (b) Aspartate

(c) Keratin (d) Tryptophan **Ans. a**

278. Tertiary structure of a protein describes

(a) The order of amino acids (b) Location of disulphide bonds

(c) Loop regions of proteins (d) The ways of protein folding

Ans. d

279. In a protein molecule the disulphide bond is not broken by

(a) Reduction (b) Oxidation

(c) Denaturation (d) X-ray diffraction **Ans. c**

280. The technique for purification of proteins that can be made specific for a given protein is

(a) Gel filtration chromotography (b) Ion exchange chromatography

(c) Electrophoresis (d) Affinity chromatography

Ans. d

281. Denaturation of proteins results in

(a) Disruption of primary structure

(b) Breakdown of peptide bonds

(c) Destruction of hydrogen bonds

(d) Irreversible changes in the molecule **Ans. c**

282. Ceruloplasmin is

(a) α_1-globulin (b) α_2-globulin

(c) β-globulin (d) None of these **Ans. a**

283. The lipoprotein with the fastest lectrophoretic mobility and the lowest triglyceride content is

(a) Chylomicron (b) VLDL

(c) IDL (d) HDL **Ans. d**

284. The lipoprotein associated with activation of LCAT is
(a) HDL (b) LDL
(c) VLDL (d) IDL **Ans. d**

285. The apolipoprotein which acts as activator of LCAT is
(a) A-I (b) A-IV
(c) C-II (d) D **Ans. a**

286. The apolipoprotein which acts as activator of extrahepatic lipoprotein is
(a) Apo-A (b) Apo-B
(c) Apo-C (d) Apo-D **Ans. c**

287. The apolipoprotein which forms the integral component of chylomicron is
(a) B-100 (b) B-48
(c) C (d) D **Ans. b**

288. The apolipoprotein which from the integral component of VLDL is
(a) B-100 (b) B-48
(c) A (d) D **Ans. a**

289. The apolipoprotein which acts as ligand for LDL receptor is
(a) B-48 (b) B-100
(c) A (d) C **Ans. b**

290. Serum LDL has been found to be increased in
(a) Obstructive jaundice (b) Hepatic jaundice
(c) Hemolytic jaundice (d) Malabsorption syndrome
Ans. a

291. A lipoprotein associated with high incidence of coronary atherosclerosis is
(a) LDL (b) VLDL
(c) IDL (d) HDL **Ans. a**

292. A lipoprotein inversely related to the incidence of coronary artherosclerosis is
(a) VLDL (b) IDL
(c) LDL (d) HDL **Ans. d**

293. The primary biochemical lesion in homozygote with familial hypercholesterolemia (type IIa) is
(a) Loss of feed back inhibition of HMG reductase
(b) Loss of apolipoprotein B
(c) Increased production of LDL from VLDL
(d) Functional deficiency of plasma membrane receptors for LDL
Ans. d

294. In abetalipoproteinemia, the biochemical defect is in
(a) Apo-B synthesis (b) Lipoprotein lipase activity
(c) Cholesterol ester hydrolase (d) LCAT activity **Ans. a**

295 Familial hypertriacylglycerolemia is associated with
(a) Over production of VLDL
(b) Increased LDL concentration
(c) Increased HDL concentration
(d) Slow clearance of chylomicrons **Ans. a**

296. For synthesis of prostaglandins, the essential fatty acids give rise to a fatty acid containing
(a) 12 carbon atoms (b) 16 carbon atoms
(c) 20 carbon atoms (d) 24 carbon atoms **Ans. c**

297. All active prostaglandins have at least one double bond between positions
(a) 7 and 8 (b) 10 and 11
(c) 13 and 14 (d) 16 and 17 **Ans. a**

298. Normal range of plasma total phospholipids is
(a) 0.2–0.6 mmol/L (b) 0.9–2.0 mmol/L
(c) 1.8–5.8 mmol/L (d) 2.8–5.3 mmol/L **Ans. c**

299. HDL_2 have the density in the range of
(a) 1.006–1.019 (b) 1.019–1.032
(c) 1.032–1.063 (d) 1.063–1.125 **Ans. d**

300. β-lipoproteins have the density in the range of
(a) 0.95–1.006 (b) 1.006–1.019
(c) 1.019–1.063 (d) 1.063–1.125 **Ans. c**

301. IDL have the density in the range of
(a) 0.95–1.006 (b) 1.006–1.019
(c) 1.019–1.032 (d) 1.032–1.163 **Ans. b**

302. Aspirin inhibits the activity of the enzyme
(a) Lipoxygenase (b) Cyclooxygenase
(c) Phospholipase A_1 (d) Phospholipase A_2 **Ans. b**

303. A 'suicide enzyme' is
(a) Cycloxygenase (B) Lipooxygenase
(c) Phospholipase A_1 (d) Phospholipase A_2 **Ans. a**

304. In adipose tissue prostaglandins decrease
(a) Lipogenesis (b) Lipolysis
(c) Gluconeogenesis (d) Glycogenolysis **Ans. a**

305. The optimal pH for the enzyme pepsin is
(a) 1.0–2.0 (b) 4.0–5.0
(c) 5.2–6.0 (d) 5.8–6.2 **Ans. a**

306. Pepsinogen is converted to active pepsin by
(a) HCl (b) Bile salts
(c) Ca^{++} (d) Enterokinase **Ans. a**

307. The optimal pH for the enzyme rennin is
(a) 2.0 (b) 4.0
(c) 8.0 (d) 6.0 **Ans. b**

308. The optimal pH for the enzyme trypsin is
(a) 1.0–2.0 (b) 2.0–4.0
(c) 5.2–6.2 (d) 5.8–6.2 **Ans. c**

309. The optimal pH for the enzyme chymotrypsin is
(a) 2.0 (b) 4.0
(c) 6.0 (d) 8.0 **Ans. d**

310. Trypsinogen is converted to active trypsin by
(a) Enterokinase (b) Bile salts
(c) HCl (d) Mg^{++} **Ans. a**

311. Pepsin acts on denatured proteins to produce
(a) Proteoses and peptones (b) Polypeptides
(c) Peptides (d) Dipeptides **Ans. a**

312. Renin converts casein to paracasein in presence of
(a) Ca^{++} (b) Mg^{++}
(c) Na^+ (d) K^+ **Ans. a**

313. An expopeptidase is

(a) Trypsin (b) Chymotrypsin

(c) Enolase (d) Elastase **Ans. d**

314. The enzyme trypsin is specific for peptide bonds of

(a) Basic amino acids

(b) Acidic amino acids

(c) Aromatic amino acids

(d) Next to small amino acid residues **Ans. a**

315. Chymotrypsin is specific for peptide bonds containing

(a) Uncharged amino acid residues

(b) Acidic amino acids

(c) Basic amino acid

(d) Small amino acid residues **Ans. a**

316. The end product of protein digestion in G.I.T. is

(a) Dipeptide (b) Tripeptide

(c) Polypeptide (d) Amino acid **Ans. d**

317. Natural L-isomers of amino acids are absorbed from intestine by

(a) Passive diffusion (b) Simple diffusion

(c) Facilitated diffusion (b) Active process **Ans. d**

318. Abnormalities of blood clotting are

(a) Haemophilia (b) Christmas disease

(c) Gout (d) Both (a) and (b) **Ans. d**

319. An important reaction for the synthesis of amino acid from carbohydrate intermediates is transamination which requires the cofactor

(a) Thiamin (b) Riboflavin

(c) Niacin (d) Pyridoxal phosphate **Ans. d**

320. The main sites for oxidative deamination are

(a) Liver and kidney

(b) Skin and pancreas

(c) Intestine and mammary gland

(d) Lung and spleen **Ans. a**

321. A positive nitrogen balance occurs
(a) In growing infant (b) Following surgery
(c) In advanced cancer (d) In kwashiorkor **Ans. a**

322. The main site of urea synthesis in mammals is
(a) Liver (b) Skin
(c) Intestine (d) Kidney **Ans. a**

323. The enzymes of urea synthesis are found in
(a) Mitochondria only
(b) Cytosol only
(c) Both mitochondria and cytosol
(d) Nucleus **Ans. c**

324. The number of ATP required for urea synthesis is
(a) 0 (b) 1
(c) 2 (d) 3 **Ans. d**

325. Most of the ammonia released from L-α-amino acids reflects the coupled action of transaminase and
(a) L-glutamate dehydrogenase (b) L-amino acid oxidase
(c) Histidase (d) Serine dehydratase **Ans. a**

326. In urea synthesis, the amino acid functioning solely as an enzyme activator
(a) N-acetyl glutamate (b) Ornithine
(c) Citrulline (d) Arginine **Ans. a**

327. The enzyme carbamoyl phosphate synthetase requires
(a) Mg^{++} (b) Ca^{++}
(c) Na^{+} (d) K^{+} **Ans. a**

328. Control of urea cycle involves the enzyme
(a) Carbamoyl phosphate synthetase
(b) Ornithine transcarbamoylase
(c) Argininosuccinase
(d) Arginase **Ans. a**

329. Transfer of the carbamoyl moiety of carbamoyl phosphate to ornithine is catalyzed by a liver mitochondrial enzyme
(a) Carbamoyl phosphate synthetase
(b) Ornithine transcarbamoylase

(c) N-acetyl glutamate synthetase
(d) N-acetyl glutamate hydrolase **Ans. b**

330. A compound serving a link between citric acid cycle and urea cycle is
(a) Malate (b) Citrate
(c) Succinate (d) Fumarate **Ans. d**

331. The 2 nitrogen atoms in urea are contributed by
(a) Ammonia and glutamate (b) Glutamine and glutamate
(c) Ammonia and aspartate (d) Ammonia and alanine **Ans. c**

332. In carcinoid syndrome the argentaffin tissue of the abdominal cavity overproduce
(a) Serotonin (b) Histamine
(c) Tryptamine (d) Tyrosine **Ans. a**

333. Tryptophan could be considered as precursor of
(a) Melanotonin (b) Thyroid hormones
(c) Melanin (d) Epinephrine **Ans. a**

334. Conversion of tyrosine to dihydroxyphenylalanine is catalyzed by tyrosine hydroxylase which requires
(a) NAD (b) FAD
(c) ATP (d) Tetrahydrobiopterin **Ans. d**

335. The rate limiting step in the biosynthesis of catecholamines is
(a) Decarboxylation of dihydroxyphenylalanine
(b) Hydroxylation of phenylalanine
(c) Hydroxylation of tyrosine
(d) Oxidation of dopamine **Ans. c**

336. The enzyme dopamine β-oxidase which catalyses conversion of dopamine to norepinephrine requires
(a) Vitamin A (b) Vitamin C
(c) Vitamin E (d) Vitamin B_{12} **Ans. b**

337. In humans the sulphur of methionine and cysteine is excreted mainly as
(a) Ethereal sulphate (b) Inorganic sulphate
(c) Sulphites (d) Thioorganic compound **Ans. b**

338. Small amount of urinary oxalates is contributed by the amino acid
(a) Glycine (b) Tyrosine
(c) Alanine (d) Serine **Ans. a**

339. The amino acid which detoxicated benzoic acid to form hippuric acid is
(a) Glycine (b) Alanine
(c) Serine (d) Glutamic acid **Ans. a**

340. The amino acids involved in the synthesis of creatin are
(a) Arginine, glycine, active methionine
(b) Arginine, alanine, glycine
(c) Glycine, lysine, methionine
(d) Arginine, lysine, methionine **Ans. a**

341. Chemical score of egg proteins is considered to be
(a) 100 (b) 60
(c) 50 (d) 40 **Ans. a**

342. Chemical score of milk protein is
(a) 70 (b) 65
(c) 60 (d) 40 **Ans. b**

343. Chemical score of proteins of bengal gram is
(a) 70 (b) 60
(c) 44 (d) 42 **Ans. c**

344. Chemical score of protein gelatin is
(a) 0 (b) 44
(c) 57 (d) 60 **Ans. a**

345. Chemical score of protein zein is
(a) 0 (b) 57
(c) 60 (d) 70 **Ans. a**

346. Biological value of egg white protein is
(a) 94 (b) 83
(c) 85 (d) 77 **Ans. b**

347. Net protein utilization of egg protein is
(a) 75% (b) 80%
(c) 91% (d) 72% **Ans. c**

348. Net protein utilization of milk protein is
(a) 75% (b) 80%
(c) 86% (d) 91% **Ans. a**

349. A limiting amino acid is an essential amino acid
(a) That is most deficient in proteins
(b) That is most excess in proteins
(c) That which increases the growth
(d) That which increases the weight gain **Ans. a**

350. The limiting amino acid of rice is
(a) Lysine (b) Tryptophan
(c) Phenylalanine (d) Tyrosine **Ans. a**

351. The limiting amino acid of fish proteins is
(a) Tryptophan (b) Cysteine
(c) Lysine (d) Threonine **Ans. a**

352. Pulses are deficient in
(a) Lysine (b) Threonine
(c) Methionine (d) Tryptophan **Ans. c**

353. A trace element deficient in the milk is
(a) Magnesium (b) Copper
(c) Zinc (d) Chloride **Ans. a**

354. A conjugated protein present in the egg yolk is
(a) Vitellin (b) Livetin
(c) Albuminoids (d) Ovo-mucoid **Ans. a**

355. The chief protein of cow's milk is
(a) Albumin (b) Vitellin
(c) Livetin (d) Casein **Ans. d**

356. A water soluble vitamin deficient in egg is
(a) Thiamin (b) Ribofalvin
(c) Ascrobic acid (d) Cobalamin **Ans. c**

357. Pulses are rich in
(a) Lysine (b) Methionine
(c) Tryptophan (d) Phenylalanine **Ans. a**

358. When net protein utilization (NPU) is low, the requirements for proteins are
(a) High (b) Moderate
(c) Low (d) Supplementary **Ans. a**

359. Protein content of human milk is about

(a) 1.4% (b) 2.4%

(c) 3.4% (d) 4.4% **Ans. b**

360. Protein content of cow's milk is about

(a) 2.5% (b) 3.5%

(c) 4.5% (d) 5.5% **Ans. b**

361. Protein content of soyabean is about

(a) 30% (b) 40%

(c) 50% (d) 60% **Ans. b**

362. Lipid content of egg white is

(a) 12% (b) 33%

(c) 10–11% (d) Traces **Ans. d**

363. The recommended daily allowance (RDA) of proteins for an adult man is

(a) 70 gms (b) 50 gms

(c) 40 gms (d) 30 gms **Ans. a**

364. The basic amino acids are

(a) Lysine (b) Bile acids

(c) Glycine (d) Alanine **Ans. a**

365. The daily caloric requirement for the normal adult female is about

(a) 1500 (b) 2100

(c) 2500 (d) 2900 **Ans. b**

366. In the total proteins, the percentage of albumin is about

(a) 20–40 (b) 30–45

(c) 50–70 (d) 80–90 **Ans. c**

367. In the total proteins percentage of α_1 globulin is about

(a) 0.2–1.2% (b) 1.2–2.0%

(c) 2.4–4.4% (d) 5.0–10.0% **Ans. c**

368. In the total proteins the percentage of γ globulin is about

(a) 2.4–4.4% (b) 10.0–21.0%

(c) 6.1–10.1% (d) 1.2–2.0% **Ans. b**

369. Most frequently the normal albumin globulin ratio (A : G) is
(a) 1.0 : 0.8 (b) 1.5 : 1.0
(c) 2.0 : 1.0 (d) 2.4 : 1.0 **Ans. c**

370. In Thymol turbidity test the protein involved is mainly
(a) Albumin (b) α_1-Globulin
(c) α_2-Globulin (d) β Globulin **Ans. d**

371. In quaternary structure, subunits are linked by
(a) Peptide bonds (b) Disulphide bonds
(c) Covalent bonds (d) Non-covalent bonds **Ans. d**

372. Molecular weight of human albumin is about
(a) 156,000 (b) 90,000
(c) 69,000 (d) 54,000 **Ans. c**

373. At isoelectric pH, an amino acid exists as
(a) Anion (b) Cation
(c) Zwitterion (d) None of these **Ans. c**

374. A disulphide bond can be formed between
(a) Two methionine residues
(b) Two cysteine residues
(c) A methionine and a cysteine residue
(d) All of these **Ans. b**

375. A coagulated protein is
(a) Insoluble (b) Biologically non-functional
(c) Unfolded (d) All of the above **Ans. d**

376. At a pH below the isoelectric point, an amino acid exists as
(a) Cation (b) Anion
(c) Zwitterion (d) Undissociated molecule **Ans. a**

377. An amino acid having a hydrophilic side chain is
(a) Alanine (b) Proline
(c) Methionine (d) Serine **Ans. d**

378. An amino acid that does not take part in α-helix formation is
(a) Histidine (b) Tyrosine
(c) Proline (d) Tryptophan **Ans. c**

379. A protein rich in cysteine is
(a) Collagen (b) Keratin
(c) Haemoglobin (d) Gelatin **Ans. b**

380. Primary structure of proteins can be determined by the use of
(a) Electrophoresis (b) Chromatography
(c) Ninhydrin (d) Sanger's reagent **Ans. d**

381. Electrostatic bonds can be formed between the side chains of
(a) Alanine and leucine (b) Leucine and valine
(c) Asparate and glutamate (d) Lysine and aspartate **Ans. d**

382. Sanger's reagent contains
(a) Phenylisothiocyanate (b) Dansyl chloride
(c) 1-Fluoro-2, 4-dinitrobenzene (d) Ninhydrin **Ans. c**

383. Folding of newly synthesized proteins is accelerated by
(a) Protein disulphide isomerase (b) Prolyl cis-trans isomerase
(c) Chaperonins (d) All of these **Ans. d**

384. Primary structure of a protein is formed by
(a) Hydrogen bonds (b) Peptide bonds
(c) Disulphide bonds (d) All of these **Ans. b**

385. α-Helix is formed by
(a) Hydrogen bonds (b) Hydrophobic bonds
(c) Electrostatic bonds (d) Disulphide bonds **Ans. a**

386. Glutelins are present in
(a) Milk (b) Eggs
(c) Meat (d) Cereals **Ans. d**

387. Aromatic amino acids can be detected by
(a) Sakaguchi reaction (b) Millon-Nasse reaction
(c) Hopkins-Cole reaction (d) Xanthoproteic reaction**Ans. d**

388. Two amino groups are present in
(a) Leucine (b) Glutamate
(c) Lysine (d) Threonine **Ans. c**

389. During denaturation of proteins, all of the following are disrupted except
(a) Primary structure (b) Secondary structure
(c) Tertiary structure (d) Quaternary structure **Ans. b**

390. All the following are branched chain amino acids except
(a) Isoleucine (b) Alanine
(c) Leucine (d) Valine **Ans. b**

391. An –OH group is present in the side chain of
(a) Serine (b) Arginine
(c) Lysine (d) Proline **Ans. a**

392. Edman's reagent contains
(a) Phenylisothiocyanate (b) 1-Fluoro-2, 4-dinitrobenzene
(c) Dansyl Chloride (d) tBOC azide **Ans. a**

393. Edman's reaction can be used to
(a) Determine the number of tyrosine residues in a protein
(b) Determine the number of aromatic amino acid residues in a protein
(c) Determine the amino acid sequence of a protein
(d) Hydrolyse the peptide bonds in a protein **Ans. c**

394. Inherited deficiency of β-glucosidase causes
(a) Tay-Sachs disease
(b) Metachromatic leukodystrophy
(c) Gaucher's disease
(d) Multiple sclerosis **Ans. c**

395. Tay-Sachs disease results from inherited deficiency of
(a) Arylsulphatase A (b) Hexosaminidase A
(c) Sphingomyelinase (d) Ceramidase **Ans. b**

396. The largest alpolipoprotein is
(a) Apo E (b) Apo B-48
(c) Apo B-100 (d) Apo A-I **Ans. c**

397. Apolipoprotein B-100 is synthesised in
(a) Adipose tissue (b) Liver
(c) Intestine (d) Liver and intestine **Ans. b**

398. Apolipoprotein B-48 is synthesized in
(a) Adipose tissue (b) Liver
(c) Intestine (d) Liver and intestine **Ans. c**

399. Apolipoproteins A-I and A-II are present in
(a) LDL only (b) LDL and VLDL
(c) HDL only (d) HDL and chylomicrons
Ans. d

400. Apolipoprotein B-48 is present in
(a) Chylomicrons (b) VLDL
(c) LDL (d) HDL **Ans. a**

401. Apolipoprotein B-100 is present in
(a) Chylomicrons (b) VLDL only
(c) LDL only (d) VLDL and LDL **Ans. b**

402. Apolipoproteins C-I, C-II and C-III are present in
(a) Chylomicrons (b) VLDL
(c) HDL (d) All of these **Ans. d**

403. Apolipoprotiens C-I, C-II and C-III are present in all of the following except
(a) Chylomicrons (b) VLDL
(c) LDL (d) HDL **Ans. c**

404. Apolipoprotein A-I acts as
(a) Enzyme activator (b) Ligand for receptor
(c) Both (a) and (b) (b) None of these **Ans. c**

405. Apolipoprotien B-100 acts as
(a) Enzyme activator (b) Ligand for receptor
(c) Both (a) and (b) (d) None of these **Ans. b**

406. Apolipoprotein C-II is an activator of
(a) Lecithin cholesterol acyl transferase
(b) Phospholipase C
(c) Extrahepatic lipoprotein lipase
(d) Hepatic lipoprotein lipase **Ans. b**

407. Nascent chylomicron receives apolipoproteins C and E from
(a) VLDL remnant (b) VLDL
(c) LDL (d) HDL **Ans. d**

408. Terminal transferase
(a) Removes nucleotides from 3' end
(b) Adds nucleotides at 3' end

(c) Removes nucleotides from 3' end
(d) Adds nucleotides at 3' end **Ans. b**

409. S_1 nuclease hydrolyses
(a) DNA of somatic cells (b) DNA of sperms
(c) Any double stranded DNA (d) Any single stranded DNA
Ans. d

410. Positive nitrogen balance is seen in
(a) Starvation (b) Wasting diseases
(c) Growing age (d) Intestinal malabsorption
Ans. c

411. Alanine can be synthesized from
(a) Glutamate and α-ketoglutarate
(b) Pyruvate and glutamate
(c) Pyruvate and α-ketoglutarate
(d) Asparate and α-ketoglutarate **Ans. b**

412. All of the following are required for synthesis of alanine except
(a) Pyruvate (b) α-ketoglutarate
(c) Glutamate (d) Pyridoxal phosphate **Ans. b**

413. All of the following statements about aspartate are true except
(a) It is non-essential amino acid
(b) It is a dicarboxylic amino acid
(c) It can be synthesized from pyruvate and glutamate
(d) It can be converted into asparagines **Ans. c**

414. Glycine can be synthesized from
(a) Serine (b) Choline
(c) Betaine (d) All of these **Ans. d**

415. All of the following are required for synthesis of glutamine except
(a) Glutamate (b) Ammonia
(c) Pyridoxal phosphate (d) ATP **Ans. c**

416. A coenzyme required for the synthesis of glycine from serine is
(a) ATP (b) Pyridoxal phosphate
(c) Tetrahydrofolate (d) NAD **Ans. c**

417. All of the following statements about proline are true except
(a) It is an imino acid
(b) It can be synthesized from glutamate
(c) It can be catabolised to glutamate
(d) Free proline can be hydroxylated to hydroxyproline **Ans. d**

418. A protein rich in hydroxyproline is
(a) Prolamin (b) Procollagen
(c) Collagen (d) Proinsulin **Ans. c**

419. All the following statement about hydroxyproline are true except
(a) There is no codon for hydroxyproline
(b) It is present in large amounts in collagen
(c) Free proline cannot be hydroxylated to hydroxyproline
(d) Hydroxylation of proline residues is catalyzed by a dioxygenase **Ans. d**

420. All of the following are required for hydroxylation of proline residues except
(a) Ascorbic acid (b) Glutamate
(c) Ferrous ions (d) Molecular oxygen **Ans. b**

421. Cysteine can be synthesized from methionine and
(a) Serine (b) Homoserine
(c) Homocysteine (d) Threonine **Ans. a**

422. Methionine is synthesized in human body from
(a) Cysteine and homoserine (b) Homocysteine and serine
(c) Cysteine and serine (d) None of these **Ans. d**

423. Hydroxylation of phenylalanine requires all of the following except
(a) Phenylalanine hydroxylase (b) Tetrahydrobiopterin
(c) NADH (d) Molecular oxygen **Ans. c**

424. Non-Protein amino acids are
(a) Ornithine (b) β-alanine
(c) α-amino butyric acid (d) All of these **Ans. a**

425. The amino acid that undergoes oxidative deamination at significant rate is
(a) Alanine (b) Aspartate
(c) Glutamate (d) Glutamine **Ans. c**

426. Allosteric inhibitor of glutamate dehydrogenase is

(a) ATP (b) ADP

(c) AMP (d) GMP **Ans. a**

427. Allsoteric activator of glutamate dehydrogenase is

(a) ATP (b) GTP

(c) ADP and GDP (d) AMP and GMP **Ans. c**

428. Free ammonia is released during

(a) Oxidative deamination of glutamate

(b) Catabolism of purines

(c) Catabolism of pyrimidines

(d) All of these **Ans. d**

429. An organ which is extremely sensitive to ammonia toxicity is

(a) Liver (b) Brain

(c) Kidney (d) Heart **Ans. b**

430. Ammonia is transported from muscles to liver mainly in the form of

(a) Free ammonia (b) Glutamine

(c) Asparagine (d) Alanine **Ans. d**

431. The major site of urea synthesis is

(a) Brain (b) Kidneys

(c) Liver (d) Muscles **Ans. b**

432. Carbamoyl phosphate required for urea synthesis is formed in

(a) Cytosol (b) Mitochondria

(c) Both (a) and (b) (d) None of these **Ans. b**

433. Cytosolic and mitochondrial carbamoyl phosphate synthetase have the following similarity

(a) Both use ammonia as a substance

(b) Both provide carbamoyl phosphate for urea synthesis

(c) Both require N-acetylglutamate as an activator

(d) Both are allosteric enzymes **Ans. c**

434. The following enzyme of urea cycle is present in cytosol

(a) Argininosuccinic acid synthetase

(b) Argininosuccinase

(c) Arginase
(d) All of these **Ans. d**

435. ATP is required in following reactions of urea cycle
(a) Synthesis of carbamoyl phosphate and citrulline
(b) Synthesis of citrulline and argininosuccinate
(c) Synthesis of argininosuccinate and arginine
(d) Synthesis of carbamoyl phosphate and argininosuccinate **Ans. c**

436. Daily excretion of nitrogen by an adult man is about
(a) 15–20 mg (b) 1.5–2 gm
(c) 5–10 gm (d) 15–20 gm **Ans. c**

437. Maple syrup urine diseases is an inborn error of metabolism of
(a) Sulphur-containing amino acids
(b) Aromatic amino acids
(c) Branched chain amino acids
(d) Dicarboxylic amino acids **Ans. c**

438. Cystinuria results from inability to
(a) Metabolise cysteine
(b) Convert cystine into cysteine
(c) Incorporate cysteine into proteins
(d) Reabsorb cystine in renal tubules **Ans. d**

439. The defective enzyme in histidinemia is
(a) Histidine carboxylase (b) Histidine decarboxylase
(c) Histidase (d) Histidine oxidase **Ans. c**

440. All the following statements about phenylketonuria are correct except
(a) Phenylalanine cannot be converted into tyrosine
(b) Urinary excretion of phenylpyruvate and phenyllactate is increased
(c) It can be controlled by giving a low phenylalanine diet
(d) It leads to decreased synthesis of thyroid hormones, catecholamines and melanin **Ans. d**

441. All the following statements about albinism are correct except
(a) Tyrosine hydroxylase (tyrosinase) is absent or deficient in melanocytes
(b) Skin is hypopigmented

(c) It results in mental retardation

(d) Eyes are hypopigmented **Ans. c**

442. Glycine is not required for the formation of

(a) Taurocholic acid (b) Creatine

(c) Purines (d) Pyrimidines **Ans. a**

443. Histamine is formed from histidine by

(a) Deamination (b) Dehydrogenation

(c) Decarboxylation (d) Carboxylation **Ans. c**

444. DOPA is an intermediate in the synthesis of

(a) Thyroid hormones (b) Catecholamines

(c) Melanin (d) Catecholamines and melanin **Ans. d**

445. All the following statements about pepsin are correct except

(a) It is smaller than pepsinogen

(b) It is formed by the action of HCl on its precursor

(c) Its optimum pH is 1.0–2.0

(d) It hydrolyses the C-terminal and N-terminal peptide bonds of proteins **Ans. d**

446. Pancreatic juice contains the precursors of all of the following except

(a) Trypsin (b) Chymotrypsin

(c) Carboxypeptidase (d) Aminopeptidase **Ans. d**

447. The only correct statement about chymotrypsin is

(a) It is formed from trypsin

(b) Carboxypeptidase converts trypsin into chymotrypsin

(c) Its optimum pH is around 7

(d) It hydrolyses peptide bonds involving basic amino acids **Ans. c**

448. The portion of the antigen molecule which is recognized by antibody is known as

(a) Hapten (b) Epitope

(c) Complement (d) Variable region **Ans. b**

449. All the following statements about haptens are true except

(a) They have high molecular weights

(b) They cannot elicit an immune response by themselves

(c) When combined with some other large molecule; they can elicit an immune response

(d) Once an immune response develops, the free hapten can be recognized by the antibody **Ans. a**

450. Antigens and haptens have the following similarity

(a) They have high molecular weights

(b) They can elicit immune response by themselves

(c) They can elicit an immune response only in association with some other large molecule

(d) Once an immune response develops, free antigen and free hapten can be recognized by the antibody **Ans. d**

451. The minimum number of polypeptide chains in an immunoglobulin is

(a) Two (b) Four

(c) Five (d) Six **Ans. b**

452. Light chains of immunoglobulins are of following types

(a) Alpha and kappa (b) Alpha and gamma

(c) Lambda and delta (d) Kappa and lambda **Ans. d**

453. Immunoglobulins are classified on the basis of

(a) Type of light chains

(b) Type of heavy chains

(c) Types of light and heavy chains

(d) Molecular weight **Ans. b**

454. The molecular weight of light chains is

(a) 10,000–15,000 (b) 20,000–25,000

(c) 25,000–50,000 (d) 50,000–75,000 **Ans. b**

455. The molecular weight of heavy chains is

(a) 20,000–25,000 (b) 25,000–50,000

(c) 50,000–70,000 (d) 70,000–1,00,000 **Ans. c**

456. Secretory component is present in

(a) IgA (b) IgG

(c) IgM (d) All of these **Ans. a**

457. The variable region of light chains is the

(a) N-terminal quarter (b) N-terminal half

(c) C-terminal quarter (d) C-terminal half **Ans. b**

458. The variable region of heavy chain is the
(a) N-terminal quarter (b) N-terminal half
(c) C-terminal quarter (d) C-terminal half **Ans. a**

459. The variable region of light chains has
(a) One hypervariable region (b) Two hypervariable regions
(c) Three hypervariable regions (d) Four hypervariable regions
Ans. c

460. The variable region of heavy chains has
(a) One hypervariable region (b) Two hypervariable regions
(c) Three hypervariable regions (d) Four hypervariable regions
Ans. d

461. The most abundant immunoglobulin in plasma is
(a) IgA (b) IgG
(c) IgM (d) IgD **Ans. b**

462. The largest immunoglobulin is
(a) IgA (b) IgG
(c) IgM (D) IgD **Ans. c**

463. The plasma concentration of IgA is
(a) 1–5 mg/dl (b) 40–200 mg/dl
(c) 60–500 mg/dl (d) 700–1,500 mg/dl **Ans. c**

464. An immunoglobulin found in exocrine secretions is
(a) IgA (b) IgG
(c) IgM (d) IgE **Ans. a**

465. Allergic reactions are mediated by
(a) IgA (b) IgG
(c) IgD (d) IgE **Ans. d**

466. An immunoglobulin which can cross the placental barrier is
(a) IgA (b) IgM
(c) IgD (d) None of these **Ans. d**

467. IgM possesses
(a) Two light chains and two heavy chains
(b) Four light chains and four heavy chains

(c) Six light chains and six heavy chains
(d) Ten light chains and ten heavy chains **Ans. d**

468. The immunoglobulin having the longest half-life is
(a) IgA (b) IgG
(c) IgM (d) IgE **Ans. b**

469. The half-life of IgG is
(a) 2–3 days (b) 5–6 days
(c) 8–10 days (d) 20–25 days **Ans. d**

470. Recognition of antigen is the function of
(a) Variable region of light chains
(b) Variable regions of light and heavy chains
(c) Constant region of heavy chains
(d) Constant regions of light and heavy chains **Ans. b**

471. The effector function of antibody is performed by
(a) Variable region of light chains
(b) Constant region of heavy chains
(c) Variable regions of light and heavy chains
(d) Constant regions of light and heavy chains **Ans. d**

472. Complement system can be activated by binding of antigen to
(a) IgA (b) IgD
(c) IgE (d) IgM **Ans. d**

473. C_1 component of classical complement pathway is made up of
(a) Complements 1q and 1r (b) Complements 1q and 1s
(c) Complements 1r and 1s (d) Complements 1q, 1r and 1s
Ans. d

474. The components of complement system are activated by
(a) Microsomal hydroxylation (b) Phosphorylation
(c) Glycosylation (d) Proteloysis **Ans. d**

475. The component system forms a membrane attack complex made up of
(a) Complements 1q, 1r and 1s (b) Complements 1, 2, 3 and 4
(d) Complements 5b, 6, 7 and 8 (d) Factors B and D **Ans. d**

476. Factors B and D are required in
(a) The classical pathway of complement fixation
(b) The alternate complement pathway
(c) Both (a) and (b)
(d) None of these **Ans. b**

477. The alternate complement pathway doesn't involve
(a) Antigen-antibody complex (b) Complement 3
(c) Factors B and D (d) Membrane attack unit **Ans. a**

478. Antibody diversity arises from
(a) Gene amplification (b) Gene re-arrangement
(c) Alternative splicing (d) All of these **Ans. b**

479. A light chain gene is constructed from the following segments
(a) Variable and constant segments
(b) Variable, joining and constant segments
(c) Variable, diversity and constant segments
(d) Variable, joining, diversity and constant segments **Ans. b**

480. In metabolic point of view, amino acids are classified as
(a) Glycogenic (b) Ketogenic
(c) Glycogenic or Ketogenic (d) All of these **Ans. d**

481. Diversity segments are present in
(a) Light chain genes (b) Heavy chain genes
(c) Light and heavy chain genes (d) None of these **Ans. b**

482. Constant segments of heavy chains are of
(a) Five types (b) Six types
(c) Seven types (d) Eight types **Ans. d**

483. Gamma heavy chains are of
(a) Two types (b) Three types
(c) Four types (d) Five types **Ans. c**

484. Gamma heavy chains are present in
(a) IgA (b) IgG
(c) IgM (d) IgD **Ans. b**

485. Heavy chains in IgD are of following type

(a) Alpha (b) Gamma

(c) Delta (d) Epsilon **Ans. c**

486. On exposure to any antigen, the first antibody to be formed is of the following class

(a) IgA (b) IgG

(c) IgM (d) IgE **Ans. c**

487. Constant segment genes of heavy chains are present in a cluster in which the first gene on side is

(a) Alpha (b) Gamma

(c) Delta (d) None of these **Ans. d**

488. Cell-mediated immunity is the function of

(a) B lymphocytes (b) T lymphocytes

(c) Plasma cells (d) Basophils **Ans. b**

489. The most abundant T cells are

(a) Cytotoxic T cells (b) Helper T cells

(c) Suppressor T cells (d) Memory T cells **Ans. b**

490. T cells can recognize

(a) Free antigens

(b) Antigens bound to cells

(c) Antigens bound to antibodies

(d) Antigens bound to MHC proteins **Ans. d**

491. MHC proteins are unique to

(a) Each cell (b) Each organ

(c) Each individual (d) Each species **Ans. c**

492. MHC class I proteins are present on the surface of

(a) B cells only (b) T cells only

(c) Macrophages only (d) All cells **Ans. d**

493. MHC class I proteins, in conjunction with antigens are recognized by

(a) Cytotoxic T cells (b) Helper T cells

(c) Suppressor T cells (d) Memory T cells **Ans. a**

494. MHC class II proteins are present on the surface of
(a) All cells
(b) B lymphocytes only
(c) Macrophages only
(d) Macrophages and B lymphocytes **Ans. d**

495. MHC Class II proteins, in conjunction with antigens, are recognized by
(a) Cytotoxic T cells (b) Helper T cells
(c) Suppressor T cells (d) Memory T cells **Ans. b**

496. CD 8 is a trans-membrane glycoprotein present in
(a) Cytotoxic T cells (b) Helper T cells
(c) Suppressor T cells (d) Memory T cells **Ans. c**

497. CD 4 is a transmembrane glycoprotein present in
(a) Cytotoxic T cells (b) Helper T cells
(c) Suppressor T cells (d) Memory T cells **Ans. a**

498. CD 3 complex and p 56lck proteins are present in
(a) Cytotoxic T cells (b) Helper T cells
(c) Both (a) and (b) (d) None of these **Ans. d**

499. Cytotoxic T cells release
(a) Perforins (b) Interleukins
(c) Colony stimulating factors (d) Tumour necrosis factor
Ans. b

500. Helper T cells release
(a) Interleukins (b) Colony stimulating factors
(c) Tumour necrosis factor (d) All of these **Ans. b**

501. MHC Class III proteins include
(a) Immunoglobulins
(b) Components of complement system
(c) T cells receptors
(d) CD4 and CD8 proteins **Ans. d**

502. Human immunodeficiency virus destroys
(a) Cytotoxic T cells (b) Helper T cells
(c) B cells (d) Plasma cells **Ans. b**

503. In allergic diseases, the concentration of the following is increased in plasma

(a) IgA (b) IgG

(c) IgD (d) IgE **Ans. d**

504. IgE has a tendency to attach to

(a) Basophils (b) Mast cells

(c) Both (a) and (b) (d) None of these **Ans. c**

505. Reaginic antibody is

(a) IgA (b) IgG

(c) IgD (d) IgE **Ans. d**

506. Active immunity can be produced by administration of

(a) Killed bacteria or viruses

(b) Live attenuated bacteria or viruses

(c) Toxoids

(d) All of these **Ans. d**

507. Passive immunity can be produced by administration of

(a) Pure antigens (b) Immunoglobulins

(c) Toxoids (d) Killed bacteria or viruses

Ans. b

508. Helper T cells release all the following except

(a) Interleukins (b) Colony stimulating factors

(c) Perforins (d) Tumour necrosis factor **Ans. c**

509. IgG cleaved by papain into

(a) Two light and two heavy chains

(b) Two F_{ab} and one F_c fragments

(c) Two pairs of one light and one heavy chain each

(d) One F_{ab} and two F_c fragments **Ans. b**

510. Bence-Jones protein is

(a) An immunoglobulin

(b) A dimer of heavy chains

(c) A dimer of light chains

(d) A dimer of one heavy and one light chain **Ans. c**

511. Bence-Jones proteins possess all the following properties except

(a) They are dimers of light chains

(b) Their amino acids sequences are identical

(c) Their N-terminal halves have variable amino acid sequences

(d) Their C-terminal halves have constant amino acid sequences

Ans. b

512. A Zwitterion is

(a) Positive ion (b) Negative ion

(c) Both (a) and (b) (d) None of these **Ans. c**

513. After accounting for SDA, the net gain of energy from 25 gm of proteins is about

(a) 70 kcal (b) 100 kcal

(c) 130 kcal (d) 200 kcal **Ans. a**

514. After accounting for SDA, the net gain of energy from 25 gm of carbohydrates is about

(a) 70 kcal (b) 95 kcal

(c) 100 kcal (d) 105 kcal **Ans. b**

515. After accounting for SDA, the net gain of energy from 100 gm of fat is about

(a) 600 kcal (b) 780 kcal

(c) 900 kcal (d) 1020 kcal **Ans. b**

516. If proteins, carbohydrates and fats are consumed together

(a) The total SDA is the sum of individual SDAs of proteins, carbohydrates and fats

(b) The total SDA is more than the sum of individual SDAs of proteins, carbohydrates and fats

(c) Carbohydrates and fats lower the SDA of proteins

(d) Proteins raise the SDA of carbohydrates and fats **Ans. c**

517. After calculating the energy requirement of a person

(a) 10% kcal are subtracted on account of SDA

(b) 10% kcal are added on account of SDA

(c) 20% kcal are subtracted on account of SDA

(d) 20% kcal are subtracted on account of SDA **Ans. b**

518. The recommended energy intake for an adult sedentary Indian man is
(a) 1,900 kcal/day (b) 2,400 kcal/day
(c) 2,700 kcal/day (d) 3,000 kcal/day **Ans. b**

519. The recommended energy intake for an adult sedentary Indian woman is
(a) 1,900 kcal/day (b) 2,200 kcal/day
(c) 2,400 kcal/day (d) 2,700 kcal/day **Ans. a**

520. During pregnancy, the following should be added to the calculated energy requirement
(a) 300 kcal/day (b) 500 kcal/day
(c) 700 kcal/day (d) 900 kcal/day **Ans. a**

521. During first six months of lactation, the following increment in energy intake is recommended
(a) 200 kcal/day (b) 300 kcal/day
(c) 550 kcal/day (d) 1,000 kcal/day **Ans. c**

522. The proximate principles of diet are
(a) Vitamins and minerals
(b) Proteins
(c) Carbohydrates and fats
(d) Carbohydrates, fats and proteins **Ans. d**

523. The limiting amino acid in wheat is
(a) Leucine (b) Lysine
(c) Cysteine (d) Methionine **Ans. b**

524. The limiting amino acid in pulses is
(a) Leucine (b) Lysine
(c) Tryptophan (d) Methionine **Ans. d**

525. Maize is poor in
(a) Lysine (b) Methionine
(c) Tryptophan (d) Lysine and tryptophan
Ans. d

526. The percentage of ingested protein/nitrogen absorbed into blood stream is known as
(a) Net protein utilisation (b) Protein efficiency ratio
(c) Digestibility coefficient (d) Biological value of protein
Ans. c

527. Biological value of a protein is

(a) The percentage of ingested protein/nitrogen absorbed into circulation

(b) The percentage of ingested protein/nitrogen in the body

(c) The percentage of ingested protein utilized for protein synthesis in the body

(d) The gain in body weight (gm) per gm of protein ingested **Ans. b**

528. Net protein utilisation depends upon

(a) Protein efficiency ratio

(b) Digestibility coefficient

(c) Digestibility coefficient and protein efficiency ratio

(d) Digestibility coefficient and biological value **Ans. d**

529. The gain in body weight (gm) per gm of protein ingested is known as

(a) Net protein utilisation (b) Protein efficiency ratio

(c) Digestibility coefficient (d) Biological value of protein

Ans. b

530. The following is considered as reference standard for comparing the nutritional quality of proteins

(a) Milk proteins (b) Egg proteins

(c) Meat proteins (d) Fish proteins **Ans. b**

531. Biological value of egg proteins is about

(a) 70 % (b) 80 %

(c) 86 % (d) 94 % **Ans. d**

532. The following has the highest protein efficiency ratio

(a) Milk proteins (b) Egg proteins

(c) Meat proteins (d) Fish proteins **Ans. b**

533. The following has the lowest protein efficiency ratio

(a) Maize proteins (b) Wheat proteins

(c) Milk proteins (d) Rice proteins **Ans. a**

534. Protein content of egg is about

(a) 10% (b) 13%

(c) 16% (d) 20% **Ans. b**

535. Protein content of meat is about

(a) 10% (b) 13%

(c) 16% (d) 20% **Ans. d**

536. Protein content of rice is about

(a) 7% (b) 12%

(c) 15% (d) 20% **Ans. a**

537. The calorific value of wheat is about

(a) 2.5 kcal/gm (b) 3.5 kcal/gm

(c) 4.5 kcal/gm (d) 5.5 kcal/gm **Ans. b**

538. For vegetarians, pulses are an important source of

(a) Carbohydrates (b) Proteins

(c) Fat (d) Iron **Ans. b**

539. The amino acids present in pulses can supplement the limiting amino acids of

(a) Cereals (b) Milk

(c) Fish (d) Nuts and beans **Ans. a**

540. Milk is a good source of

(a) Proteins, calcium and iron

(b) Proteins, calcium and ascorbic acid

(c) Proteins, lactose and retinol

(d) Proteins, lactose and essential fatty acids **Ans. c**

541. Milk is a good source of all of the following except

(a) Essential amino acids (b) Vitamin C

(c) Galactose (d) Calcium and phosphorous **Ans. b**

542. Milk is poor in

(a) Cholesterol (b) Retinol

(c) Calcium (d) Iron **Ans. d**

543. Egg is rich in all of the following except

(a) Cholesterol (b) Saturated fatty acids

(c) Ascorbic acid (d) Calcium **Ans. c**

544. A phosphoprotein present in egg is
(a) Casein (b) Albumin
(c) Ovoglobulin (b) Ovovitellin **Ans. d**

545. Consumption of raw eggs can cause deficiency of
(a) Calcium (b) Lipoic acid
(c) Biotin (d) Vitamin A **Ans. c**

546. Egg is poor in
(a) Essential amino acids (b) Carbohydrates
(c) Avidin (d) Biotin **Ans. b**

547. Cholesterol is present in all the following except
(a) Milk (b) Fish
(c) Egg white (d) Egg yolk **Ans. c**

548. Meat is rich in all of the following except
(a) Iron (b) Fluorine
(c) Copper (d) Zinc **Ans. b**

549. Kwashiorkor occurs when the diet is severely deficient in
(a) Iron (b) Calories
(c) Proteins (d) Essential fatty acids **Ans. c**

550. Clinical features of Kwashiorkor include all of the following except
(a) Mental retardation (b) Muscle wasting
(c) Oedema (d) Anaemia **Ans. a**

551. Kwashiorkor usually occurs in
(a) The post-weaning period (b) Pregnancy
(c) Lactation (d) Old age **Ans. a**

552. Marasmus occurs from deficient intake of
(a) Essential amino acids (b) Essential fatty acids
(c) Calories (d) Zinc **Ans. c**

553. Marasmus differs from Kwashiorkor in which of these following respect
(a) Mental retardation occurs in kwashiorkor but not in marasmus
(b) Growth is retarded in kwashiorkor but not in marasmus
(c) Muscle wasting occurs in marasmus but not in kwashiorkor
(d) Subcutaneous fat disappears in marasmus but not in kwashiorkor
Ans. d

554. Energy reserves of an average well-fed adult man are about
(a) 50,000 kcal (b) 100,000 kcal
(c) 200,000 kcal (d) 300,000 kcal **Ans. b**

555. During starvation, the first reserve nutrient to be depleted is
(a) Glycogen (b) Proteins
(c) Triglycerides (d) Cholesterol **Ans. a**

556. Synthesis of the following enzymes is increased during starvation
(a) Digestive enzymes (b) Gluconeogenic enzymes
(c) Urea cycle enzymes (d) Glucokinase **Ans. b**

557. In hypoparathyroidism
(a) Plasma calcium and inorganic phosphorous are low
(b) Plasma calcium and inorganic phosphorous are high
(c) Plasma calcium is low and inorganic phosphorous high
(d) Plasma calcium is high and inorganic phosphorous low **Ans. c**

558. The number of amino acid residues in calcitonin in
(a) 9 (b) 32
(c) 51 (d) 84 **Ans. b**

559. Calcitonin is synthesised in
(a) Parathyroid glands (b) Thyroid gland
(c) Pars intermedia of pituitary (d) Adrenal cortex **Ans. b**

560. Plasma calcium is lowered by
(a) Parathormone (b) Calcitonin
(c) Aldosterone (d) Deoxycorticosterone **Ans. b**

561. α-Cells of Islets of Langerhans secrete
(a) Insulin (b) Glucagon
(c) Somatostatin (d) Cholecystokinin **Ans. b**

562. A/G ratio is
(a) Strength of proteins (b) Ratio of serum proteins
(c) Ratio of ceruloplasmin (d) None of these **Ans. b**

563. Insulin is made up of
(a) A single polypeptide chain having 51 amino acid residues
(b) A single polypeptide chain having 84 amino acid residues

(c) A-chain having 21 and B-chain having 30 amino acid residues
(d) A-chain having 30 and B-chain having 21 amino acid residues
Ans. c

564. The number of amino acid residues in preproinsulin is
(a) 51 (b) 84
(c) 109 (d) 119 **Ans. c**

565. Pre-proinsulin contains a signal sequence having
(a) 9 amino acid residues (b) 19 amino acid residues
(c) 27 amino acid residues (d) 33 amino acid residues
Ans. b

566. The number of intra-chain disulphide bonds in pro-insulin
(a) One (b) Two
(c) Three (d) Four **Ans. c**

567. Pentagastrin is a
(a) Naturally occurring form of gastrin
(b) Inactive metabolite of gastrin
(c) Active metabolite of gastrin
(d) Synthetic form of gastrin **Ans. d**

568. Secretion of gastrin is evoked by
(a) Entry of food into stomach (b) Vagal stimulation
(c) Lower aliphatic alcohols (d) All of these **Ans. d**

569. Gastrin stimulates
(a) Gastric motility (b) Gastric secretion
(c) Both (a) and (b) (d) None of these **Ans. c**

570. Secretin is made up of
(a) 17 amino acids (b) 27 amino acids
(c) 37 amino acids (d) 47 amino acids **Ans. b**

571. Secretin causes all of the following except
(a) Secretion of pancreatic juice
(b) Secretion of bile juice
(c) Inhibition of gastric secretion
(d) Stimulation of intestinal motility **Ans. d**

572. All of the following statements about cholecystokinin pancreozymin are true except

(a) It is secreted by mucosa of small intestine

(b) It stimulates secretion of pancreatic juice rich in enzymes

(c) It stimulates contraction of gall bladder

(d) It inhibits gastric motility **Ans. d**

573. All of the following statements about pancreatic somatostain are true except

(a) It is secreted by δ-cells of islets of Langerhans

(b) It stimulates the secretion of gastrin

(c) It inhibits the secretion of secretin

(d) It inhibits the secretion of cholecysto-kinin-pancreozymin **Ans. b**

574. Histidine is converted into histamine by

(a) Carboxylation (b) Decarboxylation

(c) Methylation (d) Hydroxylation **Ans. b**

575. Histamine is synthesised in

(a) Brain (b) Mast cells

(c) Basophils (d) All of these **Ans. d**

576. Histamine causes all the following except

(a) Stimulation of gastric secretion

(b) Vasoconstriction

(c) Pruritus

(d) Increase in capillary permeability **Ans. b**

577. H_2-receptors are blocked by

(a) Diphenhydramine (b) Mepayramine

(c) Pyrilamine (d) Cimetidine **Ans. d**

578. Serotonin is synthesised from

(a) Serine (b) Phenylalanine

(c) Tyrosine (d) Tryptophan **Ans. d**

579. All the following statements about serotonin are true except

(a) It causes vasolidatation

(b) It causes bronchoconstriction

(c) It is metabolized by monoamine oxidase

(d) Its metabolite is 5-hydroxyindole acetic acid **Ans. a**

580. All the following statements about angiotensin are true except

(a) Its precursor is an α_2-globulin

(b) Its active form is an octapeptide

(c) It is a vasodilator

(d) It increases the secretion of aldosterone **Ans. c**

581. Methyl dopa decreases blood pressure by

(a) Inhibiting the synthesis of catecholamines

(b) Antagonising the action of aldosterone

(c) Stimulating the release of renin

(d) Inhibiting the breakdown of angiotensin **Ans. a**

582. Binding of gamma-aminobutyric acid to its receptors in brain increases the permeability of cell membrane to

(a) Cl^- (b) Na^+

(c) K^+ (d) Ca^{++} **Ans. a**

583. Binding of acetylcholine to its receptors increases the permeability of cell membrane to

(a) Ca^{++} (b) Na^+

(c) K^+ (d) Na^+ and K^+ **Ans. d**

584. All of the following are glycoproteins except

(a) Collagen (b) Albumin

(c) Transferrin (d) IgM **Ans. b**

585. Sialic acids are present in

(a) Proteoglycans (b) Glycoproteins

(c) Both (a) and (b) (d) None of these **Ans. b**

586. Hyaluronidase hydrolyses

(a) Hyaluronic acid

(b) Chondroitin sulphate

(c) Heparin

(d) Hyaluronic acid and chondroitin sulphate **Ans. d**

587. The most abundant protein in bones is
(a) Collagen type I (b) Collagen type II
(c) Collagen type III (d) Non-collagen proteins **Ans. a**

588. The most abundant collagen in cartilages is
(a) Type I (b) Type II
(c) Type III (d) Type IV **Ans. b**

589. Collagen and elastin have the following similarity
(a) Both are triple helices
(b) Both have hydroxyproline residues
(c) Both have hydrolysine residues
(d) Both are glycoproteins **Ans. a**

590. Abnormal collagen structure is seen in all of the following except
(a) I-cell disease (b) Osteogenesis imperfecta
(c) Menke's disease (d) Ehlers-Danlos syndrome
Ans. b

591. I-cell disease results from absence of the following from lysosomal enzymes
(a) Signal sequence (b) Mannose-6-phosphate
(c) Sialic acid (d) A serine residue **Ans. d**

592. In I-cell disease, lysosomal enzymes
(a) Are not synthesised (b) Are inactive
(c) Lack signal sequence (d) Cannot reach lysosomes
Ans. b

593. Renal glycosuria occurs due to
(a) Increased filtration of glucose in glomeruli
(b) Increased secretion of glucose by renal tubular cells
(c) Decreased reabsorption of glucose by renal tubular cells
(d) Increased conversion of glycogen into glucose in tubular cells
Ans. d

594. Haematuria can occur in
(a) Haemolytic anaemia (b) Mismatched blood transfusion
(c) Yellow fever (d) Stone in urinary tract **Ans. d**

595. Haematuria can occur in all of the following except
(a) Acute glomerulonephritis (b) Cancer of urinary tract
(c) Stone in urinary tract (d) Mismatched blood transfusion
Ans. d

596. Chyluria can be detected by addition of the following to the urine
(a) Sulphosalicylic acid (b) Nitric acid
(c) Acetic anhydride (d) Chloroform **Ans. d**

597. Normal range of serum urea is
(a) 0.6–1.5 mg/dl (b) 9–11 mg/dl
(c) 20–45 mg/dl (d) 60–100 mg/dl **Ans. c**

598. Normal range of serum creatinine is
(a) 0.6–1.5 mg/dl (b) 9–11 mg/dl
(c) 20–45 mg/dl (d) 60–100 mg/dl **Ans. a**

599. Standard urea clearance is
(a) 54 ml/min (b) 75 ml/min
(c) 110 ml/min (d) 130 ml/min **Ans. a**

600. Maximum urea clearance is
(a) 54 ml/min (b) 75 ml/min
(c) 110 ml/min (d) 130 ml/min **Ans. b**

601. Average creatinine clearance in an adult man is about
(a) 54 ml/min (b) 75 ml/min
(c) 110 ml/min (d) 130 ml/min **Ans. c**

602. Inulin clearance in an average adult man is about
(a) 54 ml/min (b) 75 ml/min
(c) 110 ml/min (b) 130 ml/min **Ans. d**

603. Among the following, a test of tubular function is
(a) Creatinine clearance (b) Inulin clearance
(c) PAH clearance (d) PSP excretion test **Ans. d**

604. A simple way to assess tubular function is to withhold food and water for 12 hours and, then, measure
(a) Serum urea (b) Serum creatinine
(c) Urine output in one hour (d) Specific gravity of urine **Ans. d**

605. Among the following, the most sensitive indicator of glomerular function is

(a) Serum urea (b) Serum creatinine

(c) Urea clearance (d) Creatinine clearance **Ans. d**

606. All the following statements about inulin are correct except

(a) It is completely non-toxic

(b) It is completely filtered by glomeruli

(c) It is not reabsorbed by tubular cells

(d) It is secreted by tubular cells **Ans. d**

607. Non-protein nitrogenous substances in blood include all of the following except

(a) Urea (b) Uric acid

(c) Creatinine (d) Inositol **Ans. c**

608. Non-protein nitrogenous substances in blood are raised in

(a) Starvation (b) Liver damage

(c) Renal failure (d) All of these **Ans. b**

609. Creatinine clearance is deceased in

(a) Acute tubular necrosis (b) Acute glomerulonephritis

(c) Hypertension (d) Myopathies **Ans. d**

610. Serum amylase is increased in

(a) Acute parotitis (b) Acute pancreatitis

(c) Pancreatic cancer (d) All of these **Ans. b**

611. Maximum rise in serum amylase occurs in

(a) Acute parotitis (b) Acute pancreatitis

(c) Chronic pancreatitis (d) Pancreatic cancer **Ans. b**

612. Serum lipase is increased in

(a) Acute parotitis (b) Acute pancreatitis

(c) Infective hepatitis (d) Biliary obstruction **Ans. a**

613. Which one of the following metabolites is not directly produced in the hexose monophosphate pathway?

(a) Fructose-6-phosphate (b) Dihydroxy acetone phosphate

(c) CO_2 (d) Erythrose-4-phosphate

Ans. b

614. Which one of the following statements concerning glucose-6-phosphate dehydrogenase deficiency is correct?

(a) Young R.B.Cs, particularly reticulocytes, contain the highest enzyme activity cells show less enzyme activity

(b) Glucose-6-P Dehydroglucose deficiency leads to disfuction of many tissues

(c) G-6-P Dehydroglucose deficiency is due to a single deletion of a large sequence of DNA in the G-6-PD gene

(d) G-6-PD deficiency is precipitated by ingestion of drugs such as aspirin **Ans. a**

615. The phenomenon of inhibition of glycolysis by O_2 is termed as

(a) Red drop (b) Pasteur effect

(c) Michaelis effect (d) Fischer's effect **Ans. b**

616. Seratonin is derived in the body from the following amino acid

(a) Phenylalanine (b) Histidine

(c) Tryptophan (d) Serine **Ans. c**

617. Which amino acid is a lipotropic factor?

(a) Lysine (b) Leucine

(c) Tryptophan (d) Methionine **Ans. d**

618. Which among the following is a nutritionally essential amino acid for man?

(a) Alanine (b) Glycine

(c) Tyrosine (d) Tryptophan **Ans. d**

619. The essential amino acids

(a) Must be supplied in the diet because the organism has lost the capacity to aminate the corresponding ketoacids

(b) Must be supplied in the diet because the human has an impaired ability to synthesize the carbon chain of the corresponding ketoacids

(c) Are identical in all species studied

(d) Are defined as those amino acids which cannot be synthesized by the organism at a rate adequate to meet metabolic requirements

Ans. b

620. Which among the following is an essential amino acid?
(a) Cysteine (b) Leucine
(c) Tyrosine (d) Aspartic acid **Ans. b**

621. Which among the following is a basic amino acid?
(a) Aspargine (b) Arginine
(c) Proline (d) Alanine **Ans. b**

622. This amino acid cannot have optical isomers
(a) Alanine (b) Histidine
(c) Threonine (d) Glycine **Ans. d**

623. The amino acid which contains a guanidine group is
(a) Histidine (b) Arginine
(c) Citrulline (d) Ornithine **Ans. b**

624. GABA (gama amino butyric acid) is
(a) Post-synaptic excitatory transmitter
(b) Post-synaptic inhibitor transmitter
(c) Activator of glia-cell function
(d) Inhibitor of glia-cell function **Ans. b**

625. Sulphur-containing amino acid is
(a) Glutathione (b) Chondroitin sulphate
(c) Homocysteine (d) Tryptophan **Ans. c**

626. The useful reagent for detection of amino acids is
(a) Molisch reagent (b) Dichlorophenol Indophenol
(c) Ninhydrin (d) Biuret **Ans. c**

627. The amino acid which contains an indole group is
(a) Histidine (b) Arginine
(c) Glycine (d) Tryptophan **Ans. d**

628. Sakaguchi reaction is answered by
(a) Lysine (b) Ornithine
(c) Arginine (d) Arginino succinic acid **Ans. c**

629. The pH of an amino acid depends
(a) Optical rotation (b) Dissociation constant
(c) Diffusion coefficient (d) Chain length **Ans. b**

630. When amino acids are treated with neutral formaldehyde, the pH of the mixture

(a) Is not altered (b) Increases

(c) Decreases (d) First increases then decreases

Ans. c

631. Which among the following has an imidazole group?

(a) Histidine (b) Tryptophan

(c) Proline (d) Hydroxy proline **Ans. a**

632. The amino acid exist as Zwitter ions when they are in

(a) Solid state (b) acidic solution

(c) Alkaline solution (d) neutral solution **Ans. d**

633. Plasma proteins are isolated by

(a) Salting out (b) Electrophoresis

(c) Flourimetry (d) Both (a) and (b) **Ans. d**

634. After digestion amino acids

(a) Are absorbed into portal circulation

(b) Are absorbed into lymph

(c) Are excreted to the extent of 50%

(d) Converted into glucose in the intestine **Ans. a**

635. Cysteine has the formula

(a) CH_3SH

(b) H_2N-CH_2-COOH

(c) $HS-CH_2-CH(NH_2)-COOH$

(d) $S-CH_2-CH(NH_2)-COOH$
|
$S-CH_2-CH(NH_2)-COOH$ **Ans. c**

636. The compound having the formula

$H_2N-CO-NH-CH_2-CH_2-CH_2-CH-COOH$
|
NH_2 is

(a) Lysine (b) Glutamine

(c) Serine (d) Citrulline **Ans. d**

637. An amino acid which contains a disulphide bond is

(a) Lysine (b) Methionine

(c) Homocysteine (d) Cystine **Ans. d**

638. One of the following has a phenolic group

(a) Histidine (b) Hydroxy lysine

(c) Seratonine (d) Hydroxy proline **Ans. c**

639. An amino acid not containing the usual – COOH group is

(a) Alanine (b) Tryptophan

(c) Methionine (d) Taurine **Ans. d**

640. Branched chain amino acids are

(a) Cysteine and cystine (b) Tyrosine and Tryptophan

(c) Glycine and Serine (d) Valine, Leucine and Isoleucine **Ans. d**

641. A Zwitter ion is one which has in aqueous solution

(a) One positive charge and one negative charge

(b) Two positive charges and one negative charge

(c) Two negative charges and one positive charge

(d) No electrical charges at all **Ans. a**

642. The amino acid which gives yellow colour with Ninhydrin in paper chromatography is

(a) Tyrosine (b) Proline

(c) Tryptophan (d) Alanine **Ans. b**

643. Hydroxylation of Proline and Lysine in a protein is effected by

(a) Vitamin B_1 (b) Vitamin B_2

(c) Vitamin B_6 (d) Vitamin C **Ans. d**

644. Millon's test is for identification of

(a) Tyrosine (b) Tryptophan

(c) Proline (d) Arginine **Ans. a**

645. Hopkins-Cole test is for identification of

(a) Tyrosine (b) Tryptophan

(c) Arginine (d) Cysteine **Ans. b**

646. Collagen is very rich in

(a) Glycine (b) Serine

(c) Aspartic acid (d) Glutamic acid **Ans. a**

647. All amino acids are optically active except

(a) Glycine (b) Serine

(c) Threonine (d) Tryptophan **Ans. a**

648. Out of 200 different amino acids form in nature the number of amino acids present in protein

(a) 20 (b) 25

(c) 40 (d) 35 **Ans. a**

649. Enzyme catalyzed hydrolysis of proteins produces amino acids of the form

(a) D (b) L

(c) DL (d) All of these **Ans. b**

650. The ionizable groups of amino acids are at least

(a) 1 (b) 2

(c) 3 (d) 4 **Ans. b**

651. The neutral amino acid is

(a) Lysine (b) Proline

(c) Leucine (d) Histidine **Ans. c**

652. The amino acid containing hydroxyl group

(a) Alanine (b) Isoleucine

(c) Arginine (d) Threonine **Ans. d**

653. The sulphur containing amino acid

(a) Homoserine (b) Serine

(c) Methionine (d) Valine **Ans. c**

654. The basic amino acid

(a) Glycine (b) Leucine

(c) Histidine (d) Proline **Ans. c**

655. The amino acid which synthesizes many hormones

(a) Valine (b) Phenyl alanine

(c) Alanine (d) Histidine **Ans. b**

656. Amino acids are insoluble in

(a) Acetic acid (b) Chloroform

(c) Ethanol (d) Benzene **Ans. d**

657. The major end product of protein nitrogen metabolism in man is

(a) Glycine (b) Uric acid

(c) Urea (d) NH_3 **Ans. c**

658. An amino acid not involved in urea cycle is

(a) Arginine (b) Histidine

(c) Ornithine (d) Citrulline **Ans. b**

659. NH_3 is detoxified in brain chiefly as

(a) Urea (b) Uric acid

(c) Creatinine (d) Glutamine **Ans. d**

660. In humans, NH_3 is detoxified in liver as

(a) Creatinine (b) Uric acid

(c) Urea (d) Uronic acid **Ans. c**

661. The body protein after eighteen years

(a) Remains unchanged

(b) Is decomposed only slightly at intervals of one month

(c) Is in a constant state of flux

(d) Is used only for energy requirement **Ans. c**

662. The only known physiological methylating agents in the animal organism are

(a) Choline and betaine

(b) Choline and δ-adenosyl methionine

(c) Betaine and δ-adenyosyl methionine

(d) Dimehtyl glycine and betaine **Ans. c**

663. In the synthesis of 1 molecule of urea in the Kreb's Hanseleit cycle, the number of ATPs required is

(a) 1 (b) 2

(c) 3 (d) 4 **Ans. c**

664. For biosynthesis of proteins

(a) Amino acids only are required

(b) Amino acids and nucleic acids only are required

(c) Amino acid, nucleic acids and ATP only are required

(d) Amino acids, nucleic acids, ATP, GTP, enzymes and activators are required **Ans. d**

665. Transmethylation of guanido acetic acid gives

(a) Creatine phosphate (b) Creatinine

(c) Choline (d) n-methyl nicotinamide **Ans. a**

666. The 2 energy rich compounds needed for protein biosynthesis are

(a) ATP and GTP (b) ATP and UTP

(c) ATP and CTP (d) ATP and TTP **Ans. a**

667. The following ketoacid is involved in fixing dietary NH_3 into amino acid

(a) Pyruvate (b) Oxalo acetate

(c) Oxalo succinate (d) α-keto glutarate **Ans. b**

668. The metabolite which sustains urea cycle is

(a) Ornithine (b) Citrulline

(c) Carbamoyl phosphate (d) n-acetyl glutamate **Ans. c**

669. Tetra hydroglolate can be freed from N^5-methyl tetrahydrofolate only by

(a) Nor epinephrine (b) Ethanol amine

(c) Nicotinamide (d) Vitamin B_{12} **Ans. d**

670. Neogenesis of methyl group is

(a) The availability of methyl group form adenosyl methionine

(b) The availability of methyl group from betaine

(c) Interaction between N^5, N^{10}-methylene tetra hydrofolate with a NAD^+ dependent reductase

(d) Availability of methyl group from methyl B_{12} **Ans. c**

671. More creatinine is excreted by

(a) Adult males (b) Adult females

(c) Children (d) Pregnant women **Ans. a**

672. A growing peptide in a ribosome cannot be shifted to the adjacent ribosome because

(a) It is firmly attached

(b) It will get the amino acid cleaved

(c) The gap between the ribosomes is too big for a shift
(d) The adjacent ribosomes have different composition **Ans. c**

673. The first amino acid incorporated in a polypeptide in a ribosome of a human is

(a) N-formyl methionine (b) Methionine
(c) Phenyl alanine (d) Hydroxy lysine **Ans. b**

674. The first amino acid incorporated in a polypeptide in a ribosome of a bacterium is

(a) N-formyl methionine (b) Methionine
(c) Alamine (d) Glycine **Ans. a**

675. The integrator between the TCA cycle and urea cycle is

(a) Fumarate (b) Malate
(c) Pyruvate (d) Citrate **Ans. a**

676. Bence jones proteinurial characterized by

(a) Non-heat coagulability
(b) Heat coagulability at 100°C
(c) Heat coagulability at 45 to 60°C
(d) Precipitation at 25°C **Ans. c**

677. Bence Jones proteins may be excreted in urine of patients suffering from

(a) Tuberculosis (b) Diabetes mellitus
(c) Multiple myeloma (d) Hyperthyroidism **Ans. c**

678. Xanthuric acid is an abnormal metabolite of

(a) Xanthine (b) Uric acid
(c) Tyrosine (d) Tryptophan **Ans. d**

679. Two nitrogen atoms of Urea in the urea cycle come from

(a) NH_3
(b) One from NH_3 and one from aspartate
(c) One from NH_3 and one from glutamate
(d) One from NH_3 and one from alanine **Ans. b**

680. Pyruvic acid can be obtained by transamination of alanine with

(a) α- keto glutaric acid (b) Acetoacetic acid
(c) β-OH butyric acid (d) Phosphoenol Pyruvic acid
Ans. a

681. In the synthesis of 1 molecule of urea in the Kreb's Henseleit cycle the number of AMPs formed is

(a) 1 (b) 2

(c) 3 (d) 4 **Ans. a**

682. Formation of melanin from tyrosine requires the action of

(a) Dopa decarboxylation (b) Diamine oxidase

(c) Peroxidase (d) Tyrosinase **Ans. d**

683. In one of the following the quality of the protein synthesis is affected

(a) Diabetes mellitus (b) Gont

(c) Multiple myeloma (d) Primaquine sensitivity **Ans. c**

684. Citrulline is an intermediate of

(a) TCA cycle (b) Urea cycle

(c) Pentose cycle (d) Calvin cycle **Ans. c**

685. The semialdehydes are formed under the action of enzymes characterised as

(a) Aldolases (b) Peptidyl lysyl oxidases

(c) Collagenases (d) Elastases **Ans. b**

686. Which of the following statement about the peptide bond is true?

(a) It is a carbon-carbon bond

(b) It has cis hydrogen and oxygen groups

(c) It is planar

(d) It has rotational freedom **Ans. d**

687. Isoenzymes for a given reaction

(a) Have different spedificities

(b) Have identical affinities for the same substrate

(c) Exhibit different electrophoretic motilities

(d) Contain similar ratios of different polypeptide chains **Ans. b**

688. The highest concentration of cystine can be found in

(a) Melanin (b) Chondroitin sulphate

(c) Myosin (d) Keratin **Ans. d**

689. One round of Edman degradation of the peptide: H_2N- Gly$-$Arg$-$Lys$-$Phe$-$Asp$-$ COOH would result in which of the following structures or their phenyl isothiocyanate derivatives?

(a) H_2N-Gly$-$Arg$-$COOH + H_2N-Lys$-$ Phe$-$ Asp$-$COOH

(b) H_2N-Gly$-$Arg$-$Lys$-$Phe$-$COOH + Asp

(c) $H_2N-Arg-Lys-Phe-Asp-COOH + Gly$

(d) $H_2N-Gly-Arg-Lys-COOH + H_2N-Phe-Asp-COOH$

Ans. c

690. Which of the following techniques is used to separate proteins based upon differences in their mass?

(a) Isoelectric focusing (b) Dialysis

(c) SDS-gel Electrophoresis (d) Western blotting **Ans. c**

691. The greatest buffering capacity at physiologic pH would be provided by a protein rich in which of the following amino acids?

(a) Lysine (b) Histidine

(c) Aspartic acid (d) Valine **Ans. b**

692. Which one of the amino acids could serve as the best buffer at pH 7?

(a) Glutamic acid (b) Arginine

(c) Valine (d) Histidine **Ans. d**

693. Which one of the following statements concerning glutamine is correct?

(a) Contains three tetratable groups

(b) Is classified as an acidic amino acid

(c) Contains an amide group

(d) Migrates to the cathode during electrophoresis at pH 7.0 **Ans. c**

694. One of the given example is an amino acid

(a) Oh-Lysine (b) Protein

(c) Leucine (d) Serine **Ans. b**

695. The lone pair of electrons at one of the ring nitrogens in the given amino acid makes a potential ligand, which is important in binding the iron atoms in hemoglobin

(a) Tryptophan (b) Threonine

(c) Histidine (d) Serine **Ans. c**

696. The amino acid which is not optically active is

(a) Alanine (b) Glycine

(c) Glutamine (d) Lysine **Ans. b**

697. Optically active compounds are capable of

(a) Different reactions

(b) Rotating plane of polarized light

(c) Showing same chemical properties
(d) None of these **Ans. b**

698. The reference compound for absolute configuration of optically active compound is
(a) Alanine (b) Lactic acid
(c) Glyceraldehyde (d) Dihydroxy acetone **Ans. c**

699. All the standard amino acids except the following have one chiral 'c' atom
(a) Threonine, Isoleucine (b) Isoleucine, Alanine
(c) Threonine, Alanine (d) Alanine, Glutamine **Ans. c**

700. The role of complement proteins
(a) Defense (b) Helps immunity of the body
(c) Not predictable (d) None of these **Ans. d**

701. Optical isomers that are mirror images and non superimposable are called
(a) Diastereomers (b) Enantiomers
(c) dl isomers (d) Stereomers **Ans. b**

702. Living cells have the unique ability to synthesize only ______________ optical isomer due to ________.
(a) 'd' form, stereospecific enzymes
(b) 'l' form, stereospecific enzymes
(c) 'd' form, DNA
(d) 'L' form, DNA **Ans. b**

703. Isoelectric pH of an amino acid is that pH at which it has a
(a) Positive charge (b) Negative charge
(c) No net charge (d) All of these **Ans. c**

704. Albuminoids are similar to
(a) Albumin (b) Globulin
(c) Both (a) and (b) (d) None of these **Ans. c**

705. Abnormal chain of amino acids in sickle cells anaemia is
(a) Alpha chain (b) Beta chain
(c) Gama chain (d) Delta chain **Ans. b**

706. In prehepatic jaundice, protein flocculation test is

(a) Normal/weekly positive (b) Usually positive

(c) Negative (d) None of these **Ans. a**

707. Side chains of all amino acids contain aromatic rings except

(a) Phenyl alanine (b) Alanine

(c) Tyrosine (d) Tryptophan **Ans. b**

708. In Nitroprusside test, amino acid cystein produces

(a) Blue colour complex (b) Red colour

(c) Yellow colour (d) Purple colour **Ans. b**

709. Bonds that are formed between two cysteine residues is

(a) Disulphide (b) Peptide

(c) Electrostatic (d) Hydrophobic **Ans. a**

710. The acid amide of Aspartic acid is

(a) Glutamine (b) Arginine

(c) Asparagine (d) Ornithine **Ans. c**

711. It is the only amino acid having an ionizing 'R' group with a pK' near 7 and is important in the active site of some enzymes

(a) Arginine (b) Cystein

(c) Cystine (d) Histidine **Ans. d**

712. Hemoglobin has a high content of this amino acid

(a) Proline (b) Leucine

(c) Arginine (d) Histidine **Ans. d**

713. A hexa peptide with 5 aspartic acid would have a net charge at pH 7

(a) Neutral (b) Positive

(c) Negative (d) Not predictable **Ans. c**

714. In the genetic disorder of cystinuria, the patient excretes large quantities of cystine in their urine and its low solubility causes crystalline cystine to precipitate as stones in kidneys. The remedy involves ingesting Na HCO_3. Reaction of this treatment is

(a) $NaHCO_2$ combines with cystine

(b) $NaHCO_3$ raises the pH above the isoelectric point of cystine

(c) $NaHCO_3$ prevents stone formation by hydrolysis of cystine to cysteine

(d) None of these **Ans. c**

715. Amino acids excepting histidine are not good buffering agents in cell because

(a) They exist as zwitter ions

(b) Their pk and not in the physiological pH of a cell

(c) Only Histidine has pk of its R group at 6.0 unlike the others which have at a different pH

(d) None of these **Ans. c**

716. At neutral pH Alanine has the following structure

(a) $H_2N-\overset{H}{\underset{CH_3}{C}}-COOH$ (b) $H_3^+N-\overset{H}{\underset{CH_3}{C}}-COO^-$

(c) $H_2N-\overset{H}{\underset{CH_3}{C}}-COO^-$ (d) $^+H_2N-\overset{H}{\underset{CH_3}{C}}-COO^-$ **Ans. b**

717. The amino acids in which the R groups have a net positive charge at pH 7.0 are

(a) Lysine, Arginine, Histidine (b) Lysine, Aspargine

(c) Histidine, Aspargine (d) Glutamine, Arginine **Ans. a**

718. Apolipoproteins are

(a) AI (b) AI1

(c) C1 (d) All of these **Ans. d**

719. The amino acid which has a pK near 4 and thus is negatively charged at pH 7 is

(a) Alanine (b) Glutamic acid

(c) Glutamine (d) Aspargine **Ans. b**

720. The side chain of which of the following amino acid contain sulphur atom?

(a) Methionine (b) Threonine

(c) Leucine (d) Tryptophan **Ans. a**

721. Which of the followings gives a positive test for Ninhydrin?

(a) Reducing sugars (b) Triglycerides

(c) Alpha amino acids (d) Esterified Fats **Ans. c**

722. In glutathione (a tripeptide) is present apart from Glutamic acid and cysteine

(a) Serine (b) Glycine

(c) Leucine (d) Phenyl alanine **Ans. b**

723. 2-Amino 3-OH propanoic acid is

(a) Glycine (b) Alanine

(c) Valine (d) Serine **Ans. d**

724. All amino acids have one asymmetric carbon atom, except

(a) Arginine (b) Aspargine

(c) Histidine (d) Glycine **Ans. d**

725. Number of amino acids present in the plant, animal and microbial proteins

(a) 20 (b) 80

(c) 150 (d) 200 **Ans. a**

726. Immunoglobulins are characterized by their

(a) Heavy chains (b) Molecular weight

(c) Light chains (d) Electrophoretic behavior **Ans.a**

727. The bond in proteins that is not hydrolysed under usual conditions of denaturation

(a) Hydrophobic bond (b) Hydrogen bond

(c) Disulphide bond (d) Peptide bonds **Ans. c**

728. If the amino group and a carboxylic group of the amino acid are attached to same carbon atom, the amino acid is called

(a) Alpha (b) Beta

(c) Gamma (d) Delta **Ans. a**

729. Zymogen is

(a) An intracellular enzyme

(b) Serum enzyme

(c) A complete extracellular enzyme

(d) An inactivated enzyme **Ans. d**

730. SGOT level in a adult is

(a) 5–40 units/dl (b) 1–4 units/dl

(c) 5–15 units/dl (d) 50–100 units/dl **Ans. a**

731. Activity of ceruloplasmin shown *in vitro*

(a) Reductase (b) Hydrolase
(c) Ligase (d) Oxidase **Ans. d**

732. Increased serum alanine during fasting is due to

(a) Breakdown of muscle proteins
(b) Decreased utilization of non essential amino acids
(c) Leakage of aminoacids to plasma
(d) Impaired renal function **Ans. a**

733. The following amino acids are required for completion of urea cycle except

(a) Aspartic acid (b) Arginine
(c) Ornithine (d) Glycine **Ans. d**

734. Number of amino acids present in the dietary proteins

(a) 22 (b) 23
(c) 20 (d) 19 **Ans. c**

735. Urea synthesis takes place in

(a) Blood (b) Liver
(c) Kidney (d) Heart **Ans. b**

736. All followings are ketogenic amino acids except

(a) Leucine (b) Isoleucine
(c) Phenyl alanine (d) Glycine **Ans. d**

737. The amino acid containing an indole ring

(a) Tryptophan (b) Arginine
(c) Threonine (d) Phenylalanine **Ans. a**

738. Histidine is converted to histamine through the process of

(a) Transamination (b) Decarboxylation
(c) Oxidative deamination (d) Urea cycle **Ans. b**

739. Physiologically active configuration of amino acids

(a) L
(b) D
(c) For some amino acids it is either of two
(d) Neither L nor D **Ans. a**

740. Cystine is synthesized from
(a) Cysteine (b) Methionine
(c) Arginine (d) Leucine **Ans. a**

741. The major constituent of the proteins of hair and keratin of skin
(a) Arginine (b) Cysteine
(c) Glycine (d) Arginine **Ans. b**

742. NH_3 is removed from brain mainly by
(a) Creatinine formation (b) Uric acid production
(c) Urea formation (d) Glutamine formation **Ans. d**

743. Mechanism by which NH_3 is removed from the kidneys is
(a) Urea formation (b) Uric acid formation
(c) Creatinine formation (d) None of these **Ans. d**

744. Low density plasma proteins are rich in
(a) Chylomicrons (b) Cholesterol
(c) Triglycerides (d) Phospholipids **Ans. b**

745. Transcortins are
(a) Mucoproteins (b) Glycoproteins
(c) Metalloproteins (d) Lipoproteins **Ans. a**

746. Proteins that carries Iron into different tissues is
(a) Ceruloplasmin (b) Trans cortin
(c) Mucoproteins (d) Glycoproteins **Ans. b**

747. Naturally occurring amino acids have
(a) L-Configuration (b) D-Configuration
(c) DL-Configuration (d) None of these **Ans. a**

748. Abnormal chain of aminoacids in sickle cell anemia is
(a) α-chain (b) β-chain
(c) γ-chain (d) r-chain **Ans.b**

749. A dietary deficiency of tryptophan and nicotinate leads to
(a) Beri Beri (b) Xerophthalmia
(c) Anemia (d) Pellegra **Ans. d**

750. Which one of the following is an essential amino acid?
(a) Arginine (b) Tyrosine
(c) Phenylalanine (d) Proline **Ans. c**

751. One of the following amino acid is solely ketogenic

(a) Lysine (b) Alanine

(c) Valine (d) Glutamate **Ans. a**

752. Along with CO_2, NH_3 and ATP, the amino acid that is needed in urea cycle is

(a) Alanine (b) Isoleucine

(c) Aspartate (d) Glycine **Ans. c**

753. Isoelectric pH of an amino acid is that pH at which it has a

(a) Positive charge (b) Negative charge

(c) No charge (d) None of these **Ans. c**

754. Which of the following contributes nitrogen atoms to both purine and pyrimidine rings?

(a) Aspartate (b) Carbamoyl phosphate

(c) CO_2 (d) Glutamine **Ans. a**

755. Which amino acid is a lipotropic factor?

(a) Lysine (b) Lecuine

(c) Tryptophan (d) Methionine **Ans. d**

756. Which of the following protein is rich in cysteine?

(a) Elastine (b) Collagen

(c) Fibrin (d) Keratin **Ans. d**

757. Which amino acid is present at 6th position of α-chain of Hbs instead of glutamate in HbA?

(a) Cysteine (b) Valine

(c) Aspartate (d) Glutamate **Ans. d**

758. The amino acid which contains an indole group is

(a) Histidine (b) Arginine

(c) Cystine (d) Tryptophan **Ans. d**

759. From two amino acids peptide bond formation involves removal of one molecule of

(a) Water (b) Ammonia

(c) Carbondioxide (d) Carboxylic acid **Ans. b**

760. Polymers of more than 100 amino acids are termed

(a) Proteins (b) Polypeptides

(c) Both (a) and (b) (d) None of these **Ans. a**

761. The example of globulins

(a) Leucosin (b) Tuberin

(c) Oryzenin (d) Legunelin **Ans. b**

762. The example of scleroproteins

(a) Glutamin (b) Giladin

(c) Salmine (d) Elastin **Ans. d**

763. The example of phosphoprotein

(a) Mucin (b) Ovovitellin

(c) Ovomucoid (d) Tendomucoid **Ans. b**

764. The example of metalloproteins

(a) Siderophilin (b) OREES mucoid

(c) Elastin (d) All of these **Ans. a**

765. The example of chromoprotein

(a) Salmine (b) Catalase

(c) Zein (d) Gliadin **Ans. b**

766. Deamination is ______ of amino group

(a) Removal (b) Addition

(c) Supplementation (d) None of these **Ans. a**

767. Proteins produce polypeptides from proteins by

(a) Oxidizing (b) Reducing

(c) Hydrolyzing (d) None of these **Ans. c**

768. Proteins react with biuret reagent which is suggestive of 2 or more

(a) Hydrogen bonds (b) Peptide bonds

(c) Disulphide bonds (d) Hydrophobic bonds **Ans. b**

769. The disulphide bond is not broken under the usual conditions of

(a) Filtration (b) Reduction

(c) Oxidation (d) Denaturation **Ans. d**

770. Insulin is oxidized to separate the protein molecule into its constituent polypeptide chains without affecting the other part of the molecule by the use of

(a) Performic acid (b) Oxalic acid

(c) Citric acid (d) Malic acid **Ans. a**

771. Each hydrogen bond is quite
(a) Weak (b) Strong
(c) Both (a) and (b) (d) None of these **Ans. a**

772. A coiled structure in which peptide bonds are folded in regular manner by
(a) Globular proteins (b) Fibrous proteins
(c) Both (a) and (b) (d) None of these **Ans. a**

773. In many proteins the hydrogen bonding produces a regular coiled arrangement called
(a) α-helix (b) β-helix
(c) Both (a) and (b) (d) None of these **Ans. a**

774. Many globular proteins are stable in solution although they lack in
(a) Hydrogen bonds (b) Salt bonds
(c) Non-polar bonds (d) Disulphide bonds **Ans. d**

775. Each turn of α-helix contains the number of amino acids
(a) 2.8 (b) 3.2
(c) 3.4 (d) 3.6 **Ans. d**

776. The distance travelled per turn of α-helix in nm is
(a) 0.34 (b) 0.44
(c) 0.54 (d) 0.64 **Ans. c**

777. α-helix is disrupted by certain amino acids like
(a) Proline (b) Arginine
(c) Histidine (d) Lysine **Ans. a**

778. α-helix is stabilized by
(a) Hydrogen bonds (b) Disulphide bonds
(c) Salt bonds (d) Non-polar bonds **Ans. a**

779. Foetal haemoglobin contains
(a) Two α and two γ chains (b) Two β and two γ chains
(c) Both (a) and (b) (d) None of these **Ans. a**

780. When haemoglobin takes up oxygen there is a change in the structure due to the moving closer together of
(a) β-chains (b) α, β-chains
(c) γ-chains (d) α and γ chains **Ans. a**

781. The hydrogen bonds in the secondary and tertiary structure of proteins are directly attacked by

(a) Salts (b) Alkalies

(c) Detergents (d) All of these **Ans.b**

782. The hydrogen bonds between peptide linkages are interfered by

(a) Guanidine (b) Uric acid

(c) Salicylic acid (d) Oxalic acid **Ans. a**

783. The digestability of certain denatured proteins by proteolytic enzymes

(a) Decreases (b) Increases

(c) Normal (d) None of these **Ans. a**

784. The antigenic antibody functions of proteins by denaturation are frequently

(a) Not changed (b) Changed

(c) Both (a) and (b) (d) None of these **Ans. b**

785. In case of severe denaturation of protein, there is

(a) Reversible denaturation

(b) Moderate reversible denaturation

(c) Irreversible denaturation

(d) None of these **Ans. c**

786. When egg albumin is heated till it is coagulated, the secondary and tertiary structures of the proteins are completely lost resulting in a mixture of randomly arranged

(a) Dipeptide chains (b) Tripeptide chains

(c) Polypeptide chains (d) All of these **Ans. c**

787. In glycoproteins the carbohydrate is in the form of disaccharide units, the number of units are

(a) 50–100 (b) 200–300

(c) 400–500 (d) 600–700 **Ans. d**

788. The milk protein in the stomach of the infants is digested by

(a) Pepsin (b) Trypsin

(c) Chymotrypsin (d) Rennin **Ans. d**

789. Achylia gastrica is said to be when absence of

(a) Pepsin only (b) Both pepsin and HCl

(c) HCl only (d) All of these **Ans. b**

790. The pH of gastric juice become low in
(a) Hemolytic anemia (b) Pernicious anemia
(c) Both (a) and (b) (d) None of these **Ans. b**

791. In small intestine trypsin hydrolyzes peptide linkages containing
(a) Arginine (b) Histidine
(c) Serine (d) Aspartate **Ans. a**

792. Chymotrypsin in the small intestine hydrolyzes peptide linkages containing
(a) Alanine (b) Pheynl alanine
(c) Valine (d) Methionine **Ans. b**

793. Carboxy peptidase B in the small intestine hydrolyzes peptides containing
(a) Leucine (b) Isoleucine
(c) Arginine (d) Cysteine **Ans. c**

794. The transport of amino acids regulated by active processes of different numbers
(a) 1 (b) 2
(c) 3 (d) 4 **Ans. c**

795. The third active process for amino acids transport involves
(a) Acidic amino acids
(b) Basic amino acids
(c) Neutral amino acids
(d) Sulphur containing amino acids **Ans. c**

796. The neutral amino acids for absorption need
(a) TPP (b) $B_6 - PO_4$
(c) NAD^+ (d) $NADP^+$ **Ans. b**

797. If one amino acid is fed excess, the absorption of another is
(a) Slightly accelerated (b) Moderately accelerated
(c) Highly accelerated (d) Retarded **Ans. d**

798. Under normal conditions, food proteins are generally readily digested upto the present
(a) 67 to 73 (b) 74 to 81
(c) 82 to 89 (d) 90 to 97 **Ans. d**

799. By overheating the nutritional value of cereal proteins is

(a) Increased (b) Decreased

(c) Unchanged (d) None of these **Ans. b**

800. More than half of the protein of the liver and intestinal mucosa are broken down and resynthesised in

(a) 10 days (b) 12 days

(c) 15 days (d) 18 days **Ans. a**

801. The half-life of antibody protein is about

(a) 4 weeks (b) 3 weeks

(c) 2 weeks (d) 1 week **Ans. c**

802. Protein anabolism is stimulated by

(a) ACTH (b) Testosterone

(c) Glucagon (d) Epinephrine **Ans. b**

803. The metabolism of protein is integrated with that of carbohydrate and fat through

(a) Oxaloacetate (b) Citrate

(c) Isocitrate (d) Malate **Ans. a**

804. The building up and breaking down of protoplasm are concerned with the metabolism of

(a) Carbohydrate (b) Lipid

(c) Protein (d) Minerals **Ans. c**

805. The amino acids abstracted from the liver are not utilized for repair or special synthesis but are broken down to

(a) Keto acids (b) Sulphur dioxide

(c) Water (d) Ammonia **Ans. d**

806. The unwanted amino acids abstracted from the tissues are either used up by the tissue or in the liver converted into

(a) Ammonia (b) Urea

(c) Ammonium salts (d) Uric acid **Ans. b**

807. Amino acids provide the nitrogen for the synthesis of

(a) The bases of the phospholipids

(b) Uric acid

(c) Glycolipids

(d) Chondroitin sulphates **Ans. a**

808. The metabolism of all proteins ingested over and above the essential requirements is called

(a) Exogenous metabolism (b) Endogenous metabolism
(c) Both (a) and (b) (d) None of these **Ans. a**

809. Sulphur containing amino acids after catabolism produces a substance which is excreted

(a) SO_2 (b) HNO_3
(c) H_2SO_4 (d) H_3PO_4 **Ans. c**

810. Ethereal sulphate is synthesized from the ________ amino acid

(a) Neutral (b) Acidic
(c) Basic (d) Sulphur containing **Ans. d**

811. The amino acids required for creatine formation

(a) Glycine (b) Arginine
(c) Methionine (d) All of these **Ans. d**

812. In human and other ureotelic organisms, the end product of amino acid nitrogen metabolism

(a) Bile acids (b) Ketone bodies
(c) Urea (d) Barium sulphate **Ans. c**

813. The end product of amino acid nitrogen metabolism in uricotelic organisms (reptiles and birds) is

(a) Bilirubin (b) Urea
(c) Uric acid (d) Biliverdin **Ans. c**

814. The transaminase activity needs the coenzyme

(a) ATP (b) $B_6 - PO_4$
(c) FAD^+ (d) NAD^+ **Ans. b**

815. Transamination is a

(a) Irreversible process (b) Reversible process
(c) Both (a) and (b) (d) None of these **Ans. b**

816. Most amino acids are substrates for transamination except

(a) Alanine (b) Threonine
(c) Serine (d) Valine **Ans. b**

817. Oxidative conversion of many amino acids to their corresponding - keto acids occurs in mammalian

(a) Liver and kidney (b) Adipose tissue

(c) Pancreas (d) Intestine **Ans. a**

818. The ketoacid is decarboxylated by H_2O_2 forming a carboxylic acid with one carbon atom less in the absence of the enzyme

(a) Catalase (b) Decarboxylase

(c) Deaminase (d) Phosphatase **Ans. a**

819. The activity of mammalian L-amino acid oxidase, an FMN - flavo protein, is quite

(a) Slow (b) Rapid

(c) Both (a) and (b) (d) None of these **Ans. a**

820. From dietary protein as well as from the urea present in fluids secreted into the gastrointestinal tract intestinal bacteria produce

(a) Carbondioxide (b) Ammonia

(c) Ammonium sulphate (d) Creatine **Ans. a**

821. The symptom of ammonia intoxication includes

(a) Blurring of vision (b) Constipation

(c) Mental confusion (d) Diarrhoea **Ans. a**

822. Ammonia intoxication symptoms occur when brain ammonia levels are

(a) Slightly diminished (b) Highly diminished

(c) Increased (d) All of these **Ans. c**

823. Ammonia production by the kidney is depressed in

(a) Acidosis (b) Alkalosis

(c) Both (a) and (b) (d) None of these **Ans. b**

824. Ammonia is excreted as ammonium salts during metabolic acidosis but the majority is excreted as

(a) Phosphates (b) Creatine

(c) Uric acid (d) Urea **Ans. d**

825. Synthesis of glutamine is accompanied by the hydrolysis of

(a) ATP (b) ADP

(c) TPP (d) Creatin phosphate **Ans. a**

826. In brain, the major metabolism for removal of ammonia is the formation of

(a) Glutamate (b) Aspartate
(c) Asparagine (d) Glutamine **Ans. d**

827. Carbamoyl phosphate synthetase structure is marked by change in the presence of

(a) N-Acetyl glutamate (b) N-Acetyl Aspartate
(c) Neuraminic acid (d) Oxalate **Ans. a**

828. The biosynthesis of Urea occurs mainly in the Liver

(a) Cytosol (b) Microsomes
(c) Nucleus (d) Mitochondria **Ans. d**

829. One mol of Urea is synthesized at the expense of the _______ mols of ATP

(a) 2 (b) 3
(c) 4 (d) 5 **Ans. b**

830. Urea biosynthesis occurs mainly in the liver involving the number of amino acids

(a) 3 (b) 4
(c) 5 (d) 6 **Ans. d**

831. The normal daily output of Urea through urine in grams

(a) 10 to 20 (b) 15 to 25
(c) 20 to 30 (d) 25 to 35 **Ans. c**

832. In severe acidosis, the output of urea is

(a) Decreased (b) Slightly increased
(c) Highly increased (d) Moderately increased **Ans. a**

833. Uremia occurs in

(a) Cirrhosis of the liver (b) Nephritis
(c) Diabetes mellitus (d) Coronary thrombosis **Ans. b**

834. Clinical symptom in urea cycle disorder is

(a) Mental retardation (b) Drowsiness
(c) Diarrhoea (d) Oedema **Ans. a**

835. The sparing action of methionine is

(a) Tyrosine (b) Cystine
(c) Arginine (d) Tryptophan **Ans. b**

836. NH^+_4 aminates glutamate to form glutamine requiring ATP and
(a) K^+ (b) Na^+
(c) Ca^{++} (d) Mg^{++} **Ans. d**

837. Glutathione is a
(a) Dipeptide (b) Tripeptide
(c) Polypeptide (d) None of these **Ans. b**

838. All following are conjugated proteins except
(a) Nucleoproteins (b) Proteoses
(c) Metalloproteins (d) Flavoproteins **Ans. b**

839. Number of amino acids present in plants, animals and microbial proteins
(a) 20 (b) 80
(c) 150 (d) 200 **Ans. d**

840. Hydrated density of (HD) lipoproteins is
(a) 0.94 gm/ml (b) 0.94-1.006 gm/ml
(c) 1.006-1.063 gm/ml (d) 1.063-1.21 gm/l **Ans. b**

841. The bond in proteins that is not broken under usual conditions of denaturation
(a) Hydrophobic bond (b) Hydrogen bond
(c) Disulphide bond (d) Peptide bonds **Ans. d**

842. Plasma proteins act as
(a) Buffers (b) Immunoglobulins
(c) Reserve proteins (d) All of these **Ans. d**

843. Group that reacts in the Biuret test
(a) Peptide (b) Amino group
(c) Carboxylic group (d) Aldehyde group **Ans. a**

844. In nitroprusside test, amino acid cysteine produces a
(a) Red colour (b) Blue colour
(c) Yellow colour (d) Purple colour **Ans. a**

845. Protein present in hemoglobin has the structure known as
(a) Primary (b) Secondary
(c) Tertiary (d) Quarternary **Ans. d**

846. Isoelectric pH of an amino acid is that pH at which it has a
(a) Positive charge (b) Negative charge
(c) Nil net charge (d) None of these **Ans. c**

847. Albuminoids are similar to
(a) Albumin (b) Globulin
(c) Both (a) and (b) (d) None of these **Ans. a**

848. Proteins that constitute keratin, collagen and elastin in body are
(a) Protamines (b) Phosphol proteins
(c) Scleroproteins (d) Metaproteins **Ans. c**

849. Systematic name of lysine is
(a) Amino acetic acid (b) 2,6-diaminohexanoic acid
(c) Aminosuccinic acid (d) 2-Aminopropanoic acid
Ans. b

850. Side chains of all following amino acids contain aromatic rings except
(a) Phenyl alanine (b) Alanine
(c) Tyrosine (d) Tryptophan **Ans. b**

851. Abnormal chain of amino acids in sickle cell anaemia is
(a) Alpha chain (b) Beta chain
(c) Delta chain (d) Gama chain **Ans. b**

852. Number of chains in globin part of normal Hb
(a) 1 (b) 2
(c) 3 (d) 4 **Ans. d**

853. The pH of albumin is
(a) 3.6 (b) 4.7
(c) 5.0 (d) 6.1 **Ans. b**

854. Ninhydrin reaction gives a purple colour and evolves CO_2 with
(a) Peptide bonds (b) Histamine
(c) Ergothioneine (d) Aspargine **Ans. d**

855. Denaturation of proteins involves breakdown of
(a) Secondary structure (b) Tertiary structure
(c) Quarternary structure (d) All of these **Ans. d**

856. In denaturation of proteins, the bond which is not broken

(a) Disulphide bond (b) Peptide bond

(c) Hydrogen bond (d) Ionic bond **Ans. b**

857. The purity of an isolated protein can be tested by employing various methods

(a) Solubility curve (b) Molecular weight

(c) Ultra Centrifugation (d) Immuno Reactivity

(e) All of these **Ans. c**

858. More than one break in the line or in saturation curve indicates the following quality of protein

(a) Non homogenity (b) Purity

(c) Homogenity (d) None of these **Ans. a**

859. A sharp moving boundary is obtained between the pure solvent and solute containing layer in

(a) Chromatography (b) Immuno Reactivity

(c) Ultra Centrifugation (d) Solubility curve **Ans. c**

860. The antibodies raised against a pure protein will show only one sharp spike on this technique

(a) Solubility curve

(b) Solvent precipitation

(c) Molecular weight determination

(d) Immuno electrophoresis **Ans. d**

861. This technique takes the advantage of the fact that each protein has different pH at which it is electrically neutral i.e., its isoelectric pH

(a) Isoelectric focussing (b) Immunoel Ectrophoresis

(c) Chromatography (d) HPLC **Ans. a**

862. The following technique makes use of the difference in net charges of proteins at a given pH

(a) Thin layer chromatography

(b) Ion exchange chromatography

(c) High performance liquid chromatography

(d) Paper chromatography **Ans. b**

863. The ratio of the distance moved by a compound to the distance moved by the solvent front is known as its

(a) PI value (b) Linking number

(c) R_f value (d) Gold number **Ans. c**

864. The movement of charged particles towards one of the electrodes under the influence of electrical current is

(a) Gel filtration (b) Molecular sieving

(c) Gas liquid chromatography (d) Electrophoresis **Ans. d**

865. An anion exchange resin linked to cellulose backbone is

(a) DEAE cellulose (b) CM cellulose

(c) Sephadex (d) None of these **Ans. a**

866. A cation exchange resin linked to cellulose backbone is

(a) CM-cellulose (b) DEAE cellulose

(c) Starch (d) Biogel **Ans. a**

867. The sorting out of molecules according to size and shape may be adapted to protein purification in this technique

(a) Adsorption chromatography (b) Gel filtration chromatography

(c) Paper chromatography (d) None of these **Ans. b**

868. Frequently employed materials for the adsorption chromatography of proteins include

(a) High capacity supporting gel

(b) Starch blocks

(c) Calcium phosphate gel, alumina gel and hydroxy apatite

(d) All of these **Ans. c**

869. The solubility of most proteins is lowered at high salt concentrations is called as

(a) Salting in process (b) Salting out process

(c) Isoelectric focusing (d) None of these **Ans. b**

870. Phenylalanine, ornithine and methionine are involved in the biogenesis of

(a) Lysergic acid (b) Reserpine

(c) L-Hyoscyamine (d) Papaverine **Ans. a**

871. All the following diuretics inhibit the carbonic anhydrase except
(a) Acetazolamide (b) Bumetanide
(c) Furosemide (d) Ethacrynic acid **Ans. d**

872. Protein is a polymer of
(a) Sugars (b) Phenols
(c) Amino acids (d) Carboxylic acids **Ans. c**

873. Proteinous substances which catalyze biochemical reactions are known as
(a) Activators (b) Catalysts
(c) Enzymes (d) Hormones **Ans. c**

874. Insulin is a protein which controls
(a) Blood clotting (b) Metabolic pathway
(c) Digestion (d) Kreb's cycle **Ans. b**

875. Proteins which are responsible for defence mechanism are called
(a) Antimetabolites (b) Antibodies
(c) Antimycins (d) Apoproteins **Ans. b**

876. When the net charge on an amino acid is zero, the pH is maintained as?
(a) 4.5 (b) 11.2
(c) 7.0 (d) 9.1 **Ans. c**

877. Isoelectric point of amino acids is used for
(a) Crystallisation (b) Precipitation
(c) Solubility (d) Reactivity **Ans. b**

878. Xanthoproteic test is positive in proteins containing
(a) Sulphur amino acids (b) α-Amino acids
(c) Aromatic amino acids (d) Aliphatic amino acids **Ans. c**

879. All α-amino acids give positive
(a) Million's test (b) Biurete test
(c) Xanthproteic test (d) Ninhydrin test **Ans. d**

880. N-terminal amino acids of a polypeptide are estimated by
(a) Edmann reaction (b) Sanger's reagent
(c) Formaldehyde test (d) Ninhydrin reaction **Ans. a**

881. Million's test is positive for
(a) Phenylalanine (b) Glycine
(c) Tyrosine (d) Proline **Ans. c**

882. Indole group of tryptophan responses positively to
(a) Glyoxylic acid (b) Schiff's reagent
(c) Biuret test (d) Resorcinol test **Ans. a**

883. Guanidine group of argentine gives positive test with
(a) Lead acetate (b) Sakaguchi reagent
(c) Tricholoroacetic acid (d) Molisch's reagent **Ans. b**

884. Thiol group of cysteine gives red colour with
(a) Sodium acetate (b) Lead acetate
(c) Sodium nitroprusside (d) Barfoed's reagent **Ans. c**

885. Protein deficiency disease is known as
(a) Cushing's disease (b) Fabry's disease
(c) Parkinson's disease (d) Kwashiorkor and marasmus **Ans. d**

886. A vegetable source of protein is
(a) Egg plant (b) Soyabean
(c) Tree of the Heaven (d) Devil's dung **Ans. b**

887. Oxaloacetate is converted to aspartic acid by
(a) Reductase (b) Oxidase
(c) Transminase (d) Catalase **Ans. c**

888. Deficiency of biotin results in decrease in
(a) Amino acid synthesis (b) Lipid synthesis
(c) Kidney (d) Fatty acid synthesis **Ans. d**

889. The precursor of bile salts, sex hormones and vitamin D is
(a) Diosgenin (b) Cholesterol
(c) Campesterol (d) Ergosterol **Ans. b**

890. Unsaturated fatty acids is known as
(a) Non-essential fatty acids (b) Essential fatty acids
(c) Cerebrosides (d) Phospholipids **Ans. b**

891. Biuret test is specific for
(a) Two peptide linkage (b) Phenolic group
(c) Imidazole ring (d) None of these **Ans. a**

892. Most of calcium is present in bone, but 2% present in soft tissue and the blood is called
(a) Calcinated blood (b) Solidified blood
(c) Physiological blood (d) Colloidal blood **Ans. c**

893. Calcium present with protein is known as free while in salt form is called as
(a) Bound (b) Precipitated
(c) Solid (d) Polymorphs **Ans. a**

894. The following ions help in enzymatic transfer of phosphate from ATP to pyruvic acid
(a) Sodium (d) Calcium
(c) Magnesium (d) Potassium **Ans. d**

895. International enzyme commission classifies enzymes into
(a) Three classes (b) Six classes
(c) Four classess (d) Ten classes **Ans. b**

896. Michaelis - Menten equation is used to explain the effect of substrate concentration on
(a) Carbohydrate (b) Enzyme
(c) Lipid (d) Protein **Ans. b**

897. The pH at which an enzyme has maximum activity is known as
(a) Isoelectric pH (b) Optimum pH
(c) Low pH (d) High pH **Ans. b**

898. Degradation of proteins to amino acids, glucose from carbohydrates and fatty acids from lipids is known as
(a) Anabolism (b) Metabolism
(c) Catabolism (d) Cretinism **Ans. c**

899. During glycolysis of glucose the energy liberated in the absence of oxygen is known as
(a) Oxygenesis (b) Glyconeogenesis
(c) Glycogenolysis (d) Anaerobic fermentation
Ans. d

900. Deficiency of urea cycle enzymes results into accumulation of citrulline argininosuccinate arginine in the liver resulting in increasing concentration of ....in the blood.

(a) Calcium (b) Sodium
(c) Ammonia (d) Lipid **Ans. c**

901. Accumulation of tryptophan in blood is known as

(a) Pompe's disease (b) Wilson's disease
(c) Wolman's disease (d) Hartnup's disease **Ans. d**

902. Lymphocytes are responsible for the formation of

(a) Serum (b) Plasma
(c) Antibody (d) Calcium **Ans.c**

903. Platelets contain an enzyme which has important role in clotting in blood. This enzyme is known as

(a) Cholinesterase (b) Transaminase
(c) Decarboxylase (d) Thrombokinase **Ans. d**

904. Treatment of pentoses with a concentrated mineral acid yields a cyclic aldehyde known as

(a) Pentaldehyde (b) Cyclopental
(c) Hexaldehyde (d) Furfural **Ans. d**

905. Isoelectric pH is that pH at which protein is electrically

(a) Neutral (b) Anionic
(c) Cationic (d) None of these **Ans. a**

906. About 6.25 g of haemoglobin is produced and destroyed in the body each day and the total amount of haemoglobin in a normal healthy 70 kg weighing male adult is

(a) 250 g (b) 150 g
(c) 100 g (d) 70 g **Ans. d**

907. Pancreatic juice contains all of the following except

(a) Trypsinogen (b) Lipase
(c) Cholecystokinin (d) Chymnotrypsinogen **Ans. c**

908. Carboxypeptidase, an enzyme of pancreatic juice, contains

(a) Mn (b) Zinc
(c) Magnesium (d) Manganese **Ans. b**

909. The zymogen from trypsinogen of pancreatic juice is converted to active trypsin by

(a) Pepsin (b) Enterocrinin

(c) Enterokinase (d) Rennin **Ans. c**

910. Inactive zymogens are precursors of all the following gastrointestinal enzymes except

(a) Carboxypeptidase (b) Pepsin

(c) Amino peptidase (d) Chymotrypsin **Ans. c**

911. Rennin acts on casein of milk in infants in presence of

(a) Mg^{++} (b) Zn^{++}

(c) Co^{++} (d) Ca^{++} **Ans. d**

912. All the following are true about phenylketonuria except

(a) Deficiency of phenylalanine hydroxylase

(b) Mental retardation

(c) Increased urinary excretion of p-hydroxyphenyl pyruvic acid

(d) Decrease serotonin formation **Ans. c**

913. Which of the amino acid produces a vasodilator on decarboxylation?

(a) Glutamic acid (b) Histidine

(c) Ornithine (d) Cysteine **Ans. b**

914. Neutral amino acid is

(a) Leucine (b) Lysine

(c) Aspartic acid (d) Histidine **Ans. a**

915. The amino acid containing hydroxy group

(a) Glycine (b) Isoleucine

(c) Arginine (d) Thereonine **Ans. d**

916. The amino acid which synthesizes many hormornes

(a) Valine (b) Phenylalanine

(c) Alanine (d) Histidine **Ans. b**

917. Insulin degradation of disulfide bond formation is effected by

(a) Pyruvate dehydrogenase (b) Xylitol reductase

(c) Gutathione reductase (d) Xanthine oxidase **Ans. c**

918. A protein reacts with biuret reagent which indicates 2 or more

(a) Blood clotting (b) Peptide bond

(c) Disulphide bonds (d) Hydrophobic bonds **Ans. b**

919. In many proteins the hydrogen bonding produces a regular coiled arrangement which is called as

(a) β-Helix (b) α-Helix

(c) Both (a) and (b) (d) Spiral **Ans. b**

920. The milk protein in the stomach of the infants is digested by

(a) Pepsin (b) Trypsin

(c) Chymotrypsin (d) Rennin **Ans. d**

921. Protein anabolism is stimulated by

(a) ACTH (b) Testosterone

(c) Glucagon (d) Epinephrine **Ans. b**

922. The number of helices present in a collagen molecule is

(a) 1 (b) 2

(c) 3 (d) 4 **Ans. c**

923. Which amino acids are essential building units of proteins ?

(a) α-amino acids (b) β-amino acids

(c) γ-amino acids (d) None of these **Ans. a**

924. Sakaguchi reaction is specific for

(a) Guanidine group (b) Phenolic group

(c) Carboxylic group (d) None of these **Ans. a**

925. With the exception of glycine all amino acids found in protein are

(a) Isocitrate dehydrogenase (b) Fumarase

(c) Succinate thiokinase (d) ATPase **Ans. b**

926. In protein structure the α-helix and β- pleated sheets are example of

(a) Primary structure (b) Secondary structure

(c) Tertiary structure (d) Quaternary structure **Ans. b**

927. An essential amino acid in man is

(a) Proline (b) Threonine

(c) Asparagine (d) Tyrosine **Ans. b**

928. An amino acid that does not form an α- helix is
(a) Asparagine (b) Tyrosine
(c) Tryptophan (d) Proline **Ans. d**

929. The protein present in hair is
(a) Elastin (b) Prolamine
(c) Keratin (d) Gliadin **Ans. c**

930. Plasma protein can be separated by
(a) Salting out with $(NH_4)_2SO_4$ (b) Ultracentrifugation
(c) Immuno electrophoresis (d) All of these **Ans. d**

931. RNA does not contain
(a) Uracil (b) Adenine
(c) Hydroxy methyl cytosine (d) Phosphate **Ans. c**

932. In mammalian cells, ribosomal RNA is produced mainly in the
(a) Nucleus (b) Nucleolus
(c) Ribosome (d) Golgi apparatus **Ans. b**

933. Which co-enzyme is not involved in oxidative decarboxylation of pyruvic acid?
(a) TPP (b) Mg^{++}
(c) Biotin (d) CoA-SH **Ans. c**

934. A polymeric unit of starch which has a branched structure is
(a) Glucose (b) Amylopectin
(c) Isomaltose (d) Amylose **Ans. b**

935. The repeating unit in hyaluronic acid is
(a) Glucuronic acid and Galactosamine
(b) Glucuronic acid and glucosamine
(c) Glucuronic acid and N-acetyl glucosamine
(d) Glucuronic acid and N-acetyl galactosamine **Ans. c**

936. The repeating disaccharide unit in celluslose is
(a) Sucrose (b) Maltose
(c) Dextrose (d) Cellobiose **Ans. d**

937. Using lysogeny in λ phage the lytic functions are
(a) Expressed in less amount (b) Expression is completely lost
(c) Dormant but not lost (d) Expressed besides lysogeny
Ans. c

938. During protein DNA interaction, the most important feature is
(a) The charge of protein and DNA
(b) Charge of protein only
(c) The bonds that form between them
(d) Their symmetry matching **Ans. d**

939. The expression of most genes is regulated primarily at the level of
(a) Translation (b) Transcription
(c) Transduction (d) Post translational modification
Ans. b

940. The binding of cAMP-CAP to a specific site in the promoter region
(a) Enhances the binding of RNA polymerase and the initiation of transcription
(b) Inhibits the binding of RNA polymerase
(c) Acts as a terminator during translation
(d) None of the above **Ans. a**

941. The autoregulation of λ repressor depends upon
(a) OR_1 OL_1
(b) OR_1, OR_2, OR_3
(c) OR_1, OR_2, OR_3, OL_1, OL_{2-}, PL_3
(d) OL_1, OL_{2-}, PL_3 **Ans. c**

942. The regulation coordinated with amino acid concentration is called
(a) Stringent factor (b) SOS response
(c) Stringent response (d) Autoregulation **Ans. c**

943. Role of stringent factor is
(a) To catalyze the formation of the unusual nucleotide ppG_{pp}
(b) To control the number of amino acids being added to the polypeptide chain
(c) To break the phosphodiester bond
(d) To increase the rate of translation **Ans. a**

944. What does ppG_{pp} stand for
(a) Two phosphate guanine diphosphate
(b) Guanyl tetraphosphate

(c) Unusual guanine phosphate

(d) Guanosine tetraphosphate **Ans. d**

945. The rise in the concentration of ppG_{pp} in response to amino acid saturation leads to

(i) Rise in rRNA synthesis

(ii) Reduction in rRNA and mRNA synthesis

(iii) Formation of amino acids

(iv) Reduction in rRNA synthesis only

Choose the correct answer

(a) ii, iii (b) iii

(c) iv (d) None of the above **Ans. c**

946. Nucleosomes if treated with salt on dilute acid

(a) Hydrogen bond between the bases breaks down

(b) Phosphodiester bond breaks

(c) Nucleosome changes its state

(d) Histones dissociate from the DNA **Ans. b**

947. Largest DNA molecule (41 x 10^6 kd) is bound in

(a) *Drosophila* (b) Human

(c) *Acetabularia* (d) *E. coli* **Ans. d**

948. Match the following

Organism	No. of Chromosomes
(a) *E. coli*	16
(b) *Saccharomyces cerevisae*	1
(c) *Drosophila melanogaster*	23
(d) Human	4

Ans.

949. The four types of histone proteins (H_1, H_{24}, H_{28}, H_3, H_4) exist

(a) In a variety of forms

(b) Only in single forms

(c) In different forms only in plants

(d) None of the above **Ans. a**

950. During DNA replication the major distortion takes place in
(a) H_1 Histones (b) All the four types of histones
(c) H_3 and H_{2A} histones (d) All of the above **Ans. a**

951. The variable histone/histones is/are
(a) H_1 only (b) $H_1 + H_4$
(c) $H_{2A} + H_2$ (d) All the above **Ans. a**

952. The major role in DNA compaction is played by
(a) The negative charge on DNA (b) The double helix of DNA
(c) H_1 histone protein (d) All the histone proteins **Ans. c**

953. The most compact DNA is bound in
(a) Sperm heads (b) Sperm tails
(c) *Drosophila* (d) *E. coli* **Ans. a**

954. The extreme compact state of DNA is brought about by
(a) The size of the cell (b) Histones
(c) Size of the nucleus (d) Protamines **Ans. d**

955. Hypersensitive sites are
(a) Sequences of DNA capable of undergoing mutation
(b) Sequences of DNA sensitive to DNAase I
(c) Regions in immune system responsible for hypersensitivity
(d) mRNA sequences undergoing attenuation **Ans. b**

956. Hypersensitive sites have a range of round
(a) 100-200 base pairs (b) 100-150
(c) 150-300 (d) 500-1000 **Ans. a**

957. Hypersensitive sites are found
(a) Sometimes at 3′ end or at 5′ end of the transcribed gene
(b) In between 3′ and 5′ end of mRNA
(c) Towards the 3′ end of mRNA
(d) Towards the 5′ end of transcribe gene **Ans. a**

958. Transcriptionally active regions are
i) Uncoiled ii) Deficient of histones
iii) Sensitive to DNAase I iv) All of the above

Choose the correct answer

(a) ii, iii (b) i, ii

(c) iii (d) iv **Ans. d**

959. The phage is known is prophage when it

(a) Undergoes lysis (b) Stays in the cell at lytic phase

(c) In the lysogenized state (d) Is undergoing mitosis **Ans. c**

960. Which of the following statement is not correct

(a) Lysogenic state can continue unless interrupted by DNA damage that induces SOS response

(b) Lysogeny switches to lysis with the age of the host cell

(c) Crp protein binds to the same operators as cI protein

(d) Synthesis **Ans. b**

961. In *E. coli*, the physiologic inducer is

(a) Lactose (b) Allolatose

(c) Glucose (d) Fructose **Ans. b**

962. Control of ribosomal protein synthesis is exerted at

(a) Transcription level

(b) Post transcriptional level

(c) When assembled into ribosomes

(d) Translational level **Ans. d**

963. SRP stops protein synthesis

(a) Help recognizing the stop sequences by

(b) Acting as an inhibitor

(c) Means of stearic hindrance

(d) Acting as a toxic material **Ans. c**

964. SRP is helpful to cytosolic protein because

(a) It binds to the protein and prevents premature folding on the cytosolic side

(b) It prevents premature folding in the ER lumen

(c) It helps in folding of the protein at the cytosolic side

(d) All of the above **Ans. a**

965. All integral membrane proteins of the cell are formed by the ribosomes bound to the ER except

(a) Mitochondria only (b) Mitochondria and nucleus

(c) Mitochondria and chloroplast (d) Chloroplast only **Ans. c**

966. Membrane bound ribosomes and free ribosomes are structurally

(a) Similar (b) Slightly different

(c) Totally different (d) None of the above **Ans. a**

967. Membrane bound and free ribosomes differ

(a) In their structure

(b) In their function

(c) In the type of protein synthesis

(d) All of the above **Ans. c**

968. Ubiquitin of all eukaryotes is

(a) Perform different functions (b) Essentially similar

(c) All of the above (d) None of the above **Ans. b**

969. Clathrin is a protein that forms

(a) Polyhedral lattices (b) Tetrahedral lattices

(c) Hexagonal lattices (d) Forms no lattice **Ans. a**

970. (i) All integral membrane proteins of the cell are formed by ribosomes bond ER.

(ii) All integral membrane proteins of the cell are formed by free ribosomes.

(iii) Both free and bound ribosomes participate in the formation of integral membrane proteins.

(iv) Except chloroplast and mitochondrial membranes all other integral membrane proteins of the cell are formed by bound ER.

The correct statement is/are

(a) (i), (iv) (b) (ii), (iv)

(c) (ii), (iii), (iv) (d) (iv) only **Ans. d**

971. During translocation the signal for attachment (a sequence of amino acid residues are attached) is at

(a) Carboxyl terminus

(b) Amino terminus

(c) Sometimes at carboxyl and sometimes amino terminus
(d) All of the above **Ans. b**

972. Choose the correct statement
(a) Most genes of eukaryotic proteins are encoded by single copy genes
(b) Repetitive genes for ribosomal RNAs and histones are exceptions
(c) mRNA for histones do not have poly A chain
(d) All of the above **Ans. d**

973. "Clustering is not essential for coordinated gene expression". This statement is verified by
(a) Different subunits of ribosomes are scattered but expressed together
(b) Immunoglobulins are arranged distantly but expressed simultaneously
(c) Histones genes are dispersed and scattered but expressed together
(d) Same subunit of ribosomes scattered **Ans. c**

974. The significance of the absence of poly A tail from histone mRNA is
(a) Absence of poly A tail hinders is the transfer of mRNA from nucleus to cytoplasm which is a favourable condition for histone protein
(b) Absence of poly A tail facilitates the quick transfer of histone mRNA from nucleus to cytoplasm
(c) Poly A tail inhibits with the transcription of histone mRNA
(d) Poly A tail formation is not possible with the histone mRNA **Ans. b**

975. The percentage of mammalian DNA coding for protein is
(a) 3% only (b) 4% only
(c) 2% only (d) 1% only **Ans. c**

976. The DNA of the following organism is non methylated
(a) *Xenopus* (b) *Yeast*
(c) *E. coli* (d) *Drosophila* **Ans. d**

977. Eukaryotic proteins are isoprenylated i.e.
(a) Isoprenyl group is attached at the 3′ end of the polypeptide chain.
(b) Isoprenyl group is deleted from the protein
(c) A thioester bond is formed between the isoprenyl group and cys residue of the protein
(d) An ester bond is formed between isoprenyl group and ser residue of the protein **Ans. c**

978. Example of isoprenylated protein is
(a) Proto-oncogenes (b) G protein
(c) Lamins (d) All of the above **Ans. d**

979. Amino acid sequences of H_3 and H_4 is
(a) Different in plants and animals
(b) Same in all plants and animals
(c) Same in some plants and animals and different in some
(d) Different in animals and same in all plants **Ans. b**

980. Proteins at the time of degradation becomes joined to
(a) Clathrin (b) Peptidase
(c) Ubiquitin (d) SRP **Ans. c**

981. The clathrin unit (triskelion) is a
(a) 3 legged structure (b) 4 legged structure
(c) Tetrahedral structure (d) Beaded structure **Ans. a**

982. The pH of the lysosomes is.......than endosomes
(a) Higher
(b) Equal
(c) Lower
(d) All of the above depending on the situation **Ans. c**

983. The acidic nature of the endosome leads to
(a) Killing of microbes
(b) Inactivating the interferon
(c) Separating the microbes from surface proteins
(d) Conformational change that is necessary for infection **Ans. d**

984. All protein transport except the outer membrane transport is driven by
(a) ATP hydrolysis (b) GTP hydrolysis
(c) Proton motive force (d) All of the above **Ans. c**

❑❑❑

Chapter-8

Biosynthesis of Protein

Fill in the blanks

1. RNA polymerase after binding to DNA template of *E. Coli* moves in........... direction and...........dimension.

Ans. one, one

2. RNA polymerase binds to DNA installing itself to about................base pairs.

Ans. 60

3. In the process of protein synthesis ribosomes work as.................

Ans. platform

4. During translation AUG codes for methionine at............as well as...........of a polypeptide chain.

Ans. start, interior

5. The DNA interacting antibiotic is...............

Ans. actinomycin D

6. Denaturation involves conversion of .... into.................. proteins.

Ans. globular, fibrous

7. Fibrous proteins are present in ....and ....

Ans. myosin collagen, fibroin

8. There are 20 naturally occuring amino acids. The maximum number of tripeptides that can be obtained is ....

Ans. 8000

9. Secondary structure of protein is mainly governed by ....

Ans. hydrogen bonds

10. The sequence in which amino acids are arranged in a protein is called.........

Ans. primary structure

Multiple Choice Questions

1. Ribosomes are the actual sites for the preparation of
   (a) Proteins (b) Proteins and nucleic acid
   (c) Simple proteins only (d) All of the above **Ans. a**

2. Some facts about the study of translation are given below. Match them properly

   A. Ribosomes 1. Polypeptide chain termination
   B. Endoplasmic Reticulum 2. SRP
   C. RNA polymerase 3. Transcription
   D. Rho factor 4. Polypeptide chain folding

   (a) A1, B2, C3, D4 (b) A4, B2, C3, D1
   (c) A2, B4, C1, D3 (d) A3, B1, C4, D2 **Ans. b**

3. Ribosomes were discovered by
   (a) Huxley (b) Miller
   (c) Benda (d) Palade **Ans. d**

4. Ribosomes are the centres of
   (a) Fat synthesis (b) Protein synthesis
   (c) Respiration (d) Photosynthesis **Ans. b**

5. The kinds of RNA present in any plant cell will be
   (a) 3 (b) 4
   (c) 2 (d) 5 **Ans. a**

6. *E. coli* DNA has several promoter sites. Their number lies near about
   (a) 2000 (b) 3000
   (c) 4000 (d) 1000 **Ans. a**

7. Which of the following is not the function of RNA polymerase
   (a) It searches DNA for initiation sites
   (b) It unwinds a short stretch of double helical DNA to produce a DNA template
   (c) It detects termination signals that specify where a transcript ends.
   (d) Several RNA polymerase are involved in the synthesis of a single transcript **Ans. d**

8. RNA polymerase binds to DNA installing itself to about
   (a) 60 base pairs (bps) (b) 65 bps
   (c) 75 bps (d) None of the above **Ans. a**
9. Holoenzyme binds to
   (a) Single stranded DNA (b) Double stranded DNA
   (c) Both types of DNA (d) All the above **Ans. b**
10. RNA polymerase after binding to DNA template of *E. coli* moves in
    (a) One direction and one dimension
    (b) Three dimensions
    (c) Multiple dimensions
    (d) Does not move at all **Ans. a**
11. σ subunits of an organism are
    (a) All similar in structure and function
    (b) All similar in function only
    (c) Different in structure and function
    (d) Different in structure only **Ans. c**
12. Role of Topoisomerase is
    (a) Initiate transcription (b) Induce negative supercoiling
    (c) Initiate translation (d) Induce positive supercoiling **Ans. b**
13. All promoters are stimulated by negative supercoiling
    (a) The above statement is true in all cases
    (b) The above statement is not true in all cases
    (c) The above statement is not true only for the promoters of gyrase
    (d) The above statement is not true only for the promoters of Topoisomerase **Ans. b**
14. The core enzyme binds to the DNA template
    (a) Less strongly than holoenzyme
    (b) More strongly than holoenzyme
    (c) As strongly as that of holoenzyme
    (d) Its binding depends on the type of DNA it binds **Ans. b**

15. The following does not have a nuclease activity
(a) DNA polymerase (b) RNA polymerase
(c) Endonuclease (d) None of the above **Ans. d**

16. nuSA protein is a
(a) Part of RNA polymerase
(b) Part of DNA polymerase
(c) Regulates membrane protein targating
(d) Regulates adenylate cyclase **Ans. a**

17. The mRNA transcribed from the DNA sequence 5′pCTTGGATTC-OH3′ is
(a) 5′pTGAACTAAG-OH3′ (b) 5′pUGAACUAAGU-OH3′
(c) 5′pGAATCAAGT-OH3′ (d) 5′pGAAUCAAGU-OH3′
Ans. d

18. During translation AUG codes for methionine at
(a) The end of polypeptide chain
(b) The start of a polypeptide chain
(c) The start as well as the interior of a polypeptide chain
(d) None of the above **Ans. c**

19. Following is the pictorial representation of the transcription of strand A of a DNA fragment. Which of the following assignment of the DNA strand polarities is correct?
(a) 5′– – – – – – – – – – – – – –3′
5′– – – – – – – – – – – – – –3′
(b) 3′– – – – – – – – – – – – – –5′A
5′– – – – – – – – – – – – – –3′B
(c) 3′– – – – – – – – – – – – – –5′
5′– – – – – – – – – – – – – –3′
(d) 5′– – – – – – – – – – – – – –3′A
5′– – – – – – – – – – – – – –3′B **Ans. c**

20. The DNA interacting antibiotic is
(a) Mitomycin C (b) Actinomycin D
(c) Puromycin (d) Polymyxin B **Ans. b**

21. During translocation the ribosome moves by the distance of one codon towards

(a) 3′ end of mRNA

(b) 5′ end of mRNA

(c) 3′ end of DNA

(d) The given statement is not correct **Ans. a**

22. The statement is not correct for shine Dalgarno sequence

(a) It is present only in prokaryotes

(b) It pairs with the 16S rRNA

(c) It is present several times on mRNA

(d) It is present in both prokaryotes and eukaryotes **Ans. d**

23. During elongation the uncharged tRNAs leave from the

(a) P site (b) A site

(c) A site in smaller subunit (d) A site in larger subunit **Ans. a**

24. The role of CAP in protein synthesis is

(a) It helps tRNA recognize codons on mRNA

(b) It facilitates complex formation of small and large subunit of ribosomes

(c) It facilitates formation of complex between the mRNA and the 30S ribosomal subunit

(d) It facilitates formation of complex between the mRNA and the 40S ribosomal subunit **Ans. d**

25. For the addition of each amino acid to the peptide chain requires

(a) 3GTPs are required (b) 2GDP are required

(c) 2GTPs are required (d) No GTP is required **Ans. c**

26. Formation of transcription bubble involves

(a) 12 base pairs (b) 10 base pairs

(c) 8 base pairs (d) 17 base pairs **Ans. d**

27. In prokaryotic RNA polymerase the largest subunit is

(a) α (b) β

(c) β' (d) σ **Ans. c**

28. In prokaryotic RNA polymerase the smallest subunit is

(a) α (b) β

(c) β' (d) σ **Ans. a**

29. σ subunit helps in specific initiating by

(a) Recognizing initiation codon

(b) Decreasing the affinity of RNA polymerase for general regions of DNA by a factor of 10^4

(c) It helps RNA polymerase bind to DNA template

(d) Helps in phosphodiester bond formation **Ans. b**

30. For circular DNA, negative supercoiling favours transcription because

(a) It helps RNA polymerase installation easy

(b) Sliding of RNA polymerase is facilitated

(c) It facilitates unwinding

(d) All of the above **Ans. c**

31. Fidelity of transcription as compared to translation is

(a) Lower

(b) Higher

(c) Equal

(d) Depends on the pH of the nucleoplasm **Ans. a**

32. The weakest base pairs are

(a) rC – rG (r = RNA) (b) rC – dG (d = DNA)

(c) rU – dA base pairs (d) dU – rA **Ans. c**

33. The percentage of methylated bases in prokaryotes and eukaryotes is

(a) Same

(b) More in prokaryotes

(c) More in eukaryotes

(d) Methylation of bases takes place only in eukaryotes **Ans. c**

34. hn RNA (heterogenous nuclear RNA) are precursors of mRNA and percentage of hn RNA converted into mRNA is

(a) 20%

(b) 25%

(c) 100%

(d) Depends on the requirement of mRNA **Ans. a**

35. Spliceosomes help in splicing at

(a) 5′ splice site

(b) 3′ splice site

(c) Branch site of mRNA precursors
(d) All of the above **Ans. d**

36. In *E. coli* the RNA which does not undergo modification
(a) rRNA (b) tRNA
(c) mRNA (d) None of the above **Ans. c**

37. Amino acids undergo modification during translation because
(a) Unmodified amino acid does not bind to tRNA
(b) On modified amino acid is prone to degradation
(c) The peptide bond formed by unmodified amino acid is unfavourable
(d) Unmodified amino acid cannot be accommodated in amino acyl tRNA **Ans. c**

38. The large (50) and small (30) subunit of 70s ribosome consists of 55 proteins distributed as
(a) 34 & 21 proteins (b) 34 & 22 proteins
(c) 33 & 21 proteins (d) 45 & 10 proteins **Ans. a**

39. Of all the eukaryotic initiation factors the ATP driven is/are
(a) elF_2 (b) elF_2 & elF_3
(c) elF_4 (d) elF_5 **Ans. c**

40. The GTP driven proteins are
(a) EF-2 only (b) EF-G
(c) EF-2 and eRF (d) All of the above **Ans. d**

41. Diphtheria toxin blocks protein synthesis by
(a) ADP ribosylation EF-2
(b) Phosphorylating EF-1
(c) Inhibiting peptidyl transferase activity
(d) Acting of EF-G **Ans. a**

42. Most regulatory proteins recognize................DNA most easily
(a) A (b) B
(c) A & B (d) All of the above **Ans. b**

43. Regulatory proteins recognize B-DNA most easily because
(a) It is smallest DNA
(b) They can form phosphodiester bond most readily

(c) They can slide along the DNA while searching for the specific target sequence

(d) All of the above **Ans. c**

44. The essence of attenuation is

(a) Translation is followed by transcription

(b) Transcription is followed by translation

(c) Symmetry plays an important role in the interaction of protein with DNA

(d) Transcription and translation are closely coupled **Ans. d**

45. GTP hydrolysis is protein synthesis is carried out by

(a) L7 and L12 proteins of 50S subunit

(b) L7 and L7 proteins of 40S subunit

(c) L7 and L9 proteins of 30S subunit

(d) L9 and L9 proteins of 50S subunit **Ans. a**

46. In tRNA the unusual bases are mostly

(a) Carboxylated derivatives

(b) Methylated derivatives

(c) Methylated and demethylated derivatives

(d) All of the above **Ans. c**

47. In case of protein synthesis the steps powered by GTP is/are

(a) Binding of amino acyl tRNA

(b) Movement of peptidyl tRNA from A to P site

(c) The associated movement of the mRNA to the next codon

(d) All of the above **Ans. d**

48. At the time of completion of protein synthesis the sequence in which they leave the ribosome is

(a) Polypeptide chain, mRNA, tRNA

(b) tRNA, mRNA, polypeptide chain

(c) polypeptide chain, tRNA, mRNA

(d) mRNA, polypeptide chain, tRNA **Ans. c**

❑❑❑

Chapter-9

Protein Regulation and Targeting

Fill in the blanks

1. Hypersensitive sites have a range of round.............base pairs.

Ans. 100-200

2. Control of ribosomal protein synthesis is exerted at............level.

Ans. translational

3. Clathrin is a protein that forms............

Ans. polyhedral lattices

4. The clathrin unit (triskelion) is a.................structure.

Ans. 3 legged

5. Example of isoprenylated protein is........,.........and...........

Ans. laminus, proto-oncogenes, G-protein

6. Initiator codons in protein synthesis are ....and ....

Ans. AUG, GUG

7. The two enzyme that help in DNA replication are .... and ....

Ans. DNA polymerse, DNA ligase

8. Replication occurs in the .... during the .... phase of the cell cycle.

Ans. Nucleus, S

9. The term cistron is used for a ....

Ans. structural gene

10. A eukaryotic gene consists of coding segments called .... and noncoding inserts termed ....

Ans. exons, introns

Multiple Choice Questions

1. During lysogeny in λ phage the lytic functions are
   (a) Expressed in less amount
   (b) Expression is completely lost
   (c) Dormant but not lost
   (d) Expressed besides lysogeny
   Ans. c

2. During protein-DNA interaction, the most important feature is
   (a) The charge of protein and DNA
   (b) Charge of protein only
   (c) The bonds that form between them
   (d) Their symmetry matching **Ans. d**

3. The expression of most genes is regulated primarily at the level of
   (a) Translation
   (b) Transcription
   (c) Transduction
   (d) Post translational modification
   Ans. b

4. The binding of cAMP-CAP to a specific site in the promoter region
   (a) Enhances the binding of RNA polymerase and the initiation of transcription
   (b) Inhibits the binding of RNA polymerase
   (c) Acts as a terminator during translation
   (d) None of the above **Ans. a**

5. The autoregulation of λ repressor depends upon
   (a) OR_1 OL_1
   (b) OR_1, OR_2, OR_3
   (c) OR_1, OR_2, OR_3, OL_1, OL_{2-}, PL_3
   (d) OL_1, OL_{2-}, PL_3 **Ans. c**

6. The regulation coordinated with amino acid concentration is called
   (a) Stringent factor
   (b) SOS response
   (c) Stringent response
   (d) Autoregulation **Ans. c**

7. Role of stringent factor is
   (a) To catalyze the formation of the unusual nucleotide ppG_{pp}
   (b) To control the number of amino acids being added to the polypeptide chain
   (c) To break the phosphodiester bond
   (d) To increase the rate of translation **Ans. a**

8. What does ppG_{pp} stand for

(a) Two phosphate guanine diphosphate

(b) Guanyl tetraphosphate

(c) Unusual guanine phosphate

(d) Guanosine tetraphosphate **Ans. d**

9. The rise in the concentration of ppG_{pp} in response to amino acid saturation leads to

(i) Rise in rRNA synthesis

(ii) Reduction in rRNA and mRNA synthesis

(iii) Formation of amino acids

(iv) Reduction in rRNA synthesis only

Choose the correct answer

(a) ii, iii (b) iii

(c) iv (d) None of the above **Ans. c**

10. Nucleosomes if treated with salt on dilute acid

(a) Hydrogen bond between the bases breaks down

(b) Phosphodiester bond breaks

(c) Nucleosome changes its state

(d) Histones dissociate from the DNA **Ans. b**

11. Largest DNA molecule (41×10^6 kd) is bound in

(a) *Drosophila* (b) Human

(c) *Acetabularia* (d) *E. coli* **Ans. d**

12. Match the following

Organism		No. of Chromosomes	
A.	*E. coli*	1.	16
B.	*Saccharomyces cerevisae*	2.	1
C.	*Drosophila melanogaster*	3.	23
D.	Human	4.	4

(a) A1, B3, C2, D4 (b) C1, B2, D3, A4

(c) D1, A2, C3, B4 (d) B1, C4, A2, D3 **Ans. d**

13. The four types of histone proteins (H_1, H_{24}, H_{28}, H_3, H_4) exist

(a) In a variety of forms

(b) Only in single forms

(c) In different forms only in plants
(d) None of the above **Ans. a**

14. During DNA replication the major distortion takes place in
(a) H_1 Histones
(b) All the four types of histones
(c) H_3 and H_{2A} histones
(d) All of the above **Ans. a**

15. The variable histone/histones is/are
(a) H_1 only (b) $H_1 + H_4$
(c) $H_{2A} + H_2$ (d) All the above **Ans. a**

16. The major role in DNA compaction is played by
(a) The negative charge on DNA (b) The double helix of DNA
(c) H_1 histone protein (d) All the histone proteins **Ans. c**

17. The most compact DNA is bound in
(a) Sperm heads (b) Sperm tails
(c) *Drosophila* (d) *E. coli* **Ans. a**

18. The extreme compact state of DNA is brought about by
(a) The size of the cell (b) Histones
(c) Size of the nucleus (d) Protamines **Ans. d**

19. Hypersensitive sites are
(a) Sequences of DNA capable of undergoing mutation
(b) Sequences of DNA sensitive to DNAase I
(c) Regions in immune system responsible for hypersensitivity
(d) mRNA sequences undergoing attenuation **Ans. b**

20. Hypersensitive sites have a range of round
(a) 100-200 base pairs (b) 100-150 base pairs
(c) 150-300 base pairs (d) 500-1000 base pairs **Ans. a**

21. Hypersensitive sites are found
(a) Sometimes at 3′ end or at 5′ end of the transcribed gene
(b) In between 3′ and 5′ end of mRNA
(c) Towards the 3′ end of mRNA
(d) Towards the 5′ end of transcribe gene **Ans. a**

22. Transcriptionally active regions are

(i) Uncoiled (ii) Deficient of histones

(iii) Sensitive to DNAase I (iv) All of the above

Choose the correct answer

(a) ii, iii (b) i, ii

(c) iii (d) iv **Ans. d**

23. The phage is known as prophage when it

(a) Undergoes lysis (b) Stays in the cell at lytic phase

(c) In the lysogenized state (d) Is undergoing mitosis **Ans. c**

24. Which of the following statement is not correct

(a) Lysogenic state can continue unless interrupted by DNA damage that induces SOS response

(b) Lysogeny switches to lysis with the age of the host cell

(c) Crp protein binds to the same operators as cI protein

(d) Synthesis **Ans. b**

25. In *E. coli,* the physiologic inducer is

(a) Lactose (b) Allolatose

(c) Glucose (d) Fructose **Ans. b**

26. Control of ribosomal protein synthesis is exerted at

(a) Transcription level

(b) Post transcriptional level

(c) When assembled into ribosomes

(d) Translational level **Ans. d**

27. SRP stops protein synthesis

(a) Help recognizing the stop sequences

(b) Acting as an inhibitor

(c) Means of stearic hindrance

(d) Acting as a toxic material **Ans. c**

28. SRP is helpful to cytosolic protein because

(a) It binds to the protein and prevents premature folding on the cytosolic side

(b) It prevents premature folding in the ER lumen

(c) It helps in folding of the protein at the cytosolic side

(d) All of the above **Ans. a**

29. All integral membrane proteins of the cell are formed by the ribosomes bound to the ER except

(a) Mitochondria only (b) Mitochondria and nucleus

(c) Mitochondria and chloroplast (d) Chloroplast only **Ans. c**

30. Membrane bound ribosomes and free ribosomes are structurally

(a) Similar (b) Slightly different

(c) Totally different (d) None of the above **Ans. a**

31. Membrane bound and free ribosomes differ

(a) In their structure

(b) In their function

(c) In the type of protein synthesis

(d) All of the above **Ans. c**

32. Ubiquitin of all eukaryotes is

(a) Perform different functions (b) Essentially similar

(c) All of the above (d) None of the above **Ans. b**

33. Clathrin is a protein that forms

(a) Polyhedral lattices (b) Tetrahedral lattices

(c) Hexagonal lattices (d) Forms no lattice **Ans. a**

34. (i) All integral membrane proteins of the cell are formed by ribosomes bond ER.

(ii) All integral membrane proteins of the cell are formed by free ribosomes.

(iii) Both free and bound ribosomes participate in the formation of integral membrane proteins.

(iv) Except chloroplast and mitochondrial membranes all other integral membrane proteins of the cell are formed by bound ER.

The correct statement is/are

(a) (i), (iv) (b) (ii), (iv)

(c) (ii), (iii), (iv) (d) (iv) only **Ans. d**

35. During translocation the signal for attachment (a sequence of amino acid residues are attached) is at

(a) Carboxyl terminus

(b) Amino terminus

(c) Sometimes at carboxyl and sometimes amino terminus

(d) All of the above **Ans. b**

36. Choose the correct statement

(a) Most genes of eukaryotic proteins are encoded by single copy genes

(b) Repetitive genes for ribosomal RNAs and histones are exceptions

(c) mRNA for histones do not have poly A chain

(d) All of the above **Ans. d**

37. "Clustering is not essential for coordinated gene expression". This statement is verified by

(a) Different subunits of ribosomes are scattered but expressed together

(b) Immunoglobulins are arranged distantly but expressed simultaneously

(c) Histones genes are dispersed and scattered but expressed together

(d) One subunit of ribosome is scattered but expressed with another subunit **Ans. c**

38. The significance of the absence of poly A tail from histone mRNA is

(a) Absence of poly A tail hinders is the transfer of mRNA from nucleus to cytoplasm which is a favourable condition for histone protein

(b) Absence of poly A tail facilitates the quick transfer of histone mRNA from nucleus to cytoplasm

(c) Poly A tail inhibits with the transcription of histone mRNA

(d) Poly A tail formation is not possible with the histone mRNA

Ans. b

39. The percentage of mammalian DNA coding for protein is

(a) 3% only (b) 4% only

(c) 2% only (d) 1% only **Ans. c**

40. The DNA of the following organism is non methylated

(a) *Xenopus* (b) *Yeast*

(c) *E. coli* (d) *Drosophila* **Ans. d**

41. Eukaryotic proteins are isoprenylated i.e.

(a) Isoprenyl group is attached at the 3′ end of the polypeptide chain.

(b) Isoprenyl group is deleted from the protein

(c) A thioester bond is formed between the isoprenyl group and cys residue of the protein

(d) An ester bond is formed between isoprenyl group and ser residue of the protein **Ans. c**

42. Example of isoprenylated protein is

(a) Proto-oncogenes (b) G protein

(c) Lamins (d) All of the above **Ans. d**

43. Amino acid sequences of H_3 and H_4 is

(a) Different in plants and animals

(b) Same in all plants and animals

(c) Same in some plants and animals and different in some

(d) Different in animals and same in all plants **Ans. b**

44. Proteins at the time of degradation becomes joined to

(a) Clathrin (b) Peptidase

(c) Ubiquitin (d) SRP **Ans. c**

45. The clathrin unit (triskelion) is a

(a) 3 legged structure (b) 4 legged structure

(c) Tetrahedral structure (d) Beaded structure **Ans. a**

46. The pH of the lysosomes is.......than endosomes

(a) Higher

(b) Equal

(c) Lower

(d) All of the above depending on the situation **Ans. c**

47. The acidic nature of the endosome leads to

(a) Killing of microbes

(b) Inactivating the interferon

(c) Separating the microbes from surface proteins

(d) Conformational change that is necessary for infection **Ans. d**

48. All protein transport except the outer membrane transport is driven by

(a) ATP hydrolysis (b) GTP hydrolysis

(c) Proton motive force (d) All of the above **Ans. c**

❑❑❑

Chapter-10

Lipids

Fill in the blanks

1. Lecithin in an important example of a..............

Ans. phospholipid

2. Waxes are esters of long chain........... with long chain.............alcohols.

Ans. Fatty acids, monohydric

3. Ripe fruit walls contain high percentage of...........................

Ans. pectin

4. Normal fat content of liver is about _______%.

Ans. 5

5. Obesity is accumulation of _______ in the body.

Ans. Fat

6. When the stired triacylglycerol is lipolysed in the adipose tissue blood levels of ___and__increased.

Ans. Free fatty acids (FFA), Glycerol

7. All long chain fatty acids with even number of carbon atoms are oxidized to a pool of _________ by β-oxidation.

Ans. Acetyl CoA

8. Cholesterol is excreted as such into ________.

Ans. Bile

9. Cholesterol molecule has _______ carbon atoms.

Ans. 27

10. While citrate is converted to isocitrate in the mitochondria, it is converted to _______in the cytosol.

Ans. Acetyl CoA + oxaloacetate

11. Lipase can act only at pH________.

Ans. 5–7

12. The caloric value of lipids is________.

Ans. 9.0 Kcal/g

13. When choline of lecithin is replaced by ethanolamine, the product is_________.

Ans. Cephalin

14. _____Amylase present in saliva is.

Ans. α

15. The absorption of glucose is decreased by the deficiency of____________.

Ans. Thiamine

16. Non-protein part of rhodopsin is__________.

Ans. Retinal

17. Alkaline hydrolysis of esters is known as................

Ans. saponification

18. The term lipid was given by...............

Ans. Bloor

19. Number of double bonds present in arachidonic acid is.............

Ans. 4

20. Number of essential fatty acids is..............

Ans. 3

21. Wax is ester of fatty acid with................

Ans. long chain monohydric alcohol

22. Commonly candles are prepared from..........acid and...........wax.

Ans. steric, paraffin

23. Lanolin or wood fat is a............

Ans. wax

24. Linolenic acid is essential unsaturated fatty acid. Its content is highest in............oil.

Ans. groundnut

25. Arachidonic acid is..............

Ans. non-essential fatty acid (NEFA)

26. The lipid bilayer of cell membrane exists as...............

Ans. semisolid

27. Cholesterol content is minimum in.............oil.

Ans. vegetable

28. Triglycerides become oil at about............°C.

Ans. 20

29. The polar groups of a phospholipid molecule is..................

Ans. hydrophobic

30. Waxes are esters of long chain .... with long chain .... alcohols.

Ans. fatty acids, monohydric.

Multiple Choice Questions

1. Alanine hydrolysis of a triglyceride is
   (a) Saponification (b) Esterification
   (c) Hydrogenation (d) Dehydration **Ans. a**
2. Physical property of fatty acid depends on
   (a) Length of hydrocarbon chains
   (b) Degree of unsaturation
   (c) Branching
   (d) All the above **Ans. d**
3. What are the bile salts
   (a) Charged phospholipids
   (b) Amphipathic cholesterol analogs with detergent properties
   (c) Esterified cholesterols
   (d) Hydrolyzed forms or tryacylglycerols **Ans. b**
4. The combination of an amino alcohol, fatty acid and sialic acid form
   (a) Phospholipids (b) Sulpholipids
   (c) Glycolipids (d) Aminolipids **Ans. c**
5. Prostaglandins are
   (a) Monocyclic (b) Bicyclic
   (c) Tricyclic (d) Polycyclic **Ans. a**
6. Which of the following compound is not amphipathic?
   (a) Phosphatidylcholine (b) Cholesterol
   (c) Oleic acid (d) Succinate **Ans. d**
7. The most abundant lipids in eukaryotic cell membranes are
   (a) Cholesterol (b) Glycolipids
   (c) Phospholipids (d) Lypopolysaccharides **Ans. c**
8. Most of the free fatty acids are transported in the blood
   (a) Inside the red blood cells (b) As lipoproteins
   (c) Combined with glucose (d) Bound to albumin **Ans. d**
9. Which compound is not a lipid?
   (a) Lecithin (b) Lysine
   (c) Cerebroside (d) Cephalin **Ans. b**

10. Simple lipids are

(a) Triglycerides (b) Phosphoglycerides

(c) Esters of lower acids (d) None **Ans. a**

11. The lipids which contain glycerol, fatty acid, phosphoric acid and an alcoholic base are called

(a) Phosphoglycerides (b) Sphingolipids

(c) Glycolipids (d) Lipids **Ans. a**

12. Which of the following is not a glyceride?

(a) Fat (b) Oil

(c) Phospholipid (d) Soaps **Ans. d**

13. Fat is a

(a) Lipid (b) Protein

(c) Carbohydrate (d) Amino acid **Ans. a**

14. Alkaline hydrolysis of esters is known as

(a) Esterification (b) Saponification

(c) Dehydration (d) Alkalination **Ans. b**

15. The term lipid was given by

(a) Sutherland (b) Bloor

(c) Altmann (d) Berzelius **Ans. b**

16. Each molecule of fat has

(a) One glycerol molecule and one fatty acid molecule

(b) One glycerol molecule and three fatty acid molecules

(c) Three glycerol molecules and one fatty acid molecule

(d) Three glycerol molecules and three fatty acid molecules **Ans. b**

17. The melting point of unsaturated fatty acids

(a) Increases with increase in double bonds

(b) Decreases with increase in double bonds

(c) Rises in some and falls in others

(d) These is no relationship between unsaturation and melting point **Ans. b**

18. Essential fatty acids were discovered by

(a) Evans and Burr (b) Bloor

(c) Sutherland (d) Beevers **Ans. a**

19. Number of essential fatty acids is

(a) 6 (b) 4

(c) 3 (d) 2 **Ans. c**

20. Essential fatty acids are

(a) Saturated

(b) Unsaturated

(c) Both saturated and unsaturated

(d) Cyclic **Ans. b**

21. Number of double bonds present in arachidonic acid is

(a) 4 (b) 3

(c) 2 (d) 1 **Ans. a**

22. Arachidonic acid gives rise to

(a) Terpenes (b) Prostaglandins

(c) Gangliosides (d) Cerebrosides **Ans. b**

23. Hydrogenation

(a) Increases percentage of EFA

(b) Destroys EFA

(c) Converts waxes into neutral fats

(d) Forms conjugated linoleic acid **Ans. b**

24. Major function of PUFA is

(a) Supply of essential fatty acids (b) Lowering of cholesterol level

(c) Prevent atherosclerosis (d) All the above **Ans. d**

25. Number of fatty acids present in a molecule of phospholipid is

(a) Two (b) Three

(c) One (d) None **Ans. a**

26. In contact with water, fatty acids produce

(a) Monolayer (b) Bilayer

(c) Trilayer (d) Membrane **Ans. a**

27. Wax is ester of fatty acid with

(a) Long chain dihydric alcohol

(b) Trihydric alcohol

(c) Long chain monohydric alcohol
(d) Short chain monohydric alcohol **Ans. c**

28. Bee's wax consists of
(a) Mericyl palmitate (b) Mericyl stearate
(c) Cetyl palmitate (d) Cetyl stearate **Ans. a**

29. Phosphatidic acid is
(a) Fatty acid (b) Phospholipid
(c) Ganglioside (d) Wax **Ans. b**

30. Cephalin is a
(a) Phospholipid (b) Terpenoid
(c) Sterol (d) Prostaglandin **Ans. a**

31. Lycopene, the colouring agent of Tomato, is a
(a) Sterol (b) Terpenoid
(c) Sphingolipid (d) Phospholipid **Ans. b**

32. Rubber is
(a) Monoterpene (b) Diterpene
(c) Tetraterpene (d) Polyterpene **Ans. d**

33. $-NH_2$ signifies the chemical
(a) Carbonyl (b) Amino
(c) Hydroxyl (d) Aldehyde **Ans. b**

34. A saturated fatty acid is
(a) With no double bond (b) High melting point
(c) Low melting point (d) Both a and b **Ans. d**

35. Which one is essential fatty acid
(a) Linoleic acid (b) Linolenic acid
(c) Arachiodonic acid (d) All the above **Ans. d**

36. Essential fatty acids occur in
(a) Animal fat (b) Plant oils
(c) Fish and fowl (d) Both b and c **Ans. d**

37. Deficiency of EFA causes
(a) Follicular keratosis (b) Kidney failure
(c) Sterility (d) All the above **Ans. d**

38. Cholesterol is

(a) Monosaccharide (b) Protein

(c) Sterol (d) Wax **Ans. c**

39. In contact with water, phospholipids produce

(a) Monolayer (b) Bilayer

(c) Trilayer (d) Emulsion **Ans. b**

40. A steroid that prevents fertility is

(a) Campesterol (b) Cholesterol

(c) Diosgenin (d) Testosterone **Ans. c**

41. Wax present in human blood is

(a) Mericyl palmitate (b) Cetyl palmitate

(c) Cholesterol palmitate (d) Ceryl palmitate **Ans. c**

42. A ceramide structure, galactose and sialic acid present in..............

(a) Gangliosides (b) Cerebrosides

(c) Globosides (d) Sphingomyelien **Ans. a**

43. Commonly candles are prepared from

(a) Bee's wax (b) Paraffin wax

(c) Steric acid (d) Both b and c **Ans. d**

44. Lanolin or wool fat is a

(a) Hard fat (b) Oil

(c) Wax (d) Sterol **Ans. c**

45. PUFA is

(a) Unsaturated fatty acid with one double bond

(b) Unsaturated fatty acid with two double bond

(c) Unsaturated fatty acid with more than one double bond

(d) Unsaturated fatty acid with more than two double bond **Ans. d**

46. 18C unsaturated fatty acid with three double bonds is

(a) Oleic acid (b) Linoleic acid

(c) Linolenic acid (d) Arachidonic acid **Ans. d**

47. A lipid having tetracyclic structure is

(a) Hard fat (b) Wax

(c) Phospholipid (d) Sterol **Ans. c**

48. Linolenic acid is essential unsaturated fatty acid. Its content is highest in

(a) Coconut oil (b) Sunflower oil

(c) Groundnut oil (d) Cotton oil **Ans. c**

49. Arachidonic acid is

(a) Non- essential fatty acid (NEFA)

(b) Poly unsaturated fatty acid (PUFA)

(c) Both a and b

(d) None of the above **Ans. a**

50. The lipid bilayer of cell membrane exists as

(a) Solid (b) Fibrous sheath

(c) Semi solid (d) Gel **Ans. c**

51. Phospholipids are important cell membrane constituents because they

(a) Contain glycerol

(b) Can form bilayers in H_2O

(c) Combine covalently with proteins

(d) Contain both polar & nonpolar portion **Ans. b**

52. Cholesterol content is minimum in

(a) Fish (b) Cow's skimmed milk

(c) Liver (d) Vegetable oils **Ans. d**

53. Select the odd from the following

(a) Glutamic acid (b) Capric acid

(c) Butyric acid (d) Oleic acid **Ans. c**

54. Minimum fatty acid contents is found in

(a) Peanut oil (b) Soybean oil

(c) Sunflower oil (d) Coconut oil **Ans. d**

55. Hydrolysis of fat by an alkali is known as

(a) Reduction (b) Esteritfication

(c) Condensation (d) Saponification **Ans. d**

56. Which one is most unsaturated

(a) Dalda/vegetable ghee (b) Coconut oil

(c) Mustard oil (d) All are equally unsaturated

Ans. a

57. Solid esters of long fatty acids with

(a) Lipids (b) Waxes

(c) Carotenoids (d) Steroids **Ans. a**

58. The polar groups of a phospholipid molecule is

(a) Water repellent (b) Hydrophilic

(c) Hydrophobic (d) Amphiatic **Ans. c**

59. Sudan Black B stain is used to colour

(a) Lipid (b) Protein

(c) Nucleic acid (d) Polysaccharides **Ans. c**

60. At about 20^0C, triglycerides become

(a) Gas (b) Solid

(c) Oil (d) Crystal **Ans. c**

61. Butter is soft because it contains

(a) Good quality of long chain fatty acid

(b) Good quality of short chain fatty acid

(c) Poor quality of short fatty acid

(d) Little of water and little amount of fats **Ans. a**

62. Which one is a 16C saturated fatty acid?

(a) Cholesterol (b) Palmitic acid

(c) Uridylic acid (d) Stearic acid **Ans. a**

63. Which of the following fatty acid (FAs) of similar size have higher melting point

(a) FAs having a cis double bond

(b) FAs having a trans double bond

(c) Both have same melting point

(d) Melting point does not depends upon cis and trans configuration

Ans. b

64. All of the following statements describe phosphoglycerides except

(a) They are both amphipathic and amphoteric

(b) They arise from phosphatidic acid

(c) They are found in cell membranes

(d) They are a major store of metabolic energy **Ans. d**

65. Alkaline hydrolysis of a triglyceride is

(a) Saponification (b) Esterification

(c) Hydrogenation (d) Dehydration **Ans. a**

66. Gangliosides contain

(a) A ceramide structure (b) Glucose or galactose

(c) Sialic acid (d) All the above **Ans. d**

67. Elevation of intracellular inositol triphosphate results in the release of Ca^{++} from the following organelle

(a) Peroxisome (b) Lysosome

(c) Mitochondria (d) Endoplasmic reticulum **Ans. d**

68. Which of the following not belongs to glycosphingolipids?

(a) Cerebrosides (b) Gangliosides

(c) Globosides (d) Sphingomyelin **Ans. d**

69. Which of the following statement is incorrect about fatty acid

(a) All non-essential fatty acid can be synthesized from acetyl CoA

(b) Essential fatty acid are unsaturated

(c) Most naturally occurring fatty acids contain even number of carbon atoms

(d) Unsaturated fatty acid produce more energy than saturated one of the same size when oxidized **Ans. d**

70. A ganglioside must contain in its structure

(a) N-acetyl neuraminic acid (NANA), hexoses, sphingosine, long chain fatty acid

(b) NANA, a hexose, a fatty acid, sphingosine, phosphorylcholine

(c) NANA, sphingosine, ethanolamine

(d) NANA, hexose, fatty acid, glycerol **Ans. a**

71. Which of the following statements about natural sterols is incorrect?

(a) Cholesterol is the most abundant sterol in animal tissue

(b) All the carbon atoms of cholesterol are derived from acetyl CoA

(c) β-sitosterol is the most abundant plant sterol

(d) Dietary β-sitosterol and cholesterol are absorbed to about the same extent in the intestine of normal human **Ans. d**

72. The concentrations of sphingomyellins are increased in
   (a) Gaucher's disease (b) Fabry's disease
   (c) Febrile disease (d) Niemann-Pick disease **Ans. c**

73. Cholesterol is essential for normal membrane functions because it
   (a) Cannot be made by higher organisms, e.g. mammals
   (b) Spans the thickness of the bilayer
   (c) Keeps membranes fluid
   (d) Catalyzes lipid flip-flop in the bilayer **Ans. d**

74. Assuming they all had the same number of carbon atoms, which of the following has the most C-H bonds?
   (a) An unsaturated fat (b) A polyunsaturated fat
   (c) A polysaccharide (d) A saturated fat **Ans. d**

75. Correct statements about inherited sphingolipid disorders include which of the following?
   (a) Gaucher's disease is due to a deficiency in the degradation of glucocerebrosides
   (b) Tay-Sach's disease is due to a deficiency of lysosomal hexosaminidase A
   (c) Niemann-Pick disease is due to a deficiency of sphingomyelinase
   (d) All the above **Ans. d**

76. Gangliosides have which of the following characteristics?
   (a) They contain sugars and sugar derivatives
   (b) They contain sialic acid (N-acetylneuraminic acid)
   (c) They are derived from cerebrosides
   (d) All the above **Ans. d**

77. Cholesterol can act as a precursor for each of the following compounds except
   (a) Dihydroxycholecalciferol (b) Testosterone
   (c) Glycocholic (d) Cholecystokinin **Ans. a**

78. Which one of the following processes is common to the synthesis of all steroid hormones
   (a) Conversion of presnenolone to progesterone
   (b) Cholesterol side chain cleavage
   (c) Isomerization
   (d) Dehydrogenation **Ans. d**

79. Which of the following is not a biological role of fatty acids?

(a) Fatty acid is essential component of plasma membrane

(b) Fatty acid stored as triacylglycerols in body

(c) Fatty acids as intracellular second messengers

(d) Fatty acids mainly contain odd number of carbons **Ans. c**

80. Which one will have high melting point. Fatty acid (having 15 carbons) contains

(a) Single double bond (b) More than one double bond

(c) Only single bond (d) None is correct **Ans. a**

81. Some unsaturated fat used in cooking, such as olive oil, spoil rapidly upon exposure to air at room temperature because

(a) Unsaturated fats are susceptible to oxidation by molecular oxygen

(b) Of low melting point of unsaturated fat

(c) Carboxyl group oxidized upon exposure to air

(d) None **Ans. b**

82. Arrange a triacylglycerol, a diacylglycerol and a monoacylglycerol, all containing only palmitic acid, in order of increasing solubility in water

(a) Diacylglycerol, monoacylglycerol, triacylglycerol

(b) Triacylglycerol, diacylglycerol, monoacylglycerol

(c) Monoacylglycerol, diacylglycerol, triacylglycerol

(d) All have similar solubility **Ans. b**

83. Which of the following glycerophospholipid is the major lipid of mitochondrial membrane?

(a) Phosphatidyl choline (b) Lecithin

(c) Phosphatidyl inositol (d) Cardiolipin **Ans. d**

84. Tay-Sach's disease, caused due to accumulation of GM2 gangliosides, a complex sphingolipids is a/an

(a) Autosomal dominant genetic disorder

(b) Sex chromosomal recessive genetic disorder

(c) Sex chromosomal dominant genetic disorder

(d) Autosomal recessive genetic disorder **Ans. d**

85. Which of the following lipids have a net negative charge

(a) Phosphatidylcholine (b) Cholesterol

(c) Phosphatidylserine (d) Phosphatidylethanolamine

Ans. c

86. The UV rays of sunlight is responsible for the formation of the vitamin D in the skin is from

(a) Cholesterol (b) Ergasterol

(c) Cholecalciferol (d) Calciferol **Ans. a**

87. What are bile salts?

(a) Charged phospholipids

(b) Amphipathic cholesterol analogs with detergent properties

(c) Esterified cholesterols

(d) Hydrolyzed forms of triacylglycerols **Ans. b**

88. The acetyl groups required for cytoplasmic fatty acid synthesis appear in the cytoplasm as a result of the activity of

(a) Citrate synthetase (b) Isocitrate dehydrogenase

(c) Citrate lyase (d) Thiolase **Ans. c**

89. The most important source of reducing equivalents for fatty acid synthesis in the liver is

(a) Oxidation of glucouronic acid

(b) Oxidation of acetyl coenzyme A

(c) The tricarboxylic acid cycle

(d) The pentose phosphate pathway **Ans. d**

90. Which of the following vitamins is necessary as a coenzyme in the initial step of fatty acid synthesis?

(a) Biotin (b) Thiamine

(c) Vitamin D (d) Riboflavin **Ans. a**

91. Which of the following fatty acids has the lowest melting point?

(a) Fatty acids with sites of unsaturation with cis double bonds

(b) Fatty acids with sites of unsaturation with trans double bonds

(c) Fatty acids with no sites of unsaturation

(d) Fatty acids with longer hydrophobic tails **Ans. a**

92. Bile salts act as detergents in the duodenum because

(a) Of their amphipathic structures

(b) They contain hydrophobic groups

(c) They carry an overall positive charge

(d) They are zwitter ions at the pH of the duodenum **Ans. a**

93. Which of the following cofactor is essential for the activity of acetyl CoA carboxylase?

(a) NAD^+ (b) Biotin

(c) TPP (d) Vit B_6 **Ans. b**

94. Which of the following step is not involved in cholesterol synthesis

(a) Formation of mevalonate from acetate

(b) Conversion of mevalonate into isopentenyl pyrophosphate

(c) Polymerization of six isoprene units into squalene

(d) Formation of malonyl CoA from acetyl CoA **Ans. d**

95. Which of the following statement is incorrect

(a) Fatty acid synthesis occur in cytosol of animal cell

(b) Fatty acid desaturation and elongation occurs in ER

(c) In diabetes mellitus ketone bodies production increases and results in ketoacidosis

(d) None of the above **Ans. d**

96. Which of the following compound provides the carbon atoms for cholesterol synthesis?

(a) Pregnenolone (b) Acetyl CoA

(c) Phytanic acid (d) None **Ans. b**

97. Which one of the following cofactor is utilized during the conversion of acetyl CoA to malonyl CoA?

(a) Thiamine pyrophosphate (b) Acyl carrier protein (ACP)

(c) NAD_1 (d) Biotin **Ans. d**

98. A patient has a genetic defect resulting in a deficiency of lipoprotein lipase. After eating a meal containing a large amount of fat, one would expect to see a plasma elevation of

(a) Chylomicrons (b) VLDLs

(c) LDLs (d) HDLs **Ans. a**

99. Which of the following substances represents an unsaturated fatty acid?

(a) Palmitate (b) Stearate

(c) Choline (d) Oleate **Ans. d**

100. Which of the following is a false statement regarding Thromboxane A_2 (TXA_2)

(a) Is derived from the membrane phospholipid

(b) Its production is decreased by non-steroidal anti-inflammatory drugs

(c) Causes platelet aggregation
(d) Causes vasodialation **Ans. d**

101. The following is false about cholesterol
(a) Cholesterol is the most abundant sterol found in human tissue
(b) Cholic acid is a precursor in the biosynthesis of cholesterol
(c) All the carbon atoms of cholesterol are derived from acetyl CoA
(d) HMG CoA reductase is the main regulator enzyme of cholesterol synthesis **Ans. b**

102. Ketone bodies are formed due to
(a) Breakdown in the β-oxidation pathway which would produce an excess of acetyl CoA
(b) Inhibition of fatty acid activation and transport
(c) Channelization of oxaloacetate from the citric acid cycle to gluconeogenesis
(d) Inhibition of cAMP production **Ans. c**

103. A common step involved in the biosynthesis of all steroid hormones is
(a) Conversion of cortisol to cortisterone
(b) Cholesterol side chain cleavage
(c) Aromatization
(d) Dehydrogenation **Ans. b**

104. An essential building block of phosphatidic acid and phosphatidylcholine
(a) Glycerol (b) Lysine
(c) Cholesterol (d) Glucose **Ans. a**

105. An example of a hydroxy fatty acid is
(a) Ricinoleic acid (b) Crotonic acid
(c) Butyric acid (d) Oleic acid **Ans. a**

106. An example of a saturated fatty acid is
(a) Palmitic acid (b) Oleic acid
(c) Linoleic acid (d) Erucic acid **Ans. a**

107. If the fatty acid is esterified with an alcohol of high molecular weight instead of glycerol, the resulting compound is
(a) Lipositol (b) Plasmalogen
(c) Wax (d) Cephalin **Ans. c**

108. A fatty acid which is not synthesized in the body and has to be supplied in the diet is

(a) Palmitic acid (b) Lauric acid
(c) Linolenic acid (d) Palmitoleic acid **Ans. c**

109. Essential fatty acid

(a) Linoleic acid (b) Linolenic acid
(c) Arachidonic acid (d) All these **Ans. d**

110. The fatty acid present in cerebrosides is

(a) Lignoceric acid (b) Valeric acid
(c) Caprylic acid (d) Behenic acid **Ans. a**

111. In humans, a dietary essential fatty acid is

(a) Palmitic acid (b) Stearic acid
(c) Oleic acid (d) Linoleic acid **Ans. d**

112. A lipid containing alcoholic amine residue is

(a) Phosphatidic acid (b) Ganglioside
(c) Glucocerebroside (d) Sphingomyelin **Ans. d**

113. Cephalin consists of

(a) Glycerol, fatty acids, phosphoric acid and choline
(b) Glycerol, fatty acids, phosphoric acid and ethanolamine
(c) Glycerol, fatty acids, phosphoric acid and inositol
(d) Glycerol, fatty acids, phosphoric acid and serine **Ans. b**

114. In mammals, the major fat in adipose tissues is

(a) Phospholipid (b) Cholesterol
(c) Sphingolipids (d) Triacylglycerol **Ans. d**

115. Glycosphingolipids are a combination of

(a) Ceramide with one or more sugar residues
(b) Glycerol with galactose
(c) Sphingosine with galactose
(d) Sphingosine with phosphoric acid **Ans. a**

116. The importance of phospholipids as constituent of cell membrane is because they possess

(a) Fatty acids
(b) Both polar and nonpolar groups

(c) Glycerol
(d) Phosphoric acid **Ans. b**

117. In neutral fats, the unsaponificable matter includes
(a) Hydrocarbons (b) Triacylglycerol
(c) Phospholipids (d) Cholsesterol **Ans. a**

118. Higher alcohol present in waxes is
(a) Benzyl (b) Methyl
(c) Ethyl (d) Cetyl **Ans. d**

119. Kerasin consists of
(a) Nervonic acid (b) Lignoceric acid
(c) Cervonic acid (d) Clupanodonic acid **Ans. b**

120. Gangliosides are complex glycosphingolipids found in
(a) Liver (b) Brain
(c) Kidney (d) Muscle **Ans. b**

121. Unsaturated fatty acid found in the cod liver oil and containing 5 double bonds is
(a) Clupanodonic acid (b) Cervonic acid
(c) Elaidic acid (d) Timnodonic acid **Ans. d**

122. Phospholipid acting as surfactant is
(a) Cephalin (b) Phosphatidyl inositol
(c) Lecithin (d) Phosphatidyl serine **Ans. c**

123. An oil which contains cyclic fatty acids and once used in the treatment of leprosy is
(a) Elaidic oil (b) Rapeseed oil
(c) Lanoline (d) Chaulmoogric oil **Ans. d**

124. Unpleasant odours and taste in a fat (rancidity) can be delayed or prevented by the addition of
(a) Lead (b) Copper
(c) Tocopherol (d) Ergosterol **Ans. c**

125. Gangliosides derived from glucosylceramide contain in addition one or more molecules of
(a) Sialic acid (b) Glycerol
(c) Diacylglycerol (d) Hyaluronic acid **Ans. a**

126. 'Drying oil', oxidized spontaneously by atmospheric oxygen at ordinary temperature and forms a hard water proof material is

(a) Coconut oil (b) Peanut oil

(c) Rape seed oil (d) Linseed oil **Ans. d**

127. Deterioration of food (rancidity) is due to presence of

(a) Cholesterol (b) Vitamin E

(c) Peroxidation of lipids (d) Phenolic compounds **Ans. c**

128. The number of ml of N/10 KOH required to neutralize the fatty acids in the distillate from 5 gm of fat is called

(a) Reichert-Meissel number (b) Polenske number

(c) Acetyl number (d) Non volatile fatty acid number **Ans. a**

129. Molecular formula of cholesterol is

(a) $C_{27}H_{45}OH$ (b) $C_{29}H_{47}OH$

(c) $C_{29}H_{47}OH$ (d) $C_{23}H_{41}OH$ **Ans. a**

130. The cholesterol molecule is

(a) Benzene derivative (b) Quinoline derivative

(c) Steroid (d) Straight chain acid **Ans. c**

131. Salkowski test is performed to detect

(a) Glycerol (b) Cholesterol

(c) Fatty acids (d) Vitamin D **Ans. b**

132. Palmitic, oleic or stearic acid ester of cholesterol used in manufacture of cosmetic creams is

(a) Elaidic oil (b) Lanoline

(c) Spermaceti (d) Chaulmoogric oil **Ans. b**

133. Dietary fats after absorption appear in the circulation as

(a) HDL (b) VLDL

(c) LDL (d) Chylomicron **Ans. d**

134. Free fatty acids are transported in the blood

(a) Combined with albumin

(b) Combined with fatty acid binding protein

(c) Combined with β -lipoprotein

(d) In unbound free salts **Ans. a**

135. Long chain fatty acids are first activated to acetyl-CoA in

(a) Cytosol (b) Microsomes
(c) Nucleus (d) Mitochondria **Ans. a**

136. The enzyme acyl-CoA synthase catalyses the conversion of a fatty acid of an active fatty acid in the presence of

(a) AMP (b) ADP
(c) ATP (d) GTP **Ans. c**

137. Carnitine is synthesized from

(a) Lysine and methionine (b) Glycine and arginine
(c) Aspartate and glutamate (d) Proline and hydroxyproline **Ans. a**

138. The enzymes of β-oxidation are found in

(a) Mitochondria (b) Cytosol
(c) Golgi apparatus (d) Nucleus **Ans. a**

139. Long chain fatty acids penetrate the inner mitochondrial membrane

(a) Freely
(b) As acyl-CoA derivative
(c) As carnitine derivative
(d) Requiring Na dependent carrier **Ans. c**

140. An important feature of Zellweger's syndrome is

(a) Hypoglycemia
(b) Accumulation of phytanic acid in tissues
(c) Skin eruptions
(d) Accumulation of C_{26}-C_{38} polyenoic acid in brain tissues **Ans. d**

141. An important finding of Fabry's disease is

(a) Skin rash (b) Exophthalmos
(c) Hemolytic anemia (d) Mental retardation **Ans. a**

142. Gaucher's disease is due to deficiency of the enzyme

(a) Sphingomyelinase (b) Glucocerebrosidase
(c) Galactocerbrosidase (d) β-Galactosidase **Ans. b**

143. Characteristic finding in Gaucher's disease is

(a) Night blindness (b) Renal failure
(c) Hepatosplenomegaly (d) Deafness **Ans. c**

144. An important finding in Neimann-Pick disease is
(a) Leukopenia (b) Cardiac enlargement
(c) Corneal opacity (d) Hepatosplenomegaly **Ans. d**

145. Fucosidosis is characterized by
(a) Muscle spasticity (b) Liver enlargement
(c) Skin rash (d) Kidney failure **Ans. a**

146. Metachromatic leukodystrophy is due to deficiency of enzyme
(a) α-Fucosidase (b) Arylsulphatase A
(c) Ceramidase (d) Hexosaminidase A **Ans. b**

147. A significant feature of Tangier disease is
(a) Impairment of chylomicron formation
(b) Hypotriacylglycerolmia
(c) Absence of Apo-C-II
(d) Absence of Apo-C-I **Ans. c**

148. A significant feature of Broad Beta disease is
(a) Hypocholesterolemia (b) Hypotriacylglycerolemia
(c) Absence of Apo-D (d) Abnormality of Apo-E **Ans. d**

149. Neonatal tyrosinemia improves on administration of
(a) Thiamin (b) Riboflavin
(c) Pyridoxine (d) Ascorbic acid **Ans. a**

150. Absence of phenylalanine hydroxylase causes
(a) Neonatal tyrosinemia (b) Phenylketonuria
(c) Primary hyperoxaluria (d) Albinism **Ans. d**

151. Richner-Hanhart syndrome is due to defect in
(a) Tyrosinase
(b) Phenylalanine hydroxylase
(c) Hepatic tyrosine transaminase
(d) Fumarylacetoacetate hydrolase **Ans. b**

152. Plasma tyrosine level in Richner-Hanhart syndrome is
(a) 1–2 mg/dL (b) 2–3 mg/dL
(c) 4–5 mg/dL (d) 8–10 mg/dL **Ans. c**

153. Amount of phenylacetic acid excreted in the urine in phenylketonuria is
(a) 100–200 mg/dL (b) 200–280 mg/dL
(c) 290–550 mg/dL (d) 600–750 mg/dL **Ans. c**

154. Tyrosinosis is due to defect in the enzyme
(a) Fumarylacetoacetate hydrolase
(b) p-Hydroxyphenylpyruvate hydroxylase
(c) Tyrosine transaminase
(d) Tyrosine hydroxylase **Ans. a**

155. An important finding in Histidinemia is
(a) Impairment of conversion of α-Glutamate to α-ketoglutarate
(b) Speech defect
(c) Decreased urinary histidine level
(d) Patients can not be treated by diet **Ans. b**

156. An important finding in glycinuria is
(a) Excess excretion of oxalate in the urine
(b) Deficiency of enzyme glycinase
(c) Significantly increased serum glycine level
(d) Defect in renal tubular reabsorption of glycine **Ans. d**

157. Increased urinary indole acetic acid is diagnostic of
(a) Maple syrup urine disease (b) Hartnup disease
(c) Homocystinuia (d) Phenylketonuria **Ans. b**

158. In glycinuria daily urinary excretion of glycine ranges from
(a) 100–200 mg (b) 300–500 mg
(c) 600–1000 mg (d) 1100–1400 mg **Ans. c**

159. An inborn error, maple syrup urine disease is due to deficiency of the enzyme
(a) Isovaleryl-CoA hydrogenase (b) Phenylalnine hydroxylase
(c) Adenosyl transferase (d) α-Ketoacid decarboxylase
Ans. d

160. Maple syrup urine disease becomes evident in extra uterine life by the end of
(a) First week (b) Second week
(c) Third week (d) Fourth week **Ans. a**

161. Alkaptonuria occurs due to deficiency of the enzyme
(a) Maleylacetoacetate isomerase
(b) Homogentisate oxidase
(c) p-Hydroxyphenylpyruvate hydroxylase
(d) Fumarylacetoacetate hydrolase **Ans. a**

162. An important feature of maple syrup urine disease is
(a) Patient can not be treated by dietary regulation
(b) Without treatment death, of patient may occur by the end of second year of life
(c) Blood levels of leucine, isoleucine and serine are increased
(d) Excessive brain damage **Ans. d**

163. Ochronosis is an important finding of
(a) Tyrosinemia (b) Tyrosinosis
(c) Alkaptonuria (d) Richner Hanhart syndrome
Ans. c

164. Phrynoderma is a deficiency of
(a) Essential fatty acids (b) Proteins
(c) Amino acids (d) None of these **Ans. a**

165. The percentage of linoleic acid in safflower oil is
(a) 73 (b) 57
(c) 40 (d) 15 **Ans. a**

166. The percentage of polyunsaturated fatty acids in soyabean oil is
(a) 62 (b) 10
(c) 3 (d) 2 **Ans. a**

167. The percentage of polyunsaturated fatty acids in butter is
(a) 60 (b) 37
(c) 25 (d) 3 **Ans. d**

168. Dietary fibre denotes
(a) Undigested proteins
(b) Plant cell components that cannot be digested by own enzymes
(c) All plant cell wall components
(d) All non digestible water insoluble polysaccharide **Ans. b**

169. A high fibre diet is associated with reduced incidence of

(a) Cardiovascular disease (b) C.N.S. disease

(c) Liver disease (d) Skin disease **Ans. a**

170. Dietary fibres are rich in

(a) Cellulose (b) Glycogen

(c) Starch (d) Proteoglycans **Ans. a**

171. Minimum dietary fibre is found in

(a) Dried apricot (b) Peas

(c) Bran (d) Cornflakes **Ans. b**

172. A bland diet is recommended in

(a) Peptic ulcer (b) Atherosclerosis

(c) Diabetes (d) Liver disease **Ans. a**

173. A dietary deficiency in both the quantity and the quality of protein results in

(a) Kwashiorkar (b) Marasmus

(c) Xerophtalmia (d) Liver diseases **Ans. a**

174. The deficiency of both energy and protein causes

(a) Marasmus (b) Kwashiorkar

(c) Diabetes (d) Beri-beri **Ans. a**

175. Kwashiorkar is characterized by

(a) Night blindness (b) Edema

(c) Easy fracturability (d) Xerophthalmia **Ans. b**

176. A characteristic feature of Kwashiorkar is

(a) Fatty liver

(b) Emaciation

(c) Low insulin liver

(d) Occurrence in less than 1 year infant **Ans. a**

177. A characteristic feature of marasmus is

(a) Severe hypoalbuminemia

(b) Normal epinephrine level

(c) Mild muscle wasting

(d) Low insulin and high cortisol level **Ans. d**

178. Obesity generally reflects excess intake of energy and is often associated with the development of
(a) Nervousness
(b) Non-insulin dependent diabetes mellitus
(c) Hepatitis
(d) Colon cancer **Ans. b**

179. Atherosclerosis and coronary heart diseases are associated with the diet
(a) High in total fat and saturated fat
(b) Low in protein
(c) High in protein
(d) High in carbohydrate **Ans. a**

180. Cerebrovasular disease and hypertension is associated with
(a) High calcium intake (b) High salt intake
(c) Low calcium intake (d) Low salt intake **Ans. b**

181. The normal range of total serum bilirubin is
(a) 0.2–1.2 mg/100 ml (b) 1.5–1.8 mg/100 ml
(c) 2.0–4.0 mg/100 ml (d) Above 7.0 mg/100 ml **Ans. a**

182. The normal range of direct reacting (conjugated) serum bilirubin is
(a) 0–0.1 mg/100 ml (b) 0.1–0.4 mg/100 ml
(c) 0.4–06 mg/100 ml (d) 0.5–1 mg/100 ml **Ans. b**

183. The normal range of indirect (unconjugated) bilirubin in serum is
(a) 0–0.1 mg/100 ml (b) 0.1–0.2 mg/100 ml
(c) 0.2–0.7 mg/100 ml (d) 0.8–1.0 mg/100 ml **Ans. c**

184. Jaundice is visible when serum bilirubin exceeds
(a) 0.5 mg/100 ml (b) 0.8 mg/100 ml
(c) 1 mg/100 ml (d) 2.4 mg/100 ml **Ans. c**

185. An increase in serum unconjugated bilirubin occurs in
(a) Hemolytic jaundice (b) Obstructive jaundice
(c) Nephritis (d) Glomerulonephritis **Ans. a**

186. One of the causes of hemolytic jaundice is
(a) G-6-phosphatase deficiency
(b) Increased conjugated bilirubin

(c) Glucokinase deficiency
(d) Phosphoglucomutase deficiency **Ans. a**

187. Increased urobilinogen in urine and absence of bilirubin in the urine suggests
(a) Obstructive jaundice (b) Hemolytic jaundice
(c) Viral hepatitis (d) Toxic jaundice **Ans. a**

188. A jaundice in which serum alanine transaminase and alkaline phosphatase are normal is
(a) Hepatic jaundice (b) Hemolytic jaundice
(c) Parenchymatous jaundice (d) Obstructive Jaundice **Ans. b**

189. Fecal stercobilinogen is increased in
(a) Hemolytic jaundice (b) Hepatic jaundice
(c) Viral hepatitis (d) Obstructive jaundice **Ans. b**

190. Fecal urobilinogen is increased in
(a) Hemolytic jaundice (b) Obstruction of biliary duct
(c) Extrahepatic gall stones (d) Enlarged lymphnodes **Ans. a**

191. A mixture of conjugated and unconjugated bilirubin is found in the circulation in
(a) Hemolytic jaundice (b) Hepatic jaundice
(c) Obstructive jaundice (d) Post hepatic jaundice **Ans. b**

192. Hepatocellular jaundice as compared to pure obstructive type of jaundice is characterized by
(a) Increased serum alkaline phosphate, LDH and ALT
(b) Decreased serum alkaline phosphatase, LDH and ALT
(c) Increased serum alkaline phosphatase and decreased levels of LDH and ALT
(d) Decreased serum alkaline phosphatase and increased serum LDH and ALT **Ans. d**

193. Icteric index of an normal adult varies between
(a) 1–2 (b) 2–4
(c) 4–6 (d) 10–15 **Ans. c**

194. Clinical jaundice is present with an icteric index above
(a) 4 (b) 8
(c) 10 (d) 15 **Ans. d**

195. Normal quantity of urobilinogen excreted in the feces per day is about
(a) 10–25 mg (b) 50–250 mg
(c) 300–500 mg (d) 700–800 mg **Ans. b**

196. Fecal urobilinogen is decreased in
(a) Obstruction of biliary duct (b) Hemolytic jaundice
(c) Excess fat intake (d) Low fat intake **Ans. a**

197. A complete absence of fecal urobilinogen is strongly suggestive of
(a) Obstruction of bile duct (b) Hemolytic jaundice
(c) Intrahepatic cholestasis (d) Malignant obstructive disease **Ans. d**

198. Immediate direct Vanden Bergh reaction indicates
(a) Hemolytic jaundice (b) Hepatic jaundice
(c) Obstructive jaundice (d) Megalobastic anemia **Ans. b**

199. The presence of bilirubin in the urine without urobilinogen suggests
(a) Obstructive jaundice
(b) Hemolytic jaundice
(c) Pernicious anemia
(d) Damage to the hepatic parenchyma **Ans. a**

200. Impaired galactose tolerance test suggests
(a) Defect in glucose utilisation (b) Liver cell injury
(c) Renal defect (d) Muscle injury **Ans. b**

201. Increased serum ornithine carabamoyl transferase activity is diagnostic of
(a) Myocardial infarction (b) Hemolytic jaundice
(c) Bone disease (d) Acute viral hepatitis **Ans. d**

202. The best known and most frequently used test of the detoxicating functions of liver is
(a) Hippuric acid test (b) Galactose tolerance test
(c) Epinephrine tolerance test (d) Rose Bengal dye test **Ans. a**

203. The ability of liver to remove a dye like BSP from the blood suggests a normal
(a) Excretory function (b) Detoxification function
(c) Metabolic function (d) Circulatory function **Ans. a**

204. Removal of BSP dye by the liver involves conjugation with
(a) Thiosulphate
(b) Glutamine
(c) Cystein component of glutathione
(d) UDP glucuronate **Ans. c**

205. Normal value of plasma total proteins varies between
(a) 3–4 gm/100ml (b) 6–8 gm/100ml
(c) 10–12 gm/100ml (d) 14–16 gm/100ml **Ans. b**

206. A decrease in albumin with increased production of other unidentified proteins which migrate in β, γ region suggests
(a) Cirrhosis of liver (b) Nephrotic syndrome
(c) Infection (d) Chronic lymphatic leukemia
Ans. a

207. In increase in α_2-Globulin with loss of albumin in urine suggests
(a) Primary immune deficiency (b) Nephrotic syndrome
(c) Cirrhosis of liver (d) Multiple myeloma **Ans. b**

208. The normal levels of prothrombin time is about
(a) 2 sec (b) 4 sec
(c) 14 sec (d) 10–16 sec **Ans. c**

209. In obstructive jaundice prothrombin time
(a) Remains normal
(b) Decreases
(c) Responds to vit K and becomes normal
(d) Responds to vit K and increases **Ans. c**

210. In parenhymatous liver disease the prothrombin time
(a) Remains normal (b) Increases
(c) Decreases (d) Responds to Vit K **Ans. b**

211. Urea clearance test is used to determine the
(a) Glomerular filtration rate
(b) Renal plasma flow
(c) Ability of kidney to concentrate the urine
(d) Measurement of tubular mass **Ans. a**

212. Average maximum urea clearance is

(a) 30 ml (b) 50 ml

(c) 75 ml (d) 90 ml **Ans. c**

213. The average normal value for standard urea clearance is

(a) 20 ml (b) 30 ml

(c) 40 ml (d) 54 ml **Ans. d**

214. Urea clearance is lowered in

(a) Acute nephritis

(b) Pneumonia

(c) Early stage of nephritic syndrome

(d) Benign hypertension **Ans. a**

215. Glomerular filtration rate can be measured by

(a) Endogenous creatinine clearance

(b) Para-aminohippurate test

(c) Addis test

(d) Mosenthal test **Ans. a**

216. At normal levels of creatinine in the blood, this metabolite is

(a) Filtered at the glomerulus but neither secreted nor reabsorbed by the tubule

(b) Secreted by the tubule

(c) Reabsorbed by the tubule

(d) Secreted and reabsorbed by tubule **Ans. a**

217. The normal values for creatinine clearance varies from

(a) 20–40 ml/min (b) 40–60 ml/min

(c) 70–85 ml/min (d) 95–105 ml/min **Ans. d**

218. Measurement of insulin clearance test is a measure of

(a) Glomerular filtration rate (b) Filtration factor

(c) Renal plasma flow (d) Tubular secretary mass **Ans. a**

219. The polysaccharide insulin is

(a) Filtered at the glomerulus but neither secreted nor reabsorbed by the tubule

(b) Filtered at the glomerulus and secreted by the tubule

(c) Filtered at the glomerulus and reabsorbed by the tubule

(d) Filtered at the glomerulus, secreted and reabsorbed by the tubule

Ans. a

220. Normal insulin clearance is

(a) 40 ml/1.73 sqm (b) 60 ml/1.73 sqm

(c) 80 ml/1.73 sqm (d) 120 ml/1.73 sqm **Ans. d**

221. Creatinine EDTA clearance is a test to measure

(a) Renal plasma flow (b) Filtration fraction

(c) Glomerular filtration rate (d) Tubular function **Ans. c**

222. The end products of saponification

(a) glycerol (b) acid

(c) soap (d) Both (a) and (c) **Ans. d**

223. The normal PAH clearance for a surface area of 1.73 sqm. is

(a) 200 ml/min (b) 300 ml/min

(c) 400 ml/min (d) 574 ml/min **Ans. d**

224. Para amino hippurate is

(a) Filtered at glomeruli and secreted by the tubules

(b) Filtered at glomeruli and not secreted by the tubules

(c) Filtered at glomeruli and reabsorbed completely

(d) Not removed completely during a single circulation of the blood through the kidney **Ans. a**

225. The T_m for PAH i.e. the maximal secretary capacity of the tubule for PAH can be used to gavge the

(a) Extent of tubular damage

(b) Impairment of the capacity of the tubule to perform osmotic work

(c) Impairment of renal plasma flow

(d) Glomerular filtration rate **Ans. a**

226. The normal Tm in mg/min/1.73 sqm for PAH is

(a) 20 (b) 40

(c) 60 (d) 80 **Ans. d**

227. The normal range of filtration factor in an adult is

(a) 0.10–0.15 (b) 0.16–0.21

(c) 0.25–0.30 (d) 0.35–0.40 **Ans. b**

228. The filtration factor tends to be normal in
- (a) Early essential hypertension
- (b) Malignant phase of hypertension
- (c) Glomerulonephritis
- (d) Acute nephritis **Ans. a**

229. The filtration factor is increased in
- (a) Glomerulonephritis
- (b) Malignant phase of hypertension
- (c) Early essential hypertension
- (d) Acute nephritis **Ans. b**

230. The filtration factor is decreased in
- (a) Glomerulonephritis
- (b) Early essential hypertension
- (c) Malignant phase of hypertension
- (d) Starvation **Ans. a**

231. Excretion of phenolsulphanpthalein (PSP) reflects
- (a) Glomerulonephritis
- (b) Maximaltabular excretory capacity
- (c) Filtration factor
- (d) Renal plasma flow **Ans. b**

232. Which of the following is a polyunsaturated fatty acid?
- (a) Palmitic acid
- (b) Palmitoleic acid
- (c) Linoleic acid
- (d) Oleic acid **Ans. c**

233. Which of the following is omega-3 polyunsaturated fatty acid?
- (a) Linoleic acid
- (b) α-Linolenic acid
- (c) γ-Linolenic acid
- (d) Arachidonic acid **Ans. b**

234. Triglycerides are
- (a) Heavier than water
- (b) Major constituents of membranes
- (c) Non-polar
- (d) Hydrophilic **Ans. c**

235. Cerebronic acid is present in
(a) Glycerophospholipids (b) Sphingophospholipids
(c) Galactosyl ceramide (d) Gangliosides **Ans. c**

236. Acylsphingosine is also known as
(a) Sphingomyelin (b) Ceramide
(c) Cerebroside (d) Sulphatide **Ans. b**

237. The highest phospholipids content is found in
(a) Chylomicrons (b) VLDL
(c) LDL (d) HDL **Ans. d**

238. The major lipid in chylomicrons is
(a) Triglycerides (b) Phospholipids
(c) Cholesterol (d) Free fatty acids **Ans. a**

239. Number of carbon atoms in cholesterol is
(a) 17 (b) 19
(c) 27 (d) 30 **Ans. c**

240. The lipoprotein richest in cholesterol is
(a) Chylomicrons (b) VLDL
(c) LDL (d) HDL **Ans. c**

241. The major storage form of lipids is
(a) Esterified cholesterol (b) Glycerophospholipids
(c) Triglycerides (d) Sphingolipids **Ans. c**

242. Cerebonic acid is present in
(a) Triglycerides (b) Cerebrosides
(c) Esterified cholestrol (d) Sphingomyelin **Ans. b**

243. The nitrogenous base in lecithin is
(a) Ethanolamine (b) Choline
(c) Serine (d) Betaine **Ans. b**

244. All the following are omega-6-fatty acids except
(a) Linoleic acid (b) α-Linolenic acid
(c) γ-Linolenic acid (d) Arachidonic acid **Ans. b**

245. All the following have 18 carbon atoms except
(a) Linoleic acid (b) Linolenic acid
(c) Arachidonic acid (d) Stearic acid **Ans. c**

246. A 20-carbon fatty acid among the following is
(a) Linoleic acid (b) α -Linolenic acid
(c) β -Linolenic acid (d) Arachidonic acid **Ans. d**

247. Triglycerides are transported from liver to extrahepatic tissues by
(a) Chylomicrons (b) VLDL
(c) HDL (d) LDL **Ans. b**

248. Cholesterol is transported from liver to extrahepatic tissues by
(a) Chylomicrons (b) VLDL
(c) HDL (d) LDL **Ans. d**

249. Elevated plasma level of the following projects against atherosclerosis
(a) Chylomicrons (b) VLDL
(c) HDL (d) LDL **Ans. c**

250. All the following amino acids are nonessential except
(a) Alanine (b) Histidine
(c) Cysteine (d) Proline **Ans. b**

251. Sulphydryl group is present in
(a) Cysteine (b) Methionine
(c) Both (a) and (b) (d) None of these **Ans. a**

252. Oligosaccharide-pyrophosphoryl dolichol is required for the synthesis of
(a) N-linked glycoproteins (b) O-linked glycoproteins
(c) GPI-linked glycoproteins (d) All of these **Ans. a**

253. In N-linked glycoproteins, oligosaccharide is attached to protein through its
(a) Asparagine residue (b) Glutamine residue
(c) Arginine residue (d) Lysine residue **Ans. a**

254. De novo synthesis of fatty acids occurs in
(a) Cytosol (b) Mitochondria
(c) Microsomes (d) All of these **Ans. a**

255. Acyl Carrier Protein contains the vitamin
(a) Biotin (b) Lipoic acid
(c) Pantothenic acid (d) Folic acid **Ans. c**

256. Which of the following is required as a reductant in fatty acid synthesis?
(a) NADH (b) NADPH
(c) $FADH_2$ (d) $FMNH_2$ **Ans. b**

257. Hepatic liponenesis is stimulated by
(a) cAMP (b) Glucagon
(c) Epinephrine (d) Insulin **Ans. d**

258. De novo synthesis of fatty acids requires all of the following except
(a) Biotin (b) NADH
(c) Panthothenic acid (d) ATP **Ans. d**

259. Acetyl CoA carboxylase regulates fatty acid synthesis by which of the following mechanism?
(a) Allosteric regulation (b) Covalent modification
(c) Induction and repression (d) All of these **Ans. d**

260. β-Oxidation of fatty acids requires all the following coenzymes except
(a) CoA (b) FAD
(c) NAD (d) NADP **Ans. d**

261. Which of the following can be oxidized by β-oxidation pathway?
(a) Saturated fatty acids (b) Monosaturated fatty acids
(c) Polyunsaturated fatty acids (d) All of these **Ans. d**

262. Propionyl CoA is formed on oxidation of
(a) Monounsaturated fatty acids
(b) Polyunsaturated fatty acids
(c) Fatty acids with odd number of carbon atoms
(d) None of these **Ans. c**

263. An enzyme required for the synthesis of ketone bodies as well as cholesterol is
(a) Acetyl CoA carboxylase (b) HMG CoA synthetase
(c) HMG CoA reductase (d) HMG CoA lyase **Ans. b**

264. Ketone bodies are synthesized in
(a) Adipose tissue (b) Liver
(c) Muscles (d) Brain **Ans. b**

265. All the following statements about ketone bodies are true except

(a) Their synthesis increases in diabetes mellitus

(b) They are synthesized in mitchondria

(c) They can deplete the alkali reserve

(d) They can be oxidized in the liver **Ans. d**

266. All the following statements about carnitine are true except

(a) It can be synthesised in the human body

(b) It can be synthesized from methionine and lysine

(c) It is required for transport of short chain fatty acids into mitochondria

(d) Its deficiency can occur due to haemodialysis **Ans. c**

267. Which of the following can be synthesized in the human body if precursors are available?

(a) Oleic acid (b) Palmitoleic acid

(c) Arachidonic acid (d) All of these **Ans. d**

268. All the following can be oxidized by β-oxidation except

(a) Palmitic acid

(b) Phytanic acid

(c) Linoleic acid

(d) Fatty acids having an odd number of carbon atoms **Ans. b**

269. Anti-inflammatory corticosteroids inhibit the synthesis of

(a) Leukotrienes (b) Prostaglandins

(c) Thromboxanes (d) All of these **Ans. d**

270. Diets having a high ratio of polyunsaturated : saturated fatty acids can cause

(a) Increase in serum triglycerides

(b) Decrease in serum cholesterol

(c) Decrease in serum HDL

(d) Skin lesions **Ans. b**

271. Thromboxanes cause

(a) Vasodilation (b) Bronchoconstriction

(c) Platelet aggregation (d) All of these **Ans. c**

272. Prostaglandins lower camp in
(a) Adipose tissue (b) Lungs
(c) Platelets (d) Adenohypophysis **Ans. a**

273. Slow reacting Substance of Anaphylaxis is a mixture of
(a) Prostaglandins (b) Prostacyclins
(c) Thromboxanes (d) Leukotrienes **Ans. d**

274. Dipalmitoyl lecithin acts as
(a) Platelet activating factor
(b) Second messenger for hormones
(c) Lung surfactant
(d) Anti-ketogenic compound **Ans. c**

275. Reichert-Meissl number
(a) 0.1 N KOH (b) 0.5 KOH
(c) 0.1 N NaOH (d) 0.5 NaOH **Ans. a**

276. In glycerophospholipids, a polyunsaturated fatty acid is commonly attached to which of the following carbon atom of glycerol?
(a) Carbon 1 (b) Carbon 2
(c) Both (a) and (b) (d) None of these **Ans. b**

277. Lysolecithin is formed from lecithin by removal of
(a) Fatty acid from position 1 (b) Fatty acid from position 2
(c) Phosphorylcholine (d) Choline **Ans. b**

278. Sphingosine is synthesized from
(a) Palmitoyl CoA and Choline
(b) Palmitoyl CoA and ethanolamine
(c) Palmitoyl CoA and serine
(d) Acetyl CoA and choline **Ans. c**

279. For synthesis of sphingosine, all the following coenzymes are required except
(a) Pyridoxal phosphate (b) NADPH
(c) FAD (d) NAD **Ans. d**

280. Cerebrosides contain all the following except
(a) Galactose (b) Sulphate
(c) Sphingosine (d) Fatty acid **Ans. b**

281. Niemann-Pick disease results from deficiency of
(a) Ceramidase (b) Sphingomyelinase
(c) Arylsulphatase A (d) Hexosaminidase A **Ans. b**

282. Chylomicron remnants are catabolised in
(a) Intestine (b) Adipose tissue
(c) Liver (d) Liver and intestine **Ans. c**

283. VLDL remnant may be converted into
(a) VLDL (b) LDL
(c) HDL (d) Chylomicrons **Ans. c**

284. Receptors for chylomicron remnants are
(a) Apo A specific (b) Apo B-48 specific
(c) Apo C specific (d) Apo E specific **Ans. b**

285. LDL receptor is specific for
(a) Apo B-48 and Apo B 100 (b) Apo B-48 and Apo E
(c) Apo B-100 and Apo D (d) Apo B-100 and Apo D **Ans. c**

286. Nascent HDL of intestinal origin lacks
(a) Apo A (b) Apo C
(c) Apo E (d) Apo C and Apo E **Ans. d**

287. HDL is synthesized in
(a) Adipose tissue (b) Liver
(c) Intestine (d) Liver and intestine **Ans. d**

288. Nascent HDL of intestinal origin acquires Apo C and Apo E from
(a) Chylomicrons (b) VLDL
(c) LDL (d) HDL of the hepatic origin
Ans. d

289. Heparin releasable hepatic lipase converts
(a) VLDL remnants into LDL (b) Nascent HDL into HDL
(c) HDL_2 into HDL_3 (d) HDL_3 into HDL_2 **Ans. c**

290. Activated lecithin cholesterol acyl transferase is essential for the conversion of
(a) VLDL remnants into LDL (b) Nascent HDL into HDL
(c) HDL_2 into HDL_3 (d) HDL_3 into HDL_2 **Ans. b**

291. Fatty liver may be caused by
(a) Deficiency of methionine (b) Puromycin
(c) Chronic alcoholism (d) All of these **Ans. d**

292. Alcohol dehydrogenase converts ethanol into
(a) Acetyl CoA (b) Acetaldehyde
(c) Acetate (d) CO_2 and H_2O **Ans. b**

293. Lipids are stored in the body mainly in the form of
(a) Phospholipids (b) Glycolipids
(c) Triglycerides (d) Fatty acids **Ans. c**

294. Lipid stores are mainly present in
(a) Liver (b) Brain
(c) Muscles (d) Adipose tissue **Ans. d**

295. Glycerol is converted into glycerol-3- phosphate by
(a) Thiokinase (b) Triokinase
(c) Glycerol kinase (d) All of these **Ans. c**

296. In adipose tissue, glycerol-3-phosphate required for the synthesis of triglycerides comes mainly from
(a) Hydrolysis of pre-existing triglycerides
(b) Hydrolysis of phospholipids
(c) Dihydroxyacetone phosphate formed in glycolysis
(d) Free glycerol **Ans. c**

297. Glycerol released from adipose tissue by hydrolysis of triglycerides is mainly
(a) Taken up by liver
(b) Taken up by extrahepatic tissues
(c) Reutilised in adipose tissue
(d) Excreted from the body **Ans. a**

298. Free glycerol cannot be used for triglyceride synthesis in
(a) Liver (b) Kidney
(c) Intestine (d) Adipose tissue **Ans. d**

299. Adipose tissue lacks
(a) Hormone-sensitive lipase
(b) Glycerol kinase

(c) cAMP-dependent protein kinase
(d) Glycerol-3-phosphate dehydrogenase **Ans. b**

300. A digestive secretion that does not contain any digestive enzyme is
(a) Saliva (b) Gastric juice
(c) Pancreatic juice (d) Bile **Ans. d**

301. Saliva contains a lipase which acts on triglycerides having
(a) Short chain fatty acids (b) Medium chain fatty acids
(c) Long chain fatty acids (d) All of these **Ans. a**

302. Salivary lipase hydrolyses the ester bond at
(a) Position 1 of triglycerides (b) Position 2 of triglycerides
(c) Position 3 of triglycerides (d) All of these **Ans. c**

303. Salivary lipase converts dietary triglycerides into
(a) Diglycerides and fatty acids (b) Monoglycerides and fatty acids
(c) Glycerol and fatty acids (d) All of these **Ans. a**

304. Pancreatic lipase requires for its activity
(a) Co-lipase (b) Bile salts
(c) Phospholipids (d) All of these **Ans. d**

305. Pancreatic lipase converts triacylglycerols into
(a) 2, 3-Diacylglycerol (b) 1-Monoacylglycerol
(c) 2-Monoacylglycerol (d) 3-Monoacylglycerol **Ans. c**

306. Oxidation of fatty acids occurs
(a) In the cytosol
(b) In the matrix of mitochondria
(c) On inner mitochondrial membrane
(d) On the microsomes **Ans. b**

307. Activation of fatty acids requires all the following except
(a) ATP (b) Coenzyme A
(c) Thiokinase (d) Carnitine **Ans. d**

308. Mitochondrial thiokinase acts on
(a) Short chain of fatty acids (b) Medium chain fatty acids
(c) Long chain fatty acids (d) All of these **Ans. a**

309. Carnitine is required for the transport of

(a) Triglycerides out of liver

(b) Triglycerides into mitochondria

(c) Short chain fatty acids into mitochondria

(d) Long chain fatty acids into mitochondria **Ans. d**

310. Carnitine acylcarnitine translocase is present

(a) In the inner mitochondrial membrane

(b) In the mitochondrial matrix

(c) On the outer surface of inner mitochondrial membrane

(d) On the inner surface of inner mitochondrial membrane **Ans. a**

311. Net ATP generation on complete oxidation of stearic acid is

(a) 129 (b) 131

(c) 146 (d) 148 **Ans. c**

312. Propionyl CoA formed oxidation of fatty acids having an odd number of carbon atoms is converted into

(a) Acetyl CoA (b) Acetoacetyl CoA

(c) D-Methylmalonyl CoA (d) Butyryl CoA **Ans. c**

313. α-Oxidation of fatty acids occurs mainly in

(a) Liver (b) Brain

(c) Muscles (d) Adipose tissue **Ans. b**

314. Refsum's disease results from a defect in the following pathway except

(a) Alpha-oxidation of fatty acids

(b) Beta-oxidation of fatty acids

(c) Gamma-oxidation of fatty acids

(d) Omega-oxidation of fatty acids **Ans. a**

315. The end product of omega-oxidation of fatty acids having an even number of carbon atoms is

(a) Adipic acid (b) Suberic acid

(c) Both (a) and (b) (d) None of these **Ans. c**

316. De novo synthesis of fatty acids is catalysed by a multi-enzyme complex which contains

(a) One-SH group (b) Two-SH groups

(c) Three-SH groups (D) Four-SH groups **Ans. d**

317. Fat depots are located in
(a) Intermuscular connective tissue
(b) Mesentary
(c) Omentum
(d) All of these **Ans. d**

318. Salivary lipase is secreted by
(a) Parotid glands (b) Sub-maxillary glands
(c) Dorsal surface of tongue (d) None of these **Ans. c**

319. Co-lipase is a
(a) Bile salt (b) Vitamin
(c) Protein (d) Phospholipid **Ans. c**

320. Plasma becomes milky
(a) Due to high level of HDL (b) Due to high level of LDL
(c) During fasting (d) After a meal **Ans. d**

321. Mitochondrial membrane is permeable to
(a) Short chain fatty acids (b) Medium chain fatty acids
(c) Long chain fatty acids (d) All of these **Ans. a**

322. During each cycle of β-oxidation
(a) One carbon atom is removed from the carboxyl end of the fatty acid
(b) One carbon atom is removed from the methyl end of the fatty acid
(c) Two carbon atoms are removed from the carboxyl end of the fatty acid
(d) Two carbon atoms are removed from the methyl end of the fatty acid **Ans. c**

323. Net generation of energy on complete oxidation of palmitic acid is
(a) 129 ATP equivalents (b) 131 ATP equivalents
(c) 146 ATP equivalents (d) 148 ATP equivalents **Ans. d**

324. Net generation of energy on complete oxidation of a 17-carbon fatty acid is
(a) Equal to the energy generation from a 16-carbon fatty acid
(b) Equal to the energy generation from an 18-carbon fatty acid
(c) Less than the energy generation from a 16-carbon fatty acid
(d) In between the energy generation from a 16-carbon fatty acid and an 18-carbon fatty acid **Ans. c**

325. Net energy generation on complete oxidation of linoleic acid is
(a) 148 ATP equivalents (b) 146 ATP equivalents
(c) 144 ATP equivalents (d) 142 ATP equivalents **Ans. d**

326. Extramitochondrial synthesis of fatty acids occurs in
(a) Mammary glands (b) Lungs
(c) Brain (d) All of these **Ans. d**

327. One functional sub-unit of multi-enzyme complex for de novo synthesis of fatty acids contains
(a) One –SH group (b) Two –SH groups
(c) Three –SH groups (d) Four –SH groups **Ans. b**

328. NADPH required for fatty acid synthesis can come from
(a) Hexose monophosphate shunt
(b) Oxidative decarboxylation of malate
(c) Extramitochondrial oxidation of isocitrate
(d) All of these **Ans. d**

329. Fatty liver may be prevented by all of the following except
(a) Choline (b) Betaine
(c) Methionine (d) Ethionine **Ans. d**

330. Human desaturase enzyme system cannot introduce a double bond in a fatty acid beyond
(a) Carbon 9 (b) Carbon 6
(c) Carbon 5 (d) Carbon 3 **Ans. a**

331. Which of the following lipid is absorbed actively from intestines?
(a) Glycerol (b) Cholesterol
(c) Monoacylglycerol (d) None of these **Ans. d**

332. C_{22} and C_{24}, fatty acids required for the synthesis of sphingolipids in brain are formed by
(a) De novo synthesis
(b) Microsomal chain elongation
(c) Mitochondrial chain elongation
(d) All of these **Ans. b**

333. Sphingomyelins
(a) Phospholipids (b) Nitrolipids
(c) Alcohols (d) None of these **Ans. a**

334. All of the following statements about hypoglycin are true except
(a) It is a plant toxin
(b) It causes hypoglycaemia
(c) It inhibits oxidation of short chain fatty acids
(d) It inhibits oxidation of long chain fatty acids **Ans. a**

335. Synthesis of prostaglandins is inhibited by
(a) Glucocorticoids (b) Aspirin
(c) Indomethacin (d) All of these **Ans. d**

336. Lipo-oxygenase is required for the synthesis of
(a) Prostaglandins (b) Leukotrienes
(c) Thromboxanes (d) All of these **Ans. b**

337. All of the following statements about multiple sclerosis are true except
(a) There is loss of phospholipids from white matter
(b) There is loss of sphingolipids from white matter
(c) There is loss of esterified cholesterol from white matter
(d) White matter resembles gray matter in composition **Ans. c**

338. After entering cytosol, free fatty acids are bound to
(a) Albumin (b) Globulin
(c) Z-protein (d) None of these **Ans. c**

339. Release of free fatty acids from adipose tissue is increased by all of the following except
(a) Glucagon (b) Epinephrine
(c) Growth hormone (d) Insulin **Ans. d**

340. All the following statements about brown adipose tissue are true except
(a) It is rich in cytochromes
(b) It oxidizes glucose and fatty acids
(c) Oxidation and phosphorylation are tightly coupled in it
(d) Dinitrophenol has no effect on it **Ans. c**

341. Lovastatin and mevastatin lower
(a) Serum triglycerides (b) Serum cholesterol
(c) Serum phospholipids (d) All of these **Ans. b**

342. Lovastatin is a
(a) Competitive inhibitor of acetyl CoA carboxylase
(b) Competitive inhibitor of HMG CoA synthetase
(c) Non-competitive inhibitor of HMG CoA reductase
(d) Competitive inhibitor of HMG CoA reductase **Ans. d**

343. Abetalipoproteinaemia occurs due to a block in the synthesis of
(a) Apoprotein A (b) Apoprotein B
(c) Apoprotein C (d) Cholesterol **Ans. b**

344. All of the following statements about Tangier disease are true except
(a) It is a disorder of HDL metabolism
(b) Its inheritance is autosomal recessive
(c) Apoproteins A-I and A-II are not synthesised
(d) Plasma HDL is increased **Ans. d**

345. Genetic deficiency of lipoprotein lipase causes hyperlipoproteinaemia of following type
(a) Type I (b) Type IIa
(c) Type IIb (d) Type V **Ans. a**

346. Chylomicrons are present in fasting blood samples in hyperlipoproteinaemia of following types
(a) Types I and IIa (b) Types IIa and IIb
(c) Types I and V (d) Types IV and V **Ans. c**

347. Glutathione is a constituent of
(a) Leukotriene A_4 (b) Thromboxane A_1
(c) Leukotriene C_4 (d) None of these **Ans. c**

348. Prostaglandins are inactivated by
(a) 15-Hydroxyprostaglandin dehydrogenase
(b) Cyclo-oxygenase
(c) Lipo-oxygenase
(d) None of these **Ans. a**

349. Phenylbutazone and indomethacin inhibit
(a) Phospholipase A_1 (b) Phospholipase A_2
(c) Cyclo-oxygenase (d) Lipo-oxygenase **Ans. c**

350. Prostaglandins stimulate

(a) Aggregation of platelets (b) Lipolysis in adipose tissue
(c) Bronchodilatation (d) Gastric acid secretion **Ans. c**

351. For extramitochondrial fatty acid synthesis, acetyl CoA may be obtained from

(a) Citrate (b) Isocitrate
(c) Oxaloacetate (d) Succinate **Ans. a**

352. Fluidity of membranes is increased by the following constituent except

(a) Polyunsaturated fatty acids (b) Saturated fatty acids
(c) Integral proteins (d) Cholesterol **Ans. a**

353. Transition temperature of membranes may be affected by the following constituent of membranes

(a) Peripheral proteins (b) Integral proteins
(c) Cholesterol (d) Oligosachharides **Ans. c**

354. Acetyl CoA formed from pyruvate can be used for the synthesis of all the following except

(a) Glucose (b) Fatty acids
(c) Cholesterol (d) Steroid hormones **Ans. a**

355. Which of the following can be used as a source of energy in extrahepatic tissues?

(a) Acetoacetate (b) Acetone
(c) Both (a) and (b) (b) None of these **Ans. a**

356. Anti-inflammatory corticosteroids inhibit

(a) Phospholipase A_1 (b) Phospholipase A_2
(c) Cyclo-oxygenase (d) Lipo-oxygenase **Ans. b**

357. Cyclo-oxygenase is involved in the synthesis of

(a) Prostaglandins (b) Thromboxanes
(c) Both (a) and (b) (d) None of these **Ans. c**

358. Leukotrienes cause

(a) Increase in capillary permeability
(b) Aggregation of platelets
(c) Bronchodilatation
(d) None of these **Ans. a**

359. Prostaglandins decrease all of the following except
(a) Gastric acid secretion (b) Blood pressure
(c) Uterine contraction (d) Platelet aggregation **Ans. c**

360. Hypocholesterolaemia can occur in
(a) Hyperthyroidism (b) Nephrotic syndrome
(c) Obstructive jaundice (d) Diabetes mellitus **Ans. a**

361. De novo synthesis and oxidation of fatty acids differ in the following respect
(a) Synthesis occurs in cytosol and oxidation in mitochondria
(b) Synthesis is decreased and oxidation increased by insulin
(c) NADH is required in synthesis and FAD in oxidation
(d) Malonyl CoA is formed during oxidation but not during synthesis **Ans. a**

362. Free fatty acids released from adipose tissue are transported in blood by
(a) Albumin (b) VLDL
(c) LDL (d) HDL **Ans. a**

363. β -Galactosidase is deficient in
(a) Fabry's disease
(b) Krabbe's disease
(c) Gaucher's disease
(d) Metachromatic leukodystrophy **Ans. b**

364. The enzyme deficient in metachromatic leukodystrophy is
(a) Arylsulphatase A (b) Hexosaminidase A
(c) Ceramidase (d) Sphingomyelinase **Ans. a**

365. All of the following statements about generalized gangliosidosis are true except
(a) It results from deficiency of G_{M1}-β-Gangliosidase
(b) Breakdown of G_{M1} ganglioside is impaired
(c) G_{M2} ganglioside accumulates in liver and elsewhere
(d) It leads to mental retardation **Ans. c**

366. Hexosaminidase A is deficient in
(a) Tay-Sach's disease (b) Gaucher's disease
(c) Niemann-Pick disease (d) Fabry's disease **Ans. a**

367. Mental retardation occurs in
(a) Tay-Sach's disease (b) Gaucher's disease
(c) Niemann-Pick disease (d) All of these **Ans. d**

368. The enzyme deficient in Fabry's disease is
(a) α-Galactosidase (B) β-Galactosidase
(c) α-Glucosidase (d) β-Glucosidase **Ans. a**

369. Highest protein content amongst the following is present in
(a) Wheat (b) Rice
(c) Pulses (d) Soyabean **Ans. d**

370. Daily protein requirement of an adult man is
(a) 0.5 gm/kg of body weight (b) 0.8 gm/kg of body weight
(c) 1.0 gm/kg of body weight (d) 1.5 gm/kg of body weight
Ans. c

371. Daily protein requirement of an adult woman is
(a) 0.5 gm/kg of body weight (b) 0.8 gm/kg of body weight
(c) 1.0 gm/kg of body weight (d) 1.5 gm/kg of body weight
Ans. c

372. Cysteine can partially meet the requirement of
(a) Phenylalanine (b) Threonine
(c) Methionine (d) None of these **Ans. c**

373. Invisible fat is present in
(a) Milk (b) Coconut oil
(c) Groundnut oil (d) Hydrogenated oils **Ans. a**

374. Visible fat is present in
(a) Milk (b) Pulses
(c) Coconut oil (d) Egg yolk **Ans. c**

375. Fat content of eggs is about
(a) 7% (b) 10%
(c) 13% (d) 16% **Ans. c**

376. Fat content of pulses is about
(a) 5% (b) 10%
(c) 15% (d) 20% **Ans. a**

377. Predominant fatty acids in meat are

(a) Saturated (b) Monounsaturated

(c) Polyunsaturated (d) Mono and poly-unsaturated

Ans. a

378. Oils having more than 50 % polyunsaturated fatty acids include all of the following except

(a) Groundnut oil (b) Soyabean oil

(c) Sunflower oil (d) Safflower oil **Ans. a**

379. Cholesterol is present in all of the following except

(a) Egg (b) Fish

(c) Milk (d) Pulses **Ans. d**

380. Which of the following has the highest cholesterol content?

(a) Meat (b) Fish

(c) Butter (d) Milk **Ans. c**

381. Which of the following has the highest cholesterol content?

(a) Egg yolk (b) Egg white

(c) Meat (d) Fish **Ans. a**

382. The following contains the least cholesterol

(a) Milk (b) Meat

(c) Butter (d) Cheese **Ans. a**

383. Which of the following constitutes fibre or roughage in food?

(a) Cellulose (b) Pectin

(c) Inulin (d) All of these **Ans. d**

384. The starch content of wheat is about

(a) 50% (b) 60%

(c) 70% (d) 80% **Ans. c**

385. The starch content of pulses is about

(a) 50% (b) 60%

(c) 70% (d) 80% **Ans. b**

386. A significant source of starch among vegetables is

(a) Radish (b) Spinach

(c) Potato (d) Cauliflower **Ans. c**

387. The cyclic ring present in all the steroids

(a) Cyclopentano perhydrophenanthrene
(b) Nitropentano
(c) Both (a) and (b)
(d) None of these **Ans. a**

388. In Ames' assay, addition of a carcinogen to the culture medium allows *S. typhimurium* to grow

(a) In the presence of histidine (b) In the presence of arginine
(c) In the absence of histidine (d) In the absence of arginine
Ans. c

389. In Ames' assay, liver homogenate is included in the culture medium because

(a) It converts pro-carcinogens into carcinogens
(b) Liver can metabolise histidine
(c) Salmonella mainly infects liver
(d) Liver is very susceptible to cancer **Ans. a**

390. Bile pigments are present and urobilinogen absent in urine in

(a) Haemolytic jaundice (b) Hepatocellular jaundice
(c) Obstructive jaundice (d) Crigler-Najjar syndrome
Ans. c

391. Bile pigments are absent and urobilinogen increased in urine in

(a) Haemolytic jaundice (b) Hepatocellular jaundice
(c) Obstructive jaundice (d) Rotor's syndrome **Ans. a**

392. In obstructive jaundice, urine shows

(a) Absence of bile pigments and urobilinogen
(b) Presence of bile pigments and urobilinogen
(c) Absence of bile pigments and presence of urobilinogen
(d) Presence of bile pigments and absence of urobilinogen **Ans. d**

393. In haemolytic jaundice, urine shows

(a) Absence of bile pigments and urobilinogen
(b) Presence of bile pigments and urobilinogen
(c) Absence of bile pigments and presence of urobilinogen
(d) Presence of bile pigments and absence of urobilinogen **Ans. c**

394. Serum albumin may be decreased in

(a) Haemolytic jaundice (b) Hepatocellular jaundice
(c) Obstructive jaundice (d) All of these **Ans. b**

395. Normal range of serum albumin is

(a) 2.0–3.6 gm/dl (b) 2.0–3.6 mg/dl
(c) 3.5–5.5 gm/dl (d) 3.5–5.5 mg/dl **Ans. c**

396. Serum albumin: globulin ratio is altered in

(a) Gilbert's disease (b) Haemolytic jaundice
(c) Viral hepatitis (d) Stones in bile duct **Ans. c**

397. Esterification of cholesterol occurs mainly in

(a) Adipose tissue (b) Liver
(c) Muscles (d) Kidneys **Ans. b**

398. Galactose intolerance can occur in

(a) Haemolytic jaundice (b) Hepatocellular jaundice
(c) Obstructive jaundice (d) None of these **Ans. b**

399. Prothrombin is synthesised in

(a) Erythrocytes (b) Reticulo-endothelial cells
(c) Liver (d) Kidneys **Ans. c**

400. Prothrombin time remains prolonged even after parenterals administration of vitamin K in

(a) Haemolytic jaundice (b) Liver damage
(c) Biliary obstruction (d) Steatorrhoea **Ans. b**

401. All the following statements about obstructive jaundice are true except

(a) Conjugated bilirubin in serum is normal
(b) Total bilirubin in serum is raised
(c) Bile salts are present in urine
(d) Serum alkaline phosphatase is raised **Ans. a**

402. All the following statements about obstructive jaundice are true except

(a) Prothrombin time may be prolonged due to impaired absorption of vitamin K
(b) Serum alkaline phosphatase may be raised due to increased release of the enzyme from liver cells

(c) Bile salts may enter systemic circulation due to biliary obstruction
(d) There is no defect in conjugation of bilirubin **Ans. b**

403. A test to evaluate detoxifying function of liver is
(a) Serum albumin: globulin ratio (b) Galactose tolerance test
(c) Hippuric acid test (d) Prothrombin time **Ans. c**

404. Hippuric acid is formed from
(a) Benzoic acid and alanine (b) Benzoic acid glycine
(c) Glucuronic acid and alanine (d) Glucuronic acid and glycine
Ans. b

405. An enzyme which is excreted in urine is
(a) Lactase dehydrogenase (b) Amylase
(c) Ornithine transcarbamoylase (d) None of these **Ans. c**

406. Serum gamma glutamyl transpeptidase is raised in
(a) Haemolytic jaundice (b) Myocardial infarction
(c) Alcoholic hepatitis (d) Acute cholecystitis **Ans. c**

407. Oliguria can occur in
(a) Diabetes mellitus (b) Diabetes insipidus
(c) Acute glomerulonephritis (d) Chronic glomerulonephritis
Ans. a

408. Urea clearance is the
(a) Amount of urea excreted per minute
(b) Amount of urea present in 100 ml of urine
(c) Volume of blood cleared of urea in one minute
(d) Amount of urea filtered by glomeruli in one minute **Ans. a**

409. Inulin clearance is a measure of
(a) Glomerular filtration rate (b) Tubular secretion flow
(c) Tubular reabsorption rate (d) Renal plasma flow **Ans. b**

410. Phenolsulphonephthalein excretion test is an indicator of
(a) Glomerular filtration (b) Tubular secretion
(c) Tubular reabsorption (d) Renal blood low **Ans. d**

411. Para-amino hippurate excretion test is an indicator of
(a) Glomerular filtration (b) Tubular secretion
(c) Tubular reabsorption (d) Renal plasma flow **Ans. d**

412. Renal plasma flow of an average adult man is

(a) 120–130 ml/minute (b) 325–350 ml/minute

(c) 480–52 ml/minute (d) 560–830 ml/minute **Ans. d**

413. Filtration fraction can be calculated from

(a) Standard urea clearance and PSP excretion

(b) Maximum urea clearance and PSP excretion

(c) Maximum urea clearance and PAH clearance

(d) Inulin clearance and PAH clearance **Ans. a**

414. Normal filtration fraction is about

(a) 0.2 (b) 0.4

(c) 0.6 (d) 0.8 **Ans. c**

415. Filtration fraction is increased in

(a) Acute glomerulonephritis (b) Chronic glomerulonephritis

(c) Hypertension (d) Hypotension **Ans. a**

416. Among the following, a test of Glomerular function is

(a) Urea clearance (b) PSP excretion test

(c) PAH clearance (d) Hippuric acid excretion test

Ans. d

417. Esters of fatty acids with higher alcohols other than glycerol are said to be

(a) Waxes (b) Fats

(c) Both (a) and (b) (d) None of these **Ans. a**

418. The combination of an amino alcohol, fatty acid and sialic acid form

(a) Phospholipids (b) Sulpholipids

(c) Glycolipids (d) Aminolipids **Ans. c**

419. Hydrolysis of fats by alkali is called

(a) Saponification number (b) Saponification

(c) Both (a) and (b) (d) None of these **Ans. b**

420. The number of milliliters of 0.1 N KOH required to neutralize the insoluble fatty acids from 5 gms of fat is called

(a) Acid number (b) Acetyl number

(c) Halogenation (d) Polenske number **Ans. d**

421. The rate of fatty acid oxidation is increased by
(a) Phospholipids (b) Glycolipids
(c) Aminolipids (d) All of these **Ans. a**

422. Lecithin contains a nitrogenous base named as
(a) Ethanolamine (b) Choline
(c) Inositol (d) All of these **Ans. b**

423. Lecithins contain an unsaturated fatty acid at position
(a) α (b) α and β
(c) β (d) None of these **Ans. c**

424. Lecithins are soluble in ordinary solvents except
(a) Benzene (b) Ethyl alcohol
(c) Methyl alcohol (d) Acetone **Ans. d**

425. Lecithins combine with protein to form
(a) Phosphoprotein (b) Mucoprotein
(c) Lipoprotein (d) Glycoprotein **Ans. c**

426. Instead of ester link plasmalogens possess an other link in position
(a) α- (b) β-
(c) γ- (d) None of these **Ans. b**

427. The alkyl radical in plasmalogen is an alcohol
(a) Saturated (b) Unsaturated
(c) Both (a) and (b) (d) None of these **Ans. a**

428. The concentration of sphingomyelins are increased in
(a) Gaucher's disease (b) Fabry's disease
(c) Fabrile disease (d) Niemann-Pick disease **Ans. b**

429. Sphingomyelins contain a complex amino alcohol named as
(a) Serine (b) Lysolecithin
(c) Sphingosine (d) Glycol **Ans. c**

430. The types of sphingomyelins are
(a) 1 (b) 3
(c) 4 (d) 5 **Ans. b**

431. Glycolipids contain an amino alcohol
(a) Sphingosine (b) Iso-sphingosine
(c) Both (a) and (b) (d) None of these **Ans. c**

432. Cerebrosides may also be classified as
(a) Sphingolipids (b) Sulpholipids
(c) Aminolipids (d) Glycolipids **Ans. a**

433. Gaucher's disease is characterized specially by the increase in
(a) Lignoceric acid (b) Nervonic acid
(c) Cerebomic acid (d) Hydroxynervonic acid **Ans. c**

434. Gangliosides are the glycolipids occurring in
(a) Brain (b) Liver
(c) Kidney (d) Muscle **Ans. a**

435. Lipoprotein present in cell membrane is by nature
(a) Hydrophilic (b) Hydrophobic
(c) Both (a) and (b) (d) None of these **Ans. a**

436. The density of lipoproteins increases as the protein content
(a) Increases
(b) Decreases
(c) Highly decreases
(d) Slightly and promptly decreases **Ans. a**

437. Lipoprotiens may be identified more accurately by means of
(a) Electrophoresis (b) Ultra centrifugation
(c) Centrifugation (d) Immunoelectrophoresis **Ans. d**

438. Very low density lipoproteins are also known as
(a) β-lipoproteins (b) Pre β – lipoproteins
(c) α-lipoproteins (d) None of these **Ans. b**

439. The protein moiety of lipoprotein is known as
(a) Apoprotein (b) Pre-protein
(c) Post-protein (d) Pseudoprotein **Ans. a**

440. The β-lipoprotein fraction increases in severe
(a) Diabetes Mellitus (b) Uremia
(c) Nephritis (d) Muscular dystrophy **Ans. a**

441. Δ^9 indicates a double bond between carbon atoms of the fatty acids
(a) 8 and 9 (b) 9 and 10
(c) 9 and 11 (d) 9 and 12 **Ans. b**

442. The number of carbon atoms in decanoic acid present in butter

(a) 6 (b) 8

(c) 10 (d) 12 **Ans. c**

443. The prostaglandins are synthesized from

(a) Arachidonic acid (b) Oleic acid

(c) Linoleic acid (d) Linolenic acid **Ans. a**

444. The Iodine number of essential fatty acids of vegetable oils

(a) High (b) Very high

(c) Very low (d) Low **Ans. d**

445. Cholesterol is a

(a) Animal sterol (b) M.F. $C_{27}H_{46}O$

(c) 5 methyl groups (d) All of these **Ans. d**

446. Waxes contain higher alcohols named as

(a) Methyl (b) Ethyl

(c) Phytyl (d) Cetyl **Ans. d**

447. Lieberman-Burchard reaction is performed to detect

(a) Cholesterol (b) Glycerol

(c) Fatty acid (d) Vitamin D **Ans. a**

448. Lipose present in the stomach cannot hydrolyze fats owing to

(a) Alkalinity (b) Acidity

(c) High acidity (d) Neutrality **Ans. c**

449. Fatty acids are oxidized by

(a) α -oxidation (b) β -oxidation

(c) ω -oxidation (d) All of these **Ans. d**

450. The fatty acids containing even number and odd number of carbon atoms as well as the unsaturated fatty acids are oxidized by

(a) α-oxidation (b) β-oxidation

(c) ω-oxidation (d) All of these **Ans. b**

451. Long chain fatty acids are first activated to acyl CoA in the

(a) Cytosol (b) Mitochondria

(c) Ribosomes (d) Microsome **Ans. a**

452. Long chain acyl CoA penetrates mitochondria in the presence of

(a) Palmitate (b) Carnitine
(c) Sorbitol (d) DNP **Ans. b**

453. Acyl-CoA dehydrogenase converts Acyl CoA to α - β unsaturated acyl-CoA in presence of the coenzyme

(a) NAD^+ (b) $NADP^+$
(c) ATP (d) FAD **Ans. d**

454. For the activation of long chain fatty acids the enzyme thiokinase requires the cofactor

(a) Mg^{++} (b) Ca^{++}
(c) Mn^{++} (d) K^+ **Ans. b**

455. ω-oxidation takes place by the hydroxylase in microsomes involving

(a) Cytochrome b (b) Cytochrome c
(c) Cytochrome p-4500 (d) Cytochrome a_3 **Ans. c**

456. Carboxylation of acetyl–CoA to malonyl CoA takes place in presence of

(a) FAD^+ (b) Biotin
(c) NAD^+ (d) $NADP^+$ **Ans. c**

457. Malonyl-CoA reacts with the central

(a) –SH group (b) $-NH_2$ group
(c) –COOH group (d) $-CH_2OH$ group **Ans. a**

458. Fatty acid synthesis takes place in the presence of the coenzyme

(a) NAD^+ (b) Reduced NAD
(c) $NADP^+$ (d) Reduced NADP **Ans. d**

459. Fatty acids are activated to acyl CoA by the enzyme thiokinase

(a) NAD^+ (b) $NADP^+$
(c) CoA (d) FAD^+ **Ans. c**

460. Phospholipids help the oxidation of

(a) Glycerol (b) Fatty acids
(c) Glycerophosphates (d) None of these **Ans. b**

461. The desaturation and chain elongation system of polyunsaturated fatty acids are greatly diminished in the absence of

(a) Insulin (b) Glycagon
(c) Epinephrine (d) Thyroxine **Ans. a**

462. Prostaglandins are liberated in the circulation by the stimulation of

(a) Anterior pituitary glands (b) Posterior pituitary glands

(c) Adrenal gland (d) Thyroid gland **Ans. c**

463. Prostaglandins have a common structure based on prostanoic acid which contains carbon atoms

(a) 12 (b) 16

(c) 18 (d) 20 **Ans. d**

464. The carbon chains of prostanoic acid are bonded at the middle of the chain by a

(a) 5-membered ring (b) 6-membered ring

(c) 8-membered ring (d) None of these **Ans. b**

465. All active prostaglandins have atleast one double bond between positions

(a) 7 and 8 (b) 9 and 10

(c) 11 and 12 (d) 13 and 14 **Ans. a**

466. The enzyme systems for lengthening and shortening for saturating and desaturating of fatty acids occur in

(a) Intestine (b) Muscle

(c) Kidney (d) Liver **Ans. d**

467. Which of the following are classified as essential fatty acids?

(a) Arachidonic acid (b) Oleic acid

(c) Acetic acid (d) Butyric acid **Ans. a**

468. Prostaglandins are synthesized in the body from

(a) Myristic acid (b) Arachidonic acid

(c) Stearic acid (d) Lignoceric acid **Ans. d**

469. All the following saturated fatty acids are present in buffer except

(a) Butyric acid (b) Caprylic acid

(c) Caproic acid (d) Capric acid **Ans. c**

470. Biological functions of lipids include

(a) Source of energy

(b) Insulating material

(c) Maintenance of cellular integrity

(d) All of these **Ans. d**

471. Saponification number is
(a) mg of KOH required to saponify one gm of fat or oil
(b) mg of KOH required to neutralize free fatty acids of one gms of fat
(c) mg of KOH required to neutralize the acetic acid obtained by saponification of one gm of fat after it has been acetylated
(d) None of these **Ans. c**

472. Lipids have the following properties
(a) Insoluble in water and soluble in fat solvent
(b) High energy content
(c) Structural component of cell membrane
(d) All of these **Ans. d**

473. Carbohydrate moiety in cerebrosides is
(a) Glucose (b) Sucrose
(c) Galactose (d) Maltose **Ans. b**

474. Which of the following is not an unsaturated fatty acid?
(a) Oleic acid (b) Stearic acid
(c) Linaoleic acid (d) Palmitic acid **Ans. b**

475. All the following are functions of prostaglandins except
(a) Lowering of B.P
(b) Introduction of labour
(c) Anti inflammatory
(d) Prevention of myocardial infraction **Ans. d**

476. Calorific value of lipids per gm is
(a) 4 Kcal (b) 8 Kcal
(c) 9 Kcal (d) None of these **Ans. c**

477. Fatty acid present in kerotin is
(a) Lignoceric acid (b) Cerebromic acid
(c) Nervonic acid (d) Hydroxynervonic acid **Ans. a**

478. All the following are ketones except
(a) Xylulose (b) Ribolose
(c) Erythrose (d) Fructose **Ans. c**

479. Saponification
(a) Hydrolysis of fats by alkali
(b) Hydrolysis of glycerol by liposes
(c) Esterification
(d) Reduction **Ans. a**

480. Number of ml of 0.1 N KOH required to neutralize fatty acids from 5 gms of fat
(a) Iodine number (b) Polenske number
(c) Reichert-Miessl number (d) None of these **Ans. b**

481. Hydrated density (HD) of lipoproteins is
(a) 0.94 gm/ml (b) 0.94–1.006 gm/ml
(c) 1.006–1.063 gm/ml (d) 1.063–1.21 gm/ml **Ans. d**

482. Saponification number indicates
(a) Unsaturation in fat (b) Average M.W of fatty acid
(c) Acetyl number (d) Acid number **Ans. b**

483. Acrolein Test is positive for
(a) Glycerol (b) Prostaglandins
(c) Carbohydrates (d) Proteins **Ans. a**

484. Iodine number denotes
(a) Degree of unsaturation (b) Saponification number
(c) Acid number (d) Acetyl number **Ans. a**

485. Maximum energy produced by
(a) Fats (b) Carbohydrates
(c) Proteins (d) Nucleic acids **Ans. a**

486. Lecithins are composed of
(a) Glycerol + Fatty acids + Phosphoric acid + Choline
(b) Glycerol + Fatty acids + Phosphoric acid + Ethanolamine
(c) Glycerol + Fatty acids + Phosphoric acid + Serine
(d) Glycerol + Fatty acids + Phosphoric acid + Beaine **Ans. a**

487. Sphingomyelins are composed of fatty acids, phosphoric acid and
(a) Sphingosine and choline (b) Glycerol and sphingosine
(c) Glycerol and Serine (d) Glycerol and Choline **Ans. a**

488. Depot fats of mammalian cells comprise mostly of
(a) Cholesterol (b) Cholesterol esters
(c) Triacyl glycerol (d) Phospholipids **Ans. c**

489. When choline of lecithine is replaced by ethanolamine the product is
(a) Sphingomyelin (b) Cephalin
(c) Plasmalogens (d) Lysolecithine **Ans. b**

490. Which of the following is a hydroxy fatty acid?
(a) Oleic acid (b) Ricinoleic acid
(c) Caproic acid (d) Stearic acid **Ans. b**

491. Acrolein test is answered by
(a) Cholesterol (b) Glycerol
(c) Glycosides (d) Sphingol **Ans. b**

492. The smell of fat turned rancid is due to
(a) Presence of vit E (b) Presence of quinones
(c) Phenols (d) Volatile fatty acids **Ans. d**

493. Phospholipids are important cell membrane components because
(a) They have glycerol
(b) They can form bilayers in water
(c) They have both polar and non polar portions
(d) They combine covalently with proteins **Ans. c**

494. Which one of the following is not a phospholipid?
(a) Lecithin (b) Plasmalogen
(c) Lysolecithin (d) Gangliosides **Ans. d**

495. A fatty acid which is not synthesized in human body and has to be supplied in the diet
(a) Palmitic acid (b) Oleic acid
(c) Linoleic acid (d) Stearic acid **Ans. c**

496. In cephalin, choline is replaced by
(a) Serine (b) Ethanolamine
(c) Betaine (d) Sphingosine **Ans. b**

497. The triacyl glycerol present in plasma lipoproteins are hydrolyzed by
(a) Linqual lipase (b) Pancreatic lipase
(c) Colipase (d) Lipoprotein lipase **Ans. d**

498. Amphiphatic lipids are
(a) Hydrophilic (b) Hydrophobic
(c) Both (a) and (b) (d) Lipophilic **Ans. c**

499. Which of the following is not essential fatty acid?
(a) Oleic acid (b) Linoleic acid
(c) Arachidonic acid (d) Linolenic acid **Ans. a**

500. The calorific value of lipid is
(a) 4.0 Kcal/gm (b) 6.0 Kcal/gm
(c) 9.0 Kcal/gm (d) 15 Kcal/gm **Ans. c**

501. Rancidity of butter is prevented by the addition of
(a) Vitamin D (b) Tocopherols
(c) Presence of priotin (d) Presence of 'Cu' **Ans. b**

502. Sphingomyelins on hydrolysis yields
(a) Glycerol, fatty acids, phosphoric acid and choline
(b) Glycerol, sphingosine, choline and fatty acids
(c) Sphingosine, phosphoric acid, Glycerol and inositol
(d) Sphingosine, fatty acids, phosphoric acid and choline **Ans. d**

503. Inherited deficiency of enzyme cerebrosidase produces
(a) Fabry's disease (b) Niemann pick disease
(c) Gaucher's disease (d) Tay-sach's disease **Ans. c**

504. Phosphatidic acid on hydrolysis yields
(a) Glycerol, fatty acids, phosphoric acid, choline
(b) Glycerol, fatty acids, phosphoric acid
(c) Glycerol, fatty acids, phosphoric acid, Glucose
(d) Sphingol, fatty acids, phosphoric acid **Ans. b**

505. The maximum number of double bonds present in essential fatty acid is
(a) 1 (b) 2
(c) 3 (d) 4 **Ans. d**

506. Cerebrosides are composed of
(a) Sphingosine, fatty acids, glycerol and phosphoric acid
(b) Sphingosine, fatty acids, galactose
(c) Glycerol, fatty acids, galactose
(d) Glycerol, fatty acids, galactose, sphingol **Ans. b**

507. Acetoacetic acid and β-OH butyric acid are formed as
(a) Kidneys (b) Heart
(c) Liver (d) Intestine **Ans. c**

508. Which amino acid is a lipotropic factor?
(a) Lysine (b) Leucine
(c) Tryptophan (d) Methionine **Ans. d**

509. The class of lipoproteins having a beneficial effect in atherosclerosis is
(a) Low density of lipoproteins (b) Very low density lipoproteins
(c) High density lipoproteins (d) Chylomicrons **Ans. c**

510. Cholesterol is the precursor for the biosynthesis of
(a) Fatty acid (b) Prostaglandins
(c) Bile acids (d) Sphingomyelin **Ans. c**

511. Which of the following condition is characterized by ketonuria but without glycosuria?
(a) Diabetes mellitus (b) Diabetes insipidus
(c) Prolonged starvation (d) Addison's disease **Ans. c**

512. Ketone bodies are formed in
(a) Kidney (b) Liver
(c) Heart (d) Intestines **Ans. b**

513. Changes in serum high density lipoproteins (HDL) are more truly reflected by those of
(a) HDL-1 (b) HDL-2
(c) HDL-3 (d) HDL_C **Ans. b**

514. Mitochondrial lipogenesis requires
(a) bicarbonate (b) biotin
(c) acetyl CoA carboxylase (d) NADPH **Ans. d**

515. Fatty acids having chain length of 10 carbon atoms enter the
(a) Portal ciruclation (b) Lacteals
(c) Systemic circulation (d) Colon **Ans. a**

516. A soluble system for synthesis of fatty acids have been isolated from avian liver, required for the formation of long chain fatty acids by this system is
(a) ATP (b) Acetyl CoA
(c) NADPH (d) All of these **Ans. d**

517. Most animal tissues contain appreciable amounts of lipid, when in the form of depot fat it consists largely of

(a) Cholesterol ester (b) Phosphatides
(c) Chylomicrons (d) Triacylglycerol **Ans. d**

518. A fatty acid not synthesized in man is

(a) Oleic (b) Palmitic
(c) Linoleic (d) Stearic **Ans. c**

519. The 'free fatty acids' (FFA) of plasma

(a) metabolically inert
(b) mainly bound to β-lipoproteins
(c) stored in the fat
(d) mainly bound to serum albumin **Ans. d**

520. Adipose tissue which is a store house for triacyl glycerol synthesis the same using

(a) The glycerol released by hydrolysis of triacyl glycerol
(b) The glycerol-3-phosphate obtained in the metabolism of glucose
(c) 2-phosphoglycerate
(d) 3-phosphoglycerate **Ans. b**

521. Increase in blood of this class of lipoproteins is beneficial to ward off coronary heart disease

(a) HDL (b) LDL
(c) VLDL (d) IDL **Ans. a**

522. In the extra mitochondrial synthesis of fatty acids, CO_2 is utilized

(a) To keep the system anaerobic and prevent regeneration of acetyl CoA
(b) In the conversion of malonyl to CoA hydroxybutyryl CoA
(c) In the conversion of acetyl CoA to malonyl CoA
(d) In the formation of acetyl CoA from 1 carbon intermediates **Ans. c**

523. Current concepts concerning the intestinal absorption of triacylglycerols are that

(a) They must be completely hydrolysed before the constituent fatty acids can be absorbed
(b) They are hydrolysed partially and the material absorbed consists of free fatty acids, mono and diacyl glycerols and unchanged triacyl glycerols

(c) Fatty acids with less than 10 carbon atoms are absorbed about equally via lymph and via portal blood
(d) In the absence of bile the hydrolysis of triacyl glycerols is absorbed

Ans. b

524. Main metabolic end product of cholesterol

(a) Coprosterol (b) 5-pregnenolone
(c) Bile acid (d) Glycine **Ans. c**

525. In the type II (a) hyper lipoproteinemia there is increase in

(a) Chylomicron bond (b) β
(c) Pre beta (d) α **Ans. b**

526. The first lipoprotein to be secreted by the liver is

(a) VLDL (b) nascent VLDL
(c) LDL (d) IDL **Ans. b**

527. This lipoprotein removes cholesterol from the body

(a) HDL (b) VLDL
(c) IDL (b) Chylomicrons **Ans. a**

528. The level of free fatty acids in plasma is increased by

(a) Insulin (b) Caffeine
(c) Glucose (d) Niacin **Ans. b**

529. LCAT is

(a) Lactose choline alamine transferse
(b) Lecithin cholesterol acyl transferase
(c) Lecithin carnitine acyl transferase
(d) Lanoleate carbamoyl acyl transferase **Ans. b**

530. A hydrocarbon formed in cholesterol synthesis is

(a) Mevalonate (b) HMG CoA
(c) Squalene (d) Zymosterol **Ans. c**

531. Avidin is antigonistic to

(a) Niacin (b) PABA
(c) Biotin (d) Pantothenic acid **Ans. c**

532. CTP is required for the synthesis of

(a) Fatty acids (b) Proteins
(c) Phospholipids (d) Cholesterol **Ans. c**

533. Lysolecithin is formed from lecithin by the action of

(a) Phospholipase A_1 (b) Phospholipase A_2

(c) Phospholipase C (d) Phospholipase D **Ans. b**

534. Fatty acids can not be converted into carbohydrates in the body, as the following reaction is not possible

(a) Conversion of glucose-6-phosphate into glucose

(b) Fructose 1, 6-diphosphate to fructose-6-phosphate

(c) Transformation of acetyl CoA to pyruvate

(d) Formation of acetyl CoA from fatty acids **Ans. c**

535. Cholesterol circulates in blood stream chiefly as

(a) Free cholesterol

(b) Ester cholesterol

(c) Low density lipoproteins

(d) Low density lipoproteins and high density lipoproteins **Ans. d**

536. What is the sub cellular site for the β- oxidation of fatty acids?

(a) Nucleus (b) Mitochondria

(c) Lysosome (d) Cytosol **Ans. b**

537. A diet containing this fat is helpful in lowering the blood cholesterol level

(a) Unsaturated (b) Saturated

(c) Vitamin enriched (d) Refined **Ans. a**

538. Phospholipase A_2 is an enzyme which removes a fatty acid residue from lecithin to form

(a) Lecithin fragments (b) Phosphotidic acid

(c) Glyceryl phosphate (d) Lysolecithin **Ans. d**

539. Pancreatic lipose is an enzyme which hydrolyzes facts. It acts as a

(a) peptidase (b) hydrolase

(c) carbohydrates (d) dehydrogenase **Ans. b**

540. This interferes with cholesterol absorption

(a) Lipoprotein lipase (b) Creatinase

(c) 7-dehydrocholesterol (d) β-sitosterol **Ans. d**

541. The carbon chain of fatty acids is shortened by 2 carbon atoms at a time. This involves successive reactions catalysed by 4-enzymes. These act the following order

(a) Acetyl CoA dehydrogenase, β-OH acyl CoA dehydrogenase, enoyl hydrase, thiolose

(b) Acyl CoA dehydrogenase, thiolase, enoyl hydrase, β-OH acyl CoA dehydrogenase

(c) Acyl CoA dehydrogenase, thiolose, enoyl hydrase, β-OH acyl CoA dehydrogenase

(D) Enoyl hydrase, β-OH acyl CoA dehydrogenase, acyl CoA dehydrogenase, thiolose **Ans. b**

542. Acyl carrier protein is involved in the synthesis of

(a) protein

(b) glycogen

(c) fatty acid outside the mitochondria

(d) fatty acid in the mitochondria **Ans. c**

543. 1 molecule of palmitic acid on total oxidation to CO_2 will yield molecules of ATP (as high energy bonds)

(a) 129 (b) 154

(c) 83 (d) 25 **Ans. a**

544. HMG CoA is formed in the metabolism of

(a) Cholesterol, ketones and leucine

(b) Cholesterol, fatty acid and Leucine

(c) Lysine, Lecuine and Isoleucine

(d) Ketones, Leucine and Lysine **Ans. a**

545. NADPH is produced when this enzyme acts

(a) Pyruvate dehydrogenase (b) Malic enzyme

(c) Succinate dehydrogenase (d) Malate dehydrogenase **Ans. b**

546. As a result of each oxidation a long chain fatty acid is cleaved to give

(a) An acid with 3-carbon less and propionyl CoA

(b) An acid with 2-carbon less and acetyl CoA

(c) An acid with 2-carbon and acetyl CoA

(d) An acid with 4-carbon and butyryl CoA **Ans. b**

547. Liposomes are
(a) Lipid bilayered (b) Water in the middle
(c) Carriers of drugs (d) All of these **Ans. d**

548. Long chain fatty acyl CoA esters are transported across the mitochondrial membrane by
(a) cAMP (b) Prostaglandin
(c) Carnitine (d) Choline **Ans. c**

549. The acetyl CoA formed on β-oxidation of all long chain fatty acids is metabolized under normal circumstances to
(a) CO_2 and water (b) Cholesterol
(c) Fatty acids (d) Ketone bodies **Ans. a**

550. Very low density lipoproteins are relatively rich in
(a) Cholesterol (b) Triacyl glycerol
(c) Free fatty acids (d) Phospholipids **Ans. b**

551. Neutral fat is stored in
(a) Liver (b) Pancreas
(c) Adipose tissue (d) Brain **Ans. c**

552. A pathway that requires NADPH as a cofactor is
(a) Fatty acid oxidation
(b) Extra mitochondrial denovo fatty acid synthesis
(c) Ketone bodies formation
(d) Glycogenesis **Ans. b**

553. The 'Committed step' in the biosynthesis of cholesterol from acetyl CoA is
(a) Formation of acetoacetyl CoA from acetyl CoA
(b) Formation of mevalonate from HMG CoA
(c) Formation of HMG CoA from acetyl CoA and acetoacetyl CoA
(d) Formation of squalene by squalene synthetase **Ans. b**

554. In β-Oxidation of fatty acids, which of the following are utilized as coenzymes?
(a) NAD^+ and $NADP^+$ (b) $FADH_2$ and $NADH + H^+$
(c) FAD and FMN (d) FAD and NAD^+ **Ans. d**

555. The most important source of reducing equivalents for FA synthesis on the liver is

(a) Glycolysis (b) HMP-Shunt
(c) TCA cycle (d) Uronic acid pathway **Ans. b**

556. All of the following tissue are capable of using ketone bodies except

(a) Brain (b) Renal cortex
(c) R.B.C. (d) Cardiac muscle **Ans. c**

557. The major source of cholesterol in arterial smooth muscle cells is from

(a) IDL (b) LDL
(c) HDL (d) Chylomicrons **Ans. b**

558. Ketone bodies are synthesized from fatty acid oxidation products by which of the following organs?

(a) Liver (b) Skeletal muscles
(c) Kidney (d) Brain **Ans. a**

559. Chain elongation of fatty acids occurring in mammalian liver takes place in which of the following subcellular fractions of the cell?

(a) Nucleus (b) Ribosomes
(c) Lysosomes (d) Microsomes **Ans. d**

560. Which of the following cofactors or their derivatives must be present for the conversion of acetyl CoA to malonyl CoA extramitochondrial fatty acid synthesis?

(a) Biotin (b) FAD
(c) FMN (d) ACP **Ans. a**

561. Which of the following statement regarding β-oxidation is true?

(a) Requires β-ketoacyl CoA as a substrate
(b) Forms CoA thioesters
(c) Requires GTP for its activity
(d) Yields acetyl CoA as a product **Ans. a**

562. All statements regarding 3-OH-3-methyl glutaryl CoA are true except

(a) It is formed in the cytoplasm
(b) Required in ketogenesis
(c) Involved in synthesis of Fatty acid
(d) An intermediate in cholesterol biosynthesis **Ans. b**

563. Which of the following lipoproteins would contribute to a measurement of plasma cholesterol in a normal individual following a 12 hr fast?

(a) Chylomicrons (b) VLDL

(c) Both VLDL and LDL (d) LDL **Ans. d**

564. All the following statements regarding ketone bodies are true except

(a) They may result from starvation

(b) They are formed in kidneys

(c) They include acetoacetic acid and acetone

(d) They may be excreted in urine **Ans. b**

565. In synthesis of Triglyceride from α-Glycero phosphate and acetyl CoA, the first intermediate formed is

(a) β-diacyl glycerol (b) Acyl carnitine

(c) Monoacyl glycerol (d) Phosphatidic acid **Ans. d**

566. During each cycle of β-oxidation of fatty acid, all the following compounds are generated except

(a) NADH (b) H_2O

(c) FAD (d) Acyl CoA **Ans. b**

567. The energy yield from complete oxidation of products generated by second reaction cycle of β-oxidation of palmitoyl CoA will be

(a) 5 ATP (b) 12 ATP

(c) 17 ATP (d) 34 ATP **Ans. d**

568. β-Oxidation of odd-carbon fatty acid chain produces

(a) Succinyl CoA (b) Propionyl CoA

(c) Acetyl CoA (d) Malonyl CoA **Ans. d**

569. Brown adipose tissue is characterized by which of the following?

(a) Present in large quantities in adult humans

(b) Mitochondrial content higher than white adipose tissue

(c) Oxidation and phosphorylation are tightly coupled

(d) Absent in hibernating animals

570. Ketosis in partly ascribed to

(a) Over production and Glucose

(b) Under production of Glucose

(c) Increased carbohydrate utilization

(d) Increased fat utilization **Ans. d**

571. The free fatty acids in blood are
(a) Stored in fat depots
(b) Mainly bound to β-lipoproteins
(c) Mainly bound to serum albumin
(d) Metabolically most inactive **Ans. c**

572. Carnitine is synthesized from
(a) Lysine (b) Serine
(c) Choline (d) Arginine **Ans. a**

573. A metabolite which is common to pathways of cholesterol biosynthesis from acetyl-CoA and cholecalciferol formation from cholesterol is
(a) Zymosterol (b) Lumisterol
(c) Ergosterol (d) 7 Dehydrocholesterol **Ans. d**

574. Acetyl CoA required for extra mitochondrial fatty acid synthesis is produced by
(a) Pyruvate dehydrogenase complex
(b) Citrate lyase
(c) Thiolase
(d) Carnitine-acyl transferase **Ans. b**

575. Biosynthesis of Triglyceride and Lecithine both require an intermediate
(a) Monoacyl glycerol phosphate (b) Phosphatidic acid
(c) Phosphatidyl ethanol amine (d) Phosphatidyl cytidylate **Ans. b**

576. The rage limiting step cholesterol biosynthesis is
(a) Squalene synthetase (b) Mevalonate kinase
(c) HMG CoA synthetase (d) HMG CoA reductase **Ans. d**

577. All the following are constituents of ganglioside molecule except
(a) Glycerol (b) Sialic acid
(c) Hexose sugar (d) Sphingosine **Ans. a**

578. An alcoholic amine residue is present in which of the following lipids?
(a) Phosphatidic acid (b) Cholesterol
(c) Sphingomyelin (d) Ganglioside **Ans. c**

579. Sphingosine is the backbone of all the following except
(a) Cerebroside (b) Ceramide
(c) Sphingomyelin (d) Lecithine **Ans. d**

580. Chylomicron, intermediate density lipoproteins (IDL), low density lipoproteins (LDL) and very low density lipoproteins (VLDL) all are serum lipoproteins. What is the correct ordering of these particles from the lowest to the greatest density?

(a) LDL, IDL, VLDL, Chylomicron

(b) Chylomicron, VLDL, IDL, LDL

(c) VLDL, IDL, LDL, Chylomicron

(d) Chylomicron, IDL, VLDL, LDL **Ans. b**

581. A compound normally used to conjugate bile acids is

(a) Serine (b) Glycine

(c) Glucoronic acid (d) Fatty acid **Ans. b**

582. Which of the following lipoproteins would contribute to a measurement of plasma cholesterol in a normal person following a 12 hr fast?

(a) High density lipoprotiens (b) Low density lipoproteins

(c) Chylomicron (d) Chylomicron remnants **Ans. a**

583. Which of the following products of triacylglycerol breakdown and subsequent β-Oxidation may undergo gluconeogenesis?

(a) Acetyl CoA (b) Porpionyl CoA

(c) All ketone bodies (d) Some amino acids **Ans. b**

584. Which of the following regulates lipolysis in adipocytes?

(a) Activation of fatty acid synthesis mediated by cAMP

(b) Glycerol phosphorylation to prevent futile esterification of fatty acids

(c) Activation of triglyceride lipase as a result of hormone stimulated increases in CAMP levels

(d) Activation of CAMP production by Insulin

585. Which one of the following compounds is a key intermediate in the synthesis of both triacyl glycerols and phospholipids?

(a) CDP Choline (b) Phosphatidase

(c) Triacyl glyceride (d) Phosphatidyl serine **Ans. b**

586. During each cycle of ongoing fatty acid oxidation, all the following compounds are generated except

(a) H_2O (b) Acetyl CoA

(c) Fatty acyl CoA (d) NADH **Ans. a**

587. All the following statements describing lipids are true except
(a) They usually associate by covalent interactions
(b) They are structurally components of membranes
(c) They are an intracellular energy source
(d) They are poorly soluble in H_2O **Ans. a**

588. All the following statements correctly describe ketone bodies except
(a) They may result from starvation
(b) They are present at high levels in uncontrolled diabetes
(c) They include—OH β-butyrate and acetone
(d) They are utilized by the liver during long term starvation **Ans. d**

589. Which of the following features is predicted by the Nicolson–Singer fluid mosaic model of biological membranes?
(a) Membrane lipids do not diffuse laterally
(b) Membrane lipid is primarily in a monolayer form
(c) Membrane lipids freely flip-flop
(d) Membrane proteins may diffuse laterally **Ans. d**

590. Oxidative degradation of acetyl CoA in the citric acid cycle gives a net yield of all the following except
(a) $FADH_2$ (b) 3 NADH
(c) 2 ATP (d) $2CO_2$ **Ans. c**

591. All the following correctly describe the intermediate 3-OH-3-methyl glutaryl CoA except
(a) It is generated enzymatically in the mitochondrial matrix
(b) It is formed in the cytoplasm
(c) It inhibits the first step in cholesterol synthesis
(d) It is involved in the synthesis of ketone bodies **Ans. c**

592. Intermediate in the denovo synthesis of triacyl glycerols include all the following except
(a) Fatty acyl CoA (b) CDP diacyl glycerol
(c) Glycerol-3-phosphate (d) Lysophosphatidic acid **Ans. b**

593. Mitochondrial α-ketoglutarate dehydrogenase complex requires all the following to function except
(a) CoA (b) FAD
(c) NAD^+ (d) $NADP^+$ **Ans. d**

594. Each of the following can be an intermediate in the synthesis of phosphatidyl choline except

(a) Phosphatidyl inositol (b) CDP-choline

(c) Phosphatidyl ethanolamine (d) Diacylglycerol **Ans. a**

595. High iodine value of a lipid indicates

(a) Polymerization (b) Carboxyl groups

(c) Hydroxyl groups (d) Unsaturation **Ans. d**

596. Cholesterol, bile salts, vitamin D and sex hormones are

(a) Mucolipids (b) Glycolipids

(c) Phospholipids (d) Isoprenoid lipids **Ans. d**

597. Water soluble molecular aggregates of lipids are known as

(a) Micelle (b) Colloids

(c) Sphingol (d) Mucin **Ans. a**

598. Hypoglycemia depresses insulin secretion and thus increases the rate of

(a) Hydrolysis (b) Reduction

(c) Gluconeogenesis (d) Respiratory acidosis **Ans. c**

599. The process of breakdown of glycogen to glucose in the liver and pyruvate and lacate in the muscle is known as

(a) Glyogenesis (b) Glycogenolysis

(c) Gluconeogenesis (d) Cellular degradation **Ans. b**

600. Across a membrane phospholipids act as carrier of

(a) Organic compounds (b) Inorganic ions

(c) Nucleic acids (d) Food materials **Ans. b**

601. Osteomalacia can be prevented by the administration of calcium and a vitamin

(a) A (b) B

(c) C (d) D **Ans. d**

602. Milk sugar is known as

(a) Fructose (b) Glucose

(c) Sucrose (d) Lactose **Ans. d**

603. The Instrinisic Factor (HCl and mucoproteins) present in the gastric juice help in the absorption of

(a) Vitamin B_2 (b) Tocopherols

(c) Folic acid (d) Vitmain B_{12} **Ans. d**

604. Bile is produced by
(a) Liver (b) Gall-bladder
(c) Pancreas (d) Intestine **Ans. a**

605. A pathway that requires NADPH as a cofactor is
(a) Extramitochondrial folic acid synthesis
(b) Ketone body formation
(c) Glycogenesis
(d) Gluconeogenesis **Ans. a**

606. LCAT activity is associated with which of the lipo-protein complex?
(a) VLDL (b) Chylomicrones
(c) IDL (d) HDL **Ans. d**

607. In β-oxidation of fatty acids which of the following are utilized as co-enzymes?
(a) NAD^+ and $NADP^+$ (b) $FADH_2$ and NADH + H^+
(c) FAD and FMN (d) FAD and NAD^+ **Ans. d**

608. The lipoprotein with the fastest electrophoretic mobility and lowest TG content are
(a) VLDL (b) LDL
(c) HDL (d) Chylomicrones **Ans. c**

609. The essential fatty acids retard
(a) Atherosclerosis (b) Diabetes mellitus
(c) Nepritis (d) Oedema **Ans. a**

610. The majority of absorbed fat appears in the forms of
(a) HDL (b) Chylomicrone
(c) VLDL (D) LDL **Ans. b**

611. Daily output of urea in grams is
(a) 10 to 20 (b) 15 to 25
(c) 20 to 30 (d) 35 to 45 **Ans. c**

612. Uremia occurs in
(a) Cirrohsis of liver (b) Nephritis
(c) Diabetes mellitus (d) Coronary thrombosis **Ans. b**

613. Carboxyhemoglobin is formed by
(a) CO (b) CO_2
(c) HCO_3 (d) HCN **Ans. a**

614. Methemoglobin is formed as a result of the oxidation of haemoglobin by oxidation agent

(a) Oxygen of Air (b) H_2O_2

(c) $K_4Fe(CN)_6$ (d) $KMnO_4$ **Ans. c**

615. Methemoglobin can be reduced to haemoglobin by

(a) Removal of hydrogen (b) Vitamin C

(c) Glutathione (d) Creatinine **Ans. b**

616. Fats are solids at

(a) 10°C (b) 20°C

(c) 30°C (d) 40°C **Ans. b**

617. Esters of fatty acids with higher alcohol other than glycerol are called as

(a) Oils (b) Polyesters

(c) Waxes (d) Terpenoids **Ans. c**

618. The main physiological buffer in the blood is

(a) Haemoglobin buffer (b) Acetate

(c) Phosphate (d) Bicarbonate **Ans. d**

619. All of the following substances have been used to estimate GFR except

(a) Inulin (b) Creatinine

(c) Phenol red (d) Mannitol **Ans. c**

620. Relationship between GFR and serum creatinine concentration is

(a) Non-existent (b) Inverse

(c) Direct (d) Indirect **Ans. b**

621. Urine turbidity may be caused by any of the following except

(a) Phosphates (b) Protein

(c) RBC (d) WBC **Ans. b**

622. Urine specific gravity of 1.054 indicates

(a) Excellent renal function

(b) Inappropriate secretion of ADH

(c) Extreme dehydration

(d) Presence of glucose or protein **Ans. d**

623. In hemolytic jaundice, the urinary bilirubin is

(a) Normal (b) Absent

(c) More than normal (d) Small amount is present

Ans. c

624. In obstructive jaundice, urinary bilirubin is

(a) Absent (b) Increased

(c) Present (d) Present in small amount

Ans. b

625. The pH of gastric juice of infants is

(a) 2.0 (b) 4.0

(c) 4.5 (d) 5.0 **Ans. d**

626. The pH of blood is about 7.4 when the ratio between ($NaHCO_3$) and (H_2CO_3) is

(a) 10 : 1 (b) 20 : 1

(c) 25 : 1 (d) 30 : 1 **Ans. b**

627. For the activity of amylase which of the following is required as co-factor?

(a) HCO_3 (b) Na^+

(c) K^+ (d) Cl **Ans. d**

628. Which of the following hormone increases the absorption of glucose from G.I.T?

(a) Insulin (b) Thyroid hormones

(c) Glucagon (d) FSH **Ans. b**

629. Predominant form of storage

(a) Carbohydrates (b) Fats

(c) Lipids (d) Both (b) and (c) **Ans. d**

630. Degradations of Hb takes place in

(a) Mitochondrion (b) Erythrocytes

(c) Cytosol of cell (d) R.E. cells **Ans. d**

631. Biluveridin is converted to bilirubin by the process of

(a) Oxidation (b) Reduction

(c) Conjugation (d) Decarboxylation **Ans. b**

632. Phospholipids are important cell membrane components since

(a) They have glycerol

(b) Form bilayers in water

(c) Have polar and non-polar portions

(d) Combine covalently with proteins **Ans. c**

633. Which of the following is not a phospholipids?
(a) Lecithin (b) Plasmalogen
(c) Lysolecithin (d) Gangliosides **Ans. d**

634. A fatty acid which is not synthesized in human body and has to be supplied in the diet is
(a) Palmitic acid (b) Oleic acid
(c) Linoleic acid (d) Stearic acid **Ans. c**

635. Phospholipids occur in
(a) Myelin sheath (b) Stabilizes chylomicrons
(c) Erythrocyte membrane (d) All of these **Ans. d**

636. Which of the following is not essential fatty acids?
(a) Oleic acid (b) Linoleic acid
(c) Arachidonic acid (d) Linolenic acid **Ans. a**

637. Prostaglandin synthesis is increased by activating phospholipases by
(a) Mepacrine (b) Angiotensin II
(c) Glucocorticoids (d) Indomenthacin **Ans. b**

638. Selwanof's test is positive in
(a) Glucose (b) Fructose
(c) Galactose (d) Mannose **Ans. b**

639. Spermatozoa in seminal fluid utilises the following sugar for metabolism
(a) Galactose (b) Glucose
(c) Sucrose (d) Fructose **Ans. d**

640. Depot fats of mammalian cells comprise mostly of
(a) Cholesterol (b) Phospholipid
(c) Cerebrosides (d) Triglycerol **Ans. d**

641. Which of the following is a hydroxyl fatty acid?
(a) Oleic Acid (b) Ricinoleic acid
(c) Caproic acid (d) Arachidonic acid **Ans. b**

642. Acroleic test is given by
(a) Cholesterol (b) Glycerol
(c) Glycosides (d) Sphingol **Ans. b**

❑❑❑

Chapter-11

Nucleic Acids (Part - I) DNA (Deoxyribonucleic acid)

Fill in the blanks

1. In Z-DNA, the ripening unit is....................

Ans. dinucleotide

2. A-form DNA has.................bases per turn.

Ans. 11

3. Repeating unit is dinucleotide in......... DNA.

Ans. Z

4. DNA with left-hand coiling is called..........................

Ans. Z-DNA

5. Adenine and guanine are.................

Ans. purines

6. The diameter of DNA is..................

Ans. 20Å

7. A ratio of.................occurs between purines and pyrimidine bases of DNA.

Ans. 1:1

8. 1 μm of DNA contain about.............. base pairs.

Ans. 3000

9. The prosthetic group of nucleoproteins is....................

Ans. nucleic acids

10. A nucleotide consists of three parts: a phosphate unit, a sugar unit and a.............

Ans. purine or pyrimidine base

11. A DNA molecule has...........but RNA has only...........

Ans. double strand, single strand

12. The relationship between nucleotide triplets and the α-amino acids is called the................

Ans. genetic code

13. The base which is present only in RNA but not in DNA is...............

Ans. uracil

14. ....has the unique property of replication.

Ans. DNA

15. The two strands of the double helix of DNA are held together at definite distances through.............

Ans. H-bonds

16. Type of DNA exhibiting left handed helix is........

Ans. Z only

17. Bean golden mosaic virus and maize streak virus contain................

Ans. single standard DNA

18. Nucleic acids are polymers of................

Ans. phosphorylated nucleosides

19. Nucleic acids were for the first time observed in pus cells by..............

Ans. Holfmiester

20. Frankel-Contrat and Singer experimented on TMV to show that ....is the genetic material.

Ans. RNA

21. DNase I is normally obtained from cow ....

Ans. phoshodiester bond.

22. DNA polymerase I was for the first time isolated by ....

Ans. Arthur Korenberg

23. Nucleic acid was discovered by ....

Ans. Meischer

24. DNA polymerase I is mainly involved in removing .... from okazaki fragments and in filling in the gaps so created.

Ans. RNA primers

Multiple Choice Questions

1. Which of the following statements represents the criteria of DNA being the genetic material. Because
   (a) It is found in small amounts in cell
   (b) It is a complex structure entity
   (c) It is essentially stable & is transmitted through generation to generation
   (d) It is a biomolecule with specific function **Ans. c**

2. The output of Griffith's experiment was that
   (a) RNA is a genetic material
   (b) RNA and protein is the genetic material
   (c) DNA and protein is genetic material
   (d) DNA is the genetic material **Ans. d**

3. F. Griffiths experimented to show
   (a) Transmission (b) Transformation
   (c) Conjugation (d) Transduction **Ans. b**

4. The experimental material taken by Griffith was
   (a) T-2 bacteriophage (b) *Escherichia coli*
   (c) *Diplococcus pneumonia* (d) *Penicillium notatum* **Ans. c**

5. O.T. Avery, C.M. Macleod and M. Mc Carthy
   (a) Carried a new set of experiments to show that DNA is the genetic material
   (b) Repeated Griffith's experiment and showed that DNA is the genetic material
   (c) Experimented to show that DNA undergoes mutation
   (d) Worked out the biosynthesis of DNA **Ans. b**

6. What does R and S signify in Avery, Macleod and Mc Carthy experiment
   (a) Resistant and susceptible bacterial colonies
   (b) Red and smooth bacterial colonies
   (c) Red and silver bacterial colonies
   (d) Rough and smooth bacterial colonies **Ans. d**

7. In Avery, Macleod and Mc Carthy experiment transformation occurred from
   (a) Heat killed S-III to S-III
   (b) Nonvirulent S-III to heat killed S-III
   (c) R-II to S-III
   (d) S-III to R-II **Ans. a**
8. Hershey and Chase conducted their experiment on
   (a) *Escherichia coli* (b) *Pneumococcus*
   (c) *Pseudomonas* (d) T-2 phage **Ans. d**
9. The organism containing the highest amount of DNA, is
   (a) *Saccharomyces* (Yeast) (b) *Plagusia* (Grab)
   (c) *Homo sapiens* (Humans) (d) *Triturus* (Newt) **Ans. d**
10. Arrange the following organisms in ascending order according to their DNA content (Picograms/cell)
    (i) *Drosophila* (Fruit fly) (ii) *Zea mays* (Corn)
    (iii) *Lilium* (Lily) (iv) *Homo sapiens* (Humans)
    Choose the correct answer
    (a) i, ii, iii, iv (b) iv, iii, ii, i
    (c) iii, ii, iv, i (d) iii, iv, i, ii **Ans. c**
11. Frankel- Contrat and Singer experimented on TMV to show that
    (a) DNA is the genetic material
    (b) Proteins are the genetic material
    (c) DNA is the double helical
    (d) RNA is the genetic material **Ans. d**
12. Gierer and Schramm are related with
    (a) DNA structure
    (b) RNA structure
    (c) DNA and RNA are genetic material
    (d) RNA is the genetic material **Ans. d**
13. DNA normally contains
    (i) Phosphoric acid (ii) Sulphuric acid
    (iii) Nitrogen bases (iv) Ribose sugar

Choose correct answer is

(a) i, ii (b) iii, ii, i

(c) i, iv (d) i, iii, iv **Ans. d**

14. The type of DNA found in Human genome is

(i) Z & B (ii) Z only

(iii) B only (iv) A & B

Choose correct answer

(a) i, ii, iii, iv (b) iii

(c) ii, iii (d) None of the above **Ans. b**

15. Watson, Crick and Wilkins are credited with the discovery that DNA is

(a) The carrier of genetic information

(b) Responsible for synthesis of mRNA

(c) Contains deoxyribose sugar

(d) A double stranded helix **Ans. d**

16. Both DNA and tRNA are believed to be characterized by following

(i) Both are composed of purines, pyrimidines and phosphate groups

(ii) Both are unbranched linear macro molecules

(iii) Both exists atleast partially in a helical configuration

(iv) Both contains the same group of nitrogenous bases

Choose correct answer

(a) i, ii (b) i, iii

(c) i, ii, iii (d) i, ii, iii, iv **Ans. d**

17. Type of DNA exhibiting left handed helix is/are

(a) B type (b) A+B type

(c) B & Z type (d) Z only **Ans. d**

18. Phosphates forms the backbone of all the three types of DNA but the Zig-Zag pattern of phosphate is met in with

(a) B type (b) A type

(c) Z type (d) None **Ans. c**

19. Some viruses have no DNA but contain RNA. This statement indicated that

(i) These viruses cannot replicate

(ii) These viruses have no transmittable hereditary information

(iii) RNA can act to transmit hereditary information

(iv) Their nucleic acids must combine with host DNA for virus duplication

(v) They can direct the manufacture of proteins but not of nucleic acids

Choose correct answer

(a) i, ii (b) i, iii, v

(c) iii (d) ii, iii, iv **Ans. c**

20. In the case of double stranded DNA, which of the following base ratios are always equal to one

(i) A+T/C+G (ii) A+G/C+T

(iii) C/G (iv) G+T/A+C

Choose correct answer

(a) i, ii (b) i, iii

(c) ii, iii, iv (d) i, iii, iv **Ans. c**

21. Restituted viruses can be formed by organizing

(a) A type protein + B type protein

(b) A type nucleic acid + A type protein

(c) A type DNA + B type RNA

(d) A type DNA/RNA + B type protein **Ans. d**

22. Pace and Spiegelman who worked out that RNA is the genetic material, concluded that

(a) RNA replication is independent of parental type RNA

(b) RNA replication is partially dependent on parental type

(c) RNA replication is completely dependent on parental type

(d) RNA replication depends on DNA in TMV **Ans. c**

23. Pace and Spigelman experimented to show that RNA is genetic material in viruses, they experimented on

(a) TMV (b) CMV

(c) T_2 phage (d) Qβ phage **Ans. d**

24. Dahlia mosaic virus and Cauliflower mosaic virus are unique in that they

(i) Have both DNA and RNA (ii) Have first RNA and then DNA

(iii) Have double stranded DNA (iv) Have single stranded DNA

(v) Have single stranded RNA (vi) Have double stranded RNA

Choose the correct answer

(a) i, ii, iv are correct (b) only iii is correct
(c) i, iii, v are correct (d) ii, v, vi are correct **Ans. b**

25. Bean golden mosaic virus and Maize streak virus contain

(a) DNA as genetic material (b) RNA as genetic material
(c) Single stranded DNA (d) Single stranded RNA **Ans. c**

26. There are only few viruses that contain double stranded RNA as genetic material, select the correct pair of viruses

(a) TMV and CMV
(b) Rice dwarf virus and Potato virus
(c) Wound Tumour virus and maize rough dwarf virus
(d) All the above are correct **Ans. c**

27. Nucleic acids are polymers of

(a) Nucleosides (b) Phosphorylated nucleosides
(c) Glycosides (d) Peptides **Ans. b**

28. Hereditary characters are passed on from parent to offsprings chiefly through

(a) Gametes (b) Enzymes
(c) Genes (d) Centrosome **Ans. a**

29. Particulate theory of inheritance was proposed by

(a) Weismann (b) Mendel
(c) Darwin (d) De Vries **Ans. a**

30. The difference in DNA and RNA structure is the presence of uracil in RNA instead of thymine in DNA. At molecular structural level there is only a difference of

(a) Ketone group (b) Aldehyde group
(c) Methyl group (d) Carboxylic group **Ans. c**

31. Which of the following is the physical basis of heredity

(a) DNA (b) Genes
(c) Chromosomes (d) DNA and proteins **Ans. c**

32. The sequence of three consecutive nitrogenous bases in DNA forms a/an

(a) Codon (b) Peptide
(c) Anticodon (d) Nucleotide triphosphate **Ans. a**

33. The characteristic feature(s) of genetic code is/are

(i) Being comma less

(ii) Being absolutely universal in nature

(iii) Being degenerate

(iv) Being biologically static in nature

Choose the correct answer

(a) i, ii, iii, iv are correct (b) ii, iii, iv are correct

(c) i, iii are correct (d) ii, iv are correct **Ans. c**

34. Given below are four DNA molecules, which of them is most susceptible to 'kinking' due to some biological forces

(a) ACCTGCAATCGC
TGGACGTTAGCG

(b) ACCGTTACGCT
TGGCAATGCGA

(c) CGCTTAACGA
GCGAATTGCT

(d) TATTAACGTA
ATAATTGCA **Ans. d**

35. There are large number of unusual bases in RNA, which comprises nearly of the total bases of RNA

(a) 15% (b) 25%

(c) 5% (d) 10% **Ans. c**

36. Which of the following statement is not true for Chargaff's studies of base equivalence

(a) The number of A, C, T and G is always equal

(b) The number of purines to pyrimidines is always equal

(c) The base composition of DNA varies from one species to another

(d) DNA specimen isolated from different tissues of same species have same base composition **Ans. a**

37. RNA molecules are more susceptible to alkaline and enzymatic degradation compared to DNA molecules because

(i) RNA has uracil instead of thymine

(ii) Presence of methyl group in uracil

(iii) Presence of hydroxyl group at second carbon of deoxyribose sugar

(iv) Presence of hydroxyl group at second carbon of ribose sugar

Choose the correct answer

(a) i, iv (b) iv only

(c) iv, iii, ii (d) iii only **Ans. b**

38. Nucleic acids were for the first time observed in pus cells by

(a) Weismann (b) Holfmiester

(c) Watson and Crick (d) Stanley **Ans. b**

39. Hydrolysis of RNA molecule can result in formation of

(a) Nucleosides

(b) Dinucleotides

(c) Ribonucleoside 3-5 biphosphates and 2, 3 cyclic mono phosphates

(d) Nucleosomes **Ans. c**

40. Sugar present in DNA is a typical pentose specifically named as

(a) β-deoxyribofuranose (b) α-2, D-deoxyribofuranose

(c) β-D-2 deoxyribofuranose (d) I-β, D ribofuranosyl **Ans. c**

41. Nitrogenous bases are liked to sugar (pentose) at

(a) C-1 position (b) C-5 position

(c) C-3 position (d) At any of the above positions **Ans. a**

42. Major difference between purines and pyrimidines is that

(a) Former is a single ring compound while the later is a double ring compound

(b) Former is a double ring compound while the later is a single ring compound

(c) Purines are found in DNA and pyrimidines in RNA

(d) All of the above **Ans. b**

43. DNA is made up of two antiparallel polynucleotide chains, the word antiparallel means

(a) The distance between the two strands is not constant throughout

(b) The strands have opposite polarity i.e. one is from 3′-5′ and the other is 5′-3′ directed

(c) That they are not complementary

(d) That one grows faster than the other strand **Ans. b**

44. The presence of sugar phosphate backbone in the DNA helix points out to give hydrophilic ridges with one negative charge per phosphate. This negative charge is responsible for

(i) Binding of various proteins particularly histones

(ii) Various types of staining procedures

(iii) Nucleosome formation and DNA packing

(iv) For easy template formation

Choose the correct answer

(a) i, iii (b) ii, iv

(c) i, ii, iii (d) All the above **Ans. c**

45. Native form of DNA is one

(a) That forms after DNA melting

(b) In the transition state between stable & melted DNA

(c) The ordered state of DNA present originally

(d) That represents the DNA fragments after enzymatic digestion

Ans. c

46. One Fermi is equal to

(a) 10^{-10} m (b) 10^{-12} m

(c) 10^{-14} m (d) 10^{-13} cm **Ans. d**

47. One Dalton is equal to

(a) 10^{-9} gm

(b) 10^{-12} gm

(c) Mass of hydrogen atom or 3.32×10^{-24} gm

(d) 10^{-10} gm **Ans. c**

48. A powerful denaturant is one that

(a) Destroys the DNA completely

(b) Breaks DNA into fragments

(c) Separates the two strands by breaking the H-bond

(d) Separates the strands by breaking phosphodiester bonds **Ans. c**

49. The phenomenon of hypochromism and hyperchromism is exhibited by

(i) Native DNA (ii) Denaturated DNA

(iii) Complete DNA in solution (iv) DNA bases

(v) Nucleotides in compact DNA

Choose the correct answer

(a) i, ii (b) ii, iii, iv

(c) v, iv (d) All of the above **Ans. d**

50. Nitrogenous bases of a denaturated DNA exhibit which of the following phenomenon

(a) Hypochromism (b) Amphichromism

(c) Hyperchromism (d) None of the above **Ans. c**

51. The melting temperature (T_m) of DNA depends on the amount of

(i) A, T, C, G (ii) A + T

(iii) C + G (iv) C + T

Choose the correct answer

(a) i, iii

(b) i, ii, iii

(c) ii, iv, iii

(d) none of the above, DNA is stable molecule **Ans. b**

52. Which of the given strands will have maximum and minimum T_m

(i) ACTCGTATACCG (ii) CCGGATCCGT

(iii) ATATGCCGCG (iv) CCGGCGTCGC

Choose the correct answer

(a) ii, iv

(b) i, ii

(c) iv, i

(d) All will have the same melting point **Ans. c**

53. The generalized expression to calculate melting temperature (T_m) of DNA is given by

(a) $T_m = 69.3 + 0.49\ (GC)°C$ (b) $T_m = 6.93 + 0.49\ (GC)°C$

(c) $T_m = (AT\text{-}X)\ 2.44°C$ (d) $T_m = 69.3 + 0.49\ (GC)°F$ **Ans. a**

54. Methylation of cytosine and adenine

(i) Stabilizes the DNA

(ii) Destabilizes the DNA

(iii) Former stabilizes and later destabilizes the DNA

(iv) Does not affect the stability

Choose the correct answer

(a) i, iii (b) i, ii, iii

(c) ii only (d) iii only **Ans. d**

55. The optimal level of renaturation temperature is

(a) 10-50°C below the value of T_m

(b) 15-20°C below the value of T_m

(c) 20-25°C below the value of T_m

(d) 25-30°C below the value of T_m **Ans. c**

56. Choose the correct statement

(a) Renaturation is comparatively a slower process than the denaturation

(b) Denaturation is a slower process than the renaturation

(c) Both processes occur at the same rate

(d) Totally depends on the condition **Ans. a**

57. Renaturation can be studied by

(a) Nitrocellulose filters (b) Western blotting

(c) ELISA (d) RIA **Ans. a**

58. Cot ½ of DNA is defined as

(a) The time taken by a denatured method molecule to reanneal

(b) The initial concentration (Co) in moles/litre multiplied by time (t) in seconds, taken for 50% of DNA to reanneal

(c) The concentration of DNA in a cell at any particular time

(d) Amount of cytosine bases in a single strand by DNA **Ans. b**

59. The Buoyant density of DNA is expressed as

(a) $\sigma = 1.660 + 0.098$ (GC) (b) $\sigma = 1.660 \times$ GC

(c) $\sigma = 69.3 + 0.49$ (GC) (d) $\sigma = 69.3 \times$ GC **Ans. a**

60. Which of the following enzymes was the first template dependent enzyme (polymerase) discovered in 1957

(a) RNA polymerase (b) DNA polymerase

(c) DNA polymerase I (d) DNA polymerase II **Ans. c**

61. Which of the following dye/antibiotic shows strong affinity for A:T region

(a) Comassie blue (b) Distamycin, Netropsin

(c) Actinomycin (d) Phenyl neutral red **Ans. b**

62. The G:C region specific dye/antibiotic are given in which of the following pair

(a) Actinomycin and phenyl neutral red

(b) Crystal violet and Safranin

(c) Comassie blue

(d) Silver dye **Ans. a**

63. The highest percentage of G + C is found in

(a) *Bacillus cereus* (b) *Diplococcus pneumonia*

(c) *Mycobacterium phlei* (d) T_3 phage **Ans. c**

64. C-value represents

(a) Concentration of DNA in a cell

(b) Content of DNA in the nucleus

(c) Content of DNA in a prokaryotic nucleoid or in a eukaryotic haploid nucleus

(d) Concentration of DNA in chloroplast and mitochondria **Ans. c**

65. At low pH (1 or less), the DNA and RNA will behave differently because

(a) All the H bonds will be broken down

(b) All the phosphodiester bonds break, which is accompanied by the breakage of N-glycosidic base and sugar

(c) Bases get denatured

(d) The helix gets uncoiled **Ans. b**

66. Breathing is

(a) A process of normal respiration

(b) A process where double DNA helix frequently opens and become single stranded bubbles

(c) A process of protein synthesis in intervals

(d) An unrelated term **Ans. b**

67. RNA molecule gets completely hydrolysed at

(a) pH 11 and temperature 37°C (b) pH 7 and temperature 37°C

(c) pH 11 and temperature 60°C (d) pH 5.6 and temperature 37°C

Ans. a

68. Meselson and Stahl grew *E. coli* in a medium containing $^{15}NH_4Cl$ to show that

(a) Heavy nitrogen is essential for *E. coli* growth

(b) DNA replicates semiconservatively as ^{15}N gets incorporated into DNA

(c) Heavy nitrogen degrade DNA

(d) ^{15}N stops the life cycle of *E. coli* and helps in studies **Ans. b**

69. Most of the DNA replication is

(a) Unidirectional (b) Bidirectional

(c) Multidirectional (d) Localized **Ans. b**

70. Primers synthesized during DNA replication on lagging strand are removed by

(a) DNA polymerase I

(b) Primase

(c) DNA Pol III

(d) DNA Pol I due to its 5′-3′ exonuclease activity **Ans. d**

71. SSB proteins are

(a) Specifically enzymatic proteins used in DNA replication

(b) Basic proteins that binds to single stranded DNA and stabilizes it

(c) Proteins that degrade DNA

(d) Proteins that helps in decoiling of double helix of DNA **Ans. b**

72. DNA polymerase which is multimeric with high processivity, catalytic potency and high degree of fidelity, is

(a) DNA Pol I (b) DNA Pol II

(c) DNA Pol III (d) DNA ligase **Ans. c**

73. DNA polymerase III has 8 subunits. The subunit(s) related to catalytic activity is/are

(a) α, β, ν (b) $\alpha, \varepsilon, \theta$

(c) θ, ν, ε (d) θ only **Ans. d**

74. The enzyme that stitches the broken DNA strands or replicating DNA as in okazaki fragment is

(a) DNA Pol I (b) DNA ligase

(c) DNA Pol II (d) DNA Pol III **Ans. b**

75. DNA synthesis on leading and lagging strand follows a direction of

(a) 5′-3′ and 5′-3′ (b) 3′-5′ and 5′-3′

(c) 5′-3′ and 3′-5′ (d) 5′-3′ on both strands **Ans. d**

76. The DNA repair is a/an

(a) Autocatalytic system (b) Allocatalytic system

(c) Heterocatalytic system (d) Noncatalytic system **Ans. a**

77. Damages in the DNA are essentially to be repaired because they

(a) Are harmful to the cell

(b) Are deleterious and may cause mutation or death

(c) Regulate speedy synthesis of proteins

(d) Regulate DNA replication at a very high speed where chances of error are very high **Ans. b**

78. Alkylating agents generally cause mutation because they

(a) Alkylate the phosphates of DNA

(b) Alkylate the nitrogenous bases of DNA

(c) Both (a) and (b)

(d) None of the above **Ans. c**

79. Hydrazine-a chemical, damages DNA by acting on the ring structure of cytosine to produce

(a) Adenine (b) Pyrazalone

(c) Xanthine (d) 3-aminopyrasole **Ans. d**

80. In nucleotide excision repair, the recognition and removal of damaged nucleotides is done by

(a) Endonuclease (b) DNA polymerase

(c) ABC- exinuclease (d) Dam-methylase **Ans. c**

❑❑❑

Chapter-12

Nucleic Acids (Part - II) RNA (Ribonucleic acid)

Fill in the blanks

1. tRNA was discovered by..................

Ans. Zamecknik et al.

2. The largest type of RNA is................

Ans. Ribosomal RNA

3. ....discovered mRNA.

Ans. Jacob and Monod

4. The base.............is found only in the nucleotides of RNA.

Ans. Uracil

5. In the lac operon concept..................is a protein.

Ans. repressor

6. mRNA is complementary copy of...................strand of DNA.

Ans. 32-52

7. In the biosynthesis of cDNA, the joining enzyme ligase requires.............

Ans. ATP

8. Amanitin the mushroom poison inhibits................synthesis.

Ans. mRNA

9. The anticodon region is an important part of the structure of.................

Ans. tRNA

10. Catabolism of thymidylate gives.................

Ans. β-aminoisobutyrate

11. ....is the catabolic end product of porphyrine.

Ans. Uric acid

12. cAMD is destroyed by..................

Ans. phosphodiesterase

13. The total RNA is the cell tRNA constitutes........

Ans. 10-20%

14. ....is unit of genetic information.

Ans. cistron

15. RNA synthesis requires..............

Ans. DNA template

16. The antibiotic which leaves DNA is.................

Ans. mitomycin C

17. ....gene is a tumour suppessor gene.

Ans. RB1

18. Ames assay is a rapid method for detection of.............

Ans. chemical carcinogens

19. N-formiminoglutamate is a metabolite of................

Ans. histidine

20. Homogentisic acid is formed from.............

Ans. tyrosine

21. DNA fragments upto 45 kilobases in size can be cloned in...............

Ans. cosmid

22. Fragments of DNA can be identified by the .... technique.

Ans. Southern blotting

23. An antibody probe is used in.................

And. Western blotting

24. A particular RNA in a mixture can be identified by..............

Ans. Northern blotting

25. Monoclonal antibodies are prepared by cloning..................

Ans. hybridoma cells

26. Restriction endonucleases split.............

Ans. double stranded DNA

27. Restriction endonucleases are present in................

Ans. bacteria

28. Suppressor mutations occur in..................

Ans. anticodons of tRNA

29. Mitochondrial protein synthesis is inhibited by...................

Ans. chloramphenicol

30. cis-acting elements include................

Ans. silencers

31. RNA and DNA are chiral molecules, their chirality is due to ....

Ans. D-Sugar component

32. It is estimated that human and chimpanzee genome sequences differ for only upto........................... percent of their base pairs.

Ans. 2(1–3)

33. Many human disease networks are highly conserved in...................

Ans. Fruit fly (*D. melanogaster*)

34. SNPs are single base positions in genomic DNA at which different .... occur in different individuals of a population.

Ans. nucleotides

35. The minimum genome size for a unicellular eukaryote discovered so far is .... bp for the parasite *Encephalitozoan cuniculi.*

Ans. 2.9×10^6

Multiple Choice Questions

1. Ribonucleic acids consists of
   (a) Polypeptides (b) Deoxyribonucleotides
   (c) Ribonucleotides (d) Polysaccharides **Ans. c**
2. The unique features of RNA are
   (a) Presence of uracil (b) Presence of ribose sugar
   (c) Both (a) and (b) (d) Presence of deoxyribose sugar
   Ans. c
3. RNA molecules are single stranded
   (a) In all organisms
   (b) In all plants
   (c) In all viruses
   (d) In all organisms except few viruses **Ans. d**
4. The most abundant type of RNA in the cell-pool in
   (a) mRNA (b) rRNA
   (c) tRNA (d) All are found in equal ratios
   Ans. b
5. In terms of length, the most heterogenous RNA component is
   (a) rRNA (b) tRNA
   (c) mRNA (d) All RNAs are homogenous
   Ans. c
6. Which types of RNAs are involved in protein synthesis
   (a) mRNA
   (b) tRNA
   (c) rRNA
   (d) All types of RNAs are equally involved **Ans. d**
7. Trinucleotide sequence complementary to codon is present on
   (a) tRNA (b) mRNA
   (c) DNA (d) rRNA **Ans. a**
8. Clover leaf model represents a
   (a) 3-dimensional structure of rRNA
   (b) 2-dimensional structure of rRNA

(c) 3-dimensional structure of tRNA

(d) 2-dimensional structure of rRNA **Ans. d**

9. Robert Holley is related to

(a) Structure of mRNA (b) Structure of tRNA

(c) Structure of rRNA (d) Structure of RNA **Ans. b**

10. Which of the following ion is important for tRNA stabilization

(a) Mn^{++} (b) Mg^{++}

(c) Cl^{-} (d) Na^{+} **Ans. b**

11. In a tRNA molecule unpaired CCA sequence is found at

(a) 5′ end only (b) 3′ end only

(c) 5′ and 3′ end (d) None of the above **Ans. b**

12. Three dimensional structure of tRNA was given by

(a) M. Epstein (b) Watson

(c) Jacob and Monod (d) A. Klug **Ans. d**

13. The concept of mRNA in eukaryotes as a chemical intermediate was formulated by

(a) F. Jacob and J. Monod (b) Watson and Crick

(c) A. Klug (d) Ladderberg and Tatum **Ans. a**

14. What will be the base composition of the mRNA derived from a DNA of following sequences 3′ACTCGGTACAT5′

(a) 3′UGACCAUGUA5′ (b) 5′UGAGCCAUGUA3′

(c) 3′TGAGCCATGTA5′ (d) 5′TGAGCCATGTA3′ **Ans. b**

15. Which of the following are the characteristic features of mRNA

(i) Heterogenous nature

(ii) Short biological functioning period

(iii) Polynucleotide nature

(iv) Presence of polydeoxyribo nucleotides

Choose the correct answer

(a) i, ii, iii, iv (b) ii, iii, iv

(c) i, ii, iii (d) ii, iv **Ans. c**

16. The sigma factor of RNA polymerase plays a role of

(a) Recognition of promoter sequences

(b) Elongation of growing mRNA

(c) Stabilization of unwound DNA
(d) Unwinding of double helix of DNA **Ans. a**

17. RNA polymerase synthesizes
(a) DNA from DNA (b) RNA from RNA
(c) RNA from DNA (d) DNA from RNA **Ans. c**

18. Which of the following groups represent the composition of core RNA polymerase
(a) $\sigma, \alpha_2, \beta, \beta'$ (b) α_2, β_2
(c) σ, α_2, β (d) σ, β, β' **Ans. b**

19. The start point for transcription in prokaryotes is
(a) Always 100 bases downstream
(b) 100 bases upstream (mostly)
(c) 10 bases upstream
(d) 10 bases downstream **Ans. b**

20. Pribnow box (TATAAT) is found in all
(a) Promoters (b) Repressors
(c) Eukaryotes (d) Prokaryotic promoters **Ans. d**

21. Which of the following group represents the termination codons
(a) UGA, UAA, UAT (b) UGA, UAG, UAA
(c) UAG, UAT, UUU (d) AUG, UGA, UAG **Ans. b**

22. 7-methyl guanosine cap is found in
(a) mRNA at 3′ end (b) mRNA at both 5′ and 3′ ends
(c) mRNA at 5′ end (d) tRNA at 5′ end **Ans. c**

23. The presence of 7-methyl guanosine cap and poly-A tail in mRNA is meant for
(a) Protection from enzymatic degradation
(b) Transportation from nucleus to the cytoplasm
(c) Both (a) and (b)
(d) Some unknown physio-genetic functions **Ans. c**

24. In eukaryotic system there are three types of RNA polymerases, which of the following is responsible for rRNA and tRNA synthesis
(a) RNA Pol. I (b) RNA Pol. II
(c) RNA Pol. III (d) RNA Pol. I and RNA Pol. III
Ans. d

25. The only RNA polymerase found in nucleolus is

(a) RNA Pol. I (b) RNA Pol. II

(c) RNA Pol. III (d) RNA Pol. I and III **Ans. a**

26. CAAT box is located at

(a) -80 to -60 basepairs (b) +80 to +60

(c) -60 to -40 (d) +60 to -40 **Ans. a**

27. The length of poly-A tail in human mRNA is

(a) 250 (b) 300

(c) 100 (d) 150 **Ans. a**

28. 7-methyl guanosine cap prevents the mRNA from enzymatic degradation because

(a) It degrade the cellular enzymes

(b) It blocks the free phosphate group

(c) It reacts with enzymes activity & of the mRNA selectively

(d) It inhibits enzyme activity **Ans. b**

29. There is no 7-methyl cap as well as poly-A tail, but still translation is normal, this condition is found in

(a) Haemoglobin gene (b) Histone mRNA

(c) Globulin mRNA (d) Albumin mRNA **Ans. b**

30. 7-methyl guanosine is a – – – – modification found in all – – –

(a) Pretranscriptional, Prokaryotes

(b) Pretranscriptional, eukaryotes

(c) Post-transcriptional, Prokaryotes

(d) Post-transcriptional, eukaryotes **Ans. d**

31. Splicing of mRNA is done to

(a) Ensure continuous protein synthesis

(b) Shorten the length & facilitate the easier transportation of mRNA

(c) Both (a) and (b)

(d) Exact function is still not known **Ans. c**

32. 23 S, 16S and 5S types of rRNAs are found in

(a) Prokaryotes (b) Eukaryotes

(c) Both (a) and (b) (d) Some lower eukaryotes only

Ans. a

33. Mammalian ribosomes contain 4 types of rRNAs, they are
(a) 23, 16 and 5S (b) 55, 23, 16 and 5S
(c) 55, 28, 18 and 5.8S (d) Some lower eukaryotes only
Ans. c

34. The most unique feature of rRNA is
(a) Base methylation (b) Decarboxylation
(c) Base replacement (d) De-hydrogenation **Ans. a**

35. Which of the following rRNAs are produced from a common precursor
(a) 55, 28 and 18 S (b) 28, 18 and 5.8S
(c) 28, 18 and 16S (d) 55 and 5.8S **Ans. b**

36. Consider the double helix structure for DNA. Are the base pairs?
(a) Part of the backbone structure (b) Inside the helix
(c) Outside the helix (d) None of these **Ans. b**

37. Mutation in DNA occurs due to changes in the sequence of
(a) Nitrogenous bases (b) Ribose units
(c) Phosphate units (d) None of these **Ans. a**

38. In DNA, the complementary bases are
(a) Uracil and adenine : cytosine and guanine
(b) Adenine and thymine: guanine and cytosine
(c) Adenine and thymine : guanine and uracil
(d) Adenine and guanine : thymine and cytosine **Ans. b**

39. Which substance is not present in nucleic acids?
(a) Cytosine (b) Adenine
(c) Thymine (d) Guanidine **Ans. d**

40. The backbone of DNA/RNA contain ....
(a) Nucleotides (b) Sugar-phosphate linkage
(c) Nucleosides (d) None of these **Ans. b**

41. Which of the following is responsible for hereditary character of the cell?
(a) DNA (b) RNA
(c) Proteins (d) Hormones **Ans. a**

42. Nonmineral essential elements are
(a) Carbon, hydrogen and oxygen
(b) Hydrogen, nitrogen and sulphur

(c) Hydrogen, nitrogen and fluorine
(d) Chlorine, fluorine and nitrogen **Ans. a**

43. Number of essential elements in animals is
(a) 16 (b) 18
(c) 20 (d) 24 **Ans. d**

44. Number of essential elements in plants is
(a) 10 (b) 17
(c) 20 (d) 22 **Ans. b**

45. Nitrogen is
(a) Non-mineral element (b) Mineral element
(c) Nonmetal element (d) All the above **Ans. d**

46. In relation to insulin, Zinc is required for
(a) Synthesis of insulin (b) Release of insulin
(c) Activity of insulin (d) Breakdown of insulin **Ans. b**

47. Element required for phloem transport is
(a) Copper (b) Iron
(c) Boron (d) Sodium **Ans. c**

48. The most abundant mineral of animal body is
(a) Calcium (b) Sodium
(c) Potassium (d) Iron **Ans. a**

49. Most abundant intracellular mineral is
(a) Sodium (b) Potassium
(c) Phosphate (d) Chloride **Ans. b**

50. Most abundant mineral of extracellular fluids is
(a) Calcium (b) Potassium
(c) Sodium (d) Phosphate **Ans. c**

51. Chitin is strengthened by
(a) Calcium phosphate (b) Magnesium phosphate
(c) Calcium carbonate (d) Magnesium carbonate **Ans. c**

52. Mineral forming 90% of the cations of blood plasma is
(a) Mg (b) Ca
(c) K (d) Na **Ans. d**

53. Iron is stored in the body as
(a) Haemoglobin (b) Ferritin
(c) Ferredoxin (d) Myoglobin **Ans. b**

54. Maximum content of iron is present in
(a) Ferritin (b) Myoglobin
(c) Haemoglobin (d) Cytochrome **Ans. c**

55. Purines are
(a) Single ring structures
(b) Double ring structures
(c) Open chain structures
(d) Open chain with alternate double bonds **Ans. b**

56. Purines possess nitrogen at
(a) 1, 2, 4 and 6 positions (b) 1, 3, 5 and 7 positions
(c) 1, 3, 7 and 9 positions (d) 1, 2, 6 and 8 positions **Ans. c**

57. Pseudouracil occurs in
(a) rRNA (b) mRNA
(c) tRNA (d) All the above **Ans. c**

58. Number of pyrimidines found in nucleotides is
(a) 3 (b) 2
(c) 1 (d) Numerous **Ans. a**

59. Nitrogen bases are
(a) Homocyclic (b) Heterocyclic
(c) Open chain hydrocarbons (d) All the above **Ans. b**

60. In nucleoside, nitrogen base is attached to pentose sugar at
(a) Carbon-5′ of pentose sugar (b) Carbon-1′ of pentose sugar
(c) N-1 and N-9 (d) Both (b) and (c) **Ans. d**

61. Nucleoside of uracil is
(a) Uricine (b) Deoxyuridine
(c) Uridine (d) Deoxyuricine **Ans. c**

62. Phosphate of a nucleotide is attached to carbon atom of its pentose sugar
(a) 1′ (b) 2′
(c) 3′ (d) 5′ **Ans. d**

63. Higher nucleotides found free in cytoplasm are
(a) Deoxyribonucleoside diphosphate
(b) Deoxyribonucleoside triphosphate
(c) Ribonucleoside diphosphates triphosphates
(d) All the above **Ans. c**

64. Cyclic AMP is
(a) Adenosine 1-3 monophosphate
(b) Adenosine 2-4 monophosphate
(c) Adenosine 3-5 monophosphate
(d) Adenosine 1-5 monophosphate **Ans. c**

65. Phosphate bonds present in ATP are
(a) 1 (b) 2
(c) 3 (d) 4 **Ans. c**

66. Macromolecules are produced from
(a) Polymerization (b) Micromolecules
(c) Condensation (d) All the above **Ans. d**

67. Polymerization is important in
(a) Producing new chemicals (b) Reducing osmotic influence
(c) Storage (d) All the above **Ans. d**

68. Adenine of DNA is equimolar with
(a) Uridine (b) Thymine
(c) Guanine (d) Cytosine **Ans. b**

69. DNA is a polymer of
(a) Proteins (b) Carbohydrates
(c) Guanine (d) Cytosine **Ans. d**

70. DNA is directly involved in the synthesis except that of
(a) DNA (b) Protein
(c) tRNA (d) mRNA **Ans. b**

71. DNA is unique in that it has
(a) Nitrogen bases (b) Ability to withstand heat
(c) Ability to replicate (d) Ability for replication **Ans. d**

72. DNA can be

(a) A-DNA (b) B-DNA

(c) Z-DNA (d) All the above **Ans. d**

73. A-DNA is characterized by

(a) Right handed helix with 20° bending of nitrogen bases to the axis

(b) Eleven base pairs per turn of duplex

(c) Both (a) and (b)

(d) Right hand helix with ten base pairs in each turn **Ans. c**

74. Z-DNA is characterized by

(a) Left handed helix

(b) Right handed helix

(c) Bending of nitrogen bases to helix axis at an angle other than perpendicular

(d) Both (a) and (b) **Ans. a**

75. Number of base pairs in each turn of B-DNA helix is

(a) 10 (b) 11

(c) 12 (d) Variable **Ans. a**

76. Number of base pairs in each turn of Z-DNA helix is

(a) 10 (b) 11

(c) 12 (d) 15 **Ans. c**

77. DNA differs from RNA in

(a) Nature of sugar alone

(b) Nature of purines alone

(c) Nature of sugar and pyrimidines

(d) All the above **Ans. c**

78. Difference between RNA and DNA is that of

(a) Sugar and base (b) Sugar and phosphate

(c) Phosphate and base (d) All the above **Ans. a**

79. Clover leaf structure is found in

(a) DNA (b) rRNA

(c) tRNA (d) mRNA **Ans. c**

80. Nitrogen bases of RNA are

(a) AUCG (b) AUTC

(c) ATUG (d) ATCG **Ans. a**

81. Initiation code of mRNA occurs

(a) Midway (b) At 3′ end

(c) At 5′ end (d) Submetacentric region **Ans. c**

82. Amino acid binding site of tRNA occurs at

(a) Pseudouridine loop (b) Dihydrouridine loop

(c) 5′ end (d) 3′ end **Ans. d**

83. Which one of the following is not present in RNA?

(a) Uracil (b) Thymine

(c) Ribose (d) Phosphate **Ans. B**

84. A nucleoside consists of

(a) Nitrogenous base

(b) Purine or pyrimidine base + sugar

(c) Purine or pyrimidine base + phosphorous

(d) Purine + pyrimidine base + sugar + phosphorous **Ans. b**

85. A nucleotide consists of

(a) A nitrogenous base like choline

(b) Purine + pyrimidine base + sugar + phosphorous

(c) Purine or pyrimidine base + sugar

(d) Purine or pyrimidine base + phosphorous **Ans. b**

86. A purine nucleotide is

(a) AMP (b) UMP

(c) CMP (d) TMP **Ans. a**

87. A pyrimidine nucleotide is

(a) GMP (b) AMP

(c) CMP (d) IMP **Ans. c**

88. Adenine is

(a) 6-Amino purine (b) 2-Amino-6-oxypurine

(c) 2-Oxy-4-aminopyrimidine (d) 2, 4-Dioxypyrimidine **Ans. a**

89. 2, 4-Dioxypyrimidine is

(a) Thymine (b) Cystosine

(c) Uracil (d) Guanine **Ans. c**

90. The chemical name of guanine is

(a) 2,4-Dioxy-5-methylpyrimidine (b) 2-Amino-6-oxypurine

(c) 2-Oxy-4-aminopyrimidine (d) 2, 4-Dioxypyrimidine **Ans. b**

91. Nucleotides and nucleic acids concentration are often also expressed in terms of

(a) ng (b) mg

(c) meq (d) OD at 260 nm **Ans. d**

92. The pyrimidine nucleotide acting as the high energy intermediate is

(a) ATP (b) UTP

(c) UDPG (d) CMP **Ans. c**

93. The carbon of the pentose in ester linkage with the phosphate in a nucleotide structure is

(a) C_1 (b) C_3

(c) C_4 (d) C_5 **Ans. d**

94. Uracil and ribose form

(a) Uridine (b) Cytidine

(c) Guanosine (d) Adenosine **Ans. a**

95. The most abundant free nucleotide in mammalian cells is

(a) ATP (b) NAD

(c) GTP (d) FAD **Ans. d**

96. The mean intracellular concentration of ATP in mammalian cell is about

(a) 1 mM (b) 2 mM

(c) 0.1 mM (d) 0.2 mM **Ans. a**

97. The nucleic acid base found in mRNA but not in DNA is

(a) Adenine (b) Cytosine

(c) Guanine (d) Uracil **Ans. d**

98. In RNA moleule 'Caps'

(a) Allow tRNA to be processed

(b) Are unique to eukaryotic mRNA

(c) Occur at the 3′ end of tRNA

(d) Allow correct translation of prokaryotic mRNA **Ans. b**

99. In contrast to eukaryotic mRNA, prokaryotic mRNA
(a) Can be polycistronic (b) Is synthesized with introns
(c) Can only be monocistronic (d) Has a poly A tail **Ans. a**

100. The size of small stable RNA ranges from
(a) 0–40 nucleotides (b) 40–80 nucleotides
(c) 90–300 nucleotides (d) More than 320 nucleotides **Ans. c**

101. The number of small stable RNAs per cell ranges from
(a) 10–50,000 (b) 50,000–1,00,000
(c) 1,00,000–10,00,000 (d) More than 10 lakhs **Ans. c**

102. Molecular weight of heterogenous nuclear RNA (hnRNA) is
(a) More than 10^7 (b) 10^5 to 10^6
(c) 10^4 to 10^5 (d) Less than 10^4 **Ans. a**

103. In RNA molecule guanine content does not necessarily equal its cytosine content nor does its adenine content necessarily equal its uracil content since it is a
(a) Single strand molecule
(b) Double stranded molecule
(c) Double stranded helical molecule
(d) Polymer of purine and pyrimidine ribonucleotides **Ans. a**

104. The nitrogenous base present in the RNA molecule is
(a) Thymine (b) Uracil
(c) Xanthine (d) Hypoxanthine **Ans. b**

105. RNA does not contain
(a) Uracil (b) Adenine
(c) Thymine (d) Ribose **Ans. c**

106. The sugar moiety present in RNA is
(a) Ribulose (b) Arabinose
(c) Ribose (d) Deoxyribose **Ans. c**

107. In RNA molecule
(a) Guanine content equals cytosine
(b) Adenine content equals uracil
(c) Adenine content equals guanine
(d) Guanine content does not necessarily equal its cytosine content. **Ans. d**

108. Methylated purines and pyrimidines are characteristically present in

(a) mRNA (b) hnRNA
(c) tRNA (d) rRNA **Ans. c**

109. Thymine is present in

(a) tRNA (b) Ribosomal RNA
(c) Mammalian mRNA (d) Prokaryotic mRNA **Ans. a**

110. The approximate number of nucleotides in tRNA molecule is

(a) 25 (b) 50
(c) 75 (d) 100 **Ans. c**

111. In every cell, the number of tRNA molecules is at least

(a) 10 (b) 20
(c) 30 (d) 40 **Ans. b**

112. The structure of tRNA appears like a

(a) Helix (b) Hair pin
(c) Clover leaf (d) Coil **Ans. c**

113. Although each specific tRNA differs from the others in its sequence of nucleotides, all tRNA molecules contain a base paired stem that terminates in the sequence CCA at

(a) 32 Termini (b) 52 Termini
(c) Anticodon arm (d) 32 52 -Termini **Ans. a**

114. Transfer RNAs are classified on the basis of the number of base pairs in

(a) Acceptor arm (b) Anticodon arm
(c) D arm (d) Extra arm **Ans. d**

115. In tRNA molecule D arm is named for the presence of the base

(a) Uridine (b) Pseudouridine
(c) Dihydrouridine (d) Thymidine **Ans. a**

116. The acceptor arm in the tRNA molecule has

(a) 5 Base pairs (b) 7 Base pairs
(c) 10 Base pairs (d) 20 Base pairs **Ans. b**

117. In tRNA molecule, the anticodon arm possesses

(a) 5 Base pairs (b) 7 Base pairs
(c) 8 Base pairs (d) 10 Base pairs **Ans. a**

118. The T ψ C arm in the tRNA molecule possesses the sequence
(a) T, pseudouridine and C (b) T, uridine and C
(c) T, dihydrouridine and C (d) T, adenine and C **Ans. a**

119. Double helical structure model of the DNA was proposed by
(a) Pauling and Corey (b) Peter Mitchell
(c) Watson and Crick (d) King and Wooten **Ans. c**

120. DNA does not contain
(a) Thymine (b) Adenine
(c) Uracil (d) Deoxyribose **Ans. c**

121. The sugar moiety present in DNA is
(a) Deoxyribose (b) Ribose
(c) Lyxose (d) Ribulose **Ans. a**

122. DNA rich in A-T pairs have
(a) 1 Hydrogen bond (b) 2 Hydrogen bonds
(c) 3 Hydrogen bonds (d) 4 Hydrogen bonds **Ans. b**

123. In DNA molecule
(a) Guanine content does not equal cytosine content
(b) Adenine content does not equal thymine content
(c) Adenine content equals uracil content
(d) Guanine content equals cytosine content **Ans. d**

124. DNA rich in G-C pairs have
(a) 1 Hydrogen bond (b) 2 Hydrogen bonds
(c) 3 Hydrogen bonds (d) 4 Hydrogen bonds **Ans. c**

125. The fact that DNA bears the genetic information of an organism implies that
(a) Base composition should be identical from species to species
(b) DNA base composition should charge with age
(c) DNA from different tissues in the same organism should usually have the same base composition
(d) DNA base composition is altered with nutritional state of an organism **Ans. c**

126. The width (helical diameter) of the double helix in B-form DNA in nm is
(a) 1 (b) 2
(c) 3 (d) 4 **Ans. b**

127. The number of base pair in a single turn of B-form DNA about the axis of the molecule is

(a) 4 (b) 8
(c) 10 (d) 12 **Ans. c**

128. The distance spanned by one turn of B form DNA is

(a) 1.0 nm (b) 2.0 nm
(c) 3.0 nm (d) 3.4 nm **Ans. d**

129. In a DNA molecule the thymine concentration is 30%, the guanosine concentration will be

(a) 10% (b) 20%
(c) 30% (d) 40% **Ans. b**

130. In a DNA molecule, the guanosine content is 40%, the adenine content will be

(a) 10% (b) 20%
(c) 30% (d) 40% **Ans. a**

131. An increased melting temperature of duplex DNA results from a high content of

(a) Adenine + Guanine (b) Thymine + Cytosine
(c) Cytosine + Guanine (d) Cytosine + Adenine **Ans. c**

132. A synthetic nucleotide analogue, 4-hydroxypyrazolopyrimidine is used in the treatment of

(a) Acute nephritis (b) Gout
(c) Cystic fibrosis of lung (d) Multiple myeloma **Ans. b**

133. A synthetic nucleotide analogue, used in the chemotherapy of cancer and viral infections is

(a) Arabinosyl cytosine (b) 4-Hydroxypyrazolopyrimidine
(c) 6-Mercaptopurine (d) 6-Thioguanine **Ans. a**

134. Histamine is formed from histidine by the enzyme histidine decarboxylase in the presence of

(a) NAD (b) FMN
(c) HS-CoA (d) B_6-PO_4 **Ans. d**

135. Infantile convulsions due to lesser formation of gamma amino butyric acid from glutamic acid is seen in the deficiency of

(a) Glutamate-dehydrogenase (b) Pyridoxine
(c) Folic acid (d) Thiamin **Ans. b**

136. Which of the following amino acids produce a vasoconstrictor on decarboxylation?

(a) Histidine (b) Tyrosine
(c) Threonine (d) Arginine **Ans. b**

137. The degradation of RNA by pancreatic ribonuclease produces

(a) Nucleoside 2-Phosphates
(b) Nucleoside 52 -phosphates
(c) Oligonucleosides
(d) Nucleoside 32 -phosphate and oligonucleotide **Ans. d**

138. Intestinal nucleosidases act on nucleosides and produce

(a) Purine base only
(b) Phosphate only
(c) Sugar only
(d) Purine or pyrimidine bases and sugars **Ans. d**

139. In purine biosynthesis carbon atoms at 4 and 5 position and N at 7 position are contributed by

(a) Glycine (b) Glutamine
(c) Alanine (d) Threonine **Ans. a**

140. N^{10}-formyl and N^5, N^{10}-methenyl tetrahydrofolate contributes purine carbon atoms at position

(a) 4 and 6 (b) 4 and 5
(c) 5 and 6 (d) 2 and 8 **Ans. d**

141. In purine nucleus nitrogen atom at 1 position is derived from

(a) Aspartate (b) Glutamate
(c) Glycine (d) Alanine **Ans. a**

142. The key substance in the synthesis of purine, phosphoribosyl pyrophosphate is formed by

(a) α-D-ribose 5-phosphate (b) 5-phospho β-D-ribosylamine
(c) D-ribose (d) Deoxyribose **Ans. a**

143. In purine biosynthesis ring closure in the molecule formyl glycinamide ribosyl-5-phosphate requires the cofactors

(a) ADP (b) NAD
(c) FAD (d) ATP and Mg^{++} **Ans. d**

144. Ring closure of formimidoimidazole carboxamide ribosyl-5-phosphate yields the first purine nucleotide

(a) AMP (b) IMP

(c) XMP (d) GMP **Ans. b**

145. The cofactors required for synthesis of adenylosuccinate are

(a) ATP, Mg^{++} (b) ADP

(c) GTP, Mg^{++} (d) GDP **Ans. c**

146. Conversion of inosine monophosphate to xanthine monophosphate is catalysed by

(a) IMP dehydrogenase

(b) Formyl transferase

(c) Xanthine-guanine phosphoribosyl transferase

(d) Adenine phosphoribosyl transferase **Ans. a**

147. Phosphorylation of adenosine to AMP is catalysed by

(a) Adenosine kinase (b) Deoxycytidine kinase

(c) Adenylosuccinase (d) Adenylosuccinate synthetase **Ans. a**

148. The major determinant of the overall rate of denovo purine nucleotide biosynthesis is the concentration of

(a) 5-phosphoribosyl 1-pyrophosphate

(b) 5-phospho β-D-ribosylamine

(c) Glycinamide ribosyl-5-phosphate

(d) Formylglycinamide ribosyl-5-phosphate **Ans. a**

149. An enzyme which acts as allosteric regulator and sensitive to both phosphate concentration and to the purine nucleotides is

(a) PRPP synthetase

(b) PRPP glutamyl midotransferase

(c) HGPR Tase

(d) Formyl transferase **Ans. a**

150. PRPP glutamyl amidotransferase, the first enzyme uniquely committed to purine synthesis is feed back inhibited by

(a) AMP (b) IMP

(c) XMP (d) CMP **Ans. a**

151. Conversion of formylglycinamide ribosyl-5-phosphate to formylglycinamide ribosyl-5-phosphate is inhibited by

(a) Azaserine (b) Diazonorleucine
(c) 6-Mercaptopurine (d) Mycophenolic acid **Ans. a**

152. In the biosynthesis of purine nucleotides the AMP feed back regulates

(a) Adenylosuccinase (b) Adenylosuccinate synthetase
(c) IMP dehydrogenase (d) HGPR Tase **Ans. b**

153. 6-Mercapto purine inhibits the conversion of

(a) IMP $\rightarrow$ XMP
(b) Ribose 5 phosphate $\rightarrow$ PRPP
(c) PRPP $\rightarrow$ 5-phospho $\rightarrow$ β -D-ribosylamine
(d) Glycinamide ribosyl 5-phosphate $\rightarrow$ formylglycinamide ribosyl-5-phosphate **Ans. a**

154. Purine biosynthesis is inhibited by

(a) Aminopterin (b) Tetracyclin
(c) Methotrexate (d) Chloramphenicol **Ans. a**

155. Pyrimidine and purine nucleoside biosynthesis share a common precursor

(a) PRPP (b) Glycine
(c) Fumarate (d) Alanine **Ans. a**

156. Pyrimidine biosynthesis begins with the formation from glutamine, ATP and CO_2, of

(a) Carbamoyl aspartate (b) Orotate
(c) Carbamoyl phosphate (d) Dihydroorotate **Ans. c**

157. The two nitrogen of the pyrimidine ring are contributed by

(a) Ammonia and glycine
(b) Asparate and carbamoyl phosphate
(c) Glutamine and ammonia
(d) Aspartate and ammonia **Ans. b**

158. A cofactor in the conversion of dihydroorotate to orotic acid, catalysed by the enzyme dihydroorotate dehydrogenase is

(a) FAD (b) FMN
(c) NAD (d) NADP **Ans. c**

159. The first true pyrimidine ribonucleotide synthesized is

(a) UMP (b) UDP
(c) TMP (d) CTP **Ans. a**

160. UDP and UTP are formed by phosphorylation from

(a) AMP (b) ADP
(c) ATP (d) GTP **Ans. c**

161. Reduction of ribonucleotide diphosphates (NDPs) to their corresponding deoxy ribonucleotide diphosphates (dNDPs) involves

(a) FMN (b) FAD
(c) NAD (d) NADPH **Ans. d**

162. Conversion of deoxyuridine monophosphate to thymidine monophosphate is catalysed by the enzyme

(a) Ribonucleotide reductase (b) Thymidylate synthetase
(c) CTP synthetase (d) Orotidylic acid decarboxylase **Ans. b**

163. d-UMP is converted to TMP by

(a) Methylation (b) Decarboxylation
(c) Reduction (d) Deamination **Ans. a**

164. UTP is converted to CTP by

(a) Methylation (b) Isomerisation
(c) Amination (d) Reduction **Ans. c**

165. Methotrexate blocks the synthesis of thymidine monophosphate by inhibiting the activity of the enzyme

(a) Dihydrofolate reductase
(b) Orotate phosphoribosyl transferase
(c) Ribonucleotide reductase
(d) Dihydroorotase **Ans. a**

166. A substrate for enzymes of pyrimidine nucleotide biosynthesis is

(a) Allopurinol (b) Tetracylin
(c) Chloramphenicol (d) Puromycin **Ans. a**

167. An enzyme of pyrimidine nucleotide biosynthesis sensitive to allosteric regulation is

(a) Aspartate transcarbamoylase
(b) Dihydroorotase

(c) Dihydroorotate dehydrogenase
(d) Orotidylic acid decarboxylase **Ans. a**

168. An enzyme of pyrimidine nucleotides biosynthesis regulated at the genetic level by apparently coordinate repression and derepression is
(a) Carbamoyl phosphate synthetase
(b) Dihydroorotate dehydrogenase
(c) Thymidine kinase
(d) Deoxycytidine kinase **Ans. a**

169. The enzyme aspartate transcarbamoylase of pyrimidine biosynthesis is inhibited by
(a) ATP (b) ADP
(c) AMP (d) CTP **Ans. d**

170. In humans end product of purine catabolism is
(a) Uric acid (b) Urea
(c) Allantoin (d) Xanthine **Ans. a**

171. In humans purine are catabolised to uric acid due to lack of the enzyme
(a) Urease (b) Uricase
(c) Xanthine oxidase (d) Guanase **Ans. b**

172. In mammals other than higher primates uric acid is converted by
(a) Oxidation to allantoin (b) Reduction to ammonia
(c) Hydrolysis to ammonia (d) Hydrolysis to allantoin **Ans. a**

173. The correct sequence of the reactions of catabolism of adenosine to uric acid is
(a) Adenosine→hypoxanthine→xanthine→uric acid
(b) Adenosine→xanthine→inosine→uric acid
(c) Adenosine→inosine→hypoxanthine→ xanthine uric acid
(d) Adenosine→xanthine→inosine→hypoxanthine uric acid **Ans. c**

174. Gout is a metabolic disorder of catabolism of
(a) Pyrimidine (b) Purine
(c) Alanine (d) Phenylalanine **Ans. b**

175. Gout is characterized by increased plasma levels of
(a) Urea (b) Uric acid
(c) Creatine (d) Creatinine **Ans. b**

176. Lesch-Nyhan syndrome, the sex linked recessive disorder is due to the lack of the enzyme

(a) Hypoxanthine-guanine phosphoribosyl transferse
(b) Xanthine oxidase
(c) Adenine phosphoribosyl transferase
(d) Adenosine deaminase **Ans. a**

177. Lesch-Nyhan syndrome, the sex linked, recessive absence of HGPRTase, may lead to

(a) Compulsive self destructive behaviour with elevated levels of urate in serum
(b) Hypouricemia due to liver damage
(c) Failure to thrive and megaloblastic anemia
(d) Protein intolerance and hepatic encephalopathy **Ans. a**

178. The major catabolic product of pyrimidines in human is

(a) β-Alanine (b) Urea
(c) Uric acid (d) Guanine **Ans. a**

179. Orotic aciduria type I reflects the deficiency of enzymes

(a) Orotate phosphoribosyl transferase and orotidylate decarboxylase
(b) Dihydroorotate dehydrogenase
(c) Dihydroorotase
(d) Carbamoyl phosphate synthetase **Ans. a**

180. Orotic aciduria type II reflects the deficiency of the enzyme

(a) Orotate phosphoribosyl transferase
(b) Orotidylate decarboxylase
(c) Dihydroorotase
(d) Dihydroorotate dehydrogenase **Ans. b**

181. An autosomal recessive disorder, xanthinuria is due to deficiency of the enzymes

(a) Adenosine deaminase (b) Xanthine oxidase
(c) HGPRTase (d) Transaminase **Ans. b**

182. Enzymic deficiency in β-aminoisobutyric aciduria is

(a) Adenosine deaminase (b) Xanthine oxidase
(c) Orotidylate decarboxylase (d) Transaminase **Ans. d**

183. Polysomes lack in

(a) DNA (b) mRNA

(c) rRNA (d) tRNA **Ans. a**

184. Genetic information flows from

(a) DNA to DNA (b) DNA to RNA

(c) RNA to cellular proteins (d) DNA to cellular proteins **Ans. b**

185. Genetic code is

(a) Collection of codon

(b) Collection of amino acids

(c) Collection of purine nucleotide

(d) Collection of pyrimidine nucleotide **Ans. a**

186. Degeneracy of genetic code implies that

(a) Codons do not code for specific amino acid

(b) Multiple codons must decode the same amino acids

(c) No anticodon on tRNA molecule

(d) Specific codon decodes many amino acids **Ans. b**

187. Genetic code is

(a) Overlapping (b) Non-overlapping

(c) Not universal (d) Ambiguous **Ans. b**

188. mRNA is complementary to the nucleotide sequence of

(a) Coding strand (b) Ribosomal RNA

(c) tRNA (d) Template strand **Ans. a**

189. In DNA replication the enzyme required in the first step is

(a) DNA directed polymerase (b) Unwinding proteins

(c) DNA polymerase (d) DNA ligase **Ans. b**

190. The smallest unit of DNA capable of coding for the synthesis of a polypeptide is

(a) Operon (b) Repressor gene

(c) Cistron (d) Replicon **Ans. c**

191. Termination of the synthesis of the RNA molecule is signaled by a sequence in the template strand of the DNA molecule, a signal that is recognized by a termination protein, the

(a) Rho (ρ) factor (b) σ factor

(c) δ factor (d) ε factor **Ans. a**

192. After termination of the synthesis of RNA molecule, the core enzymes separate from the DNA template. The core enzymes then recognize a promoter at which the synthesis of a new RNA molecule commences, with the assistance of

(a) Rho (ρ) factor (b) δ factor

(c) β factor (d) σ factor **Ans. d**

193. In the process of transcription in bacterial cells

(a) Initiation requires rho protein

(b) RNA polymerase incorporates methylated bases in correct sequence

(c) Both the sigma unit and core enzymes of RNA polymerase are required for accurate promotor site binding

(d) Primase is necessary for initiation **Ans. c**

194. The correct statement concerning RNA and DNA polymerases is

(a) RNA polymerase use nucleoside diphosphates

(b) RNA polymerase require primers and add bases at 5′ end of the growing polynucleotide chain

(c) DNA polymerases can add nucleotides at both ends of the chain

(d) All RNA and DNA polymerases can add nucleotides only at the 3′ end of the growing polynucleotide chain **Ans. d**

195. The eukaryotic nuclear chromosomal DNA

(a) Is a linear and unbranched molecule

(b) Is not associated with a specific membranous organelle

(c) Is not replicated semiconservatively

(d) Is about of the same size as each prokaryotic chromosomes **Ans. a**

196. The function of a repressor protein in an operon system is to prevent synthesis by binding to

(a) The ribosome

(b) A specific region of the operon preventing transcription of structural genes

(c) The RNA polymerase
(d) A specific region of the mRNA preventing translation to protein

Ans. b

197. All pribnow boxes are variants of the sequence

(a) 5′ -TATAAT -3′ (b) 5′ -GAGCCA -3′
(c) 5′ -UAACAA -3′ (d) 5′ -TCCTAG -3′ **Ans. a**

198. 5′-Terminus of mRNA molecule is capped with

(a) Guanosine triphosphate
(b) 7-Methylguanosine triphophate
(c) Adenosine triphosphate
(d) Adenosine diphosphate **Ans. b**

199. The first codon to be translated on mRNA is

(a) AUG (b) GGU
(c) GGA (d) AAA **Ans. a**

200. AUG, the only identified codon for methionine is important as

(a) A releasing factor for peptide chains
(b) A chain terminating codon
(c) Recognition site on tRNA
(d) A chain initiating codon **Ans. d**

201. In biosynthesis of proteins the chain terminating codons are

(a) UAA, UAG and UGA (b) UGG, UGU and AGU
(c) AAU, AAG and GAU (d) GCG, GCA and GCU **Ans. a**

202. The formation of initiation complex during protein synthesis requires a factor

(a) IF-III (b) EF-I
(c) EF-II (d) IF-I **Ans. a**

203. The amino terminal of all polypeptide chain at the time of synthesis in *E. coli* is tagged to the amino acid residue

(a) Methionine (b) Serine
(b) N-formyl methinine (d) N-formal serine **Ans. c**

204. Initiation of protein synthesis begins with binding of

(a) 40S ribosomal unit on mRNA
(b) 60S ribosomal unit

(c) Charging of tRNA with specific amino acid
(d) Attachment of aminoacyl tRNA on mRNA **Ans. a**

205. Initiation of protein synthesis requires

(a) ATP (b) AMP
(c) GDP (d) GTP **Ans. d**

206. The enzyme amino acyl tRNA synthetase is involved in

(a) Dissociation of discharged tRNA from 80S ribosome
(b) Charging of tRNA with specific amino acids
(c) Termination of protein synthesis
(d) Nucleophilic attack on esterified carboxyl group of peptidyl tRNA **Ans. b**

207. In the process of activation of amino acids for protein synthesis, the number of high energy phosphate bond equivalent utilised is

(a) 0 (b) 1
(c) 2 (d) 4 **Ans. c**

208. Translation results in a product known as

(a) Protein (b) tRNA
(c) mRNA (d) rRNA **Ans. a**

209. In the process of elongation of chain binding of amino acyl tRNA to the A site requires

(a) A proper codon recognition (b) GTP
(c) EF-II (d) GDP **Ans. a**

210. The newly entering amino acyl tRNA into A site requires

(a) EF-II (b) Ribosomal RNA
(c) mRNA (d) EF-I **Ans. d**

211. The α-amino group of the new amino acyl tRNA in the A site carries out a nucleophilic attack on the esterified carboxyl group of the peptidyl tRNA occupying the P site. This reaction is catalysed by

(a) DNA polymerase (b) RNA polymerase
(c) Peptidyl transferase (d) DNA ligase **Ans. c**

212. The nucleophilic attack on the esterified carboxyl group of the peptidyl-tRNA occupying the P site and the α-amino group of the new amino acyl tRNA, the number of ATP required by the amino acid on the charged tRNA is

(a) Zero (b) One

(c) Two (d) Four **Ans. a**

213. Translocation of the newly formed peptidyl tRNA at the A site into the empty P site involves

(a) EF-II, GTP (b) EF-I, GTP

(c) EF-I, GDP (d) Peptidyl transferase, GTP

Ans. a

214. In eukaryotic cells

(a) Formylated tRNA is important for initiation of translation

(b) Cyclohexamide blocks elongation during translation

(c) Cytosolic ribosomes are smaller than those found in prokaryotes

(d) Erythromycin inhibits elongation during translation **Ans. a**

215. The mushroom poison amanitin is an inhibitor of

(a) Protein synthesis (b) mRNA synthesis

(c) DNA synthesis (d) Adenosine synthesis **Ans. b**

216. Tetracylin prevents synthesis of polypeptide by

(a) Blocking mRNA formation from DNA

(b) Releasing peptides from mRNA-tRNA complex

(c) Competing with mRNA for ribosomal binding sites

(d) Preventing binding of aminoacyl tRNA **Ans. d**

217. In prokaryotes, chloramphenicol

(a) Causes premature release of the polypeptide chain

(b) Causes misreading of the mRNA

(c) Depolymerises DNA

(d) Inhibits peptidyl transferase activity **Ans. a**

218. Streptomycin prevents synthesis of polypeptide by

(a) Inhibiting initiation process

(b) Releasing premature polypeptide

(c) Inhibiting peptidyl transferase activity

(d) Inhibiting translocation **Ans. a**

219. Erythromycin acts on ribosomes and inhibit
(a) Formation of initiation complex
(b) Binding of aminoacyl tRNA
(c) Peptidyl transferase activity
(d) Translocation **Ans. d**

220. The binding of prokaryotic DNA dependent RNA polymerase to promoter sites of genes is inhibited by the antibiotic
(a) Puromycin (b) Rifamycin
(c) Terramycin (d) Streptomycin **Ans. b**

221. The gene which is transcribed during repression is
(a) Structural (b) Regulator
(c) Promoter (d) Operator **Ans. b**

222. The gene of lac operon which has constitutive expression is
(a) i (b) c
(c) z (d) p **Ans. a**

223. The minimum effective size of an operator for lac repressor binding is
(a) 5 base pairs (b) 10 base pairs
(c) 15 base pairs (d) 17 base pairs **Ans. d**

224. To commence structural gene transcription the region which should be free on lac operation is
(a) Promoter site (b) Operator locus
(c) Y gene (d) A gene **Ans. b**

225. In the lac operon concept, a protein molecule is
(a) Operator (b) Inducer
(c) Promoter (d) Repressor **Ans. d**

226. The catabolite repression is mediated by a catabolite gene activator protein (CAP) in conjunction with
(a) AMP (b) GMP
(c) cAMP (d) cGMP **Ans. c**

227. The enzyme DNA ligase
(a) Introduces superhelical twists
(b) Connects the end of two DNA chains
(c) Unwinds the double helix
(d) Synthesises RNA primers **Ans. b**

228. Restriction endonucleases
(a) Cut RNA chains at specific locations
(b) Excise introns from hnRNA
(c) Remove Okazaki fragments
(d) Act as defensive enzymes to protect the host bacterial DNA from DNA of foreign organisms **Ans. d**

229. The most likely lethal mutation is
(a) Substitution of adenine for cytosine
(b) Insertion of one nucleotide
(c) Deletion of three nucleotides
(d) Substitution of cytosine for guanine **Ans. b**

230. In the following partial sequence of mRNA, a mutation of the template DNA results in a change in codon 91 to UAA. The type of mutation is

88	89	90	91	92	93	94
GUC	GAC	CAG	UAG	GGC	UAA	CCG

(a) Missene (b) Silent
(c) Nonsense (d) Frame shit **Ans. b**

231. Restriction endonucleases recognize and cut a certain sequence of
(a) Single stranded DNA (b) Double stranded DNA
(c) RNA (d) Protein **Ans. b**

232. Positive control of induction is best described as a control system in which an operon functions
(a) Unless it is switched off by a derepressed repressor protein
(b) Only after a repressor protein is inactivated by an inducer
(c) Only after an inducer protein, which can be inactivated by a corepressor, switches it on
(d) Only after an inducer protein, which is activated by an inducer, switch it on **Ans. d**

233. Interferon
(a) Is virus specific
(b) Is a bacterial product
(c) Is a synthetic antiviral agent
(d) Requires expression of cellular genes **Ans. d**

234. Repressor binds to DNA sequence and regulate the transcription. This sequence is called

(a) Attenuator (b) Terminator

(c) Anti terminator (d) Operator **Ans. d**

235. Okazaki fragment is related to

(a) DNA synthesis (b) Protein synthesis

(c) mRNA formation (d) tRNA formation **Ans. a**

236. The region of DNA known as TATA BOX is the site for binding of

(a) DNA polymerase

(b) DNA topoisomerase

(c) DNA dependent RNA polymerase

(d) Polynucleotide phosphorylase **Ans. c**

237. Reverse transcriptase is capable of synthesizing

(a) RNA → DNA (b) DNA → RNA

(c) RNA → RNA (d) DNA → DNA **Ans. a**

238. A tetrovirus is

(a) Polio virus (b) HIV

(c) Herpes virus (d) Tobacco mosaic virus **Ans. b**

239. Peptidyl transferase activity is located in

(a) Elongation factor (b) A charged tRNA molecule

(c) Ribosomal protein (d) A soluble cytosolic protein

Ans. c

240. Ultraviolet light can damage a DNA strand causing

(a) Two adjacent purine residue to form a covalently bounded dimer

(b) Two adjacent pyrimidine residues to form covalently bonded dimer

(c) Disruption of phosphodiesterase linkage

(d) Disruption of non-covalent linkage **Ans. b**

241. Defective enzyme in Hurler's syndrome is

(a) α-L-diuronidase (b) Iduronate sulphatase

(c) Arylsulphatase B (d) C-acetyl transferase **Ans. a**

242. Presence of arginine can be detected by

(a) Sakaguchi reaction (b) Million-Nasse reaction

(c) Hopkins-Cole reaction (d) Gas chromatography **Ans. a**

243. A nitrogenous base that does not occur in mRNA is
(a) Cytosine (b) Thymine
(c) Uracil (d) All of these **Ans. a**

244. In nucleotides, phosphate is attached to sugar by
(a) Salt bond (b) Hydrogen bond
(c) Ester bond (d) Glycosidic bond **Ans. c**

245. Cyclic AMP can be formed from
(a) AMP (b) ADP
(c) ATP (d) All of these **Ans. c**

246. A substituted pyrimidine base of pharmacological value is
(a) 5-Iododeoxyuridine (b) Cytisine arabinoside
(c) 5-Fluorouracil (d) All of these **Ans. c**

247. The 'transforming factor' discovered by Avery, McLeod and McCarty was later found to be
(a) mRNA (b) tRNA
(c) DNA (d) None of these **Ans. c**

248. In DNA, the complementary base of adenine is
(a) Guanine (b) Cytosine
(c) Uracil (d) Thymine **Ans. d**

249. In DNA, three hydrogen bonds are formed between
(a) Adenine and guanine (b) Adenine and thymine
(c) Guanine and cytosine (d) Thymine and cytosine **Ans. c**

250. Left handed double helix is present in
(a) Z-DNA (b) A-DNA
(c) B-DNA (d) None of these **Ans. a**

251. Nuclear DNA is present in combination with
(a) Histones (b) Non-histones
(c) Both (a) and (b) (d) None of these **Ans. c**

252. Number of guanine and cytosine residues is equal in
(a) mRNA (b) tRNA
(c) DNA (d) None of these **Ans. c**

253. Alkali cannot hydrolyse

(a) mRNA (b) tRNA

(c) rRNA (d) DNA **Ans. d**

254. Codons are present in

(a) Template strand of DNA (b) mRNA

(c) tRNA (d) rRNA **Ans. b**

255. Amino acid is attached to tRNA at

(a) 5′-End (b) 3′-End

(c) Anticodon (d) DHU loop **Ans. b**

256. In prokaryotes, the ribosomal subunits are

(a) 30 S and 40 S (b) 40 S and 50 S

(c) 30 S and 50 S (d) 40 S and 60 S **Ans. c**

257. Ribozymes are

(a) Enzymes present in ribosomes

(b) Enzymes which combine the ribosomal subunits

(c) Enzymes which dissociate

(d) Enzymes made up of RNA **Ans. d**

258. The smallest RNA among the following is

(a) rRNA (b) hnRNA

(c) mRNA (d) tRNA **Ans. d**

259. The number of adenine and thymine bases is equal in

(a) DNA (b) mRNA

(c) tRNA (d) rRNA **Ans. a**

260. The number of hydrogen bonds between adenine and thymine in DNA is

(a) One (b) Two

(c) Three (d) Four **Ans. b**

261. The complementary base of adenine in RNA is

(a) Thymine (b) Cystosine

(c) Guanine (d) Uracil **Ans. d**

262. Extranuclear DNA is present in

(a) Ribosomes (b) Endoplasmic reticulum

(c) Lysosomes (d) Mitochondria **Ans. d**

263. Mitochondrial DNA is present in
(a) Bacteria (b) Viruses
(c) Eukaryotes (d) All of these **Ans. c**

264. Ribothymidine is present in
(a) DNA (b) tRNA
(c) rRNA (d) hnRNA **Ans. b**

265. Ten base pairs are present in one turn of the helix in
(a) A-DNA (b) B-DNA
(c) C-DNA (d) Z-DNA **Ans. b**

266. Transfer RNA transfers
(a) Information from DNA to ribosomes
(b) Information from mRNA to cytosol
(c) Amino acids from cytosol to ribosomes
(d) Proteins from ribosomes to cytosol **Ans. c**

267. Ceramidase is deficient in
(a) Fabry's disease (b) Farber's disease
(c) Krabbe's disease (d) Tay-Sachs disease **Ans. b**

268. Ceramide is present in all of the following except
(a) Plasmalogens (b) Cerebrosides
(c) Sulphatides (d) Sphingomyelin **Ans. a**

269. Nucleotides required for the synthesis of nucleic acids can be obtained from
(a) Dietary nucleic acids and nucleotides
(b) De novo synthesis
(c) Salvage of pre-existing bases and nucleosides
(d) De novo synthesis and salvage **Ans. d**

270. De novo synthesis of purine nucleotide occurs in
(a) Mitochondria (b) Cytosol
(c) Microsmes (d) Ribosomes **Ans. b**

271. The nitrogen atoms for de novo synthesis of purine nucleotides are provided by
(a) Aspartate and glutamate
(b) Aspartate and glycine

(c) Aspartate, glutamine and glycine
(d) Aspartate, glutamate and glycine **Ans. c**

272. For de novo synthesis of purine nucleotides, glycine provides
(a) One nitrogen atom
(b) One nitrogen and one carbon atom
(c) Two carbon atoms
(d) One nitrogen and two carbon atoms **Ans. d**

273. For de novo synthesis of purine nucleotides, aspartate provides
(a) Nitrogen 1 (b) Nitrogen 3
(c) Nitrogen 7 (b) Nitrogen 9 **Ans. a**

274. In the purine nucleus, carbon 6 is contributed by
(a) Glycine (b) CO_2
(c) Aspartate (d) Glutamine **Ans. b**

275. 5-Phosphoribosyl-1-pyrophosphate is required for the synthesis of
(a) Purine nucleotides (b) Pyrimidine nucleotides
(c) Both (a) and (b) (d) None of these **Ans. c**

276. Inosine monophophate is an intermediate during the de novo synthesis of
(a) AMP and GMP (b) CMP and UMP
(c) CMP and TMP (d) All of these **Ans. a**

277. Xanthosine monophosphate is an intermediate during de novo synthesis of
(a) TMP (b) CMP
(c) AMP (d) GMP **Ans. d**

278. In the pathway of de novo synthesis of purine nucleotides, all the following are allosteric enzymes except
(a) PRPP glutamyl amido transferase
(b) Adenylosuccinate synthetase
(c) IMP dehydrogenase
(d) Adenylosuccinase **Ans. d**

279. All of the following enzymes are unique to purine nucleotide synthesis except
(a) PRPP synthetase
(b) PRPP glutamyl amido transferase

(c) Adenylosuccinate synthetase
(d) IMP dehydrogenase **Ans. a**

280. PRPP synthetase is allosterically inhibited by
(a) AMP (b) ADP
(c) GMP (d) All of these **Ans. d**

281. An allosteric inhibitor of PRPP glutamyl amido transferase is
(a) AMP (b) ADP
(c) GMP (d) All of these **Ans. c**

282. An allosteric inhibitor of adenylosuccinate synthetase is
(a) AMP (b) ADP
(c) GMP (d) GDP **Ans. a**

283. An allosteric inhibitor of IMP dehydrogenase is
(a) AMP (b) ADP
(c) GMP (d) GDP **Ans. c**

284. GMP is an allosteric inhibitor of all the following except
(a) PRPP synthetase
(b) PRPP glutamyl amido synthetase
(c) IMP dehydrogenase
(d) Adenylosuccinate synthetase **Ans. d**

285. AMP is an allosteric inhibitor of
(a) PRPP synthetase (b) Adenylosucciante synthetase
(c) Both (a) and (b) (d) None of these **Ans. c**

286. The first reaction unique to purine nucleotide synthesis is catalysed by
(a) PRPP synthetase
(b) PRPP glutamyl amido transferase
(c) Phosphoribosyl glycinamide synthetase
(d) Formyl transferase **Ans. b**

287. Free purine bases which can be salvaged are
(a) Adenine and guanine
(b) Adenine and hypoxanthine
(c) Guanine and hypoxanthine
(d) Adenine, guanine and hypoxanthine **Ans. d**

288. The enzyme required for salvage of free purine bases is
(a) Adenine phosphoribosyl transferase
(b) Hypoxanthine guanine phosphoribosyl transferase
(c) Both (a) and (b)
(d) None of these **Ans. c**

289. Deoxycytidine kinase can salvage
(a) Adenosine
(b) Adenosine and deoxyadenosine
(c) Adenosine and guanosine
(d) Adenine and adenosine **Ans. d**

290. Adenosine kinase can salvage
(a) Adenosine
(b) Adenosine and deoxyadenosine
(c) Adenosine and guanosine
(d) Adenine and adenosine **Ans. b**

291. Salvage of purine bases is regulated by
(a) Adenosine phosphoribosyl transferase
(b) Hypoxanthine guanine phosphoribosyl transferase
(c) Availability of PRPP
(d) None of these **Ans. c**

292. The available PRPP is used preferentially for
(a) De novo synthesis of purine nucleotides
(b) De novo synthesis of pyrimidine nucleotides
(c) Salvage of purine bases
(d) Salvage of pyrimidine bases **Ans. c**

293. The end product of purine catabolism in man is
(a) Inosine (b) Hypoxanthine
(c) Xanthine (d) Uric acid **Ans. d**

294. The enzyme common to catabolism of all the purines is
(a) Adenosine deaminase
(b) Purine nucleoside phosphorylase
(c) Guanase
(d) None of these **Ans. b**

295. Uric acid is the end product of purine as well as protein catabolism in

(a) Man (b) Fish

(c) Birds (d) None of these **Ans. c**

296. Daily uric acid excretion in adult men is

(a) 2–6 mg (b) 20–40 mg

(c) 150–250 mg (d) 40–600 mg **Ans. d**

297. Dietary purines are catabolised in

(a) Liver (b) Kidneys

(c) Intesitnal mucosa (d) All of these **Ans. c**

298. De novo synthesis of pyrimidine nucleotides occurs in

(a) Mitochondria (b) Cytosol

(c) Microsomes (d) Ribosomes **Ans. b**

299. An enzyme common to de novo synthesis of pyrimidine nucleotides and urea is

(a) Urease

(b) Carbamoyl phosphate synthetase

(c) Aspartate transcarbamoylase

(d) Argininosuccinase **Ans. b**

300. The nitrogen atoms of pyrimidine nucleus are provided by

(a) Glutamate (b) Glutamate and aspartate

(c) Glutamine (d) Glutamine and aspartate

Ans. d

301. The carbon atoms of pyrimidine nucleus are provided by

(a) Glycine and aspartate (b) CO_2 and aspartate

(c) CO_2 and glutamate (d) CO_2 and glutamine **Ans. b**

302. Nitrogen at position 1 of pyrimidine nucleus comes from

(a) Glutamine (b) Glutamate

(c) Glycine (d) Aspartate **Ans. d**

303. Nitrogen at position 3 of pyrimidine nucleus comes from

(a) Glutamine (b) Glutamate

(c) Glycine (d) Aspartate **Ans. a**

304. The carbon atom at position 2 of pyrimidine nucleus is contributed by
(a) CO_2 (b) Glycine
(c) Aspartate (d) Glutamine **Ans. a**

305. Aspartate contributes the following carbon atoms of the pyrimidine nucelus
(a) C_2 and C_4 (b) C_5 and C_6
(c) C_2, C_4 and C_6 (d) C_4, C_5 and C_6 **Ans. d**

306. The first pyrimidine nucleotide to be formed in de novo synthesis pathway is
(a) UMP (b) CMP
(c) CTP (d) TMP **Ans. a**

307. Conversion of uridine diphosphate into deoxyuridine diphosphate requires all the following except
(a) Ribonucleotide reductase (b) Thioredoxin
(c) Tetrahydrobiopterin (d) NADPH **Ans. c**

308. Amethopterin and aminopterin decrease the synthesis of
(a) TMP (b) UMP
(c) CMP (d) All of these **Ans. a**

309. For synthesis of CTP and UTP, the amino group comes from
(a) Amide group of Asparagine (b) Amide group of glutamine
(c) α-Amino group of glutamine (d) α-Amino group of glutamate
Ans. b

310. CTP synthetase forms CTP from
(a) CDP and inorganic phosphate (b) CDP and ATP
(c) UTP and glutamine (d) UTP and glutamate **Ans. c**

311. For the synthesis of TMP from dump, a coenzyme is required which is
(a) N^{10}- Formyl tetrahydrofolate
(b) N^5- Methyl tetrahydrofolate
(c) N^5, N^{10}- Methylene tetrahydrofolate
(d) N^5- Formimino tetrahydrofolate **Ans. c**

312. All the enzymes required for de novo synthesis of pyrimidine nucleotides are cytosolic except
(a) Carbamoyl phosphate synthetase
(b) Aspartate transcarbamoylase

(c) Dihydro-orotase
(d) Dihydro-orotate dehydrogenase **Ans. d**

313. During de novo synthesis of pyrimidine nucleotides, the first ring compound to be formed is
(a) Carbamoyl aspartic acid (b) Dihydro-orotic acid
(c) Orotic acid (d) Orotidine monophosphate
Ans. b

314. Tetrahydrofolate is required as a coenzyme for the synthesis of
(a) UMP (b) CMP
(c) TMP (d) All of these **Ans. c**

315. All of the following statements about thioredoxin reductase are true except
(a) It requires NADH as a coenzyme
(b) Its substrates are ADP, GDP, CDP and UDP
(c) It is activated by ATP
(d) It is inhibited by dADP **Ans. a**

316. De novo synthesis of pyrimidine nucleotides is regulated by
(a) Carbamoyl phosphate synthetase
(b) Aspartate transcarbamoylase
(c) Both (a) and (b)
(d) None of these **Ans. c**

317. Cytosolic carbamoyl phosphate synthetase is inhibited by
(a) UTP (b) CTP
(c) PRPP (d) TMP **Ans. a**

318. Cytosolic carbamoyl phosphate synthetase is activated by
(a) Glutamine (b) PRPP
(c) ATP (d) Aspartate **Ans. b**

319. Aspartate transcarbamoylase is inhibited by
(a) CTP (b) PRPP
(c) ATP (d) TMP **Ans. a**

320. The following cannot be salvaged in human beings
(a) Cytidine (b) Deoxycytidine
(c) Cytosine (d) Thymidine **Ans. c**

321. β -Aminoisobytyrate is formed from catabolism of

(a) Cytosine (b) Uracil

(c) Thymine (d) Xanthine **Ans. c**

322. Free ammonia is liberated during the catabolism of

(a) Cytosine (b) Uracil

(c) Thymine (d) All of these **Ans. d**

323. β -Alanine is formed from catabolism of

(a) Thymine (b) Thymine and cytosine

(c) Thymine and uracil (d) Cytosine and uracil **Ans. d**

324. The following coenzyme is required for catabolism of pyrimidine bases

(a) NADH (b) NADPH

(c) $FADH_2$ (d) None of these **Ans. b**

325. Inheritance of primary gout is

(a) Autosomal recessive (b) Autosomal dominant

(c) X-linked recessive (d) X-linked dominant **Ans. c**

326. The following abnormality in PRPP synthetase can cause primary gout

(a) High V_{max}

(b) Low K_m

(c) Resistance to allosteric inihbition

(d) All of these **Ans. d**

327. All the following statements about primary gout are true except

(a) Its inheritance is X-linked recessive

(b) It can be due to increased activity of PRPP synthetase

(c) It can be due to increased activity of hypoxanthine guanine phosphoribosyl transferase

(d) De novo synthesis of purines is increased in it **Ans. c**

328. All of the following statements about uric acid are true except

(a) It is a catabolite of purines

(b) It is excreted by the kidneys

(c) It is undissociated at pH above 5.8

(d) It is less soluble than sodium urate **Ans. c**

329. In inherited deficiency of hypoxanthine guanine phosphoribosyl transferase
(a) De novo synthesis of purine nucleotides is decreased
(b) Salvage of purines is decreased
(c) Salvage of purines is increased
(d) Synthesis of uric acid is decreased **Ans. b**

330. All of the following statements about uric acid are true except
(a) It can be formed from allantoin
(b) Formation of uric acid stones in kidneys can be decreased by alkalinisation of urine
(c) Uric acid begins to dissociate at pH above 5.8
(d) It is present in plasma mainly as monosodium urate **Ans. a**

331. All of the following statements about primary gout are true except
(a) Uric acid stones may be formed in kidneys
(b) Arthritis of small joints occurs commonly
(c) Urinary excretion of uric acid is decreased
(d) It occurs predominantly in males **Ans. c**

332. All of the following statements about allopurinol are true except
(a) It is a structural analogue of uric acid
(b) It can prevent uric acid stones in the kidneys
(c) It increases the urinary excretion of xanthine and hypoxanthine
(d) It is a competitive inhibitor of xanthine oxidase **Ans. a**

333. Orotic aciduria can be controlled by
(a) Oral administration of orotic acid
(b) Decreasing the dietary intake of orotic acid
(c) Decreasing the dietary intake of pyrimidines
(d) Oral administration of uridine **Ans. d**

334. All of the following occur in orotic aciduria except
(a) Increased synthesis of pyrimidine nucleotides
(b) Increased excretion of orotic acid in urine
(c) Decreased synthesis of cytidine triphosphate
(d) Retardation of growth **Ans. a**

335. Inherited deficiency of adenosine deaminase causes

(a) Hyperuricaemia and gout (b) Mental retardation

(c) Immunodeficiency (d) Dwarfism **Ans. c**

336. Complete absence of hypoxanthine guanine phospharibosyl transferase causes

(a) Primary gout (b) Immunodeficiency

(c) Uric acid stones (d) Lesh-Nyhan syndrome **Ans. d**

337. Increased urinary excretion of orotic acid can occur in deficiency of

(a) Orotate phosphoribosyl transferase

(b) OMP decarboxylase

(c) Mitochondrial ornithine transcarbamoylase

(d) Any of the above **Ans. d**

338. All of the following can occur in Lesch- Nyhan syndrome except

(a) Gouty arthritis (b) Uric acid stones

(c) Retarted growth (d) Self-mutiliating behavior

Ans. c

339. Inherited deficiency of purine nucleoside phosphorylase causes

(a) Dwarfism (b) Mental retardation

(c) Immunodeficiency (d) Gout **Ans. c**

340. Deoxyribonucleotides are formed by reduction of

(a) Ribonucleosides

(b) Ribonucleoside monophosphates

(c) Ribonucleoside diphosphates

(d) Ribonucleoside triphosphates **Ans. c**

341. An alternate substrate for orotate phosphoribosyl transferase is

(a) Allopurinol (b) Xanthine

(c) Hypoxanthine (d) Adenine **Ans. d**

342. Mammals other than higher primates do not suffer from gout because they

(a) Lack xanthine oxidase

(b) Lack adenosine deaminase

(c) Lack purine nucleoside phosphorylase

(d) Possess uricase **Ans. d**

343. Hypouricaemia can occur in

(a) Xanthine oxidase deficiency (b) Psoriasis

(c) Leukaemia (d) None of these **Ans. a**

344. Synthesis of DNA is also known as

(a) Duplication (b) Replication

(c) Transcription (d) Translation **Ans. b**

345. Replication of DNA is

(a) Conservative (b) Semi-conservative

(c) Non-conservative (d) None of these **Ans. b**

346. Direction of DNA synthesis is

(a) $5' \rightarrow 3'$ (b) $3' \rightarrow 5'$

(c) Both (a) and (b) (d) None of these **Ans. a**

347. Formation of RNA primer

(a) Precedes replication (b) Follows replication

(c) Precedes transcription (d) Follows transcription **Ans. a**

348. Okazaki pieces are made up of

(a) RNA (b) DNA

(c) RNA and DNA (d) RNA and proteins **Ans. c**

349. Okazaki pieces are formed during the synthesis of

(a) mRNA (b) tRNA

(c) rRNA (d) DNA **Ans. d**

350. After formation of replication fork

(a) Both the new strands are synthesized discontinuously

(b) One strand is synthesized continuously and the other discontinuously

(c) Both the new strands are synthesized continuously

(d) RNA primer is required only for the synthesis of one new strand

Ans. b

351. An Okazaki fragment contains about

(a) 10 Nucleotides (b) 100 Nucleotides

(c) 1,000 Nucleotides (d) 10,000 Nucleotides **Ans. c**

352. RNA primer is formed by the enzyme
(a) Ribonuclease (b) Primase
(c) DNA polymerase I (d) DNA polymerase III **Ans. b**

353. In RNA, the complementary base of adenine is
(a) Cytosine (b) Guanine
(c) Thymine (d) Uracil **Ans. d**

354. During replication, the template DNA is unwound
(a) At one of the ends (b) At both the ends
(c) At multiple sites (d) Nowhere **Ans. c**

355. During replication, unwinding of double helix is initiated by
(a) DNAA protein (b) DnaB protein
(c) DNAC protein (d) Rep protein **Ans. b**

356. For unwinding of double helical DNA,
(a) Energy is provided by ATP
(b) Energy is provided by GTP
(c) Energy can be provided by either ATP or GTP
(d) No energy is required **Ans. a**

357. Helicase and DNAB protein cause
(a) Rewinding of DNA and require ATP as a source of energy
(b) Rewinding of DNA but do not require any source of energy
(c) Unwinding of DNA and require ATP as a source of energy
(d) Unwinding of DNA but do not require any source of energy **Ans. c**

358. The unwound strands of DNA are held apart by
(a) Single strand binding protein (b) Double strand binding protein
(c) Rep protein (d) DNAA protein **Ans. a**

359. Deoxyribonucleotides are added to RNA primer by
(a) DNA polymerase I
(b) DNA polymerase II
(c) DNA polymerase III holoenzyme
(d) All of these **Ans. c**

360. Ribonucleotides of RNA primer are replaced by deoxyribonucleotides by the enzyme

(a) DNA polymerase I

(b) DNA polymerase II

(c) DNA polymerase III holoenzyme

(d) All of these **Ans. a**

361. DNA fragments are sealed by

(a) DNA polymerase II (b) DNA ligase

(c) DNA gyrase (d) DNA topoisomerase II **Ans. b**

362. Negative supercoils are introduced in DNA by

(a) Helicase

(b) DNA ligase

(c) DNA gyrase

(d) DNA polymerase III holoenzyme **Ans. c**

363. Reverse transcriptase activity is present in the eukaryotic

(a) DNA polymerase α (b) DNA polymerase γ

(c) Telomerase (d) DNA polymerase II **Ans. c**

364. DNA polymerase III holoenzyme possesses

(a) Polymerase activity

(b) $3' \rightarrow 5'$ Exonuclease activity

(c) $5' \rightarrow 3'$ Exonuclease and polymerase activities

(d) $3' \rightarrow 5'$ Exonuclease and polymerase activities **Ans. d**

365. DNA polymerase I possesses

(a) Polymerase activity (b) $3' \rightarrow 5'$ Exonuclease activity

(c) $5' \rightarrow 3'$ Exonuclease activity (d) All of these **Ans. d**

366. $3' \rightarrow 5'$ Exonuclease activity of DNA polymerase I

(a) Removes ribonucleotides

(b) Adds deoxyribonucleotides

(c) Corrects errors in replication

(d) Hydrolyses DNA into mononucleotides **Ans. c**

367. All of the following statements about RNA-dependent DNA polymerase are true except

(a) It synthesizes DNA using RNA as a template

(b) It is also known as reverse transcriptase

(c) It synthesizes DNA in 5′→3′ direction
(d) It is present in all the viruses **Ans. d**

368. Reverse transcriptase catalyses
(a) Synthesis of RNA (b) Breakdown of RNA
(c) Synthesis of DNA (d) Breakdown of DNA **Ans. c**

369. DNA A protein can bind only to
(a) Positively supercoiled DNA (b) Negatively supercoiled DNA
(c) Both (a) and (b) (d) None of these **Ans. b**

370. DNA topoisomerase I of *E. coli* catalyses
(a) Relaxation of negatively supercoiled DNA
(b) Relaxation of positively supercoiled DNA
(c) Conversion of negatively supercoiled DNA into positively supercoiled DNA
(d) Conversion of double helix into supercoiled DNA **Ans. a**

371. In mammalian cell cycle, synthesis of DNA occurs during
(a) S phase (b) G_1 phase
(c) Mitotic Phase (d) G_2 phase **Ans. a**

372. Melting temperature of DNA is the temperature at which
(a) Solid DNA becomes liquid
(b) Liquid DNA evaporates
(c) DNA changes from double helix into supercoiled DNA
(d) Native double helical DNA is denatured **Ans. d**

373. Melting temperature of DNA is increased by its
(a) A and T content (b) G and C content
(c) Sugar content (d) Phosphate content **Ans. b**

374. Buoynat density of DNA is increased by its
(a) A and T content (b) G and C content
(c) Sugar content (d) None of these **Ans. b**

375. Relative proportions of G and C versus A and T in DNA can be determined by its
(a) Melting temperature (b) Buoyant density
(c) Both (a) and (b) (d) None of these **Ans. c**

376. Some DNA is present in mitochondria of

(a) Prokaryotes (b) Eukaryotes

(c) Both (a) and (b) (d) None of these **Ans. b**

377. Satellite DNA contains

(a) Highly repetitive sequences

(b) Moderately repetitive sequences

(c) Non-repetitive sequences

(d) DNA-RNA hybrids **Ans. a**

378. Synthesis of RNA and a DNA template is known as

(a) Replication (b) Translation

(c) Transcription (d) Mutation **Ans. c**

379. Direction of RNA synthesis is

(a) $5' \rightarrow 3'$ (b) $3' \rightarrow 5'$

(c) Both (a) and (b) (d) None of these **Ans. a**

380. DNA-dependent RNA polymerase is a

(a) Monomer (b) Dimer

(c) Trimer (d) Tetramer **Ans. d**

381. DNA-dependent RNA polymerase requires the following for its catalytic activity:

(a) Mg^{++} (b) Mn^{++}

(c) Both (a) and (b) (d) None of these **Ans. c**

382. The initiation site for transcription is recognized by

(a) α–Subunit of DNA-dependent RNA polymerase

(b) β–Subunit of DNA-dependent RNA polymerase

(c) Sigma factor

(d) Rho factor **Ans. c**

383. The termination site for transcription is recognized by

(a) α–Subunit of DNA-dependent RNA polymerase

(b) β–Subunit of DNA-dependent RNA polymerase

(c) Sigma factor

(d) Rho factor **Ans. d**

384. Mammalian RNA polymerase I synthesizes

(a) mRNA (b) rRNA

(c) tRNA (d) hnRNA **Ans. b**

385. Mammalian RNA polymerase III synthesizes

(a) rRNA (b) mRNA

(c) tRNA (d) hnRNA **Ans. c**

386. In mammals, synthesis of mRNA is catalysed by

(a) RNA polymerase I (b) RNA polymerase II

(c) RNA polymerase III (d) RNA polymerase IV **Ans. b**

387. Heterogeneous nuclear RNA is the precursor of

(a) mRNA (b) rRNA

(c) tRNA (d) None of these **Ans. a**

388. Post-transcriptional modification of hnRNA involves all of the following except

(a) Addition of 7-methylguanosine triphosphate cap

(b) Addition of polyadenylate tail

(c) Insertion of nucleotides

(d) Deletion of introns **Ans. c**

389. Newly synthesized tRNA undergoes posttranscriptional modifications which include all the following except

(a) Reduction in size

(b) Methylation of some bases

(c) Formation of pseudouridine

(d) Addition of C-C-A terminus at 5′ end **Ans. d**

390. Post-transcriptional modification does not occur in

(a) Eukaryotic tRNA (b) Prokaryotic tRNA

(c) Eukaryotic hnRNA (d) Prokaryotic mRNA **Ans. d**

391. A consensus sequence on DNA, called TATA box, is the site for attachment of

(a) RNA-dependent DNA polymerase

(b) DNA-dependent RNA polymerase

(c) DNA-dependent DNA polymerase

(d) DNA topoisomerase II **Ans. b**

392. Polyadenylate tail is not present in mRNA synthesizing
(a) Globin (b) Histone
(c) Apoferritin (d) Growth hormone **Ans. b**

393. Introns are present in DNA of
(a) Viruses (b) Bacteria
(c) Man (d) All of these **Ans. c**

394. A mammalian DNA polymerase among the following is
(a) DNA polymerase α (b) DNA polymerase I
(c) DNA polymerase II (d) DNA polymerase IV **Ans. a**

395. Mammalian DNA polymerase γ is located in
(a) Nucleus (b) Nucleolus
(c) Mitochondria (d) Cytosol **Ans. c**

396. Replication of nuclear DNA in mammals is catalysed by
(a) DNA polymerase α (b) DNA polymerase β
(c) DNA polymerase γ (d) DNA polymerase III **Ans. a**

397. Primase activity is present in
(a) DNA polymerase II (b) DNA polymerase α
(c) DNA polymerase β (d) DNA polymerase δ **Ans. b**

398. The mammalian DNA polymerase involved in error correction is
(a) DNA polymerase α (b) DNA polymerase β
(c) DNA polymerase γ (d) DNA polymerase δ **Ans. b**

399. Novobicin inhibits the synthesis of
(a) DNA (b) mRNA
(c) tRNA (d) rRNA **Ans. a**

400. Ciprofloxacin inhibits the synthesis of
(a) DNA (b) mRNA
(c) tRNA (d) rRNA **Ans. a**

401. Ciprofloxacin inhibits
(a) DNA topisomerase II (b) DNA polymerase I
(c) DNA polymerase III (d) DNA gyrase **Ans. d**

402. Rifampicin inhibits
(a) Unwinding of DNA (b) Initiation of replication
(c) Initiation of translation (d) Initiation of transcription
Ans. d

403. Actinomycin D binds to

(a) Double stranded DNA (b) Single stranded DNA

(c) Single stranded RNA (d) DNA-RNA hybrid **Ans. a**

404. DNA contains some palindromic sequences which

(a) Mark the site for the formation of replication forks

(b) Direct DNA polymerase to turn back to replicate the other strand

(c) Are recognized by restriction enzymes

(d) Are found only in bacterial DNA **Ans. c**

405. Introns in genes

(a) Encode the amino acids which are removed during post-translational modification

(b) Encode signal sequences which are removed before secretion of the proteins

(c) Are the non-coding sequences which are not translated

(d) Are the sequences that intervene between two genes **Ans. c**

406. All of the following statements about post-transcriptional processing of tRNA are true except

(a) Introns of some tRNA precursors are removed

(b) CCA is added at 3′ end

(c) 7-Methylguanosine triphosphate cap is added at 5′ end

(d) Some bases are methylated **Ans. c**

407. α-Amanitin inhibits

(a) DNA polymerase II of prokaryotes

(b) DNA polymerase α of eukaryotes

(c) RNA polymerase II of eukaryotes

(d) RNA-dependent DNA polymerase **Ans. c**

408. Ciprofloxacin inhibits the synthesis of

(a) DNA in prokaryotes

(b) DNA in prokaryotes and eukaryotes

(c) RNA in prokaryotes

(d) RNA in prokaryotes and eukaryotes **Ans. a**

409. All of the following statements about bacterial promoters are true except

(a) They are smaller than eukaryotic promoters

(b) They have two consensus sequences upstream from the transcription star site

(c) TATA box is the site for attachment of RNA polymerase

(d) TATA box has a high melting temperature **Ans. d**

410. All of the following statements about eukaryotic promoters are true except

(a) They may be located upstream or down stream from the structural gene

(b) They have two consensus sequences

(c) One consensus sequence binds RNA polymerase

(d) Mutations in promoter region can decrease the efficiency of transcription of the structural gene **Ans. a**

411. In sanger's method of DNA sequence determination, DNA synthesis is stopped by using

(a) 1′ , 2′ - Dideoxyribonucleoside triphosphates

(b) 2′ , 3′ - Dideoxyribonucleoside triphosphates

(c) 2′ , 4′ - Dideoxyribonucleoside triphosphates

(d) 2′ , 5′ - Dideoxyribonucleoside triphosphates **Ans. b**

412. tRNA genes have

(a) Upstream promoters (b) Downstream promoters

(c) Intragenic promoters (d) No promoters **Ans. c**

413. All of the following statements about tRNA are true except

(a) It is synthesized as a large precursor

(b) It is processed in the nucelolus

(c) It has no codons or anticodons

(d) Genes for rRNA are present in single copies **Ans. d**

414. Anticodons are present on

(a) Coding strand of DNA (b) mRNA

(c) tRNA (d) rRNA **Ans. c**

415. Codons are present on

(a) Non-coding strand of DNA (b) hnRNA

(c) tRNA (d) None of these **Ans. b**

416. Nonsense codons are present on

(a) mRNA (b) tRNA

(b) rRNA (d) None of these **Ans. a**

417. Genetic code is said to be degenerate because
(a) It can undergo mutations
(b) A large proportion of DNA is non-coding
(c) One codon can code for more than one amino acids
(d) More than one codons can code for the same amino acids **Ans. d**

418. All the following statements about genetic code are correct except
(a) It is degenerate (b) It is unambigous
(c) It is nearly universal (d) It is overlapping **Ans. d**

419. All of the following statements about nonsense codons are true except
(a) They do not code for amino acids
(b) They act as chain termination signals
(c) They are identical in nuclear and mitochondrial DNA
(d) They have no complementary anticodons **Ans. c**

420. A polycistronic mRNA can be seen in
(a) Prokaryotes (b) Eukaryotes
(c) Mitochondria (d) All of these **Ans. a**

421. Non-coding sequence are present in the genes of
(a) Bacteria (b) Viruses
(c) Eukaryotes (d) All of these **Ans. c**

422. Non-coding sequences in a gene are known as
(a) Cistrons (b) Nonsense codons
(c) Introns (d) Exons **Ans. c**

423. Splice sites are present in
(a) Prokaryotic mRNA (b) Eukaryotic mRNA
(c) Eukaryotic hnRNA (d) All of these **Ans. c**

424. The common features of introns include all the following except
(a) The base sequence begins with GU
(b) The base sequence ends with AG
(c) The terminal AG sequence is preceded by a purine rich tract of ten nucleotides
(d) An adenosine residue in branch site participates in splicing **Ans. c**

425. A splice some contains all the following except

(a) hnRNA (b) snRNAs

(b) Some proteins (d) Ribosome **Ans. d**

426. Self-splicing can occur in

(a) Some precursors of rRNA (b) Some precursors of tRNA

(c) hnRNA (d) None of these **Ans. a**

427. Pribnow box is present in

(a) Prokaryotic promoters (b) Eukaryotic promoters

(c) Both (a) and (b) (d) None of these **Ans. a**

428. Hogness box is present in

(a) Prokaryotic promoters (b) Eukaryotic promoters

(c) Both (a) and (b) (d) None of these **Ans. b**

429. CAAT box is present in

(a) Prokaryotic promoters 10 bp upstream of transcription start site

(b) Prokaryotic promoters 35 bp upstream of transcription start site

(c) Eukaryotic promoters 25 bp upstream of transcription start site

(d) Eukaryotic promoters 70–80 bp upstream of transcription start site **Ans. d**

430. Eukaryotic promoters contain

(a) TATA box 25bp upstream of transcription start site

(b) CAAT box 70-80 bp upstream of transcription start site

(c) Both (a) and (b)

(d) None of these **Ans. c**

431. All the following statements about tRNA are correct except

(a) A given tRNA can be charged with only one particular amino acid

(b) The amino acid is recognized by the anticodon of tRNA

(c) The amino acid is attached to end of tRNA

(d) The anticodon of tRNA finds the complementary codon on mRNA **Ans. b**

432. All the following statements about charging of tRNA are correct except

(a) It is catalysed by amino acyl tRNA synthetase

(b) ATP is converted into ADP and Pi in this reaction

(c) The enzyme recognizes the tRNA and the amino acid

(d) There is a separate enzyme for each tRNA **Ans. b**

433. All the following statements about recognition of a codon on mRNA by an anticodon on tRNA are correct except

(a) The recognition of the third base of the codon is not very precise

(b) Imprecise recognition of the third base results in wobble

(c) Wobble is partly responsible for the degeneracy of the genetic code

(d) Wobble results in incorporation of incorrect amino acids in the protein **Ans. d**

434. The first amino acyl tRNA which initiates translation in eukaryotes is

(a) Mehtionyl tRNA (b) Formylmethionyl tRNA

(c) Tyrosinyl tRNA (d) Alanyl tRNA **Ans. a**

435. The first amino acyl tRNA which initiates translation in prokaryotes is

(a) Mehtionyl tRNA (b) Formylmethionyl tRNA

(c) Tyrosinyl tRNA (d) Alanyl tRNA **Ans. b**

436. In eukaryotes, the 40 S pre-initiation complex contains all the following initiation factors except

(a) eIF-1A (b) eIF-2

(c) eIF-3 (d) eIF-4 **Ans. d**

437. Eukaryotic initiation factors 4A, 4B and 4F bind to

(a) 40 S ribosomal subunit (b) 60 S ribosomal subunit

(c) mRNA (d) Amino acyl tRNA **Ans. c**

438. The codon which serves as translation start signal is

(a) AUG (b) UAG

(c) UGA (d) UAA **Ans. a**

439. The first amino acyl tRNA approaches S ribosomal subunit in association with

(a) eIF-1A and GTP (b) eIF-2 and GTP

(c) eIF-2C and GTP (d) eIF-3 and GTP **Ans. b**

440. eIF-1A and eIF-3 are required

(a) For binding of amino acyl tRNA to 40 S ribosomal subunit

(b) For binding of mRNA to 40 S ribosomal subunit

(c) For binding of 60 S subunit to 40 S subunit

(d) To prevent binding of 60 S subunit to 40 S subunit **Ans. d**

441. eIF-4 A possesses
(a) ATPase activity (b) GTPase activity
(c) Helicase activity (d) None of these **Ans. a**

442. eIF-4 B
(a) Binds to 3′ chain initiation codon on mRNA
(b) Binds to 3′ end of mRNA
(c) Binds to 5′ end of mRNA
(d) Unwinds mRNA near its 5′ end **Ans. d**

443. Peptidyl transferase activity is present in
(a) 40 S ribosomal subunit (b) 60 S ribosomal subunit
(c) eEF-2 (d) Amino acyl tRNA **Ans. b**

444. After formation of a peptide bond, mRNA is translocated along the ribosome by
(a) eEF-1 and GTP (b) eEF-2 and GTP
(c) Peptidyl transferase and GTP (d) Peptidyl transferase and ATP
Ans. b

445. Binding of formylmethionly tRNA to 30 S ribosomal subunit of prokaryotes is inhibited by
(a) Streptomycin (b) Chloramphenicol
(c) Erythromycin (d) Mitomycin **Ans. a**

446. Tetracyclines inhibit binding of amino acyl tRNAs to
(a) 30 S ribosomal subunits (b) 40 S ribosomal subunits
(c) 50 S ribosomal subunits (d) 60 S ribosomal subunits **Ans. a**

447. Peptidyl transferase activity of 50 S ribosomal subunits is inhibited by
(a) Rifampicin (b) Cycloheximide
(c) Chloramphenicol (d) Erythromycin **Ans. c**

448. Erythromycin binds to 50 S ribosomal subunit and
(a) Inhibits binding of amino acyl tRNA
(b) Inhibits Peptidyl transferase activity
(c) Inhibits translocation
(d) Causes premature chain termination **Ans. c**

449. Puromycin causes premature chain termination in
(a) Prokaryotes (b) Eukaryotes
(c) Both (a) and (b) (d) None of these **Ans. c**

450. Diphtheria toxin inhibits

(a) Prokaryotic EF-1
(b) Prokaryotic EF-2
(c) Eukaryotic EF-1
(d) Eukaryotic EF-2 **Ans. d**

451. The proteins destined to be transported out of the cell have all the following features except

(a) They possess a signal sequence
(b) Ribosomes synthesizing them are bound to endoplasmic reticulum
(c) After synthesis, they are delivered into Golgi apparatus
(d) They are tagged with ubiquitin **Ans. d**

452. SRP receptors involved in protein export are present on

(a) Ribosomes
(b) Endoplasmic reticulum
(c) Golgi appartus
(d) Cell membrane **Ans. b**

453. The signal sequence of proteins is cleaved off

(a) On the ribosomes immediately after synthesis
(b) In the endoplasmic reticulum
(c) During processing in Golgi apparatus
(d) During passage through the cell membrane **Ans. b**

454. The half-life of a protein depends upon its

(a) Signal sequence
(b) N-terminus amino acid
(c) C-terminus amino acid
(d) Prosthetic group **Ans.b**

455. Besides structural genes that encode proteins, DNA contains some regulatory sequences which are known as

(a) Operons
(b) Cistrons
(c) Cis-acting elements
(d) Trans-acting factors **Ans. c**

456. Inducers and repressors are

(a) Enhancer and silencer elements respectively
(b) Trans-acting factors
(c) Cis-acting elements
(d) Regulatory proteins **Ans. b**

457. cis-acting elements include

(a) Steroid hormones
(b) Calcitriol
(c) Histones
(d) Silencers **Ans. d**

458. Silencer elements

(a) Are trans-acting factors

(b) Are present between promoters and the structural genes

(c) Decrease the expression of some structural genes

(d) Encode specific repressor proteins **Ans. c**

459. trans-acting factors include

(a) Promoters (b) Repressors

(c) Enhancers (d) Silencers **Ans. b**

460. Enhancer elements have all the following features except

(a) They increase gene expression through a promoter

(b) Each enhancer activates a specific promoter

(c) They may be located far away from the promoter

(d) They may be upstream or downstream from the promoter **Ans. b**

461. Amplification of dihydrofolate reductase gene may be brought about by

(a) High concentrations of folic acid

(b) Deficiency of folic acid

(c) Low concentration of thymidylate

(d) Amethopterin **Ans. d**

462. Proteins which interact with DNA and affect the rate of transcription possess the following structural motif

(a) Helix-turn-helix motif (b) Zinc finger motif

(c) Leucine zipper motif (d) All of these **Ans. d**

463. Lac operon is a cluster of genes present in

(a) Human beings (b) *E. coli*

(c) Lambda phage (d) All of these **Ans. b**

464. Lac operon is a cluster of

(a) Three structural genes

(b) Three structural genes and their promoter

(c) A regulatory gene, an operator and a promoter

(d) A regulatory gene, an operator, a promoter and three structural genes **Ans. d**

465. The regulatory i gene of lac operon
(a) Is inhibited by lacotse
(b) Is inhibited by its own product, the repressor protein
(c) Forms a regulatory protein which increases the expression of downstream structural genes
(d) Is constitutively expressed **Ans. d**

466. RNA polymerase holoenzyme binds to lac operon at the following site
(a) i gene (b) z gene
(c) Operator locus (d) Promoter region **Ans. d**

467. Trancription of z, y and a genes of lac operon is prevented by
(a) Lactose (b) Allo-lactose
(c) Repressor (d) cAMP **Ans. c**

468. Transcription of structural genes of lac operon is prevented by binding of the repressor tetramer to
(a) i gene (b) Operator locus
(c) Promoter (d) z gene **Ans. b**

469. The enzymes encoded by z, y and a genes of lac operon are inducible, and their inducer is
(a) Lactose
(b) Allo-lactose
(c) Catabolite gene activator protein
(d) All of these **Ans. b**

470. Binding of RNA polymerase holoenzyme to the promoter region of lac operon is facilitated by
(a) Catabolite gene activator protein (CAP)
(b) cAMP
(c) CAP-cAMP complex
(d) None of these **Ans. c**

471. Lactose or its analogues act as positive regulators of lac operon by
(a) Attaching to i gene and preventing its expression
(b) Increasing the synthesis of catabolite gene activator protein
(c) Attaching to promoter region and facilitating the binding of RNA polymerase holoenzyme
(d) Binding to repressor subunits so that the repressor cannot attach to the operator locus **Ans. d**

472. Expression of structural genes of lac operon is affected by all the following except

(a) Lactose or its analogues (b) Repressor tetramer
(c) cAMP (d) CAP-cAMP complex **Ans. c**

473. The coding sequences in lac operon include

(a) i gene
(b) i gene, operator locus and promoter
(c) z, y and a genes
(d) i, z, y and a genes **Ans. d**

474. Mutations can be caused by

(a) Ultraviolet radiation (b) Ionising radiation
(c) Alkylating agents (d) All of these **Ans. c**

475. Mutations can be caused by

(a) Nitrosamine (b) Dimethyl sulphate
(c) Acridine (d) All of these **Ans. d**

476. Nitrosamine can deaminate

(a) Cytosine to form uracil
(b) Adenine to form xanthine
(c) Guanine to form hypoxanthine
(d) All of these **Ans. a**

477. Exposure of DNA to ultraviolet radiation can lead to the formation of

(a) Adenine dimers (b) Guanine dimers
(c) Thymine dimers (d) Uracil dimers **Ans. c**

478. Damage to DNA caused by ultraviolet radiation can be repaired by

(a) uvr ABC excinuclease (b) DNA polymerase I
(c) DNA ligase (d) All of these **Ans. d**

479. Xeroderma pigmentosum results from a defect in

(a) uvr ABC excinuclease (b) DNA polymerase I
(c) DNA ligase (d) All of these **Ans. a**

480. All the following statements about xeroderma pigmentosum are true except

(a) It is a genetic disease
(b) Its inheritance is autosomal dominant

(c) uvr ABC excinuclease is defective in this disease
(d) It results in multiple skin cancers **Ans. b**

481. Substitution of an adenine base by guanine in DNA is known as
(a) Transposition (b) Transition
(c) Transversion (d) Frameshift mutation **Ans. b**

482. Substitution of a thymine base by adenine in DNA is known as
(a) Transposition (b) Transition
(c) Transversion (d) Frameshift mutation **Ans. c**

483. A point mutation results from
(a) Substitution of a base (b) Insertion of a base
(c) Deletion of a base (d) All of these **Ans. a**

484. Substitution of a base can result in a
(a) Silent mutation (b) Mis-sense mutation
(c) Nonsense mutation (d) All of these **Ans. d**

485. A silent mutation is most likely to result from
(a) Substitution of the first base of a codon
(b) Substitution of the third base of a codon
(c) Conversion of a nonsense codon into a sense codon
(d) Conversion of a sense codon into a nonsense codon **Ans. b**

486. The effect of a mis-sense mutation can be
(a) Acceptable (b) Partially acceptable
(c) Unacceptable (d) All of these **Ans. d**

487. Amino acid sequence of the encoded protein is not changed in
(a) Silent mutation
(b) Acceptable mis-sense mutation
(c) Both (a) and (b)
(d) None of these **Ans. a**

488. Haemoglobin S is an example of a/an
(a) Silent mutation
(b) Acceptable mis-sense mutation
(c) Unacceptable mis-sense mutation
(d) Partially acceptable mis-sense mutation **Ans. d**

489. If the codon UAC on mRNA changes into UAG as a result of a base substitution in DNA, it will result in

(a) Silent mutation (b) Acceptable mis-sense mutation
(c) Nonsense mutation (d) Frameshift mutation **Ans. c**

490. Insertion of a base in a gene can cause

(a) Change in reading frame
(b) Garbled amino acid sequence in the encoded protein
(c) Premature termination of translation
(d) All of these **Ans. d**

491. A frameshift mutation changes the reading frame because the genetic code

(a) Is degenerate (b) Is overlapping
(c) Has no punctuations (d) Is universal **Ans. c**

492. Suppressor mutations occur in

(a) Structural genes (b) Promoter regions
(c) Silencer elements (d) Anticodons of tRNA **Ans. d**

493. Suppressor tRNAs can neutralize the effects of mutations in

(a) Structural genes (b) Promoter regions
(c) Enhancer elements (d) All of these **Ans. a**

494. Mutations in promoter regions of genes can cause

(a) Premature termination of translation
(b) Change in reading frame of downstream structural gene
(c) Decreased efficiency of transcription
(d) All of these **Ans. c**

495. Mitochondrial protein synthesis is inhibited by

(a) Cycloheximide (b) Chloramphenicol
(c) Diptheria toxin (d) None of these **Ans. b**

496. All of the following statements about puromycin are true except

(a) It is an alanyl tRNA analogue
(b) It causes premature termination of protein synthesis
(c) It inhibits protein synthesis in prokaryotes
(d) It inhibits protein synthesis in eukaryotes **Ans. a**

497. Leucine zipper motif is seen in some helical proteins when leucine residues appear at every

(a) 3rd position (b) 5th position

(c) 7th position (d) 9th position **Ans. c**

498. Zinc finger motif is formed in some proteins by binding of zinc to

(a) Two cysteine residues

(b) Two histidine residues

(c) Two arginine residues

(d) Two cysteine and two histidine residues or two pairs of two cysteine residues each **Ans. d**

499. Restriction endonucleases are present in

(a) Viruses (b) Bacteria

(c) Eukaryotes (d) All of these **Ans. b**

500. Restriction endonucleases split

(a) RNA (b) Single stranded DNA

(c) Double stranded DNA (d) DNA-RNA hybrids **Ans. c**

501. Restriction endonucleases can recognise

(a) Palindromic sequences (b) Chimeric DNA

(c) DNA-RNA hybrids (d) Homopolymer sequences **Ans. a**

502. All of the following statements about restriction endonucleases are true except

(a) They are present in bacteria

(b) They act on double stranded DNA

(c) They recognize palindromic sequences

(d) They always produce sticky ends **Ans. d**

503. Which of the following is a palindromic sequence

(a) 5′- ATGCAG -3′ (b) 3′- TACGTC - 5′

(c) 5′- CGAAGC - 3′ (d) 3′- GCTTCG - 5′ **Ans. c**

504. In sticky ends produced by restriction endonucleases

(a) The 2 strands of DNA are joined to each other

(b) The DNA strands stick to the restriction endonuclease

(c) The ends of a double stranded fragment are overlapping

(d) The ends of a double stranded fragment are non overlapping **Ans. c**

505. All of the following may be used as expression vectors except
(a) Plasmid (b) Bacteriophage
(c) Baculovirus (d) *E. coli* **Ans. d**

506. A plasmid is a
(a) Single stranded linear DNA (b) Single stranded circular DNA
(c) Double stranded linear DNA (d) Double stranded circular DNA **Ans. d**

507. Fragments of DNA can be identified by the technique of
(a) Western blotting (b) Eastern blotting
(c) Northern blotting (d) Southern blotting **Ans. d**

508. A particular RNA in a mixture can be identified by
(a) Western blotting (b) Eastern blotting
(c) Northern blotting (d) Southern blotting **Ans. c**

509. A radioactive isotope labeled cDNA probe is used in
(a) Southern blotting (b) Northern blotting
(c) Both (a) and (b) (d) None of these **Ans. c**

510. An antibody probe is used in
(a) Southern blotting (b) Northern blotting
(c) Western blotting (d) None of these **Ans. c**

511. A particular protein in a mixture can be detected by
(a) Southern blotting (b) Northern blotting
(c) Western blotting (d) None of these **Ans. c**

512. The first protein synthesized by recombinant DNA technology was
(a) Streptokinase (b) Human growth hormone
(c) Tissue plasminogen activator (d) Human insulin **Ans. d**

513. For production of eukaryotic protein by recombinant DNA technology in bacteria, the template used is
(a) Eukaryotic gene (b) hnRNA
(c) mRNA (d) All of these **Ans. c**

514. Monoclonal antibodies are prepared by cloning
(a) Myeloma cells (b) Hybridoma cells
(c) T-Lymphocytes (d) B-Lymphocytes **Ans. b**

515. Myeloma cells are lacking in

(a) TMP synthetase (b) Formyl transferase

(c) HGPRT (d) All of these **Ans. c**

516. Hybridoma cells are selected by culturing them in a medium containing

(a) Adenine, guanine, cytosine and thymine

(b) Adenine, guanine, cytosine and uracil

(c) Hypoxanthine, aminopterin and thymine

(d) Hypoxanthine, aminopterin and thymidine **Ans. d**

517. Myeloma cells and lymphocytes can be fused by using

(a) Calcium chloride (b) Ethidium bromide

(c) Polyethylene glycol (d) DNA polymerase **Ans. c**

518. Trials for gene therapy in human beings were first carried out, with considerable success, in a genetic disease called

(a) Cystic fibrosis

(b) Thalasemia

(c) Adenosine deaminase deficiency

(d) Lesch-Nyhan syndrome **Ans. c**

519. Chimeric DNA

(a) Is found in bacteriophages (b) Contains unrelated genes

(c) Has no restriction sites (d) Is palindromic **Ans. b**

520. Which of the following may be used as cloning vector?

(a) Prokaryotic plasmid (b) Lambda phage

(c) Cosmid (d) All of these **Ans. d**

521. The plasmid pBR322 has

(a) Ampicillin resistance gene (b) Tetracycline resistance gene

(c) Both (a) and (b) (d) None of these **Ans. c**

522. Lambda phage can be used to clone DNA fragments of the size

(a) Upto 3 kilobases (b) Upto 20 kilobases

(c) Upto 45 kilobases (d) Upto 1,000 kilobases **Ans. b**

523. DNA fragments upto 45 kilobases in size can be cloned in

(a) Bacterial plasmids (b) Lambda phage

(c) Cosmids (d) Yeast artificial chromosomes

Ans. c

524. A cosmid is a
(a) Large bacterial plasmid (b) Viral plasmid
(c) Hybrid of plasmid and phage (d) Yeast plasmid **Ans. c**

525. Polymerase chain reaction can rapidly amplify DNA sequences of the size
(a) Upto 10 kilobases (b) Upto 45 kilobases
(c) Upto 100 kilobases (d) Upto 1,000 kilobases **Ans. a**

526. The DNA polymerase commonly used in polymerase chain reaction is obtained from
(a) *E. coli* (b) Yeast
(c) *T. aquaticus* (d) Eukaryotes **Ans. c**

527. Base sequence of DNA can be determined by
(a) Maxam-Gilbert method (b) Sanger's dideoxy method
(c) Both (a) and (b) (d) None of these **Ans. c**

528. From a DNA-RNA hybrid, DNA can be obtained by addition of
(a) DNA B protein and ATP (b) Helicase and ATP
(c) DNA topoisomerase I (d) Alkali **Ans. d**

529. Optimum temperature of DNA polymerase of *T. aquaticus* is
(a) 30°C (b) 37°C
(c) 54°C (d) 72°C **Ans. d**

530. In addition to Taq polymerase, polymerase chain reaction requires all of the following except
(a) A template DNA
(b) Deoxyribonucleoside triphosphates
(c) Primers
(d) Primase **Ans. d**

531. DNA polymerase of *T. aquaticus* is preferred to that of *E. coli* in PCR because
(a) It replicates DNA more efficiently
(b) It doesn't require primers
(c) It is not denatured at the melting temperature of DNA
(d) It doesn't cause errors in replication **Ans. c**

532. Twenty cycles of PCR can amplify DNA

(a) 2^{20} fold (b) 20^2 fold

(c) 20 x 2 fold (d) 20 fold **Ans. a**

533. Transgenic animals may be prepared by introducing a foreign gene into

(a) Somatic cells of young animals

(b) Testes and ovaries of animals

(c) A viral vector and infecting the animals with the viral vector

(d) Fertilised egg and implanting the egg into a foster mother **Ans. d**

534. Yeast artificial chromosome can be used to amplify DNA sequences of the size

(a) Upto 10 kb (b) Upto 45 kb

(c) Upto 100 kb (d) Upto 1,000 kb **Ans. d**

535. DNA finger printing is based on the presence in DNA of

(a) Constant number of tandem repeats

(b) Varibale number of tandem repeats

(c) Non-repititive sequences in each DNA

(d) Introns in eukaryotic DNA **Ans. b**

536. All the following statements about restriction fragment length polymorphism are true except

(a) It results from mutations in restriction sites

(b) Mutations in restriction sites can occur in coding or non-coding regions of DNA

(c) It is inherited in Mendelian fashion

(d) It can be used to diagnose any genetic disease **Ans. d**

537. Inborn errors of urea cycle can cause all the following except

(a) Vomiting (b) Ataxia

(c) Renal failure (d) Mental retardation **Ans. d**

538. Hyperammonaemia type I results from congenital absence of

(a) Glutamate dehydrogenase

(b) Carbamoyl phosphate synthetase

(c) Ornithine transcarbamoylase

(d) None of these **Ans. b**

539. Congenital deficiency of ornithine transcarbamoylase causes
(a) Hyperammonaemia type I (b) Hyperammonaemia type II
(c) Hyperornithinaemia (d) Citrullinaemia **Ans. c**

540. A ketogenic amino acid among the following is
(a) Leucine (b) Serine
(c) Threonine (d) Proline **Ans. a**

541. Carbon skeleton of the following amino acid can serve as a substance for gluconeogenesis
(a) Cysteine (b) Aspartate
(c) Glutamate (d) All of these **Ans. d**

542. N-Formiminoglutamate is a metabolite of
(a) Glutamate (b) Histidine
(c) Tryptophan (d) Methionine **Ans. b**

543. Methylmalonyl CoA is a metabolite of
(a) Valine (b) Leucine
(c) Isoleucine (d) All of these **Ans. a**

544. Homogentisic acid is formed from
(a) Homoserine (b) Homocysteine
(c) Tyrosine (d) Tryptophan **Ans. c**

545. Maple syrup urine disease results from absence or serve deficiency of
(a) Homogentisate oxidase
(b) Phenylalanine hydroxylase
(c) Branched chain amino acid transaminase
(d) None of these **Ans. d**

546. Which of the following is present as a marker in lysosomal enzymes to direct them to their destination?
(a) Glucose-6-phosphate (b) Mannose-6-phosphate
(c) Galactose-6-phosphate (d) N-Acetyl neuraminic acid
Ans. c

547. Marfan's syndrome results from a mutation in the gene coding
(a) Collagen (b) Elastin
(c) Fibrillin (d) Keratin **Ans. b**

548. All the following statements about fibronectin are true except
(a) It is glycoprotein
(b) It is a triple helix
(c) It is present in extra cellular matrix
(d) It binds with integrin receptors of cell **Ans. a**

549. Fibronectin has binding sites for all of the following except
(a) Glycophorin (b) Collagen
(c) Heparin (d) Integrin receptor **Ans. c**

550. Fibronectin is involved in
(a) Cell adhension (b) Cell movement
(c) Both (a) and (b) (d) None of these **Ans. b**

551. Glycoproteins are marked for destruction by removal of their
(a) Oligosaccharide prosthetic group
(b) Sialic acid residues
(c) Mannose residues
(d) N-terminal amino acids **Ans. d**

552. Glycophorin is present in cell membranes of
(a) Erythrocytes (b) Platelets
(c) Neutrophils (d) Liver **Ans. a**

553. Selectins are proteins that can recognize specific
(a) Carbohydrates (b) Lipids
(c) Amino acids (d) Nucleotides **Ans. a**

554. Hunter's syndrome results from absence of
(a) Hexosaminidase A (b) Iduronate sulphatase
(c) Neuraminidase (d) Arylsulphatase B **Ans. b**

555. A cancer cell is characterized by
(a) Uncontrolled cell division (b) Invasion of neighbouring cells
(c) Spread to distant sites (d) All of these **Ans. d**

556. If DNA of a cancer cell is introduced into a normal cell, the recipient cell
(a) Destroys the DNA (b) Loses its ability to divide
(c) Dies (d) Changes into a cancer cell
Ans. d

557. A normal cell can be transformed into a cancer cell by all of the following except

(a) Ionising radiation (b) Mutagenic chemicals

(c) Oncogenic bacteria (d) Some viruses **Ans. c**

558. Proto-oncogenes are present in

(a) Oncoviruses (b) Cancer cells

(c) Healthy human cells (d) Prokaryotes **Ans. c**

559. All the following statements about protooncogenes are true except

(a) They are present in human beings

(b) They are present in healthy cells

(c) Proteins encoded by them are essential

(d) They are expressed only when a healthy cell has been transformed into a cancer cell **Ans. d**

560. Various oncogenes may encode all of the following except

(a) Carcinogens

(b) Growth factors

(c) Receptors for growth factors

(d) Signal transducers for growth factors **Ans. a**

561. Ras proto-oncogene is converted into oncogene by

(a) A point mutation

(b) Chromosomal translocation

(c) Insertion of a viral promoter upstream of the gene

(d) Gene amplification **Ans. a**

562. Ras proto-oncogene encodes

(a) Epidermal growth factor (EGF)

(b) Receptor for EGF

(c) Signal transducer for EGF

(d) Nuclear transcription factor **Ans. c**

563. P 53 gene

(a) A proto-oncogene (b) An oncogene

(c) A tumour suppressor gene (d) None of these **Ans. c**

564. Retinoblastoma can result from a mutation in
(a) ras proto-oncogene (b) erbB proto-oncogene
(c) p 53 gene (d) RB 1 gene **Ans. d**

565. All the following statements about retino blastoma are true except
(a) At least two mutations are required for its development
(b) One mutation can be inherited from a parent
(c) Children who have inherited one mutation develop retinoblastoma at a younger age
(d) RB 1 gene promotes the development of retinoblastoma **Ans. d**

566. Ames assay is a rapid method for detection of
(a) Oncoviruses (b) Retroviuses
(c) Chemical carcinogens (d) Typhoid **Ans. c**

567. Amplification of dihydrofolate reductase gene in a cancer cell makes the cell
(a) Susceptible to folic acid deficiency
(b) Less malignant
(c) Resistant to amethopterin therapy
(d) Responsive to amethopterin therapy **Ans. c**

568. Conversion of a procarcinogen into a carcinogen often requires
(a) Proteolysis
(b) Microsomal hydroxylation
(c) Exposure to ultraviolet radiation
(d) Exposure to X-rays **Ans. b**

569. The only correct statement about oncoviruses is
(a) All the oncoviruses are RNA viruses
(b) Reverse transcriptase is present in all oncoviruses
(c) Viral oncogenes are identical to human protooncogens
(d) Both DNA and RNA viruses can be oncoviruses **Ans. d**

570. RB 1 gene is
(a) A tumour suppressor gene (b) Oncogene
(c) Proto-oncogene (d) Activated proto-oncogene
Ans. a

571. Cancer cells may become resistant to amethopterin by
(a) Developing mechanisms to destroy amethopterin
(b) Amplification of dihydrofolate reducatse gene
(c) Mutation in the dihydrofolate reductase gene so that the enzyme is no longer inhibited by amethopterin
(d) Developing alternate pathway of thymidylate synthesis **Ans. b**

572. The major source of NH_3 produced by the kidney is
(a) Leucine (b) Glycine
(c) Alanine (d) Glutamine **Ans. d**

573. Which of these methyl donors is not a quanternary ammonium compound?
(a) Methionine (b) Choline
(c) Betain (d) Betainaldehyde **Ans. a**

574. L-glutamic acid is subjected to oxidative deaminition by
(a) L-amino acid dehydrogenase (b) L-glutamate dehydrogenase
(c) Glutaminase (d) Glutamine synthetase **Ans. b**

575. A prokaryotic ribosome is made up of ________ sub units.
(a) 20 S and 50S (b) 30S and 50S
(c) 30S and 60S (d) 20S and 50S **Ans. b**

576. An Eukaryotic ribosome is made up of ________ sub unit.
(a) 40S and 60S (b) 40S and 50S
(c) 40S and 80S (d) 60S and 80S **Ans. a**

577. GTP is not required for
(a) Capping L of mRNA
(b) Fusion of 40S and 60S of ribosome
(c) Accommodation of tRNA amino acid
(d) Formation of tRNA amino acid complex **Ans. d**

578. The antibiotic which inhibits DNA dependent RNA polymerase is
(a) Mitomycin C (b) Actinomycin d
(c) Streptomycin (d) Puromycin **Ans. b**

579. The antibiotic which cleaves DNA is
(a) Actinomycin d (b) Streptomycin
(c) Puromycin (d) Mitomycin C **Ans. d**

580. The antibiotic which has a structure similar to the amino acyl end of tRNA tyrosine is

(a) Actinomycin d (b) Streptomycin
(c) Puromycin (d) Mitomycin c **Ans. c**

581. ATP is required for

(a) Fusion of 40S and 60S of ribosome
(b) Accommodation tRNA amino acid in a site of ribosome
(c) Movement of ribosome along mRNA
(d) Formation of tRNA amino acid complex **Ans. d**

582. What is the subcellular site for the biosynthesis of proteins?

(a) Chromosomes (b) Lymosomes
(c) Ribosomes (d) Centrosomes **Ans. c**

583. An animal is in negative nitrogen balance when

(a) Intake exceeds output
(b) New tissue is being synthesized
(c) Output exceeds intake
(d) Intake is equal to output **Ans. c**

584. When NH_3 is perfused through a dog's liver ______ is formed, while ______ is formed in the birds liver.

(a) Urea, Uric acid (b) Urea, allantoin
(c) Uric acid, creatinine (d) Uric acid, Urea **Ans. a**

585. Aspartate amino transferase uses the following for transamination

(a) Glutamic acid and pyruvic acid
(b) Glutamic acid and oxaloacetic acid
(c) Aspartic acid and pyruvic acid
(d) Aspartic acid and keto adipic acid **Ans. b**

586. Which among the following compounds is not a protein?

(a) Insulin (b) Heparin
(c) Mucin (d) Pepsin **Ans. d**

587. Almost all the urea is formed in this tissue

(a) Kidney (b) Urethra
(c) Uterus (d) Liver **Ans. d**

588. A polyribosome will have about _______ individual ribosomes.

(a) 20 (b) 10

(c) 5 (d) 2 **Ans. c**

589. Progressive transmethylation of ethanolamine gives

(a) Creatinine (b) Choline

(c) Methionine (d) N-methyl nicotinamide **Ans. c**

590. Genetic information originates from

(a) Cistron of DNA (b) Codons of mRNA

(c) Anticodons of tRNA (d) Histones of nucleoproteins

Ans. a

591. The genetic code operates through

(a) The protein moiety of DNA

(b) Cistron of DNA

(c) Nucleotide sequence of m RNA

(d) The anticodons of tRNA **Ans. c**

592. DNA synthesis in laboratory was first achieved by

(a) Watson and crick (b) Khorana

(c) A. Kornberg (d) Ochoa **Ans. c**

593. Among the different types of RNA, which one has the highest M.W.?

(a) mRNA (b) rRNA

(c) yeast RNA (d) tRNA **Ans. b**

594. From DNA the genetic message is transcribed into this compound

(a) Protein (b) mRNA

(c) tRNA (d) rRNA **Ans. b**

595. This compound has a double helical structure

(a) Deoxyribonucleic acid

(b) RNA

(c) Flavine-adenine dinucleotide

(d) Nicotinamide adenine dinucleotide **Ans. a**

596. The structural stability of the double helix of DNA is ascribed largely to

(a) Hydrogen bonding between adjacent purine bases

(b) Hydrophobic bonding between staked purine and pyrimidine nuclei

(c) Hydrogen bonding between adjacent pyrimidine bases
(d) Hydrogen bonding between purine and pyrimidine bases **Ans. d**

597. Which of the following statements about nucleic acid is most correct?
(a) Both pentose nucleic acid and deoxypentose nucleic acid contain the same pyrimidines
(b) Both pentose nucleic acid and deoxypentose nucleic acid contain the same purines
(c) RNA contains cytosine and thymine
(d) DNA and RNA are hydrolysed by weak alkali **Ans. b**

598. Acid hydrolysis of ribonucleic acid would yield the following major products
(a) d- deoxyribose, cytosine, adenine
(b) d-ribose, thymine, Guanine
(c) d-ribose, cytosine, uracil, thymine
(d) d-ribose, uracil, adenine, guanine, cytosine **Ans. d**

599. RNA does not contain
(a) adenine (b) OH methyl cytosine
(c) d-ribose (d) Uracil **Ans. b**

600. Which of the following statements is correct?
(a) A nucleo protein usually contain deoxy sugars of the hexose type
(b) Nucleoproteins are usually absent from the cytoplasm
(c) Nucleoproteins usually are present in the nucleus only
(d) Nucleoproteins usually occur in the nucleus and cytoplasm **Ans. d**

601. Which of the following compound is present in RNA but absent from DNA?
(a) Thymine (b) Cytosine
(c) Uracil (d) Guanine **Ans. c**

602. Nucleic acids can be detected by means of their absorption maxima near 260 nm. Their absorption in this range is due to
(a) Proteins (b) Purines and pyrimidines
(c) Ribose (d) Deoxyribose **Ans. b**

603. Which of the following contains a deoxy sugar?
(a) RNA (b) DNA
(c) ATP (d) UTP **Ans. b**

604. DNA is

(a) Usually present in tissues as a nucleo protein and cannot be separated from its protein component

(b) A long chain polymer in which the internucleotide linkages are of the diester type between C-3′ and C-5′

(c) Different from RNA since in the latter the internucleotide linkages are between C-2′ and C-5′

(d) Hydrolyzed by weak alkali (pH 9 to 100°C) **Ans. b**

605. Nobody is the name given to

(a) Ribosome (b) Microsome

(c) Centrosome (d) Nucleosome **Ans. c**

606. Transcription is the formation of

(a) DNA from a parent DNA (b) mRNA from a parent mRNA

(c) pre mRNA from DNA (d) protein through mRNA **Ans. c**

607. Translation is the formation of

(a) DNA from DNA (b) mRNA from DNA

(c) Protein through mRNA (d) mRNA from pre mRNA

Ans. b

608. Sigma and Rho factors are required for

(a) Replication (b) Transcription

(c) Translation (d) Polymerisation **Ans.c**

609. The genine of φ×174 bacteriophage is interesting in that if contains

(a) No DNA (b) DNA with uracil

(c) Single stranded DNA (d) Triple standard DNA **Ans. c**

610. Okazaki fragments are small bits of

(a) RNA (b) DNA

(c) DNA with RNA heads (d) RNA with DNA heads **Ans. c**

611. In addition to the DNA of nucleus, there is also a DNA of

(a) Mitochondrian (b) Endoplasmic reticulum

(c) Golgi apparatus (d) Plasma membrane **Ans. a**

612. The mitochondrial DNA is

(a) Like the nuclear DNA in structure

(b) Single stranded, linear

(c) Double stranded, circular
(d) Single stranded, circular **Ans. c**

613. A synthetic RNA having the sequence of UUUUUU (Poly U) will give a protein having poly ______.
(a) Alanine (b) Phenyl alanine
(c) Glycine (d) Methionine **Ans. b**

614. Lac operon of *E. coli* contains _______ in continuity.
(a) Regulator and operator genes only
(b) Operator and structural genes only
(c) Regular and structural genes only
(d) Regulator, operator and structural genes **Ans. d**

615. A mRNA of eukaryotes can code for
(a) Only one polypeptide (b) Two polypeptides
(c) Three polypeptides (d) Five polypeptides **Ans. a**

616. mRNA of prokaryotes can code for
(a) More than one polypeptide (b) Only one polypeptide
(c) Many exons and introns (d) Introns only **Ans. a**

617. DNA directed RNA polymerase is
(a) Replicase (b) Transcriptase
(c) Reverse transcriptase (d) Polymerase III **Ans. b**

618. RNA directed DNA polymerase is
(a) Replicase (b) Transcriptase
(c) Reversetranscriptase (d) Polymerase–III **Ans. c**

619. RNA synthesis requires
(a) RNA primer (b) RNA template
(c) DNA template (d) DNA primer **Ans. c**

620. The mRNA ready for protein synthesis has the ________ cap.
(a) ATP (b) CTP
(c) GTP (d) UTP **Ans. c**

621. mRNA ready for protein synthesis has the poly _______ tail.
(a) G (b) A
(c) U (d) C **Ans. b**

622. The codon for phenyl Alanine is
(a) AAA (b) CCC
(c) GGG (d) UUU **Ans. d**

623. Blue print for genetic information residues in
(a) mRNA (b) tRNA
(c) rRNA (d) DNA **Ans. d**

624. Genes are
(a) RNA (b) DNA
(c) Lipoproteins (d) Chromoproteins **Ans. b**

625. Codons are in
(a) DNA (b) mRNA
(c) tRNA (d) rRNA **Ans. b**

626. The genetic code operates via
(a) The protein moiety of DNA
(b) The base sequences of DNA
(c) The nucleotide sequence of mRNA
(d) The base sequence of tRNA **Ans. c**

627. Urine bases with methyl substituents occurring in plants are
(a) Caffeine (b) Theophylline
(b) Theobromine (d) All of these **Ans. d**

628. Genetic information in human beings is stored in
(a) DNA (b) RNA
(c) Both (a) and (b) (d) None of these **Ans. a**

629. All following are naturally occurring nucleotides except
(a) Cyclic AMP (b) ATP
(c) DNA (d) Inosine monophosphate **Ans. c**

630. If the amino group and a carboxylic group of the amino acid are attached to same carbon atom, the amino acid is called as
(a) Alpha (b) Beta
(c) Gamma (d) Epsilon **Ans. a**

631. If in a nucleic acid there are more than 8000 nucleotides it is most likely
(a) RNA (b) DNA
(c) Both (a) and (b) (d) None of these **Ans. b**

632. Genetic information in human beings is stored in
(a) RNA (b) DNA
(c) Both (a) and (b) (d) mRNA **Ans. b**

633. In RNA, apart from ribose and phosphate, all following are present except
(a) Adenine (b) Guanine
(c) Thymine (d) Cytosine **Ans. c**

634. Which of the following gives a positive Ninhydrin test?
(a) Reducing sugar (b) Triglycerides
(c) α-amino acids (d) Phospholipids **Ans. c**

635. A Gene is
(a) A single protein molecule
(b) A group of chromosomes
(c) An instruction for making a protein molecule
(d) A bit of DNA molecule **Ans. d**

636. In DNA, genetic information is located in
(a) Purine bases (b) Pyrimidine bases
(c) Purine and pyrimidine bases (d) Sugar **Ans. c**

637. Which one of the following is not a constituent of RNA?
(a) Deoxyribose (b) Uracil
(c) Adenine (d) Thymine **Ans. a**

638. Which of the following are nucleo proteins?
(a) Protamines
(b) Histones
(c) Deoxy and Ribo nucleo proteins
(d) All of these **Ans. d**

639. In the total RNA of cell, tRNA constitutes
(a) 1–10% (b) 10–20%
(c) 30–50% (d) 50–80% **Ans. b**

640. Unit of genetic information
(a) DNA (b) RNA
(c) Cistron (d) None of these **Ans. c**

641. Anticodon sequence are seen in
(a) tRNA and transcribed DNA strand
(b) tRNA and complementary DNA strand
(c) mRNA
(d) mRNA and complementary DNA strand **Ans. a**

642. cAMD is destroyed by
(a) Adenylate cyclase (b) Phosphodiesterase
(c) Synthetase phosphatase (d) Synthetase kinase **Ans. b**

643. Restriction enzymes have been found in
(a) Humans (b) Birds
(c) Bacteria (d) Bacteriophage **Ans. c**

644. Sulphur is not present in
(a) Thiamine (b) Lipic acid
(c) Thymine (d) Biotin **Ans. c**

645. Which one of the following binds to specific nucleotide sequences?
(a) RNA polymerase (b) Repressor
(c) Inducer (d) Restriction **Ans. a**

646. Using written conversion which one of the following sequences is complimentary to TGGCAGCCT?
(a) ACC GTC GGA (b) ACC GUC GGA
(c) AGG CTG CCA (d) TGG CTC GGA **Ans. a**

647. Ribosomes similar to those of bacterial found in
(a) Plant nuclei (b) Cardiac muscle cytoplasm
(c) Liver endoplasmic reticulum (d) Neuronal cytoplasm **Ans. a**

648. The mechanism of synthesis of DNA and RNA are similar in all the following ways except
(a) They involve release of pyrophosphate from each nucleotide added
(b) They require activated nucleotide precursor and Mg^{2+}
(c) The direction of synthesis is $5' \rightarrow 3'$
(d) They require a primer **Ans. a**

649. Template-directed DNA synthesis occurs in all the following except
(a) The replication fork
(b) Polymerase chain reaction

(c) Growth of RNA tumor viruses

(d) Expression of oneogenes **Ans. c**

650. Which one of the following statements correctly describes eukaryotic DNA?

(a) They involve release of pyrophosphate from each nucleotide precussor and Mg^{2+}

(b) The direction of synthesis is $5' \rightarrow 3'$

(c) They require a primer

(d) None of these **Ans. c**

651. Which one of the following causes frame shift mutation?

(a) Transition

(b) Transversion

(c) Deletion

(d) Substitution of purine to pyrimidine **Ans. c**

652. Catabolism of thymidylate gives

(a) α-alanine (b) β-alanine

(c) α-aminoisobutyrate (d) β-aminoisobutyrate **Ans. d**

653. Glycine gives __________ atoms of purine.

(a) C_2, C_3 (b) C_4, C_5 and N_7

(c) C_4, C_5 and N_9 (d) C_4, C_6 and N_7 **Ans. b**

654. A common substrate of HGPRTase, APRTase and PRPP glutamyl amidotransferase is

(a) Ribose 5 phosphate

(b) Phosphoribosyl pyrophosphate

(c) Hypoxanthine

(d) Adenosine **Ans. b**

655. Carbon 6 of purine skeleton comes from

(a) Atmospheric CO_2 (b) 1 carbon carried by folate

(c) Betoine (d) Methionine **Ans. a**

656. Uric acid is the catabolic end product of

(a) Porphyrine (b) Purines

(c) Pyrimidines (d) Pyridoxine **Ans. a**

657. Diphenylamine method is employed in the quantitation of

(a) Nucleic acid (b) RNA
(c) DNA (d) Proteins **Ans. c**

658. Orcinol method is employed in the quantitation of

(a) Nucleic acid (b) DNA
(c) RNA (d) Proteins **Ans. c**

659. Nucleic acid show strong absorption at one of the wavelength

(a) 280 nm (b) 220 nm
(c) 360 nm (d) 260 nm **Ans. d**

660. tRNA has

(a) Clover leaf structure (b) Anticodon arm
(c) Poly 'A' tail at 3′ end (d) Cap at 5′ end **Ans. a**

661. Which one of the following contributes nitrogen atoms to both purine and pyrimidine rings?

(a) Aspartate (b) Carbanoyl phosphate
(c) Carbondioxide (d) Tetrahydrofolate **Ans. a**

662. The four nitrogen atoms of purines are derived from

(a) Urea and NH_3
(b) NH_3, Glycine and Glutamate
(c) NH_3, Asparate and Glutamate
(d) Aspartate, Glutamine and Glycine **Ans. d**

663. A drug which prevents uric acid synthesis by inhibiting the enzyme Xanthine oxidase is

(a) Aspirin (b) Allopurinal
(c) Colchicine (d) Phenyl benzoate **Ans. b**

664. Glycine contributes to the following C and N of purine nucleus

(a) C_1, C_2 and N_7 (b) C_8, C_8 and N_9
(c) C_4, C_5 and N_7 (d) C_4, C_5 and N_9 **Ans. c**

665. Insoinic acid is the biological precursor of

(a) Cytosine and Uric acid
(b) adenylic acid and Glycine folic acid
(c) Orotic acid and Uridylic acid
(d) Adenosine acid Thymidine **Ans. b**

666. The probable metabolic defect in gents is
(a) A defect in excretion of uric acid by kidney
(b) An overproduction of pyrimidines
(c) An overproduction of uric acid
(d) Rise in calcium leading to deposition of calcium urate **Ans. c**

667. In humans, the principal break down product of purines is
(a) NH_3 (b) Allantoin
(c) Alanine (d) Uric acid **Ans. d**

668. A key substance in the committed step of pyrimidines biosynthesis is
(a) Ribose-5-phosphate (b) Carbamoyl phosphate
(c) ATP (d) Glutamine **Ans. b**

669. In humans, the principal metabolic product of pyrimidines is
(a) Uric acid (b) Allantoin
(c) Hypoxanthine (d) β-alanine **Ans. d**

670. In most mammals, except primates, uric acid is metabolized by
(a) Oxidation to allantoin (b) Reduction to NH_3
(c) Hydrolysis to allantoin (d) Hydrolysis to NH_3 **Ans. a**

671. Two nitrogen of the pyrimidines ring are obtained from
(a) Glutamine and Carbamoyl-P (b) Asparate and Carbamoyl-P
(c) Glutamate and NH_3 (d) Glutamine and NH_3 **Ans. b**

672. All are true about lesch-nyhan syndrome except
(a) Produces self-mutilation
(b) Genetic deficiency of the enzyme
(c) Elevated levels of uric acid in blood
(d) Inheritance is autosomal recessive **Ans. b**

673 Synthesis of GMP and IMP requires the following
(a) NH_3, NAD^+, ATP (b) Glutamine, NAD^+, ATP
(c) NH_3, GTP, $NADP^+$ (d) Glutamine, GTP, $NADP^+$
Ans. b

674. Which pathway is correct for catabolism of purines to form uric acid?
(a) Guanylate → Adenylate → Xanthine → hypoxanthine → Uric acid
(b) Guanylate → inosinate → Xanthine → hypoxanthine → Uric acid

(c) Adenylate → Inosinate → Xanthine→ hypoxanthine → Uric acid
(d) Adenylate → Inosinate → hypoxanthine → Xanthine → Uric acid
Ans. d

675. Polysomes do not contain
(a) Protein (b) DNA
(c) mRNA (d) rRNA **Ans. b**

676. The formation of a peptide bond during the elongation step of protein synthesis results in the splitting of how many high energy bonds?
(a) 1 (b) 2
(c) 3 (d) 4 **Ans. b**

677. Translocase is an enzyme required in the process of
(a) DNA replication (b) RNA synthesis
(c) Initiation of protein synthesis (d) Elongation of peptides **Ans. d**

678. Nonsense codons bring about
(a) Amino acid activation
(b) Initiation of protein synthesis
(c) Termination of protein synthesis
(d) Elongation of polypeptide chains **Ans. c**

679. Which of the following genes of the *E. coli* "Lac operon" codes for a constitutive protein?
(a) The 'a' gene (b) The 'i' gene
(c) The 'c' gene (d) The 'z' gene **Ans. b**

680. In the process of transcription, the flow of genetic information is from
(a) DNA to DNA (b) DNA to protein
(b) RNA to protein (d) DNA to RNA **Ans. d**

681. The anticodon region is an important part of the structure of
(a) rRNA (b) tRNA
(c) mRNA (d) hrRNA **Ans. b**

682. The region of the Lac operon which must be free from structural gene transcription to occur is
(a) The operator locus (b) The promoter site
(c) The 'a' gene (d) The 'i' gene **Ans. a**

683. Another name for reverse transcriptase is
(a) DNA dependent DNA polymerase
(b) DNA dependent RNA polymerase
(c) RNA dependent DNA polymerase
(d) RNA dependent RNA polymerase **Ans. c**

684. In the 'lac operon' concept, which of the following is a protein?
(a) Operator (b) Repressor
(c) Inducer (d) Vector **Ans. b**

685. Degeneracy of the genetic code denotes the existence of
(a) Base triplets that do not code for any amino acids
(b) Codons consisting of only two bases
(c) Codons that include one or more of the unusual bases
(d) Multiple codons for a single amino acid **Ans. b**

686. The normal function of restriction endonucleases is to
(a) Excise introns from hrRNA
(b) Polymerize nucleotides to form RNA
(c) Remove primer from okazaki fragments
(d) Protect bacteria from foreign DNA **Ans. d**

687. In contrast to Eukaryotic mRNA, prokaryotic mRNA is characterized by
(a) Having 7-methyl guanosine triphosphate at the 5′ end
(b) Being polycystronic
(c) Being only monocystronic
(d) Being synthesized with introns **Ans. a**

688. DNA ligase of *E. coli* requires which of the following co-factors?
(a) FAD (b) NAD^+
(c) $NADP^+$ (d) NADH **Ans. b**

689. Which of the following is transcribed during repression?
(a) Structural gene (b) Promoter gene
(c) Regulator gene (d) Operator gene **Ans. c**

690. mRNA is complementary copy of
(a) 5′ -3′ strand of DNA (b) 3′ -5′ strand of DNA
(c) Antisense strand of DNA (d) tRNA **Ans. b**

691. Synthesis of RNA molecule is terminated by a signal which is recognised by

(a) α-factor (b) β-factor

(c) δ-factor (d) ρ–factor **Ans. d**

692. The binding of prokaryotic DNA dependent RNA polymerase to promoter sits of genes is inhibited by the antibiotic

(a) Streptomycin (b) Rifamcin

(c) Aueromycin (d) Puromycin **Ans. b**

693. In *E. coli* the chain initiating amino acid in protein synthesis is

(a) N-formyl methionine (b) Methionine

(c) Serine (d) Cysteine **Ans. a**

694. How many high-energy phosphate bond equivalents are required for amino acid activation in protein synthesis?

(a) One (b) Two

(c) Three (d) Four **Ans. b**

695. Translation results in the formation of

(a) mRNA

(b) tRNA

(c) rRNA

(d) A protein molecule **Ans. d**

696. Elongation of a peptide chain involves all the following except

(a) mRNA

(b) GTP

(c) Formyl-Met-tRNA

(d) Tu, TS and G factors **Ans. c**

697. The 'rho' (ρ) factor is involved

(a) To increase the rate of RNA synthesis

(b) In binding catabolite repressor to the promoter region

(c) In proper termination of transcription

(d) To allow proper initiation of transcription **Ans. c**

698. In the biosynthesis of c-DNA, the joining enzyme ligase requires

(a) GTP (b) ATP

(c) CTP (d) UTP **Ans. b**

699. Which of the following is not correct about RNA ?

(a) RNA molecules ordinarily consist of a single polyribonucleotide strand

(b) RNA was, most likely, the first genetic material it was much later replaced by DNA

(c) Phosphodiester bonds are formed between ribonucleotides in the same manner as in the case of deoxyribonucleotides

(d) RNA molecules contain uracil in the place of cytosine **Ans. d**

❑❑❑

Chapter-13

Replication, Transcription and Translation

Fill in the blanks

1. Enzymes called......................couple each amino acid to its appropriate tRNA molecule to create an...................... molecule.

Ans. aminoacyl-tRNA synthetases, aminoacyl-tRNA

2. A..............contains two binding sites for tRNA molecules; the.............., or P-site, holds the tRNA molecule that is linked to the growing end of the polypeptide chain, and the............, or A-site, holds the incoming tRNA molecule that is charged with an amino acid.

Ans. ribosome, peptidyl-tRNA-binding site, aminoacyl-tRNA-binding site

3. Peptide bond formation is catalyzed by............, a catalytic activity that is thought to be mediated by the major............molecule in the large ribosomal subunit.

Ans. peptidyl transferase, rRNA

4. Proteins called release factors bind to................ codons in the A-site on the ribosome, causing peptidyl trnasferase to hydrolyze the bond that links the nascent polypeptide to the tRNA molecule.

Ans. stop

5. An RNA sequence can be translated in any of three............ each of which will specify a completely different polypeptide chain.

Ans. reading frames

6. The initiation process of protein synthesis is complicated, involving a number of steps catalyzed by proteins called.............

Ans. initiation factors (IFs)

7. In all cells, a special...........molecule, recognizing the.......codon AUG and carrying the amino acid............... provides the amino acid that begins a protein chain.

Ans. initiator tRNA, start, methionine

8. Most spontaneous changes in DNA are quickly erased by a correction process called...............; only rarely do DNA maintenance procedures fail and allow a permanent sequence change, which is called a ....

Ans. DNA repair, mutation

9. The genes encoding the.........., which are discarded during blood clotting, accumulate mutations without being selected against.

Ans. fibrinopeptides

10. Two common spontaneous changes in DNA are................, which results from disruption on the N-glycosyl linkages of adenine and guanine to deoxyribose, and............., which converts cytosine to uracil.

Ans. depurination, deamination

11. DNA repair involves three steps: recognition and removal of the altered portion of the DNA strand by enzymes called.................., resynthesis of the excised region by..........., and sealing the remaining nick by the enzyme.........

Ans. DNA repair nucleases, DNA polymerase, DNA ligase

12. One major repair pathway, called............., involves a battery of enzymes called..............., each of which recognizes an altered base in DNA and catalyzed its removal.

Ans. base excision repair, DNA glycosylases

13. The..............pathway in capable of removing almost any type of DNA damage that creates a large change in the DNA double helix.

Ans. nucleotide excision repair

14. In *E. coli* any block to DNA replication caused by DNA damage produces a signal that induces the............., which allows replication through the block, hence giving the cell a chance for survival.

Ans. SOS response

15. The enzyme responsible for DNA synthesis in both replication and repair is called..................

Ans. DNA polymerase

16. The active region of a chromosome involved in replication is a Y-shaped structure called a............

Ans. DNA replication fork

17. The enzyme that seals nicks in the DNA helix during DNA synthesis and repair is called................

Ans. DNA ligase

18. During DNA replication, the daughter strand that is synthesized continuously is called the............., and the strand that is synthesized discontinuously is known as the...............

Ans. leading strand, lagging strand

19. If DNA polymerase adds an incorrect nucleotide to the 3′ terminus, a separate catalytic domain containing a 3′- to 5′......................activity removes the mismatched base.

Ans. proofreading exonuclease

20. Initiation of DNA synthesis on the lagging strand requires short.......... made by an enzyme called............, which uses ribonucleotide triphosphates as substrates.

Ans. RNA primers, DNA primase

21. The unwinding of the DNA helix at the replication fork is catalyzed by a.............., which uses the energy from ATP hydrolysis to move unidirectionally along DNA.

Ans. DNA helicase

22. ...., which aid DNA unwinding, bind to single-stranded DNA in such a way that the bases are still available for templating reactions.

Ans. single-stranded DNA-binding (SSB) proteins

23. The DNA primase molecule is linked directly to the DNA helicase to form a unit called a.........., which moves down the lagging strand at the fork, synthesizing RNA primers as it goes.

Ans. primosome

24. If DNA polymerase makes a mistake, thus creating a pair of incorrectly hydrogen-bonded bases, the mistake is corrected by a special ....system that uses methylation to distinguish new strands from old.

Ans. mismatch proofreading (mismatch repair)

25. For bacteria and yeasts, as well as for several viruses that grow in eukaryotic cells, replication bubbles have been shown to form at special DNA sequences called.............

Ans. replication origins

26. ....can be viewed as reversible nucleases that create either a transient single-strand break (type I) or a transient double-strand break (type II).

Ans. DNA topoisomerases

27. In..............., genetic exchange occurs between homologous DNA sequences, most commonly between two copies of the same chromosome.

Ans. general recombination

28. At the site of exchange, a strand of one DNA molecule has become base-paired to a strand of the second DNA molecule to create a...............between the two double helices.

Ans. heteroduplex joint (staggered joint)

29. Two single-stranded, complementary DNA molecules come together to form a fully double-stranded helix by..............., which is thought to start with a slow............step.

Ans. DNA renaturation (hybridization), helix nucleation

30. The..............is required for chromosome pairing in *E. coli*; it binds to single-stranded DNA and promotes its pairing with homologous, double-stranded DNA.

Ans. RecA protein

31. Once synapsis has occurred, the short heteroduplex region where strands from two different DNA molecules have begun to pair is enlarged through a reaction called................

Ans. branch migration

32. A central intermediate in general recombination is the............., which is also called a................, after its discover.

Ans. cross-strand exchange, Holiday junction

33. Occasionally during recombination between two slightly different copies of the same gene (alleles), one allele is replaced by the other in a process known as................

Ans. gene conversion

34. Mobile DNA sequences and some viruses enter and leave a target chromosome by.....................genetic recombination.

Ans. site-specific

Multiple Choice Questions

1. Nobel Prize for one gene one enzyme hypothesis was given to
   (a) Watson and Crick (b) Sutton and Boveri
   (c) Avery et al (d) Beadle and Tatum **Ans. d**

2. *E. coli* fully labeled with ^{15}N medium. The two strands of DNA molecule of the first generation bacteria have
   (a) Different density and do not resemble parent DNA
   (b) Different density but resemble parent DNA
   (c) Same density and resemble parent DNA
   (d) Same density but do not resemble parent DNA **Ans. a**

3. DNA polymerase helps in
   (a) Splitting of two DNA strands (b) Proof reading of DNA
   (c) Renaturation of DNA (d) Joining monomers of DNA
   Ans. d

4. DNA having labeled thymidine is allowed to replicate in medium having non-radioactive thymidine. After three duplication, the DNA molecules having labeled thymidine shall be
   (a) One molecule (b) Two molecule
   (c) Four molecules (d) Eight molecules **Ans. b**

5. Transcription involves
   (a) Synthesis of RNA over DNA
   (b) Joining of amino acids in a polypeptide
   (c) Synthesis of RNA over ribosome
   (d) Synthesis of DNA **Ans. a**

6. DNA replication requires
   (a) DNA polymerase
   (b) RNA polymerase and translocase
   (c) DNA ligase
   (d) DNA polymerase and DNA ligase **Ans. d**

7. Ligase is an enzyme required for
   (a) Breaking of DNA (b) Joining of DNA bits
   (c) Renaturation of DNA (d) Proof reading **Ans. b**

8. Polypeptide chain is initiated by
   (a) Glycine (b) Leucine
   (c) Methionine (d) Lysine **Ans. c**

9. Process of protein synthesis in a cell is
   (a) Translocation (b) Translation
   (c) Transcription (d) Transduction **Ans. b**

10. The one which is capable of self replication is
   (a) DNA (b) RNA
   (c) Enzyme (d) Protein **Ans. a**

11. Degeneracy of genetic code was discovered by
   (a) Khorana (b) Ochoa
   (c) McClintock (d) Bernfield and Nirenberg **Ans. d**

12. Usual method of DNA replication is
   (a) Semiconservative (b) Conservative
   (c) Dispersive (d) Non conservative **Ans. a**

13. *E. coli* with completely radioactive DNA was allowed to replicate in non-radioactive medium for two generations. Percentage of bacteria with radioactive DNA is
   (a) 100% (b) 50%
   (c) 25% (d) 12.5% **Ans. b**

14. DNA replication in eukaryotes commences
   (a) From both ends of a chrosome simultaneously
   (b) Several sites along DNA of a chromosome simultaneously
   (c) From centromere to either end
   (d) From one end of chromosome to the other **Ans. b**

15. *In vitro* synthesis of RNA and DNA was carried out respectively first by
   (a) Kornberg and Nirenberg (b) Ochoa and Kornberg
   (c) Ochoa and Nirenberg (d) Nirenberg and Khorana **Ans. b**

16. Genetic code consists of
   (a) Adenine and guanine (b) Cytosine and uracil
   (c) Cytosine and guanine (d) All of the above **Ans. d**

17. Genetic code determines

(a) Sequence of amino acids in protein chain

(b) Variations

(c) Constancy of morphological traits

(d) Structural protein **Ans. a**

18. First deciphering of genetic code through trinucleotide synthesis was performed by

(a) Beadle and Tatum (b) Watson and Crick

(c) Nirenberg (d) Ochoa **Ans. c**

19. Initiation of polypeptide chain is through

(a) Lysine (b) Glycine

(c) Leucine (d) Methionine **Ans. d**

20. Codon of mRNA and anticodon of tRNA is made of

(a) A set of three out of U, A, C and G

(b) A set of three and two respectively

(c) A set of two nitrogen base

(d) Three and one nitrogen bases respectively **Ans. a**

21. Antibiotic inhibiting translation in eukaryotes is

(a) Tetracycline (b) Puromycin

(c) Penicillin (d) Chloromycetin **Ans. b**

22. Who established that RNA can be genetic material?

(a) Griffith (b) Conrat

(c) Lederberg (d) Nirenberg and Mathaei **Ans. b**

23. Nirenberg synthesized an RNA having 34 adenine residues (AAAA...) and obtained a polypeptide of 11 lysine residues. It proved that genetic code for lysine is

(a) Adenine (b) AA

(c) AAA (d) AAAA...A **Ans. c**

24. Khorana et al. synthesized RNA with copolymer of UGUGUGUGUG... It produced a peptide with alternate cytosine and valine. The codons for the two are

(a) UGU and GUG (b) UUG and GGU

(c) GUG and UGU (d) UGG and GUU **Ans. a**

25. Who proposed that genes control production of enzymes?
(a) R.D. Kornberg (b) Beadle and Tatum
(c) A.E. Garrod (d) T.H. Morgan **Ans. c**

26. Genes are formed of
(a) RNA (b) DNA
(c) DNA and RNA (d) DNA and Proteins **Ans. b**

27. Gene is a
(a) Segment of DNA
(b) Segment of chromosome
(c) Functional unit of DNA
(d) DNA segment capable of crossing over **Ans. c**

28. Regulator gene controls chemical synthesis (Operon concept) by
(a) Inhibiting transcription of mRNA
(b) Inhibiting enzymes
(c) Inhibiting passage of mRNA
(d) Inhibiting substrate enzyme reaction **Ans. c**

29. In operon concept, the operator gene combines with
(a) Regulator protein to switch off structural gene transcription
(b) Regulator protein to switch on structural gene transcription
(c) Inducer to switch off structural gene transcription
(d) Regulator gene to switch off structural gene transcription **Ans. a**

30. The terms cistron, recon and muton were proposed by
(a) Johannsen (b) Morgan
(c) Lederberg (d) Benzer **Ans. d**

31. Smallest segment of DNA capable of undergoing crossing over and recombination is
(a) Muton (b) Cistron
(c) Recon (d) Intron **Ans. c**

32. Biologically functional tyrosine suppressor tRNA of *E. coli* synthesized by Khorana in 1979 had
(a) 333 nucleotides (b) 312 nucleotide pairs
(c) 207 nucleotide pairs (d) 77 nucleotide pairs **Ans. c**

33. In operon model, RNA polymerase binds to
(a) Structural gene (b) Promoter gene
(c) Regulator (d) Operator gene **Ans. b**

34. A gene which synthesizes a repressor protein is
(a) Operator gene (b) Structural gene
(c) Promoter gene (d) Regulator gene **Ans. d**

35. The term gene was coined by
(a) McClintock (b) Morgan
(c) Johannsen (d) De Duve **Ans. c**

36. Khorana first deciphered the triplet codons of
(a) Serine and isoleucine (b) Cystiene and valine
(c) Tyrosine and tryptophan (d) Phenylalanine and methionine
Ans. b

37. In the genetic dictionary, there are 64 codons as
(a) 64 amino acids are to be coded
(b) 64 types of tRNAs are present
(c) There are 44 nonsense codons and 20 sense codons
(d) Genetic code is triplet **Ans. d**

38. An anticodon of tRNA can recognize more than one codon of mRNA. It is
(a) Wobble hypothesis (b) Gene flow hypothesis
(c) Template hypothesis (d) Richmond and Lang effect
Ans. a

39. Wobble hypothesis establishes
(a) Peptide chain formation (b) Initiation of peptide chain
(c) Termination of peptide chain (d) Economy in tRNA molecules
Ans. d

40. Synthesis of DNA over RNA template/teminism occurs in
(a) TMV (b) Retrovirus
(c) Rous Sarcoma Virus (d) T_2 **Ans. c**

41. Experimental material in the study of DNA replication has been
(a) *E. coli* (b) *Neurospora crassa*
(c) *Pneumococcus* (d) *Drosophila melanogaster*
Ans. a

42. Site of tRNA that binds to mRNA molecule is

(a) 3′ end (b) 5′ end

(c) Codon (d) Anticodon **Ans. d**

43. Find out the correct answers out of the following discoveries

1. Griffith-Transformation
2. Gamow-Triplet code
3. Miescher-Nucleic acid

(a) 1, 2 and 3 correct (b) 1 and 2 correct, 3 false

(c) 1 correct, 2 and 3 false (d) 1 and 3 correct, 2 false **Ans. d**

44. Nucleotide arrangement in DNA can be seen by

(a) X-ray crystallography (b) Electron microscope

(c) Ultracentrifuge (d) Light microscope **Ans. a**

45. A DNA with unequal nitrogen bases would most probably be

(a) Single stranded (b) Double stranded

(c) Triple stranded (d) Four stranded **Ans. a**

46. Pneumococcus experiment proved that

(a) Bacteria do not reproduce asexually

(b) Bacteria undergo binary fission

(c) DNA is genetic material

(d) RNA may sometimes control production of DNA and proteins

Ans. c

47. Select correct/incorrect functions

1. Ligase-Joins short DNA segments
2. DNA polymerase-Cuts DNA a specific sites
3. Helicase-Breaks H-bonds between complementary DNA strands

(a) 1, 2 and 3 correct (b) 1 correct, 2 and 3 false

(c) 1 and 3 correct, 2 false (d) 1 and 2 correct, 3 false **Ans. c**

48. Because most of the amino acids are represented by more than one codon, the genetic code is

(a) Overlapping (b) Wobbling

(c) Degenerate (d) Generate **Ans. c**

49. Who proved that DNA is basic genetic material?

(a) Griffith (b) Watson

(c) Boveri and Sutton (d) Hershey and Chase **Ans. d**

50. Information transfer from RNA to DNA is

(a) Transcription (b) Translation

(c) Replication (d) Reverse transcription **Ans. d**

51. Double stranded DNA virus with 20,000 base pairs has nucleotides

(a) 20,000 (b) 10,000

(c) 666 (d) 40,000 **Ans. d**

52. A DNA nucleotide chain has AGCTTCGA sequence. The nucleotide sequence of other chain would be

(a) TCGAAGCT (b) GCTAAGCT

(c) TAGCATAT (d) GATCCTAG **Ans. a**

53. Eukaryotes differ from prokaryotes in mechanism of DNA replication due to

(a) Different enzymes for opening of strands

(b) DNA primers instead of RNA primers

(c) Unidirectional rather than bidirectional

(d) Discontinuous rather than semi continuous **Ans. d**

54. Protein helping in opening of DNA double helix in front of replication fork is

(a) DNA gyrase (b) DNA polymerase I

(c) DNA ligase (d) DNA topoisomerase **Ans. a**

55. Which is not involved in protein synthesis?

(a) Transcription (b) Initiation

(c) Elongation (d) Termination **Ans. d**

56. Characters are transferred from parents to progeny through

(a) DNA (b) RNA

(c) Protein (d) Phospholipid **Ans. a**

57. Circular DNA is present in

(a) ER and ribosomes (b) Ribosomes and chloroplasts

(c) Ribosomes and mitochondria (d) Mitochondria and chloroplasts

Ans. d

58. A functional unit in synthesis of protein is
(a) Lysosome (b) Peroxisome
(c) Polysome (d) Dictyosome **Ans. c**

59. Arrangement of three successive bases in genetic code signifies
(a) Protein (b) Amino acid
(c) Nucleic acid (d) Plasmid **Ans. b**

60. Genes connected with cancer are
(a) Cancer genes (b) Carcinoma genes
(c) Oncogenes (d) Sex-linked genes **Ans. c**

61. Smallest genes affected by mutation is
(a) Muton (b) Cistron
(c) Recon (d) Exon **Ans. a**

62. Gene regulate growth and differentiation through
(a) Transcription and translation (b) Transduction and translation
(c) Transformation (d) Translation **Ans. a**

63. In *E. coli,* lac operon is induced by
(a) Lactose (b) Promoter gene
(c) b-galactose (d) I-gene **Ans. a**

64. Reverse transcriptase is
(a) RNA dependent RNA polymerase
(b) DNA dependent RNA polymerase
(c) DNA dependent DNA polymerase
(d) RNA dependent DNA polymerase **Ans. d**

65. In some viruses RNA is present instead of DNA indicating that
(a) Their nucleic acid must combine with host DNA before replication
(b) They cannot replicate
(c) There is no hereditary information
(d) RNA can act to transfer heredity **Ans. d**

66. Lac operon is
(a) Arabinose operon (b) Repressible operon
(c) Inducible operon (d) Overlapping operon **Ans. c**

67. Wild type *E. coli* growing on medium having glucose is transferred to lactose containing medium. Which change occurs?
(a) Lac operon is induced (b) Lac operon is suppressed
(c) All operons are induced (d) The bacterium stops dividing
Ans. a

68. In split genes, the coding sequences are called
(a) Exons (b) Introns
(c) Cistrons (d) Operons **Ans. a**

69. The number of base substitution possible in amino acid codons is
(a) 261 (b) 264
(c) 535 (d) 549 **Ans. d**

70. Watson and Crick are known for their discovery that DNA is
(a) Single stranded (b) Double stranded
(c) Having deoxyribose only (d) Template for rRNA synthesis
Ans. b

71. Base sequence on one strand of DNA is GCATG. Replicated complementary strand has
(a) GCATG (b) GCATC
(c) CGTAC (d) ATGCG **Ans. c**

72. Double hydrogen bonds occur in DNA between
(a) Adenine and Thymine (b) Uracil and Thymine
(c) Adenine and Guanine (d) Thymine and Cytosine **Ans. a**

73. Nucleic acids are made of
(a) Nucleotides (b) Nucleosides
(c) Amino acids (d) Proteins **Ans. a**

74. Hydrogen between cytosine and guanine are
(a) 1 (b) 2
(c) 3 (d) 4 **Ans. c**

75. Environmental agent triggering transcription from an operon is
(a) Inducer (d) Regulator
(c) Derepressor (d) Controlling factor **Ans. a**

76. Nuclear DNA sends information for protein synthesis through
(a) tRNA (b) mRNA
(c) rRNA (d) All of the above **Ans. b**

77. Coded information in nucleic acid depends upon
(a) Arrangement of nucleotides (b) Position of nucleotides
(c) Number of nucleotides (d) All of the above **Ans. a**

78. RNA polymerase (Ochon 1953) has polypeptide chains
(a) 1 (b) 2
(c) 4 (d) 5 **Ans. d**

79. DNA strand with nitrogen base sequence ATTGCC will have sequence in mRNA
(a) UAACGG (b) ATCGCC
(c) ATTGCA (d) UGGACC **Ans. a**

80. DNA sequence of ATTCGATG is transcribed as
(a) AUUCGAUG (b) UAAGCUAC
(c) CAUCGAAU (d) GUAGCUUA **Ans. b**

81. Cytosine base inserted in the beginning of DNA codons ATGATGATG will produce
(a) CATGATGATG (b) CATGATGATG
(c) CATGATGATG (d) Nonsense mutation **Ans. b**

82. Virus can survive in
(a) Soil (b) Air
(c) Water (d) Living tissue **Ans. d**

83. Central dogma of modern genetics is
(a) RNA $\rightarrow$ RNA $\rightarrow$ DNA $\rightarrow$ Protein
(b) DNA $\rightarrow$ DNA $\rightarrow$ RNA $\rightarrow$ Protein
(c) DNA $\rightarrow$ RNA $\rightarrow$ Protein
(d) RNA $\rightarrow$ DNA $\rightarrow$ RNA $\rightarrow$ Protein **Ans. b**

84. Okazaki segments are formed during
(a) Transduction (b) Transcription
(c) Replication (d) Translation **Ans. c**

85. The enzyme taking part in joining two ends of DNA is
(a) Ligase (b) Polymerase
(c) Gyrase (d) Helicase **Ans. a**

86. Which one is not true?

(a) A = T, G = C (b) A = G, T = C
(c) C = G, T = A (d) G = C, A = T **Ans. b**

87. Who proposed central dogma?

(a) Watson and Crick (b) Beadle and Tatum
(c) Klug (d) Crick **Ans. d**

88. Nucleotide base present in DNA and not in RNA is

(a) Cytosine (b) Uracil
(c) Thymine (d) Guanine **Ans. c**

89. Hereditary information is indicated by

(a) Number of nucleic acids (b) Position of nucleic acids
(c) Sequence of nucleic acids (d) All of the above **Ans. d**

90. What is correct for protein synthesis?

(a) Code transfer on mRNA
(b) Code transfer on tRNA
(c) Coding is done by DNA strands
(d) DNA coding takes place in antiparallel fashion **Ans. a**

91. Triplet codon refers to sequence of three bases on

(a) mRNA (b) tRNA
(c) rRNA (d) All of the above **Ans. a**

92. Genetic code was discovered by frameshift mutation by

(a) Crick (b) Watson
(c) Franklin (d) Khorana **Ans. a**

93. Who synthesized the first artificial gene?

(a) Morgan (b) Nirenberg
(c) Khorana (d) Mendel **Ans. c**

94. Products of one gene required to activate another gene are called

(a) Catalases (b) Coenzymes
(c) Episomes (d) Transcription factors **Ans. d**

95. Operator gene of lac-operon is turned on when lactose molecules bind to

(a) Promoter site (b) mRNA
(c) Operator gene (d) Regulator gene **Ans. c**

96. Point mutation is

(a) Loss of gene (b) Change in base of gene
(c) Addition of a gene (d) Deletion of a segment of gene

Ans. b

97. A mutation at one base of first codon of a gene forms a nonfunctional protein. It is

(a) Nonsense mutation (b) Reverse mutation
(c) Frames shift mutation (d) Mis-sense mutation **Ans. d**

98. Taylor demonstrated chromosome replication to be semiconservative in

(a) Vicia faba (b) he La cell
(c) Pea (d) Mouse liver cell **Ans. a**

99. Process of altering code in DNA where message is reversed is

(a) Deletion (b) Insertion
(c) Inversion (d) Translocation **Ans. c**

100. Which step of translation does not consume high energy phosphate bond?

(a) Translocation
(b) Peptidyl transferase reaction
(c) Amino acid activation
(d) Aminoacyl tRNA binding to A-site **Ans. b**

101. DNA synthesis can be measured by estimating incorporation of radio-labelled

(a) Uracil (b) Ribose sugar
(c) Thymidine (d) Adenine **Ans. c**

102. DNA duplex shows

(a) Right handed coiling and parallel
(b) Right handed coiling and antiparallel
(c) Left handed coiling and antiparallel
(d) Left handed coiling and parallel **Ans. b**

103. mRNA is polymer of

(a) Ribonucleotides (b) Deoxyribonucleotides
(c) Ribonucleosides (d) Deoxyribonucleosides **Ans. a**

104. tRNA recognizes ribosome by

(a) T ψ C loop (b) DHU loop
(c) Anticodon (d) AA-site **Ans. a**

105. Enzyme required for removing RNA primer during DNA replication is

(a) Primase (b) Ligase

(c) DNA polymerase I (d) DNA polymerase III **Ans. c**

106. Which one is involved in DNA repair?

(a) Ligase (b) Primase

(c) DNA polymerase III (d) DNA polymerase I **Ans. d**

107. Double chained DNA strand is made radioactive in both its chains. It is allowed to replicate twice in non-radioactive medium. The result would be

(a) All strands have radioactivity

(b) Half strands have radioactivity

(c) Three strands have radioactivity

(d) Radioactivity is absent in all strands **Ans. b**

108. Who made discovery for genetic engineering?

(a) Morgan (b) Watson

(c) Crick (d) Khorana **Ans. d**

109. Experimental evidence supporting concept of triplet genetic code was first provided by

(a) Beadle and Tatum (b) Watson

(c) Crick (d) Michaelis and Menten **Ans. c**

110. Hargobind Khorana is known for

(a) Discovery of DNA (b) Discovery of DNA ligase

(c) Discovery of tRNA (d) Synthesis of proteins **Ans. b**

111. Amino acid binding site of tRNA is

(a) 5′ end (b) Anticodon loop

(c) DHU loop (d) –CCA 3′ end **Ans. d**

112. In a DNA molecule cytosine is 18%. Percentage of adenine would be

(a) 32% (b) 64%

(c) 36% (d) 18% **Ans. a**

113. Which one is DNA nitrogen base with single ring structure?

(a) Uracil (b) Thymine

(c) Adenine (d) Guanine **Ans. b**

114. Genetic code translates the language of

(a) Amino acids into that of RNA (b) RNA into that of protein
(c) Protein into that of DNA (d) RNA into that of DNA **Ans. b**

115. Restriction endonuclease is employed for cutting

(a) A single stranded DNA (b) Double stranded DNA
(c) RNA fragment (d) mRNA **Ans. b**

116. Which enzyme is useful in genetic engineering?

(a) DNA-ase (b) Amylase
(c) Lipase (d) Restriction endonuclease
Ans. d

117. It is now possible to breed plants and animals of desired characters through

(a) Tissue culture (b) Genetic engineering
(c) Ikebana technique (d) Chromosome engineering
Ans. b

118. Genetic engineering is

(a) Plastic surgery (b) Addition or removal of genes
(c) Study of extra nuclear genes (d) All of the above **Ans. b**

119. Advancement in genetic engineering has been possible due to discovery of

(a) Oncogenes (b) Transposons
(c) Restriction endonuclease (d) Exonucleases **Ans. c**

120. Genetically engineered bacteria are being used in commercial production of

(a) Melatonin (b) Testosterone
(c) Human insulin (d) Thyroxine **Ans. c**

121. Transgenic plants are developed by

(a) Introducing foreign genes
(b) Introducing gene mutations
(c) Deleting certain chromosome parts
(c) Stopping spindle formation **Ans. a**

122. Bacterial plasmid contains

(a) RNA (b) RNA + Protein
(c) DNA (d) Photosynthetic structures
Ans. c

123. Structure involved in genetic engineering is

(a) Plastid (b) Restriction endonuclease

(c) DNA polymerase I (d) Prochrosome **Ans. b**

124. Plasmids are vectors for gene cloning because they

(a) Self replicate in bacterial cells

(b) Replicate freely outside bacterial cells

(c) Can be multiplied in culture

(d) Can be multiplied in laboratories using enzymes **Ans. a**

125. Two bacteria most useful in genetic engineering are

(a) *Rhizobium* and *Azotobacter* (b) *Escherichia* and *Agrobacterium*

(b) *Rhizobium* and *Diplococcus* (d) *Nitrosomonas* and *Klebsiella*

Ans. b

126. Genetic engineering has been made possible due to

(a) Observation of DNA under electron microscope

(b) We can break DNA at specific points by DNA-ases

(c) Availability of restriction endonucleases in purified form

(d) Knowledge of transduction **Ans. c**

127. Cloning is means to

(a) Replace original genotype

(b) Preserve genotype

(c) Production of HGH gene in *E. coli*

(d) None of the above **Ans. b**

128. DNA finger printing is useful for study of

(a) Forensics (b) Polymorphism

(c) Identity and relationships (d) All of the above **Ans. d**

129. The first codon discovered by Nirenberg and Mathaei was

(a) GGG (b) AAA

(c) UUU (d) CCC **Ans. c**

130. The process of transfer of genetic information from DNA to RNA/ formation of RNA from DNA is

(a) Transversion (b) Transcription

(c) Translocation (d) Translation **Ans. b**

131. One gene one enzyme hypothesis was proposed by
(a) Watson and Crick (b) Beadle and Tatum
(c) Jacob and Monod (d) Garrod and Jensen **Ans. b**

132. Initiation codon of protein synthesis (in eukaryotes) is
(a) CCA (b) GCA
(c) GUA (d) AUG **Ans. d**

133. Genetic code was discovered by
(a) Holley and Ochoa
(b) Nirenberg and Holley
(c) Nirenberg and Mathaei
(d) Holley, Nirenberg and Khorana **Ans. d**

134. Semiconservative DNA/chromosome replication using ^{15}N was demonstrated by
(a) Meselson (b) Hershey and Chase
(c) Meselsona and Stahl (d) Taylor **Ans. c**

135. Central dogma of modern biology is
(a) DNA $\leftrightarrow$ RNA $\rightarrow$ Proteins (b) RNA $\rightarrow$ Proteins $\rightarrow$ DNA
(c) RNA $\rightarrow$ DNA $\rightarrow$ Proteins (d) DNA $\rightarrow$ RNA $\rightarrow$ Proteins
Ans. d

136. DNA duplication or multiplication is known as
(a) Replication (b) Translation
(c) Transduction (d) Transcription **Ans. a**

137. The two strands of DNA are held together by
(a) Peptide bond (b) S-S bonds
(c) Hydrogen bonds (d) Phosphodiester bonds **Ans. c**

138. DNA replication is
(a) Semiconservative and discontinuous
(b) Semiconservative and semidiscontinuous
(c) Conservative and discontinuous
(d) Conservative **Ans. b**

139. Genetic code is said to be degenerate because
(a) One codon codes for more than one amino acid
(b) One amino acid is coded by more than one codon

(c) Codons degenerate very quickly

(d) None of the above **Ans. b**

140. Functional unit of gene that specifies synthesis of one polypeptide is

(a) Codon (b) Cistron

(c) Muton (d) Recon **Ans. b**

141. Characteristic unique in DNA is

(a) Denaturation and renaturation

(b) Polymer complex

(c) Replication

(d) Resistance to temperature change **Ans. c**

142. A solution contains DNA polymerase I, Mg^{2+} salts of dATP, dGTP, dCTP, and dTTP, and an appropriate buffer. Which of the following DNA molecules would serve as a template for DNA synthesis when added to this solution

(a) A single-stranded closed circle

(b) A single-stranded closed circle base-paired to a shorter linear strand with 3′-terminal hydroxyl

(c) A single-stranded closed circle base-paired to a shorter linear strand with 3′-terminal phosphate

(d) A single-stranded closed circle **Ans. b**

143. Which of the following activities is typical of histones during replication of eukaryotic DNA?

(a) They remain bound to the DNA throughout replication

(b) They undergo proteolysis and are replaced by newly synthesized protein on each daughter strand

(c) They are both conserved and associated with the leading daughter strand and newly synthesized and associated with the lagging daughter strand

(d) They are not synthesized for cell division but are retained from generation to generation **Ans. c**

144. Correct statements describing DNA polymerase I include which of the following?

(a) It can add deoxyribonucleoside triphosphates to the 3′-hydroxyl end of a DNA primer chain

(b) It can act as an endonuclease in DNA repair

(c) It can act as a 5′ to 3′-exonuclease

(d) It is the major polymerizing enzyme in the replication process

Ans. c

145. Approximately how many Okazaki fragments are synthesized during one round of replication of the *E. coli* genome?

(a) 5000 to 10000 (b) 4×10^6

(c) 2500 to 5000 (d) 2 **Ans. c**

146. DNA replication under going in an *E. coli* cell resulted into two circular DNAs interlocked with each other; this may be due to a defective gene encoding

(a) Primase (b) DNA polymerase I

(c) DNA polymerase (d) DNA topoisomerase II **Ans. d**

147. A bacterial culture growing in a medium containing $^{15}NH_4Cl$ is switched to a medium containing $^{14}NH_4Cl$ for three generations (resulting into eight fold increase in its population), what is the molar ratio of hybrid DNA (^{15}N-^{14}N) to light DNA (^{14}N-^{14}N) at this point?

(a) 2/6 or 0.33 (b) 3/5 or 0.60

(c) 4/4 or 1.00 (d) 1/7 or 0.14 **Ans. a**

148. The fact that DNA polymerase I from *E.coli* has a 5′-3′ exonuclease activity

(a) Implies that the enzyme has multiple subunits

(b) Implies that DNA polymerase I can use both DNA and RNA as primers

(c) Makes the enzyme able to detect thymine dimers in double-stranded DNA

(d) Enables the enzyme to play an important role in DNA replication

Ans. d

149. Which of the following features of transcription is similar to that of replication?

(a) No primer is requires for polymerization

(b) Polymerization does not have proof reading activity

(c) Newly synthesized strand starts falling off the template before complete RNA is synthesized

(d) RNA synthesis requires DNA topoisomerase action **Ans. d**

150. In prokaryotes, the lagging primers are removed by

(a) 3′ to 5′ exonuclease (b) DNA ligase

(c) DNA polymerase I (d) DNA polymerase III **Ans. c**

151. The essential initiator protein at the *E. coli* origin of replication is

(a) DnaA (b) DnaB

(c) DnaC (d) DnaE **Ans. a**

152. Which phase would a cell enter if it was starved of mitogens before the R point?

(a) G_1 (b) S

(c) G_2 (d) G_0 **Ans. d**

153. Prokaryotic plasmids can replicate in yeast cells if they contain a cloned yeast

(a) ORC (b) CDK

(c) ARS (d) RNA **Ans. c**

154. During DNA replication, the synthesis of the leading strand occurs in the 5′→3′; direction along with the movement of the replication fork. Which one of the following statements is true about the synthesis of the lagging strand?

(a) DNA polymerase II, which has an additional activity of polymerization in 3′→5′ direction, is utilized

(b) DNA polymerase I is utilized, which can carry out DNA synthesis in the absence of any primer

(c) Synthesis of the lagging strand occurs in the small fragments by DNA polymerase in 5′→3′ direction using a de novo synthesized DNA primer

(d) Synthesis of the lagging strand occurs in small fragments in 5′→3′ direction utilizing RNA primers **Ans. d**

155. Okazaki fragments are used to elongate

(a) The leading strand towards the replication fork

(b) The lagging strand towards the replication fork

(c) The leading strand away from the replication fork

(d) The lagging strand away from the replication fork **Ans. d**

156. *E. coli* cells that had been cultured in a light medium containing ^{14}N-ammonium chloride as the sole nitrogen source were transferred to a heavy medium containing ^{15}N-ammonium chloride. After three generations of growth in the heavy medium, what will be the proportion of dual light strands (LL), dual heavy strands (HH), and hybrid light-plus-heavy strands (LH) in the DNA duplexes?

(a) 33 LH to 1 HH (b) 3 HH to 1 LH

(c) 3 HH to 1 LH/4LL (d) 15 LL to 1 LH **Ans. b**

157. Which of the following statements about DNA polymerase I is correct?

(a) It functions as a DNA repair enzyme but is not involved in the DNA replication process

(b) It requires a template and a primer to polymerize deoxyribonucleoside triphosphates

(c) It joins together Okazaki fragments to complete the lagging strand during DNA replication

(d) It produces Okazaki fragments linked to RNA primer chains

Ans. b

158. DNA ligase of *E. coli* requires which of the following cofactors?

(a) Flavin-adenine dinucleotide as an electron acceptor

(b) Nicotinamide-adenine dinucleotide phosphate as a phosphate donor

(c) Nicotinamide-adenine dinucleotide (NAD^+) to form an active adenyl enzyme

(d) NAD^+ as an electron acceptor **Ans. c**

159. Which enzyme removes the RNA primer and fills in deoxyribonucleotides in prokaryotic replication?

(a) DNA polymerase III (b) DNA polymerase II

(c) DNA polymerase I (d) Exonuclease MFI **Ans. c**

160. Which of the sequence listed below best describes the order in which the following enzymes participate in the replication of DNA in bacteria?

(i) DNA polymerase I (ii) $5' \rightarrow 3'$ Exonuclease

(iii) DNA polymerase III (iv) DNA ligase

(a) iii, i, ii, iv (b) iii, ii, i, iv

(c) iii, iv, ii, i (d) iii, ii, iv, i **Ans. d**

161. Which of the following statements regarding the supercoiling of duplex DNA is correct?

(a) DNA replication promotes positive supercoiling downstream from the replication fork

(b) DNA polymerase III can replicate positively supercoiled duplex DNA

(c) DNA topoisomerase unwind duplex DNA into single-stranded DNA chains

(d) DNA topoisomerases unwind positively supercoiled duplex DNA without cutting the phosphosugar backbone **Ans. a**

162. Short chains of nucleic acid can be isolated from cells in which DNA is undergoing replication. These segments, known as Okazaki fragments, have which of the following properties?

(a) They are double-stranded

(b) They contain covalently linked RNA and DNA

(c) They are DNA-RNA hybrids

(d) They arise from the nicking of the sugar-phosphate backbone of the parental DNA chain **Ans. b**

163. *E. coli* grown on ^{15}N medium are transferred to ^{14}N medium and allowed to grow for two generations (two cell replications). DNA extracted from these cells is centrifuged. What density distribution of DNA would you expect in this experiment? Explain your answer.

(a) One high density and one low density band

(b) One intermediate density band

(c) One high density and one intermediate density band

(d) One low density and one intermediate density band **Ans. d**

164. The bacterial enzyme that changes positively supercoiled DNA into negatively supercoiled DNA is

(a) DNA helicase (b) DNA gyrase

(c) Single strand binding protein (d) Polymerase **Ans. b**

165. Radioactive thymidine when added to the medium surrounding living mammalian cells gets incorporated into the newly synthesized DNA. Which of the following types of chromatin is expected to become radioactive if cells are exposed to radioactive thymidine as soon as they enter the S-phase?

(a) Neither heterochromatin nor euchromatin but only the nucleus

(b) Heterochromatin

(c) Euchromatin

(d) Both heterochromatin and euchromatin **Ans. c**

166. DNA polymerization requires the presence of template and of a free 3′-OH end. This end may be generated by

P. synthesis of an RNA primer at the origin or at Okazaki fragment start points

Q. nicking one strand of duplex DNA followed by strand displacement

R. by loop formation at the 3′ end (self-priming)

S. binding of a terminal nucleotide-carrying protein to the 3′ end of template

(a) P,Q (b) P,R

(c) R,S (d) P,Q,R,S **Ans. d**

167. Which of the following property not associated with DNA polymerase I

(a) $5' \rightarrow 3'$ exonuclease activity (b) $3' \rightarrow 5'$ exonuclease activity

(c) $5' \rightarrow 3'$ endonuclease activity (d) $5' \rightarrow 3'$ polymerase activity

Ans. c

168. Isotopes used for proving semiconservative replication of DNA are

(a) N^{14} and P^{31} (b) N^{14} and N^{15}

(c) N^{14} and C^{14} (d) C^{14} and P^{31} **Ans. b**

169. The exonuclease activity of the DNA polymerases functions to

(a) Remove the RNA primer sequences

(b) Proofread the new DNA strand and remove inappropriate nucleotides

(c) Maximize the fidelity of DNA replication

(d) All the above **Ans. d**

170. In prokaryotes, the lagging strand primers are removed by

(a) 3′ to 5′ exonuclease (b) DNA ligase

(c) DNA polymerase I (d) DNA polymerase III **Ans. c**

171. T_4 DNA ligase

(a) Requires ATP

(b) Joins double-stranded DNA fragments with an adjacent 3′-phosphate and 5′-OH

(c) Requires NADH

(d) Joins single-stranded DNA **Ans. a**

172. How does the DNA of the circular *E. coli* chromo-some replicate?

(a) Unidirectionally: it starts at one point on the chromosome and replicates in one direction

(b) Bidirectionally: it starts at one point on the chromosome and replicates in both directions

(c) Multidirectionally: it starts at several points on the chromosome proceeding in both directions

(d) All the above are true **Ans. b**

173. In genetics, an origin is
(a) A protein that binds to DNA at the point where replication begins
(b) A protein that binds to RNA at the point where replication begins
(c) The point in DNA where transcription starts
(d) The point in DNA where replication starts **Ans. d**

174. Which of the following is used in rolling circle DNA replication but not in normal cellular DNA replication?
(a) An endonuclease (b) An exonuclease
(c) Primase (d) DNA ligase **Ans. a**

175. In eukaryotes, RNA primers are removed after replication by
(a) DNA Pol δ (b) Helicase
(c) FEN I (d) DNA Pol α **Ans. c**

176. Which of the following DNA Pol has proofreading property in animal cells
(a) DNA Pol α (b) DNA Pol δ
(c) DNA Pol β (d) All **Ans. b**

177. Which of the following statement is correct?
(a) Meselson and Stahl used radioactive isotope of ^{15}N
(b) Klenow fragment in DNA Pol I has $5'\rightarrow3'$ exonuclease property
(c) DNA replication is semi conservative as well as semi discontinuous
(d) All the above are correct **Ans. c**

178. Choose the incorrect statement?
(a) Mitochondrial DNA replicates by D-loop formation
(b) Rolling circle replication is also known as σ-replication
(c) DNA Pol I is made up of single polypeptide
(d) DNA Pol III has ability of nick translation **Ans. d**

179. Which of the following DNA Pol in eukaryote has $5' \rightarrow 3'$ exonuclease activity?
(a) DNA Pol α (b) DNA Pol β
(c) DNA Pol δ (d) None **Ans. d**

180. Choose the correct statement
(a) Both circular and linear DNA chromosomes after replication are separated by type II topoisomerase
(b) In *E. coli*, all DNA Pols have proofreading properties

(c) Telomerase is chemically ribonucleo protein

(d) All the statements are correct **Ans. d**

181. Mitochondrial DNA is replicated from

(a) A single ori site bidirectionally

(b) Two different ori sites in the same direction

(c) Two different ori sites at different times in opposite directions

(d) Many sites bidirectionally, like nuclear chromosomes **Ans. c**

182. Chloroplast DNA is replicated from

(a) A single ori site bidirectionally

(b) Two different ori sites in the same direction

(c) Two different ori sites at different times in opposite directions

(d) Many sites bidirectionally, like nuclear chromosomes **Ans. c**

183. Which of the following statements about topoisomerases is incorrect?

(a) They alter the linking numbers of topoisomers

(b) They break and reseal phosphodiester bonds

(c) They require NAD^+ as a cofactor to supply the energy to drive the conversion of a supercoiled molecule to its relaxed form

(d) They form covalent intermediates with their DNA substrates

Ans. c

184. During replication of DNA, which one of the following enzymes polymerizes the Okazaki fragments?

(a) DNA polymerase I (b) DNA polymerase II

(c) DNA polymerase III (d) RNA polymerase I **Ans. c**

185. All of the following cell types contain the enzyme telomerase which protects the length of telomerase at the end of chromosomes, except

(a) Germinal (b) Somatic

(c) Haemopoetic (d) Tumour **Ans. b**

186. *E. coli* DNA ligase uses NAD^+ as a cofactor. During the actual reaction (joining of the 5′ phosphate and 3′ hydroxyl to form a phosphor-diester bond), the leaving group is

(a) NAD (b) NMN

(c) ATP (d) AMP **Ans. d**

187. The synthesis of DNA was shown to be template dependent and semi-conservative. This would predict that

(a) No organism can have a single-stranded genome

(b) RNA cannot serve as the genome of any organism

(c) All single stranded genomes would be synthesized via a double-stranded intermediate

(d) Template-independent DNA polymerases do not exist in nature

Ans. c

188. Reverse transcriptases are encoded by retro-viruses. The only enzyme from the mammalian cells with reverse transcriptase like activity is

(a) Telomerase (b) Ribonuclease

(c) Ribozyme (d) Caspase **Ans. a**

189. What is the role of gyrase in DNA replication?

(a) To relax or prevent supercoiling of the double helix

(b) To separate the two template strands of DNA at each replication fork

(c) To synthesize a short RNA primer for each new DNA strand

(d) To protect single-stranded regions of DNA from damage **Ans. a**

190. The large (Klenow) fragment of *E. coli* DNA polymerase I contains which of the following activities?

(a) Reverse transcriptase activity and nick-translation activity

(b) Polymerase activity and 3′-5′-exonuclease activity

(c) Polymerase activity and nick-translation activity

(d) 5′-3′-exonuclease activity **Ans. b**

191. The replicative polymerase in *E. coli* is

(a) DNA polymerase I (b) DNA polymerase II

(c) DNA polymerase III (d) DNA polymerase IV **Ans. c**

192. The function of *E. coli* DNA polymerase II in the cell is to

(a) Initiate replication at the origins

(b) Carry out 'fill in' reaction at the Okazaki fragments after RNA primer removal

(c) Synthesis of leading strand only

(d) Replication restart at the stalled replication fork **Ans. d**

193. Eukaryotic DNA synthesis is inhibited by

(a) Aphidicoline (b) Cyclohexamide

(c) Chlorammphenicol (d) Ampicillin **Ans. a**

194. Action of topoisomerase leads to changes in

(a) Linking number of single-stranded linear DNA

(b) Linking number of double-stranded linear DNA

(c) Linking number of closed circular single-stranded linear DNA

(d) Linking number of closed circular double-stranded linear DNA

Ans. d

195. In a reaction containing dNTPs and DNA polymerase in buffer, on which of the following molecules DNA synthesis would occur

(a) A single-stranded closed circular DNA

(b) A double-stranded closed circular DNA

(c) A single-stranded closed circular DNA paired with a short oligonucleotide with a 3′terminal phosphate

(d) A double-stranded closed circular DNA paired with a short oligonucleotide with a 3′terminal phosphate **Ans. d**

196. Okazaki fragments

(a) Require the activity of only a DNA polymerase for synthesis

(b) Require only RNA polymerase activity for synthesis

(c) Are made when DNA is exposed to UV radiation

(d) Are composed of both DNA and RNA **Ans. d**

197. Replication of DNA in eukaryotes commences from

(a) One end of chromatid extending to the other end

(b) Both ends of the chromatid simultaneously

(c) The centromere to either of the ends of chromatids

(d) Several sites of the DNA of the chromatid simultaneously **Ans.d**

198. In the Meselson and Stahl experiment, *E. coli* cells grown on heavy nitrogen were transferred to light nitrogen. What % of DNA can be expected to be constituted of light nitrogen after 3 generation of multiplication?

(a) 25 (b) 50

(c) 75 (d) 100 **Ans. c**

199. Which of the following statements about topoisomerases is incorrect?

(a) They alter the linking numbers of topoisomers

(b) They break and reseal phosphodiester bonds

(c) They require NAD^+ as a cofactor to supply the energy to drive the conversion of a supercoiled molecule to its relaxed from

(d) They form covalent intermediates with their DNA substrates

Ans. c

200. *E. coli* DNA ligase uses NAD^+ as a cofactor. During the actual ligation reaction (joining of the 5′ phosphate and 3′ hydroxyl to form a phosphodiester bond), the leaving group is

(a) NAD (b) NMN

(c) ATP (d) AMP **Ans. d**

201. Mark the correct statement

(a) Transcription is a process that yields, semi-conservatively, two identical strands of DNA

(b) DNA-dependent DNA polymerases are multi-subunit enzymes responsible for DNA transcription

(c) Bacterial transcripts (mRNAs) are polygenic

(d) Sigma factors guide the posttranscriptional modification of eukaryotic hnRNAs to mRNAs **Ans. c**

202. The DNA-binding protein that initiates the transcription of bacterial genes is called a/an

(a) Operator (b) Promoter

(c) Repressor (d) Sigma factor **Ans. d**

203. Binding of a given sigma factor depends on the

(a) Recognition of the length and spacing of promoter consensus sequences

(b) Interaction with core enzyme

(c) Length of the transcription unit

(d) Distance of the translational start codon **Ans. a**

204. Mark all true statements

P. exons are found in the same order in the genome

Q. all cells in the human body have the same set of genes

R. all cells in the human body express the same set of genes

S. all cells in the human body splice mRNAs for each gene in the same way

(a) P and S (b) P and Q
(c) R and S (d) Q and R **Ans. b**

205. Alternative splicing can give rise to transcripts that differ in
(a) The 5′ untranslated region of the mRNA
(b) The coding region of the mRNA
(c) The 3′ untranslated region of the mRNA
(d) All the above **Ans. b**

206. Consider the following statement
(1) A factor called 'sigma factor' is involved in the transcription process during protein synthesis. This factor
(2) Interacts with the core enzyme RNA-polymerase and forms the complete active enzyme
(3) Recognizes the start signal on DNA and directs the binding of the enzyme to the promoter region
(4) Along with the rho factor helps RNA to dissociate from RNA-polymerase and thus helps to terminate transcription
Which of these statements are correct?
(a) 1 and 2 (b) 1 and 3
(c) 2 and 3 (d) 1, 2 and 3 **Ans. a**

207. Sigma about of RNA polymerase
(a) Specifies the site for transcription
(b) Initiate replication
(c) Provides contact between ribonucleotides and DNA template
(d) Terminate transcription **Ans. a**

208. Synthesis of mRNA on DNA template is
(a) Unidirectional
(b) Bidirectional
(c) Bidirectional with help of primer
(d) Unidirectional with the help of primer **Ans. a**

209. How many different types of polypeptide subunits are present in RNA polymerase of *E. coli*?
(a) Two (b) Five
(c) Three (d) Four **Ans. d**

210. Which of the following statement about transcription is correct?

(a) RNA synthesis occurs in the 3′ to 5′ direction

(b) The RNA polymerase enzyme moves along the sense strand of the DNA in a 5′ to 3′ direction

(c) The RNA polymerase enzyme moves along the template strand of the DNA in a 5′ to 3′ direction

(d) The transcribed RNA is complementary to the non template strand

Ans. b

211. Pribnow box located at

(a) +10 position (b) -10 position

(c) -35 position (d) +35 position **Ans. b**

212. Splicing of pre-mRNA occurs in the

(a) Nucleus (b) Cytoplasm

(c) Internuclear space (d) Lysosome **Ans. a**

213. If the sequence of bases in one strand of DNA is 5′ATTGCGCA3′. What will be the complementary bases in mRNA?

(a) 5′UAACGCGU3′ (b) 5′TAACGCGT3′

(c) 5′TUUCGCGT3′ (d) 5′UGCGCAAY3′ **Ans. d**

214. Spliceosome, a complex of snRNA, protein and pre-mRNA found in

(a) Eukaryotic cell

(b) Prokaryotic cell

(c) Both eukaryotic and prokaryotic cell

(d) Only in plant cell **Ans. a**

215. If a segment of an mRNA molecule has the sequence 5′GUACCGAUCG3′, which of the following could have been the template DNA molecule?

(a) 5′GCUAGCCAUG3′ (b) 5′GUACCGAUCG3′

(c) 5′CATGGCTAGC3′ (d) 5′CGATCGGTAC3′ **Ans. d**

216. Alternate splicing means that

(a) The same gene can code for several different proteins

(b) Several different genes can code for the same protein

(c) Gene expression can be regulated at the level of transcription

(d) Pieces of DNA can move around within the genome **Ans. a**

217. Most eukaryotic pre-mRNAs are matured by which of the following modifications to their ends?

(a) Capping at the 3′-end, cleavage and polyadenylation at the 5′-end

(b) Addition of a GMP to the 5′-end, cleavage and polyadenylation to create the 3′-end

(c) Addition of a guanine residue to the 5′-end, cleavage and polyadenylation to create the 3′-end

(d) Addition of a GMP to the 5′-end, polyadenylation, then cleavage to create the 3′-end **Ans. b**

218. Which of the following statements are incorrect?

(a) There are two α subunits in the *E. coli* RNA polymerase

(b) There is one σ subunit in the *E. coli* RNA polymerase

(c) *E. coli* has 3 different RNA polymerase

(d) The α subunit of *E. coli* RNA polymerase is inhibited by rifampicin **Ans. c**

219. What is the role of the promoter region of a gene?

(a) It is the site where the ribosome binds to the mRNA

(b) It is the site where RNA polymerase binds to the DNA

(c) It is the site where DNA polymerase binds to the DNA

(d) It is the site where Sine-Dalgarno sequence present **Ans. b**

220. Which of the following processes does not have a mechanism that checks to make sure that the right monomer is incorporated into the polymer?

(a) Protein synthesis

(b) Double stranded rolling circle DNA synthesis

(c) RNA synthesis

(d) Excision repair DNA synthesis **Ans. c**

221. Rho-dependent terminators

(a) Require the RNA involved to take on a specific secondary structure before termination can occur

(b) Require that a specific protein bind to the RNA and cause termination

(c) Require that the RNA be cut before transcription terminates further downstream

(d) Only occurs on tRNA genes in eukaryotes **Ans. b**

222. Which of the following is true about sigma factors?

(a) They are made by eukaryotes to regulate synthesis of different mRNAs

(b) They are necessary for prokaryotic RNA polymerase to recognize promoters

(c) They are necessary for prokaryotic RNA polymerase to elongate RNA, but not for promoter recognition

(d) They are involved in initiation of prokaryotic translation **Ans. b**

223. In *E. coli* rho-factor is required for

(a) Initiation of transcription at specific site

(b) Initiation of transcription at any site

(c) Termination of transcription at specific site

(d) Termination of transcription at any site **Ans. c**

224. Which one of the following regions in the mature mRNA is not found in the gene for the mRNA?

(a) Start codon (b) Leader sequence

(c) Poly A tail (d) Exon **Ans. c**

225. The prokaryotic RNA polymerase holoenzyme has the subunit structure

(a) $\alpha\beta\beta'\sigma$ (b) $\alpha\beta\sigma$

(c) $\alpha\beta$ (d) $\alpha\beta\beta'$ **Ans. a**

226. A new antibiotic was discovered which strongly inhibited mRNA precursor transcripts and snRNA transcripts. This antibiotic was predicted to be an inhibitor of

(a) RNA polymerase I (b) RNA polymerase II

(c) RNA polymerase III (d) Helicase **Ans. b**

227. There are three kinds of RNA polymerases (I, II, III) in eukaryotic cells, each specific for one class of RNA molecule. Which of the following is a correct match?

(a) RNA Pol I-rRNA; RNA Pol II-tRNA

(b) RNA Pol II-mRNA; RNA Pol III-rRNA

(c) RNA Pol I-rRNA; RNA Pol II-mRNA

(d) RNA Pol I-tRNA; RNA Pol III-rRNA **Ans. c**

228. In which of the following splicing events, the intermediates do not form the lariat or the branched structure?

(a) Trans splicing (b) Group I intron splicing

(c) Group II intron splicing (d) hnRNA splicing **Ans. b**

229. Mode of action of rifampicin in *E. coli* is through

(a) Cell division

(b) Initiation of transcription

(c) RNA polymerase binding to DNA template

(d) Inhibition of the oxidation potential **Ans. b**

230. Rho-dependent and rho-independent transcription termination mechanisms operate in prokaryotes. Rho independent termination mechanism involves

(a) Binding of the rho protein upstream of the termination element

(b) No protein factors and only RNA secondary structure and run of 'U's

(c) Presence of UGA or UAA stop codons

(d) Binding of accessory factors at termination signal **Ans. b**

231. In eukaryotes the ribosomal RNA genes are transcribed by

(a) Reverse transcriptase

(b) RNA dependent RNA polymerase

(c) RNA polymerase I

(d) RNA polymerase III **Ans. c**

232. Which of these substrates is best suited for measuring RNA synthesis by RNA polymerase

(a) γ P^{32} ATP (b) α P^{32} d ATP

(c) γ TP^{32} d ATP (d) α P^{32} UTP **Ans. a**

233. DNA contains sequences of bases known as promoters which

(a) Are recognized by the core enzyme of RNA polymerase

(b) Occur in the chain just before the message sequence that defines the messenger RNA

(c) Are single-stranded regions of the duplex DNA

(d) Are transcribed into RNA sequences **Ans. b**

234. RNA polymerase recognizes sites on DNA known as promoters. All of the following statements regarding these sites are correct except

(a) A sequence of DNA base-pairs known as the pribnow box is located 5-10 bases upstream from the first base to be copied into RNA

(b) All pribnow box sequences are variants of the sequence TATAATG

(c) A second sequence, the -35 sequence, in many promoters, is located about 35 bases downstream from the pribnow box

(d) The σ subunit of prokaryotic RNA polymerase is involved in recognition of the promoter sequence **Ans. c**

235. During the processing of introns, a single snRNP complex catalyzes both the cleavage of the RNA and the joining of the cut ends. What would be the consequence if these two processes were catalyzed by separate enzymes not associated in a single complex?

(a) The rate of RNA processing would be much faster

(b) The cell would be unable to identify the correct cleavage sites

(c) The exons might not be joined in the correct sequence

(d) Exons instead of introns would be cleaved from the RNA molecule

Ans. c

236. Which of the following agents inhibits the initiation of RNA synthesis in prokaryotes by binding to the b subunit of RNA polymerase?

(a) α-amanitin (b) Actinomycin D

(c) Rifamycin SV (d) Puromycin **Ans. c**

237. Which of the following statements regarding the termination of transcription of DNA is correct?

(a) The rho protein binds tightly to RNA polymerase and inactivates the polymerization activity

(b) Termination sites of DNA contain an inverted-repeat base sequence that results in the RNA formed adopting a "stem and loop" configuration

(c) Each termination site contains a unique base where transcription stops

(d) All prokaryote termination sites contain the "stem and loop" promoting sequence and also require the participation of the rho protein to halt RNA synthesis **Ans. b**

238. The colinearity of gene and product clearly seen in prokaryote system is obscured in eukaryotes because

(a) The DNA sequences coding for proteins is interrupted by noncoding sequences that are not transcribed

(b) The sequence of bases in the mRNA is modified in the cytoplasm prior to interaction with the ribosomes

(c) Ribosomes in eukaryotes only translate exon regions of the mRNA

(d) The coding sequence in DNA is interrupted by DNA sequences that are transcribed and then spliced out of the mRNA **Ans. d**

239. All of the following statements about tRNA molecules are true except that

(a) They have extensive intrachain hydrogen bonding

(b) They contain a number of base that have been modified post-transcriptionally

(c) The amino acid is attached to the 2′ or 3′ position of the 3′ terminal adenosine

(d) They contain an anticodon triplet of bases at the 5′ end of the triplet codon on mRNA **Ans. d**

240. In *E. coli*, transcription of tRNA genes is promoted by

(a) RNA polymerase I (b) RNA polymerase II

(c) RNA polymerase III (d) RNA polymerase **Ans. d**

241. Mature eukaryotic mRNAs have a 5′ cap that is residue of

(a) 2-methylguanosine (b) 3-methylguanosine

(c) 5-methylguanosine (d) 7-methylguanosine **Ans. d**

242. RNA polymerase II is responsible for synthesis of

(a) rRNA precursors (b) mRNA precursors

(c) tRNA precursors (d) 5S rRNA and tRNA precursors **Ans. b**

243. The TATA box plays a key role in assembling active transcription complexes by

(a) Binding with TATA box-binding protein

(b) Binding with DNA protein

(c) Binding with RNA protein

(d) It does not bind with any protein **Ans. a**

244. Changes in RNA polymerase III promoters result in changes in the sequence of the gene product. This is because

(a) RNA polymerase III initiates transcription at various positions

(b) The promoter elements are located within the coding region of the gene

(c) Promoters are located far upstream of the gene transcribed

(d) Polymerase is unable to bind to the promoter **Ans. b**

245. Which one of the following statements about the primary transcripts in eukaryotes is incorrect?

(a) They are usually longer than the functional RNA

(b) They contain no modified bases

(c) They contain nucleotide sequences which are complementary to the DNA template

(d) They contain the TATA box sequence **Ans. d**

246. Which one of the following transcription factors bind to TATA box?

(a) TFIID (b) TFII B

(c) TFII A (d) TFII E **Ans. a**

247. Which one of the following is an inhibitor of bacterial transcription?

(a) Streptomycin (b) Kanamycin

(c) Rifampicin (d) Nalidixic acid **Ans. c**

248. Site of DNA where RNA polymerase binds to initiate transcription is called

(a) Terminator (b) Operator

(c) ORF (d) Promoter **Ans. d**

249. Which ribonucleases are involved in producing mature tRNA in *E. coli*?

(a) RNases A, D, E and F (b) RNases D, E, F and H

(c) RNases D, E, F and P (d) RNases A, D, H and P **Ans. c**

250. Most eukaryotic pre-mRNAs are matured by which of the following modifications to their ends

(a) Capping at the 3′ end, cleavage and polyadenylation at the 5′ end

(b) Addition of a GMP to the 5′ end, cleavage and polyadenylation to create the 3′ end

(c) Addition of a guanine residue to the 5′ end, cleavage and polyadenylation to create the 3′ end

(d) Addition of a GMP to the 5′ end, polyadenylation, then cleavage to create the 3′ end **Ans. b**

251. Which one of the following statements correctly describes the splicing process undergone by most eukaryotic pre-mRNAs?

(a) In a two-step reaction, the spliceosome removes the exon as lariat and joins the two introns together

(b) Splicing requires conserved sequences, which are the 5′ splice site, the 3′ splice site, the branch point and the polypurine tract

(c) The U1 snRNP initially binds to the 5′ splice site, U2 to the branch point sequence and then the three snRNP, U4, U5 and U6 can bind

(d) In the first step of splicing the G at the 3′ end of the intron is joined to the 2′ hydroxyl group of the a residue of the branch point sequence to create a lariat **Ans. c**

252. Which of the following statements is true?

(a) RNA Pol II only transcribes protein-coding genes

(b) The TATA box has a role in transcription efficiency but not in positioning the start of transcription

(c) TBP binds to the TATA box

(d) Enhancers lie 100-200 bps upstream from the start of transcription

Ans. c

253. Which of the following statements about general transcription factors is false?

(a) TFIID binds to the TATA box

(b) TFIID is a multiprotein complex consisting of TBP and TAF

(c) TBP is a common factor in transcription by RNA Pol I, RNA Pol II and RNA Pol III

(d) TFIIH dephosphorylates the carboxyl-terminal domain (CTD)

Ans. d

254. Which one of the following statements about eukaryotic RNA polymerases I, II and III is false?

(a) RNA Pol II is very sensitive to α-amanitin

(b) RNA Pol II is located in the nucleoplasm

(c) RNA Pol III transcribes the genes for tRNA

(d) The carboxyl end of RNA Pol II contains a short sequence of only seven amino acids which is called the CTD and which may be phosphorylated **Ans. d**

255. Which one of the following statements about *E. coli* RNA polymerase is false?

(a) The holoenzyme includes the sigma factor

(b) The core enzyme includes the sigma factor

(c) It requires Mg^{2+} for its activity

(d) It requires Zn^{2+} for its activity **Ans. b**

256. Which one of the following statements is incorrect?

(a) There are two α subunits in the *E. coli* RNA polymerase

(b) There in one β subunits in the *E. coli* RNA polymerase

(c) *E. coli* has 3 RNA polymerase

(d) The β subunits of *E. coli* RNA polymerase is inhibited by rifampicin **Ans. c**

257. ....is the untranslated sequence at the 5′ end of mRNA

(a) Leader sequence (b) Nuclear localization sequence

(c) Poly-A-tail (d) Enhancer sequence **Ans. a**

258. Which one of the following is correct descending order based on their size?

(a) Gene, cistron, recon, muton (b) Cistron, gene, recon, muton

(c) Gene, cistron, muton, recon (d) Cistron, gene, muton, recon **Ans. a**

259. According to operon concept, in the tryptophan operon the regulator gene forms

(a) An inducer (b) A repressor

(c) An aporepressor (d) General inhibitor **Ans. c**

260. Transcription termination of mRNA genes in eukaryotes occurs

(a) At polyadenylation sites by the action of a termination factor

(b) By the formation of a strong hairpin structure in the vicinity of polyadenylation site

(c) Termination factor bound to the termination site in the vicinity of polyadenylation site

(d) At pause sites following the polyadenylation sites **Ans. d**

261. A transesterification reaction

(a) Requires on ATP

(b) Breaks one bond and forms one bond

(c) Involves the nucleophilic attack of an OH group on the sugar phosphate backbone

(d) All the above **Ans. d**

262. Mark all true statements about snRNAs

(a) snRNAs are strictly localized to the nucleus

(b) Most snRNAs are very abundant

(c) Some snRNAs can base pair with intron consensus sequences

(d) All the above **Ans. d**

263. Chaperone proteins help in

(a) Protein folding and assembly (b) Protein stability

(c) Both (a) and (b) (d) None **Ans. c**

264. How many energy bonds are expended in the formation of a peptide bond?

(a) 2 (b) 4

(c) 3 (d) 6 **Ans. b**

265. Which of the following statements is correct in case of chaperone proteins?

(a) These do not prevent aggregation

(b) They cleave incorrect S-S bonds

(c) Act on fully synthesized polypeptides

(d) Are involved in the transport of proteins across mitochondria and endoplasmic reticulum **Ans. d**

266. Diphtheria toxin inhibits protein synthesis by

(a) Causing formation of ADP-EF2 complex

(b) Release of peptidyl-t-RNA from the 'P' site

(c) Binding to factor eEF-2

(d) Inhibiting peptide bond formation **Ans. a**

267. An amino-acyl-tRNA synthetase is responsible for

(a) Formation of a peptide bond

(b) Binding of mRNA to ribosomes

(c) Attaching an amino group to an organic acid

(d) Joining an amino acid to tRNA **Ans. d**

268. The term protein splicing refers to

(a) Removal of intervening sequences between the genes

(b) Splicing out of introns from RNA

(c) Removal of intervening protein sequences from the translated protein

(d) Joining (splicing) of two different gene products to generate a novel protein **Ans. c**

269. 'Kozak' is associated with

(a) Transcription (b) DNA replication

(c) DNA repair (d) Translation **Ans. d**

270. Formyl methionine tRNA is present in

(a) Yeast (b) Bacteria

(c) Insects (d) Mammals **Ans. b**

271. The genetic code refers to the

(a) Number of chromosomes in the diploid cells of a species

(b) Nucleotide sequences that correspond to common amino acids

(c) Amino acid sequence of cellular proteins

(d) Ratio of Mendelian inheritance **Ans. b**

272. The wobble hypothesis refers to the less stringent base pairing specificity of the

(a) 5′ end base of the codon (b) 3′ end base of the anticodon

(c) Middle base of the anticodon (d) 5′ end base of the anticodon

Ans. d

273. All of the following statements about amino acid activation are correct except

(a) Aminoacyl-tRNA synthetases transfer the amino acids to the A on the 3′ end of tRNA via an aminoacyl-adenosine monophosphate intermediate

(b) The aminoacyl-tRNA synthetase is specific for a given amino acid

(c) There may be more than one kind of tRNA for a given amino acid

(d) The activation step requires the hydrolysis of the terminal phosphate group of guanosine triphosphate **Ans. d**

274. All of the following factors are required to initiate translation of mRNA in prokaryotes except

(a) Formylmethionine-tRNA (b) 30S ribosomal subunits

(c) Guanosine triphosphate (d) EF-Tu **Ans. d**

275. Which of the following statements about chain termination in peptide synthesis is true?

(a) The formation of peptide bonds by the ribosomal mRNA complex continues until a stop codon on mRNA is reached

(b) The formation of peptide bonds by the ribosomal mRNA complex continues until a tRNA with an anticodon for UAA, UAG, or UGA interacts with the A site on the ribosome

(c) In prokaryotes, peptide bond formation ceases when a formylmethionine-tRNA, interacts with the A site on the ribosome

(d) Peptide bond formation ceases when the ribosome reaches the 5′ end of the mRNA **Ans. a**

276. Which of the following is involved in the majority of the ATP-dependent cytosolic degradation of proteins in eukaryotes?

(a) Cathepsins (b) Calpains

(c) Lysosome (d) 26S proteosomes **Ans. d**

277. If a genetic code is degenerate, it means that

(a) A given base triplet can code for more than one amino acid

(b) There is no punctuation in the code sequences

(c) The third base in a codon is not important in coding

(d) A given amino acid can be coded for by more than one base triplet **Ans. d**

278. What is the E-site of the ribosome?

(a) The site where eukaryotic mRNA is processed

(b) An exciting term made up by your instructor

(c) The site where tRNA exits the prokaryotic ribosome

(d) The site where the endonuclease EcoRI restrict the ribosome **Ans. c**

279. Which bases(s) in the anticodon contribute to the degeneracy in codon reading?

(a) First (b) Second

(c) Third (d) First and second **Ans. a**

280. The number of aminoacyl-tRNA synthetase is

(a) Determined by the number of existing tRNAs

(b) Determined by the number of translatable mRNA codons

(c) Determined by the number of utilized amino acids

(d) Determined by the individual organisms species **Ans. c**

281. Which statement about the codon-anticodon interaction is false?

(a) It is antiparallel and can include nonstandard base pairs

(b) Inosine in the 5′-anticodon position can pair with A, C or U in the 3′-codon position

(c) Inosine in the 3′-anticodon position can pair with A, C or U in the 5′-codon position

(d) A is never found in the 5′-anticodon as it is modified by anticodon deaminase **Ans. c**

282. Which one of the following statements correctly describes initiation of protein synthesis in *E. coli*?

(a) The initiator tRNA binds to the Shine-Dalgarno sequence

(b) Three initiation factors are involved and IF2 binds to GTP

(c) The intermediate containing IF1, IF2, IF3 initiator tRNA and mRNA is called the 30S initiation complex

(d) Binding of the 50S subunit releases IF1, IF2, GMP and PPi **Ans. b**

283. *E. coli* release factor1 (RF1) recognizes which codons?

(a) UAA only (b) UGA only

(c) UGA and UAA (d) UAG and UAA **Ans. d**

284. ....is a reading frame sequence not interrupted by a stop codon

(a) Regulatory sequence (b) Open reading frame

(c) Putative gene sequence (d) None **Ans. b**

285. ....is the complete collection of proteins that an organism produces

(a) Poly proteins (b) Protein complex

(c) Proteome (d) Expression tags **Ans. c**

286. All aminoglycoside antibiotics inhibit protein synthesis by

(a) Binding to small ribosomal subunit

(b) Binding to large ribosomal subunit

(c) Binding to both small and large ribosomal subunit

(d) Inactivating eEF-2 **Ans. a**

287. In the protein synthesis, tRNA carrying the initiating amino acid enters in which site of ribosome?

(a) 'A' site (b) 'P' site
(c) Anticodon site (d) Recognition site **Ans. b**

288. GUG codes for valine in both prokaryotes & eukaryotes but when GUG is initiation codon, this codes

(a) Methionine (b) Valine
(c) Tryptophan (d) None **Ans. a**

289. An antibiotic which inhibits translation in both prokaryote and eukaryotes is

(a) Chloromycetin (b) Puromycin
(c) Actinomycin-D (d) Tetracycline **Ans. b**

290. In the case of humans lysine is an essential acid because

(a) It is present in all proteins
(b) It is highly nutritive
(c) It is not formed in the body and had to be supplied the diet
(d) It is required for protein synthesis **Ans. c**

291. N→C polarity of a polypeptide is collinear with the

(a) 3′→5′ polarity of its mRNA (b) 3′→5′ polarity of its tRNA
(c) 5′→3′ polarity of its tRNA (d) 5′→3′ polarity of its mRNA
Ans. d

292. Which of the following statements about the diphtheria toxin are correct?

(a) It is cleaved on the surface of susceptible eukaryotic cells into two fragments, one of which enters the cytosol
(b) It binds to peptidyl transferase and inhibits protein synthesis
(c) It reacts with ATP to phosphorylate eIF2 & prevent the insertion of the Met-tRNAi into the P site
(d) It reacts with NAD^+ to add ADP-ribose to eEF2 and prevent movement of the peptidyl-tRNA from the A to the P site **Ans. d**

293. The wobble hypothesis explains

(a) How multiple different types of tRNA are charged with a single type of amino acid
(b) How multiple stop codons can associate with the same release factor
(c) How multiple tRNAs can associate with the same codon
(d) How one tRNA can associate with multiple codons **Ans. d**

294. Which of the following group of amino acids have maximum codons

(a) Methionine, tryptophan

(b) Leusine, serine, valine

(c) Leusine, serine, Arginine

(d) Tryptophan, methionine, Arginine **Ans. c**

295. Which site of ribosome first fill up by tRNA carrying initiating amino acid?

(a) Peptidyl site (b) Amino aceyl site

(c) Either P-site or A-site (d) None **Ans. a**

296. Shine-Dalgarno sequence of mRNA is helpful in

(a) Recognition of 50S ribosomal subunit

(b) Recognition of 60S ribosomal subunit

(c) Recognition of 50S rRNA

(d) Recognition of 30S subunit of ribosome **Ans. d**

297. In genetic code is tetraplet then what is the possible number of codons which code 20 types of amino acids?

(a) 261 (b) 64

(c) 256 (d) 43 **Ans. c**

298. Initiation of protein synthesis from most eukaryotic mRNA requires

(a) A 3′poly-A tail (b) A 5′ cap

(c) A 3′ cap (d) Both a poly-A tail and a 5′ cap **Ans. d**

299. Which of the following amino acids has the greatest number of codons?

(a) Proline (b) Leucine

(c) Tryptophan (d) Aspartic acid **Ans. b**

300. In a heteropolymer experiment using 1/2C:1/4A:1/4G, how many different triplets will occur in the synthetic RNA molecule and also how often will the most frequemt triplet occur?

(a) 27 triplets, most frequent will be CAG

(b) 9 triplets, most frequent will be CCC

(c) 8 triplets, most frequent will be CAG

(d) 27 triplets, most frequent will be CCC **Ans. d**

❑❑❑

Chapter-14

Chromosomal Basis of Inheritance

Fill in the blanks

1. Cell division is favoured by............

Ans. cytokinins

2. Meiotic divisions are present in..........

Ans. spore mother cells

3. Exchange of sections of chromatid between synapsing chromosomes during meiosis is termed as.................

Ans. crossing over

4. ....theory of inheritance was proposed by W.S. Sutton and T. Boveri.

Ans. Chromosomal

5. The segregation of homologous chromosomes during anaphase I of meiosis is called..........

Ans. disjunction

6. ....is the term used for the division of cytoplasm.

Ans. Cytokinesis

7. The movement of intermediary chiasmata towards the ends of chromosomes occurs during.........stage.

Ans. terminalisation

8. In meiosis, pairing of homologous maternal and paternal chromosomes occurs during.........stage.

Ans. zygotene

9. Meiosis was first observed in plants by..........

Ans. Strasburger

10. The cells which undergo meiosis are known as............

Ans. meiocytes

11. In the life cycle of lower plants...........undergoes meiosis.

Ans. zygote

12. ....of plants are the best material to observe meiosis.

Ans. Anthers

13. Meiosis cannot occur in the cells with........set of genome.

Ans. one

14. Prophase I refers to the prophase of...........

Ans. meiosis

15. Spindle is not formed in.........division.

Ans. amitotic

16. The enzymes.........and....help in the breaking and joining of chromatids.

Ans. endonuclease, ligase

17. Spindle fibres which extend from pole to pole are called...............

Ans. continuous fibres

18. Nuclear division without the breakdown of nuclear envelope, and chromosome separating while attached to nuclear envelope is called...........

Ans. dinomitosis

19. Replication of DNA occurs in the........phase of cell cycle.

Ans. S

20. Division of nucleus not followed by cytokinesis is called...............

Ans. free nuclear division

21. Uncontrolled mitosis leads to...........

Ans. cancer

22. Regeneration in organisms take place by...........division.

Ans. mitotic

23. The fibres appearing between separating chromatids are called..............

Ans. interzonal fibres

24. The shape of phragmoplast is like a...........

Ans. barrel

25. 2,4 dinitrophenol prevents mitosis by inhibiting..........phosphorylation.

Ans. oxidative

26. The future plane of cell division in plants is marked by a.............

Ans. preprophase band

27. Replication of chromosomes without division of the nucleus is called................

Ans. endomitosis

28. The ski configuration results due to action of the......on mitotic metaphase and anaphase.

Ans. colchicine

29. Karyochorisis is found in the group of organism called.............

Ans. fungi

30. The terms 'heterotype' and 'homotype' were coined by............

Ans. Flemming

31. The word 'Meiosis I' and 'Meiosis II' were used by...........

Ans. Gregoire

32. The chromosome is attached to nucleus, at leptotene, by the structure called...............

Ans. attachement plaque

33. The synaptonemal complex was first described by...............

Ans. Moses

34. During pachytene, ...and .... chromosomes do not synapse fully.

Ans. X, Y

35. In pachytene, the crossing over is mediated by...............

Ans. recombination nodule

36. Chiasmata are the expression of.............

Ans. crossing over

37. The spindle in meiosis II is oriented at an angle of..........to the spindle in meiosis I.

Ans. 90^0

38. ....division provided cytological basis for Mendel's laws of inheritance.

Ans. Meiotic

39. The decision for cell division occurs at..........stage.

Ans. G_1

40. The passage of cells through the restriction point is triggered by............

Ans. U-protein

41. Equal distribution of extranuclear material of a cell is brought about by....................

Ans. cytokinesis

42. The minimum number of chiasmata in a bivalent is.....................

Ans. one

43. During mitotic anaphase, the chromatids move towards the pole they face at the speed of..........per minute.

Ans. 1 μm

44. In..............part of seeds, mitosis occurs without cytokinesis.

Ans. endosperm

45. ....is longest phase of meiosis I.

Ans. Prophase I

46. The transfer of genetic message from DNA to mRNA is described as............

Ans. transcription

47. Puffs in giant chromosomes are called.................

Ans. Balbiani rings

48. True nucleus is absent in.............

Ans. prokaryotes

49. Shape of the chromosome is determined by the position of................

Ans. centromere

50. Number of chromosomes in man is............

Ans. 2n = 46

51. ....is the chief constituent of chromatin and chromosomes.

Ans. DNA

52. Chromatids of the same chromosomes are termed as...................

Ans. sister chromatids

53. ....prevent the chromosomes from sticking to each other.

Ans. Telomere

54. Chromosomes with diffused centrameric activity are called................

Ans. polycentric

55. Replication of chromosomes not followed by separation of chromatids results in the formation of..................chromosomes.

Ans. polytene

56. DNA content is measured in..........unit.

Ans. picogram

57. Chromosomes with morphologically and genetically identical arms are called..............

Ans. isochromosomes

58. Chromatin material showing positive heteropycnosis is called ....

Ans. heterochromatin

59. A banding in chromosomes is obtained by using the chemical ....

Ans. Quinacine mustard

60. The chromosomes concerned with somatic characters only are called.................

Ans. autosomes

61. The term chromonema was coined by............

Ans. Vejdovsky

62. The position of.................determine the shape of the chromosome.

Ans. centromere

63. The division of nucleus not followed by cytokinesis is called ....division.

Ans. free nuclear

64. The cell which undergo meiosis are known as.....................

Ans. reproductive cells

65. ....is the term used for the division of cytoplasm.

Ans. Meiocytes

Multiple Choice Questions

1. Term chromosome was coined by
   (a) Hofmeister (b) Sutton
   (c) Boveri (d) Waldeyer **Ans. d**
2. Chromosomes were first seen by
   (a) Hofmeister (b) Waldeyer
   (c) Strasbuger (d) Flemming **Ans. a**
3. Chromosomes found in the salinity glands of *Drosophila* are
   (a) Lampbrush (b) Polytene
   (c) Supernumerary (d) B- chromosomes **Ans. b**
4. A giant chromosome with a number of chromonemata is
   (a) Lampbrush chromosome (b) Heterochromosome
   (c) Supernumerary chromosomes (d) Polytene chromosome **Ans. d**
5. Lampbrush chromosomes occur in
   (a) Salivary glands (b) Lymph glands
   (c) Cancer cells (d) Oocytes **Ans. d**
6. Chromosome ends are called
   (a) Satellites (b) Telomeres
   (c) Centromeres (d) Kinetochores **Ans. b**
7. Chromatid is
   (a) One half of chromosome (b) Haploid chromosome
   (c) Complete chromosome (d) Duplicate chromosome **Ans. a**
8. Centromere is that part of chromosome where
   (a) Nucleoli are formed (b) Crossing over takes place
   (c) Chromatids are attached (d) Nicking occurs **Ans. c**
9. A chromosome with sub-terminal centromere is
   (a) Acentric (b) Acrocentric
   (c) Metacentric (d) Telocentric **Ans. b**
10. A chromosome with centromere near the middle is called
   (a) Metacentric (b) Submetacentric
   (c) Acrocentric (d) Telocentric **Ans. b**

11. Puffs or balbiani rings in salivary gland chromosomes are sites of
(a) DNA replication (b) DNA duplication
(c) RNA synthesis (d) Protein synthesis **Ans. c**

12. Chromosome theory of inheritance was proposed by
(a) Sutton (b) Boveri
(c) Both (a) and (b) (d) Correns **Ans. c**

13. Chromatin fibres condense to form chromosomes during
(a) G_1-phase (b) S – phase
(c) Metaphase (d) Prophase **Ans. d**

14. The stage for counting of chromosomes is
(a) Early prophase (b) Mid prophase
(c) Late prophase (d) Metaphase **Ans. d**

15. More than 200 chromosomes occur in
(a) Chicken (b) Dog
(c) Amoeba (d) Gorilla **Ans. c**

16. The genes which are present on the same chromosome are called
(a) Identical genes (b) Allelomorphs
(c) Linked genes (d) Dominant genes **Ans. c**

17. *Drosophila* has four homologous pairs of chromosomes. What is the number of linkage groups in this animal?
(a) Four (b) Two
(c) Eight (d) Uncertain **Ans. a**

18. Gene for colour blindness in man is located on
(a) X-chromosome only (b) Y-chromosome only
(c) Either X or Y chromosome (d) Both X and Y chromosomes
Ans. a

19. A colour blind daughter may be born if the
(a) Father is normal and mother is colour blind
(b) Father is colour blind and mother is normal
(c) Father is normal and mother is a carrier
(d) Father is colour blind and mother is a carrier **Ans. d**

20. A somatic cell in human male contains
(a) No gene on sex chromosomes
(b) Only one sex linked gene for each character
(c) Two genes for every sex linked character
(d) Genes only on sex chromosomes **Ans. b**

21. A normal woman is married to a colour blind man. The children are expected to be
(a) All normal
(b) 50% sons are colour blind
(c) All daughters normal but carrier whereas all sons are normal phenotypically as well as genotypically
(d) 50% daughters are colour blind **Ans. c**

22. Which of the following disease is sex linked?
(a) Colour blindness (b) Malignancy
(c) Hepatitis (d) Leukemia **Ans. a**

23. The genes for the eye colour and size of the wing in Drosophila are located on the same chromosome. They can be separated by
(a) Non-disjunction (b) Crossing over
(c) Hybridization (d) Not be separated at any stage
Ans. b

24. A colour blind son is born to normal parents. It shows that
(a) The father was heterozygous for colour blindness
(b) The mother was genotypically homozygous
(c) The mother was heterozygous for colour blindness
(d) Both parents carried a recessive gene for the disorder **Ans. c**

25. Chromosomal constitution in human female can best be written as
(a) 46 (b) 44 + 2
(c) 44A + XX (d) 44A + XY **Ans. c**

26. The sex linked characters are those
(a) Which are related to sexual physiology
(b) The genes of which are present on the sex chromosomes
(c) Which appear either in male or in female
(d) Which are controlled by sex hormones **Ans. b**

27. The chromosomes other than sex chromosomes are called

(a) Autosomes (b) Heterosomes

(c) Karyosomes (d) None of the above **Ans. a**

28. The points where the crossing over occurs between the homologous chromosomes of a pair are called

(a) Centromeres (b) Chromomeres

(c) Chromocentres (d) Chiasmata **Ans. d**

29. If the crossing over had occurred at two-strand stage in *Neurospora*, the ascospores would be arranged in

(a) 1 - 1 position (b) 2 - 2 position

(c) 4 - 4 position (d) None of these **Ans. c**

30. Complete linkage is found in

(a) Birds (b) Snakes

(c) Female *Drosophilla* (d) Male *Drosophila* **Ans. d**

31. In most of the higher unisexual animals there is one chromosomal pair which is not identical in two sexes. These are called

(a) Non-homologous chromosomes

(b) Non-identical chromosomes

(c) Non-compatible chromosomes

(d) Sex chromosomes **Ans. d**

32. The two diverse disciplines of cytology and genetics were co-related by

(a) Muller (b) Morgan

(c) Bridges (d) Tschermak **Ans. c**

33. A linkage group is defined as

(a) All the linked genes of a chromosomal pair

(b) Different groups of genes present on different chromosomes

(c) All the genes located on the same chromosome

(d) None of the above **Ans. a**

34. Linkage in *Drosophila* was reported by

(a) Mendel (b) Correns

(c) Morgan (d) None of these **Ans. c**

35. A phenomenon which works opposite to the linkage is

(a) Independent assortment (b) Crossing over

(c) Segregation (d) Mutation **Ans. b**

36. When two gametes are situated very close to one another at a chromosome
- (a) The percentage of crossing over between them is very high
- (b) Hardly any cross-overs are produced
- (c) No crossing over can take place
- (d) Only double cross overs can occur between them **Ans. b**

37. Greater is the distance between the two genes on a chromosome
- (a) Greater is the linkage strength
- (b) Lesser is the linkage strength
- (c) Linkage strength remains unchanged
- (d) There is no relationship between the two **Ans. b**

38. Crossing over occurs at
- (a) Single-strand stage of chromosomes during prophase
- (b) Two-strand stage during zygotene
- (c) Four-strand stage of pachytene
- (d) Metaphase II of meiosis **Ans. c**

39. The term coupling and repulsion signify the same phenomenon which is termed
- (a) Synapsis (b) Disjunction
- (c) Linkage (d) Non-disjunction **Ans. c**

40. The linked characters would always inherit together till they are
- (a) Delinked due to segregation (b) Marked by dominance
- (c) Mutated (d) Separated due to crossing over **Ans. d**

41. During meiosis, the crossing over which occurs between the homologous chromosomes is of
- (a) No genetic significance
- (b) Importance to bring about recombination of characters
- (c) Importance because it produces nutrition
- (d) Importance because it causes reorganization of cytoplasm **Ans. b**

42. A white eyed (recessive) and long winged male *Drosophila* was crossed to a vestigial winged (recessive) red-eyed female. The two characters are linked in this animal. When a female F_1 was test crossed, the F_2 generation produced
- (a) Mostly white eyed with long wings
- (b) Mostly red eyed with vestigial wings

(c) Mostly white eyed with long wings and red eyed with vestigial winged

(d) All white eyed and vestigial winged **Ans. c**

43. A haemophilic man marries a normal woman. What is the probability of the disease in the grand children in case the daughter marries a normal man?

(a) All grandsons would be haemophilic, and grand daughters normal

(b) 50% of grandsons are haemophilic. Other grand children are normal

(c) Only grand daughters would be haemophilic

(d) All grand children would be haemophilic **Ans. b**

44. A colour blind man marries a normal woman whose father was colour blind. What percentage of children is expected to be colour blind?

(a) 25% (b) 50%

(c) 75% (d) 100% **Ans. b**

45. Mendel did not get any linkage in his experiments on pea. One of the reasons was that

(a) He did not keep an exact record

(b) There is no linkage in pea

(c) He did not have means to detect linkage

(d) All the seven characters selected by him were present on different chromosomes or showed 50% cross-overs **Ans. d**

46. The genes of different traits located on different loci on the same chromosome are

(a) Alleles (b) Linked

(c) Pleiomorphic (d) Mutated **Ans. b**

47. In Four O'clock plant, normal leaves (A) and variegated leaves (B) male is crossed with (A) female, the hybrid has normal leaves, but when (B) female is crossed with (A) male, the hybrid has variegated leaves. It is a case of

(a) Mutation (b) Cytoplasmic inheritance

(c) Complementary genes (d) Supplementary genes **Ans. b**

48. Crossing over occurs between

(a) Two non-sister chromatids of a homologous pair of chromosomes

(b) Two chromatids of any chromosomes

(c) Two chromatids of same chromosomes

(d) All the foregoing **Ans. a**

49. *Neurospora crassa* is widely used in genetical studies because of all the following except one

(a) It is a haploid plant and mutations can be easily detected

(b) It can be easily cultured

(c) Life-cycle short

(d) Spores are not affected by mutagens **Ans. d**

50. The frequency of crossing over would be higher if

(a) Two genes are located closely

(b) Two genes are far apart on a chromosome

(c) Two genes are not located on the same chromosome

(d) None of the above **Ans. b**

51. Mendelian recombinations are due to

(a) Mutation

(b) Linkage

(c) Crossing over

(d) Independent assortment of characters **Ans. d**

52. The blue green algae and bacteria contain

(a) One linkage group (b) Two linkage groups

(c) Three linkage groups (d) None of the above **Ans. a**

53. Which one of the following character in man is controlled by a recessive gene?

(a) Colour blindness (b) Wooly hair

(c) Brachydactyly (d) Curly hairs **Ans. a**

54. A normal plant suddenly started reproducing parthenogenitically. The number of chromosomes of the second generation as compared to the parent will be

(a) One half (b) One fourth

(c) Same (d) Double **Ans. c**

55. Plotting of specific genes on the chromosome is known as

(a) Chromosome map (b) Linkage map

(c) Genetic map (d) All the above **Ans. d**

56. Crossing over may result into

(a) Gene mutation (b) Genomatic mutation

(c) Genetic recombination (d) None of the above **Ans. c**

57. Maize has 10 pairs of chromosomes. How many linkage groups does it have?

(a) 20 (b) 10
(c) 5 (d) 40 **Ans. b**

58. Linkage in *Drosophila* was first discovered by

(a) Bridges (b) Mendel
(c) Bateson and Punnet (d) Morgan **Ans. d**

59. In flowering plants meiosis occurs at the time of

(a) Germination of seed (b) Formation of endosperm
(c) Formation of embryo (d) Formation of pollens **Ans. d**

60. A child is born with an extra chromosome in each of its cell. This condition is the result of

(a) Synapsis (b) Crossing over
(c) Non-disjunction (d) Disjunction **Ans. c**

61. The number of Barr bodies in Turner syndrome is

(a) 0 (b) 1
(c) 2 (d) 3 **Ans. a**

62. In gene mutation, adenine is replaced by guanine. It is known as

(a) Substitution (b) Point mutation
(c) Transition (d) Transversion **Ans. c**

63. In transversion

(a) Purine is replaced by another purine
(b) Purine is replaced by pyrimidine or vice versa
(c) Some nitrogen bases are eliminated from a gene
(d) Some nitrogen bases are added in a gene **Ans. b**

64. Alkaptonurics excrete excess of

(a) Urine (b) Albumen
(c) Malonylacetic acid (d) Homogentisic acid **Ans. d**

65. Colour blindness is more likely to occur in males than females because

(a) Genes for this character are located on the X-chromosomes
(b) The trait is dominant in males and recessive in females
(c) Some males suffer from deficiency of vitamin A
(d) The Y-chromosome of males have the genes for distinguishing colours **Ans. a**

66. Suppose that out of 1000 offspring in the F_2 generation of a dihybrid cross, 748 are tall and hairy, 6 are tall and smooth, 4 are short and hairy, 242 are short and smooth. The reason is that the genes controlling these characters are

(a) Lethal genes (b) Complementary genes
(c) Linked genes (d) Epistatic genes **Ans. c**

67. An animal where male carries half the chromosomes present in female is

(a) *Amoeba* (b) Gorilla
(c) Honey Bee (d) Geometrid Moth **Ans. c**

68. The males of roundworms have

(a) One Y-chromosome
(b) One chromosome less than female
(c) Two similar sex chromosomes
(d) Distinct sex chromosomes **Ans. b**

69. XX-XO chromosome complement occurs in

(a) Cockroach (b) Honey Bee
(c) Human beings (d) Chimpanzee **Ans. a**

70. An example of complete linkage is

(a) Human beings (b) *Neurospora*
(c) Female *Drosophila* (d) Male *Drosophila* **Ans. d**

71. Which one is homogametic?

(a) Human child (b) Human embryo
(c) Human male (d) Human female **Ans. d**

72. Crossing over occurs in

(a) Four strand stage (b) Three strand stage
(c) Two strange stage (d) Single strand stage **Ans. a**

73. In *Neurospora*, both recessive and dominant alleles express their effect because

(a) Two genes control each character
(b) There are two alleles for each character
(c) The organism contains only one allele for a gene
(d) Each gene has only one allele **Ans. c**

74. In human males, some recessive genes express their effect because they have

(a) Only one Y-chromosome (b) Single genome

(c) Only two sex chromosomes (d) Only one X-chromosome

Ans. d

75. Genes located on X-chromosomes are called

(a) Sex influenced (b) Sex limited

(c) Sex linked (d) Mantel genes **Ans. c**

76. The term mutation was coined by

(a) Morgan (b) Beadle and Tatum

(c) Hugo de Vries (d) H.J. Muller **Ans. c**

77. Most of the mutations remain unexpressed because they are

(a) Too small (b) Very significant

(c) Very few (d) Mostly recessive **Ans. d**

78. Mutations were first reported by De Vries in

(a) Pea (b) Datura

(c) *Oenothera lamarckiana* (d) None of the above **Ans. c**

79. An exchange of segments between the two non-homologous chromosomes is called

(a) Polyploidy (b) Chromosomal aberration

(c) Translocation (d) Inversion **Ans. c**

80. Who induced mutations by X-ray irradiation in fruit fly for the first time?

(a) De Vries (b) H.J. Miller

(c) Morgan (d) Bateson **Ans. b**

81. If a part of a chromosome gets separated and reattached in reverse position to the same chromosome, the mutation is called

(a) Inversion (b) Transversion

(c) Translocation (d) Gene mutation **Ans. a**

82. If a part of the chromosome gets detached and lost during the cell division, the mutation so produced is called

(a) Deletion (b) Euploidy

(c) Inversion (d) Transcription **Ans. a**

83. Mutation is

(a) A change in phenotype of an organism caused by some environmental factor

(b) An inheritable change in genetic material

(c) A temporary change in structure of the nucleus

(d) Any one of the above **Ans. b**

84. A sudden or spontaneous change in the structure and action of a particular gene is called

(a) Linkage (b) Variation

(c) Mutation (d) Allelomorph **Ans. c**

85. In case of an abnormal crossing over, the two homologous chromosomes happen to exchange unequal length of segments. The phenomenon is called

(a) Transition (b) Duplication

(c) Inversion (d) Gene mutation **Ans. b**

86. 'Illegitimate crossing over' is another term for

(a) Transition (b) Transversion

(c) Reciprocal translocation (d) None of the above **Ans. c**

87. Mutations do not result in

(a) Death of organism

(b) Better progenies

(c) Change in genetic constitution of the cell

(d) Hybrid vigour **Ans. d**

88. Gene mutations are those which involve

(a) The change in nature and sequence of base triplets of DNA

(b) The change in genome

(c) The change in all the genes

(d) The disappearance of certain part of chromosome **Ans. a**

89. A genomatic mutation is the mutation involving

(a) Change in gene

(b) Change in chromosomal structure

(c) Change in the number of chromosomes

(d) All the above **Ans. c**

90. Aneuploidy is the term applied for the

(a) Gene mutation

(b) Chromosomal mutation

(c) Chromosomal mutations involving the addition or loss of one or more chromosome

(d) Chromosomal mutation involving the addition of one or more complete set of chromosomes **Ans. c**

91. What term is applied to the gene mutation where a base is replaced by another base?

(a) Duplication (b) Aneuploidy

(c) Euploidy (d) Substitution **Ans. d**

92. A monosomic organism can best be represented as

(a) 2n + 1 (b) 2n + 2

(c) n + 1 (d) 2n – 1 **Ans. d**

93. When a purine is replaced by a pyrimidine in a part of DNA, the mutation so produced is called

(a) Transition (b) Transversion

(c) Deletion (d) Reversal **Ans. b**

94. A flower was fertilized by pollen grains having the constitution (n-1). If the ovules are normal, which of the following is applicable to the seeds so produced?

(a) Allopoluploidy (b) Monosomy

(c) Nullisomy (d) Both (a) and (b) **Ans. b**

95. Nullisomy is the term used for the condition when an organism has

(a) An additional chromosome

(b) One chromosome less than normal

(c) A complete set of chromosomes except one homologous pair

(d) None of the above **Ans. c**

96. A mutation changes the original base sequence of DNA, GATAGCCG to a new sequence GGTAGCCG. What is the type of mutation?

(a) Transition (b) Transversion

(c) Translocation (d) Inversion **Ans. a**

97. A classical example of allopolyploidy is

(a) *Brassica* (b) *Raphano-brassica*

(c) *Raphanus* (d) All the above **Ans. b**

98. The mutation which returns to the original state is called

(a) Reversible mutation (b) Lethal mutation

(c) Backward mutation (d) Abnormal mutation **Ans. a**

99. The first man made cereal Triticale has been developed from a cross between

(a) Wheat and Oat (b) Wheat and Maize

(c) Maize and Rice (d) Wheat and Rye **Ans. d**

100. Mutations are

(a) Always useful (b) Seldom useful

(c) Always harmful (d) May be useful or harmful

Ans. d

101. Mutations were first induced in *Drosophila* by the use of

(a) X-ray (b) DDT

(c) High temperature (d) UV-rays **Ans. a**

102. Euploidy is the term applied for those mutations in which

(a) There is multiplication of chromosomal number

(b) Involves addition of complete set of chromosomes

(c) There is loss of a few chromosomes

(d) None of the above **Ans. b**

103. Which is not applicable to mutations?

(a) They are inheritable genetic changes

(b) Many of them remain undetected

(c) They can be induced artificially

(d) They are possible only in diploid cells **Ans. d**

104. Autopolypolids are those which have

(a) More than one set of chromosomes

(b) Only one additional set of chromosomes

(c) The same basic set of chromosomes

(d) All the above **Ans. c**

105. Which of the following is a case of autoallopolyploidy?

(a) AA, AA, A (b) AA, AA, AB

(c) AA-1 (d) 2n-2 **Ans. b**

106. The frequency of mutations in nature is one in

(a) 1×10^5 (b) 1×10^{12}

(c) 1×10^{-5} (d) 1×10^{-12} **Ans. a**

107. When chromosome sets are present in multiple of 'n' the condition is termed

(a) Diploidy (b) Haploidy

(c) Euploidy (d) Aneuploidy **Ans. c**

108. Which of the following is not an aneuploid?

(a) Monoploid (b) 2n – 1

(c) Trisomic (d) 2n + 2 **Ans. a**

109. The reason of genetic variability in sexually organisms is

(a) Temperature (b) Mutations

(c) Certain chemicals (d) Evolution **Ans. b**

110. Most of the high yielding varieties of wheat are

(a) Imported (b) Monosomic

(c) Trisomic (d) Mutants **Ans. d**

111. The smallest segment of a gene which can undergo mutation is called

(a) Muton (b) Recon

(c) Cistron (d) Interferon **Ans. a**

112. The X-rays were used to induce mutations in *Drosophila* by

(a) Hooker (b) Morgan

(c) Muller (d) Khurana **Ans. c**

113. The mutations which prove fatal for the organism are called

(a) Spontaneous (b) Induced

(c) Deleterious (d) Lethal **Ans. d**

114. Mutation theory of evolution by Hugo de Vries was based on his studies on

(a) *Pisum sativum* (b) *Oenothera lamarckiana*

(c) *Acetabularia* (d) *Lathyrus odoratus* **Ans. b**

115. Most of the mutations are

(a) Forward (b) Backward

(c) Reversible (d) Lethal **Ans. a**

116. Mutagens are
(a) The genes which can mutate
(b) The organism which show mutation
(c) Genes that regulate mutations
(d) The agents which cause mutations **Ans. d**

117. The mutation which occur in germ cells can be detected phenotypically in
(a) Parents (b) Next generation
(c) Third generation (d) None of the above **Ans. b**

118. The frequency of mutation in a species can be increased by the use of
(a) X-rays (b) UV-rays
(c) Nitrous acid (d) All the above **Ans. d**

119. A pleiotropic mutation is the one which
(a) Affects one character at a time
(b) Affects many characters at a time
(c) Is always lethal
(d) Easily gets preserved in natural conditions **Ans. b**

120. Down's syndrome is a typical case of
(a) Trisomy (b) Nullisomy
(c) Gene mutation (d) Monosomy **Ans. a**

121. It is a common observation that some insects have developed resistance to DDT. The reason may be attributed to
(a) Poor quality of DDT (b) Genetic recombination
(c) Genetic mutation (d) Hybrid vigour **Ans. c**

122. Datura is a classical example of study of
(a) Monosomics (b) Trisomics
(c) Triploids (d) Nullisomics **Ans. b**

123. Which one is a hexaploid species of wheat?
(a) *Triticum monococcum* (b) *T. aestivum*
(c) *T. durum* (d) *T. turgidum* **Ans. b**

124. When chromosome sets are present in multiple of 'n', the condition is called as
(a) Diploidy (b) Aneuploidy
(c) Haploidy (d) Euploidy **Ans. d**

125. A trisomic individual has

(a) One extra-chromosome (b) One less chromosome

(c) Two extra chromosomes (d) One pair of extra chromosomes

Ans. a

126. X-rays cause mutations by

(a) Breaking spindle

(b) Rupturing of nuclear membrane

(c) Changing the chromosome morphology

(d) Inducing karyokinesis **Ans. c**

127. Colchicines interferes in

(a) Spindle organization (b) DNA replication

(c) Chromosome condensation (d) Polyploidy **Ans. a**

128. Gene mutations occur in

(a) Chromosome (b) DNA

(c) Nucleus (d) RNA **Ans. b**

129. The frequency of mutations

(a) Can be increased by recombination

(b) Varies in the character and the organism

(c) Is affected by environmental factors

(d) Can be increased by X-rays **Ans. d**

130. In Sweet Pea, the flower colour changes from Red to White and seed coat colour from grey to white. This is example of

(a) Spontaneous mutation (b) Pleiotropic mutation

(c) Reverse mutation (d) Induced mutation **Ans. b**

131. Haploids are used for researches because

(a) They contain only one chromosome

(b) They contain two sets of chromosomes

(c) They contain three sets of chromosomes

(d) They contain only one set of chromosomes **Ans. d**

132. Male sterile lines were first discovered in

(a) Wheat (b) Maize

(c) Onion (d) Sunflower **Ans. b**

133. The scientist who first discovered cytoplasmic inheritance was
(a) Correns (b) Rhoades
(c) Mendel (d) Morgan **Ans. a**

134. Shell coiling in Limnaea is an example of
(a) Material inheritance (b) Biparental inheritance
(c) Predetermination (d) Dauermodiication **Ans. c**

135. Kappa particles were discovered by
(a) Correns (b) Sonneborn
(c) Rhoades (d) Bycott et al **Ans. b**

136. Kappa particles are present in
(a) *Mirabilis jalapa* (b) *Zea mays*
(c) *Limnaea peregra* (d) *Paramecium Aurelia* **Ans. d**

137. Kappa particles make an animal killer when their number in an individual is
(a) 6 (b) 60
(c) 400 (d) 150 **Ans. c**

138. Kappa particles are present in
(a) Plastids (b) Mitochondria
(c) Plasmids (d) Cytoplasm **Ans. d**

139. The killer chemical secreted by kappa particles is
(a) Secretin (b) Paramecin
(c) Plasmon (d) Poky **Ans. b**

140. Male sterility was discovered by
(a) Rhoades (b) Sonneborn
(c) Bycott et al (d) Correns **Ans. a**

141. A sinistral shelled female snail having Dd gene complement breeds with dextral shelled male snail having dd gene component. What type of shall be present in the progeny?
(a) Dextral (b) Sinistral
(c) Lateral coiling (d) Vertical coiling **Ans. a**

142. A pollen from green branch fertilizes an ovum of pale type in *Mirabilis jalapa*. What shall be the progeny?
(a) Green (b) Variegated
(c) Pale (d) All the above **Ans. c**

143. Cytoplasmic inheritance differs from nuclear inheritance in the absence of
(a) Similarity of reciprocal crosses
(b) Biparental contribution
(c) Effect on backcrossing
(d) All the above **Ans. d**

144. Cytoplasmic inheritance is due to
(a) Plastids (b) Mitochondria
(c) Cytoplasmic particles (d) All the above **Ans. d**

145. Cytoplasmic inheritance is also called maternal inheritance because it is
(a) Without sexual reproduction
(b) Only female parent takes part in multiplication
(c) Most of the cytoplasm of the zygote is provided by ovum
(d) All the above **Ans. c**

146. Length of X-chromosome is
(a) 8.5 - 9.5 μm (b) 7.5 - 8.0 μm
(c) 6.5 - 7.5 μm (d) 5.0 - 5.5 μm **Ans. d**

147. Length of Y-chromosome is
(a) 2.0 μm (b) 3.0 μm
(c) 4.0 μm (d) 5.0 μm **Ans. a**

148. X-chromosome is
(a) Telocentric (b) Metacentric
(c) Submetacentric (d) Acentric **Ans. b**

149. Y-chromosome is
(a) Acrocentric (b) Telocentric
(c) Submetacentric (d) Acentric **Ans. a**

150. Theory of heterogamesis for sex determination was proposed by
(a) Morgan (b) Darwin
(c) Correns (d) Bridges **Ans. c**

151. Percentage of colour blindness in white male population is
(a) 1.5% (b) 2.5%
(c) 5.9% (d) 8% **Ans. d**

152. Percentage of colour blindness in white female population is
(a) 4.5% (b) 2.3%
(c) 0.5% (d) Zero% **Ans. c**

153. XY chromosomes are
(a) Homomorphic (b) Heteromorphic
(c) Heterologous (d) Both (b) and (c) **Ans. b**

154. Role of chromosomes in sex determination was proposed by
(a) Sutton and Boveri (b) Henking
(c) Mc Clung (d) Morgan **Ans. c**

155. Chromosome theory of XY sex determination was proposed by
(a) Henking (b) Wilson and Stevens
(c) Johannsen (d) Punnet **Ans. b**

156. In colour blindness red, green and other colours appear
(a) White (b) Yellow
(c) Grey (d) Pink **Ans.c**

157. Heterogamesis is
(a) Formation of two types of gametes
(b) Morphological distinction of male and female gametes
(c) Formation of two types of gametes by one sex and one type by other sex
(d) Formation of two types of gametes by both the sexes **Ans. c**

158. Who discovered sex-linked inheritance?
(a) Morgan (b) Bridges
(c) Mc clung (d) Wilson and Stevens **Ans. a**

159. A character is transmitted from father to daughter and from there to grandson. It is
(a) Holandric inheritance (b) Hologenic inheritance
(c) Criss cross inheritance (d) Dominant inheritance **Ans. c**

160. Colour blindness in which all colours appear grey is
(a) Monochromatism (b) Dichromatism
(c) Protanopia (d) Deuteronopia **Ans. a**

161. In protanopia, a person cannot distinguish
(a) Green colour (b) Red colour
(c) Blue colour (d) Blue and green colour **Ans. b**

162. Females seldom become bald as they lack
(a) The gene for baldness (b) Y-chromosome
(c) Male sex hormone (d) All the above **Ans. c**

163. Beard appears only after attaining maturity as the gene
(a) Expression is delayed
(b) Expression occurs only in presence of male hormones
(c) Remains dominant in childhood
(d) All the above **Ans. b**

164. Traits controlled by genes present on X-chromosome are called
(a) Sex limited (b) Sex influenced
(c) Autosomal (d) Sex-linked **Ans. d**

165. Phenylketonuria is
(a) Sex linked dominant trait (b) Sex linked recessive trait
(c) Autosomal dominant trait (d) Autosomal recessive trait
Ans. d

166. Colour blindness is disease in which the factor is usually transmitted to children by woman. It is because the factor is located on
(a) An autosome (b) X-chromosome
(c) Y-chromosome (d) Cytoplasm **Ans. b**

167. Gene for sickle cell anaemia is carried by
(a) Sex cells (b) Sex chromosomes
(c) Autosomes (d) Bone cells **Ans. c**

168. A red-green colour blind woman marries to a normal man. The progeny shall be
(a) 50% colour blind and 50% normal sons
(b) Normal daughters and colour blind sons
(c) Carrier daughters and colour blind sons
(d) All children colour blind **Ans. c**

169. A normal woman whose father was haemophilic, marries a normal man. The offspring shall be
(a) All normal (b) All sons haemophilic
(c) All daughters haemophilic (d) Some sons are haemophilic
Ans. d

170. A late acting dominant disorder is
(a) Tay-Sach's disease (b) Polydactyly
(c) Huntington's chorea (d) Phenylketonuria **Ans. c**

171. Huntington's chorea is characterized by
(a) Incongruant muscle movement
(b) Disordered muscle movement and metal mental deterioration
(c) Weak eye sight and hearing power
(d) Inability to speak **Ans. b**

172. Huntington's chorea appears at the age of
(a) 25 – 55 years (b) 15 – 25 years
(c) 50 – 60 years (d) 10 – 15 years **Ans. a**

173. Down's syndrome or Mongolism is due to
(a) A missing chromosome (b) An extra chromosome
(c) A dominant gene (d) A recessive gene **Ans. b**

174. In Down's syndrome the chromosome number in each cell of body is
(a) 45 (b) 47
(c) 48 (d) 49 **Ans. b**

175. Down's syndrome is caused by the presence of extra chromosome number
(a) X (b) Y
(c) 21 (d) 22 **Ans. c**

176. The risk of Down's syndrome offspring is more to mothers at the age of
(a) 20 years (b) 25 years
(c) 30 years (d) 35 years **Ans. d**

177. The child afflicted with Down's syndrome has
(a) Flattened nasal bridge, open mouth with protruding tongue
(b) Small forehead, bulging eyes and raised nasal bridge
(c) Habitually open mouth with long protruding tongue, bulging eyes and small forehead
(d) Large forehead, raised nasal bridge and long included tongue
Ans. a

178. Persons with Klinefelter's syndrome have chromosomes
(a) XX (b) XY
(c) XXY (d) XYY **Ans. c**

179. Klinefelter's syndrome is characterized by chromosomes
(a) 46 (b) 45
(c) 47 (d) 48 **Ans. c**

180. A person having Klinefelter's syndrome is characterized by
(a) Male with some secondary sexual characters of female
(b) Female with some secondary sexual characters of male
(c) Having both male and female sex organs
(d) Female internal sex organs and male external sex organs **Ans. a**

181. In *Drosophila* the XXY constitution determines
(a) Maleness (b) Femaleness
(c) Intersex (d) Both (a) and (c) **Ans. b**

182. Turner's syndrome is characterized by chromosomes
(a) 46 (b) 45
(c) 47 (d) 48 **Ans. b**

183. The sex chromosome complement of Turner's syndrome is
(a) XXY (b) XYY
(c) XO (d) YO **Ans. c**

184. Individual with Turner's syndrome is
(a) Normal female
(b) Normal male
(c) A female with rudimentary ovaries and undeveloped breasts
(d) A male with rudimentary testes and undeveloped penis **Ans. c**

185. A supermale has a genetic constitution of
(a) XY (b) XXY
(c) XXYY (d) XYY **Ans. d**

186. A supermale XYY is characterized by
(a) Under production of sex hormones
(b) Overproduction of sex hormones
(c) Reduced intelligence but aggressive nature
(d) Both (a) and (c) **Ans. d**

187. In Huntington's chorea, limb movements are
(a) Rhythmic (b) Arrhythmic
(c) Slow and hardly noticeable (d) Absent **Ans. b**

188. Huntington's disease is

(a) Autosomal dominant disease (b) Autosomal recessive disorder
(c) Sex-linked recessive disorder (d) Sex-linked dominant disease

Ans. a

189. Dancing gait and bizarre grimacing is characteristics of

(a) Schizophrenia (b) Huntington's disease
(c) Alzheimer's disease (d) Paralysis agitans **Ans. b**

190. Alzheimer's disease is due to

(a) Poor neurotransmission (b) Degeneration of neurons
(c) Muscular dystrophy (d) Both (a) and (b) **Ans. d**

191. In Alzheimer's disease brain cells do not metabolise

(a) Glucose (b) Amyloid β-peptide
(c) GABA (d) Acetylcholine **Ans. b**

192. Cystic fibrosis is caused by

(a) Recessive autosomal allele (b) Dominant autosomal allele
(c) Recessive sex linked allele (d) Dominant sex linked allele

Ans. a

193. Allele for cystic fibrosis occurs over

(a) 21 chromosome (b) 14 chromosome
(c) 7 chromosome (d) 4 chromosome **Ans. c**

194. In cystic fibrosis there is

(a) Failure of chloride ion transport
(b) Mucous clogging of lungs
(c) Defective functioning
(d) All the above **Ans. d**

❑❑❑

Chapter-15

Genetic Basis of Inheritance

Fill in the blanks

1. The abundant genotypic and phenotypic variety of sexually reproducing organisms is largely due to..........................

Ans. recombinations

2. The term genetics was coined by...........

Ans. Bateson

3. The more closely related the sexual partners are, the................are the variations between their offsprings.

Ans. lesser

4. The expression of a character of a distant ancestor is called.................

Ans. atavism

5. Mendel began his scientific investigations on inheritance in.................

Ans. 1856

6. The process of removal of male organs from a flower is called..................

Ans. emasculation

7. Alternative forms of the same factor responsible for determining contrasting characters are called....................

Ans. allele

8. The condition in which only one allele of a pair is present is called.....................

Ans. hemizygous

9. Punnett square was discovered by.................

Ans. R.C. Punnett

10. The term allelomorph was coined by.................

Ans. Bateson

11. The cross of a hybrid with either of the parental type is known as..................

Ans. back cross

12. The type of gametes found in F_1 of a dihybrid cross is..................

Ans. 4

13. Genotypic ratio of a dihybrid cross is..................

Ans. 1 : 2 : 1 : 2 : 4 : 2 : 1 : 4 : 1

14. Mendelian recombinations are due to..............

Ans. independent assortment

15. The genes (or factors) which occupy same position on homologous chromosomes are called........................

Ans. alleles

16. When tall heterozygous plant is crossed with homozygous dwarf, the ratio of tall vs. dwarf would be..................

Ans. 1 : 1

17. Inheritance of a single pair of alleles is studied under.................cross.

Ans. monohybrid

18. The twins arising from a single zygote are...........of each other.

Ans. clones

19. Each parent has..............factors for each character in their cells.

Ans. two

20. ....is the diploid condition where different alleles are present.

Ans. Heterozygosity

21. Mendelian inheritance is not applicable to..................organisms.

Ans. haploid

22. A dihybrid cross between F_1 heterozygous for two factors, and the recessive parent yields the ratio....................

Ans. 1 : 1 : 1 : 1

23. A sudden or spontaneous change in the structure and action of a particular gene is called as................

Ans. Mutation

24. Biochemical nature of gene was established by ....

Ans. Avery, Mcleod, McCarty

25. Crossing over takes place at.................stage.

Ans. Diplotene-Diakinesis

26. .... are removed in emasculation.

Ans. Anthers

27. Polyploidy discovered by ....

Ans. Lutz

28. Inheritance of kernal colour in wheat is an example of .... inheritance.

Ans. Polygenic

29. Mangolism in human beings is due to the presence of .... chromosome.

Ans. an extra

30. A mutant organism incapable to grow on minimal medium but could grow when such medium supplemented by some growth factor are known as.....................

Ans. Auxotrop

31. Clone is term refers to individuals derived by vegetative propagation from.............. individual.

Ans. single

32. Linkage of hereditary unit was discovered by ________.

Ans. W. Bateson and R.C. Punnett

33. One gene one enzyme theory was proposed by...........................

Ans. G.W. Beadle and E.L. Tatum

34. ....proposed the chromosome theory of inheritance.

Ans. Sutton & Boveri

35. Process in which genetic material is transferred from one bacterium to another through the agency of virus called ....

Ans. Transduction

36. Harmful recessive characters carried by ....individuals are not eliminated from gene pool.

Ans. heterozygous

37. A colour blind man marries a woman with normal vision whose father was colour blind. The percentage of their children expected to be colour blind is ....

Ans. 50%

38. Linked gene may be separated by the process of .... during meiosis.

Ans. crossing over

39. A substance, usually a protein that stimulate the production of antibodies when introduced into living organism is known as ....

Ans. Antigen

40. Linkage was discovered by......................

Ans. Bateson and Punnett

41. The abundant genotypic and phenotypic variety of sexually reproducing organisms is largely due to.........

Ans. recombination

42. Mendelian recombinants are due to.............

Ans. crossing over

43. The killer strains of Paramoecia contain cytoplasmic particles called...................

Ans. Kappa articles

44. A plant having red flowers is crossed with one having yellow flowers. If their offsprings produce orange flowers, the colour change is probably due to................

Ans. incomplete dominance

45. The more the distance between the genes, the more are the chances of..............

Ans. crossing over

46. Crossing over occurs at.............stranded stage.

Ans. four

47. Bateson and Punnett coined the term.............. for the tendency of two alleles to stay together.

Ans. coupling

48. The percentage frequency with which a gene successfully manifests its phenotypic effect is called..................

Ans. penetrance

49. The kappa particles are, in fact, the bacteria called................

Ans. *Caedobacter taeniospirales*

50. Inheritance of skin colour in man is an example of..............inheritance.

Ans. quantitative

51. Coiling of shells in snails is an example of................

Ans. maternal influence

52. Sum total of genes in reproductive gametes of a population is called........

Ans. gene pool

53. If the dominant alleles of two liked genes are present on same chromosome, and their recessive alleles on the homologous chromosomes, the arrangement of genes is called........... arranegement.

Ans. cis

54. If the chances of separation of two genes are nil, the linkage is said to be.....................

Ans. complete

55. A genetic clone is.................by asexual means.

Ans. plants produced

56. A pleiotropic gene is................

Ans. Hbs

57. Siblings are................of same parents.

Ans. sons and daughters

58. ....converted Mendel's conclusions into principles of heredity.

Ans. Carl correns

59. Mendel was lucky and could discover a law of heredity because he selected traits which................

Ans. crossed independently

60. A mechanism that can cause a gene to more from one linkage group to another is.................

Ans. translocation

61. In a................individual the number of chromosome is 2n+1.

Ans. trisomic

Multiple Choice Questions

1. The ratio of phenotypes in F_2 of a monohybrid cross is
   (a) 3:1 (b) 1:2:1
   (c) 9:3:3:1 (d) 2:1 **Ans. a**

2. A pure tall Pea was crossed with a pure dwarf Pea. All the plants of F_1 were found to be tall. This is due to
   (a) Dominance
   (b) Disappearance of factor for dwarfness in F_1 generation
   (c) Segregation of factors
   (d) Coordination **Ans. a**

3. The monohybrid genotypic ratio 1:2:1 in F_2 generation indicates
   (a) Segregation (b) Independent assortment
   (c) Dominance (d) Incomplete dominance **Ans. a**

4. A monohybrid cross is the one in which
   (a) Only a single plant is involved for the experiment
   (b) A single pair of contrasting characters is considered for the genetic results
   (c) A hybrid is crossed to a homozygous
   (d) None of the above **Ans. b**

5. The F_2 generation of a cross produced identical phenotypic and genotypic ratio. It is not an excepted Mendelian result, and can be attributed to
   (a) Independent assortment (b) Linkage
   (c) Incomplete dominance (d) None of the above **Ans. c**

6. The law of independent assortment can be related to one of the following
   (a) Both homologous chromosomes of each pair are received by a single gamete
   (b) The genes present on the same chromosome get randomly redistributed
   (c) The non-homologous chromosomes show random distribution during anaphase-1 of meiosis
   (d) The exchange of segments between the non-homologous chromosomes **Ans. c**

7. The law of segregation of characters postulated by Mendel can be related to
   (a) The presence of two genes for each character in a somatic cell
   (b) A gamete receiving only one of the two homologous chromosomes during meiosis
   (c) Presence of both genes on the same chromosome
   (d) None of the above **Ans. b**

8. The factors which represents the contrasting pairs of characters are called
   (a) Dominant and recessive (b) Alleles
   (c) Homologous pairs (d) Determinants **Ans. b**

9. The back cross is
   (a) A cross between F_1 individual and F_2 individual
   (b) A cross between an F_1 individual with another F_1 individual
   (c) Cross between F_1 and one of the two parents
   (d) Cross between F_2 with one of the parents **Ans. c**

10. Mendel selected Pea as material for his experiments because
   (a) It is an annual with comparatively short life cycle
   (b) The flowers are self-pollinated
   (c) The number of seeds produced is quite large
   (d) All the above **Ans. d**

11. One of the following did not constitute the seven contrasting pairs of characters noticed by G.J. Mendel in Pea
   (a) Height of the plants (b) Shape of the leaves
   (c) Shape of pod (d) Colour of pod **Ans. b**

12. In genetics the term test cross means
   (a) The crossing of F_1 individual with homozygous recessive
   (b) Crossing an F_1 individual with either of the two parents
   (c) Crossing an F_1 individual with another F_1 individual
   (d) Crossing F_1 individual with that of F_2 **Ans. a**

13. The first work on genetics was done by
   (a) Lamarck (b) Hugo de Vries
   (c) Mendel (d) Darwin **Ans.c**

14. According to the law of Independent Assortment in a dihybrid was done by

(a) There are four genotypes in F_2

(b) F_2 contains 16 phenotypes

(c) There is a single individual which is homozygous recessive for both the characters

(d) It is not possible to forecast the different phenotypes **Ans. c**

15. The discipline which deals with the study of inheritance of characters is

(a) Darwinism (b) Cytology

(c) Genetics (d) Evolution **Ans. c**

16. Who is regarded as the 'Father of Genetics'?

(a) Gregor Johann Mendel (b) Morgan

(c) Lamarck (d) Hugo de Vries **Ans. a**

17. The number of different type of gametes produced from a plant with genotype Aa Bb Cc is

(a) 2 (b) 8

(c) 4 (d) 16 **Ans. b**

18. How many types of gametes are excepted from the organism with genotype AABBCC?

(a) One (b) Two

(c) Four (d) Eight **Ans. a**

19. How many types of gametes are expected from the plants with genotypes AA, BB, aa, bb?

(a) 7 (b) 4

(c) 8 (d) 16 **Ans. b**

20. A haploid set of all the genes present in a gamete is called

(a) Genotype (b) Phenotype

(c) Genome (d) Linkage group **Ans. c**

21. A plant with a genotype Aa Bb is crossed with a plant having the genotype aa bb. The genotype of F_1 would be

(a) AaBb, AABB (b) aabb, aaBb

(c) aaBB, AAbb (d) AaBb, Aabb, aaBb, abab

Ans. d

22. Mendel was born in

(a) Australia (b) Heizendorf

(c) Maravia (d) Brunn **Ans. b**

23. How many types of gametes are expected from a plant with genotype Pp Qq, provided there occurs independent assortment?

(a) 4 (b) 8

(c) 2 (d) 1 **Ans. a**

24. How would you test a Pea plant whether it is a pure or hybrid for tallness?

(a) Cross it with another tall Pea plant of unknown genotype

(b) Cross it with a pure tall Pea plant

(c) Cross with a homozygous dwarf Pea

(d) Cross it with any Pea plant **Ans. c**

25. The Mendelian principle which has always stood true is

(a) The law of independent assortment

(b) The law of segregation

(c) The law of dominance

(d) All the above **Ans. b**

26. Which of the following crosses would produce a genotypic ratio of 1:2:1 in F_2?

(a) AB x AB (b) Ab x ab

(c) Ab x Ab (d) Ab x ab **Ans. c**

27. In *Mirabilis jalapa* when two F_1 pink flowered plants were crossed with each other, the F_2 generation produced 40 red, 80 pink and 40 white flowering plants. This is a case of

(a) Duplicate genes (b) Lethal genes

(c) Incomplete dominance (d) Epistasis **Ans. c**

28. The term genetics was coined by

(a) Mendel (b) Bateson

(c) Muller (d) Morgan **Ans. b**

29. Genotype-phenotype concept was first proposed by

(a) Bateson (b) Johannsen

(c) Sutton and Boveri (d) Punnet **Ans. b**

30. At which stage is the fate of genetic constitution of gametes finally decided?

(a) Metaphase - I (b) Anaphase - I

(c) Anaphase - II (d) Interkinesis **Ans. c**

31. The phenomenon which defines the independent assortment is

(a) Segregation (b) Crossing over

(c) Dominance (d) Linkage **Ans. d**

32. A tobacco plant heterozygous for albinism (a recessive character) is self pollinated and 1200 seeds are subsequently germinated. How many seedlings would have the parental phenotype?

(a) 900 (b) 600

(c) 1200 (d) 300 **Ans. b**

33. Mendel's law were rediscovered by

(a) Lamarck, de Vries and Correns

(b) Hugo de Vries, Correns and Tschermak

(c) Morgan, Beadle and Tatum

(d) Hugo de Vries, Morgan and Correns **Ans. b**

34. When a yellow mouse was crossed to another yellow mouse, the F_1 generation produces yellow and brown-black mice in the ratio 2:1. The yellow mice are never homozygous. The reason is

(a) Homozygous yellow cannot survive due to lethal effect of genes

(b) Yellow mice are not very suitable to live

(c) There is no formation of zygotes with homozygous yellow constitution

(d) None of the above **Ans. a**

35. A pea with whit flowers was crossed to another pea which is also white flower plant. When selfed the F_2 generation produced purple and white in the ratio 9:7. The reason for the result is that

(a) It is typical monohybrid Mendelian ratio

(b) Purple flower colour is dominant over the white

(c) It is a complementary factor

(d) None of the above **Ans. c**

36. Multiple alleles are present

(a) At different loci in the same chromosome

(b) In different chromosomes

(c) At the same locus in one type of chromosomes

(d) None of the above **Ans. c**

37. An example of the quantitative trait in man is

(a) Hair colour (b) Colour of eye

(c) Skin colour (d) Shape of nose Ans. c

38. The law of segregation of characters is also called the law of purity of gametes because

(a) Gametes have only one of the two alleles for each character

(b) Gametes cannot be contaminated

(c) Gametes are very different type of cells

(d) It was just another name adopted accidentally **Ans. a**

39. A dwarf pea plant was treated with GA. The plant became tall. The treated plant was then crossed with a homozygous tall pea. The results in F_2 are expected to be

(a) All tall (b) Tall and dwarf in 3:1 ratio

(c) 50% tall (d) All dwarf **Ans. b**

40. Number of characters studies by Mendel in Pea was

(a) 5 (b) 7

(c) 6 (d) 4 **Ans. b**

41. The genes for same trait present on nonhomologous chromosomes are

(a) Alleles (b) Linked genes

(c) Multiple alleles (d) None of these **Ans. d**

42. Mendel observed red flowers in F_1 when he crossed red and white because of

(a) Dominance

(b) Recessive gene

(c) Law of independent assortment

(d) Law of segregation **Ans. a**

43. The genotypic ratio of a monohybrid cross will be

(a) 3:1 (b) 1:1

(c) 1:2:1 (d) 2:1 **Ans. c**

44. Mendel formulated the laws of heredity considering seven pairs of contrasting characters in the pea plant. If he had studied an eighth pair, the law which would have been altered is

(a) Law of segregation

(b) Law of dominance

(c) Law of independent assortment

(d) Law of unit characters **Ans. c**

45. Mendel was successful in formulating the laws of inheritance whereas his predecessors were not because

(a) He studied one clear-cut character at a time

(b) The characters studied by him were present on separate chromosomes

(c) Of the right choice of material

(d) He kept accurate records of his experiments **Ans. a**

46. Appearance of hidden character in some progeny of F_2 population indicates

(a) Law of purity of gametes

(b) Law of independent assortment

(c) Law of dominance

(d) None of the above **Ans. a**

47. Mendel's law of segregation is based upon the F_2 ratio of

(a) 1:2 (b) 9:3:3:1

(c) 1:2:1 (d) 3:1 **Ans. d**

48. Mendel is popular for postulating

(a) Origin of species (b) Cell theory

(c) Linkage theory (d) Laws of inheritance **Ans. d**

49. 'Like begets like' an important and universal phenomenon of life, is due to

(a) Eugenics (b) Inheritance

(c) Dominance (d) Crossing over **Ans. b**

50. Why were pea plants more suitable than dogs for Mendelian experiments?

(a) Pea plants can be self-pollinated

(b) All pea plants are diploid

(c) There were no pedigree records of dogs
(d) Dogs have many genetics traits **Ans. a**

51. A cross between unlike, organisms is called
(a) Test-cross (b) Back-cross
(c) Heterosis (d) Hybrid **Ans. d**

52. Genes do not occur in pairs in
(a) Zygote (b) Somatic cell
(c) Endosperm cells (d) Gametes **Ans. d**

53. If dwarf pea plant was treated with GA_3 it grew as tall as the pure tall pea plant. If this treated plant is crossed with a pure tall plant the phenotypic ratio F_1 is likely to be
(a) All tall (b) All dwarf
(c) 50% tall and 50% dwarf (d) 75% tall and 25% dwarf **Ans. a**

54. A modified dihybrid mendelian ratio of 9:3:4 indicates
(a) Supplementary genes (b) Complementary genes
(c) Lethal genes (d) Epistatic genes **Ans. a**

55. Segregation of genes take place during
(a) Metaphase (b) Anaphase
(c) Prophase (d) Embryo formation **Ans. b**

56. A cross between hybrid and a parent is known as
(a) Test cross (b) Back cross
(c) Monohybrid cross (d) Reciprocal cross **Ans. b**

57. A cross between offspring and recessive parent is
(a) Monohybrid cross (b) Back cross
(c) Test cross (d) Reciprocal cross **Ans. c**

58. In an experiment on pea plant, pure plants with yellow round seeds (YYRR) were crossed with plants producing green wrinkled seeds (yyrr). What will be phenotypic ratio of F_1 progeny?
(a) 9 yellow round : 3 round green : 3 wrinkled yellow : 1 green wrinkled
(b) All yellow round
(c) 1 round yellow : 1 round green : 1 wrinkled yellow : 1 wrinkled green
(d) All wrinkled green **Ans. b**

59. Genetics deals with

(a) Heredity and variations

(b) Heredity

(c) Mutations

(d) Nuclear and cytoplasmic inheritance **Ans. a**

60. When a wheat variety of red kernels (homozygous for two nonallelic and independent dominant genes) is crossed with white kenelled wheat (homozygous for two recessive nonallelic independent genes), the phenotypic ratio in F_2 generation would be

(a) 9 : 7 (b) 1 : 10 : 4 : 1

(c) 1 : 4 : 6 : 4 : 1 (d) 1 : 2 : 4 : 2 : 4 : 2 : 1 **Ans. c**

61. ABO blood grouping in humans is an example of

(a) Polygenic inheritance (b) Multifactor inheritance

(c) Pleiotropic gene (d) Multiple alleles **Ans. d**

62. In multiple allele system a gamete possesses

(a) Two alleles (b) Three alleles

(c) One allele (d) Several allele **Ans. c**

63. ABO blood grouping shows

(a) Codominant genes

(b) Polygenes

(c) Dominant-recessive genes

(d) Both codominant and dominant recessive genes **Ans. d**

64. A pleiotropic gene is one which

(a) Affects one character

(b) Affects more than one character

(c) Supplements the effect of another gene

(d) Requires another gene for expression **Ans. b**

65. Sum total of all the genetic information in the breeding members of a population at a given time is known as

(a) Gene pool (b) Genetic clone

(c) Genome (d) Genetic drift **Ans. a**

66. The spread of genes from one breeding population to another by migration which may result in changes in gene frequency is called

(a) Genetic drift (b) Gene flow

(c) Gene frequency (d) None of the above **Ans. b**

67. Hb^A and Hb^S alleles of normal and sickle celled RBC are

(a) Dominant-recessive alleles (b) Polygenic alleles
(c) Codominant alleles (d) Multiple alleles **Ans. c**

68. A genetic clone is

(a) Plants produced by asexual means
(b) Hybrid produced by sexual means
(c) Homozygous plant produced by sexual means
(d) Heterozygous plant produced by sexual means **Ans. a**

69. A pure tall plant is reared in a soil poor in nutrition and reached the size of dwarf plant. If this plant is selfed, the phenotype in the F_1 generation is most likely to be

(a) All tall plants (b) 50% tall and 50% dwarf
(c) All dwarf (d) Data insufficient **Ans. a**

70. A pleiotropic gene is

(a) I^A (b) Hb^S
(c) Hb^A (d) I^B **Ans. b**

71. Dominant gene for tallness is T and for yellow colour is Y. A plant heterozygous for both the traits is selfed, then the ratio of pure homozygous dwarf and green offspring would be

(a) 1/4 (b) 4/16
(c) 3/16 (d) 1/16 **Ans. d**

72. A gene that shows its effect on more than one character is

(a) Polygene (b) Pleiotropic gene
(c) Multifactor gene (d) Multiple gene **Ans. b**

73. A dihybrid ratio of 1:4:6:4:1 is obtained instead of 9:3:3:1. This is an example of

(a) Complementary genes (b) Supplementary genes
(c) Polygenic inheritance (d) Pleiotropic genes **Ans. c**

74. In a dihybrid cross, F_2 phenotypic ratio is 13:3. It is case of

(a) Complementary genes (b) Epistatic genes
(c) Multigenic inheritance (d) Incomplete dominance **Ans. b**

75. Epistatic gene differs from dominant gene in

(a) Epistatic gene is nonallelic
(b) Epistatic gene never expresses itself independently

(c) Epistatic and hypostatic genes are present at different loci
(d) All the above **Ans. d**

76. Mendel was
(a) Plant breeder (b) Cytologist
(c) Physiologist (d) Taxonomist **Ans. a**

77. Mendel published his research under the title of
(a) Laws of heredity
(b) Experiments in plant hybridization
(c) Hybridization experiments on Pea
(d) My experiments on particulate inheritance **Ans. b**

78. The scientist who proposed particulate concept of inheritance was
(a) Darwin (b) Galton
(c) Mendel (d) Garrod **Ans. c**

79. Hugo de Vries, a rediscoverer of Mendel's work belonged to
(a) Holland (b) Austria
(c) Germany (d) England **Ans. a**

80. Tschermak-Seysenegg, a rediscoverer of Mendel's work belonged to
(a) USA (b) Spain
(c) Austria (d) Australia **Ans. c**

81. Carl Correns, a rediscoverer of Mendel's work, was
(a) American (b) German
(c) Austrian (d) Spanish **Ans. b**

82. Siblings are
(a) Sons and daughters of same parents
(b) Individuals formed through asexual means
(c) Individuals from interspecific cross
(d) Mutants **Ans. a**

83. Offspring are individuals developed as a result of
(a) Vegetative multiplication (b) Asexual reproduction
(c) Sexual reproduction (d) All the above **Ans. c**

84. Checkerboard method of calculations was developed by
(a) Mendel (b) Bateson
(c) Punnet (d) Morgan **Ans. c**

85. An individual having similar unit factors of a character is
(a) Heterozygote (b) Homozygote
(c) Dominant (d) Recessive **Ans. b**

86. Punnet square is used to know
(a) Outcome of a cross (b) Probable result of a cross
(c) Types of gametes (d) Result of meiosis **Ans. b**

87. Allele is
(a) Segment of gene (b) Form of a gene
(c) Special kind of gene (d) A muton **Ans. b**

88. Repeated selfing produces
(a) Heterozygosity
(b) Homozygosity
(c) Homozygosity in some and heterozygosity in other traits
(d) Pure hybrids **Ans. b**

89. YyRR is crossed with yyRR. The progeny will be
(a) 1 YyRR : 1 yyRR (b) 3 YyRR : 1 yyRR
(c) 1 YyRR : 3 yyRR (d) YyRR only **Ans. a**

90. AaBb individual produces 2 million gametes. How many of them would carry both the recessive alleles (ab)
(a) 1.5 million (b) 1.0 million
(c) 0.5 million (d) 0.25 million **Ans. c**

91. A double homozygous yellow round plant of Pea is crossed with green wrinkled plant. The offspring shall be of
(a) One type (b) Two types
(c) Four types (d) Several types **Ans. a**

92. A dihybrid cross is made between YYrr and yyRR. In F_2 generation the ratio of parental to recombinant phenotype is
(a) 9 : 7 (b) 6 : 10
(c) 10 : 6 (d) 7 : 9 **Ans. b**

93. Percentage of pure breeding F_2 individuals of a monohybrid cross would be

(a) 75% (b) 50%

(c) 25% (d) 12.5% **Ans. b**

94. Double homozygous individuals in F_2 generation of a dihybrid cross would be

(a) 1/16 (b) 2/16

(c) 6/16 (d) 9/16 **Ans. b**

95. Ratio between completely homozygous dominant and homozygous recessive individuals of a dihybrid cross is

(a) 1 : 1 (b) 2 : 2

(c) 6 : 10 (d) 10 : 6 **Ans. a**

96. Ratio of parental and recombinant phenotypes in a dihybrid cross would be

(a) 8 : 8 (b) 6 : 10

(c) 10 : 6 (d) 9 : 7 **Ans. c**

97. Which mendelian principle will not operate if two genes under study are close together

(a) Paired unit factors (b) Dominance

(c) Segregation (d) Independent assortment

Ans. d

98. In a monohybrid cross the ratio of F_2 true breeding dominant and true breeding recessive would be

(a) 50 : 50 (b) 25 : 25

(c) 75 : 25 (d) 25 : 75 **Ans. b**

99. A single heterozygous yellow wrinkled seeded Pea plant shall produce gametes

(a) YR only (b) yr only

(c) Yr and yr (d) YR and yR **Ans. c**

100. A couple with curly haired husband and straight haired wife have all their children curly haired because

(a) Both are heterozygous

(b) Husband is homozygous and wife is heterozygous

(c) Husband is heterozygous while wife is homozygous

(d) Both are homozygous **Ans. d**

101. Genes P and Q are both required in dominant state for normal hearing. A deaf couple has all children with normal hearing. The probable genotype for the couple is

(a) PPqq x ppQQ (b) PPqq x PPqq

(c) PpQq x ppqq (d) PPqq x ppQq **Ans. a**

102. Name the scientist who converted Mendel's conclusions into principles of heredity

(a) De Vries (b) Tschermak-seysenegg

(c) Carl Correns (d) T.H. Morgan **Ans. c**

103. Pure tall pea plant can be differentiated from hybrid tall pea plant by

(a) Selfing and finding the progeny tall

(b) Test cross with dwarf plant and finding the progeny tall

(c) Selfing and test crossing to find the progeny of both tall and dwarf plants

(d) Both a and b **Ans. d**

104. F_1 plants crossed with dominant individuals will yield a progeny of

(a) All recessive

(b) All dominants

(c) Dominant and recessive in the ratio of 1 : 1

(d) Dominant and recessive in the ratio of 3 : 1 **Ans. b**

105. Round seed trait (R) is dominant over wrinkled (r) seed trait in Pea. Heterozygous round seeded plant (Rr) is crossed with wrinkled seeded plant (rr). What is the possibly progeny?

(a) 302 round : 102 wrinkled (b) 210 round : 95 wrinkled

(c) 105 round : 99 wrinkled (d) 103 round : 315 wrinkled

Ans. c

106. A dihybrid yellow round seeded plant (YyRr) is crossed to another yellow round seeded plant. The progeny is also yellow round seeded plant. What is the genotype of the second plant?

(a) YyRr (b) YYRR

(c) YyRR (d) yyRr **Ans. b**

107. A yellow round seeded pea plant is crossed with green wrinkled seeded pea plant. Four phenotypes appeared in the progeny in the ratio of 1 : 1 : 1 : 1. The genotype of the two are

(a) YyRr and yyrr (b) YYRR and yyrr
(c) YYRr and yyrr (d) YyRR and yyrr **Ans. a**

108. The ratio of 1 : 1 : 1 : 1 is obtained in case of

(a) Monohybrid cross (b) Monohybrid test cross
(c) Dihybrid cross (d) Dihybrid test cross **Ans. d**

109. Position of a gene on chromosome is called

(a) Locus (b) Factor
(c) Cistron (d) Nucleosome **Ans. a**

110. A recessive character in Pea is

(a) Red flower (b) Round seed
(c) Green cotyledons (d) Tall plant **Ans. c**

111. Mendel was lucky and could discover a law of heredity because he selected traits which

(a) Possessed linkage (b) Crossed independently
(c) Had complete dominance (d) Had incomplete dominance
Ans. b

112. Phenotype is influenced by

(a) Environment (b) Development
(c) Ageing (d) All the above **Ans. d**

113. Some gene loci on two homologous chromosomes produce different phenotypes. They bear

(a) Homozygous alleles (b) Heterozygous alleles
(c) Two different genes (d) Pleiotropic genes **Ans. b**

114. In a dihybrid test cross, the proportion of individuals showing recessive phenotypes would be

(a) 1/16 (b) 7/16
(c) 1/4 (d) 3/4 **Ans. d**

115. The number of genotypes produced by gametes Y and y would be

(a) 1 (b) 12
(c) 3 (d) 4 **Ans. c**

116. A phenotypic ratio not obtained by Mendel was

(a) 3 : 1 (b) 1 : 2 : 1

(c) 1 : 1 : 1 : 1 (d) 9 : 3 : 3 : 1 **Ans. b**

117. Pea plant with double hybrid yellow round seeds (YyRr) is crossed with plant having single hybrid green round seeds (yyRr). The progeny shall be

(a) 3 : 3 : 1 : 1 (b) 1 : 1 : 1 : 1

(c) 9 : 3 : 3 : 1 (d) 3 : 1 : 3 : 1 **Ans. d**

118. Trihybrid ratio is

(a) 27 : 9 : 9 : 9 : 3 : 3 : 3 : 1 (b) 27 : 9 : 9 : 6 : 6 : 3 : 3 : 1

(c) 1 : 6 : 15 : 20 : 15 : 6 : 1 (d) 36 : 6 : 6 : 6 : 3 : 3 : 3 : 1 **Ans. a**

119. Number of gamete types produced by genotype Aa Bb Cc Dd will be

(a) 4 (b) 8

(c) 16 (d) 32 **Ans. c**

120. First generation after a cross is

(a) First filial generation (b) F_1 generation

(c) Second filial generation (d) Both (a) and (b) **Ans. d**

121. F_2 generation is produced as a result of

(a) Crossing F_1 individuals with dominant individuals

(b) Crossing F_1 individuals with recessive individuals

(c) Crossing F_1 individuals amongst themselves

(d) All the above **Ans. c**

122. Which is incorrect in Mendelian characters?

	Character	Dominant	Recessive
(a)	Pod colour	Green	Yellow
(b)	Seed shape	Round	Wrinkled
(c)	Flower position	Terminal	Axillary
(d)	Shape of pod	Full	Constricted

Ans. c

123. F_2 generation of a dihybrid cross possesses one or both the dominant traits in proportion of

(a) 1/16 (b) 6/16

(c) 9/16 (d) 15/16 **Ans. d**

124. What is incorrect in the following Mendelian traits

(a) Height Tall Dwarf
(b) Seed colour Green Yellow
(c) Flower position Axillary Terminal
(d) Flower colour Violet White **Ans. b**

125. In Shepherd's Purse, the fruit shape is controlled by

(a) Supplementary genes (b) Complementary genes
(c) Duplicate genes (d) Polymeric genes **Ans. c**

126. A pure breeding triangular fruit bearing plant of shepherd's Purse is crossed with ovoid fruit bearing plant. The ratio of two types of plants in F_2 generation would be

(a) 3 : 1 (b) 1 : 2 : 1
(c) 13 : 3 (d) 15 : 1 **Ans. d**

127. 9 : 6 : 1 F_2 generation ratio is obtained in case of

(a) Polymeric genes (b) Pleiotropic genes
(c) Supplementrary genes (d) Recessive epistasis **Ans. a**

128. Two plants of Summer Squash both having circular fruits are crossed. F_1 plants had discoid fruits. F_2 generation has 3 types of fruits, discoid, circular and long in the ratio of

(a) 9 : 3 : 4 (b) 9 : 6 : 1
(c) 12 : 3 : 1 (d) 7 : 6 : 3 **Ans. b**

129. Recessive epistasis is shown by

(a) Flower colour in Sweat Pea (b) Fruit colour of Summer Singh
(c) Coat colour Mice (d) Shape of comb in poultry
Ans. c

130. The ratio of 2 : 1 is observed in case of

(a) Suppressor gene (b) Dominant-recessive epistasis
(c) Complementary gene (d) Lethal gene **Ans. d**

131. Mendel's law of segregation, as applied to the behavior of chromosome in meiosis, means that

(a) Paring of homologs will convert one allele into the other, leading to separation of the types
(b) Allels of a gene separate from each other when homologs separate in meiosis I

(c) Genes on the same chromosome will show 50% recombination

(d) Alleles of a gene will be linked and passed on together through meiosis **Ans. b**

132. The phenomenon of independent assortment refers to

(a) Expression at the same stage of development

(b) Unlinked transmission of genes in crosses resulting from being located on different chromosomes

(c) Association of an RNA and a protein implying related function

(d) Independent location of genes from each in an interphase cell

Ans. b

133. One of Mendel's pure strains of pea plants had green peas. How many different kinds of egg could such a plant produce with regard to pea colour?

(a) One (b) Two

(c) Three (d) Four **Ans. a**

134. List all the different gametes produced by the following individuals

(1) AA BB Cc (2) aa Bb Cc

(a) 1. ABC, ABc ; 2. aBC, aBc , abC , abc

(b) 1. AbC, ABc ; 2. ABC, aBc, abC, aBC

(c) 1. ABC, ABC ; 2. aBC, aBc, abc, aBC

(d) 1. ABC, ABc ; 2. aBC, ABc, abc, abC **Ans. a**

135. If a diploid cell contains six chromosomes, how many possible random arrangements of homologous could occur during Metaphasse-I?

(a) 4 (b) 8

(c) 6 (d) 64 **Ans. b**

136. Determine the frequency of only one genotype, BBLI ,in the offspring of dihybrid parents (BbLI).

(a) 1/4 (b) 1/8

(c) 1/16 (d) 1/2 **Ans. b**

137. If an individual of genotypes AaBbCcDd is test crossed, how many different phenotypes can appear in the progeny?

(a) 16 (b) 8

(c) 32 (d) 64 **Ans. a**

138. If individuals of genotypes AaBbCc are intercrossed, how many different phenotypes can appear their offspring?

(a) 16 (b) 8
(c) 32 (d) 64 **Ans. b**

139. If individual of genotypes AaBbCc are intercrossed, how many different genotypes can occur in their progeny?

(a) 8 (b) 64
(c) 27 (d) 9 **Ans. c**

140. If individual of genotypes AaBbCc are intercrossed, how many different F_2 phenotypes can appear assuming complete codominance at all loci?

(a) 8 (b) 64
(c) 27 (d) 9 **Ans. c**

141. A man is heterozygous B/b for one autosomal gene, and he carries a recessive X-linked allele d. What proportion of his sperm will be bd?

(a) 1/4 (b) 1/8
(c) 1/16 (d) 1/2 **Ans. a**

142. A trihybrid cross is made between two yeast, both with genotype A/a, B/b, C/c. What proportion of the offspring will be of genotype a/a, b/b, c/c?

(a) 1/8 (b) 1/32
(c) 1/64 (d) 1 /74 **Ans. c**

143. When heterozygous black pigs are intercrossed then what is the chance of the first two offspring being black?

(a) 46 (b) 1/32
(c) 1/64 (d) 1/74 **Ans. b**

144. Medical geneticists usually abbreviate the normal beta- globin gene as b, and the abnormal gene (in this case) as b^0 .Neither of your patient's parents has beta –thalassemia. Which of these described the most likely genotypes of both parents?

(a) One is b^0 and one in b (b) One is b^0 b^0 and one is bb
(c) Both are b^0b^0 (d) Both are bb^0 **Ans. d**

145. If a man of blood group AB marries a woman of blood group A whose father was of blood group O, to what different blood group can this man and woman expect their children to belong?

(a) A, AB, B (b) A, AB
(c) AB, O (d) A, O, B **Ans. a**

146. In the human ABO blood system, the alleles A and B are dominant to O, what will be the number of different possible genotype?

(a) 4 (b) 8

(c) 6 (d) 12 **Ans. c**

147. A brown mouse is mated with two female blackmice. When each female has produced several litters of young, the first female has had 48 black and the second female has 14 black and 11 brown young. Deduce the pattern of inheritance coat colour and the genotype of all the parents (B-Black ; b - brown)

(a) Male Bb; female 1, BB; female 2 Bb

(b) Male BB; female 1, BB; female 2 Bb

(c) Male Bb; female 1, BB; female 2 bb

(d) None of the above **Ans. d**

148. Three genes (P, Q and R) are found of three different chromosome .For the following diploid genotypes, describe all of the possible gamete combinations and their predicted ratios.

(P) AABbCC (Q) AaBBCc

(R) Aabbcc

(a) P. 1ABC: 1AbC Q. 1ABC: 1ABc: 1aBC: 1aBc R. 1Abc: 1abc

(b) P. 1ABC: 1Abc Q. 1abc : 1Abc: 1aBC: 1aBc R. 1abc: 1abc

(c) P. 1ABC: 1Abc Q. 1ABC: 1aBc: 1aBC: R. 1Abc: 1abc

(d) P. 1ABC: 1Abc Q. 1ABC: 1Abc: 1aBC: R. 1Abc: 1Abc

Ans. a

149. A true -breeding pea plant with round and green seeds is correct is crossed to a true -breeding plant with wrinkled and yellow seeds .Round and yellow seeds are the dominant traits .The F_1 plants are allowed to self - fertilize. What are the following probabilities for the F_2 generation?

(P) An F_2 plant with wrinkled, yellow seeds

(Q) An F_2 plant with round, yellow seeds

(a) P. 2/16 Q. 9/16

(b) P. 3/16 Q. 7/16

(c) P. 3/16 Q. 9/16

(d) Cannot be determined from the information **Ans. c**

150. In Guinea pig, black coat colour is a dominant trait and white is recessive trait. A black female is test crossed, producing six black offspring. The probability that heterozygous black would do this by chance alone is approximately

(a) 50%

(b) 25%

(c) 1%

(d) cannot be determined from the information **Ans. a**

151. What are the minimum progeny population sizes allowing for random union of all kinds of gametes from AaBbCc parent?

(a) 64 (b) 32

(c) 16 (d) 8 **Ans. a**

152. A red -flowered tall parent plant (P_1) was crossed to a true breeding red-flowered dwarf plant (P_2) and half of the progenies obtained was red and tall and the other half red and dwarf. In the next generation, half of all these progenies segregated only for flower colour and the other half segregated only for height . The genotype of the P_1 is

(a) Heterozygous for colour and heterozygous for height

(b) Homozygous for colour and heterozygous for height

(c) Heterozygous for colour and homozygous for height

(d) Homozygous for colour and homozygous for height **Ans. a**

153. According to classical genetics, Which of the following statements is true?

(a) Recessive alleles are detected by the phenotype of the F_1 generation

(b) The closer two genes are, the more frequently they recombine

(c) Genes on different autosomes segregate independently

(d) Gene on sex chromosomes segregate with the same pattern as autosomal genes **Ans. c**

154. Assuming the comparable chromosomes in different individuals are genetically dissimilar because of different alleles. How many unique zygotic combinations are possible following fertilization in an organism where n = 3 (Assuming that no crossing over occurs) ?

(a) 8 (b) 16

(c) 64 (d) 216 **Ans. c**

155. The colour of flowers of an annual species of plants is controlled by a single locus with two alleles R and r, and the genotypes RR, Rr and rr and red, pink and white respectively. A large number of seeds from individuals with pink flowers were collected and planted on an island. Where because of absence of pollinators, only self -pollination is possible. What will be the most likely outcome after 25 years?

(a) About 50% plants with red flowers 50 % with white flowers

(b) Almost 100% plants with pink flowers

(c) Red, pink, and white flowered plants in a ratio of 1:2:1

(d) Red, pink and white flowered plants in equal proportion **Ans. c**

156. Coat colour of dogs depend upon the action of at least two genes. At one locus a dominant epistatic inhibitor of colour pigment (A-) prevents the expression of colour at another independently assorting locus, producing white coat colour. When the recessive condition exists at the inhibitor locus (aa) the alleles of the hypostatic locus may be expressed, aaB- producing black and aabb producing brown. When dihybrid white dogs are together, determine the phenotypic proportions expected in the progeny.

(a) 9 : 7 (b) 9 : 3 : 4

(c) 12 : 3 : 1 (d) 13 : 3 **Ans. c**

157. Short hair in rabbits is governed by a dominant gene L and long hair by its recessive allele l black hair results from the action of the dominant genotype B and brown from the recessive genotype bb.

(a) In crosses between dihybrid short-haired, black and homozygous short-haired, brown rabbits, genotypic ratio will be 3 : 1 : 1 : 9

(b) In crosses between dihybrid short-haired, black and homozygous short-haired, brown rabbits, phenotypic ratio will be 1 : 1

(c) Expected phenotypic ratio from the cross LIBb x LIbb will be 6 : 6 : 3 : 1

(d) Expected genotypic ratio of LIBb will be ¾ **Ans. b**

158. In a cross between two individuals with the genotypes AaBbccDdEeFf and AaBbCCDDeeff, the probability that an offspring will be heterozygous at all these loci is

(a) 0 (b) 1/16

(c) 1/32 (d) 1/64 **Ans. c**

159. Two pink-flowered four-o'clocks are crossed to each other. Flower colour is incompletely dominant and giving phenotypic ratio if 1 red : 2 pink : 1 white. What are the following probabilities?

(P) The first three plants with white flower

(Q) A plant with either white or pink flower

(a) P. 1/64 Q. 3/4 (b) P.3/16 Q. 1/64

(c) P.3/64 Q.9/16 (d) P. 1/64 Q.1/64 **Ans. a**

160. Which of the following is a mismatch?

(a) Phenotypic F_2 ratio of 1 : 4 : 6 : 4 : 1-polygenic inheritance

(b) Phenotypic F_2 ratio of 9 : 3 : 4- recessive epistasis

(c) Phenotypic F_2 ratio of 1 : 2 : 1-codominance

(d) Phenotypic F_2 ratio of 3 : 1-partial dominance **Ans. d**

161. Why would you predict that half of the human babies born will be males and half will be females?

(a) Because of the segregation of the X and Y chromosomes during male meiosis

(b) Because of the segregation of the X chromosomes during female meiosis

(c) Because all eggs contain an X chromosome

(d) Because, on average, one- half of all eggs produce females **Ans. a**

162. Asymmetry between reciprocal crosses is seen in

(a) Pleiotropy (b) Sex-linked inheritance

(c) Interaction of genes (d) Autosomal inheritance **Ans. b**

163. Which of the following is not true autosomal dominant traits?

(a) Every affected person should have at least one affected parent

(b) Males and females should be equally often affected

(c) An affected person has a 50% chance of transmitting the dominant allele to each offspring

(d) All the daughters of an affected male will be affected but none of the sons **Ans. d**

164. Albinism is a recessive human trait. If a normal couple produces an albino child, what is the probability that their next child will be albino?

(a) 1/4 (b) 1/8

(c) 1/16 (d) 1/64 **Ans. a**

165. When a man with hypertrichosis marries a normal woman, what %of their sons would be expected to have hairy ears?

(a) 50% (b) 100%

(c) 0% (d) 25% **Ans. b**

166. A man with a certain disease marries a normal woman. They have 8 children (4 boys and 4 girls); all of the girls have their father's disease, but none of the boys do. What inheritance is suggested?

(a) Autosomal recessive (b) Autosomal dominant

(c) Y-linked (d) X-linked **Ans. d**

167. Cystic fibrosis is a hereditary disease that affects the respiratory and digestive systems. Cystic fibrosis occurs when two recessive genes (cc) are present. A person with one allele for Cystic fibrosis is called a carrier (Cc) of the disease. If the mother is a carrier of the disease and the father is homozygous dominant, what are chances that their child will be carrier of Cystic fibrosis?

(a) 25% (b) 50%

(c) 75% (d) 100% **Ans. b**

168. Which of the following is/are correct about sex - linked recessive inheritance?

(i) Most affected individuals are male

(ii) Affected females come from affected father and affected or carrier mothers

(iii) The sons of affected females should be affected

(iv) The trait does not skip generations

(a) i and ii (b) ii only

(c) i, ii, and iii (d) iv only **Ans. c**

169. Which of the following is/are not feature (s) of quantitative trait?

(a) Characters of degree (b) Continuous variation

(c) Polygenic control (d) Discontinuous variation

Ans. d

170. Which of the following are correct?

(a) Classical Mendelian traits are qualitative in nature

(b) Qualitative traits show continuous variations

(c) Qualitative traits are polygenic traits

(d) Qualitative traits are referred to as metric traits **Ans. a**

171. Wheat grain colour is controlled by four loci and following qualitative inheritance.

(P) Calculate the number of different F_2 phenotypes in a certain stock of wheat

(Q) Calculate the number of F_2 as extreme as one parent or the other

(a) P-9, Q-1/16 (b) P-16, Q-1/256
(c) P-8, Q-1/32 (d) P-9, Q-1/256 **Ans. d**

172. Two pure lines of corn have mean cob lengths of 9 and 3 inches, respectively. The polygenes involved in this trait all exhibit additive gene action. Crossing these two lines is expected to produce a progeny with mean cob length (in inches) of

(a) 12.0 (b) 7.5
(c) 6.0 (d) 2.75 **Ans. c**

173. In a plant, height varies from 6 to 36 cm. When 6 cm and 36 cm plants were crossed all F_1plants were 21 cm. In the F_2 generation, a continuous range of heights was observed. Most were around 21 cm, and 3 of 200 were 6 cm. Find out how many gene pairs are involved in this mode of inheritance?

(a) 3 (b) 4
(c) 2 (d) 5 **Ans. a**

174. If four chromosomes synapse into a cross-shaped configuration during meiotic prophase, the organism is heterozygous for a

(a) Pericentric inversion (b) Delection
(c) Translocation (d) Paracentric inversion **Ans. c**

175. Philadelphia chromosome is generated by translocation between

(a) Chromosome 18 and chromosome 6
(b) Chromosome 22 and chromosome 9
(c) Chromosome 22 and chromosome 3
(d) Chromosome 16 and chromosome 4 **Ans. b**

176. Retinoblastoma is caused by loss of both copies of the RB gene in the chromosome band

(a) 13p11 (b) 13q11
(c) 13q14 (d) 21q14 **Ans. c**

177. Position effect is the result of

(a) Mutations (b) Deletions
(c) Inversions (d) Transversions **Ans. c**

178. The bridge-fragment configuration at anaphase-1 is characteristic of
(a) Translocation heterozygote
(b) Paracentric inversion heterozygote
(c) Pericentric inversion heterozygote
(d) Duplication heterozygote **Ans. b**

179. A mechanism that can cause a gene to move from one linkage group to another is
(a) Translocation (b) Inversion
(c) Crossing over (d) Duplication **Ans. a**

180. Pseudodominance may be observed in heterozygotes for
(a) A deletion (b) A duplication
(c) A reciprocal translocation (d) More than one of the above
Ans. a

181. Which of the following chromosomal changes is usually the most damaging when in the homozygous condition?
(a) Deletion (b) Duplication
(c) Translocation (d) Inversion **Ans. a**

182. In a trisomic individual the number of chromosome is
(a) 2n-1 (b) 2n+2
(c) 2n+3 (d) 2n+1 **Ans. d**

183. A person with Klinefelter syndrome is considered a
(a) Monosomic (b) Triploid
(c) Trisomic (d) Delection heterozygote **Ans. c**

184. If the garden pea has 14 chromosomes in its diploid complement, how many double trisomics could theoretically exist?
(a) 6 (b) 9
(c) 16 (d) 21 **Ans. d**

185. The condition in which there is too many or one too few chromosomes is called
(a) Aneuploidy (b) Polytene
(c) Polyploidy (d) Monoploidy **Ans. a**

186. Most cases of Down syndrome are caused by the presence of a third copy chromosome 21 associated with the chromosome 21 pair. This genetic condition, known as trisomy 21, is caused by

(a) A frame-shift mutation (b) Chromosome nondisjunction

(c) Fragile X syndrome (d) Chromosome transloction

Ans. b

187. Which statement about Down's syndrome are correct?

(a) The frequency increases dramatically in mothers over the age of 40

(b) The cause is a non-disjunction when chromosomes do not separate during the first meiotic division

(c) Affected individuals have an extra autosome

(d) The long time lag between onset of meiosis in ovarian tissue and its completion (at ovulation) is most likely the reason for increased incidence in older mothers **Ans. all**

188. People with klinefelter syndrome have 47 chromosome, including three sex chromosome (XXY).What is the term to describe the aberration that occurs during meiosis that results in abnormal chromosome number

(a) Crossing over

(b) Non - disjunction

(c) Independent assortment

(d) Pairing of homologous chromosome **Ans. b**

189. The genes abcde are determined to be closely linked on the *E.coli* chromosome. Three random short deletions are created in the region, resulting in the removal of various genes are shown below. Which of the following gene orders is possible based on the deletion analysis?

Deletion 1: genes bde are lost

Deletion 2: genes ac are lost

Deletion 3: genes abd are lost

(a) abcde (b) acded

(c) bdeac (d) cabde **Ans. d**

190. Datura plants have been regenerated from anther cultures, endosperm culture and embryo culture. Their respective ploidy levels will be

(a) n,2n and 2n (b) n,3n and 2n

(c) n,2n and 3n (d) 2n, 2n and 2n **Ans. b**

191. Diagnosis of chromosome aneuploidy of unborn children is normally done by a combination of amniocentesis, cell culture, and

(a) Enzyme assay (b) RRLP analysis
(c) Pedigree analysis (d) Karyotyping **Ans. d**

192. The Hardy- Weinberg law describes

(a) Genotype frequencies of a population when evolutionary forces are not acting
(b) How sexual reproduction would change the relative gene frequencies in a population
(c) How mutations occur and balance each other
(d) Genotype frequencies of a population when evolutionary forces are acting **Ans. a**

193. What are the assumptions of Hardy-Weinberg equilibrium?

(a) Small population size, random mating, no selection, no migration, no mutation
(b) Large population size, random mating, no selection, no migration, no mutation
(c) Large population size, random mating, heterozygotes survive the best, no migration, no mutation
(d) Large population size, random mating, no migrants enter from other populations,no mutation **Ans. b**

194. Genetic diversity is required for natural selection to act, but natural selection can reduce or eliminate diversity. What process can restore genetic diversity to a population?

(a) Genetic drift (b) Mutation
(c) Sexual selection (d) Stabilizing selection **Ans. b**

195. Deviation from the Hardy-Weinberg assumption of infinitely large population size results in

(a) Genetic lethal (b) Hetrozygote advantage
(c) Consanguinity (d) Genetic drift **Ans. d**

196. If two population are merged, each with different frequencies of an allele at a locus, and randomly mating occurs immediately, how long will it take to achieve a hardy –Weinberg equilibrium in the new population ?

(a) One generation
(b) Ten generations

(c) Variable-depends on allele frequencies

(d) The population will never achieve equilibrium **Ans. a**

197. In a population that is in equilibrium, the proportion of individuals showing the domain trait at a given locus having two alleles is 84%. The frequency of the recessive allele in the population is

(a) 0.4 (b) 0.3

(c) 0.2 (d) 0.16 **Ans. a**

198. How many A and a alleles are present in a sample of organisms consisting of 10 AA, 15Aa and 4aa individuals? What are the allele frequencies in the sample?

(a) A =0.6; a=0.40 (b) A =0.4; a=0.60

(c) A =0.6; a=0.6 (d) A=0.4; a=0.4 **Ans. a**

199. In a diploid organism, what is the maximum number of alleles that can exist in a population for any given gene?

(a) 1 (b) 2

(c) 4 (d) Unlimited **Ans. d**

200. Cystic fibrosis is an autosomal recessive inherited disease. ITS frequency in newborns is 1 in 1700. One of the following indicates the frequency of the disease carriers

(a) 0.024 (b) 0.047

(c) 0.976 (d) 0.148 **Ans. b**

201. Which statement about allele frequencies in not true?

(a) The sum of any set of allele frequency is always 1

(b) If there are two alleles at a locus and we know the frequency of one of them, we can obtain the frequency of the other by subtraction

(c) If an allele is missing from a population, its frequency is 0

(d) If two populations have the same gene pool for a locus, they will have the same proportion of homozygotes at that locus **Ans. d**

202. Mutation generally produces

(a) Recessive genes (b) Lethal genes

(c) Polygenes (d) Dominant genes **Ans. a**

203. Linkage was first observed in............

(a) Field pea (b) Sweet pea

(c) Pea (d) Grass pea **Ans. b**

204. Which one of the following is not a mutagen?

(a) 5 bromouracil (b) Acetic acid

(c) Nitrous acid (d) Gamma radiation **Ans. b**

205. Down's syndrome is basically a..........

(a) Sex linked abnormally (b) Autosomal abnormality

(c) Neurological abnormality (d) Physiological abnormality

Ans. b

❑❑❑

Chapter-16

Signal Transduction

Fill in the blanks

1. Regardless of the nature of the signal, the target cell responds by means of a specific protein called a..............., which binds the signaling molecules and then initiates a responses in the target cell.

Ans. receptor

2. Signaling molecules that a cell secretes may acts as.............., affecting only cells in the immediate environment of the signaling cell, in a process called......................

Ans. local mediators, paracrine signaling

3. When a nerve impulse reached the nerve terminals at the end of the axon, it stimulates the terminals to secrete a chemical signal called a............., which is delivered rapidly to the postsynaptic target cell in a process termed......................

Ans. neurotransmitter, synaptic signaling

4. ....cells secrete their signaling molecules, called............., into the bloodstream, which carries the signal to target cells throughout the body.

Ans. Endocrine, hormones

5. Cells can send signals to other cells of the same type and to themselves in a process called.................

Ans. autocrine signaling

6. ...., which are derived from fatty acids (mainly arachidonic acid), are released to the cell exterior, where they influence cells in their immediate neighborhood.

Ans. Elcosanoids

7. ....are specialized cell-cell junctions that can form between closely apposed plasma membranes, directly connecting the cytoplasma of the joined cells.

Ans. Gap junctions

8. Acetylcholine indirectly causes smooth muscle cells in blood vessel walls to relax by inducing the epithelial cells to make and release the gas............., which then signals the smooth muscle cells to relax.

Ans. Nitric oxide (NO)

9. Steroid hormones, thyroid hormones, and retinoids diffuse across the plasma membrane and bind to specific receptors that are structurally related and constitute the..........

Ans. intracellular receptor superfamily

10. ...., hormones are all made from cholesterol.

Ans. Steroid

11. The...........hormones, which are made from the amino acid tyrosine, act to increase metabolism in a wide variety of cell types.

Ans. thyroid

12. The.............., which are made from vitamin A, play important roles as local mediators in vertebrate development.

Ans. retinoids

13. ....receptors are involved in rapid synaptic signaling between electrically excitable cells.

Ans. Ion-channel-linked

14. ....receptors, are indirectly to regulate the activity of a separate plasma-membrane-bound target protein, which can be an enzyme or an ion channel.

Ans. G-protein-linked

15. ....receptors, when activated, either function directly as enzymes or are associated with enzymes.

Ans. Enzyme-linked

16. ....functionally couple G-protein-linked receptors to their target enzymes or ion channels in the plasma membrane.

Ans. Trimeric GTP-binding proteins (G proteins)

17. ....is synthesized from ATP by a plasma-membrane-bound enzyme........., and it is rapidly and continuously destroyed by one or more.................

Ans. Cyclic AMP, adenylyl cyclase, cyclic AMP phosphodiesterase

18. A trimeric G protein that activates adenylyl cyclase is called a............

Ans. stimulatory G protein (G_5)

19. ...., which mediate some of the actions of adrenaline and noradrenaline, are coupled by G proteins to the activation of adenylyl cyclase.

Ans. β-adrenergic receptors

20. ...., which is responsible for the disease symptoms of cholera, is an enzyme that catalyzed the transfer of ADP ribose from intracellular NAD^+ to the α chain of a stimulatory G protein.

Ans. Cholera toxin

21. While the β-andrenergic receptors are functionally coupled to adenylyl cyclase by G_s, the α_2-adrenergic receptors are coupled to this enzyme by an................

Ans. inhibitory G protein (G_1)

22. Thimann isolated IAA from........

Ans. fungus

23. Flowers open due to.............

Ans. epinasty

24. Gibberellins were first isolated from a plant source i.e. immature seeds of.................

Ans. *Phaseolus cocienus*

25. High concentration of cytokinin in nutrient media induced.............in callus mass.

Ans. shoot buds

26. Hormone required for flower formation is..............

Ans. florigen

27. Florigen was discovered by..............

Ans. Cajlachjan

28. Flowering in some biennial plants can be induced by...................

Ans. Bolting

29. Mesobiotic are the seeds which can remain viable for................ days.

Ans. 3-15

30. Tomato seeds fail to germinate due to a germination inhibitor in the pulp of fruit. It is....................

Ans. ferulic acid

31. ....light is necessary for the germination of photoblastic seeds.

Ans. Red

32. ....hormone is break dormancy of potato tuber.

Ans. Gibberellin

33. Bolting does not require.............

Ans. cold nights

34. Cytokinin synthesis is maximum in..................

Ans. roots

35. Hormones responsible for ageing is..................

Ans. ABA

Multiple Choice Questions

1. Which plant will flower when the photoperiod is below a critical period?
   (a) Rice (b) Wheat
   (c) Oat (d) *Bryophyllum* **Ans. a**

2. The receptor in plants which detects photoperiod is a
   (a) Phospholipid (b) Protein
   (c) Carbohydrate (d) K^+ **Ans. b**

3. Certain flowers open during day and closing at night. These sleep movements are due to
   (a) Autonomic movement of growth
   (b) Paratonic movement of growth
   (c) Autonomic movement of variation
   (d) Paratonic movement of variation **Ans. d**

4. Nodulation and mycorrhizal formation is influenced by
   (a) Ethylene (b) Phosphon-D
   (c) Riboflavin and violaxanthin (d) Cytokinins **Ans. d**

5. Lodging is prevented by the use of
   (a) NAA (b) NAAM
   (c) 2, 4, 5-T (d) IBA **Ans. b**

6. Ethaphon, a commercial compound is used for
   (a) Increase malt production
   (b) Increase size of apple
   (c) Promote flowering in pineapple
   (d) Ripening of fruits **Ans. d**

7. ABA promotes flowering in
   (a) SDP (b) LDP
   (c) SLDP (d) LSDP **Ans. a**

8. When P_{FR} of phytochrome accumulates in plants, it is inhibitory for flowering in
   (a) *Xanthium* (b) Maize
   (c) *Bryophyllum* (d) Carrot **Ans. a**

9. 'Agent orange' was used by USA in Vietnam War for defoliation of forests. It was a mixture of

(a) Auxins and cytokinins (b) Auxins and gibberellins

(c) Two types of gibberellins (d) Different type of auxins **Ans. d**

10. Thimann isolated IAA from

(a) Corn germ oil (b) Human urine

(b) Mint leaf (d) Fungus **Ans. d**

11. In storage sprouting of potato can be prevented by

(a) Zeatin (b) TIBA

(c) MH (d) CCC **Ans. c**

12. Excurrent habit of Pinus is due to

(a) Apical dominance

(b) Acropetal movement of auxins

(c) Non polar movement of auxins

(d) High C/A ratio **Ans. a**

13. Which hormone stimulates growth of root hair in plants?

(a) Dormin (b) Ethylene

(c) 2, 4-D (d) CK **Ans. c**

14. Chemotactic movement is shown by

(a) Male gametes of Bryophyte

(b) Opening of flower

(c) Growth of pollen tube towards female gamete

(d) Both a and c **Ans. a**

15. Opening of flower, bud and leaves is mostly because of differential growth of their surface. This type of movement is called

(a) Hyponasty (b) Epinasty

(c) Thigmonasty (d) Seismonasty **Ans. b**

16. Protoplasm of staminal hair of *Tradescantia* shows

(a) Amoeboid movement (b) Rotation

(c) Circulation (d) Paratonic movement **Ans. c**

17. The minimum time taken by a plant to capture a geotropic stimulus while rotating on the clinostat is called

(a) Relaxation time (b) Perception time

(c) Presentation time (d) Reaction time **Ans. b**

18. According to Melchers, the site of vernalisation in biennials and perennials is

(a) Root apex (b) Shoot apex

(c) Seed stage (d) Embryo **Ans. b**

19. Which of the following fungus secretes ABA?

(a) *Albuga candida* (b) *Alternaria solani*

(c) *Cercospora personaia* (d) *Cercospora resicola* **Ans. d**

20. TIBA is

(a) Synthetic auxin (b) Natural auxin

(c) Antiauxin (d) Antigibberellin **Ans. c**

21. By exogenous application of which hormone biennials can be induced to behave as annuals?

(a) Auxins (b) Gibberellins

(c) Cytokinins (d) Ethylene **Ans. b**

22. Which kind of geotropic behavior is shown by lateral branches of stem?

(a) +ve orthogeotropic (b) –ve orthogeotropic

(c) Plageotropic (d) Apogeotropic **Ans. c**

23. According to Statolith theory the gravitational stimulus is perceived by

(a) Auxins (b) Large starch grains

(c) Small starch grains (d) CK **Ans. b**

24. Vernalisation stimulus perceived by

(a) Meristems (b) Embryos

(c) Leaves (d) More than one is correct

Ans. d

25. Glycine max is a

(a) SDP (b) LDP

(b) Day neutral plant (d) SLDP **Ans. d**

26. Match the following

A.	Indole –3- acetic acid	i.	Antigibberellin
B.	6-furfuryl amino purine	ii.	Natural auxin
C.	NPA	iii.	Synthetic gibberellin
D.	AMO – 1618	iv.	Antiauxin
		v.	Synthetic cytokinin

(a) A-v, B-iv, C-ii, D-iii (b) A-ii, B-v, C-iv, D-i
(c) A-ii, B-v, C-i, D-iv (d) A-ii, B-i, C-iii, D-v **Ans. b**

27. Bud dormancy can be broken by
(a) Alternating low and high temperature
(b) Mechanical scarification
(c) Chemical scarification
(d) Light **Ans. a**

28. Match the following

A.	GA + Anthesins	i.	True florigen
B.	2, 4-D + 2, 4, 5-T	ii.	Agent orange
C.	GA + CK	iii.	Pomalin
D.	ABA + Kinetin	iv.	Germination controller

(a) A-ii, B-iii, C-i, D-iv (b) A-i, B-ii, C-iiii, D-iv
(c) A-iv, B-iii, C-ii, D-i (d) A-iv, B-i, C-iii, D-ii **Ans. b**

29. Match the following

A.	Ethylene	i.	Pigment
B.	Cytokinin	ii.	Growth inhibitor
C.	Phytochrome	iii.	Flowering hormone
D.	Florigen	iv.	Phytohormone

(a) A-iv, B-iii, C-ii, D-i (b) A-iii, B-iv, C-ii, D-i
(c) A-ii, B-iv, C-i, D-iii (d) A-ii, B-i, C-iii, D-iv **Ans. c**

30. Flowers open due to
(a) Nutational movements (b) Epinasty
(c) Seismonasty (d) Increase in Turgor pressure
Ans. b

31. A biennial plant can be made annual be treating the plant with a plant hormone called
(a) Auxin (b) Zeatin
(c) 6 FAP (d) GA **Ans. d**

32. A long day plant can flower under short day condition if it is
(a) Allowed to posses a certain number of leaves
(b) Allowed to attain a particular height

(c) Treated with florigen

(d) Treated with GA **Ans. d**

33. Winter wheat is a/an

(a) Annual plant (b) Polycarpic plant

(c) C_4 plant (d) Biennial plant **Ans. d**

34. Gibberellins were first isolated from a plant source i.e. immature seeds of............by

(a) *Gibberella fujikuroi* (b) *Phaseolus cocienus*

(c) *Avena sativa* (d) *Zea mays* **Ans. b**

35. Phenomenon of phototropism is verified by

(a) Korper kappe theory (b) Went cholodny theory

(c) Avena curvature test (d) Apical cell theory **Ans. b**

36. Triple response is bioassay for

(a) Zeatin (b) Ethylene

(c) Auxin (d) Abscisic acid **Ans. b**

37. Richmond Land effect is associated with the growth hormone called

(a) Cytokinins (b) Gibberellic acid

(c) Ethylene (d) Indole 3-acetic acid **Ans. a**

38. Opening of flowers of Crocus and Tulip is an example of

(a) Nystinastic movements (b) Thermonastic movement

(c) Photonastic movement (d) Haptonastic movement

Ans. b

39. Day neutral plant among following is

(a) *Cestrum nocturum* (b) *Secale cereale*

(c) *Cucumis sativaus* (d) *Xanthium pensylvanicum*

Ans. c

40. Critical photoperiod for Henbane plant is

(a) 15.45 hrs (b) 11 hrs

(c) 10 hrs (d) 12 hrs **Ans. b**

41. On which of the following plants no work has been done in plant physiology

(a) *Phalaris canariensi* (b) *Avena sativa*

(c) *Saccharum officinarum* (d) *Solanum tuberosum* **Ans. d**

42. Pick the odd one out

(a) NAA (b) 2, 4-D

(c) TIBA (d) PAA **Ans. c**

43. Ethylene was recognized as growth regulator by

(a) Addicot and Lynch (b) Gallston and Davis

(c) Slatyr and Taylor (d) Skoog and Miller **Ans. b**

44. High concentration of cytokinin in nutrient media induced............in callus mass

(a) Adventitious root (b) Shoot buds

(c) Gall formation (d) Tumour formation **Ans. b**

45. Phototropism phenomenon can be demonstrated using

(a) Clinostat (b) Luxmeter

(c) Heliotropic chamber (d) Photocalorimeter **Ans. c**

46. Rotation and circulation are a kind of

(a) Autonomic movements (b) Diurnal rhythm

(c) Paratonic movements (d) Circadian rhythm **Ans. a**

47. It is more appropriate to use the term........for sensitive plant's droping of leaves

(a) Thigmotropism (b) Haptonastic

(c) Hyponasty (d) Seismotropic **Ans. b**

48. Most pronounced avena curvature was observed under the effect of

(a) Auxanolonic acid (b) Auxintriolic acid

(b) Heteroauxin (d) None of these **Ans. c**

49. Opening of flowers is called

(a) Anthology (b) Anthesis

(c) Horticulture (d) Olericulture **Ans. b**

50. Graviception is a property exhibited by

(a) Sand crystals (b) Statoliths

(c) Druses (d) None of these **Ans. b**

51. Flowering in some biennial plants can be induced by

(a) Etiolation (b) Bolting

(c) Cauliflory (d) Pruning **Ans. b**

52. Hormone required for flower formation is

(a) Vernalin (b) Anthesin

(c) Florigen (d) Phytochrome **Ans. b**

53. Traumatropism occur in......of plant

(a) Growing regions (b) Scenescing regions

(c) Wounded regions (d) Zone of maturation **Ans. c**

54. Which of the following is bioassay for ethylene?

(a) Pea curvature test (b) Triple pea test

(c) Chlorophyll retention test (d) Barley endosperm test **Ans. b**

55. Wound hormone is

(a) Jasmonic acid (b) Traumatic acid

(c) Morphactins (d) Lunularic acid **Ans. b**

56. Why is chilling treatment necessary to accelerate flowering in biennial plants?

(a) To produce florigen for flowering

(b) To defunct inhibitors which inhibits flowering

(c) To activate meristematic activity

(d) To activate enzymes to promote flowering **Ans. b**

57. Florigen was discovered by

(a) Cajlachjan (b) Butler

(c) Melchers (d) Garner **Ans. a**

58. Prosopis is known for being

(a) A phreatophyte

(b) A drought enduring xerophytes

(c) A drought resisting xerophytes

(d) Both (a) and (b) **Ans. d**

59. Which of the following, added to a medium will support not only growth of tobacco callus but also induce the formation of roots on it?

(a) Kinetin ((b) 0 ppm)

(b) Indole 3-acetic acid ((b) 0 ppm)

(c) Kinetin ((b) 0 ppm) + IAA (0.2 ppm)

(d) Kinetin (0.2 ppm) + IAA ((b) 0 ppm) **Ans. a**

60. What does a higher carbohydrate/nitrogen ratio do in plant?
(a) Less vegetative growth
(b) Inhibits flowering
(c) Promotes flowering and fruiting
(c) Both (a) and (c) **Ans. d**

61. Select the incorrect match
(a) Auxin - Stimulates sprouting of potato
(b) Gibberellin - Promotes flowering in LDP
(c) Ethylene - Promotes lateral growth
(d) Cytokinin - Prevents senescence **Ans. a**

62. The movement of the peduncle of sunflower during daytime is an example of
(a) Phototaxis (b) Photonasty
(c) Phototropism (d) Both (b) and (c) **Ans. c**

63. Growth of plants differs from growth of animals in being
(a) Localized (b) Diffused
(c) Localized and diffused (d) Limited **Ans. a**

64. Distribution of growth in a root by marking it with India in at equal intervals was first studied by
(a) Sachs (b) Strasburger
(c) Went (d) Paal **Ans. b**

65. The establishment of localized qualitative differences in biochemicals, metabolic processes and structural organization in cells is known as
(a) Growth (b) Log phase of growth
(c) Differentiation (d) Development **Ans. c**

66. The growing zones of root and shoot are respectively
(a) 5 cm and 5-30 cm (b) 4.5 cm and 4-20 cm
(c) 2 cm and 3-18.5 cm (d) 1 cm and 2-30 cm **Ans. d**

67. Auxanometer is required for
(a) Studying rate of transpiration
(b) Measuring rate of respiration
(c) Finding out rate of photosynthesis
(d) Calculating rate of growth in length **Ans. d**

68. The phase of growth where new cells are formed by the division of meristematic cells is known as

(a) Formative phase (b) Phase of elongation
(c) Phase of differentiation (d) Phase of amitosis **Ans. a**

69. The rate of respiration of cell in formative phase is

(a) Higher (b) Lower
(c) Equal to differentiating cells (d) Equal to elongating cells **Ans. a**

70. The classical experiments in growth were performed by

(a) Darwin and Lamarck (b) Lamarck and Boysen-Jensen
(c) Boysen-Jensen and Darwin (d) due Vries and Paal **Ans. c**

71. Enlargement of cells in phase of elongation is in......direction

(a) Linear (b) Longitudinal
(c) Along volume (d) May be in any direction **Ans. a**

72. Cell enlargement is due to

(a) Florigen (b) Traumatin
(c) Gibberellin (d) Auxin **Ans. d**

73. The phase of growth where synthesis of more of protoplasm, formation of a central vacuole and addition and extension of wall material takes place is known as

(a) Phase of cell formation (b) Phase of elongation
(c) Phase of differentiation (d) Primary growth **Ans. b**

74. The differentiation which continues through all phases of growth is known as

(a) Structural differentiation (b) Physiological differentiation
(c) Growth differentiation (d) Total differentiation **Ans. b**

75. The differentiation which is restricted to the later stages of growth of a cell is known as

(a) Physiological differentiation (b) Structural differentiation
(c) Later differentiation (d) Post differentiation **Ans. b**

76. The respiratory activity in "cell maturation" phase as compared to other phases is

(a) Equal (b) Less
(c) More (d) Dependent upon temperature **Ans. b**

77. Maximum growth in root occurs
(a) At its apex (b) In the root hair zone
(c) Behind the apex (d) In the presence of soil **Ans. c**

78. When growth is going on at a slow rate, this phase is called as
(a) Lag period (b) Log period
(c) Period of diminishing growth (d) Linear phase of growth **Ans. a**

79. In plant, optimum growth occur in
(a) Blue light (b) Red light
(c) White light (d) Green light **Ans. c**

80. Maximum rate of growth is achieved during
(a) Lag period (b) Log period
(c) Steady state (d) Senescent phase **Ans. b**

81. Cell differentiation is accompanied during period of
(a) Lag period (b) Log period
(c) Diminishing growth (d) Senescence **Ans. c**

82. The period of steady state is never reached in the organs of
(a) Determinate structure (b) Indeterminate structure
(c) Reproductive structure (d) All of the above **Ans. b**

83. When the rate of maximum growth is maintained for sometime, it is known as
(a) J-shaped phase of growth (b) Linear phase of growth
(c) S-shaped phase of growth (d) All of the above **Ans. b**

84. The total period of growth is termed as
(a) Total period of growth (b) Grand period of growth
(c) Graceful period of growth (d) Magnified period of growth
Ans. b

85. Who studied the growth systematically for the first time?
(a) Krebs (b) Sachs
(c) Lysenko (d) Blackman **Ans. b**

86. If total growth is plotted against time the curve obtained is called
(a) S-shaped curve (b) G-shaped curve
(c) T-shaped curve (d) U-shaped curve **Ans. a**

87. Double sigmoid curve is found in the growth of

(a) Fruits (b) Flowers

(c) Inflorescence (d) In all organs showing growth

Ans. a

88. In plant, light is required for

(a) Photosynthesis

(b) Controlling distribution of hormones

(c) Controlling growth and movements

(d) All of the above **Ans. d**

89. Maximum growth rate recorded in the stripe equivalent to 5 mm/min has been recorded in

(a) *Dictyota dichotoma* (b) *Dendrocalamus giganteous*

(c) *Dicanthium* (d) *Dictyophora phalloidea*

Ans. d

90. Two seeds are germinated, A in light and B in dark. Which grows taller in initial stage?

(a) A>B (b) B>A

(c) A=B (d) No growth in either case

Ans.b

91. Plastids of etiolated plant possess

(a) Phycobilins (b) Carotenoids and xanthophylls

(c) Chlorophylls and carotenoids (d) Chlorophyll and carotenes

Ans. c

92. According to Blackman (1919) the magnitude of relative growth has been named

(a) Efficiency index (b) Growth index

(c) Grand period of growth (d) Phase of differentiation **Ans. a**

93. Mobilization of stored food in germinating seed is triggered by

(a) IAA (b) GA

(c) MH (d) Florigen **Ans. b**

94. Viability of seed can be tested by

(a) Tetrazolium dye (b) Germination test

(c) Both (a) and (b) (d) None of these **Ans. c**

95. A condition when seed fails to germinate because of absence of favourable environmental conditions, is called

(a) Dormancy (b) Quiescence
(c) Viability (d) Abscission **Ans. b**

96. The seeds which are not influenced by the presence or absence of light are called

(a) Photoblastic (b) Nonphotoblastic
(c) Light hard seeds (d) Light sensitive seeds **Ans. b**

97. The seeds which are affected by the presence of light at the time of germination are called

(a) Non photoblastic (b) Photoblastic
(c) Light hard seeds (d) Light sensitive seeds **Ans. b**

98. The seeds of tomato can not germinate in the presence of light and hence are called

(a) Negatively photoblastic (b) Non photoblastic
(c) Photoblastic (d) Light sensitive seeds **Ans. a**

99. Microbiotic are the seeds which can remain viable for

(a) Less than a month (b) Less than a week
(c) Less than a year (d) Less than three years **Ans. d**

100. Mesobiotic are the seeds which can remain viable for

(a) A month (b) A year
(c) Two years (d) 3-15 years **Ans. d**

101. The seeds which can remain viable for more than 15 years are called

(a) Microbiotic (b) Mesobiotic
(c) Macrobiotic (d) None of these **Ans. c**

102. The germination where hypocotyl grows actively bringing the seed above the soil is called

(a) Epigeal (b) Hypogeal
(c) Semi epigeal (d) Semi hypogeal **Ans. a**

103. The germination where epicotyl elongates and becomes curved keeping the cotyledons underground is called

(a) Epigeal (b) Hypogeal
(c) Semi epigeal (d) Semi hypogeal **Ans. b**

104. The process when the embryo of the seed continues growth while the latter is attached to the parent plant is called
(a) Epigeal germination (b) Hypogeal germination
(c) Vivipary (d) None of these **Ans. c**

105. Vivipary is known to be present in
(a) Rhizophora (b) Grasses
(c) Bamboos (b) Hydrophytes **Ans. a**

106. Evergreen trees remain green throughout the year on account of
(a) Absence of leaf fall
(b) Leaves falling in small number at intervals
(c) Supply of moisture throughout the year
(d) Cold climate **Ans. b**

107. Correct sequence of cellular growth stage is
(a) Division-Differentiation – Elongation
(b) Division-elongation-differentiation
(c) Differentiation-division-elongation
(d) Elongation-differentiation-division **Ans. a**

108. A sigmoid curve is obtained for the growth of
(a) An organ (b) An organism
(c) A population (d) All of these **Ans. d**

109. Excess amount of CO_2
(a) Retards growth (b) Accelerates growth
(c) Affects growth slowly (d) Does not affect growth **Ans. a**

110. Legume seeds exhibit dormancy due to
(a) Undeveloped embryo (b) Absence of cytokinins
(c) Hard seed coat (d) Absence of GA_3 **Ans. c**

111. A rhythmic phenomenon that occur every 24 hours is a
(a) Circadian rhythm (b) Biological clock
(c) Photoperiodism (d) Daily phase cycle **Ans. a**

112. If the tip of seedling is cut off, growth as well as bending ceases because it hampers
(a) Respiration (b) Photosynthesis
(c) Perception of light stimulus (d) Transpiration **Ans. c**

113. What does a higher C/N ratio in a plant?

(a) Improves vegetative growth (b) Inhibits flowering and fruiting
(c) Promote flowering (d) Both (a) and (b) **Ans. b**

114. In an old leaf, preparing to fall, the colour change is due to

(a) Conversion of chloroplasts into leucoplasts
(b) Chromoplasts being converted into chloroplast
(c) Chloroplast being converted into chromoplasts
(d) Chromoplasts is converted into elioplast **Ans. c**

115. During leaf fall, the conversion of chlorophyll into chromoplasts is due to

(a) Degradation of carotenoids
(b) Synthesis of carotenoids
(c) Degradation of chlorophyll and dominance of carotenoids
(d) Predominance of prochlorophyll over carotenoids **Ans. c**

116. The direct and dominating factor of growth is

(a) Light (b) Temperature
(c) Soil (d) Wind **Ans. a**

117. Crescograph, an instrument devised by J.C. Bose, was used to

(a) Investigate sleep movement of bean leaves
(b) Study the role of phytochrome in seed germination
(c) Record the progress of plant growth accurately
(d) Distinguish between SDP and LDP **Ans. c**

118. Tomato seeds fail to germinate due to a germination inhibitor in the pulp of fruit. It is

(a) Abscisic acid (b) Ferulic acid
(c) Coumarin (d) Thiourea **Ans. b**

119. Plant growth is

(a) Limited (b) Unlimited
(c) Unlocalized (d) Diffused **Ans. b**

120. Phytotron is device by which

(a) Electrons are bombarded
(b) Protons are liberated

(c) Plants are grown in controlled environment
(d) Mutations are produced in plants **Ans. c**

121. A green plant will turn yellow if placed in the dark. This is called
(a) Phototropism (b) Dormancy
(c) Chlorosis (d) Etiolation **Ans. d**

122. The law of compound interest for growth does not justify one of the following factors, that
(a) The growth is dependent upon the initial size
(b) The growth is dependent upon the rate of growth
(c) The growth is dependent upon the time interval
(d) The growth is dependent upon the location of meristem **Ans. d**

123. By which method, testa is made permeable to water
(a) Stratification (b) Denudation
(c) Vernalisation (d) Scarification **Ans. d**

124. The activity of α-amylase in the endosperm of barley germinating seeds is induced by
(a) Ethylene (b) Cytokinin
(c) IAA (d) Gibberellin **Ans. d**

125. The usual shape of growth curve is
(a) Linear (b) Inverted bell shaped
(c) Sigmoidal (d) Zig-zag **Ans. c**

126. Which light is necessary for the germination of photoblastic seeds?
(a) Red (b) Green
(c) Far red (d) Blue **Ans. a**

127. The hormone responsible for apical dominance is
(a) IAA (b) GA
(c) ABA (d) Florigen **Ans. a**

128. If a growing plant is decapitated, then
(a) Axillary buds are inactivated
(b) Axillary buds are activated
(c) Leaves become yellow and fall off
(d) Growth stops **Ans. b**

129. Growth substance that stimulates nodule formation in leguminous plant is
(a) NAA (b) IAA
(c) IBA (d) ABA **Ans. b**

130. Avena curvature test was first of all done by
(a) J.C. Bose (b) F.W. Went
(c) Thimann (d) Skoog **Ans. b**

131. The phenomenon of apical dominance can be overcome by exogenous application of
(a) Auxins (b) Gibberellins
(c) Cytokinins (d) Ethylene **Ans. c**

132. Bakanae disease is associated with discovery of
(a) 2, 4-D (b) IAA
(c) GA (d) ABA **Ans. c**

133. Bolting does not require
(a) Short days
(b) Long days
(c) Internode elongation in rosette plants
(d) Cold nights **Ans. a**

134. Gibberellin was first extracted from a species of a fungus
(a) *Fusarium* (b) *Gelidim*
(c) *Gracelaria* (d) *Aspergillus* **Ans. a**

135. Genetic dwarfness can be controlled by treating plants with
(a) Cytokinin (b) Gibberellic acid
(c) Auxin (d) Antigibberellin **Ans. b**

136. Hormone that break dormancy of potato tuber is
(a) IAA (b) ABA
(c) Zeatin (d) Gibberellin **Ans. d**

137. Hormone primarily connected with cell division is
(a) IAA (b) NAA
(c) Cytokine/zeatin (d) Gibberellic acid **Ans. c**

138. Which of the following is a cytokinin
(a) Phytochrome (b) Leucine
(c) Ethylene (d) Zeatin **Ans. d**

139. Cytokinin synthesis is maximum in
(a) Roots (b) Leaves
(c) Shoot tip (d) Fruit **Ans. a**

140. The gaseous fruit ripening phytohormone is
(a) Ethylene (b) Kinetin
(c) GA (d) ABA **Ans. a**

141. Which is not a phytohormone?
(a) IAA (b) GA
(c) Phytochrome (d) Florigen **Ans. c**

142. Ethylene gas
(a) Is saturated hydrocarbon (b) Slow down ripening of apples
(c) Retards ripening of tomatoes (d) Speed up maturation of fruits **Ans. d**

143. Hormones responsible for ageing is
(a) GA (b) IAA
(c) ABA (d) Cytokinin **Ans. c**

144. Which is a stress hormone preventing seeds during drought?
(a) Benzyl aminopurine (b) Dichlorophenoxy acetic acid
(c) Ethylene (d) Abscisic acid **Ans. d**

145. Hormones differs from enzymes in that they are
(a) Found in living and dead cells both
(b) Found only in plants
(c) Used up in the metabolism
(d) Not used up in the process **Ans. c**

146. The existence of growth regulating chemicals in plants was first suggested by
(a) Darwin (b) Went
(b) Sachs (d) Paal **Ans. a**

147. A hormone used to induce root formation in horticulture
(a) IBA (b) NAA
(c) ABA (d) 2, 4-D **Ans. a**

148. Parthenocarpic fruits can be produced by the application of which of the following auxin
(a) IAA (b) IBA
(c) 2, 4-D (d) All of the above **Ans. b**

149. Removal of apical bud makes
- (a) Plant bushy
- (b) Available more auxins to lateral buds
- (c) Formation of lateral branches
- (d) All of the above **Ans. d**

150. Substance formed at one place and express their effect at a distant place are called as
- (a) Hormone (b) Enzyme
- (c) Pheromone (d) Vitamin **Ans. a**

151. High concentration of auxin is found in
- (a) Root apex (b) Shoot apex
- (c) Petiole base (d) Node **Ans. b**

152. Differentiation of shoot is controlled by
- (a) High gibberellin : auxin ratio
- (b) High gibberellin : cytokinin ratio
- (c) High auxin : cytokinin ratio
- (d) High cytokinin : auxin ratio **Ans. d**

153. Seedless watermelons are obtained by
- (a) Triploidy (b) GA treatment
- (b) Haploidy (d) Vegetative propagation **Ans. b**

154. Which one prevents premature fall of fruits?
- (a) NAA (b) Ethylene
- (c) GA_3 (d) Zeatin **Ans. a**

155. Which of the following type of phytohormones resemble the nucleic acids in some structural aspects
- (a) Cytokinins (b) Auxins
- (c) Gibberellins (d) Abscisic acid **Ans. a**

156. A plant hormone used for inducing morphogenesis in tissue culture is
- (a) Abscisic acid (b) Gibberellins
- (c) Cytokinins (d) Ethylene **Ans. c**

157. Male hormone in plant is
- (a) IAA (b) GA
- (c) CK (d) Ethylene **Ans. b**

158. Enzyme diastase during seed germination converts

(a) Starch → glycogen

(b) Starch → simple sugars

(c) Starch → maltose

(d) Reserve food into sugars and fats **Ans. b**

159. Which combination is used in fruit ripening?

(a) 80% CH_4+20% O_2 (b) 80% C_2H_4+20% CO_2

(c) 80% CO_2+20% CH_4 (d) 80% C_2H_4+20% O_2 **Ans. b**

160. Which of the following cannot be isolated from plants?

(a) Vitamin B_{12} (b) Vitamin C

(c) Niacin (d) Riboflavins **Ans. a**

161. Pickup the correct explanation

(a) Xanthium – Long day plant (b) Sunflower – short day plant

(c) Wheat – short day plant (d) Tomato – day neutral plant **Ans. d**

162. The pigment that absorbs red and far red light in plants is

(a) Cytochrome (b) Xanthophylls

(c) Phytochrome (d) Carotene **Ans. c**

163. Which one prevents photo-oxidation and pigment destruction?

(a) Phytochrome (b) Phytohormone

(c) Phycocyanin (d) Phycoerythrin **Ans. a**

164. Phytochrome becomes biologically active in

(a) Green light (b) Blue light

(c) Red light (d) None of the above **Ans. c**

165. If a tree flowers thrice in a year in October, January and July in Northern India, the plant is

(a) Photo- and thermo-sensitive

(b) Photo- and thermo-insensitive

(c) Photo-sensitive and thermo-insensitive

(d) Thermo-sensitive and photo-insensitive **Ans. b**

166. Flash of light in dark inhibits flowering in

(a) SDP (b) LDP

(c) LSDP (d) DNP **Ans. a**

167. "Skototropic movements" are those, induced by

(a) Light (b) Touch

(c) Night (d) Heat **Ans. c**

168. Peristome teeth of moss show

(a) Hydochasy (b) Xerochasy

(c) Hydrotropism (d) Chemotropism **Ans. b**

169. Tendrils and climbers show which type of movement?

(a) Taxis (b) Tropism

(c) Nasty (d) Epitaxis **Ans. b**

170. The response of plants of light was first of all discovered by

(a) Lamarck (b) Darwin

(c) Thimann (d) Went **Ans. b**

171. Geotropic response is perceived by

(a) Mature roots

(b) Root cap

(c) Root hairs

(d) Elongating cells of quiescent centre **Ans. b**

172. Glyconcogenesis is regulated by

(a) Cartisol (b) GH

(c) Adrenalin and Glycogen (d) All of the above **Ans. d**

173. Glycogenolysis is the process of conversion of

(a) Glycogen to glucose

(b) Glucose to glycogen

(c) Glycogen to glucose 6-phosphate

(d) None of these **Ans. c**

174. α-adrenergic signals of hypopthalmic physical unit-supresses

(a) PSH (b) LH

(c) PRL (d) GH release **Ans. c**

175. Parathyroid gland act through β-adrenergic receptors by medicine

(c) cGMP (b) Ca^{+}

(c) cAMP (d) Na-K ATPase **Ans. c**

176. Which of the following is/are tropic hormone(s)?

(a) LH & FSH (b) TSH, ACTH, LH

(c) TSH, ACTH, FSH, LH (d) All of the above **Ans. c**

177. Gr RH induces the synthesis and secretion of

(a) FSH only (b) LH only

(c) PRL only (d) FSH and LH **Ans. c**

178. ACTH under the influence of corticotrophin releasing hormone or stimulate the synthesis of

(a) Cortisol (b) Glucocorticoids

(c) Minerallogcorticoids (d) Cartisol and Androgens **Ans. d**

179. The feedback regulation is found for which of the following hormones?

(a) TSH (b) ACTS

(c) FSH and LH (d) All of the above **Ans. d**

180. What is the sequence of estrogens synthesis?

(a) Acetate → Cholesterol → Pregnenolone → Progesterone → Androgens

(b) Acetate → Cholesterol → Progesterone → Estrogens

(c) Cholesterol → Pregnenolone → Progesterone → Androgens → Estrogens

(d) None of these **Ans. d**

181. Chain of the PSH, LH, HCG and TSH is

(a) α - chain (b) β - chain

(c) π - chain (d) Both α and β chain **Ans. a**

182. Hormone → receptor → membrane enzymes → cAMP → phosphorylase is above sequence which one is the first messenger

(a) cAMP (b) Membrane enzyme

(c) Hormone (d) Receptor **Ans. c**

183. Galactorrhea is a symptom caused by the deficiency of

(a) FSH (b) LH

(c) GH (d) PRL **Ans. d**

184. $\text{ATP} \xrightarrow[\text{cyclase}]{\text{Adenylite}} \text{cAMP} \xrightarrow{\text{Phosphodiestergic}} \text{AMP}$ which one is more responsible for cell physiology?

(a) ATP (b) cAMP

(c) AMP (d) ATP + cAMP **Ans. d**

185. Pituitary adenohypophysis contains which of the following hormone groups

(a) MSH, ACTH, FSH, LH, TSH, HCG

(b) PRL, GH, HCG, TSH, FSH, LH, ACTH

(c) MSH, FSH, LH, TSH, HCG

(d) ACTH, FSH, LH, GH, MSH **Ans. b**

186. Which of the hormone are glycoprotein in nature?

(a) ACTH, FSH, LH, PRL (b) LH, HCG, PRL, TSH

(c) FSH, LH, TSH, HCG (d) All of the above **Ans. c**

187. The site for action for LH is

(a) Ovary (b) Follicular cells

(c) Corpous Luteum (d) Egg **Ans. c**

188. FSH get by increasing intracellular cAMP is

(a) Testicular cells (b) Ovarian cells

(c) Thecal cells (d) Theca and granulose cells **Ans. d**

189. Which hormone is an indicator for pregnancy?

(a) FSH in blood (b) LH in blood

(c) HCG in blood (d) HCG in urine **Ans. d**

190. Ovulation in vertebrate is induced by the injection of

(a) Gonodotropin hormones (b) FSH

(c) LH (d) HCG **Ans. a**

191. The secretion of FSH and LH is regulated by feed mechanism process which hormones is/are responsible for

(a) Progesterone (b) Testosterone

(c) Estrogen (d) Estrogen + Androgen **Ans. c**

192. Myxeciema is a symptom cause due to

(a) Eypothyrodism (b) Hyperthyroidism

(c) Hypo GH (d) Hypo G + H **Ans. a**

193. Secondary sexual characters are regulated by

(a) Pituitary hormones

(b) Testis or ovarian hormones

(c) Adrenal hormones

(d) Cumulative effects of all these endocrine hormones **Ans. d**

194. Evshing's syndrome is the abnormality of

(a) Adrenal gland tumor

(b) Pituitary tumor

(c) Thyroid tumor

(d) Pituitary achenohypophysis tumor **Ans. a**

195. Autoimmune thyroditis disorder is also know as

(a) Goits (b) Grave's disease

(c) Hashimoto disease (d) Metabolic disorder **Ans. c**

196. Adrenal secretes which of the following group of hormones

(a) ACTH, MSH, Cortisol, Estrogens

(b) Adrenelin, Noradrenelin, Gluco-minerallo-corticoids

(c) Epineptrim, Nonepineprin, Glucocorticoids, minerallocorticoids, Estrogens and Androgens

(d) None of these **Ans. c**

197. Klinefelter's syndrome is caused due to?

(a) Pituitary hormones (b) Gonadotropic hormones

(c) Uvanims hormones (d) FSH and LH **Ans. d**

198. Hilar cells are located in

(a) Testes (b) Ovary

(c) Thyroid (d) Adrenal **Ans. b**

199. FSH induces the follicular growth by secretion of

(a) Estrogens (b) Testosterone

(c) Androgen (d) Estradiol – 17β **Ans. d**

200. LH level is highest in too ovary duns

(a) Follicular growth (b) Oocyte maturation

(c) Ovulation (d) Pregnancy **Ans. c**

201. Which are the hormones regulates Lactatron?

(a) Oxytocin

(b) PRL

(c) Oxytocin + PRL

(d) Oxytocin + Vassopression + PRL **Ans. c**

202. Pancreas secretes hormones which are
(a) Insulin from β cells, Glucagen from α cells
(b) Insulin – β, Glycogen – β, Somatostatron –α cells
(c) Glycogen – α, Insulin –β and Somatostatron
(d) All of the above **Ans. c**

203. The important function of pheromones are
(a) Behavioral pattern
(b) Reproductive pattern
(c) Co-ordination of behavior or physiology
(d) Socialism **Ans. c**

204. Most of the pheromones are secreted by
(a) Endocrine gland (b) Mesocrine gland
(c) Glandviar cells (d) Saliva-sweating and urines
Ans. d

205. Tendrils coil around substratum due to
(a) Geotropism (b) Hapotropism
(c) Thigmotropism (d) Both (b) and (c) **Ans. d**

206. Approdiside pheromones are responsible for
(a) Mimicry (b) Coloration
(c) Alarming (d) Copulation **Ans. d**

207. Auxanometer is needed for
(a) Study of plant movements
(b) Measurement of plant growth
(c) Geotropic growth study of roots
(d) Calculation of plant nutrition in relation to growth **Ans. b**

208. If total growth is plotted against time, the curve obtained is
(a) S – shaped (b) Parabolic
(c) T – shaped (d) U – shaped **Ans. a**

209. When growth is going on at slow rate in the beginning, this phase is called
(a) Log phase (b) Lag phase
(c) Senescence phase (d) Linear phase **Ans. b**

210. Leaf fall occurs when content of
(a) Gibberellin decreases (b) Auxin decreases
(c) Auxin increases (d) Abscisic acid decreases **Ans. b**

211. Apical dominance is caused by
(a) GA (b) IAA
(c) ABA (d) Florigen **Ans. b**

212. Which of the following is precursor of IAA synthesis in plants?
(a) Tyrosine (b) Acetyl CoA
(c) Tryptophan (d) Phenylalanine **Ans. c**

213. The growth regulator which delays ageing of plant organs is
(a) Auxin (b) Cytokinin
(c) Gibberellin (d) 2,4-D **Ans. b**

214. Gibberellins were first isolated from
(a) An alga (b) A fungus
(c) An angiosperm (d) Bacteria **Ans. b**

215. Bolting of genetic dwarfs is caused by
(a) Gibberellins (b) Traumatin
(c) ABA (d) Cytokinins **Ans. a**

216. Which of the following shows hormonal activity and used as weedicide?
(a) Zeatin (b) Florigen
(c) 2, 4, 5-T (d) Ethylene **Ans. c**

217. Photoreceptor pigment in plants for photomorphogenetic reactions is
(a) Phytochrome (b) Kinetin
(c) Florigen (d) Vernalin **Ans. a**

218. IAA was isolated by
(a) J.C. Bose (b) F.W. Went
(c) Kogl et al (d) Skoog and Miller **Ans. c**

219. In absence of light, increase takes place in
(a) Uptake of minerals (b) Uptake of water
(c) Elongation of internodes (d) Ascent of sap **Ans. c**

220. Herbicide DCMU kills small weeds by
(a) Destroying chlorophyll (b) Inhibiting PsII
(c) Inhibiting PsI (d) Blocking protein synthesis
Ans. b

221. The fruit ripening plant growth regulator is
(a) Ethylene (b) Kinetin
(c) ABA (d) Gibberellin **Ans. a**

222. A detached leaf can be made to remain green for long by keeping it in
(a) Gibberellins (b) Cytokinins
(c) Auxins (d) ABA **Ans. b**

223. Which is known as 'stress hormone'?
(a) IAA (b) GA
(c) Zeatin (d) ABA **Ans. d**

224. Phytochrome is involved in
(a) Flowering (b) Seed germination
(c) Organogenesis (d) All of the above **Ans. d**

225. Highest concentration of auxin exists in
(a) Leaves (b) Growing tips
(c) Petioles (d) Lower part of plant organs
Ans. b

226. Bananas can be prevented from one ripening by keeping them
(a) At room temperature (b) In refrigerator
(c) In ABA (d) In ascorbic acid solution
Ans. d

227. Physiologically active phytochrome is
(a) P_{730} (b) P_{660}
(c) P_{700} (d) P_{680} **Ans. a**

228. Effect of duration of light on the growth and flowering of plants is called
(a) Vernalization (b) Photoperiodism
(c) Morphogenesis (d) Photomorphogenesis **Ans. b**

229. Abscisic acid controls
(a) Stem elongation
(b) Cell expansion and cell wall plasticity
(c) Cell division
(d) Leaf fall and dormancy **Ans. d**

230. Which of the following is a phytohormone acting like growth inhibitor?
(a) IAA (b) NAA
(c) Ethylene (d) Gibberellic acid-3 **Ans. c**

231. Induction of flowering by treating with low temperature is called

(a) Photoperiodism (b) Temperature induction

(c) Vernalization (d) Enhanced flowering **Ans. c**

232. Zinc is essential for

(a) IAA biosynthesis (b) Gibberellin synthesis

(c) Chlorophyll synthesis (d) β- oxidation of fat **Ans. a**

233. When the dark period (night) is interrupted by a short exposure of light, a short day plant like sugarcane will

(a) Flower immediately resulting into loss of stem girth

(b) Not flower resulting into increase in stem girth

(c) Give more vigorous flowering resulting into decrease in sugar product

(d) Behave like a long day plant and flower after a brief period **Ans. b**

234. Sugarcane will accumulate more sugar if it is provided

(a) 2, 4. 5-T (b) 2, 4-D

(c) NAA (d) IBA **Ans. b**

235. The hormone which can changes sex of a unisexual plant is

(a) Cytokinin (b) ABA

(c) Ethylene (d) Auxin **Ans. d**

236. The hormone which can substitute low temperature treatment is

(a) Auxins (b) Gibberellins

(c) Cytokinins (d) Abscisic acid **Ans. b**

237. The hormone known to promote maleness flowering plant is

(a) GA (b) IAA

(c) Kinetin (d) ABA **Ans. a**

238. Pollen tube growth in the style of the carpel is

(a) Thigmotropic (b) Chemotactic

(c) Chemonastic (d) Chemotropic **Ans. d**

239. An example of movement of variation is shown by

(a) *Desmodium gyrans* (b) *Drosera*

(c) *Mimosa pudica* (d) All of the above **Ans. d**

240. Secondary roots are

(a) Positively geotropic (b) Plagiogeotropic

(c) Plagiophototropic (d) Ageotropic **Ans. b**

241. Sprouting of potato can be prevented by

(a) Gibberellins

(b) IAA

(c) Cytokinins

(d) Maleic hydrazide and IBA or NAA **Ans. d**

242. Premature falling of leaves and fruits can be prevented by the application of

(a) Auxins (b) Gibberellins

(c) Cytokinins (d) All of the above **Ans. a**

243. Photoperiodism is supported to be due to the synthesis of

(a) Cytokinins (b) Plorigen

(c) Gibberellins (d) Auxins **Ans. b**

244. Gibberellins are known to induce flowering more effectively in case of

(a) Long day plants (b) Short day plants

(c) Day neutral plants (d) All plants **Ans. a**

245. Which of the following is an established long day plant?

(a) Soyabean (b) Tobacco

(c) Wheat (d) Xanthium **Ans. c**

246. Bioassay of IAA is

(a) Avena cuvaeture test (b) Callus test

(c) Tobacco pith test (d) Meristematic growth **Ans. a**

247. The phase of growth where new cells are produced by cell divisions is called

(a) Formative phase (b) Protoplasmic phase

(c) Phase of differentiation (d) Phase of elongation **Ans. a**

248. When grasses are mown, it causes

(a) Excessive lateral growth

(b) Control of excessive growth

(c) Control of flowering and resulting vegetative growth

(d) None of these **Ans. a**

249. Universal natural auxin of plants is

(a) ABA (b) IAA

(c) NAA (d) IBA **Ans. b**

250. In nurseries, the rooting hormone used is

(a) IAA (b) 2, 4-D

(c) Benzoic acid (d) IBA **Ans. d**

251. Movement of male gametes towards archegonia in bryophytes and pteridophytes is an example of

(a) Chemotripsin

(b) Chemonasty

(c) Chemotactic movement

(d) Autonomic movement of locomotion **Ans. c**

252. Nyctinasty is

(a) Opening and closing of flowers due to differential growth on the two sides of sepals and petals induced by temperature or light

(b) Closing up of leaves or petals during evening due to loss of turgor in cells

(c) Autonomic sleeping movements in leaves of some plants

(d) Movement of variation induced by shock or touch **Ans. b**

253. IAA was first isolated from

(a) Corn germ oil (b) Gibberella

(c) Human urine (d) Rhizopus **Ans. a**

254. Which phytohormone overcomes dwarfism?

(a) Auxins (b) GA

(c) CK (d) Both (b) and (c) **Ans. b**

255. Chemotropic movement is shown by

(a) Male gametes of Bryophytes (b) Male gametes of Pteridophytes

(c) Pollen tube (d) All of the above **Ans. b**

256. Cocklebur is

(a) SDP (b) LDP

(c) SLDP (d) LSDP **Ans. a**

257. Went demonstrated that when the tip of Avena coleoptile was removed

(a) Coleoptile growth stopped

(b) Coleoptile bend towards source of light

(c) Coleoptile bend away from source of light

(d) Coleoptile grow lengthwise **Ans. a**

258. Under following condition

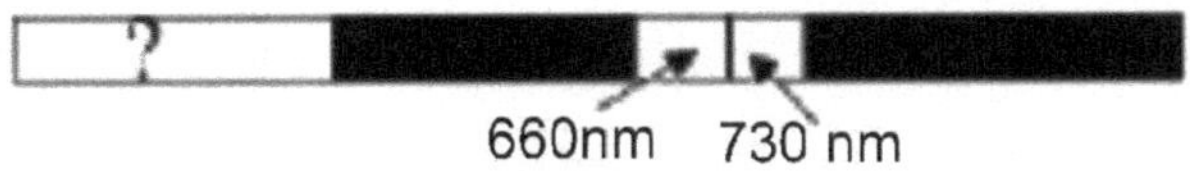

(a) LDP will flower but SDP will show vegetative growth

(b) SDP will flower but LDP will show vegetative growth

(c) Both SDP and LDP will flower

(d) LDP will flower but SDP will show vegetative growth **Ans. a**

259. Which hormone is isolated and crystallized from milky endosperm of corn by Letham?

(a) Zeatin (b) Kinetin

(c) Heteroauxin (d) Auxin – b **Ans. a**

260. In *Ginkgo biloba*, seed dormancy is due to

(a) Germinaton inhibitors

(b) Immature embryo

(c) Very hard seed coat

(d) Seed coat impermeable to water **Ans. b**

261. By exogenous application of which hormone, biennials can be induced to behave as annuals?

(a) NAA (b) IBA

(c) IPA (d) GA **Ans. d**

262. Scarification is method for

(a) Inducing flowering in a plant (b) Breaking seed dormancy

(c) Devernalization of plant (d) None of these **Ans. b**

263. Progressive senescence of older leaves and lateral organs which may occur at any time of the year is called

(a) Sequential senescence (b) Simultaneous senescence

(c) Synchronous senescence (d) Both (b) and (c) **Ans. a**

264. Which auxin is used to prevent the sprouting of potato tuber?

(a) NAA (b) IBA

(c) 2, 4-D (d) 2, 3, 6-trichlorobenzoic acid

Ans. a

265. What is not correct?

(a) The exogenous supply of auxins increases formation of male flowers

(b) Plants having excurrent habit have greater apical dominance

(c) Auxins show polar transport

(d) Foliar spray with NAA causes flowering in litchi **Ans. a**

266. 'Pomalin', particularly effective in increasing apple size is a combination of

(a) Auxins and GA (b) Auxins and CK

(c) CK and GA (d) Letham **Ans. c**

267. Who was first to declare ethylene as a phytohormone?

(a) Crocker (b) Miller

(c) Addicot (d) Letham **Ans. a**

268. Which hormone promotes transverse growth and inhibits longitudinal growth?

(a) IPA (b) Kinetin

(c) Ethylene (d) ABA **Ans. c**

269. Thigmonasty is

(a) Paratonic movement of growth

(b) Paratonic movement of variation

(c) Paratonic movement of locomotion

(d) Autonomic movement of growth **Ans. b**

270. In Acacia, the leaflets spread horizontally during the day and fold together and droop down night. This type of movement is called

(a) Phototactic (b) Phototropic

(c) Nyctinasty (d) Autonomic movement **Ans. c**

271. During nutations

(a) The stem exhibits a kind of nodding movement in two directions

(b) More growth occurs on upper side of sepals and petals

(c) Less growth occurs on lower side of sepals and petals

(d) Cytoplasm moves around a vacuole in a cell **Ans. a**

272. The plant organs (secondary roots) which are placed obliquely to the force of gravity are called?

(a) Orthogeotropic (b) Plageotropic

(c) Diageotropic (d) Apogeotropic **Ans. b**

273. Bioassay of cytokinin is
- (a) Expansion of excised radish cotyledons
- (b) Delay in senescence
- (c) Root growth inhibition test
- (d) More than one is correct **Ans. d**

274. Which of the following phase has constantly dividing cells and is restricted to the apical meristem?
- (a) Formative phase (b) Phase of elongation
- (c) Phase of maturation (d) Phase of differentiation **Ans. a**

275. Which temperature coagulates and damages the protoplasm and hinders growth?
- (a) 20°C - 25°C (b) 28°C - 30°C
- (c) 35°C - 40°C (d) Above 45°C **Ans. d**

276. Natural Ck 'Zeatin' was isolated from
- (a) Autoclaved herring sperm DNA
- (b) Corn Kernels
- (c) Coconut milk
- (d) Both (b) and (c) **Ans. d**

277. Find out incorrect match
- (a) Cosmos - SDP
- (b) Sugar beat - LDP
- (c) Tomato - Day Neutral Plant
- (d) Cucumber - SDP **Ans. d**

278. Precursor of IAA is
- (a) Methionine (b) Tryptophan
- (c) Mevalonic acid (d) Leucine **Ans. b**

279. Which light is favourable for seed germination?
- (a) Red (b) Far red
- (c) Blue (d) Yellow **Ans. a**

280. Which of the following is negative photoblastic seed?
- (a) Tobacco (b) Onion
- (c) Lettuce (d) Mistletoe **Ans. b**

281. Who proposed the existence of hypothetical substance, the vernalin?

(a) Hendricks (b) Borthwick

(c) Melchers (d) Chailakhyan **Ans. c**

282. Accumulation of which form of phytochrome inhibit flowering in short day plant?

(a) P_r (b) P_{fr}

(c) P_{660} (d) Both (a) and (b) **Ans. b**

283. Which hormone prevent premature fruit drop in apples?

(a) IAA (b) GA_3

(c) IPA (d) CK **Ans. a**

284. Epigeal germination is shown by

(a) Castor (b) Mango

(c) Pea (d) Gram **Ans. a**

285. In which plant reproduction occurs in second season?

(a) Wheat (b) Radish

(c) *Convolvulus* (d) *Fragaria* **Ans. b**

286. Which is not a natural plant hormone?

(a) GA_2 (b) GA_3

(c) IAA (d) 2, 4-D **Ans. d**

287. Optimum temperature for seed germination is

(a) 10 °C – 15 °C (b) 15 °C - 30 °C

(c) 30 °C – 40 °C (d) 40 °C - 42 °C **Ans. b**

288. Which hormone is used for defoliation of forest trees?

(a) TIBA (b) Phosphon – D

(c) AMO – 1618 (d) 2, 4 – D **Ans. d**

289. Which of the following is a photomorphoegnetic process?

(a) Seed germination (b) Flowering

(c) Chlorophyll formation (d) All of the above **Ans. d**

290. Abscisic acid cause closure of stomata because it cause

(a) Rapid movement of water in the guard cells

(b) Slow movement of water out of the guard cells

(c) Rapid movement of K^+ out the guard cells
(d) Rapid movement of K^+ into the guard cells **Ans. c**

291. According to Went the movement of auxin is
(a) Polar and basipetal (b) Polar and acropetal
(c) Non polar (d) Not fixed **Ans. a**

292. Tentacles of Drosera bend in response to the stimulus of toush by insects. This type of movement is called
(a) Thigmotropic (b) Haptotropic
(c) Thigmonastic (d) Nyctinasty **Ans. c**

293. Internodal elongation is stimulated by
(a) Auxin (b) Cytokinin
(c) Gibberellin (d) Phenol **Ans. c**

294. Reversal of apical dominance and a bushy growing habit of a plant owes to
(a) Deficiency of calcium (b) Excess of auxin in shoot
(c) Deficiency of potassium (d) Low concentration of cytokinin **Ans. c**

295. A point in growth curve at which exponential phase ends is said to be
(a) Inflexion point (b) Decreation phase
(c) Stationary phase (d) Senescent phase **Ans. a**

296. A hormone that has direct implication on water economy in plants is
(a) Auxin (b) Gibberellin
(c) Ethylene (d) Abscisic acid **Ans. d**

297. A plant hormone that promotes phloem transport is
(a) Auxin (b) Gibberellin
(c) Cytokinin (d) Ethylene **Ans. c**

298. A hormone that changes monopodial growing habit to bushy habit of a plant is
(a) Cytokinin (b) Auxin
(c) ABA (d) Gibberellins **Ans. a**

299. Which of the following is not a cytokinin?
(a) Benzoic acid (b) Benzimidazole
(c) Benzyladenine (d) 6-furfuryl aminopurine **Ans. a**

300. Latex extraction from rubber plant is enhanced by use of a hormone

(a) Auxin (b) Gibberellin

(c) Cytokinin (d) Ethylene **Ans. d**

301. Physiologically active form of phytochrome is the one that

(a) Absorbs red light (b) Absorbs far red light

(c) Absorbs blue light (d) Both (a) and (b) **Ans. b**

302. Fruit traders use a chemical to prevent 'in transit' ripening of bananas. This chemical is

(a) Cytokinin (b) Ethylene

(c) Abscisic aicd (d) Ascorbic acid **Ans. d**

303. The phytochrome that acts as antheridiogen factor in ferns is

(a) IPA (b) IAA

(c) GA (d) TIBA **Ans. c**

304. A growth pigment that regulates flowering in most plants is

(a) Phytochrome (b) Florigen

(c) Auxin (d) Cytokinin **Ans. a**

305. Action spectrum of chloroplast exhibits a parallelism to absorption spectrum of

(a) Chlorophyll a (b) Chlorophyll b

(c) Both (a) and (b) (d) Chlorophyll and carotenoids **Ans. a**

306. Specific role of growth hormones at cellular level is in

(a) Enhanced release of Ca^{++} by ER

(b) Enhanced intake of Ca^{++} through plasma membrane

(c) Enhanced elasticity of cell wall

(d) All of the above **Ans. d**

307. Photonastic movement is observed in

(a) *Nymphaea* (b) Silene

(c) Tulip (d) Victoria **Ans. c**

308. Find odd one (w.r.t. photoperiodic response)

(a) Four O' clock plant (b) Spinach

(c) *Crysanthenum* (d) Sugar beet **Ans. a**

309. Photoperiodic stimulus in plants is perceived by
(a) Stem apex (b) Root
(c) Leaf (d) Flowers **Ans. c**

310. Tomato seed never show vivipary in their pulp due to
(a) ABA (b) Ferulic acid
(c) Ethylene (d) Absence of oxygen **Ans. b**

311. Find the odd one (w.r.t germination)
(a) Pea (b) Maize
(c) *Sonneratia* (d) *Vicia faba* **Ans. c**

312. Mimosa is very quick to respond to a shock stimulus. How much time it takes to start a reaction?
(a) 1 minute (b) 1 second
(c) 1/10 second (d) 1/1000 second **Ans. c**

313. Chemotropic movement is shown by
(a) Pollen tube (b) Male gametes of bryophytes
(c) Coiling of young leaves (d) Drooping of *Mimosa pudica*
Ans. a

314. Wheat plant shows
(a) Whole plant senescence (b) Sequential senescence
(c) Simultaneous senescence (d) Synchronous senescence
Ans. a

315. Which type of movement is shown by insectivorous plant *Drosera* during capturing of prey?
(a) Haptonastic (b) Haptotropic
(c) Thigmotropic (d) Diageotropic **Ans. a**

316. Which of the following reduce transpiration by reducing stomatal opening for period of few weeks without affecting metabolic processes?
(a) PMA (b) Waxes
(c) Silicon emulsions (d) CCC **Ans. a**

317. Unicellular alga *Chlamydomonas* shows
(a) Autonomic movement of locomotion
(b) Paratonic movement of locomotion
(c) Both (a) and (b)
(d) Movement of variation **Ans. c**

318. A method of breaking dormancy in which moist seeds are subjected for variable periods of low or high temperature in the presence of oxygen, is called

(a) Scarification (b) Impaction

(c) Stratification (d) Any of these **Ans. c**

319. Vernalization was first studied in Europe on......varieties of...........

(a) Summer, Grasses (b) Winter, Cereals

(c) Summer, Fruits (d) Winter, vegetables **Ans. b**

320. Nutation is autonomic growth movement during which

(a) Alternate more growth on one side and slow growth on the other side

(b) More growth occurs at the base of thalamus

(c) Stem shows spiral coiling around a support

(d) Root grows towards source of water **Ans. a**

321. Which of the following is a photomorphogenetic process?

(a) Flowering (b) Chlorophyll formation

(c) Seed germination (d) All of the above **Ans. d**

322. A short day plant will flower under which condition?

(a) 

(b) 

(c) 

(d) 

Ans. a

323. Following diagram is showing

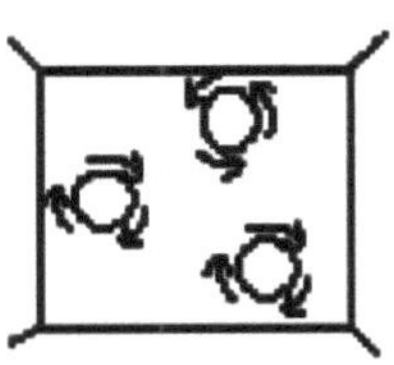

(a) Heliotropism

(b) Circulation

(c) Rotation

(d) Paratonic movement of locomotion **Ans. b**

324. Went demonstrated that when the tip of Oat caoleoptile was removed

(a) Coleoptile bend towards source of light

(b) Coleoptile bend away from source of light

(c) Coleoptile grow lengthwise

(d) Growth stopped **Ans. d**

325. Flowering hormone florigen is produced in.......and......... is translocated through

(a) Floral bud, xylem

(b) Floral bud, phloem

(c) Leaves, phloem

(d) Vegetative bud, intercellular spaces **Ans. d**

326. Which of the following plant flower after a period of vegetative growth, regardless of the photoperiod?

(a) Radish (b) Tomato

(c) Barley (d) Rice **Ans. b**

327. Match the following

A. Kinetin i. Gibbane skeleton

B. GA_3 ii. Root hair initiation

C. Ethylene iii. Purine derivative

D. ABA iv. Parthenocarpy in Rosa

(a) A-iv, B-iii, C-ii, D-i (b) A-iii, B-iv, C-ii, D-i

(c) A-iii, B-i, C-ii, D-iv (d) A-ii, B-i, C-iv, D-iiii **Ans. c**

328. If mica plate is inserted on shaded side of the coleoptile tip. What will happen?

(a) Coleptile tip will bend towards light

(b) Coleptile tip will show nutation

(c) Results in circum nutation

(d) No curvature will be formed **Ans. d**

329. Ethylene and MH have antagonistic function as

(a) Promote flowering in SDP

(b) Increase sheft life to cut vegetables

(c) Promote apical dominance

(d) Breaking dormancy of storage organs **Ans. d**

330. A photoblastic seed is given following treatment. In which case germination of seeds will be maximum?

(a) Seeds + R
(b) Seeds + R + F_r
(c) Seeds + R + F_r + R
(d) Seeds + R + F_r + R + F_r **Ans. c**

331. Seed dormancy in pulses is due to

(a) Immature embryo
(b) Hard seed coat
(c) Germination inhibitors
(d) Dormant embryo **Ans. b**

332. Which of the following hormone is isolated and crystallized from milky endospaerm of corn by Letham?

(a) Kinetin
(b) Zeatin
(c) Gibberellin A_3
(d) Dormin **Ans. b**

333. What is correct for simultaneous senescence?

(a) It is progressive senescence of older leaves which may occur at any time of the year

(b) It occurs in all annuals

(c) It occurs in monocarpic plants

(d) It is controlled by environmental factors rather than internal factors **Ans. d**

334. Naphthalene acetic acid is used for

(a) Inducing rooting in cuttings

(b) Inducing bud dormancy

(c) Substitution of cold treatment

(d) Increasing shelf life of cut flowers **Ans. a**

335. Maximum concentration of gibberellins is found in

(a) Young leaves
(b) Mature leaves
(c) Root tip
(d) Stem tip **Ans. a**

336. There is a non-indole phenylacetic acid compound, which plays a role similar to IAA. Which of the following plant is rich in it?

(a) Tomato
(b) Tobacco
(c) Potato
(d) Papaya **Ans. a**

337. Male gametes of Dryopteris show
(a) Chemotactic movement
(b) Chemotropic movement
(c) Autonomic movement of locomotion
(d) Ciliary movement **Ans. a**

338. Branches arising from the main stem usually place themselves at an angle 45^0 from the main axis and are
(a) Plagiotropic (b) Diageotropic
(c) Orthogeotropic (d) Apogeotropic **Ans. a**

339. Nutations are shown by
(a) Leaf lamina (b) Petals
(c) Sepals (d) Twiners **Ans. d**

340. Which of the following biochemical reactions is most commonly utilized by living cells to propagate intracellular signals?
(a) Acylation (b) Phosphorylation
(c) Methylation (d) Decarboxylation **Ans. b**

341. Identify the correct order in which the extra-cellular signals are transmitted
(a) Adenylate cyclase-cAMP-protein kinase A
(b) Protein kinase A-adenylate cyclase-cAMP
(c) Protein kinase A- cAMP-adenylate cyclase
(d) cAMP-adenylate cyclase-protein kinase A **Ans. a**

342. Which of the following hormone passes through the cell membrane, binds to its intracellular receptors and activates it?
(a) Estrogen (b) Non-epinephrine
(c) Acetylcholine (d) Glucagons **Ans. a**

343. In eukaryotes, which of the following is not a second messenger?
(a) 1,2-diacyl glycerol (DAG) (b) Cyclic AMP
(c) Inositol 1,4,5-triphosphate (d) Cyclins **Ans. d**

344. Which one of the following is involved in cell signaling process
(a) Cytochrome P450 (b) Protein kinase C
(c) Zinc fingers (d) All the above are involved
Ans. b

345. In most cases protein kinases
(a) Stimulate adenylyl cyclase
(b) Add phosphate groups to proteins
(c) Bind cGMP
(d) Hydrolyze proteins **Ans. b**

346. In the cAMP pathway, the G protein stimulates
(a) Phospholipase C (b) Adenylyl cyclase
(c) The endoplasmic reticulum (d) Calmodulin **Ans. b**

347. Find the correct statement
(1) Ras is a monomeric GTP binding switch protein
(2) Phosphodiesterase converts cGMP into GMP
(3) Receptor tyrosine kinase dimerise after binding with ligand
(4) Receptor tyrosine kinase after activation stimulates Ras-MAP kinase pathway
(a) 1 and 2 (b) 1 and 3
(c) 1,3 and 4 (d) 1,2,3 and 4 **Ans. d**

348. Choose the incorrect statement
(a) Phospholipase C generates DAG and IP3 from PI 4,5-bisphosphate
(b) Phospholipase C is not membrane bound enzyme
(c) DAG activates protein kinase C under increased Ca^{2+} level
(d) IP3 induces the opening of Ca^{2+} channel present on ER **Ans. b**

349. Which subunit of trimeric G-protein acts as GTPase switch protein subunit?
(a) α (b) β
(c) γ (d) $\alpha\gamma$ **Ans. a**

350. Synaptic signaling involves
(a) Endocrine signals (b) Paracrine signals
(c) Autocrine signals (d) Neurotransmitters **Ans. d**

351. Characteristics of intracellular receptors that regulate gene transcription include all of the following except
(a) A DNA binding site
(b) An extra cellular binding site
(c) A transcription activating domain
(d) May be signaled by lipid soluble molecules **Ans. b**

352. Cell surface receptors may be any of the following except

(a) G protein linked

(b) Enzymatic receptors

(c) Single-pass transmembrane proteins for neurotransmitters

(d) Ligand-gated ion channels **Ans. c**

353. The mechanism of signal transduction by steroid hormones differs from amine and peptide hormones because

(a) Steroids use small, water soluble second messengers

(b) They bind with specific receptor proteins on target-cell plasma membranes

(c) They bind to cytoplasmic or nuclear receptors and affect gene expression

(d) They are secreted from exocrine glands **Ans. c**

354. Match each enzyme column A with the product column B that it produces

	Column A		Column B
1.	Guanylcyclase	A.	cAMP
2.	Adenylylcyclase	B.	DAG and IP_3
3.	Phospholipase-C	C.	cGMP
4.	Phosphatidylinositol	D.	Phosphatidylinosi-tol 4-phosphate 4,5-bisphosphate (PIP) kinase (PIP_2)

(a) 1-C, 2-A, 3-B, 4-D (b) 1-B, 2-C, 3-A, 4-D

(c) 1-D, 2-C, 3-B, 4-A (d) 1-D, 2-A, 3-D, 4-C **Ans. a**

355. Which of the following is incorrect?

(a) Endocrine signaling molecules acts on target cell distant from their site of synthesis

(b) Paracrine signaling molecules affect target cells only in close proximity

(c) Autocrine signaling molecules affect the same cells that release

(d) None **Ans. d**

356. The receptor for nitric oxide is

(a) Intercellular (b) Intracellular

(c) Extracellular (d) Transcellular **Ans. b**

357. The mechanism of signal transduction by steroid hormones differs from amine and peptide hormones because

(a) Steroids use small, water soluble second messengers

(b) They bind with specific receptor proteins on target-cell plasma membranes

(c) They bind to cytoplasmic or nuclear receptors and affect gene expression

(d) They are secreted from exocrine glands **Ans. c**

358. All of the following statements apply to G proteins except

(a) G proteins transmit a signal from the cell surface to the interior of the cell

(b) G proteins do not use second messengers but transmit the signal directly into the nucleus

(c) All G proteins have a similar structure

(d) G proteins act to amplify the signal creating a cascade response in the cell **Ans. b**

359. Calmodulin activates protein kinases in response to a transient increase in

(a) cAMP (b) Calcium ions

(c) DAG (d) NO **Ans. b**

360. The signaling system that operates over the shortest range in

(a) Contact-dependent cell signaling, as in the inhibition of neighbors by differentiating nerve cells

(b) Paracrine secretion, as in the secretion of histamines by mast cells

(c) Neutral signaling, as in the transmission of a nerve impulse

(d) None **Ans. a**

361. Cyclin-dependent kinase (Cdk) activity increases steadily during G_2 as a result of

(a) Phosphorylation of mitotic cyclins by Cdks

(b) Activation of mitotic Cdk-cyclins through the activity of a phosphatase

(c) Phosphorylation of Cdks at a point not located inside the catalytic site of the Cdk-cyclin complex

(d) Transient increase in the cytosolic GTP concentration **Ans. a**

362. Entry into M phase is not allowed if

(a) The cell is not bid enough

(b) Sufficient nutrients are not available

(c) If DNA replication is not complete

(d) Mitotic cyclin is over expressed **Ans. c**

363. Which of the following is a component in the signaling pathway stimulated by receptor tyrosine kinases?

(a) Adenylate cyclase (b) Adaptor proteins

(c) Autophosphorylating receptor (d) Ras activating protein **Ans. d**

364. Signal transduction describes the ways in which external molecules cause internal changes in cells. These processes

(P) always involve the movement of ligands across the membrane

(Q) amplify the strength of the external stimulus

(R) rely on cell surface receptors that can interact with intracellular proteins

(S) can trigger the production of a second messenger inside the cell

(a) P,Q (b) P,R

(c) Q,R,S (d) P,Q,R,S **Ans. c**

365. Receptor tyrosine kinases

(P) have a cytoplasmic kinase domain

(Q) dimerize upon ligand binding

(R) change conformation of their cytoplasmic domains upon ligand binding

(S) become catalytically active after the conformational change

(a) P,Q (b) P,R

(c) Q,R,S (d) P,Q,R,S **Ans. d**

366. MAP kinases

(P) are activated by the Ras pathway

(Q) phosphorylated their targets on tyrosine residues

(R) act to stimulate or repress the transcription of target genes through the phosphorylation of transcription factors

(S) become active when phosphorylated by MEK kinases

(a) P,Q (b) P,R

(c) Q,R,S (d) P,Q,R,S **Ans. c**

367. Which of the following is not true about NO
(a) Acts as intracellular signaling molecule as well as neurotransmitter
(b) Regulates the vasodilation
(c) Synthesized from L-arginine
(d) Induces adenylate cyclase, which catalyzes cAMP formation
Ans. d

368. IP_3 causes calcium to be released into the cytoplasm from
(a) Lysosomes (b) Golgi bodies
(c) Endoplasmic reticulum (d) Mitochondria **Ans. c**

369. Which one of the following compounds does not acts second messenger during signaling process?
(a) cAMP (b) Calcium ions
(c) Inositol 3,4,5-triphosphate (d) Triacylglycerol **Ans. d**

370. Which one of the following bacterial toxins does not have ADP-ribosyl transferase activity
(a) Cholera toxin (b) *Diphtheria* toxin
(c) Exotoxin A (d) *S. aureus*-toxin **Ans. d**

371. Diacylglycerol activates
(a) Protein kinase A (b) Protein kinase C
(c) MAP kinase (d) Tyrosine kinase **Ans. b**

372. The receptor for which of the following hormones is a transcription factor?
(a) Insulin (b) Glucagon
(c) Estradiol (d) Adrenalin **Ans. c**

373. Which of the following statements is false for the nitric oxide gas?
(a) An intracellular signaling molecule
(b) Deamination of histidine results into nitric oxide production
(c) Stimulates guanylyl cyclase to produce cGMP
(d) Can be produced by activated neutrophils **Ans. b**

374. G-proteins are involved in relaying signals through G-protein linked receptors. Which of the following forms of G-protein is considered to be in active state?
(a) G-protein-ADP (b) G-protein-ATP
(c) G-protein-GDP (d) G-protein-GTP **Ans. d**

375. Permeability of protoplasm is accelerated by.........

(a) Na (b) K

(c) P (d) Ca **Ans. b**

376. The essential element for the synthesis of auxin is

(a) Sulphur (b) Phosphorus

(c) Zn (d) Potassium **Ans. c**

377. Coater potential in the leaf tissue is positive (+) during

(a) Excessive transpiration (b) Low transpiration

(c) Excessive absorption (d) Guttation **Ans. b**

378. Metal component of oxygen evolving complex of chloroplast is

(a) Fe (b) Mg

(c) Mn (d) Cu **Ans. b**

379. Richmond long effect is shown by

(a) Auxins (b) Gibberellins

(c) Kinetin (d) Sugar **Ans. c**

380. The growth is maximum in the zone of

(a) Cell division (b) Cell elongation

(c) Cell maturation (d) All of these **Ans. b**

❑❑❑

Chapter-17

Enzymes

Fill in the blanks

1. The enzyme ....was crystallized by Sumner.

Ans. urease

2. Feedback inhibition of enzyme occurs due to the accumulation of.............

Ans. end products

3. Enzymes catalyzing similar reactions but differing slightly in their structures are called.............................

Ans. isozymes

4. ....coined the term enzyme.

Ans. Kuhne

5. Synthesis of enzymes occurs on.................

Ans. ribosomes

6. ....are the enzymes which act in the same cell where they are synthesized.

Ans. Endoenzymes

7. The number of molecules of substrates acted upon by an enzyme in one minute are called.................

Ans. turn over number

8. Non-protein part of conjugated enzyme is known as................

Ans. prosthetic group

9. The enzymes which catalyze the removal of groups from substrates are classified as........................

Ans. lyases

10. The poisonous substances which cause permanent inactivity in enzymes are called...............inhibitors.

Ans. non competitive

11. ....is an enzyme which is not proteinaceous.

Ans. Ribozyme

12. Amount of enzyme catalyzing the transformation of 1 μ mole of substrate per unit is called................

Ans. enzyme unit

13. Enzymes are................proteins.

Ans. globular

14. The enzyme ....is used for coagulation of milk.

Ans. renin

15. Association of Zn^{++} is required for the activity of....................

Ans. carbonic anhydrase

16. Association of molybdenum is required for the activity of..............

Ans. dinitrogenase

17. The macromolecules shown to have capacity of catalysis are.................

Ans. proteins

18. Multienzyme complexes are bound to..................

Ans. plasma membrane

19. ....suggested that enzymes are proteins.

Ans. Sumner

20. The enzyme requiring metal ions are called..................

Ans. metalloenzyme

21. The enzyme ribozyme was discovered in the organism..................

Ans. *Tetrahymena thermophile*

22. The rate of enzyme controlled reactions almost................for energy rise of 10°C.

Ans. doubles

23. The substrate concentration which gives one half of the V_{max} ($V_{max}/2$) is............

Ans. Michaelis constant

24. ....inhibition often involves allosteric inhibition.

Ans. Feedback

25. Precursors of enzymes are called...........

Ans. zymogens

26. At boiling temperatures an enzyme is................

Ans. denatured

27. The proteinaceous part of an enzyme is called...........

Ans. apoenzyme

28. Enzyme urease was crystallised by.............

Ans. Sumner

29. Formation of peptide bond is catalysed by..............not by peptidyl transferase.

Ans. 23 sRNA

30. Out of the total enzyme present in the cell, mitochondria alone have.............percent.

Ans. 70%

31. Blocking activity site of an enzyme is a kind of................inhibition.

Ans. competitive

32. ....ions are required for the activity of enzyme urease.

Ans. Ni^{++}

33. ....inhibitor inactivates an enzyme by changing shape of the enzyme.

Ans. Irreversible

34. The best measure of catalytic efficiency of an enzyme is................

Ans. K_{cat}/K_m

35. Most of the enzyme of the cell are located in...................

Ans. mitochondria

36. Q_{10} for the enzyme catalysed reaction is................

Ans. 2

37. ....amino acids constitute the active site of an enzyme.

Ans. 3-12

38. The enzyme code of enzyme 2.8.1.1 refers to the...............main group.

Ans. transferase

39. Enzymes are absent in...............

Ans. viruses

40. Feedback inhibition of an enzymatic reaction is caused by.................

Ans. end product

41. The lower value of K_m means.........................substrate affinity of enzyme.

Ans. higher

42. Epimerases and mutases are...................type of enzymes.

Ans. isomerases

43. Enzymes useful for hydrolysing fats and lipid are called..............

Ans. esterases

44. Proenzyme is the...............precursor of an enzyme.

Ans. inactive

45. Control of bacterial pathogens has been effected through ....inhibition.

Ans. competitive

46. Competitive inhibitor is not connected by.............pathway catalysed by the enzyme.

Ans. metabolic

47. Catalytic anibodies are called ....

Ans. Abzymes

48. Cofactors and coenzymes may be covalently or noncovalently attached to the protein molecule called............................

Ans. Apoenzyme

49. Coenzymes and cofactors generally contribute to enzyme activity and ....

Ans. stability

50. Usudly, membrane reactors are used in the case of .... enzyme.

Ans. soluable

51. An example of .... synzyme is provided by cyclodextrins.

Ans. nonprotein (non traditional)

52. .... enzyme is used for production of aspartame.

Ans. Thermolysin

Multiple Choice Questions

1. In plants, enzymes are present
   (a) Only in leaves
   (b) Only in storage organs
   (c) Only in flowers
   (d) In all the living cells of the plant body **Ans. d**
2. Enzymes are useful because they
   (a) Catalyse biochemical reactions in the plant body
   (b) Supply energy
   (c) Enhance the absorption of water
   (d) Are important structural components **Ans. a**
3. The enzymes which act normally within cells are called
   (a) Endoenzymes (b) Exoenzymes
   (c) Apoenzymes (d) Ferment **Ans. a**
4. The scientists associated with the study of enzymes include
   (a) Buchner (b) Went
   (c) Sumner (d) Both (a) and (c) **Ans. d**
5. 'Enzymes are proteins', was suggested by
   (a) Pasteur (b) Leeuwenhoek
   (c) Miller (d) Sumner **Ans. d**
6. Noble Prize for discovering enzyme was given to
   (a) Fischer (b) Altmann
   (c) Fleming (d) Buchner **Ans. d**
7. Which of the following enzymes was first isolated and purified in the form of crystal
   (a) Amylase (b) Ribonuclease
   (c) Urease (d) Pepsin **Ans. c**
8. Urease, the first enzyme crystallized, was crystallized by
   (a) J. Northrop (b) E. Buchner
   (c) Louis Pasteur (d) J.B. Sumner **Ans. d**
9. Number of enzymes known now-a-days is approximately
   (a) 1,000 (b) 2,000
   (c) 3,000 (d) 4,000 **Ans. b**

10. Enzymes are basically made up of
(a) Proteins (b) Vitamins
(c) Carbon (d) Fats **Ans. a**

11. Which of the following enzyme transfers a phosphate group from ATP to another molecule
(a) Phosphatase (b) Phospho diesterase
(c) Kinase (d) Esterase **Ans. c**

12. An enzyme can be synthesized by chemically bonding together the molecules of
(a) Carbohydrates (b) Amino acids
(c) Lipase (d) CO_2 **Ans. b**

13. Enzymes are polymers of
(a) Amino acids (b) Hexose carbon
(c) Fatty acids (d) Inorganic phosphate **Ans. a**

14. Which one of the following statement is true for the enzyme
(a) All enzymes are proteins (b) All proteins are enzymes
(c) All enzymes are not proteins (d) All enzymes are vitamins
Ans. c

15. Which enzyme is not proteinaceous
(a) Isozyme (b) Ribozyme
(c) Holozyme (d) Trypsin **Ans. b**

16. Ribozyme is
(a) RNA without sugar (b) RNA without phosphate
(c) RNA having enzyme activity (d) RNA with extra phosphate
Ans. c

17. The non-protein part of an enzyme is called
(a) Holoenzyme (b) Prosthetic group
(c) Apoenzyme (d) None of the above **Ans. b**

18. The protein part of an enzyme is called
(a) Holoenzyme (b) Prosthetic group
(c) Apoenzyme (d) All of the above **Ans. c**

19. When coenzyme combines with apoenzymes, it is called
(a) Holoenzyme (b) Cofactor
(c) Isoenzyme (d) Prosthetic group **Ans. a**

20. Coenzyme is
(a) Always a protein
(b) Often a vitamin
(c) Always an inorganic compound
(d) Often a metal **Ans. b**

21. Cytochrome oxidase contains
(a) Iron (b) Magnesium
(c) Cobalt (d) Mercury **Ans. a**

22. Some metals known as cofactors of enzymes are
(a) Ca^{++}, Zn^{++}, Mn^{++} (b) Zn^{++}, Ca^{++}
(c) K^{+}, Co^{++} (d) All the above **Ans. d**

23. What will happen to an enzyme when apoenzyme is separated from its metal component
(a) Activity will be increased
(b) Activity will be lost
(c) Activity will be decreased
(d) There will be no change in the activity **Ans. b**

24. NADP is
(a) An enzyme (b) A part of soluble RNA
(c) A part of transfer RNA (d) A coenzyme **Ans. d**

25. FAD or FMN is a coenzyme; which vitamin is incorporated in its structure
(a) Vitamin B_1 (b) Vitamin B_2
(c) Vitamin B_6 (d) Vitamin C **Ans. b**

26. The nature of coenzyme is
(a) Non proteinaceous (b) Proteinaceous
(c) Both (a) and (b) (d) None of the above **Ans. a**

27. During enzyme activity, the coenzyme
(a) Acts as a donor or acceptor of atoms which are added to or removed from the substrates
(b) Are important in oxidation-reduction reactions
(c) Both (a) and (b)
(d) None of the above **Ans. c**

28. Which of the following are coenzymes
(a) NAD, NADP, FAD, FMN (b) Vitamin, Fe, Cu
(c) $NADPH_2$, Ca, Co (d) NAD, K, CoA **Ans. a**

29. Out of total enzymes present in the cell, mitochondrion alone has
(a) 95% (b) 20%
(c) 64% (d) 70% **Ans. d**

30. Which of the following is iron porphyrin coenzyme or cofactor
(a) Coenzyme A (b) Cytochrome
(c) NAD (d) FAD **Ans. b**

31. The ratio of enzyme : substrate molecules can be as high as
(a) 1:1000 (b) 1:10,000
(c) 1:50,000 (d) 1:100,000 **Ans. d**

32. The turnover number of the fastest enzyme is
(a) 10^5 (b) 10^4
(c) 18×10^4 (d) 4×10^6 **Ans. d**

33. Which among the following has highest turnover number
(a) Urease (b) Carbonic anhydrase
(c) Catalase (d) Pepsin **Ans. c**

34. Which among the following has highest catalytic efficiency
(a) Catalse (b) Urease
(c) Carbonic anhydrase (d) Pepsin **Ans. c**

35. The number of enzyme units per mg of protein is known as
(a) Molecular activity (b) Specific activity
(c) Turnover number (d) Molar activity **Ans. b**

36. Amount of enzyme transforming 1 µmole of substrate per minute at 25°C under optimal conditions of measurement is called
(a) Specific activity (b) Unit of enzyme (activity)
(c) Catalytic centre activity (d) Enzyme purity **Ans. b**

37. Which of the following statements is incorrect
(a) At optimum pH, the activity of enzyme is maximum
(b) Enzymes are not affected by hydrogen ion concentration
(c) The optimum pH for pepsin is 2.0
(d) The optimum pH for trypsin is 8.8 **Ans. b**

38. At temperature below the freezing point, an enzyme is
(a) Slightly activated (b) Killed
(c) Inactivated (d) Unaffected **Ans. c**

39. How is the rate of enzyme-catalyzed reactions affected by every 10°C rise of temperature
(a) Halves (b) Doubles
(c) Becomes four times (d) Remains unchanged **Ans. b**

40. At which temperature, the enzyme activity would be maximum
(a) 35°C (b) 45 °C
(c) 55 °C (d) 60 °C **Ans. a**

41. Most of the enzymes are inactivated at temperature above
(a) 25 °C (b) 45 °C
(c) 55 °C (d) 80 °C **Ans. b**

42. At boiling temperature, an enzyme is
(a) Killed (b) Denatured
(c) Inactivated (d) Unaffected **Ans. b**

43. On boiling/treating with strong acid or alkali/treating with pepsin, an enzyme loses its catalytic activity because
(a) It is a protein (b) It burns
(c) It is a functional substances (d) It is a lipoprotein **Ans. a**

44. Which part of an active enzyme is denatured by heat
(a) Apoenzyme (b) Holoenzyme
(c) Coenzyme (d) Activator **Ans. a**

45. Enzymes catalyses a reaction in
(a) Forward direction (b) Backward direction
(b) Both ways (d) Either ways **Ans. d**

46. Enzymes differ from inorganic catalysts in
(a) Having a high diffusion rate
(b) Working at high temperature
(c) Not being used up in reactions
(d) Being proteinaceous in nature **Ans. d**

47. What is true about enzymes
(a) All act best at pH 7.0 (b) All are amino acids
(c) All are not proteins (d) All act best at 0 °C **Ans. c**

48. Which of the following is not an attribute of enzymes
    (a) These are proteinaceous in nature
    (b) These speed up the rate of biochemical reaction
    (c) These are specific in nature
    (d) These are used up in reaction **Ans. d**

49. Enzymes are sensitive to
    (a) Light (b) Wind velocity
    (c) Change in pH (d) Rainfall **Ans. c**

50. Dry seeds can endure higher temperature than the germinating seeds because
    (a) Dry seeds are hard
    (b) Seedlings are tender
    (c) Dry seeds have more reserve food
    (d) Hydration makes the enzyme more sensitive to temperature **Ans. d**

51. Zymogens are
    (a) Enzyme acting upon starch (b) Groups of zymase enzymes
    (c) Inactive enzyme precursors (d) None of the above **Ans. c**

52. Enzymes which are slightly different in molecular structure, but can perform identical activities are called
    (a) Isoenzymes (b) Homoenzymes
    (c) Apoenzymes (d) Coenzymes **Ans. a**

53. Lactic dehydrogenase (LDH) which catalyses pyruvate to lactate is an example of
    (a) Apoenzyme (b) Antienzyme
    (c) Isoenzyme (d) Coenzyme **Ans. c**

54. An enzyme which catalyses the rearrangement of molecular structure and forms a compound of same molecular weight is called
    (a) Ligase (b) Oxidoreductase
    (c) Isomerase (d) Hydrolase **Ans. c**

55. The transfer of a group from a donor molecule to an acceptor molecule is catalysed by
    (a) Transferase (b) Isomerase
    (c) Protease (d) Hydrolytic enzymes **Ans. a**

56. Transferases are involved in the transfer of
(a) Methyl group (b) Amino group
(c) Phosphate group (d) All of these **Ans. d**

57. Oxidative enzymes occur in
(a) Lysosomes (b) Golgi bodies
(c) Mitochondria (d) Ribosomes **Ans. c**

58. Hydrolases are involved in the hydrolysis of
(a) Esters (b) Lipids
(c) Proteins (d) All of these **Ans. d**

59. Ligases are involved in the synthesis of
(a) C-C bonds (b) C-N bonds
(c) C-O bonds (d) All of these **Ans. d**

60. Esterase belongs to
(a) Oxidoreductases (b) Carboxylases
(c) Hydrolases (d) Transferases **Ans. c**

61. Which of the following is a hydrolytic enzyme
(a) Esterase (b) Carbohydrase
(c) Protease (d) All of these **Ans. d**

62. The enzyme converting starch into maltose is
(a) Invertase (b) Hydrogenase
(c) Maltase (d) Diastase **Ans. d**

63. Diastase enzyme helps in the digestion of
(a) Starch (b) Proteins
(c) Fats (d) Amino acids **Ans. a**

64. Which of the following enzymes acts upon fatty acids
(a) Amylase (b) Ligase
(c) Trypsin (d) Peptidase **Ans. b**

65. Lipase enzyme acts upon
(a) Water (b) Protein
(c) Fat (d) Sugar **Ans. c**

66. Which enzyme is required to digest the reserve food material (lipid) in castor seeds

(a) Lipase (b) Diastase

(c) Amylase (d) Protease **Ans. a**

67. In seeds, digestion is made possible at relatively low temperature by

(a) Auxins

(b) Proteins

(c) Enzymes

(d) Nitrogenous complex substances **Ans. c**

68. In the cell, digestive enzymes are mostly located in

(a) Lysosomes (b) Cell wall

(c) Chromosomes (d) Ribosome **Ans. a**

69. An enzyme acts by

(a) Reducing activation energy (b) Increasing activation energy

(c) Increasing reaction time (d) Decreasing reaction time

Ans. a

70. Which of the following is correct in an enzyme-controlled reaction (E=enzyme; S=Substrate, P=Product)

(a) $E + S \leftrightarrow E + P$ (b) $E + S \leftrightarrow ES \leftrightarrow E + P$

(c) $E + S \leftrightarrow ES \leftrightarrow E$ (d) $E + S \leftrightarrow P \leftrightarrow E + P$ **Ans. b**

71. Turnover number of enzyme depends upon

(a) Molecular weight of enzyme

(b) Size of enzyme molecule

(c) Active sites of enzyme molecule

(d) Concentration of substrate molecule **Ans. c**

72. Which of the following is the best evidence for template theory of enzyme action

(a) Compounds similar in structure to the substrate inhibit the reaction

(b) Enzymes speed up reaction by definite amount

(c) Enzymes determine the direction of a reaction

(d) Enzymes are found in living organisms and increase the rate

Ans. a

73. The enzyme is said to be working at maximum efficiency

(a) When substrate concentration is increased to point of saturation

(b) When substrate concentration is low

(c) When substrate coming in contact with active site are negligible

(d) None of the above is true **Ans. a**

74. When the action of the enzyme is inhibited in the presence of a substance which closely resembles the substrate molecule, then the inhibition is known as

(a) Feedback inhibition (b) Non-competitive inhibition

(c) Allosteric inhibition (d) Competitive inhibition **Ans. d**

75. In competitive inhibition, which of the following is true

(a) $E + I \leftrightarrow EI$ (b) $E + I \leftrightarrow EI + S \leftrightarrow ESI$

(c) $S + I \leftrightarrow SI$ (d) $ES + I \leftrightarrow ESI$ **Ans. a**

76. Non-competitive inhibitors are those which

(a) Alter the structure of protein molecule

(b) Get attached on the active site

(c) Activate the enzyme

(d) Break the bonds which are responsible for the formation of active sites **Ans. a**

77. The product of an enzyme-catalyzed reaction can act as inhibitor of the reaction. This mechanism of control is known as

(a) Feedback inhibition (b) Competitive inhibition

(c) Metabolic antagonism (d) Repression **Ans. a**

78. In feedback inhibition, a metabolic pathway is switched off by

(a) Competitive inhibition (b) Denaturation

(c) Accumulation of end product (d) Allosteric inhibition **Ans. d**

79. Which of the following statements is incorrect

(a) Enzymes hasten the completion of a reaction

(b) The two terms 'substrate' and 'product' signify the starting and ending materials of a reaction

(c) Enzymes are affected by the reactions they catalyse

(d) Enzyme exhibit specificity for the reactions there catalyse **Ans. c**

80. Enzymes, vitamins and hormones can be classified in a single category of biological chemicals because all of them

(a) Are proteins

(b) Are synthesized in organisms

(c) Enhance the oxidative metabolism

(d) Aid in regulating metabolism **Ans. d**

81. Which enzymes can digest plant protein

(a) Pepsin (b) Erepsin

(c) Rennin (d) All the above **Ans. a**

82. Enzymes as they exist inside the cell are

(a) In solid form (b) In crystalline form

(c) In solution form (d) In colloidal form **Ans. d**

83. Feedback inhibition is affected by

(a) End product (b) First product

(c) Enzymes (d) External factor **Ans. a**

84. K_m value refers to

(a) Maximum reaction velocity

(b) Near maximum reaction velocity

(c) One half of the maximum reaction velocity

(d) Threshold value **Ans. c**

85. The substrate concentration at which an enzyme attains half its maximum velocity is called

(a) Threshold value (b) Half life

(c) Michaelis Menton constant (d) Concentration coefficient

Ans. c

86. For the enzyme action

(a) Value of K_m is low (b) Value of K_i is low

(c) Value of K_m is high (d) Value of K_i is high **Ans. b**

87. The function of an enzyme is

(a) To cause chemical reactions which would not occur otherwise

(b) To change the rates of chemical reactions

(c) To control the equilibrium points of reactions

(d) To change the direction of reactions **Ans. b**

88. The organic compounds which have transient association with apoenzymes are called

(a) Holoenzyme (b) Coenzyme

(c) Prosthetic group (d) None of these **Ans. b**

89. Iron is a cofactor responsible for the catalytic action of

(a) Sucrose (b) Lipase

(c) Catalase (d) Cellulose **Ans. c**

90. The fastest acting enzyme is

(a) Peroxidase (b) Amylase

(c) Carbonic anhydrase (d) Phosphoglyceromutase **Ans. c**

91. The enzyme bromelain is obtained from

(a) Papaya (b) Pineapple

(c) Saliva (d) Witches broom **Ans. b**

92. The evolution of oxygen from H_2O_2 in a living tissue is catalysed by

(a) MnO_2 (b) Peroxidase

(c) SnO_2 (d) Invertase **Ans. b**

93. The enzyme sucrase acts on

(a) Sucrose only (b) Any disaccharide

(c) Starch and sucrose (d) Any organic molecule **Ans. a**

94. The active site of an enzyme is formed by some of its

(a) R groups on the amino acids

(b) Amino groups of amino acids

(c) Carboxyl groups of the amino acids

(d) Exposed sulphur bonds **Ans. a**

95. Upon binding the substrate at one site, other sites on an enzyme become more reactive. This is called

(a) Allosteric inhibition (b) Specificity

(c) Co-operativity (d) Activation **Ans. c**

96. Which of these inactivates an enzyme by changing enzyme shape

(a) Allosteric inhibitor (b) Competitive inhibitor

(c) Irreversible inhibitor (d) Multienzyme complex **Ans. a**

97. Which of these inactivates an enzyme by denaturing it

(a) Allosteric inhibitor (b) Competitive inhibitor

(c) Irreversible inhibitor (d) Multienzyme complex **Ans. c**

98. The specificity of an enzyme is governed by the tertiary structure. In forming a tertiary structure, bonds develop between certain amino acids. The resulting shape provides for the substrate

(a) Spatial fit (b) Bonding fit

(c) Spatial and bonding fit both (d) Neither spatial nor bonding fit

Ans. c

99. Only a substrate with a particular shape is help by the enzyme. It is due to

(a) Spatial fit of the active site

(b) Bonding fit of the active site

(c) Both spatial and bonding fit of the active site

(d) Neither spatial nor bonding fit of the active site **Ans. a**

100. Bonding fit of an enzyme is the presence of certain active sites

(a) Within the grooves in the surface of tertiary structure

(b) Outside the grooves on the surface of tertiary surface

(c) Both within and outside the grooves in the surface of tertiary structure

(d) Formed as a result of quaternary structure **Ans. a**

101. Specificity of the bonding fit for a particular reaction is the result of

(a) R groups on the amino acids

(b) Amino groups of amino acids

(c) Carboxyl groups of the amino acids

(d) Exposed sulphur bonds **Ans. a**

102. Which of the following statements is correct

(a) Enzymes can substitute one another

(b) No other enzyme can substitute for a missing enzyme

(c) A few enzymes can substitute a given enzyme

(d) While some enzymes can be substituted by other enzymes, there is no substitute for most of the enzymes **Ans. b**

103. All the enzymes of a multienzyme complex are arranged together in

(a) Solution of ATP (b) Membrane

(c) Quaternary protein (d) Coenzyme **Ans. b**

104. During feedback inhibition, the high concentrations of the end product of a metabolic pathway

(a) Slow down the enzymes which control earlier steps in the pathway

(b) Switch off the enzymes which control earlier steps in the pathway

(c) Slow down or switch off the enzymes which control earlier steps in the pathway

(d) Denature the enzymes which control earlier steps in the pathway

Ans. c

105. Feedback inhibition quite often involves

(a) Competitive inhibition (b) Irreversible inhibition

(c) Allosteric inhibition (d) All the above **Ans. c**

106. Glutamate pyruvate transaminase enzyme is an example of

(a) Qxidoreductases (b) Transferases

(c) Lyases (d) Ligases **Ans. b**

107. Most of the digestive enzymes belongs to the category

(a) Qxidoreductases (b) Transferases

(c) Hydrolases (d) Lyases **Ans. c**

108. Histidine decarboxylase enzyme belongs to the category

(a) Qxidoreductases (b) Isomerases

(c) Lyases (d) Ligases **Ans. c**

109. Which enzyme require energy for their action and obtain it from the hydrolysis of ATP

(a) Qxidoreductases (b) Transferases

(c) Ligases (d) Lyases **Ans. c**

110. The enzyme pyruvate carboxylase combines pyruvic acid and CO_2 to form oxaloacetic acid at the expense of ATP. The enzyme is

(a) An isomerase (b) A ligase

(c) A lyase (d) A transferase **Ans. b**

111. The activity of an enzyme declines

(a) Above the optimum temperature

(b) Below the optimum temperature

(c) Both above and below the optimum temperature

(d) Below the minimum and above the maximum temperature **Ans. c**

112. Pepsinogen (inactive form of enzyme pepsin) is converted into its active form by

(a) Isomerism (b) Hydrolysis

(c) Oxidation (d) Reduction **Ans. b**

113. Most intracellular enzymes function best

(a) At neutral pH

(b) In acidic conditions

(c) In basic conditions

(d) Either neutral or acidic conditions **Ans. a**

114. The protein-digesting enzyme trypsin functions best is

(a) pH 7.0 (b) pH 6.5

(c) pH 7.5 (d) pH 8.5 **Ans. d**

115. On reaching the stomach, the enzyme salivary amylase is inactivated by

(a) Pepsinogen (b) Pepsin

(c) Hydrochloric acid (d) Mucous **Ans. c**

116. The inhibition of succinic acid dehydrogenase by malonic acid is an example of

(a) Competitive inhibition (b) Allosteric inhibition

(c) Feedback inhibition (d) Irreversible inhibition **Ans. a**

117. Sulpha drugs are used for the control of bacterial pathogens because they cause

(a) Competitive inhibition of folic acid synthesis

(b) Allosteric inhibition of folic acid synthesis

(c) Feedback inhibition of folic acid synthesis

(d) Irreversible inhibition of folic acid synthesis **Ans. a**

118. Cyanide kills an organism by inhibiting

(a) Hexokinase (b) Cytochrome oxidase

(c) Succinate dehydrogenase (d) Histidine decarboxylase **Ans. b**

119. How many categories of enzymes have been recognized by IUB

(a) 5 (b) 6

(c) 7 (d) 8 **Ans. b**

120. Which ions are toxic for enzyme activity

(a) Mn^{++} (b) K^{+}

(c) Na^{++} (d) Hg^{++} **Ans. d**

121. The enzyme Hexokinase which catalyses glucose to Glucose-6-phosphate in glycolysis is inhibited by Glucose-6-phosphate. This is an example of

(a) Feedback allosteric inhibition (b) Positive feedback inhibition

(c) Competitive inhibition (d) Non-competitive inhibition

Ans. a

122. If temperature is increased from 30°C to 45°C, the rate enzyme activity will

(a) Only increase

(b) Only decrease

(c) Initially increase then decrease

(d) Initially decrease then increase **Ans. c**

123. The compound which has the lowest density is

(a) Chylomicron (b) β-Lipoprotein

(c) α-Lipoprotein (d) pre β-Lipoprotein **Ans. a**

124. Non steroidal anti inflammatory drugs, such as aspirin act by inhibiting the activity of the enzyme

(a) Lipoxygenase (b) Cyclooxygenase

(c) Phospholipase A_2 (d) Lipoprotein lipase **Ans. b**

125. From arachidonate, synthesis of prostaglandins is catalysed by

(a) Cyclooxygenase (b) Lipoxygenase

(c) Thromboxane synthase (d) Isomerase **Ans. a**

126. A Holoenzyme is

(a) Functional unit (b) Apo enzyme

(c) Coenzyme (d) All of these **Ans. d**

127. Gaucher's disease is due to the deficiency of the enzyme

(a) α -Fucosidase (b) β -Galactosidase

(c) β -Glucosidase (d) Sphingomyelinase **Ans. c**

128. Neimann-Pick disease is due to the deficiency of the enzyme

(a) Hexosaminidase A and B (b) Ceramidase

(c) Ceramide lactosidase (d) Sphingomyelinase **Ans. d**

129. Krabbe's disease is due to the deficiency of the enzyme

(a) Ceramide lactosidase (b) Ceramidase

(c) β -Galactosidase (d) GM1 β -Galactosidase **Ans. c**

130. Fabry's disease is due to the deficiency of the enzyme
(a) Ceramide trihexosidase (b) Galactocerebrosidase
(c) Phytanic acid oxidase (d) Sphingomyelinase **Ans. a**

131. Farber's disease is due to the deficiency of the enzyme
(a) α -Galactosidase (b) Ceramidase
(c) β -Glucocerebrosidase (d) Arylsulphatase A **Ans. b**

132. A synthetic nucleotide analogue, used in organ transplantation as a suppressor of immunologic rejection of grafts is
(a) Theophylline
(b) Cytarabine
(c) 4-Hydroxypyrazolopyrimidine
(d) 6-Mercaptopurine **Ans. d**

133. Example of an extracellular enzyme is
(a) Lactate dehydrogenase (b) Cytochrome oxidase
(c) Pancreatic lipase (d) Hexokinase **Ans.c**

134. Enzymes, which are produced in inactive form in the living cells, are called
(a) Papain (b) Lysozymes
(c) Apoenzymes (d) Proenzymes **Ans. d**

135. An example of ligases is
(a) Succinate thiokinase (b) Alanine racemase
(c) Fumarase (d) Aldolase **Ans. a**

136. An example of lyases is
(a) Glutamine synthetase (b) Fumarase
(c) Cholinesterase (d) Amylase **Ans. b**

137. Activation or inactivation of certain key regulatory enzymes is accomplished by covalent modification of the amino acid
(a) Tyrosine (b) Phenylalanine
(c) Lysine (d) Serine **Ans. d**

138. The enzyme which can add water to a carbon-carbon double bond or remove water to create a double bond without breaking the bond is
(a) Hydratase (b) Hydroxylase
(c) Hydrolase (d) Esterase **Ans. a**

139. Fischer's 'lock and key' model of the enzyme action implies that

(a) The active site is complementary in shape to that of substance only after interaction

(b) The active site is complementary in shape to that of substance

(c) Substrates change conformation prior to active site interaction

(d) The active site is flexible and adjusts to substrate **Ans. b**

140. From the Lineweaver-Burk plot of Michaelis-Menten equation, K_m and V_{max} can be determined when V is the reaction velocity at substrate concentration S, the X-axis experimental data are expressed as

(a) 1/V (b) V

(c) 1/S (d) S **Ans. c**

141. A sigmoidal plot of substrate concentration ([S]) verses reaction velocity (V) may indicate

(a) Michaelis-Menten kinetics (b) Co-operative binding

(c) Competitive inhibition (d) Non-competitive inhibition

Ans. b

142. The K_m of the enzyme giving the kinetic data as below is

(a) -0.50 (b) -0.25

(c) +0.25 (d) +0.33 **Ans. d**

143. The kinetic effect of purely competitive inhibitor of an enzyme

(a) Increases K_m without affecting V_{max}

(b) Decreases K_m without affecting V_{max}

(c) Increases V_{max} without affecting K_m

(d) Decreases V_{max} without affecting K_m **Ans. a**

144. Which type of enzyme catalyses the following reaction

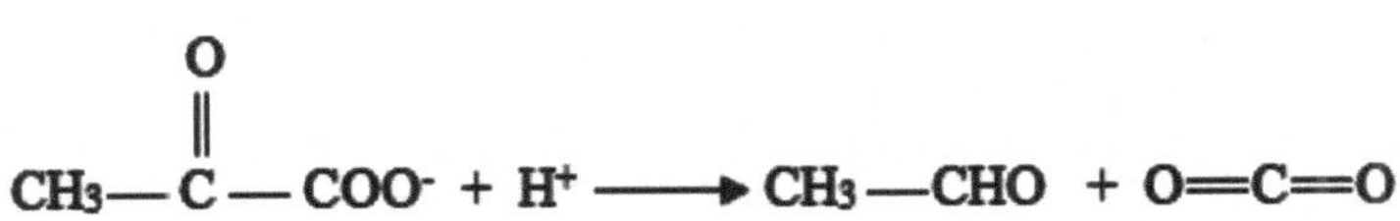

(a) Lyase (b) Ligase

(c) Isomerase (d) Oxidoreductase **Ans. a**

145. An inducer is absent in the type of enzyme:

(a) Allosteric enzyme (b) Constitutive enzyme

(c) Co-operative enzyme (d) Isoenzyme **Ans. b**

146. A demonstrable inducer is absent in

(a) Allosteric enzyme (b) Constitutive enzyme
(c) Inhibited enzyme (d) Co-operative enzyme **Ans. b**

147. In reversible non-competitive enzyme activity inhibition

(a) V_{max} is increased
(b) K_m is increased
(c) K_m is decreased
(d) Concentration of active enzyme is reduced **Ans. d**

148. In reversible non-competitive enzyme activity inhibition

(a) Inhibitor bears structural resemblance to substrate
(b) Inhibitor lowers the maximum velocity attainable with a given amount of enzyme
(c) K_m is increased
(d) K_m is decreased **Ans. b**

149. In competitive enzyme activity inhibition

(a) The structure of inhibitor generally resembles that of the substrate
(b) Inhibitor decreases apparent K_m
(c) K_m remains ineffective
(d) Inhibitor decreases V_{max} without affecting K_m **Ans. a**

150. In enzyme kinetics V_{max} reflects

(a) The amount of an active enzyme
(b) Substrate concentration
(c) Half the substrate concentration
(d) Enzyme substrate complex **Ans. a**

151. In enzyme kinetics K_m implies

(a) The substrate concentration that gives one half V_{max}
(b) The dissociation constant for the enzyme substrate complex
(c) Concentration of enzyme
(d) Half of the substrate concentration required to achieve V_{max} **Ans. b**

152. In competitive enzyme activity inhibition

(a) Apparent K_m is decreased (b) Apparent K_m is increased
(c) V_{max} is increased (d) V_{max} is decreased **Ans. b**

153. In non competitive enzyme activity inhibition, inhibitor

(a) Increases K_m (b) Decreases K_m

(c) Does not effect K_m (d) Increases K_m **Ans. c**

154. An enzyme catalyzing oxidoreduction, using oxygen as hydrogen acceptor is

(a) Cytochrome oxidase (b) Lactate dehydrogenase

(c) Malate dehydrogenase (d) Succinate dehydrogenase

Ans. a

155. The enzyme using some other substance, not oxygen as hydrogen acceptor is

(a) Tyrosinase (b) Succinate dehydrogenase

(c) Uricase (d) Cytochrome oxidase **Ans. b**

156. An enzyme which uses hydrogen acceptor as substrate is

(a) Xanthine oxidase (b) Aldehyde oxidase

(c) Catalase (d) Tryptophan oxygenase **Ans. c**

157. Enzyme involved in joining together two substrates is

(a) Glutamine synthetase (b) Aldolase

(c) Guanine deaminase (d) Arginase **Ans. a**

158. The pH optima of most of the enzymes is

(a) Between 2 and 4 (b) Between 5 and 9

(c) Between 8 and 12 (d) Above 12 **Ans. b**

159. Coenzymes are

(a) Heat stable, dialyzable, non protein organic molecules

(b) Soluble, colloidal, protein molecules

(c) Structural analogue of enzymes

(d) Different forms of enzymes **Ans. a**

160. An example of hydrogen transferring coenzyme is

(a) CoA (b) NAD^+

(c) Biotin (d) TPP **Ans. b**

161. An example of group transferring coenzyme is

(a) NAD^+ (b) $NADP^+$

(c) FAD (d) CoA **Ans. d**

162. Cocarboxylase is

(a) Thiamine pyrophosphate (b) Pyridoxal phosphate

(c) Biotin (d) CoA **Ans. c**

163. A coenzyme containing non aromatic hetero ring is

(a) ATP (b) NAD

(c) FMN (d) Biotin **Ans. d**

164. A coenzyme containing aromatic hetero ring is

(a) TPP (b) Lipoic acid

(c) Coenzyme Q (d) Biotin **Ans. a**

165. Isoenzymes are

(a) Chemically, immunologically and electrophoretically different forms of an enzyme

(b) Different forms of an enzyme similar in all properties

(c) Catalysing different reactions

(d) Having the same quaternary structures like the enzymes **Ans. a**

166. Isoenzymes can be characterized by

(a) Proteins lacking enzymatic activity that are necessary for the activation of enzymes

(b) Proteolytic enzymes activated by hydrolysis

(c) Enzymes with identical primary structure

(d) Similar enzymes that catalyse different reaction **Ans. b**

167. The isoenzymes of LDH

(a) Differ only in a single amino acid

(b) Differ in catalytic activity

(c) Exist in 5 forms depending on M and H monomer contents

(d) Occur as monomers **Ans. c**

168. The normal value of CPK in serum varies between

(a) 4–60 IU/L (b) 60–250 IU/L

(c) 4–17 IU/L (d) > 350 IU/L **Ans. a**

169. Factors affecting enzyme activity

(a) Concentration (b) pH

(c) Temperature (d) All of these **Ans. d**

170. The normal serum GOT activity ranges from

(a) 3.0–15.0 IU/L (b) 4.0–17.0 IU/L

(c) 4.0–60.0 IU/L (d) 0.9–4.0 IU/L **Ans. b**

171. The normal GPT activity ranges from

(a) 60.0–250.0 IU/L (b) 4.0–17.0 IU/L

(c) 3.0–15.0 IU/L (d) 0.1–14.0 IU/L **Ans. c**

172. The normal serum acid phosphatase activity ranges from

(a) 5.0–13.0 KA units/100 ml (b) 1.0–5.0 KA units/100 ml

(c) 13.0–18.0 KA units/100 ml (d) 0.2–0.8 KA units/100 ml

Ans. b

173. The normal serum alkaline phosphatase activity ranges from

(a) 1.0–5.0 KA units/100 ml (b) 5.0–13.0 KA units/100 ml

(c) 0.8–2.3 KA units/100 ml (d) 13.0–21.0 KA units/100 ml

Ans. b

174. In early stages of myocardial ischemia the most sensitive indicator is the measurement of the activity of

(a) CPK (b) SGPT

(c) SGOT (d) LDH **Ans. a**

175. Serum acid phosphatase level increases in

(a) Metastatic carcinoma of prostate

(b) Myocardial infarction

(c) Wilson's disease

(d) Liver diseases **Ans. a**

176. Serum alkaline phosphatase level increases in

(a) Hypothyroidism (b) Carcinoma of prostate

(c) Hyperparathyroidism (d) Myocardial ischemia **Ans. c**

177. Serum lipase level increases in

(a) Paget's disease (b) Gaucher's disease

(c) Acute pancreatitis (d) Diabetes mellitus **Ans. c**

178. Serum ferroxidase level decreases in

(a) Gaucher's disease (b) Cirrhosis of liver

(c) Acute pancreatitis (d) Wilson's disease **Ans. d**

179. The isoenzymes LDH_5 is elevated in

(a) Myocardial infarction (b) Peptic ulcer

(c) Liver disease (d) Infectious diseases **Ans. c**

180. On the third day of onset of acute myocardial infarction the enzyme elevated is

(a) Serum AST (b) Serum CK

(c) Serum LDH (d) Serum ALT **Ans. c**

181. LDH_1 and LDH_2 are elevated in

(a) Myocardial infarction (b) Liver disease

(c) Kidney disease (d) Brain disease **Ans. a**

182. The CK isoenzymes present in cardiac muscle is

(a) BB and MB (b) MM and MB

(c) BB only (d) MB only **Ans. b**

183. In acute pancreatitis, the enzyme raised in first five days is

(a) Serum amylase (b) Serum lactic dehydrogenase

(c) Urinary lipase (d) Urinary amylase **Ans. a**

184. Acute pancreatitis is characterised by

(a) Lack of synthesis of zymogen enzymes

(b) Continuous release of zymogen enzymes into the gut

(c) Premature activation of zymogen enzymes

(d) Inactivation of zymogen enzymes **Ans. c**

185. An example of functional plasma enzyme is

(a) Lipoprotein lipase (b) Amylase

(c) Aminotransferase (d) Lactate dehydrogenase **Ans. a**

186. A non-functional plasma enzyme is

(a) Pseudocholinesterase

(b) Lipoprotein lipase

(c) Proenzyme of blood coagulation

(d) Lipase **Ans. d**

187. The pH optima for salivary amylase is

(a) 6.6–6.8 (b) 2.0–7.5

(c) 7.9 (d) 8.6 **Ans. a**

188. The pH optima for pancreatic analyse is
(a) 4.0 (b) 7.1
(c) 7.9 (d) 8.6 **Ans. d**

189. The pH optima for sucrase is
(a) 5.0–7.0 (b) 5.8–6.2
(c) 5.4–6.0 (d) 8.6 **Ans. a**

190. The pH optima for maltase is
(a) 1.0–2.0 (b) 5.2–6.0
(c) 5.8–6.2 (d) 5.4–6.0 **Ans. c**

191. The pH optima for lactase is
(a) 1.0-2.0 (b) 5.4–6.0
(c) 5.0–7.0 (d) 5.8–6.2 **Ans. b**

192. The substrate for amylase is
(a) Cane sugar (b) Starch
(c) Lactose (d) Ribose **Ans. b**

193. The ion which activates salivary amylase activity is
(a) Chloride (b) Bicarbonate
(c) Sodium (d) Potassium **Ans. a**

194. The pancreatic amylase activity is increased in the presence of
(a) Hydrochloric acid (b) Bile salts
(c) Thiocyanate ions (d) Calcium ions **Ans. b**

195. A carbohydrate which can not be digested in human gut is
(a) Cellulose (b) Starch
(c) Glycogen (d) Maltose **Ans. a**

196. The sugar absorbed by facilitated diffusion and requiring Na independent transporter is
(a) Glucose (b) Fructose
(c) Galactose (d) Ribose **Ans. b**

197. In the intestine the rate of absorption is highest for
(a) Glucose and galactose (b) Fructose and mannose
(c) Fructose and pentose (d) Mannose and pentose **Ans. a**

198. Glucose absorption is promoted by
(a) Vitamin A (b) Thiamin
(c) Vitamin C (d) Vitamin K **Ans. b**

199. The hormone acting directly on intestinal mucosa and stimulating glucose absorption is

(a) Insulin (b) Glucagon

(c) Thyroxine (d) Vasopressin **Ans. c**

200. Given that the standard free energy change ($\Delta G°$) for the hydrolysis of ATP is –7.3 K cal/mol and that for the hydrolysis of Glucose 6-phosphate is –3.3 Kcal/mol, the $\Delta G°$ for the phosphorylation of glucose is Glucose + ATP $\rightarrow$ Glucose 6–Phosphate + ADP.

(a) –10.6 Kcal/mol (b) –7.3 Kcal/mol

(c) –4.0 Kcal/mol (d) +4.0 Kcal/mol **Ans. c**

201. At low blood glucose concentration, brain but not liver will take up glucose. It is due to the

(a) Low K_m of hexokinase (b) Low K_m of glucokinase

(c) Specificity of glucokinase (d) Blood brain barrier **Ans. a**

202. In the reaction below, NuTP stands for NuTP + Glucose $\rightarrow$ Glucose 6–Phosphate + NuDP.

(a) ATP (b) CTP

(c) GTP (d) UTP **Ans. a**

203. In a Michaelis - Menten enzyme mechanism, what substrate concentration (relative to K_m) is needed for the reaction rate to be 0.5 V_{max}

(a) 1/9 K_m (b) 1/3 K_m

(c) K_m (d) ¼ K_m **Ans. c**

204. The enzyme of the glycolic pathway, sensitive to inhibiton by fluoride ions is

(a) Hexokinase (b) Aldolase

(c) Enolase (d) Pyruvate kinase **Ans. c**

205. In glycolytic pathway, iodoacetate inhibits the activity of the enzyme

(a) Phosphotriose isomerase

(b) Glyceraldehyde-3-phosphate dehydrogenase

(c) Pyruvate kinase

(d) Phosphofructokinase **Ans. b**

206. In the glycolytic pathway, enolpyruvate is converted to ketopyruvate by

(a) Pyruvate kinase (b) Phosphoenolpyruvate

(c) Pyruvate dehydrogenase (d) Spontaneously **Ans. d**

207. In erythrocytes, 2, 3-biphosphoglycerate is derived from the intermediate:

(a) Glyeraldehyde-3-phosphate (b) 1, 3-Biphosphoglycerate

(c) 3-Phosphoglycerate (d) 2-Phosphoglycerate **Ans. b**

208. 2, 3 -Biphosphoglycerate in high concentrations, combines with hemoglobin, causes

(a) Displacement of the oxyhemoglobin dissociation curve to the left

(b) Displacement of the oxyhemoglobin dissociation curve to the right

(c) No change in oxyhemoglobin dissociation curve

(d) Increased affinity for oxygen **Ans. b**

209. Erythrocytes under normal conditions and microorganisms under anaerobic conditions may accumulate

(a) NADPH (b) Pyruvate

(c) Phosphoenolpyruvate (d) Lactate **Ans. d**

210. Enzymes leading to the high energy phosphorylation of substrates during glycolysis include which of the following

(a) Phosphoglycerate kinase

(b) Enolase

(c) Pyruvate Kinase

(d) Glyceraldehyde-3-phosphate dehydrogenase **Ans. d**

211. Lineweaver - Burk double reciprocal plot is related to

(a) Substrate concentration (b) Enzyme activity

(c) Temperature (d) Both (a) and (b) **Ans. d**

212. Phosphofructokinase key enzyme in glycolysis is inhibited by

(a) Citrate and ATP (b) AMP

(c) ADP (d) TMP **Ans. a**

213. One of the enzymes regulating glycolysis is

(a) Phosphofructokinase

(b) Glyceraldehyde-3-phosphate dehydrogenase

(c) Phosphotriose isomerase

(d) Phosphohexose isomerase **Ans. a**

214. Hexokinase is inhibited in an allosteric manner by

(a) Glucose-6-Phosphate (b) Glucose-1-Phosphate

(c) Fructose-6-phosphate (d) Fructose-1, 6-biphosphate

Ans. a

215. A reaction which may be considered as isomerisation is

(a) Glucose 6-Phosphate → fructose 6-phosphate

(b) 3-Phosphoglycerate → 2-phosphoglycerate

(c) 2-phosphoglycerate → phosphoenolpyruvate

(d) Pyruvate → Lactate **Ans. a**

216. The net number of ATP formed per mole of glucose in anaerobic glycolysis is

(a) 1 (b) 2

(c) 6 (d) 8 **Ans. b**

217. Pyruvate dehydrogenase a multienzyme complex is required for the production of

(a) Acetyl-CoA (b) Lactate

(c) Phosphoenolpyruvate (d) Enolpyruvate **Ans. a**

218. Dietary deficiency of thiamin inhibits the activity of the enzyme

(a) Pyruvate kinase (b) Pyruvate dehydrogenase

(c) Phosphofructokinase (d) Enolase **Ans. b**

219. Pyruvate dehydrogenase activity is inhibited by

(a) Mercury (b) Zinc

(c) Calcium (d) Sodium **Ans. a**

220. In the normal resting state of humans, most of the blood glucose burned as fuel is consumed by

(a) Liver (b) Adipose tissue

(c) Muscle (d) Brain **Ans. a**

221. All the enzymes of glycolysis pathway are found in

(a) Extramitochondrial soluble fraction of the cell

(b) Mitochondria

(c) Nucleus

(d) Endoplasmic reticulum **Ans. a**

222. Most major metabolic pathways are considered mainly either anabolic or catabolic. Which of the following pathway is most correctly considered to be amphibolic

(a) Citric acid cycle (b) Gluconeogenesis

(c) Lipolysis (d) Glycolysis **Ans. a**

223. The enzymes of the citric acid cycle are located in
(a) Mitochondrial matrix
(b) Extramitochondrial soluble fraction of the cell
(c) Nucleus
(d) Endoplasmic reticulum **Ans. a**

224. The initial step of the citric acid cycle is
(a) Conversion of pyruvate to acetyl-CoA
(b) Condensation of acetyl-CoA with oxaloacetate
(c) Conversion of citrate to isocitrate
(d) Formation of α-ketoglutarate catalysed by isocitrate dehydrogenase **Ans. b**

225. The substance which may be considered to play a catalytic role in citric acid cycle is
(a) Oxaloacetate (b) Isocitrate
(c) Malate (d) Fumarate **Ans. a**

226. An enzyme of the citric acid cycle also found outside the mitochondria is
(a) Isocitrate dehydrogenase
(b) Citrate synthetase
(c) α-Ketoglutarate dehydrogenase
(d) Malate dehydrogenase **Ans. c**

227. The reaction catalysed by α-ketoglutarate dehydrogenase in the citric acid cycle requires
(a) NAD (b) NADP
(c) ADP (d) ATP **Ans. a**

228. If all the enzymes, intermediates and cofactors of the citric acid cycle as well as an excess of the starting substrate acetyl-CoA are present and functional in an organelle free solution at the appropriate pH, which of the following factors of the citric acid cycle would prove to be rate limiting
(a) Molecular oxygen (b) Half life of enzyme
(c) Turnover of intermediates (d) Reduction of cofactors **Ans. d**

229. In TCA cycle, oxalosuccinate is converted to α-ketoglutarate by the enzyme
(a) Fumarase (b) Isocitrate dehydrogenase
(c) Aconitase (d) Succinase **Ans. b**

230. The enzyme α-ketoglutarate dehydrogenase in the citric acid cycle requires

(a) Lipoate (b) Folate

(c) Pyridoxine (d) Inositol **Ans. a**

231. The example of generation of a high energy phosphate at the substrate level in the citric acid cycle is the reaction

(a) Isocitrate → α-Ketoglutarate (b) Succinate → α -fumarate

(c) Malate → α -oxaloacetate (d) Succinyl CoA → α -Succinate **Ans. d**

232. Fluoroacetate inhibits the reaction of citric acid cycle

(a) Isocitrate → α -Ketoglutarate (b) Fumarate→ α -Malate

(c) Citrate → α -cis-aconitate (d) Succinate → α -fumarate **Ans. c**

233. Formation of succinyl- CoA from α-Ketoglutarate is inhibited by

(a) Fluoroacetate (b) Arsenite

(c) Fluoride (d) Iodoacetate **Ans. b**

234. The number of ATP molecules generated for each turn of the citric acid cycle is

(a) 8 (b) 12

(c) 24 (d) 38 **Ans. b**

235. Oxidation of one molecule of glucose yields

(a) 12 ATP (b) 24 ATP

(c) 36 ATP (d) 38 ATP **Ans. d**

236. Which of the following intermediates of metabolism can be both a precursor and a product of glucose

(a) Lactate (b) Pyruvate

(c) Alanine (d) Acetyl-CoA **Ans. a**

237. Mitochondrial membrane is freely pereamble to

(a) Pyruvate (b) Malate

(c) Oxaloacetate (d) Fumarate **Ans. b**

238. The reaction of Kreb's cycle which does not require cofactor of vitamin B group is

(a) Citrate → isocitrate (b) α –Ketoglutarate → succinate

(c) Malate → oxaloacetate (d) Succinate → fumarate **Ans. a**

239. The coenzyme not involved in the formation of acetyl-CoA from pyruvate is

(a) TPP (b) Biotin
(c) NAD (d) FAD **Ans. b**

240. A carrier molecule in the citric acid cycle is

(a) Acetyl-CoA (b) Citrate
(c) Oxaloacetate (d) Malate **Ans. c**

241. A specific inhibitor for succinate dehydrogenase is

(a) Arsenine (b) Arsenite
(c) Citrate (d) Fluoride **Ans. b**

242. The rate of citric acid cycle is controlled by the allosteric enzyme

(a) Aconitase (b) Fumarase
(c) Fumarase (d) Malate dehydrogenase **Ans. c**

243. In the erythrocytes, the net production of ATP molecules by the Rapoport-Leubering pathway is

(a) 0 (b) 2
(c) 4 (d) 8 **Ans. a**

244. The ratio that most closely approximates the number of net molecules of ATP formed per mole of glucose utilized under aerobic conditions to the net number formed under anaerobic conditions is

(a) 4:1 (b) 13:1
(c) 18:1 (d) 24:1 **Ans. c**

245. The pathway of glycogen biosynthesis involves a special nucleotide of glucose. In the reaction below, NuDP stands for

NuDP Glucose + $Glycogen_n \rightarrow$ NuDP + $Glycogen_{n+1}$

(a) ADP (b) GDP
(c) UDP (d) CDP **Ans. c**

246. Glucose 6-phosphate is converted to glucose 1-phosphate in a reaction catalyzed by the enzyme phosphoglucomutase, which is

(a) Phosphorylated
(b) Dephosphorylated
(c) Phosphorylated-dephosphorylated
(d) Phosphorylated- dephosphorylated- rephosphorylated **Ans. d**

247. The glycogen content of the liver is upto

(a) 6% (b) 8%

(c) 10% (d) 12% **Ans. a**

248. In glycogenesis a branch point in the molecule is established by the enzyme

(a) Amylo[1→4][1→6] transglucosidase

(b) α [1→4] α [1→4] Glucan transferase

(c) Amylo [1→6] glucosidase

(d) Glycogen synthase **Ans. a**

249. In glycogenolysis, the enzyme which transfers a trisaccharide unit from one branch to the other exposing 1→6 branch point is

(a) Phosphorylase

(b) α -[1→4] → α -[1→4] → Glucan transferase

(c) Amylo [1→6] glucosidase

(d) Amylo[1→4] → [1→6] transglucosidase **Ans. b**

250. In the synthesis of glycogen from glucose the reversible step is

(a) Glucose → Glucose 6-phosphate

(b) Glucose 6-phosphate → Glucose 1-phosphate

(c) Glucose 1-phosphate → UDP glucose

(d) UDP glucose → Glycogen **Ans. b**

251. The enzyme glucose-6-phosphatase which catalyses the conversion of glucose 6-phosphate to glucose is not found in

(a) Liver (b) Muscle

(c) Intestine (d) Kidney **Ans. b**

252. Allosteric activator of glycogen synthase is

(a) Glucose (b) Glucose-6-phosphate

(c) UTP (d) Glucose-1-phosphate **Ans. b**

253. Action of glycogen synthase is inhibited by

(a) Insulin (b) Glucose

(c) Mg^{2+} (d) Cyclic AMP **Ans. c**

254. The hormone activating the glycogen synthase activity is

(a) Insulin (b) Glucagon

(c) Epinephrine (d) ACTH **Ans. a**

255. Characteristic features of active site are
(a) Flexible in nature (b) Site of binding
(c) Acidic (d) Both (a) and (b) **Ans. d**

256. Von Gierke's disease is characterized by the deficiency of
(a) Glucose-6-phosphatase (b) α -1 $\rightarrow$ 4 Glucosidase
(c) 1$\rightarrow$ 6 Glucosidase (d) Liver phosphorylase **Ans. a**

257. Cori disease (Limit dextrinosis) is caused due to absence of
(a) Branching enzyme (b) Debranching enzyme
(c) Glycogen synthase (d) Phosphorylase **Ans. b**

258. Mc Ardle's syndrome is characterized by the absence of
(a) Liver phosphorylase (b) Muscle phosphorylase
(c) Branching enzyme (d) Debranching enzyme **Ans. b**

259. Pompe's disease is caused due to deficiency of
(a) Lysosomal α -1$\rightarrow$4 and 1$\rightarrow$6-glucosidase
(b) Glucose-6-phosphatase
(c) Glycogen synthase
(d) Phosphofructokinase **Ans. a**

260. Amylopectinosis is caused due to absence of
(a) Debranching enzyme (b) Branching enzyme
(c) Acid maltase (d) Glucose-6-phosphatase **Ans. b**

261. Her's disease is characterized by deficiency of
(a) Muscle phosphorylase (b) Liver phosphorylase
(c) Debranching enzyme (d) Glycogen synthase **Ans. b**

262. Tarui disease is characterized by the deficiency of the enzyme
(a) Liver phosphorylase
(b) Muscle phosphorylase
(c) Muscle and erythrocyte phosphofructokinase
(d) Lysosomal acid maltase **Ans. c**

263. The hexose monophosphate pathway includes the enzyme
(a) Maltase dehydrogenase
(b) Hexokinase
(c) α -Ketoglutarate dehydrogenase
(d) Glucose-6-phosphate dehydrogenase **Ans. d**

264. The hydrogen acceptor used in pentose phosphate pathway is

(a) NAD (b) NADP

(c) FAD (d) FMN **Ans. b**

265. The enzymes of the pentose phosphate pathway are found in the

(a) Cytosol (b) Mitochondria

(c) Nucleus (d) Endoplasmic reticulum **Ans. a**

266. In pentose phosphate pathway, D-ribulose-5-phosphate is converted to D-ribose-5-phosphate by the enzyme

(a) Fumarase (b) Ketoisomerase

(c) G-6-PD (d) Epimerase **Ans. b**

267. The transketolase enzyme in the pentose phosphate pathway requires the B vitamin

(a) Pantothenic acid (b) Thiamin

(c) Riboflavin (d) Nicotinic acid **Ans. b**

268. Xylulose-5-phosphate serves as a donar of active glycolaldehyde, the acceptor is

(a) Erythrose 4-phosphate (b) Ribose 5-phosphate

(c) Glyceraldehyde 3-phosphate (d) Sedoheptulose 7-phosphate

Ans. a

269. Pentose phosphate pathway is of significance because it generates

(a) NADPH for reductive synthesis

(b) Regenerates glucose 6-phosphate

(c) Generates fructose 6-phosphate

(d) Forms glyceraldehyde 3-phosphate **Ans. a**

270. The pentose phosphate pathway protects erythrocytes against hemolysis by assisting the enzyme

(a) Superoxide dismutase (b) Catalase

(c) Glutathionic peroxidase (d) Cytochrome oxidase **Ans. c**

271. Hemolytic anemia is caused by the deficiency of certain enzymes of the pentose phosphate pathway, the principal enzyme involved is

(a) Glucose-6-phosphate dehydrogenase

(b) Aldolase

(c) Fructose 1, 6-bisphosphatase
(d) Phosphohexose isomerase **Ans. a**

272. The sites for gluconeogenesis are
(a) Liver and kidney (b) Skin and pancreas
(c) Lung and brain (d) Intestine and lens of eye
Ans. a

273. An enzyme involved in gluconeogenesis is
(a) Pyruvate kinase (b) Pyruvate carboxylase
(c) Hexokinase (d) Phosphohexose isomerase
Ans. b

274. The enzyme pyruvate carboxylase is present in
(a) Cytosol (b) Mitochondria
(c) Nucleus (d) Golgi bodies **Ans. b**

275. The enzyme phosphoenolpyruvate carboxykinase catalyses the conversion of oxaloacetate to phosphoenolpyruvate requires
(a) ATP (b) ADP
(c) AMP (d) GTP **Ans. d**

276. The enzyme glucose 6-phosphatase is present in
(a) Liver (b) Muscle
(c) Adipose tissue (d) Brain **Ans. a**

277. In gluconeogensis, an allosteric activator required in the synthesis of oxaloacetate from bicarbonate and pyruvate, which is catalyzed by the enzyme pyruvate carboxylase is
(a) Acetyl CoA (b) Succinate
(c) Isocitrate (d) Citrate **Ans. a**

278. The number of ATP molecules required to convert 2 molecules of lactate into glucose in mammalian liver is
(a) 2 (b) 4
(c) 5 (d) 6 **Ans. d**

279. For conjugation with many endogenous and exogenous substances before elimination in urine, the uronic acid pathway provides
(a) Active glucuronate (b) Gluconate
(c) Xylulose (d) Xylitol **Ans. a**

280. UDP glucose is converted to UDP glucuronate, a reaction catalysed by UDP glucose dehydrogenase requires

(a) NAD^+ (b) FAD

(c) NADP (d) FMN **Ans. a**

281. Pentosuria is a rare hereditary disease is characterized by increased urinary excretion of

(a) L-xylulose (b) Xylitol

(c) Xylulose 5-phosphate (d) Ribose 5-phosphate **Ans. a**

282. The enzyme involved in essential pentosuria is

(a) Reductase (b) Hydroxylase

(c) Isomerase (d) Racemase **Ans. a**

283. Galactose is synthesized from glucose in

(a) Mammary gland (b) Intestine

(c) Kidney (d) Adipose tissue **Ans. a**

284. Galactose is readily converted to glucose in

(a) Liver (b) Intestine

(c) Kidney (d) Adipose tissue **Ans. a**

285. Galactose 1-phosphate is converted to uridine diphosphate galactose, the reaction is catalysed by the enzyme

(a) Galactokinase

(b) Galactose 1-phosphate uridyl transferase

(c) Uridine diphospho galactose 4-epimerase

(d) UDP glucose pyrophosphorylase **Ans. b**

286. The best known cause of galactosemia is the deficiency of

(a) Galactose 1-phosphate and uridyl transferase

(b) Phosphoglucomutase

(c) Galactokinase

(d) Lactose synthase **Ans. a**

287. Conversion of fructose to sorbitol is catalysed by the enzyme

(a) Sorbitol dehydrogenase (b) Aldose reductase

(c) Fructokinase (d) Hexokinase **Ans. a**

288. A specific fructokinase present in liver has a very high affinity for its substrate because

(a) K_m for fructose is very high (b) K_m for fructose is very low

(c) Activity is affected by fasting (d) Activity is affected by insulin

Ans. b

289. Insulin has no effect on the activity of the enzyme

(a) Glycogen synthetase (b) Fructokinase

(c) Pyruvate kinase (d) Pyruvate dehydrogenase

Ans. b

290. The pathogenesis of diabetic cataract involves accumulation of

(a) Galactose (b) Mannitol

(c) Sorbitol (d) Pyruvate **Ans. c**

291. Hereditary fructose intolerance involves the absence of the enzyme

(a) Aldolase B (b) Fructokinase

(c) Triokinase (d) Phosphotriose isomerase

Ans. a

292. Essential fructosuria is characterized by the lack of the hepatic enzyme

(a) Phosphohexose isomerase (b) Aldolase A

(c) Aldolase B (d) Fructokinase **Ans. d**

293. In normal individuals glycosuria occurs when the venous blood glucose concentration exceeds

(a) 5–6 mmol/L (b) 7–8 mmol/L

(c) 8.5–9 mmol/L (d) 9.5–10 mmol/L **Ans. d**

294. Phlorizin inhibits

(a) Renal tubular reabsorption of glucose

(b) Glycolysis

(c) Gluconeogenesis

(d) Glycogenolysis **Ans. a**

295. Renal glycosuria is characterized by

(a) Hyperglycemia

(b) Hyperglycemia with glycosuria

(c) Normal blood glucose level with glycosuria

(d) Hyperglycemia with ketosis **Ans. c**

296. Acute hemolytic anemia in persons sensitive to the Fava beans is due to the deficiency of the enzyme

(a) Pyruvate dehydrogenase (b) G-6-PD
(c) Aconitase (d) Transketolase **Ans. b**

297. Acute hemolytic episode after administration of antimalarial, primaquin, is due to deficiency of the enzyme

(a) 6-Phosphogluconate dehydrogenase
(b) Glucose-6-phosphate dehydrogenase
(c) Epimerase
(d) Transketolase **Ans. b**

298. The pH optima of gastric lipase is

(a) 3.0–6.0 (b) 1.0–2.0
(c) 8.0 (d) 8.6 **Ans. a**

299. The optimum pH of pancreatic lipase is

(a) 2.0 (b) 4.0
(c) 6.0 (d) 8.0 **Ans. c**

300. Gastric lipase is activated in the presence of

(a) Bile salts (b) Cu^{++}
(c) K^+ (d) Na^+ **Ans. a**

301. An example of enzyme inhibition

(a) Reversible inhibition (b) Irreversible inhibition
(c) Allosteric inhibition (d) All of these **Ans. d**

302. The formation of Δ^2-trans-enoyl-CoA from acyl-CoA requires the enzyme

(a) Acyl-CoA synthetase
(b) Acyl-CoA dehydrogenase
(c) 3-Hydroxy acyl-CoA dehydrogenase
(d) Thiolase **Ans. b**

303. In β-oxidation 3-ketoacyl-CoA is splitted at the 2, 3 position by the enzyme

(a) Hydratase (b) Dehydrogenase
(c) Reducatse (d) Thiolase **Ans. d**

304. Fatty acids with odd number of carbon atoms yield acetyl-CoA and a molecule of

(a) Succinyl-CoA (b) Propionyl-CoA
(c) Malonyl-CoA (d) Acetoacetyl-CoA **Ans. b**

305. For each of the first 7-acetyl-CoA molecules formed by α -oxidation of palmitic acid, the yield of high energy phosphates is

(a) 12 (b) 24
(c) 30 (d) 35 **Ans. d**

306. The net gain of ATP/mol of palmitic acid on complete oxidation is

(a) 88 (b) 105
(c) 129 (d) 135 **Ans. c**

307. ω-oxidation is normally a very minor pathway and is brought by hydroxylase enzymes involving

(a) Cytochrome a (b) Cytochrome b
(c) Cytochrome c (d) Cytochrome p-450 **Ans. c**

308. α -Oxidation i.e., the removal of one carbon at a time from the carboxyl end of the molecule has been detected in

(a) Brain tissue (b) Liver
(c) Adipose tissue (d) Intestine **Ans. a**

309. In β-oxidation, the coenzyme for acyl- CoA dehydrogenase is

(a) FMN (b) NAD
(c) NADP (d) FAD **Ans. d**

310. The coenzyme involved in dehydrogenation of 3-hydroxy acyl-CoA is

(a) FAD (b) FMN
(c) NAD (d) NADP **Ans. c**

311. The concentration of ketone bodies in the blood does not normally exceed

(a) 0.2 mmol/L (b) 0.4 mmol/L
(c) 1 mmol/L (d) 2 mmol/L **Ans. a**

312. In humans under normal conditions loss of ketone bodies via urine is usually less than

(a) 1 mg/24 hr (c) 4 mg/24 hr
(c) 8 mg/24 hr (d) 10 mg/24 hr **Ans. a**

313. The structure which appears to be the only organ to add significant quantities of ketone bodies to the blood is

(a) Brain (b) Erythrocytes
(c) Liver (d) Skeletal muscle **Ans. c**

314. The starting material for ketogenesis is
(a) Acyl- CoA (b) Acetyl- CoA
(c) Acetoacetyl-CoA (d) Malonyl-CoA **Ans. c**

315. Enzymes responsible for ketone body formation are associated mainly with the
(a) Mitochondria (b) Endoplasmic reticulum
(c) Nucleus (d) Golgi apparatus **Ans. a**

316. The synthesis of 3-hydroxy-3-methylglutaryl-CoA can occur
(a) Only in mitochondria of all mammalian tissues
(b) Only in the cytosol of all mammalian tissue
(c) In both cytosol and mitochondria
(d) In lysosomes **Ans. c**

317. In the pathway leading to biosynthesis of acetoacetate from acetyl-CoA in liver, the immediate precursor of aceotacetate is
(a) Acetoacetyl-CoA
(b) 3-Hydroxybutyryl-CoA
(c) 3-Hydroxy-3-methyl-glutaryl-CoA
(d) 3-Hydroxybutyrate **Ans. a**

318. Ketone bodies serve as a fuel for
(a) Extrahepatic tissues (b) Hepatic tissues
(c) Erythrocytes (d) Mitochondria **Ans. a**

319. In extra hepatic tissues, one mechanism for utilization of acetoacetate involves
(a) Malonyl-CoA (b) Succinyl-CoA
(c) Propionyl-CoA (d) Acetyl-CoA **Ans. b**

320. Ketosis reflects
(a) Increased hepatic glucose liberation
(b) Increased fatty acid oxidation
(c) Increased carbohydrate utilisation
(d) Increased gluconeogenesis **Ans. b**

321. Ketosis is associated with the disease
(a) Nephritis (b) Diabetes mellitus
(c) Edema (d) Coronary artery diseases
Ans. b

322. The main pathway for de novo synthesis of fatty acids occur in
(a) Cytosol (b) Mitochondria
(c) Microsomes (d) Nucleus **Ans. a**

323. Chain elongation of fatty acids in mammalian liver occurs in
(a) Nucleus (b) Ribosomes
(c) Lysosomes (d) Microsomes **Ans. d**

324. Acetyl- CoA is the principal building block of fatty acids. It is produced within the mitochondria and does not diffuse readily into cytosol. The availability of acetyl CoA involves
(a) Carnitine acyl transferase (b) Pyruvate dehydrogenase
(c) Citrate lyase (d) Thiolase **Ans. c**

325. The synthesis of fatty acids is often termed reductive synthesis
(a) $NADP^+$ (b) NADH
(c) $FADH_2$ (d) NADPH **Ans. d**

326. The protein, which is in fact a multifunctional enzyme complex in higher organism is
(a) Acetyl transacylase
(b) Malonyl transacylase
(c) 3-Hydroxy acyl-ACP dehydratase
(d) Fatty acid synthase **Ans. d**

327. The fatty acid synthase complex catalyses
(a) 4 sequential enzymatic steps (b) 6 sequential enzymatic steps
(c) 7 sequential enzymatic steps (d) 8 sequential enzymatic steps
Ans. b

328. The main source of reducing equivalents (NADPH) for lipogenesis is
(a) Pentose phosphate pathway (b) Citric acid cycle
(c) Glycolysis (d) Glycogenolysis **Ans. a**

329. In fatty acids synthase of both bacteria and mammals, ACP (acyl carrier protein) contain the vitamin
(a) Thiamin (b) Pyridoxine
(c) Riboflavin (d) Pantothenic acid **Ans. d**

330. Carboxylation of acetyl-CoA to malonyl- CoA requires the enzyme
(a) Acetyl-CoA carboxylase (b) Pyruvate carboxylase
(c) Acetyl transacylase (d) Acyl CoA-synthetase **Ans. a**

331. The rate limiting reaction in the lipogenic pathway is

(a) Acetyl-CoA carboxylase step (b) Ketoacyl synthase step

(c) Ketoacyl reductase step (d) Hydratase step **Ans. a**

332. Conversion of fatty acyl-CoA to an acyl-CoA derivative having 2 more carbon atoms involves as acetyl donar

(a) Acetyl-CoA (b) Succinyl-CoA

(c) Propionyl-CoA (d) Malonyl-CoA **Ans. d**

333. A cofactor required for the conversion of acetyl-CoA to malonyl-CoA in extramitochondrial fatty acid synthesis is

(a) Biotin (b) FMN

(c) NAD (d) NADP **Ans. a**

334. The glycerol for fatty acid esterification in adipocytes is

(a) For the most part, derived from glucose

(b) Obtained primarily from phosphorylation of glycerol by glycerol kinase

(c) Formed from gluconeogenesis

(d) Formed from glycogenolysis **Ans. a**

335. In the biosynthesis of triglycerides from glycerol 3-phosphate and acyl-CoA, the first intermediate formed is

(a) 2-Monoacylglycerol (b) 1, 2-Diacylglycerol

(c) Lysophosphatidic acid (d) Phosphatidic acid **Ans. d**

336. The enzyme glycerol kinase has low activity in

(a) Liver (b) Kidney

(c) Intestine (d) Adipose tissue **Ans. d**

337. The common precursor in the biosynthesis of triacylglycerol and phospholipids is

(a) 1, 2-Diacylglycerol phosphate (b) 1-Acylglycerol 3-phosphate

(c) Glycerol 3-phosphate (d) Dihydroxyacetone phosphate

Ans. a

338. Synthesis of polyunsaturated fatty acids involves the enzyme systems

(a) Acyl transferase and hydratase

(b) Desaturase and elongase

(c) Ketoacyl-CoA reductase and hydratase
(d) Dihydroxyacetone phosphate **Ans. a**

339. The desaturation and chain elongation system of polyunsaturated fatty acid are enhanced by
(a) Insulin (b) Glucagon
(c) Epinephrine (d) Thyroxine **Ans. a**

340. Higher rate of lipogenesis is associated with
(a) High proportion of carbohydrate in diet
(b) Restricted caloric intake
(c) High fat diet
(d) Deficiency of insulin **Ans. a**

341. Example of enzyme specificity
(a) Stereo specificity (b) Reaction specificity
(c) Substrate specificity (d) All of these **Ans. d**

342. Phospholipase C attacks the ester bond liberating 1, 2-diacylglycerol and a phosphoryl base at position
(a) 1 (b) 2
(c) Both (a) and (b) (d) 3 **Ans. d**

343. Synthesis of phosphatidylinositol by transfer of inositol to CDP diacylglycerol is catalysed by the enzyme
(a) CTP phosphatidate cytidyl transferase
(b) Phosphatidate phosphohydrolase
(c) CDP-diacylglycerol inositol transferase
(d) Choline kinase **Ans. c**

344. Synthesis of sphingosine requires the cofactor
(a) NAD (b) NADP
(c) $NADPH^+$ (d) ATP **Ans. c**

345. Ceramide is formed by the combination of sphingosine and
(a) Acetyl-CoA (b) Acyl-CoA
(c) Malonyl-CoA (d) Propionyl-CoA **Ans. b**

346. The amino alcohol sphingosine is synthesized in
(a) Mitochondria (b) Cytosol
(c) Nucleus (d) Endoplasmic reticulum **Ans. d**

347. The output of free fatty acids from adipose tissue is reduced by

(a) Insulin (b) Glucagon

(c) Growth hormone (d) Epinephrine **Ans. a**

348. The principal action of insulin in adipose tissue is to inhibit the activity of the

(a) Hormone sensitive lipoprotein lipase

(b) Glycerol phosphate acyltransferase

(c) Acetyl-CoA carboxylase

(d) Pyruvate dehydrogenase **Ans. a**

349. In non shivering thermogenesis

(a) Glucose is oxidized to lactate

(b) Fatty acids uncouple oxidative phosphorylation

(c) Ethanol is formed

(d) ATP is burned for heat production **Ans. b**

350. Brown adipose tissue is

(a) A prominent tissue in human

(b) Characterised by high content of mitochondria

(c) Associated with high activity of ATP synthase

(d) Characterised by low content of cytochromes **Ans. b**

351. Fatty liver is caused due to accumulation of

(a) Fatty acids (b) Cholesterol

(c) Phospholipids (d) Triacylglycerol **Ans. d**

352. A lipotropic factor is

(a) Choline (b) Palmitic acid

(c) Calcium (d) Vitamin C **Ans. a**

353. Fatty liver is also caused by

(a) CH_3Cl (b) CCl_4

(c) Na_2SO_4 (d) Riboflavin **Ans. b**

354. All the enzymes involved in the synthesis of cholesterol are found in

(a) Mitochondria

(b) Golgi apparatus

(c) Nucleus

(d) Endoplasmic reticulum and cytosol **Ans. d**

355. The source of all the carbon atoms in cholesterol is

(a) Acetyl-CoA (b) Bicarbonate

(c) Propionyl-CoA (d) Succinyl-CoA **Ans. a**

356. Two molecules of acetyl-CoA condense to form acetoacetyl-CoA catalysed by

(a) Thiolase (b) Kinase

(c) Reductase (d) Isomerase **Ans. a**

357. Acetoacetyl-CoA condenses with one more molecule of acetyl-CoA to form

(a) Mevalonate

(b) Acetoacetate

(c) β-Hydroxybutyrate

(d) 3-Hydroxy 3-methyl-glutaryl-CoA **Ans. d**

358. HMG-CoA is converted to mevalonate by reduction catalysed by

(a) HMG-CoA synthetase (b) HMG-CoA reductase

(c) Mevalonate kinase (d) Thiolase **Ans. b**

359. For reduction enzyme HMG-CoA reductase requires cofactor

(a) NADPH (b) NADP

(c) NAD (d) FAD **Ans. a**

360. In the biosynthesis of cholesterol, the step which controls the rate and locus of metabolic regulation is

(a) Geranyl pyrophosphate farnesyl pyrophosphate

(b) Squalene → lanosterol

(c) HMG CoA → mevalonate

(d) Lanosterol → 1, 4-desmethyl lanosterol **Ans. c**

361. The cyclisation of squalene in mammals results in the direct formation of the sterol

(a) Cholesterol (b) Lanosterol

(c) Sistosterol (d) Zymosterol **Ans. b**

362. In the biosynthesis of cholesterol, the rate limiting enzyme is

(a) Mevalonate kinase (b) HMG-CoA synthetase

(c) HMG-CoA reductase (d) Cis-prenyl transferase **Ans. c**

363. Cholesterol by a feed back mechanism inhibits the activity of

(a) HMG-CoA synthetase (b) HMG-CoA reductase

(c) Thilase (d) Mevalonate kinase **Ans. b**

364. The activity of HMG-CoA reductase is inhibited by

(a) A fungal inhibitor mevastatin (b) Probucol

(c) Nicotinic acid (d) Clofibrate **Ans. a**

365. Hypolipidemic drugs reduce serum cholesterol and triacylglycerol. The effect of clofibrate is attributed to

(a) Block in absorption from G.I.T.

(b) Decrease in secretion of triacylglycerol and cholesterol containing VLDL by liver

(c) Block in the reabsorption of bile acids

(d) Decreased synthesis of cholesterol **Ans. b**

366. In biosynthesis of cholesterol triparanol inhibits the activity of the enzyme

(a) Δ^{24} Reductase

(b) Oxidosqualene-lanosterol cyclase

(c) Isomerase

(d) Squalene epoxidase **Ans. a**

367. HMG-CoA reductase activity is increased by administration of the hormone

(a) Insulin (b) Glucagon

(c) Epinephrine (d) Glucocorticoids **Ans. a**

368. The principal sterol excreted in feces is

(a) Coprostanol (b) Zymosterol

(c) Lanosterol (d) Desmosterol **Ans. a**

369. The principal rate limiting step in the biosynthesis of bile acids is at the

(a) 7-Hydroxylase reaction (b) 12 α -Hydroxylase reaction

(c) Conjugation reaction (d) Deconjugation reaction **Ans. a**

370. Hypercholesterolemia is found in

(a) Xanthomatosis (b) Thyrotoxicosis

(c) Hemolytic jaundice (d) Malabsorption syndrome

Ans. a

371. Hypocholesterolemia is found in
(a) Thyrotoxicosis (b) Diabetes mellitus
(c) Obstructive jaundice (d) Nephrotic syndrome **Ans. a**

372. The major source of extracellular cholesterol for human tissue is
(a) Very low density lipoprotein (b) High density lipoprotein
(c) Low density lipoprotein (d) Albumin **Ans. c**

373. Correct ordering of lipoprotein molecules from lowest to the greater density is
(a) LDL, IDL, VLDL, chylomicron
(b) Chylomicron, VLDL, IDL, LDL
(c) VLDL, IDL, LDL, chylomicron
(d) LDL, VLDL, IDL, chylomicron **Ans. b**

374. In Hurler's syndrome, urine shows the presence of
(a) Keratan sulphate I
(b) Chondroitin sulphate
(c) Dermatan sulphate and heparan sulphate
(d) Keratan sulphate II **Ans. c**

375. Defective enzyme in Hunter's syndrome is
(a) α -L-iodouronidase (b) Iodouronate sulphatase
(c) Arylsulphatase B (d) C-acetyl transferase **Ans. b**

376. In Hunter's syndrome
(a) There is progressive corneal opacity
(b) Keratan sulphate is excreted in the urine
(c) Enzyme defective is arylsulphatase B
(d) Hearing loss is perceptive **Ans. d**

377. An important feature of Von-Gierke's disease is
(a) Muscle cramps (b) Cardiac failure
(c) Hypoglycemia (d) Respiratory alkalosis **Ans. c**

378. The affected organ in Mc Ardle's syndrome is
(a) Liver (b) Kidney
(c) Liver and Heart (d) Skeletal muscle **Ans. d**

379. Refsum's disease is due to deficiency of the enzyme

(a) Pytantate- α -oxidase (b) Glucocerebrosidase

(c) Galactocerebrosidase (d) Ceramide trihexosidase **Ans. a**

380. An important finding in Refsum's disease is

(a) Accumulation of ceramide trihexoside in the kidney

(b) Accumulation of phytanic acid in the blood and tissues

(c) Accumulation of gangliosides in brain and spleen

(d) Skin eruptions **Ans. b**

381. α -Galactosidase enzyme is defective in

(a) Tay-sach's disease (b) Refsum's disease

(c) Sandhoff's disease (d) Fabry's disease **Ans. d**

382. The hypothesis to explain enzyme–substrate complex formation

(a) Lock and key model (b) Induced fit theory

(c) Proenzyme theory (d) Both (a) and (b) **Ans. c**

383. An important finding in Tay-sach's disease is

(a) Renal failure

(b) Accumulation of gangliosides in brain and spleen

(c) Cardiac failure

(d) Anemia **Ans. b**

384. The enzyme deficient in Krabbe's disease is

(a) Hexosaminidase A (b) Arylsuphatase A

(c) β-Galactosidase (d) α -Fucosidase **Ans. c**

385. The enzyme ceramidase is deficient in

(a) Farber's disease (b) Fabry's disease

(c) Sandhoff's disease (d) Refsum's disease **Ans. a**

386. Niemann-Pick disease is due to deficiency of the enzyme

(a) Ceramidase (b) Glucocerebrosidase

(c) Galactocerebrosidase (d) Sphingomyelinase **Ans. d**

387. Wolman's disease is due to deficiency of

(a) Cholesteryl ester hydrolase (b) Hexosaminidase A

(c) α -Fucosidase (d) Arylsulphatase A **Ans. a**

388. The enzyme deficient in Sandhoff's disease is
(a) α -Fucosidase (b) Hexosaminidase A and B
(c) β-Galactosidase (d) β-Glucosidase **Ans. b**

389. Jamaican vomiting sickness is due to inactivation of the enzyme
(a) Pyruvate carboxylase (b) Acyl-Co-A synthetase
(c) Acyl-Co-A dehydrogense (d) Thiolase **Ans. c**

390. Zellweger's syndrome is due to inherited absence of
(a) Peroxisomes (b) Phospholipase A_1
(c) Acyl-Co-A dehydrogenase (d) Thiolase **Ans. a**

391. Bassen-Kornzweig syndrome is due to
(a) Absence of Apo-C-II (b) Defect in Apo-B synthesis
(c) Absence of Apo-E (d) Absence of Apo-D **Ans. b**

392. Enzyme deficient in Hyperammonemia type II is
(a) Glutamine synthetase
(b) Glutaminase
(c) Ornithine transcarbamoylase
(d) Carbamoylphosphate synthetase **Ans. c**

393. An important finding in Hyperammonemia type II is
(a) Increased serum gluatmine level
(b) Enlarged liver
(c) Mental retardation
(d) Increased carbamoyl phosphate synthetase level **Ans. c**

394. Absence of the enzyme argininosuccinate synthetase causes
(a) Argininosuccinic aciduria (b) Hyperargininemia
(c) Tricorrhexis nodosa (d) Citrullinemia **Ans. a**

395. Tricorrhexis nodosa is a characteristic finding of
(a) Argininosuccinic aciduria (b) Citrullinemia
(c) Phenylketonuria (d) Hyperargininemia **Ans. d**

396. Elevated blood argininosuccinate level is found in
(a) Hyperargininemia (b) Argininosuccinic aciduria
(c) Citrullinemia (d) Tyrosinosis **Ans. a**

397. Hyperargininemia, a defect in urea synthesis develops due to deficiency of the enzyme

(a) Ornithine transcarbamoylase (b) Argininosuccinase
(c) Arginase (d) Argininosuccinate synthetase

Ans. b

398. Albinism is due to deficiency of the enzyme

(a) Phenylalanine hydroxylase
(b) Tyrosinase
(c) p-Hydroxyphenylpyruvic acid oxidase
(d) Tyrosine dehydrogenase **Ans. c**

399. Neonatal tyrosinemia is due to deficiency of the enzyme

(a) p-Hydroxyphenylpyruvate hydroxylase
(b) Fumarylacetoacetate hydrolase
(c) Phenylalanine hydroxylase
(d) Tyrosine dehydrogenase **Ans. b**

400. Which of the following is a substratespecific enzyme

(a) Hexokinase (b) Thiokinase
(c) Lactase (d) Aminopeptidase **Ans. c**

401. Coenzymes combine with

(a) Proenzymes (b) Apoenzymes
(c) Holoenzymes (d) Antienzymes **Ans. b**

402. Coenzymes are required in which of the following reactions

(a) Oxidation-reduction (b) Transamination
(c) Phosphorylation (d) All of these **Ans. d**

403. Which of the following coenzyme takes part in hydrogen transfer reactions

(a) Tetrahydrofolate (b) Coenzyme A
(c) Coenzyme Q (d) Biotin **Ans. c**

404. Which of the following coenzyme takes part in oxidation-reduction reactions

(a) Pyridoxal phosphate (b) Lipoic acid
(c) Thiamin diphosphate (d) None of these **Ans. b**

405. In conversion of glucose to glucose-6-phsophate, the coenzyme is
(a) Mg^{++} (b) ATP
(c) Both (a) and (b) (d) None of these **Ans. b**

406. A coenzyme required in transamination reactions is
(a) Coenzyme A (b) Coenzyme Q
(c) Biotin (d) Pyridoxal phosphate **Ans. d**

407. Coenzyme A contains a vitamin which is
(a) Thiamin (b) Ascorbic acid
(c) Pantothenic acid (d) Niacinamide **Ans. c**

408. Cobamides contain a vitamin which is
(a) Folic acid (b) Ascorbic acid
(c) Pantothenic acid (d) Vitamin B_{12} **Ans. d**

409. A coenzyme required in carboxylation reactions is
(a) Lipoic acid (b) Coenzyme A
(c) Biotin (d) All of these **Ans. c**

410. Which of the following coenzyme takes part in tissue respiration
(a) Coenzyme Q (b) Coenzyme A
(c) NADP (d) Cobamide **Ans. a**

411. The enzyme hexokinase is a
(a) Hydrolase (b) Oxidoreductase
(c) Transferase (d) Ligase **Ans. c**

412. Which of the following is a proteolytic enzyme
(a) Pepsin (b) Trypsin
(c) Chymotrypsin (d) All of these **Ans. d**

413. Enzymes which catalyse binding of two substrates by covalent bonds are known as
(a) Lyases (b) Hydrolases
(c) Ligases (d) Oxidoreductases **Ans. c**

414. The induced fit model of enzyme action was proposed by
(a) Fischer (b) Koshland
(c) Mitchell (d) Markert **Ans. b**

415. Allosteric inhibition is also known as
(a) Competitive inhibition (b) Non-competitive inhibition
(c) Feedback inhibition (d) None of these **Ans. c**

416. An allosteric enzyme is generally inhibited by
(a) Initial substrate of the pathway
(b) Substrate analogues
(c) Product of the reaction catalysed by allosteric enzyme
(d) Product of the pathway **Ans. d**

417. When the velocity of an enzymatic reaction equals V_{max} substrate concentration is
(a) Half of K_m (b) Equal to K_m
(c) Twice the K_m (d) Far above the K_m **Ans. d**

418. In Lineweaver-Burk plot, the y-intercept represents
(a) V_{max} (b) K_m
(c) K_m (d) $1/K_m$ **Ans. b**

419. In competitive inhibition, the inhibitor
(a) Competes with the enzyme
(b) Irreversibly binds with the enzyme
(c) Binds with the substrate
(d) Competes with the substrate **Ans. d**

420. Competitive inhibitors
(a) Decrease the K_m (b) Decrease the V_{max}
(c) Increase the K_m (d) Increase the V_{max} **Ans. c**

421. Competitive inhibition can be relieved by raising the
(a) Enzyme concentration (b) Substrate concentration
(c) Inhibitor concentration (d) None of these **Ans. b**

422. Physostigmine is a competitive inhibitor of
(a) Xanthine oxidase (b) Cholinesterase
(c) Carbonic anhydrase (d) Monoamine oxidase **Ans. b**

423. Carbonic anhydrase is competitively inhibited by
(a) Allopurinol (b) Acetazolamide
(c) Aminopterin (d) Neostigmine **Ans. b**

424. Serum lactate dehydrogenase rises in
(a) Viral hepatitis (b) Myocardial infarction
(c) Carcinomatosis (d) All of these **Ans. d**

425. Which of the following serum enzyme rises in myocardial infarction
(a) Creatine kinase (b) GOT
(c) LDH (d) All of these **Ans. d**

426. From the following myocardial infarction, the earliest serum enzyme to rise is
(a) Creatine Kinase (b) GOT
(c) GPT (d) LDH **Ans. a**

427. Proenzymes
(a) Chymotrysinogen (b) Pepsinogen
(c) Both (a) and (b) (d) None of these **Ans. b**

428. Alkaline phosphatase is present in
(a) Liver (b) Bones
(c) Placenta (d) All of these **Ans. d**

429. Which of the following isoenzyme of lactate dehydrogenase is raised in serum in myocardial infarction
(a) LD_1 (b) LD_2
(c) LD_1 and LD_2 (d) LD_5 **Ans. c**

430. Enzymes which are always present in an organism are known as
(a) Inducible enzymes (b) Constitutive enzymes
(c) Functional enzymes (d) Apoenzymes **Ans. b**

431. Inactive precursors of enzymes are known as
(a) Apoenzymes (b) Coenzymes
(c) Proenzymes (d) Holoenzymes **Ans. c**

432. Which of the following is a proenzyme
(a) Carboxypeptidase (b) Aminopeptidase
(c) Chymotrypsin (d) Pepsinogen **Ans. d**

433. Allosteric enzymes regulate the formation of products by
(a) Feedback inhibition (b) Non-competitive inhibition
(c) Competitive inhibition (d) Repression-derepression
Ans. a

434. Regulation of some enzymes by covalent modification involves addition or removal of

(a) Acetate (b) Sulphate

(c) Phosphate (d) Coenzyme **Ans. c**

435. Covalent modification of an enzyme generally requires a

(a) Hormone (b) cAMP

(c) Protein kinase (d) All of these **Ans. d**

436. An inorganic ion required for the activity of an enzyme is known as

(a) Activator (b) Cofactor

(c) Coenzyme (d) None of these **Ans. b**

437. The first enzyme found to have isoenzymes was

(a) Alkaline Phosphatase (b) Lactate dehydrogenase

(c) Acid Phosphatase (d) Creatine kinase **Ans. b**

438. Lactate dehydrogenase is located in

(a) Lysosomes (b) Mitochondria

(c) Cytosol (d) Microsomes **Ans. c**

439. Lactate dehydrogenase is a

(a) Monomer (b) Dimer

(c) Tetramer (d) Hexamer **Ans. c**

440. Ceruloplasmin is absent in

(a) Cirrhosis of liver (b) Wilson's disease

(c) Menke's disease (d) Copper deficiency **Ans. b**

441. Ceruloplasmin oxidizes

(a) Copper (b) Iron

(c) Both (a) and (b) (d) None of these **Ans. b**

442. Creatine kinase is present in all of the following except

(a) Liver (b) Myocardium

(c) Muscles (d) Brain **Ans. a**

443. Alkaline phosphatase is present in

(a) Liver (b) Bones

(c) Intestinal mucosa (d) All of these **Ans. d**

444. All of the following are zinc-containing enzymes except

(a) Acid Phosphatase (b) Alkaline Phosphatase

(c) Carbonic anhydrase (d) RNA polymerase **Ans. a**

445. All of the following are iron-containing enzymes except

(a) Carbonic anhydrase (b) Catalase

(c) Peroxidase (d) Cytochrome oxidase **Ans. a**

446. Biotin is a coenzyme for

(a) Pyruvate dehydrogenase

(b) Pyruvate carboxylase

(c) PEP carboxykinase

(d) Glutamate pyruvate transminase **Ans. b**

447. Enzymes accelerate the rate of reactions by

(a) Increasing the equilibrium constant of reactions

(b) Increasing the energy of activation

(c) Decreasing the energy of activation

(d) Decreasing the free energy change of the reaction **Ans. c**

448. Kinetics of an allosteric enzyme are explained by

(a) Michaelis-Menten equation (b) Lineweaver-Burk plot

(c) Hill plot (d) All of these **Ans. c**

449. Covalent modification of an enzyme usually involves phosphorylation /dephosphorylation of

(a) Serine residue (b) Proline residue

(c) Hydroxylysine residue (d) Hydroxyproline residue

Ans. a

450. V_{max} of an enzyme may be affected by

(a) pH (b) Temperature

(c) Non-competitive inhibitors (d) All of these **Ans. d**

451. In enzyme assays, all the following are kept constant except

(a) Substrate concentration (b) Enzyme concentration

(c) pH (d) Temperature **Ans. b**

452. If the substrate concentration is much below the K_m of the enzyme, the velocity of the reaction is

(a) Directly proportional to substrate concentration

(b) Not affected by enzyme concentration
(c) Nearly equal to V_{max}
(d) Inversely proportional to substrate concentration **Ans. a**

453. Enzymes requiring NAD as co-substrate can be assayed by measuring change in absorbance at
(a) 210 nm (b) 290 nm
(c) 340 nm (d) 365 nm **Ans. c**

454. Different isoenzymes of an enzyme have the same
(a) Amino acid sequence (b) Michaelis constant
(c) Catalytic activity (d) All of these **Ans. c**

455. From the pentapeptide, phe-ala-leu-lys-arg-phenylalanine residue is split off by
(a) Trypsin (b) Chymotrypsin
(c) Aminopeptidase (d) Carboxypeptidase **Ans. c**

456. A high-energy phosphate among the following is
(a) Glucose-6-phosphate (b) Glucose-1-phosphate
(c) 1, 3-Biphoglycerate (d) All of these **Ans. c**

457. The highest energy level is present amongst the following in
(a) 1, 3-Biphosphoglycerate (b) Creatine phosphate
(c) Carbamoyl phosphate (d) Phosphoenol pyruvate **Ans. d**

458. Daily urinary urobilinogen excretion in adult men is
(a) 0–4 mg (b) 5–8 mg
(c) 9–12 mg (d) 13–20 mg **Ans. a**

459. In obstructive jaundice, faecal urobilinogen is
(a) Absent (b) Decreased
(c) Increased (d) Normal **Ans. a**

460. Acetyl-CoA can be formed from
(a) Pyruvate (b) Fatty acids
(c) ketone bodies (d) All of these **Ans. d**

461. Pyruvate is converted into acetyl-CoA by
(a) Decarboxylation (b) Dehydrogenation
(c) Oxidative decarboxylation (d) Oxidative deamination **Ans. c**

462. Conversion of pyruvate into acetyl CoA is catalysed by
(a) Pyruvate dehydrogenase
(b) Didrolipoyl acetyl transferase
(c) Dihydrolipoyl dehydrogenase
(d) All the 3 acting in concert **Ans. d**

463. Pyruvate dehydrogenase complex is located in
(a) Cytosol (b) Lysosomes
(c) Mitochondria (d) Endoplasmic reticulum **Ans. c**

464. A flavoprotein in pyruvate dehydrogenase complex is
(a) Pyruvate dehydrogenase (b) Didrolipoyl acetyl transferase
(c) Dihydrolipoyl dehydrogenase (d) None of these **Ans. c**

465. Pyruvate dehydrogenase complex is regulated by
(a) Covalent modification (b) Allosteric regulation
(c) Both (a) and (b) (d) None of these **Ans. c**

466. An allosteric inhibitor of pyruvate dehydrogenase is
(a) Acetyl CoA (b) ATP
(c) NADH (d) Pyruvate **Ans. a**

467. Ribozymes
(a) RNA enzyme (b) Non-protein enzymes
(c) Catalyst function (d) All of these **Ans. d**

468. In citric acid cycle, NAD is reduced in
(a) One reactions (b) Two reactions
(c) Three reactions (d) Four reactions **Ans. c**

469. Among citric acid cycle enzymes, a flavoprotein is
(a) Malate (b) Fumarase
(c) Succinate dehrogenase (d) Isocitrate dehrogenase **Ans. c**

470. In citric acid cycle, GDP is phosphorylated by
(a) Succinate dehydrogenase (b) Aconitase
(c) Succinate thiokinase (d) Fumarse **Ans. c**

471. Malonate is an inhibitor of
(a) Malate dehydrogenase
(b) α-Ketoglutarate dehydrogenase

(c) Succinate dehydrogenase
(d) Isocitrate dehydrogenase **Ans. c**

472. Isocitrate dehydrogenase is allosterically inhibited by
(a) Oxalosuccinate (b) α-Ketoglutarate
(c) ATP (d) NADH **Ans. c**

473. All of the following are allosteric enzymes except
(a) Citrate synthetase (b) α-Ketoglutarate dehdrogenase
(c) Succinate thiokinase (d) Succinate dehydrogenase **Ans. c**

474. All of the following are intermediates of citric acid cycle except
(a) Oxalosuccinate (b) Oxaloacetate
(c) Pyruvate (d) Fumarate **Ans. c**

475. All of the following intermediates of citric acid cycle can be formed from amino acids except
(a) α-Ketoglutarate (b) Fumarate
(c) Malate (d) Oxaloacetate **Ans. c**

476. Glycolytic pathway is located in
(a) Mitochondria (b) Cytosol
(c) Microsomes (d) Nucleus **Ans. b**

477. End product of aerobic glycolysis is
(a) Acetyl CoA (b) Lactate
(c) Pyruvate (d) CO_2 and H_2O **Ans. c**

478. During fasting, glucose is phosphorylated mainly by
(a) Hexokinase (b) Glucokinase
(c) Both (a) and (b) (d) None of these **Ans. a**

479. Glucokinase is found in
(a) Muscles (b) Brain
(c) Liver (d) All of these **Ans. c**

480. In anaerobic glycolysis, energy yield from each molecule of glucose is
(a) 2 ATP equivalents (b) 8 ATP equivalents
(c) 30 ATP equivalents (d) 38 ATP equivalents **Ans. a**

481. Which of the following is an allosteric enzyme
(a) Phosphohexose isomerase (b) Phosphotriose isomerase
(c) Lactate dehydrogenase (d) Phosphofructokinase **Ans. d**

482. Glycolysis is anaerobic in
(a) Liver (b) Brain
(c) Kidneys (d) Erythrocytes **Ans. d**

483. Phosphofructokinase is allosterically inhibited by
(a) Fructose-1, 6-biphosphate (b) Lactate
(c) Pyruvate (d) Citrate **Ans. d**

484. Glucose-6-phosphate is an allosteric inhibitor of
(a) Glucokinase (b) Hexokinase
(c) Phosphohexose isomerase (d) None of these **Ans. b**

485. ATP is a co-substrate as well as an allosteric inhibitor of
(a) Phosphofructokinase (b) Hexokinase
(c) Glucokinase (d) None of these **Ans. a**

486. Complete oxidation of one molecule of glucose into CO_2 and H_2O yields
(a) 8 ATP equivalents (b) 15 ATP equivalents
(c) 30 ATP equivalents (d) 38 ATP equivalents **Ans. d**

487. A unique by-product of glycolysis in erythrocytes is
(a) Lactate (b) 1, 3-Biphosphoglycerate
(c) 2, 3-Biphosphoglycerate (d) All of these **Ans. c**

488. Which of the following enzymes incorporates inorganic phosphate into the substrate
(a) Phosphoglycerate kinase
(b) Glyceraldehyde-3-phosphate dehydrogenase
(c) Pyruvate kinase
(d) Enolase **Ans. a**

489. Rapoport-Luebering cycle is located in
(a) Liver (b) Muscles
(c) Brain (d) Erythrocytes **Ans. d**

490. Glycerol can enter glycolytic pathway via
(a) Dihydroxyacetone phosphate (b) 1, 3-Biphospoglycerate
(c) 3-Phosphoglycerate (d) 2-Phosphoglycerate **Ans. a**

491. HMP shunt is present in
(a) Erythrocytes (b) Liver
(c) Testes (d) All of these **Ans. d**

492. Glucose-6-phosphate dehydrogenase is induced by
(a) 6-Phosphogluconolactone (b) Glucose-6-phosphate
(c) Ribose-5-phosphate (d) Insulin **Ans. d**

493. The decarboxylation reaction in HMP shunt is catalysed by
(a) Gluconolactone hydrolase
(b) 6-Phosphogluconate dehydrogenase
(c) 6-Phosphogluconate decarboxylase
(d) Transaldolase **Ans. c**

494. The first pentose formed in HMP shunt is
(a) Ribose-5-phosphate (b) Ribulose-5-phosphate
(c) Xylose-5-phosphate (d) Xylulose-5-phosphate **Ans. b**

495. The regulatory enzyme in HMP shunt is
(a) Glucose-6-phosphate dehydrogenase
(b) 6-Phosphogluconate dehydrogenase
(c) Both (a) and (b)
(d) None of these **Ans. c**

496. The rate of HMP shunt reactions is
(a) Increased by Insulin (b) Increased in diabetes mellitus
(c) Increased by glucagons (d) Increased in starvation **Ans. a**

497. Glycogenesis requires
(a) GTP (b) CTP
(c) UTP (d) None of these **Ans. c**

498. Glycogen synthetase catalyses the formation of
(a) α –1, 4-Glycosidic bonds (b) α –1, 6-Glycosidic bonds
(c) Both (a) and (b) (d) None of these **Ans. a**

499. Glycogenoloysis is increased by
(a) Glucagon (b) Insulin
(c) Epinephrine (d) cAMP **Ans. b**

500. Hepatic glycogenoloysis is increased by

(a) Insulin (b) Glucagon

(c) Epinephrine (d) Glucocorticoids **Ans. b**

501. Glycogen phosphorylase liberates the following from glycogen

(a) Glucose (b) Glucose-6-phosphate

(c) Glucose-1-phosphate (d) Maltose **Ans. c**

502. After the action of phosphorylase, glycogen is converted into

(a) Amylopectin (b) Dextrin

(c) Amylose (d) Maltose **Ans. b**

503. Glucose-1-phosphate liberated from glycogen cannot be converted into free glucose in

(a) Liver (b) Kidneys

(c) Muscles (d) Brain **Ans. c**

504. A coenzyme present in phosphorylase is

(a) NAD (b) Pyridoxal phosphate

(c) Thiamin pyrophosphate (d) Coenzyme A **Ans. b**

505. If glucose-1-phosphate formed by glycogenoloysis in muscles is oxidized to CO_2 and H_2O, the energy yield will be

(a) 2 ATP equivalents (b) 3 ATP equivalents

(c) 4 ATP equivalents (d) 8 ATP equivalents **Ans. b**

506. A molecule of phosphorylase kinase is made up of

(a) 4 subunits (b) 8 subunits

(c) 12 subunits (d) 16 subunits **Ans. d**

507. Cyclic AMP binds to

(a) Catalytic subunits of protein kinase

(b) Regulatory subunits of protein kinase

(c) Catalytic subunits of phosphorylase kinase

(d) Regulatory subunits of phosphorylase kinase **Ans. b**

508. Glucose is the only source of energy for

(a) Myocardium (b) Kidneys

(c) Erythrocytes (d) Thrombocytes **Ans. c**

509. Glycerol-3-phosphate for the synthesis of triglycerides in adipose tissue is derived from

(a) Phosphatidic acid (b) Diacylglycerol
(c) Glycerol (d) Glucose **Ans. d**

510. Gluconeogenesis does not occur in

(a) Brain (b) Kidneys
(c) Muscles (d) Liver **Ans. c**

511. Glucose cannot be synthesized from

(a) Glycerol (b) Lactate
(c) Alanine (d) Leucine **Ans. d**

512. Coenzyme for phosphoenolpyruvate carboxykinase is

(a) ATP (b) ADP
(c) GTP (d) GDP **Ans. c**

513. Therapeutic enzymes

(a) Streptokinase (b) Asparaginase
(c) Riboflavinase (d) Both (a) and (b) **Ans. d**

514. A gluconeogenic enzyme among the following is

(a) Phosphofructokinase
(b) Pyruvate kinase
(c) Phosphoenol pyruvate carboxykinase
(d) Glucokinase **Ans. c**

515. Glucose-6-phosphatase and PEP carboxy kinase are regulated by

(a) Covalent modification (b) Allosteric regulation
(c) Induction and repression (d) All of these **Ans. c**

516. The maximum possible chain length of fatty acids formed in the pathway of de novo synthesis is

(a) 16 Carbon atoms (b) 18 Carbon atoms
(c) 20 Carbon atoms (d) 24 Carbon atoms **Ans. a**

517. Acetyl CoA required for de novo synthesis of fatty acids is obtained from

(a) Breakdown of existing fatty acids
(b) Ketone bodies
(c) Acetate
(d) Pyruvate **Ans. d**

518. Formation of acetyl CoA from pyruvate for de novo synthesis of fatty acids requires
(a) Pyruvate dehydrogenase complex
(b) Citrate synthetase
(c) ATP citrate lyase
(d) All of these **Ans. d**

519. The major site for elongation of medium chain fatty acids is
(a) Mitochondria (b) Cytosol
(c) Microsomes (d) All of these **Ans. c**

520. β-oxidation of fatty acids is inhibited by
(a) $NADPH^+$ (b) Acetyl CoA
(c) Malonyl CoA (d) None of these **Ans. c**

521. The enzyme regulating extra-mitochondrial fatty acid synthesis is
(a) Thioesterase (b) Acetyl CoA carboxylase
(c) Acyl transferase (d) Multi-enzyme complex **Ans. b**

522. Acetyl CoA carboxylase is activated by
(a) Citrate (b) Insulin
(c) Both (a) and (b) (d) None of these **Ans. c**

523. All the following statements about acetyl CoA carboxylase are true except:
(a) It is activated by citrate
(b) It is inhibited by palmitoyl CoA
(c) It can undergo covalent modification
(d) Its dephosphorylated form is inactive **Ans. d**

524. All the following statements about acetyl CoA carboxylase are true except
(a) It is required for de novo synthesis of fatty acids
(b) It is required for mitochondrial elongation of fatty acids
(c) It is required for microsomal elongation of fatty acids
(d) Insulin converts its inactive form into its active form **Ans. b**

525. Both Acyl carrier protein (ACP) of fatty acid synthetase and coenzyme (CoA) are
(a) Contain reactive phosphorylated
(b) Contain thymidine
(c) Contain phosphopantetheine reactive groups
(d) Contain cystine reactive groups **Ans. c**

526. Which one of the following transfers acyl groups

(a) Thiamine pyrophosphate (b) Lipomide
(c) ATP (d) NADH **Ans. b**

527. Which one of the following cofactors must be utilized during the conversion of acetyl CoA to malonyl CoA

(a) TPP (b) ACP
(c) NAD^+ (d) Biotin **Ans. d**

528. Coenzymes derived from the vitamin shown below are required by enzymes involved in the synthesis of which of the following

(a) ATP (b) UTP
(c) CTP (d) NADH **Ans. a**

529. Coenzymes derived from the vitamin shown below are required by which of the following enzymes

(a) Lactate dehydrogenase (b) Glutamate dehydrogenase
(c) Pyruvate dehydrogenase (d) Malate dehydrogenase **Ans. c**

530. All the following are coenzymes except

(a) Ubiquinone (b) CoA
(c) Pyruvate dehydrogenase (d) Lipoic acid **Ans. c**

531. Which of the following is not a cofactor

(a) Mg (b) Iron
(c) Cu (d) Methylcobalamine

532. All the following compounds are members of the electron transport chain except

(a) Ubiquinone (b) Carnitine
(c) NAD (d) FAD **Ans. b**

533. Thiamine is essential for

(a) Pyruvate dehydrogenase (b) Isocitrate dehydrogenase
(c) Succinate dehydrogenase (d) Acetyl CoA synthetase **Ans. b**

534. Adenylate cyclase is activated by

(a) Insulin (b) Glucagon
(c) Prostaglandin E_1 (d) Ca^{2+} ions **Ans. b**

535. Maximum enzyme activity is observed at

(a) Acidic pH (b) Neutral pH
(c) Basic pH (d) Optimum pH **Ans. d**

536. Which of the following is known as bone forming enzyme
(a) Alkaline phosphatase (b) Acid phosphatase
(c) Leucine aminopeptidase (d) γ-glutamyl transpeptidase
Ans. a

537. Conversion of pepsinogen to pepsin is
(a) Intra molecular rearrangement
(b) Breaking of hydrogen bonds
(c) Covalent modification
(d) Polymerisation **Ans. c**

538. Which of the following is not having an apoenzyme and coenzyme
(a) Lactate dehydrogenase (b) Succinate dehydrogenase
(c) Malate dehydrogenase (d) Pepsin **Ans. d**

539. Pyruvate dehydrogenase is a/an
(a) Isomerase (b) Lyase
(c) Ligase (d) Oxido reductase **Ans. d**

540. Homogentisic oxidase is an
(a) Oxidase (b) Monooxygenase
(c) Dioxygenase (d) Anaerotic dehydrogenase
Ans. c

541. Isocitrate dehydrogenase can use __________ as a cofactor.
(a) NAD^+ only (b) $NADP^+$ only
(c) NAD^+ or $NADP^+$ (d) FMN and FAD **Ans. c**

542. The rate of most enzyme catalyzed reactions changes with pH. As the pH increases, this rate
(a) Reaches at minimum, then increases
(b) Reaches at maximum, then decreases
(c) Increases
(d) Decreases **Ans. b**

543. A substrate for the enzyme aldolase is
(a) Galactose-6-phosphate (b) Isocitric acid
(c) Glucose-1-phosphate (d) Fructose 1, 6-diphosphate
Ans. d

544. Decarboxylation of α-keto acids requires
(a) Thiamine pyrophosphate, FAD, NAD^+
(b) Flavin mononucleotide
(c) $NADP^+$
(d) NAD^+ only **Ans. a**

545. Coenzyme A contains the vitamin
(a) Riboflavin (b) Pantothenic acid
(c) Pyridoxine (d) Thiamine **Ans. b**

546. Which of the following is not a component of coenzyme A
(a) Adenylic acid (b) Pantothenic acid
(c) β-mercaptoethylamine (d) Deoxyadenylic acid **Ans. d**

547. Malic enzyme converts malic acid, in the presence of $NADP^+$ to Pyruvic acid. This reaction is a/an
(a) Decarboxylation
(b) Decarboxylation and Dehydrogenation
(c) Dehydrogenation
(d) Oxidation **Ans. b**

548. The following reaction is characteristic of what type of enzymes
$2H_2O_2 \rightarrow 2H_2O + O_2$
(a) Peroxides (b) Catalase
(c) Dehydrogenase (d) Copper containing oxidases
Ans. b

549. Of which warburg's yellow enzyme contains as a prosthetic group
(a) Thiamine pyrophosphate (b) Biotin
(c) NAD^+ (d) Riboflavin-5-phosphate **Ans. d**

550. Dehydrogenases utilize, as coenzymes, all of the following except
(a) NAD^+ (b) $NADP^+$
(c) FAD (d) FH_4 **Ans. d**

551. Urea is produced physiologically by the action of the enzyme
(a) Urease (b) Glutaminase
(c) Arginase (d) None of these **Ans. c**

552. Urease is a
(a) Lyase (b) Ligase
(c) Isomerase (d) Hydrolase **Ans. d**

553. Velocity maximum for an enzyme at half the substrate concentration gives

(a) The molecular weight of the enzyme

(b) K_m value

(c) Isoelectric pH

(d) Pk_a value **Ans. b**

554. Which of the following amino acid has been shown as one of the active site of phosphoglucomutase

(a) Lysine (b) Tyrosine

(c) Serine (d) Histidine **Ans. c**

555. The inhibition of succinate dehydrogenase by malonate by

(a) Competitive inhibition (b) Non-competitive inhibition

(c) Uncompetitive inhibition (d) Feedback inhibition **Ans. a**

556. Cobamide coenzymes are

(a) Vitamin B_1 (b) Riboflavin

(c) Pyridoxine (d) Vitamin B_{12} **Ans. d**

557. The isozyme CK-MB is specifically increased in the blood of patients who had

(a) Skeletal muscle disease (b) Recent myocardial infarction

(c) Infective hepatitis (d) Myxoedema **Ans. b**

558. FAD containing enzyme, catalyzing formation of α, β unsaturated fatty acyl CoA derivative

(a) Acyl CoA dehydrogenase

(b) Enoyl hydrase

(c) β-OH acyl CoA dehydrogenase

(d) Thiolase **Ans. a**

559. Immobilized enzymes

(a) Potentiation of activity (b) Presentation of activity

(c) Preparation of activity (d) All of these **Ans. b**

560. This catalyzes formation of CoA derivatives from fatty acid, CoA and ATP

(a) Acyl CoA dehydrogenase

(b) Enol hydrase

(c) β-OH acyl CoA dehydrogenase
(d) Thio kinase **Ans. d**

561. Fructose 2, 3 biphosphate is a powerful allosteric activator of
(a) Fructose 1, 6 diphosphatase (b) Phosphofructokinase
(c) Hexokinase (d) Fructokinase **Ans. b**

562. 'Clearing factor' is
(a) Lipoprotein lipase (b) Crotonase
(c) 7-dehydro cholesterol (d) β-sitosterol **Ans. a**

563. Maltase attacks only
(a) α-glucosides (b) β-glucosides
(c) Starch (d) Dextrins **Ans. a**

564. Pepsin is
(a) Exo-peptidase (b) Endo-peptidase
(c) Carboxy peptidase (d) Amino peptidase **Ans. b**

565. An enzyme in saliva which hydrolyzes starch is
(a) Pepsinogen (b) Chymotrysin
(c) α -Amylase (d) Malate **Ans. c**

566. If a coenzyme is required in an enzyme reaction, the former usually has the function of
(a) Acting as an acceptor for one of the cleavage products of the substrate
(b) Enhancing the specificity of the apo enzyme
(c) Increasing the number of receptor sites of the apoenzyme
(d) Activating the substrate **Ans. a**

567. The Michaelis-Menten hypothesis
(a) Postulates the formation of an enzyme substrate complex
(b) Enables us to calculate the isoelectric point of an enzyme
(c) States that the rate of a chemical reaction may be independent of substrate concentration
(d) States that the reaction rate is proportional to substrate concentration
Ans. a

568. Schardinger's enzyme is
(a) Lactate dehydrogenase (b) Xanthine dehydrogenase
(c) Uric oxidase (d) L-amino acid dehydrogenase
Ans. b

569. Tryptophan pyrolase is currently known as

(a) Tryptophan deaminase (b) Tryptophan dioxygenase

(c) Tryptophan mono oxygenase (d) Tryptophan decarboxylase

Ans. b

570. An enzyme which brings about lysis of bacterial cell wall is

(a) Amylase (b) Lysozyme

(c) Trypsin (d) Lipase **Ans. b**

571. Trypsin has no action on

(a) Hemoglobin (b) Albumin

(c) Histone (d) DNA **Ans. d**

572. Which one of the following coenzymes pairs correctly with the group transferred

(a) CoA - electron (b) $NADP^+$ - phosphoryl

(c) Thiamine - pyrophosphate (d) Biotin - carbondioxide **Ans. d**

573. In non-competitive enzyme action

(a) V_{max} is increased

(b) Apparent K_m is increased

(c) Apparent K_m is decreased

(d) Concentration of active enzyme molecule is reduced **Ans. c**

574. An allosteric enzyme influences the enzyme activity by

(a) Competiting for the catalytic site with the substrate

(b) Changing the specificity of the enzyme for the substrate

(c) Changing the conformation of the enzyme by binding to a site other than catalytic site

(d) Changing the nature of the products formed **Ans. c**

575. Which of the following regulatory reactions involves a reversible covalent modification of an enzyme

(a) Phosphorylation of serine OH on the enzyme

(b) Allosteric modulation

(c) Competitive inhibition

(d) Non-competitive inhibition **Ans. a**

576. A competitive inhibitor of an enzyme has which of the following properties

(a) It is frequently a feedback inhibitor

(b) It becomes covalently attached to an enzyme

(c) It decreases the V_{max}

(d) It interferes with substrate binding to the enzyme **Ans. d**

577. When [S] is equal to K_m, which of the following conditions exist

(a) Half the enzyme molecules are bound to substrate

(b) The velocity of the reaction is equal to V_{max}

(c) The velocity of the reaction is independent of substrate concentration

(d) Enzyme is completely saturated with substrate **Ans. a**

578. Which of the following statements about an enzyme exhibiting allosteric kinetics with cooperative interaction is false

(a) A plot of V-Vk [S] has a sigmoidal shape

(b) An inhibitor may increase the apparent K_m

(c) Line weaver Berk plot is useful for determining K_m and V_{max}

(d) Removal of allosteric inhibitor may result in hyperbolic V-S [S] plot **Ans. d**

579. Pantothenic acid acts on

(a) NADP (b) NADPH

(c) FAD (d) CoA **Ans. b**

580. Vitamin deficiency that causes fatty liver includes all except

(a) Vitamin E (b) Pyridoxine

(c) Retionic acid (d) Pantothenic acid **Ans. c**

581. In which of the following types of enzymes an inducer is not required

(a) Inhibited enzyme (b) Cooperative enzyme

(c) Allosteric enzyme (d) Constitutive enzyme **Ans. d**

582. In which of the following types of enzyme water may be added to a C—C double bond without breaking the bond

(a) Hydrolase (b) Hydratase

(c) Hydroxylase (d) Esterase **Ans. b**

583. 'Lock' and 'Key' model of enzyme action proposed by Fisher implies that

(a) The active site is flexible and adjusts to substrate

(b) The active site requires removal of PO_4 group

(c) The active site is complementary in shape to that of the substrate

(d) Substrates change conformation prior to active site interaction

Ans. c

584. In competitive inhibition of enzyme action

(a) The apparent K_m is decreased

(b) The apparent K_m is increased

(c) V_{max} is decreased

(d) Apparent concentration of enzyme molecules decreased **Ans. b**

585. In competitive inhibition which of the following kinetic effect is true

(a) Decreases both K_m and V_{max}

(b) Increases both K_m and V_{max}

(c) Decreases K_m without affecting V_{max}

(d) Increases K_m without affecting V_{max} **Ans. d**

586. Enzymes increase the rates of reactions by

(a) Increasing the free energy of activation

(b) Decreasing the energy of activation

(c) Changing the equilibrium constant of the reaction

(d) Increasing the free energy change of the reaction **Ans. b**

587. The most useful test for the diagnosis of acute hemorrhagic pancreatitis during the first few days is

(a) Urinary lipase test (b) Serum calcium

(c) Urinary amylase (d) Serum amylase **Ans. d**

588. The best test for acute pancreatitis in the presence of mumps is

(a) A serological test for mumps (b) Serum amylase

(c) Urinary amylase (d) Serum lipase **Ans. d**

589. The slow moving fraction of LDH is typically increased in pancreas with

(a) Cerebrovascular accidents (b) Acute myocardial infarction

(c) Acute pancreatitis (d) Acute viral hepatitis **Ans. d**

590. Which of the following enzyme typically elevated in alcoholism

(a) Serum ALP (b) Serum GOT

(c) Serum γ-GT (d) Serum acid phosphatase

Ans. c

591. Patients with hepatocellular jaundice, as compared to those with purely obstructive jaundice tend to have

(a) Lower serum ALP, LDH and AST activity

(b) Lower serum ALP, Higher LDH and AST activity

(c) Higher serum ALP, LDH and AST activity

(d) Higher serum ALP, Lower LDH and AST activity **Ans. b**

592. If results of the serum bilirubin, serum ALP, LDH and AST determinations suggest obstructive jaundice, the best confirmatory test would be the estimation of

(a) Serum ALT (b) Serum 5′ nucleotidase

(c) Serum Pseudo cholinesterase (d) None of these **Ans. b**

593. Which enzyme estimation will be helpful in differentiating the elevated serum ALP found in obstructive jaundice as well as bone disorders

(a) Serum AST (b) Serum ALT

(c) Serum LDH (d) Serum γ-GT **Ans. d**

594. Cardiac muscle contains which of the following CK osoenzyme

(a) BB only (b) MM and BB only

(c) MM, BB and MB (d) MM and MB only **Ans. d**

595. Liver and skeletal measle disorders are characterized by on disk proportionate increase in which of the LDH isoenzyme fraction

(a) LDH-1 (b) LDH-1 and LDH-2

(c) LDH-3 and LDH-4 (d) LDH-2 and LDH-3 **Ans. c**

596. On the third day following onset of acute myocardial infarction, which enzyme estimation will have the best predictive value

(a) Serum AST (b) Serum CK

(c) Serum ALT (d) Serum LDH **Ans. d**

597. Serum AST activity is not characteristically elevated as the result of

(a) Myocardial infarction (b) Passive congestion of liver

(c) Muscular dystrophies (d) Peptic ulcer **Ans. d**

598. On which day following acute myocardial infarction the estimation of serum AST will be of greatest significance

(a) First day (b) Second day

(c) Third day (d) Fourth day **Ans. b**

599. In which diseases of the following organs, isoenzymes LDH-1 and LDH-2 will be released in plasma

(a) Kidney, R.B.C and Liver (b) Heart, Kidney and R.B.C

(c) Heart, Kidney and Liver (d) Heart, Lungs and Brain **Ans. b**

600. Plasma non-functional enzymes are

(a) Totally absent

(b) Low concentration in plastic

(c) Important for diagnosis of several disease

(d) All of these **Ans. d**

601. Pyruvate dehydrogenase contains all except

(a) Biotin (b) NAD

(c) FAD (d) CoA **Ans. a**

602. An increase in LDH-5 enzyme is seen in the following except

(a) Acute hepatitis (b) Muscular distrophies

(c) Breast carcinoma (d) Pulmonary embolism **Ans. d**

603. Diastase can be used for the hydrolysis can be used for the hydrolysis of

(a) Sucrose (b) Starch

(c) Cellulose (d) Maltose **Ans. b**

604. Which of the following statements is true

(a) Enzymes have names ending -ase

(b) Enzymes are highly specific in their action

(c) Enzymes are living organisms

(d) Enzymes get activated on heating **Ans. b**

605. Enzymes activity is controlled by

(a) pH of the solution (b) Temperature

(c) Concentration of the enzyme (d) Concentration of the substrate

Ans. c

606. Which of the following is not true regarding enzymes

(a) They catalyze only a particular type of reaction

(b) They remain active even after separation from the source

(c) They are destroyed after the completion of the reaction they catalyse
(d) They are irreversibly destroyed at high temperature **Ans. c**

607. The number of enzymes known is about
(a) 10,000 (b) 100
(c) 50 (d) 26 **Ans. c**

608. Nicotine present in tobacco is a/an
(a) Alkaloid (b) Terpene
(c) Steroid (d) Protein **Ans. a**

609. The poisonous alkaloid present in the oil of hemlock is
(a) Cocaine (b) Nicotine
(c) Quinine (d) Morphine **Ans. a**

610. Alkaloids are usually purified by extraction with
(a) Ether (b) Dilute HCl
(c) NaOH (d) Chloroform **Ans. b**

611. The number of N-MC groups in alkaloids is best estimate with the help of
(a) HI (b) H_2SO_4
(c) $(CH_3CO)_2CO$ (d) CH_3MgI **Ans. a**

612. A competitive inhibitor of an enzyme
(a) Increases K_m without affecting V_{max}
(b) Decreases K_m without affecting V_{max}
(c) Increases V_{max} without affecting K_m
(d) Decreases both V_{max} and K_m **Ans. a**

613. The Michaelis constant, K_m is
(a) Numerically equal to ½ V_{max}
(b) Dependent on the enzyme concentration
(c) Independent of pH
(d) Numerically equal to the substrate concentration that gives half maximal velocity **Ans. d**

614. The rate of an enzyme catalyzed reaction was measured using several substrate concentrations that were much lower than K_m, the dependence of reaction velocity on substrate concentration can best be described as
(a) Independent of enzyme concentration
(b) A constant fraction of V_{max}

(c) Equal to K_m

(d) Proportional to the substrate concentration **Ans. c**

615. The presence of a non competitive inhibitor

(a) Leads to both an increase in the V_{max} of a reaction and an increase in K_m

(b) Leads to a decrease in the observed V_{max}

(c) Leads to a decrease in K_m and V_{max}

(d) Leads to an increase in K_m without affecting V_{max} **Ans. b**

616. Which one of the following statements is not characteristic of allosteric enzymes

(a) They frequently catalyze a committed step early in a metabolic pathway

(b) They are often composed of subunits

(c) They follow Michaelis-Menten kinetics

(d) They frequently show cooperativity for substrate binding **Ans. c**

617. The abnormal isoenzyme need not

(a) Be an oxidoreductase (b) Have any coenzyme

(c) Require ATP (d) Be localized intra-cellularly

Ans. a

618. LDH assays are most useful in diagnosing diseases of the

(a) Heart (b) Pancreas

(c) Brain (d) Kidney **Ans. a**

619. The chemical forces that bind most coenzymes and substrates to enzymes such as LDH are

(a) Hydrogen bonds (b) Peptide bonds

(c) Coordinate bonds (d) Covalent bonds **Ans. d**

620. How many different proteins may be present in normal LDH

(a) One (b) Two

(c) Three (d) Four **Ans. d**

621. All the isoenzymes function with the coenzyme

(a) $NADP^+$ (b) FAD

(c) Lipoate (d) NAD^+ **Ans. d**

622. 'Lock' and 'Key' theory was proposed by

(a) Sorenson (b) Fischer

(c) Mehler (d) Sanger **Ans. d**

623. Which of the following forms part of a coenzyme

(a) Zn^{2+} (b) Lipase

(c) Vitamin B_2 (d) Lysine **Ans. c**

624. The shape of an enzyme and consequently its activity can be reversibly altered from moment to moment by

(a) Heat (b) Amino acid substrate

(c) Allosteric subunits (d) Sulfur substitutions **Ans. c**

625. Which one of the following regulatory actions involves a reversible covalent modification of the enzyme

(a) Phosphorylation of ser-OH on the enzyme

(b) Allosteric modulation

(c) Competitive inhibition

(d) Non-competitive inhibition **Ans. a**

626. An enzyme is a

(a) Carbohydrate (b) Lipid

(c) Protein (d) Nucleic acid **Ans. c**

627. An enzyme promotes a chemical reaction by

(a) Lowering the energy of activation

(b) Causing the release of heat which acts as a primer

(c) Increasing molecular motion

(d) Changing the free energy difference between substrate and product **Ans. a**

628. In most metabolic pathways, all needed enzymes are arranged together in a multienzyme complex within a

(a) Solution of ATP (b) Membrane

(c) Quanternary protein (d) Coenzyme **Ans. b**

629. An enzyme catalyzes the conversion of an aldose sugar to a ketose sugar would be classified as one of the

(a) Transferases (b) Isomerases

(c) Oxido reductases (d) Hydrolases **Ans. b**

630. Binding energy released in enzyme-substrate complex formation causes the

(a) Decreases in the free energy of product

(b) Decrease in the free energy of substrate

(c) Increase in the activation energy

(d) Decrease in the activation energy **Ans. d**

631. In which of the following types of enzymes, water may be added to a C—C double bond without breaking the bond

(a) Hydrolase (b) Hydratase

(c) Hydroxylase (d) Oxygenase **Ans. a**

632. Enzymes increases the rate of reactions by

(a) Increasing the free energy of activation

(b) Decreasing the energy of activation

(c) Changing the equilibrium constant of the reaction

(d) Increasing the free energy change of the reaction **Ans. b**

633. Which of the following is a false statement

(a) Coenzymes may be derived from vitamins

(b) Coenzymes may be prosthetic groups

(c) Coenzymes increase the activation energy of the enzymatic reaction

(d) All statements are correct **Ans. c**

634. Allosteric enzymes contain

(a) Multiple subunits (b) Single chain

(c) Two chains (d) Three chains **Ans. a**

635. Isoenzymes of lactate dehydrogenase are useful for the diagnosis of

(a) Heart disease (b) Kidney disease

(c) Liver disease (d) Both (a) and (c) **Ans. d**

636. An enzyme catalysing the reaction Glycerate-3-phosphate ↔ Glycerate-2-phosphate would have an EC number with first digit

(a) 2 (b) 4

(c) 5 (d) none **Ans. c**

637. The first enzyme isolated, purified and crystallized from Jack bean (*Canavalia*) by summer in 1926 was

(a) Urease (b) Insulin

(c) Ribonuclease (d) Zymase **Ans. a**

638. Enzyme glutathione peroxidase, catalyses destruction of H_2O_2, contains

(a) Zn (b) Fe

(c) Se (d) Mo **Ans. d**

639. Lyases, a class of enzyme catalyse

(a) Hydrolysis reaction

(b) Group transfer reaction

(c) Oxidation and reduction reaction

(d) Addition of groups to double bond and vice versa **Ans. d**

640. The enzyme that converts glucose to glucose-6-phosphate is

(a) Phosphatase (b) Hexokinase

(c) Phosphorylase (d) Glucose synthetase **Ans. b**

641. Enzymes are required in traces because they

(a) Have high turnover number

(b) Remain unused at the end of reaction and can be used

(c) Show cascade effect

(d) All correct **Ans. d**

642. An organic substance bound to an enzyme and essential for the activity of enzyme is called

(a) Holoenzyme (b) Apoenzyme

(c) Coenzyme (d) Isoenzyme **Ans. c**

643. Enzyme catalysed reactions occur in

(a) Pico seconds (b) Micro seconds

(c) Milli seconds (d) None of these **Ans. c**

644. An enzyme can accelerate a reaction up to

(a) 10^{10} times (b) 10^{1} times

(c) 10^{100} times (d) 10 times **Ans. a**

645. Abzymes are

(a) Enzymes that are highly specific like antibodies

(b) Antibodies that have catalytic activities

(c) Also referred to as zymogens

(d) Enzymes that hydrolyse antibodies **Ans. b**

646. Zymogen is a

(a) Vitamin (b) Enzyme precursor

(c) Modulator (d) Hormone **Ans. b**

647. Cofactor (Prosthetic group) is a part of holoenzyme, it is

(a) Inorganic part loosely attached

(b) Accessory non-protein substance attached firmly

(c) Organic part attached loosely

(d) None of these **Ans. b**

648. A protein having both structural and enzymatic traits is

(a) Myosin (b) Collagen

(c) Trypsin (d) Actin **Ans. a**

649. Enzymes are different from catalysts in

(a) Being proteinaceous

(b) Not used up in reaction

(c) Functional at high temperature

(d) Having high rate of diffusion **Ans. a**

650. Enzymes, vitamins and hormones are common in

(a) Being proteinaceous

(b) Being synthesized in the body of organisms

(c) Enhancing oxidative metabolism

(d) Regulating metabolism **Ans. d**

651. Which of the following statement(s) is/are correct about coenzymes

P. Tetrahydrofolate transfer $-CH_3$, $-CH_2$, -CHO and -CHNH group

Q. S-adenosylmethionine transfer $-CH_3$ group

R. Biotin transfer $-CO_2$ group

S. TPP transfer acyl group

(a) P, Q and R (b) P and R

(c) R and S (d) P, Q, R and S **Ans. a**

652. A typical saturation curve, for an enzyme - catalysed reaction is shown in the following figure. The parameters, which can be derived from this equation, is indicated by A, B and C. Choose the correct parameter that corresponds to B

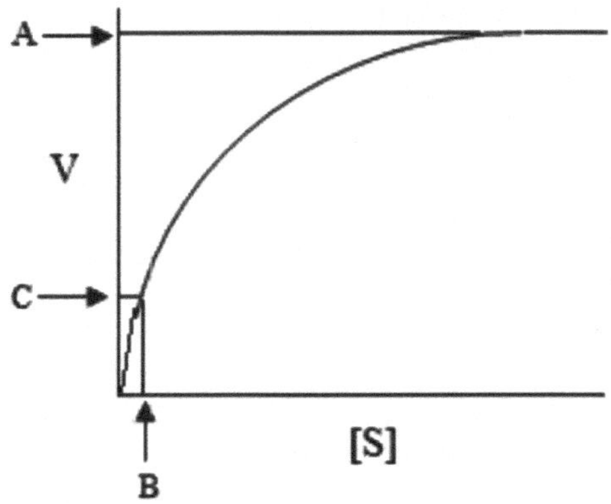

(a) K_m (b) V_{max}
(c) K_i (d) K_d **Ans. a**

653. Template/lock and key theory of enzyme action is supported by
(a) Enzymes speed up reaction
(b) Enzymes occur in living beings and speed up certain reactions
(c) Enzymes determine the direction of reaction
(d) Compounds similar to substrate inhibit enzyme activity **Ans. d**

654. Combination of apoenzyme and coenzyme produces
(a) Prosthetic group (b) Holoenzyme
(c) Enzyme substrate complex (d) Enzyme product complex
Ans. b

655. Enzyme inhibition caused by a substance resembling substrate molecule is
(a) Competitive inhibition (b) Non-competitive inhibition
(c) Feedback inhibition (d) Allosteric inhibition **Ans. a**

656. An enzyme brings about
(a) Decrease in reaction time (b) Increase in reaction time
(c) Increase in activation energy (d) Reduction in activation energy
Ans. d

657. Which of the following is an Eadie-Hofstee plot showing enzyme kinetics

(a) 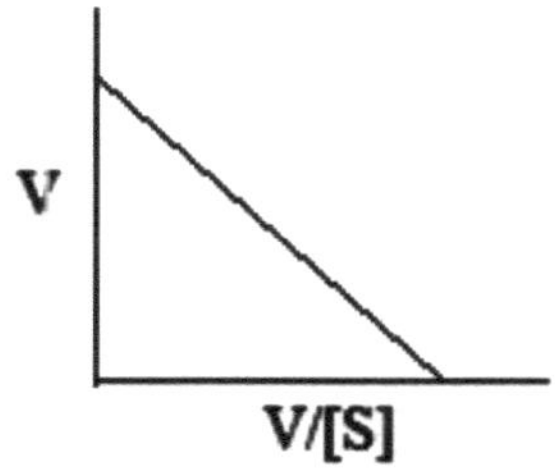

(b) 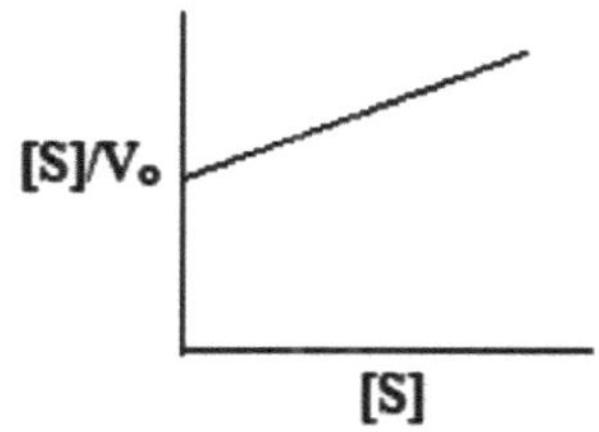

(c) 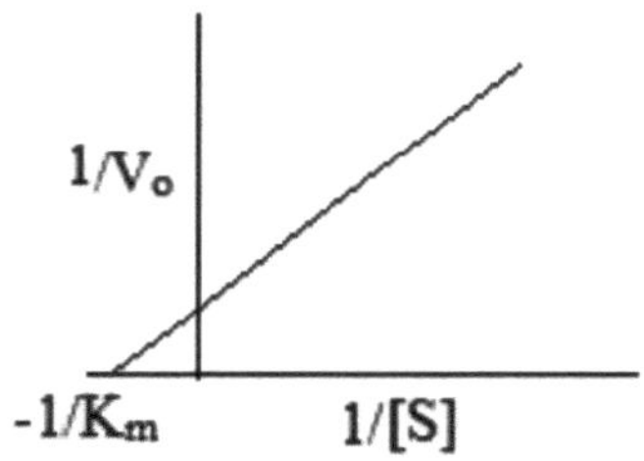

(d) 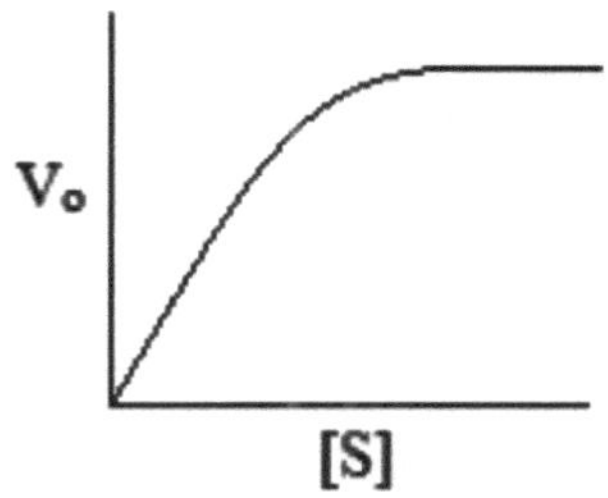

Ans. a

658. Coenzyme is

(a) Often a vitamin

(b) Always an inorganic compound

(c) Always a protein

(d) Often a metal **Ans. a**

659. Genetic engineering requires enzyme

(a) DNAase (b) Amylase

(c) Lipase (d) Restriction endonuclease

Ans. d

660. Which is not true about inorganic catalysts and enzymes

(a) They are specific

(b) Inorganic catalysts require specific not needed by enzymes

(c) They are sensitive to pH

(d) They speed up the rate of chemical reaction **Ans. b**

661. Key and lock hypothesis of enzyme action was given by

(a) Fischer (b) Koshland

(c) Buchner (d) Kuhne **Ans. a**

662. An example of feedback inhibition is
(a) Allosteric inhibition of hexokinase by glucose-6-phosphate
(b) Cyanide action on cytochrome
(c) Sulpha drug on folic acid synthesizer bacteria
(d) Reaction between succinic dehydrogenase and succinic acid **Ans. a**

663. Feedback term refers to
(a) Effect of substrate on rate of enzymatic reaction
(b) Effect of end product on rate reaction
(c) Effect of enzyme concentration on rate of reaction
(d) Effect of external compound on rate of reaction **Ans. b**

664. Allosteric inhibition
(a) Makes active site unifit for substrate
(b) Controls excess formation and end product
(c) Both (a) and (b)
(d) None of these **Ans. c**

665. The ratio of enzyme to substrate molecules can be as low as
(a) 1 : 100,000 (b) 1 : 500,000
(c) 1 : 10,000 (d) 1 : 1,000 **Ans. a**

666. Vitamin B_2 is component of coenzyme
(a) Pyridoxal phosphate (b) TPP
(c) NAD (d) FMN/FAD **Ans. d**

667. K_m value of enzyme is substrate concentration at
(a) ½ V_{max} (b) 2 V_{max}
(c) ½ V_{max} (d) 4 V_{max} **Ans. d**

668. Part of enzyme which combines with nonprotein part to form functional enzyme is
(a) Apoenzyme (b) Coenzyme
(c) Prosthetic group (d) None of these **Ans. c**

669. Who got Nobel Prize in 1978 for working on enzymes
(a) Koshland (b) Arber and Nathans
(c) Nass and Nass (d) H.G. Khorana **Ans. a**

670. Site of enzyme synthesis in a cell is

(a) Ribosomes (b) RER

(c) Golgi bodies (d) All of these **Ans. b**

671. The fruit when kept is open, tastes bitter after 2 hours because of

(a) Loss of water from juice

(b) Decreased concentration of fructose in juice

(c) Fermentation by yeast

(d) Contamination by bacterial enzymes **Ans. d**

672. Hexokinase (Glucose + ATP $\rightarrow$ Glucose-6–P + ADP) belongs to the category

(a) Transferases (b) Lysases

(c) Oxidoreductases (d) Isomerases **Ans. c**

673. For the enzyme catalysed reactions S $\leftrightarrow$ P, Keq = 5, which of the following curves best represents the change in [S] that occurs with time when S is added to a solution containing the appropriate enzyme

(a) 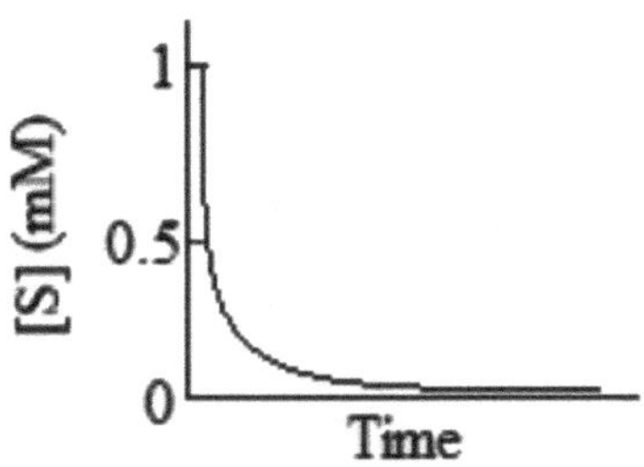

(b) 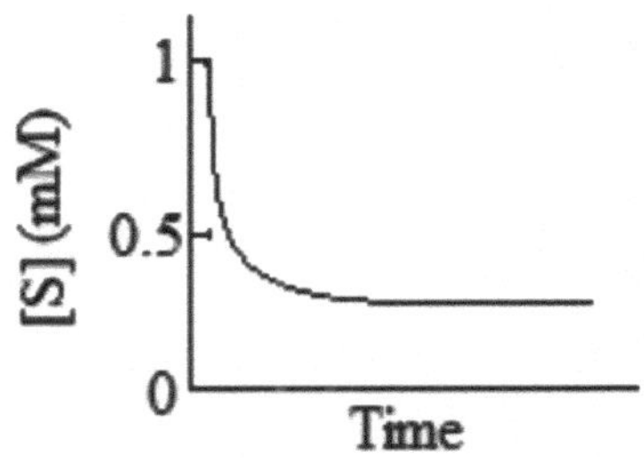

(c) 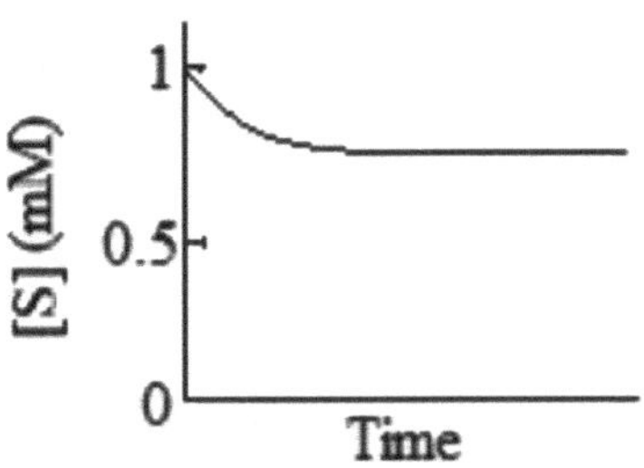

(d) 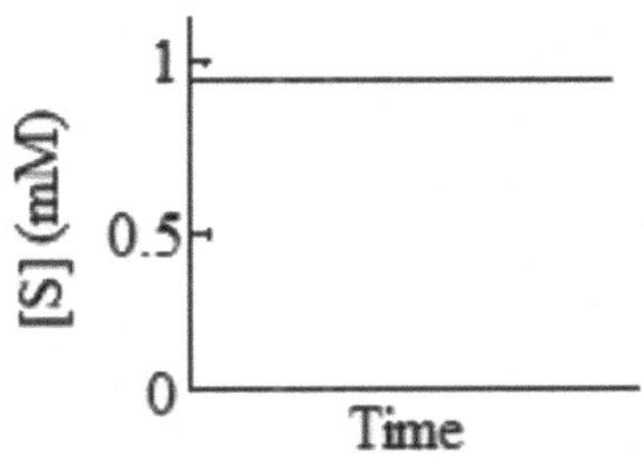

Ans. b

674. The best example of extracellular enzymes (exoenzyme) is

(a) Nucleases (b) Digestive enzymes

(c) Succinic dehydrogenase (d) None of these **Ans. c**

675. Which mineral element controls the activity of Nitrate reductase

(a) Fe (b) Mo

(c) Zn (d) Ca **Ans. a**

676. Name the enzyme that acts both as carboxylase at one time and oxygenase at another time

(a) PEP carboxylase (b) RuBP carboxylase

(c) Carbonic anyhdrase (d) None of these **Ans. b**

677. A metabolic pathways is a

(a) Route taken by chemicals

(b) Sequence of enzyme facilitated chemical reactions

(c) Route taken by an enzyme from one reaction to another

(d) Sequence of origin of organic molecules **Ans. b**

678. The energy required to start an enzymatic reaction is called

(a) Chemical energy (b) Metabolic energy

(c) Activation energy (d) Potential energy **Ans. c**

679. Out of the total enzymes present in a cell, a mitochondrion alone has

(a) 4% (b) 70%

(c) 95% (d) 50% **Ans. b**

680. Creatine phosphokinase isoenzyme is a marker for

(a) Kidney disease (b) Liver disease

(c) Myocardial infarction (d) None of these **Ans. c**

681. Which inactivates an enzyme by occupying its active site?

(a) Competitive inhibitor (b) Allosteric inhibitor

(c) Non-competitive inhibitor (d) All of these **Ans. a**

682. Which one is coenzyme?

(a) ATP (b) Vitamin B and C

(c) CoQ and CoA (d) All of these **Ans. d**

683. What is the turnover number (i.e. the K_{cat} or enzymatic activity) for an enzyme catalysed reaction which has a maximum velocity of 3.47 moles per liter^{-1}min^{-1} when the enzyme concentration is 6.85 x 10^{-1} g liter^{-1}? Molecular weight of the enzyme is 12,800 g mol^{-1}

(a) 6.48 x 104 min^{-1} (b) 5.258 x 10-3 min^{-1}

(c) 1.579 x 105 min^{-1} (d) 1.27 x 10-5 min^{-1} **Ans. a**

684. Carbonic anhydrase enzyme has maximum turn over number (36 million). Minimum turn over number for an enzyme

(a) DNA polymerase (b) Lysozyme
(c) Penicillase (d) Lactase dehydrogenase **Ans. b**

685. 677. Which of the following is/are true when DIPF inhibits chymotrypsin activity

P. $[E]_T$ decrease
Q. Apparent Vmax decrease
R. K_m decrease
S. V_{max} increase

(a) P (b) R
(c) P and Q (d) S **Ans. b**

686. Substrate concentration at which an enzyme attains half its maximum velocity is

(a) Threshold value (b) Michaelis-Menton constant
(c) Concentration level (d) None of these **Ans. b**

687. Aspartic protease

P. Possesses two aspartic acid residues at the active site
Q. Includes enzymes like pepsin, cathepsin and renin
R. Involves formation of intermediate amide dehydrate
S. Present in AIDS virus HIV-1

(a) P and S (b) Q, R and S
(c) P, Q, R and S (d) P, Q, R but not S **Ans. c**

688. Enzymes functional in cell or mitochondria are

(a) Endoenzymes (b) Exoenzymes
(c) Apoenzymes (d) Holoenzymes **Ans. c**

689. The enzymes present in the membrane of mitochondria are

(a) Flavoproteins and cytochromes
(b) Fumarase and lipase
(c) Enolase and catalase
(d) Hexokinase and zymase **Ans. a**

690. A mitochondrial marker enzyme is

(a) Aldolase (b) Amylase
(c) Succinic dehydrogenase (d) Pyruvate dehydrogenase
Ans. c

691. The enzyme used in polymerase chain reaction (PCR) is

(a) Taq polymerase (b) RNA polymerase
(c) Ribonuclease (d) Endonuclease **Ans. d**

692. Which of the following is a microsomal enzyme inducer

(a) Indomethacin (b) Clofibrate
(c) Tolbutamide (d) Glutethamide **Ans. d**

693. Identify the correct molecule which controls the biosynthesis of proteins in living organisms

(a) DNA (b) RNA
(c) Purines (d) Pyrimidines **Ans. a**

694. The tear secretion contains an antibacterial enzyme known as

(a) Zymase (b) Diastase
(c) Lysozyme (d) Lipase **Ans. c**

695. Identify one of the canbonic anhydrase inhibitor that inhibit only luminal carbonic anhydrase enzyme

(a) Methazolamide (b) Acetazolamide
(c) Dichlorphenamide (d) Benzolamide **Ans. b**

696. If the substrate concentration in an enzyme reaction is equal to ½ K_m, the initial reaction velocity will be

(a) 0.25 V_{max} (b) 0.33 V_{max}
(c) 0.50 V_{max} (d) 0.67 V_{max} **Ans. b**

697. The co-enzyme containing an automatic hetero ring in the structure is

(a) Biotin (b) TPP
(c) Sugar Phosphate (d) Co-enzyme **Ans. c**

698. The example of hydrogen transferring Co-enzyme is

(a) B_6-PO_4 (b) $NADP^+$
(c) TPP (d) ATP **Ans. d**

699. Enzyme catalyzed hydrolysis of proteins produces amino acid of the form

(a) D (b) DL
(c) L (d) Racemic **Ans. c**

700. Transaminase activity needs the Coenzyme

(a) ATP (b) B_6-PO_4
(c) FADT (d) NAD^+ **Ans. b**

701. The biosynthesis of urea occurs mainly in the liver

(a) Cytosol (b) Mitochondria

(c) Microsomes (d) Nuclei **Ans. b**

702. Bile salts make emulsification with fat for the action of

(a) Amylose (b) Lipase

(c) Pepsin (d) Trypsin **Ans. b**

703. All of the following compounds are intermediates of TCA cycle except

(a) Malate (b) Pyruvate

(c) Oxaloacetate (d) Fumarate **Ans. b**

704. In conversion of lactic acid to glucose, three reactions of glycolytic pathway are circumvented, which of the following enzymes do not participate

(a) Pyruvate carboxylase

(b) Phosphoenol pyruvate carboxy kinase

(c) Pyruvate kinase

(d) Glucose-6-phosphatase **Ans. b**

705. In the normal resting state of human most of the blood glucose burnt as fuel is consumed by

(a) Liver (b) Brain

(c) Adipose tissue (d) Muscles **Ans. b**

706. A regulator of the enzyme glucogen synthase is

(a) Citric Acid (b) Pyruvate

(c) Glucose-6-PO_4 (d) GTP **Ans. c**

707. A specific inhibitor for succinate dehydrogenase is

(a) Arsenite (b) Malonate

(c) Citrate (d) Fluoride **Ans. b**

708. During the functioning of biosensor, which of the following sequences of event occurs

(a) Enzymatic/cellular reaction → detector → transducer

(b) Enzymatic/cellular reaction → transducer → detector

(c) Enzymatic/cellular reaction → pressure gauge → time

(d) Enzymatic/cellular reaction → vibrator → mechanical signal

Ans. b

709. An immobilized enzyme being used in continuous plug flow reactor exhibits an effectiveness factor (η) of 1.2. The value of η being greater than one could be apparently due to one of the following reasons. Identify the correct reasons

(a) The enzyme follows substrate inhibited kinetics with intern pore diffusion initiation

(b) The enzyme experiences external film diffusion limitation

(c) The enzyme follows sigmoid kinetic

(d) The immobilized enzyme is operationally unstable **Ans. c**

710. The degree of inhibition for non-competitive inhibition of an enzyme catalyzed reaction

(a) Increase with increase in substrate concentration

(b) Reaches a maxima with increase in substrate concentration and then decreases

(c) Is independent of substrate concentration

(d) Decreases with increase in substrate concentration **Ans. b**

711. An enzyme following Michaelis-Menten kinetics with V_m = 2.5 mmol $m^{-3}s^{-1}$ and K_m = 5.0 mM was used to carry out the reaction in a batch stirred reactor. Starting with an initial substrate concentration of 0.1 M, the time required for 50% conversion of the substrate will be about

(a) 01 hr (b) 06 hr

(c) 02 hr (d) 12 hr **Ans. b**

712. The maximum reaction velocity (V_m) for an enzyme catalyzed reaction was experimentally measured at two different temperatures of following results were obtained

Temperature, ^{0}C -	27	37
V_m, mmolm^{-3}s^{-1}	2.25	4.50

The energy of activation for the reaction is

(a) 12834 cal mol^{-1} (b) 25668 cal mol^{-1}

(c) 6417 cal mol^{-1} (d) 19251 cal mol^{-1} **Ans. a**

713. Inversion of sucrose by immobilized invertase follows substrate inhibited kinetics. The reaction rate (V) in mol mhr^{-1} can be expressed as

$$V = 800 \frac{(S)}{(400 + 50\,(S) + (S)^2)}$$

Where, (S) is sucrose concentration

The immobilized invertase preparation is used in a CSTR with 100 mol m^{-3} sucrose concentration in the feed stream. If the reaction velocity passes through a maxima at (S) = 20 mol m^{-3} the feed flow rate for a reactor volume of 1 m^3 to get the maximum productivity from reactor should be

(a) 0.11 $m^{-3}hr^{-1}$ (b) 1.10 $m^{-3}hr^{-1}$

(c) 5.05 $m^{-3}hr^{-1}$ (d) None of these **Ans. a**

714. Enzyme papin is used with success to

(a) Increase meat production (b) Leaven bread

(c) Ripen papaya fruit (d) Tenderize meat **Ans. d**

715. Which one of the following reactions used for the purpose of recycling enzymes in bioprocesses

(a) Isomerization (b) Immobilization

(c) Phosphorylation (d) Polymerization **Ans. b**

716. Match the industrial application of the following enzymes:

A. Penicillinase 1. Pharmaceutical

B. Pectinase 2. Leather

C. Trypsin 3. Wine

D. Rennin 4. Dairy

Codes

	A	B	C	D
(a)	4	3	1	2
(b)	1	3	2	4
(c)	1	2	3	4
(d)	4	2	3	1

Ans. b

717. Which one of the following techniques is NOT ideal for immobilized cell-free enzyme

(a) Physical entrapment by encapsulation

(b) Covalent surface bonding to surface carriers

(c) Physical bonding by flocculation

(d) Covalent chemical bonding by cross-linking the precipitate **Ans. c**

718. The enzyme where catalysis involves transfer of electrons are named as

(a) Isomerase (b) Transferases

(c) Oxidoreductase (d) Lyases **Ans. c**

719. The graph shows a Lineweaver-Burke plot for an enzyme catalyzed reaction

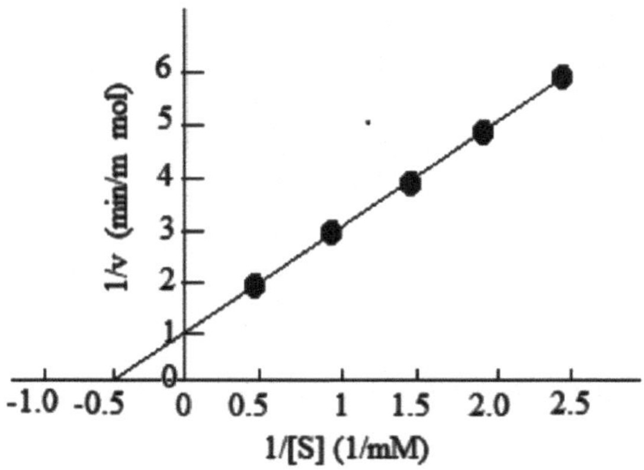

Which of the following statements is correct

(a) V_{max} is 5 M mol/min and with competitive inhibition V_{max} remains unchanged

(b) K_m is 2M mol/min and with competitive inhibition both K_m and V_{max} both decreases

(c) K_m is 0.5 mM and with competitive inhibition V_{max} increases but K_m remain unchanged

(d) K_m is 2.0 mM and with competitive inhibition K_m increases but V_{max} remains unchanged **Ans. d**

720. The kinetic data for an enzymatic in the presence and absence of inhibitors are plotted in the following figure, which line represents kinetics without inhibitors

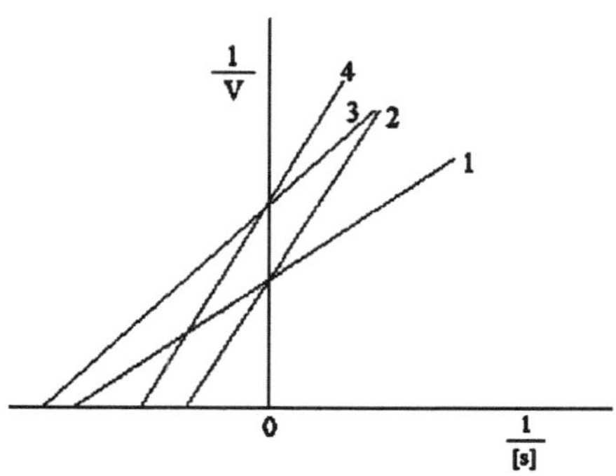

(a) Line 1 (b) Line 2

(c) Line 3 (d) Line 4 **Ans. b**

721. The kinetic data for enzymatic in the presence and absence of inhibitors are plotted in the following figure. Which line represents kinetics of non competitive inhibition

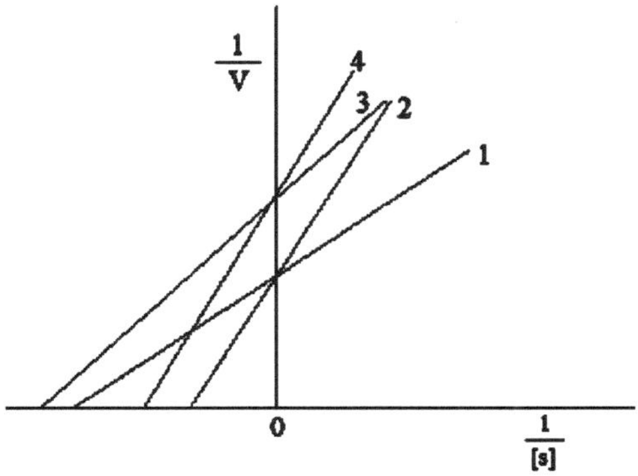

(a) Line 1 (b) Line 2

(c) Line 3 (d) Line 4 **Ans. d**

722. Using Hill equation for an enzyme $[s]_0 = (v_0 K_m/V_{max} - v_0)^{1/n}$ and the plot of $\log_{10} (v_0/V_{max} - v_0)$ vs $\log_{10} [s]_0$ one can find out

P - V_{max} from the intercept on the ordinate.

Q - K_m from the intercept on the ordinate.

R - 'n' from the slope.

S - K_m from the intercept on the abscissa.

(a) P, Q (b) Q, R

(c) R, S (d) P, S **Ans. c**

723. The Michaelis-Menton constant K_m is a measure of

(a) Rate of the reaction

(b) Affinity of the enzyme for substrate

(c) Concentration of enzyme - substrate [ES] intermediate

(d) None of these **Ans. b**

724. The Michaelis - Menton constant, K_m is

(a) Numerically equal to ½ V_{max}

(b) Dependent on enzyme concentration

(c) Independent of pH

(d) Numerically equal to the substrate concentration that gives half-maximal velocity **Ans. d**

725. Most industrial enzymes are obtained from

(a) Plants (b) Animal tissues

(c) Microbes (d) Insects **Ans. c**

726. The turnover number of chymotrypsin is $100s^{-1}$ and for DNA polymerase it is 15 S^{1}. This means that

(a) Chymotrypsin binds to its substrates with higher affinity than does DNA polymerase

(b) The velocity of chymotrypsin reaction is always greater than that of DNA polymerase

(c) The velocities of reaction catalysed by both enzymes of saturatic substrate levels could be equal if 6.7 times more DNA polymerase than chymotrypsin where used

(d) The velocities of chymotrypsin reactions at particular enzyme concentration and saturating substrate levels is lower than that of DNA polymerase reaction under the same condition **Ans. c**

727. Which of the following statement is true for enzyme catalysis of chemical reaction

(a) Increases the forward and reverse reaction rates

(b) Decreases $\Delta G'$ so that the reaction can proceed spontaneously

(c) Increases the energy of transition state

(d) Decreases the entropy of reactions **Ans. a**

728. The factor likely to increase the rate of reaction catalyzed by a surface immobilized enzyme is/are

(a) Increase agitation of the bulk liquid containing the substrate

(b) Continued replacement of the bulk liquid containing the substrate

(c) Increased concentration of the substrate in the bulk liquid

(d) All of the above **Ans. d**

729. Which of the following cases are likely to lead to faster rates of catalyzing by an enzyme immobilized on a negatively charged support

(a) A positively charged substrate and negatively charged product

(b) A negatively charged substrate and positively charged product

(c) A positively charged substrate and positively charged product

(d) None of these **Ans. a**

730. On doubling the enzyme concentration, the kinetic parameters that changes are

(a) K_m (b) V_{max}

(c) K_{cat} (d) Both V_{max} and K_{cat} **Ans. d**

731. An enzyme does the following in catalyzing a reaction

(a) Stabilizes the substrate

(b) Decreases the equilibrium constant

(c) Increases the forward reaction rate

(d) Hasten the approach of rate equilibrium **Ans. b**

732. K_i indicates

(a) Competitive inhibition (b) Denaturation of enzyme

(c) Reaction velocity (d) All the above **Ans. a**

733. K_m value of enzyme is substrate concentration at

(a) ¼ V_{max} (b) 2 V_{max}

(c) ½ V_{max} (d) 4 V_{max} **Ans. c**

734. At what [S] is the velocity (V_0) of an enzyme-catalyzed reaction is 25% of the V_{max}

(a) ¾ K_m (b) 4 K_m

(c) 1/3 K_m (d) ¼ K_m **Ans. c**

735. K_m which specified affinity of enzyme with substrate when

(a) $K_2 >> K_{-1}$ (b) $K_2 << K_{-1}$

(c) $K_2 = K_{-1}$ (d) No effect of K_{-1} **Ans. b**

736. Turn-over number of an enzyme is dependent upon

(a) Size of enzyme (b) Active site

(c) Molecular weight of enzyme (d) Concentration of substrate

Ans. c

737. Competitive inhibition is due to

(a) Protein poison

(b) Substrate analogue

(c) Non-availability of activation energy

(d) Short wave radiation **Ans. b**

738. End product inhibition is called

(a) Substrate regulation (b) Feed back regulation
(c) Irreversible inhibition (d) Non competitive inhibition
Ans. b

739. Non competitive inhibition often results in

(a) Changing in enzyme structure
(b) Blocking of active site
(c) Non synthesis of enzyme
(d) Non availability of cofactor **Ans. a**

740. Template theory of enzyme action was given by

(a) Fischer (b) Koshland
(c) Moned *et al.* (d) Michaelis and Menton **Ans. a**

741. Enzyme are immobilized through

(a) Covalently attaching to solid support
(b) Cross-linking
(c) Entrapping in gel
(d) All of the above **Ans. d**

742. Enzyme used in detergents are

(a) Amylases (b) Lipases
(c) Proteases (d) Glucoisomerase **Ans. c**

743. Enzyme TPA or tissue Plasminogen activator is used for

(a) Dissolving blood clots
(b) Maintaining plasma contents
(c) Clearing turbidity of juices
(d) Stimulating thromoboplastin production **Ans. a**

744. Enzyme immobilization is

(a) Conversion of active enzyme into inactive enzyme form
(b) Provide enzyme with protective covering
(c) Changing soluble enzyme into insoluble state
(d) None of the above **Ans. b**

745. Which of the following is the molecular weight of mRNA

(a) 15,000 to 30,000 (b) 20,000 to 35,000
(c) 25,000 to 40,000 (d) 30,000 to 50,000 **Ans. d**

746. α-helix in protein can be labeled

(a) 3.6_{13} helix (b) 4.4_{16} helix

(c) 2.6_{12} helix (d) None of these **Ans. c**

747. When a protein is denaturated the freedom of rotation about bonds in both the polypeptide backbone and the side chain is

(a) Increased

(b) Decreased

(c) Increased in the backbone only

(d) Increased in the side chain only **Ans. a**

748. The term milimol is written as

(a) M.M (b) V.M

(c) uuM (d) mM **Ans. d**

749. The phenomenon by which the synthesis of a set of enzyme leading to a product from outside is known as

(a) Repression (b) Depression

(c) Suppression (d) None of these **Ans. a**

750. Autolytic function is shown by

(a) Enzymes (b) DNA

(c) Hormones (d) RNA **Ans. a**

751. K_m is

(a) The dissociation constant for the enzyme substrate complex

(b) Equal to half the substrate concentration required to achieve V_{max}

(c) Identical for all isozymes of an enzyme

(d) The substrate concentration that gives one-half V_{max} **Ans. d**

752. Di-isopropyl fluorophosphates (DEF) reacts with serine proteases stoichiometrically and irreversibly and therefore is a

(a) Competitive inhibitor (b) Non-competitive inhibitor

(c) Uncompetitive inhibitor (d) Repressor **Ans. b**

753. In non-competitive inhibition

(a) The concentration of active enzyme molecules is reduced

(b) V_{max} is increased

(c) The concentration of active enzyme molecules is unchanged

(d) The apparent K_m is increased **Ans. a**

754. In an enzyme assay in which substrate concentration is much lower then K_m, the rate

(a) Approaches V_{max}

(b) Shows zero-order kinetics

(c) It proportional to substrate concentration

(d) Is independent of enzyme concentration **Ans. c**

755. Isoenzymes

(a) Are enzymes that exist in more than one amino acid sequence in the same species

(b) Cannot be distinguished in a given species except immunologically

(c) By definition must have the same amino acid composition

(d) Are single polypeptide chains that differ by an amino acid replacement **Ans. a**

756. Which of the following is not a component of coenzyme A

(a) Adenylic acid (b) Pantothenic acid

(c) Cystamine (d) Acetic acid **Ans. d**

757. As a coenzyme, pyruvate decarboxylase requires

(a) Coenzyme A (b) NAD^+

(c) FMN (d) Thiamine pyrophosphate **Ans. d**

758. Dehydrogenases use as coenzymes all of the following except

(a) NAD^+ (b) NADP

(c) FAD (d) Ferriprotoporphyrin **Ans. d**

759. Which of the following is an essential cofactor in carboxylation reactions

(a) Coenzyme A (b) Biotin

(c) CTP (d) Lipoic acid **Ans. b**

760. The enzyme that catalyses the reaction $2H_2O \rightarrow 2H_2O + O_2$ is a

(a) Dehydrogenase (b) Peroxidase

(c) Catalase (d) Hydrolase **Ans. c**

761. Which enzyme will cleave leucyl-glycyl-proline to leucine and glycyl-proline

(a) Carboxypeptidase (b) Glycylglycyl peptidase

(c) Aminopeptidase (d) Chymotrypsin **Ans. c**

762. The isoenzymes of lactate dehydrogenase

(a) Demonstrate the evolutionary development of this enzyme

(b) Range from monomers to tetramers

(c) Differ only in a single amino acid

(d) Exist in 5 forms depending upon the content of M and H monomers

Ans. d

763. The nerve gas, DFP, has been a useful reagent in enzyme chemistry. At the active site of many hydrolytic enzymes of DFP combines with

(a) Histidine (b) Serine

(c) Lysine (d) Aspartate **Ans. b**

764. The isocitrate dehydrogenase reaction is analogous to (performs the same type of chemical reaction)

(a) Pyruvate dehydogenase

(b) α-ketoglutarate dehydrogenase

(c) β-hydroxyacyl-CoA dehydrogenase

(d) 6-phosphogluconate dehydrogenase **Ans. d**

765. Antimetabolites act by

(a) Competitive inhibition when they combine irreversibly with an enzyme

(b) Competitive inhibition when they combine reversibly with an enzyme

(c) Non-competitive inhibition if there is no relation between degree of inhibition and substrate concentration

(d) Lactate dehydrogenase **Ans. b**

766. K_m and V_{max} can be determined from the Lineweaver - Burk plot of the Michaelis - Menton equation shoen below. Where V is the reaction velocity at substrate concentrations S, the X-axis experimental data are expressed as

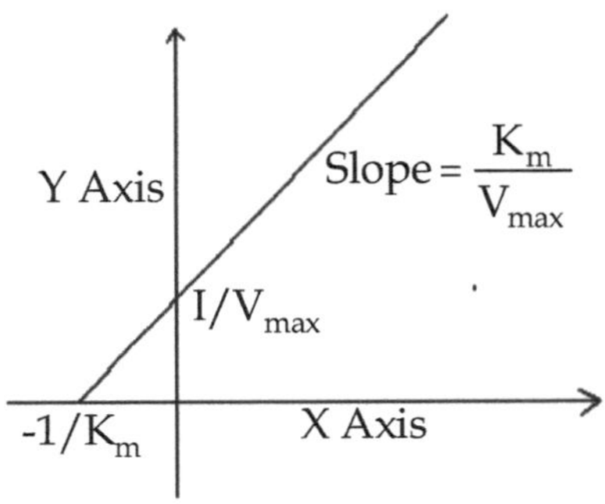

(a) 1/V (b) V
(c) 1/S (d) S **Ans. c**

767. All the following gastrointestinal enzymes are secreted as inactive zymogens (proenzymes) except

(a) Ribonuclease (b) Pepsin
(c) Trypsin (d) Chymotrypsin **Ans. a**

768. Which enzyme has the greatest specificity for peptide bonds on the carboxyl side of a cationic amino acid side chain

(a) Carboxypeptidase (b) Trypsin
(c) Rennin (d) Pepsin **Ans. b**

769. Which of the following statements about myosin is true

(a) It is a spherically symmetric molecule
(b) It is low in α-helix content
(c) It is a zinc-requiring enzyme
(d) It is an actin-binding protein **Ans. d**

770. If an enzyme behaves according to classic Michaelis-Menton kinetics, from a double reciprocal plot of velocity versus substrate concentration, the value for the Michaelis constant (K_m) of the substrate can be determined graphically as the

(a) Slope of curve
(b) Point of infection of the curve
(c) Absolute value of the intercept of the curve with the X-axis
(d) Reciprocal of the absolute value of the intercept of the curve with the Y-axis **Ans. d**

771. Which of the following oxidation-reduction systems has the highest redox potential

(a) Fumarate/succinate (b) NAD^+/NADH
(c) Fe^{+++} Cytochrome a/Fe^{++} (d) Fe^{+++} Cytochrome b/Fe^{++}
Ans. c

772. Dinitrophenol would be most likely to inhibit cell function by disrupting

(a) TCA cycle (b) Glycolysis
(c) Hepatic gluconeogenesis (d) Oxidative phosphorylation
Ans. d

773. Polymyxin is unique among chemotherapeutic agents because it is bactericidal in the absence of cell growth. It exerts its effect by

(a) Detergent-like disruption of membranes

(b) Binding of DNA polymerase

(c) Binding to DNA as an insertion its effect by

(d) Binding to polysome-bound mRNA **Ans. a**

774. Cycloserine inhibits transpeptidation in the formation of the peptidoglycan cell wall network of gram positive organisms. This action of cycloserine is competitively inhibited by

(a) L-lysine (b) D-alanine

(c) D-glutamine (d) D-serine **Ans. b**

775. Some antibiotics act as ionophores, which means that they

(a) Increase cell membrane permeability to specific ions

(b) Inhibit both translation and transcription

(c) Inhibit only translation

(d) Interfere directly with bacterial cell-wall synthesis **Ans. a**

776. De novo synthesis of an enzyme, promoted by the substrate on which it acts, is characterized by the term

(a) Induction (b) Gratuity

(c) Activation (d) Derepression **Ans. a**

777. Control of metabolic pathway may be exerted by enzyme repression or induction. In vertebrates, this form of enzyme control occurs primarily in the

(a) Heart (b) Brain

(c) Liver (d) Skeletal muscle **Ans. c**

778. Which of the following statements concerning trypsinogen and chymotrypsinogen are false

(a) They have considerable homologies in primary sequence

(b) They are secreted by exocrine cells in the pancreas

(c) They are exopeptidases

(d) They can be converted into active enzymes by limited digestion with trypsin **Ans. c**

779. It is true of most enzymes that they

(a) Increase the rapidity of the reaction they catalyze

(b) Are specific for the substrate as well as the substrate as well as the reaction catalyzed
(c) Are large polypeptides with a molecular weight greater than 5,000 daltons
(d) All of the above **Ans. d**

780. Proteins that contain a porphyrin ring include
(a) Catalase (b) Cytochrome
(c) Haemoglobin (d) All the above **Ans. d**

781. Which of the following proteins contain iron
(a) Haemoglobin (b) Cytochrome C
(c) Peroxidase (d) All of the above **Ans. d**

782. Which of the following statements about sickle cell anaemia are true
(a) It is widely distributed in areas of high malaria fatality
(b) It results from a single amino acid change in haemoglobin
(c) It is seen in homozygous individuals only
(d) All of these **Ans. d**

783. Heavy chains of I_gG antibody may be separated from light chains with
(a) Ethanolamine (b) Pepsin
(c) Papain (d) Mercaptoethanol **Ans. d**

784. NADP is
(a) Enzymes activator (b) Electron acceptors
(c) Ion carrier (d) All living cells of plant body
Ans. b

785. The first enzyme that reduces nitrates to nitrites and ammonia in plant is
(a) Glutamine synthetase (b) Glutamate dehydrogenase
(c) Nitrite reductase (d) Nitrate reductase **Ans. d**

786. A huge amount of starch is stored in potatoes which are underground. This is made possible by
(a) Synthesis of sugar in the potatoes
(b) Activity of enzymes which convert starch into sugar and sugar back into starch after it has of starch the potato
(c) Migration of starch from aerial parts
(d) Migration of starch from the soil **Ans. b**

787. Blocking of active site of an enzyme is a kind of
(a) Feedback inhibition (b) Non competitive inhibition
(c) Allosteric inhibition (d) Competitive inhibition **Ans. d**

788. The plant proteinases or endopeptidases enzyme is
(a) Trypsin (b) Papain
(c) Pepsin (d) Urease **Ans. b**

789. T. Cech and S. Altman got noble prize in 1989 for
(a) Viral infection
(b) Mechanism of DNA synthesis
(c) Catalytic role of RNA (ribozyme)
(d) All the above **Ans. c**

790. Enzyme taking part in converting dihydroxy acetone phosphate to glyceraldehydes phosphate belongs to the type of
(a) Ligases (b) Transferases
(c) Isomerases (d) Hydrolases **Ans. c**

791. The enzyme used for alcohol formation by fermentation is
(a) Invertase (b) Lipase
(c) Amylase (d) Zymase **Ans. d**

792. Feedback inhibition is caused by blocking of enzymes by
(a) Chemicals produced by hormones
(b) Hormones
(c) Competitive inhibition
(d) Accumulated end products **Ans. d**

793. A substance unrelated to substrate reversibly changes the activity of an enzyme. It is
(a) Allosteric subunit (b) Competitive inhibitor
(c) Allosteric modulator (d) Non competitive inhibitor
Ans. b

794. The transfer of a group from a donor molecule to an acceptor molecule is catalyzed by
(a) Protease (b) Isomerase
(c) Transferase (d) Hydrolytic enzymes **Ans. c**

795. Activation energy is required by
(a) Non catalyzed reactions (b) Catalyzed reaction
(c) Both (a) and (b) (d) None of the above **Ans. c**

796. Many vitamins are necessary to the cell

(a) To increase immunity (b) To act as coenzymes

(c) To provide energy (d) To help in digestion **Ans. b**

797. Coenzyme functions in association with a

(a) Apoenzyme (b) Protein

(c) Holoenzyme (d) Vitamin **Ans. a**

798. The enzymes catalyzing breakdown without addition of water are called

(a) Ligases (b) Lysases

(c) Hydrolases (d) Oxidoreductases **Ans. b**

799. Amylopectin acts upon

(a) Polysaccharide in any medium

(b) Polysaccharide in acidic medium

(c) Polysaccharide in alkaline medium

(d) Polysaccharide in neutral medium **Ans. c**

800. A coenzyme is

(a) Same enzyme that occurs in different tissues such as heart and muscle

(b) One that shares the function of other enzyme

(c) Organic or inorganic in nature and helps activate metabolic enzymes

(d) Organic non protein in nature and helps to activate metabolic enzymes **Ans. d**

801. Enzymes are the polymers of

(a) Hexose carbon (b) Fatty acids

(c) Amino acids (d) Inorganic phosphate **Ans. c**

802. The release of adenyl cyclase from the cell membrane changes

(a) ATP into ADP (b) ADP into ATP

(c) cAMP into ATP (d) ATP into cAMP **Ans. d**

803. Proteinaceous nature of the enzyme was established by obtaining crystalline form of urease from Jack beans in 1926 by

(a) Buchner (b) Summer

(c) Pouling (d) Northrop **Ans. b**

804. Which has coenzyme activity

(a) Purine (b) Pyrimidine

(c) Urease (d) Both a and b **Ans. c**

805. Which one is not a simple enzyme

(a) Amylase (b) Pepsin

(c) Urease (d) Dehydrogenase **Ans. d**

806. UbQ (ubiquinone) is

(a) Activator (b) Protein cofactor

(c) Non protein coenzyme (d) Protein coenzyme **Ans. c**

807. The enzyme used to dissolve blood clot in coronary artery is

(a) Thrombokinase (b) Rennin

(c) Streptokinase (d) Tyrosinase **Ans. c**

808. The metallic ion acting as activator of ATPase in cell membrane

(a) Fe^{++} (b) K^{+}

(c) Na^{+} (d) Mg^{++} **Ans. c**

809. Enzyme responsible for RNA directed DNA synthesis is

(a) DNA polymerase (b) Helicase

(c) Reverse transcriptase (d) Topoisomerase **Ans. c**

810. Which of the following enzyme is functional in maize grain during germination

(a) Maltase (b) Diastase

(c) Urease (d) Zymase **Ans. b**

811. Cytochrome oxidase is a marker enzyme for

(a) Circular DNA of mitochondria

(b) Outer membrane of mitochondria

(c) Inner membrane of mitochondria

(d) Matrix of mitochondria **Ans. c**

812. The enzymes are highly specific in their action. This specificity of an enzyme if due to

(a) RNA (b) Specific coenzymes

(c) Arrangement of amino acids (d) Active site **Ans. c**

813. The stoppage of an enzymatic (biological) reaction by the end product of reaction is called

(a) End product inhibition (b) Feed back inhibition

(c) Negative inhibition (d) All of the above **Ans. d**

814. Coenzyme is a part of enzyme

(a) Vitamin A

(b) Inorganic metal activator

(c) Non protein organic part attached firmly

(d) Non protein organic part attached loosely **Ans. d**

815. In competitive inhibition

(a) Inhibitor block site

(b) Inhibitor resembles the substrate in molecular structure

(c) Inhibitor has no effect on allosteric site

(d) All are correct **Ans. d**

816. If the prosthetic group or non protein part of an conjugation enzyme (Holoenzyme) is removed then

(a) Enzyme becomes inactive

(b) K_m constant increases

(c) Activity of enzyme decreases

(d) Enzyme starts inhibiting the reaction **Ans. a**

817. Where does reaction occur between enzyme and substrate

(a) Allosteric site (b) Prosthetic group

(c) Active site (d) None of the above **Ans. c**

818. Induced fit theory of Koshland states that

(a) Active sites are inactive and do not participate

(b) Active sites are static like a key

(c) Active sites undergo geometrical conformational changes

(d) None of these **Ans. c**

819. Cyanide kill an animal by inhibiting cytochrome oxidase (enzyme of respiratory chain). Is does not bind with active site this is an examples of

(a) Feedback inhibition (b) Noncompetitive inhibition

(c) Competitive inhibition (d) Allosteric inhibition **Ans. b**

820. The spoilage of food can be prevented by keeping it in cold storage. It is due to

(a) More availability of pure oxygen in cold storage

(b) Reduced respiration

(c) Reduced enzyme activity of bacteria

(d) Reduced enzyme activity in substrate molecules **Ans. c**

821. Which is not an attribute of enzymes

(a) Regulator (b) Specificity

(c) Procaine (d) Used in reaction **Ans. d**

822. Sometimes enzymes are named according to the source from which they are obtained such as papain from papaya. Name of the enzymes obtained from pineapple

(a) Maltose (b) Bromelain

(c) Sucrose (d) Pineollelase **Ans. b**

823. The Q_{10} for enzyme activity is

(a) 3-4 (b) 2-3

(c) 1-2 (d) 1-3 **Ans. b**

824. According to IUB system, isomerases belong to which class

(a) IV (b) V

(c) III (d) I **Ans. b**

825. Why heat is used to sterilized non living object in tissue culture

(a) Proteins lose their tertiary structure due to break down of bonds

(b) Proteins are denatured at temperature above 55°C

(c) Both correct

(d) Only a is correct **Ans. c**

826. A high fever is dangerous to a human because

(a) BMR is lowered (b) Fats are oxidised

(c) Proteins are used up quickly (d) Enzymes are denatured **Ans. d**

827. Which enzyme was isolated by Buchner

(a) Zymase (b) Ferment

(c) Zeamase (d) Zeathinase **Ans. a**

828. Vitamins, hormones and enzymes are nothing but

(a) Proteins (b) Biological regulators

(c) Growth regulators (d) Digestive juices **Ans. b**

829. The inorganic part of enzyme is known as

(a) Activator (b) Apoenzyme

(c) Coenzyme (d) Holoenzyme **Ans. a**

830. Who prepared pure crystals of enzyme pepsin and trypsin
(a) Northrop (b) Sumner
(c) Pasteur (d) Buchner **Ans. a**

831. Curd making, gobargas and wine making involve
(a) Digestion (b) Oxidation
(c) Acrtation (d) Fermentation **Ans. d**

832. The enzyme which combines with a non protein prosthetic group to form a functional enzyme is called
(a) Coenzyme (b) Proenzyme
(c) Holoenzyme (d) Apoenzyme **Ans. d**

833. Cl, Zn, Fe, Mn are
(a) Cofactor (b) Metabolic poisons
(c) Activators of enzymes (d) Inhibitors **Ans. b**

834. The enzyme in insectivorous plants for digesting the insects is
(a) Pepsinogen (b) Steapsin
(c) Amylopsin (d) Trypsin **Ans. d**

835. Coenzyme II is
(a) $FADH_2$ (b) FAD
(c) NAD (d) NADP **Ans. d**

836. Which enzyme converts ATP into cyclic AMP
(a) ATPase (b) Hydrolase
(b) Adenyl cyclase (d) Adenyl synthetase **Ans. c**

837. Which one is coenzyme or part of coenzyme
(a) CoQ and CoA (b) Vitamin B and C
(c) ATP (d) All of these **Ans. d**

838. The turn over number of enzymes is related with its
(a) Efficiency (b) Protein nature
(c) Catalytic nature (d) Specificity **Ans. a**

839. Enzymes have high molecular weight and therefore, cannot move from cell to cell. Every cell synthesized its own enzymes. Who is responsible for the synthesis of enzymes in each cell
(a) mRNA (b) Ribosomes
(c) Intrinsic proteins (d) Cytoplasm **Ans. a**

840. Enzymes are absent in

(a) Viruses (b) Bacteria

(c) Cyanobacteria (d) RBC **Ans. a**

841. The lower value of K_m means

(a) Lower the affinity of enzyme with substrate

(b) Higher substrate affinity of enzyme

(c) Higher enzyme activity

(d) No effect on reaction **Ans. b**

842. Turn over number of an enzyme means

(a) Number of substrate molecules acted upon by an enzyme per second

(b) Number of substrate molecules acted upon by one molecule of an enzyme per minute

(c) Number of enzyme molecules acting on one molecule of substrate per minute

(d) Number of molecules of end product produced by an enzyme in one minute **Ans. b**

843. Enzymes have 3-D structure minimum number of amino acids to make an enzyme is

(a) 1000 (b) 150

(c) 50 (d) 10000 **Ans. b**

844. Which enzyme do not obey K_m kinetics

(a) Zymogens (b) Exoenzymes

(c) Allosteric enzymes (d) Proenzymes **Ans. c**

845. In a dilute starch solution, α salivary amylase is added at pH 1.6 and kept at 35°C for half an hour and then iodine solution is added. What would be the result

(a) The solution will be sweet

(b) There will be a black solution

(c) There will be no black colour

(d) Solution will be clear and colourless **Ans. b**

846. The value of K_m varies from 10^{-1} to 10^{-6} m but for allosteric enzymes, there is no constant K_m value. This K_m is

(a) End product concentration at which reaction attains ($1/2\ V_{max}$)

(b) Substrate concentration at which the enzymatic velocity ($1/2\ V_{max}$)

(c) Enzyme concentration at which reaction attains (1/2 V_{max})

(d) None of these **Ans. b**

847. Allosteric inhibition

(a) Makes active sites of enzymes unfit for substrate

(b) Controls excess formation of end product

(c) Both (a) and (b)

(d) None of these **Ans. c**

848. Rice or bread taste sweet on prolonged chewing because of the breakdown of starch in them. The enzyme in the saliva which takes part in this reaction is

(a) Pepsin (b) Rennin

(c) Amylase (d) Invertase **Ans. c**

849. The best example of extra cellular enzymes (exoenzyme) is

(a) Succinic dehydrogenase (b) Digestive enzymes

(c) Nucleases (d) Rubisco **Ans. b**

850. Enzymes were discovered in

(a) Virus (b) Bacteria

(c) Yeast (d) Jack bean (*Canavalia*) **Ans. c**

851. Specificity of a protein in enzymatic property depends upon

(a) Linear sequence of amino acids

(b) Active sites

(c) K_m value

(d) Turn over number **Ans. a**

852. Who showed that besides living yeast cell, even their extract (zymase) could bring about fermentation

(a) Kuhne (b) Buchner

(c) Sumner (d) Pasteur **Ans. b**

853. Most of the enzymes when secreted are in (zymogens) otherwise they will destroy

(a) Cell DNA (b) Cell protein

(c) Cell mitochondria (d) Cell wall and membrane

Ans. b

854. An enzyme acts as catalyst because
(a) Enzyme catalysed reactions are reversible
(b) It accelerates the chemical reaction
(c) It remains unchanged
(d) All of the above **Ans. d**

855. Which determines specificity of enzymes
(a) Inorganic factor (b) Coenzyme
(c) Protein part (d) Full holoenzyme **Ans. c**

856. Modulators are substances which are
(a) Inhibitors (-) of enzyme activity
(b) Activators (+) of enzyme activity
(c) Both (a) and (b)
(d) Coenzymes **Ans. a**

857. Enzyme enolase required element
(a) Mn (b) Mg
(c) B (d) Cu **Ans. b**

858. The monopoly of proteins to form enzymes is now not universal. Certain, catalytic RNA enzymes are differ from protein enzymes in
(a) That they are exactly similar to enzyme
(b) That they catalyse only phosphoprotein molecules
(c) That they catalyse only nucleic acid molecules
(d) That they catalyse only RNA molecules **Ans. d**

859. Which of the following is not a part of enzyme but it activates more than 40 enzymes
(a) Mg (b) Mn
(c) K (d) Zn **Ans. c**

860. Which mineral element is essential for aconitase
(a) Cl (b) Mn
(c) Sn (d) Fe **Ans. d**

861. Enzyme Rubisco is found in storing of chloroplast. It is the most abundant protein on this earth. It constitutes....% of chloroplast proteins
(a) 26% (b) 36%
(c) 16% (d) 44% **Ans. c**

862. Enzyme kinase requires

(a) Mn (b) Mg

(c) K (d) $Mg^{++}K$ **Ans. b**

863. Enzyme responsible for biological nitrogen fixation is

(a) RUBP carboxylase (b) PEP carboxylase

(c) Oxygenase (d) Nitrogenase **Ans. d**

864. Which vitamin in incorporated into the structure of NAD/NADP

(a) Nicotinic acid (b) Vitamins B_7

(c) Vitamin PP (d) All correct **Ans. d**

865. Some enzymes when secreted are in inactive state. Such enzymes in inactive state are called

(a) Zymogens (b) Apoenzymes

(c) Coenzymes (d) Isoenzymes **Ans. a**

866. Apoenzymes and coenzymes collectively produce

(a) Cofactor (b) Enzyme product complex

(c) Holoenzyme (d) Prosthetic group **Ans. c**

867. Vitamin B_1 is a constituent of a coenzyme

(a) TPP (b) CoA

(c) NAD (d) FMN **Ans. a**

868. Competitive inhibitors are those which

(a) Act as coenzyme for reaction

(b) Alter the structure of enzyme molecule

(c) Compete for same active site of enzyme

(d) Both (a) and (c) **Ans. c**

869. Isoenzymes are structural variants of same enzyme. LDH (Lactic dehydrogenase) has 5 isomers in liver. α amylase in wheat endosperm has how many isoenzymes

(a) 4 (b) 5

(c) 16 (d) None of these **Ans. c**

870. The activator of carbonic anhydrase which is fastest enzyme RBC in of our body is

(a) Cobalt (b) Zn

(c) Fe^{++} (d) Cl^- **Ans. b**

871. Which is best evidence for Lock and Key theory (Template theory)

(a) Non competitive inhibition (b) Allosteric inhibition

(c) Competitive inhibition (d) Feedback inhibition **Ans. c**

872. Which of the following enzyme requires molybdenum as the cofactor

(a) Carboxylase (b) Hexokinase

(c) Nitrate reductase (d) Flavin oxidase **Ans. c**

873. Which of the following enzymes are located in the cytoplasm

A. Enzymes of glycolysis

B. Enzymes of hexose monophosphate shunt

C. Enzymes of oxidative phosphorylation

D. Select the correct answer using the codes given below

(a) A, B and C (b) A and C

(c) B and C (d) A and B **Ans. d**

874. The enzyme which converts glucose into glucose-6-phosphate is

(a) Phosphorylase (b) Phosphatase

(c) Glucosynthetase (d) Hexokinase **Ans. d**

875. Who got the Noble prize for working on enzymes in the year 1978

(a) W. Arber and D. Nathans (b) Nass and Nass

(c) R. Misra (d) H.G. Khorana **Ans. a**

876. No cell could live without

(a) Phytochrome (b) Enzymes

(c) Chloroplast (d) Proteins **Ans. d**

877. The enzyme which converts glucose to ethyl alcohol

(a) Diastase (b) Maltase

(c) Zymase (d) Invertase **Ans. c**

878. The lock and key theory of an enzyme structure and function was proposed by

(a) Morgan (b) Robertson

(c) Brown (d) Fischer **Ans. d**

879. Enzymes concerned with transfer of electrons are

(a) Hydrolase (b) Dehydrogenase

(c) Transaminase (d) Demolase **Ans. c**

880. Which one of the following enzymes is inactivated by oxygen

(a) Dehydrogenase (b) Nitrogenase

(c) Phosphatase (d) Urease **Ans. b**

881. Which one of the following enzyme is used to disintegrate the prokaryotic cell wall

(a) Amylase (b) Protease

(c) Lipase (d) Lysozyme **Ans. d**

882. Nickel is component of enzyme

(a) Nitrogenase (b) Nitrate reductase

(c) Urease (d) None of these **Ans. c**

883. Which enzyme is most abundantly found on earth?

(a) Invertase (b) Catalase

(c) Rubisco (d) Nitrogenase **Ans. c**

884. Identify the protein that does not contain any metal

(a) Cytochrome (b) Glycoprotein

(c) Ferritin (d) Phytochrome **Ans. b**

885. Enzymes are sensitive to

(a) Cold (b) Cell wall

(c) Heat (d) Pressure **Ans. c**

886. Allosteric enzymes have modulators for

(a) Activation only (b) Inhibition only

(c) Both activation and inhibition (d) Reduction in activation energy **Ans. c**

887. Enzyme aldolase which helps in containing dihydroxyacetone belongs to the category of

(a) Transferases (b) Lyases

(c) Hydrolases (d) Ligases **Ans. a**

888. An allosteric enzyme has

(a) Active site and two allosteric sites

(b) One active site

(c) One active site and two allesteric sites

(d) Two types of active sites **Ans. a**

889. Which one is constituent of coenzyme

(a) B_2 (b) Sucrase

(c) Lipase (d) Ascorbic acid **Ans. a**

890. Esterases belong to the category of

(a) Ligases (b) Lyases

(c) Hydrolases (d) Transferases **Ans. c**

891. Elisa test is used to

(a) Purify proteins (b) Separate viral RNA

(c) Identify specific proteins (d) Isolate DNA sequences **Ans. c**

892. Biocatalysts were found accidently in yeast extract by

(a) Pasteur (b) Kuhne

(c) Buchner (d) Sumner **Ans. c**

893. Enzyme amylase belongs to

(a) Oxido-reductase (b) Hydrolases

(c) Isomerase (d) Transferase **Ans. b**

894. Enzymes functional in cell are

(a) Inducible only

(b) Repressible only

(c) Inducible and repressible

(d) Inducible, constitutive and repressible **Ans. d**

895. Kinetics

(a) Reaction velocity (b) Competitive inhibition

(c) Denaturation of enzymes (d) All of the above **Ans. b**

896. Isoenzymes are

(a) Different in maximum activity

(b) Different in substrate affinity

(c) Different in molecular forms

(d) All of the above **Ans. d**

897. Enzyme urease, first crystallized by Sumner, was obtained from

(a) *Thalictnum* (b) *Pancratium*

(c) *Canavalia* (d) Human urine **Ans. c**

898. Constitutive enzymes are

(a) Operational all the time (b) House keeping enzyme
(c) Alloenzymes (d) Both a and b **Ans. d**

899. Sulfonamides were used to treat microbial infection because they inhibit microbial growth by

(a) Inhibiting union of ribosome subunits and preventing protein synthesis
(b) Inhibiting wall formation
(c) Breaking naked DNA of microbes
(d) Competing with PABA required for synthesis of folic acid **Ans. d**

900. Excess of ATP inhibits respiration but inhibiting one of the following enzymes

(a) Aldolase (b) Pyruvic decarboxylase
(c) Phosphofructokinase (d) Kinase **Ans. c**

901. Nerve gas (DFP) inhibits neurotrans mission as it reacts with

(a) Norad renaline
(b) Acetylcholine
(c) Serine of acetylcholine esterase
(d) Choline of acetylcholine esterase **Ans. d**

902. The enzyme used in the beverage industry to flavor soft drinks and in the baking industry to sweeten the biscuits and cakes

(a) Glycoamylase (b) Amylase
(c) Proteases (d) Glucoamylase **Ans. b**

903. Enzymes belong to which class of compounds?

(a) Polysaccharides
(b) Polypeptides
(c) Polynitro heterocyclic compounds
(d) Hydrocarbons **Ans. b**

904. Which of the following chemical unit is certainly to be found in an enzyme?

(a) 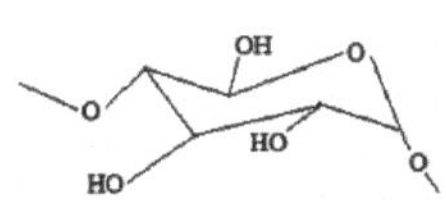

(b) 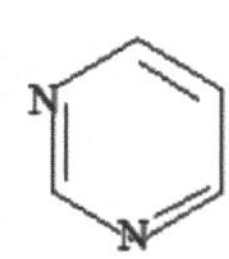

(c)  (d) 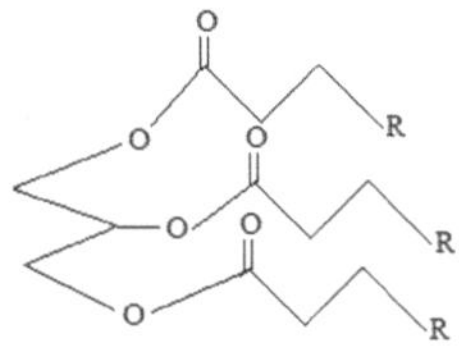 **Ans. c**

905. The function of enzymes in the living system is to

(a) Transport oxygen

(b) Provide immunity

(c) Catalyse biochemical reactions

(d) Provide energy **Ans. c**

906. Enzymes in the living systems

(a) Provide energy (b) Provide immunity

(c) Transport oxygen (d) Catalyse biological reactions

Ans. d

907. The effect of an enzyme on a biochemical reaction is that the

(a) Rate of forward reaction is increased but the rate of backward reaction is not altered.

(b) Rate of backward reaction is decreased but the rate of forward reaction is not altered.

(c) Rate of backward reaction and the rate of forward reaction are altered by the same factor so that their ratio does not change.

(d) Neither the rate of forward reaction nor the backward reaction is altered. **Ans. a**

908. The best solvent for removing butter stain from cloth is

(a) $CHCl_3$ (b) C_2H_5OH

(c) $C_2H_5OC_2H_5$ (d) H_2O **Ans. a**

909. The number of tripeptides formed by three different amino acids are

(a) Three (b) Four

(c) Five (d) Six **Ans. d**

910. Which of the following enzymes hydrolyses triglycerides to fatty acids and glycerol?

(a) Amylase (b) Maltase

(c) Lipase (d) Pepsin **Ans. c**

911. Magnesium is required for enzymes connected with
(a) Oligosaccharide formation (b) ATP utilizing reactions
(c) Glycoprotein formation (d) All the above **Ans. b**

912. A substance unrelated to substrate reversibly changes the activity of an enzyme. It is
(a) Competitive inhibitor (b) Allosteric subunit
(c) Allosteric modulator (d) None of the above **Ans. c**

913. Malonate functions as substrate analogue and inhibits enzyme
(a) Succinate dehydrogenase (b) Pyruvate oxidase
(c) Fumarase (d) Isocitrate dehydrogenase
Ans. a

914. Cyanide causes irreversible inhibition of cytochrome oxidase. It
(a) Combines with an amino acid (b) Destroys tertiary structure
(c) Attaches to copper (d) All the above **Ans. c**

915. Enzyme action comes to a stop when hydration decreases in maturing seeds to
(a) 50-60% (b) 30-45%
(c) 25-30% (d) 10-20% **Ans. d**

916. The disease phenylketoneurea is due to the deficiency of
(a) Vitamin A (b) Carbohydrates
(c) Phenylalanine hydroxylase (d) Lysine **Ans. c**

❑❑❑

Chapter-18

Hormones and Growth Substances

Fill in the blanks

1. Dormancy induction is due to ....

Ans. ABA

2. ....is a naturally occurring cytokinin.

Ans. 6-furfuryl aminopurine

3. Cell division is favoured by.................

Ans. Cytokinins

4. Gibberellins causes............................

Ans. internode elongation

5. 2, 4-D is a ....

Ans. weedicide

6. Spiral growth shown by shoot tips is called ....

Ans. circumnutation

7. Genetically dwarf plants get elongated if sprayed with ....

Ans. gibberellins

8. Cytokinins are derivatives of.................

Ans. purines

9. ....is a hormone which acts as growth inhibitor

Ans. abscisic acid

10. 2, 4, 5-T is used as a ....

Ans. weedicide

11. α-amylase activity in germinating barley seeds is due to the activity of.................

Ans. gibberellins

12. Apical dominance is maximum in............

Ans. palms

13. Sleep disease in flowers is caused by ....

Ans. ethylene

14. ....is the motivative force for cell elongation.

Ans. Turgor pressure

15. Apical dominance is due to..................

Ans. auxins (IAA)

16. Synthetic compounds which interfere with gibberellins in plant body are called......................

Ans. antigibberellins

17. Elongation of shoot just before flowering is called........................

Ans. bolting

18. Morphactins are derivatives of...............

Ans. fluorence carboxylic

19. Genetically dwarf plants are deficient in a gene responsible for the synthesis of....................

Ans. gibberellins

20. Coconut milk is used in tissue culture experiments because it contains..............

Ans. cytokinin

21. The tissues infected by *Agrobacterium tumefaciens* are rich in hormone ....

Ans. cytokinins

22. Leaf fall occurs when the content of auxin........................

Ans. decreases

23. Avena curvature test is used for bioassay of.........................

Ans. auxins

24. Cell division in vascular cambium are promoted by.......................

Ans. auxins

25. First naturally occurring cytokinin to be chemically identified was......................

Ans. zeatin

26. Leaves dipped in..............stay green longer than the control leaves.

Ans. cytokinins

27. ....is a stress plant hormone.

Ans. ABA

28. Phytochrome, present throughout the plant, is more concentrated at ....

Ans. growing tips

29. ....light inhibits seed germination.

Ans. Far red

30. Chemically, phytochrome is a..................

Ans. protein

31. The half life of glucagons is...........

Ans. ~5

32. Normal serum free testosterone in adult men varies between..................

Ans. 10–30 ng/dl

33. Normal serum free testosterone in adult women varies between................

Ans. 0.3–2 ng/dl

34. The prepubertal total serum testosterone is.....................

Ans. <100 ng/100 ml

35. The total serum testosterone in adult men is.....................

Ans. 300–1000 ng/100 ml

36. The total serum testosterone in adult women is.....................

Ans. 20–80 ng/100 ml

37. The serum estradiol level in men is.....................

Ans. 24–68 pg/ml

38. The serum estradiol level in women during 21–30 days of menstrual cycle is....................

Ans. 73-149 pg/ml

39. Serum progesterone level during luteal phase is.....................

Ans. 750 ng/ml

40. The production of progesterone by corpus luteum cell is stimulated by......................

Ans. LH

41. The enzyme catalyzing conversion of androstenedione to testosterone is a......................

Ans. dehydrogenase

42. Conversion of testosterone to estradiol requires the enzyme......................

Ans. aromatase

43. The precursor of testosterone is......................

Ans. pregnenolone

44. The hormone measured in urine to test pregnancy is......................

Ans. choroinic gonadotropin

45. G-proteins are......................

Ans. trimers

46. The nucleotide binding site of G-proteins is present on their......................

Ans. α-subunit

47. Insulin receptor is a......................

Ans. tetramer

48. Protein kinase C is activated by......................

Ans. diacyl glycerol

49. Melatonin is synthesised in......................

Ans. pineal gland

50. Diabetes insipidus is caused by deficient secretion of......................

Ans. vasopressin

51. Somatotropin is secreted by......................

Ans. anterior pituitary

52. Precursor of ACTH is......................

Ans. pro-opiomelanocortin

53. ACTH is a polypeptide made up of......................

Ans. 39 amino acids

54. CRH is a polypeptide made up of......................

Ans. 41 amino acids

55. In males, luteinising hormone acts on.....................

Ans. leyding cells

56. Secretion of luteinising hormone is increased by.....................

Ans. GnRH

57. Stature is increased in.....................

Ans. gigantism

58. An enzyme required for the synthesis of thyroid hormones is.....................

Ans. thyroperoxidase

59. Thyroid gland takes up circulating iodine by.....................

Ans. active uptake

60. The most powerful thyroid hormone is.....................

Ans. T_3

61. The most abundant thyroid hormone in blood is..................... bound to ....

Ans. T_4, TBG

62. Secretion of PTH is regulated by..................... effect of ....

Ans. feedback, plasma calcium

63. Second messenger for glucagons is.....................

Ans. cyclic AMP

64. Number of amino acid residues in glucagons is.....................

Ans. 29

65. Glucagon secretion increases when blood glucose is ....

Ans. low

66. Blood brain barrier can be crossed by.....................

Ans. Dopa

67. The major mineralcorticoid is.....................

Ans. aldactone A

68. A common intermediate in the synthesis of all the steroid hormones is.....................

Ans. pregnenolone

69. A common intermediate in the synthesis of estrogens is.....................

Ans. andostenedione

70. Secretion of mineralcorticoids is increased by.....................

Ans. angiotensin

71. The metabolites of androgens are.....................

Ans. 17-Ketosteroids

72. An androgen which is more powerful than testosterone is.....................

Ans. dihydrotestosterone

73. The major metabolite of progesterone is.....................

Ans. pregnanediol

74. A hormone used for detection of pregnancy is.....................

Ans. chorionic gonadotropin

75. Vasoconstrictor effect of ADH is mediated by.....................

Ans. protein kinase C

76. Urinary excretion of vanillyl madelic acid is increased in.....................

Ans. phaeochromocytoma

77. Iodide uptake by thyroid gland is decreased by.....................

Ans. thicyanate

78. STAT proteins are.....................

Ans. Signal transduction and activators of transcription

79. Sweat chlorides are increased in.....................

Ans. cystic fibrosis

80. Radioactive iodine uptake by thyroid gland 24 hours of a test dose is.....................

Ans. 20–40% of the test done

81. Normal range of total thyroxine in serum is.....................

Ans. 5–12 μg/dl

82. Retention of sodium in the body leads to a retention of.....................

Ans. water

83. Protein bound iodine is .... bound to protein.

Ans. thyroid hormones

84. There is polyuria without glycosuria in this disorder.....................

Ans. diabetes insipidus

85. In hyperpara thyroidism there is......................

Ans. hypophatemia

86. Tyrosine of thyroglobulin is acted upon by .... to give mono and diiodotyrosines.

Ans. higher valency state of iodine (I^+)

87. Protein bound iodine in blood is present to the extent of .... / dL

Ans. 3–8 mg

88. Prolactin release inhibiting hormone is believed to be......................

Ans. dopanine

89. Somatostatin is produced by...................... and ....

Ans. Hypothalamus, pancreas

90. ....hormone is a single chain polypeptide having 32 amino acids with molecular weight of 3,600.

Ans. Calcitonine

91. The secondary sexual characters in females is effected by......................

Ans. estrogens

92. Gastric Secretion is regulated by the hormone......................

Ans. gastrin

93. An essential agent for converting glucose to glycogen in liver is...................

Ans. UTP

94. Anticodon region is an important part of the structure of......................

Ans. t-RNA

95. Thyroid function is determined by the use of isotopes......................

Ans. I^{131}

96. Pernicious anaemia is diagnosed by the radio active substance.................

Ans. CO^{60}

97. 2, 4-D stands for ....

Ans. 2, 4 Dichlorophenoxy acetic acid

98. .... for the first time isolated cytokinin.

Ans. Miller

99. Internode elongation is promoted by ....

Ans. gibberellic acid

100. Physiologically active form of phytochrome is ....

Ans. P_{750}

101. .... hormone concerned with cell division.

Ans. Kinin (Zeatin)

102. Phytochrome is involved in ....

Ans. photoperiodism

103. Gibberellin was first extracted from fungi ....

Ans. *Gibberella fujikuroi*

104. It will not be of any advantage to produce seedless fruits in ....

Ans. pomegranate

105. Ethylene gas speed up...................... of fruits.

Ans. ripening

106. Plants which daily require shorter period of darkness for flowering are called ....

Ans. long day plants

107. Kinetin was isolated from ....

Ans. coconut milk

108. Effect of day length on flowering is called ....

Ans. photoperiodism

109. Avena curvature test was first of all doene by ....

Ans. F.W. Went

110. Cell division resulting in plant growth is ....

Ans. mitosis

111. Growth can be defined as a .... increase in weight volume or size.

Ans. permanent

112. The three components of growth in plants are cell division, elongation and ....

Ans. differentiation

113. The graph of growth against time is of .... type.

Ans. sigmoid

114. The Red (R) and far red (FR) responses are not independent of ....

Ans. GA

115. The growth regulator that retards ageing of plant organ is ....

Ans. cytokinin

Multiple Choice Questions

1. Which of the following is a gaseous plant growth regulator?
   (a) Ethylene (b) Kinetin
   (c) GA (d) ABA **Ans. a**
2. Ripening of banana is accompanied with
   (a) Sudden rise in cytokinin (b) Sudden rise in auxin
   (c) Sudden rise in ethylene (d) Sudden rise in gibberellin
   Ans. c
3. Ethylene gas
   (a) Slow down ripening of apples (b) Speed up maturation of fruits
   (c) Is saturated hydrocarbon (d) Retards ripening of tomatoes
   Ans. b
4. The ripening of fruits can be accelerated by
   (a) Artificially adding ethylene gas to the atmosphere surrounding them
   (b) Warming up the surroundings artificially
   (c) Reducing the supply of water to plant when fruits are maturing
   (d) Increasing the supply of nitrogen to the atmosphere surrounding them **Ans. a**
5. The plant hormone controlling fruit ripening is
   (a) IAA (b) Kinetin
   (c) Ethylene (d) GA **Ans. c**
6. Which one of the following responses of plants to growth hormones is true for ethylene
   (a) Increase in cell elongation
   (b) Decrease in the formation of female flowers
   (c) Increase in ripening of fruits
   (d) Decrease in abscission of flowers **Ans. c**
7. Hormone used in early ripening of fruits is
   (a) Auxin (b) ABA
   (c) Ethylene gas (d) Cytokinin **Ans. c**
8. Sleep disease of flowers is caused by
   (a) IAA (b) Ethylene
   (c) ABA (d) IBA **Ans. b**

9. Which of the following is growth inhibitor
   (a) Gibberellic acid (b) Auxin
   (c) Cytokinin (d) Ethylene **Ans. d**
10. Which one is used for artificial ripening of banana
   (a) Cytokinin (b) Ethylene
   (c) Auxin (d) Coumarin **Ans. b**
11. The term bioassay implies
   (a) Writing an essay on living organisms
   (b) Using living organisms in testing the biologically active chemicals
   (c) Description of microorganisms
   (d) Description of useful plant parts **Ans. b**
12. Pineapple can be made to flower in off-season by
   (a) Zeatin (b) Ethylene
   (c) Temperature (d) Short days **Ans. b**
13. To prevent over ripening of banana, these
   (a) Should be kept in refrigerator
   (b) Should be kept at room temperature
   (c) Should be treated with ascorbic acid
   (d) None of the above **Ans. a**
14. Which of the following combinations induces shoot differentiation in tobacco callus
   (a) High cytokinin and low auxin
   (b) High auxin and low cytokinin
   (c) Equal proportion of auxin and cytokinin
   (d) Only auxin and no cytokinin **Ans. a**
15. Which one of the following plant hormones is known as a stress hormone
   (a) Gibberellin (b) Kinetin
   (c) Auxin (d) Abscisic acid **Ans. d**
16. In unisexual plants, sex can be changed by application of
   (a) Auxins (or Ethylene) (b) ABA
   (c) Ethanol (d) Cytokinins **Ans. a**

17. The following is a naturally occurring growth inhibitor

(a) NAA (b) GA_3

(c) ABA (d) IAA **Ans. c**

18. Abscisic acid controls

(a) Shoot elongation

(b) Cell elongation and cell wall formation

(c) Cell division

(d) Leaf fall and dormancy **Ans. d**

19. Abscisic acid (ABA) promotes

(a) Budding (b) Cell division

(c) Leaf fall (d) Germination **Ans. c**

20. Early leaf fall in plants can be induced by

(a) IAA (b) GA_3

(c) ABA (d) Kinetin **Ans. c**

21. Abscisic acid treatment results in

(a) Leaf expansion (b) Stem elongation

(c) Stomatal closure (d) Root elongation **Ans. c**

22. Green organs are turned yellow by application of

(a) IAA (b) IBA

(c) 2, 4-D (d) ABA **Ans. d**

23. Photomorphogenetic pigment in plant is (or the pigment responsible for photomorphogenetic response is)

(a) Chlorophyll (b) Phytochrome

(c) Xanthophylls (d) Carotene **Ans. b**

24. Phytochrome is found in

(a) Algae (b) Fungi

(c) Gymnosperms (d) Angiosperms **Ans. d**

25. Phytochrome is used in

(a) Seed germination only

(b) Transpiration only

(c) Flowering only

(d) All physiological processes exhibited by the plants such as seed germination, flowering, stem elongation and transpiration **Ans. d**

26. Pigment involved in photoperception in flowering is

(a) Cytochrome (b) Carotene

(c) Lycopene (d) Phytochrome **Ans. d**

27. Phytochrome is involved in

(a) Photoperiodism (b) Geotropism

(c) Phototropism (d) Photorespiration **Ans. a**

28. One of the following shows inter conversion due to red-far-red light

(a) Chlorophylls (b) Carotenoids

(c) Cytochromes (d) Phytochrome **Ans. d**

29. Which one of the following statements is true for the phytochrome

(a) Phytochrome is a phytochrome

(b) Phytochrome is a photosynthetic pigment

(c) Phytochrome is a pigment that controls growth, photomorphogenesis and development of many plants

(d) Phytochrome is a regulatory protein that controls several dark-dependent development processes **Ans. c**

30. Which of the following pigments is present in two forms

(a) Cytochrome (b) Florigen

(c) Phytochrome (d) Vernalin **Ans. c**

31. Phytochrome is concerned with

(a) Absorption of blue light by leaves

(b) Absorption of P_R and P_{FR} in reversible manner

(c) Absorption of red light with wavelength 660 μm

(d) Absorption of far red light with wavelength 740 μm **Ans. b**

32. The pigment which absorbs red and far-red light is

(a) Phytochrome (b) Xanthophyll

(c) Cytochrome (d) Carotene **Ans. a**

33. In the red (P_R) and Far-red (P_{FR}) responses of plants, it is the

(a) First treatment that counts

(b) Last treatment that counts

(c) Middle treatment that counts

(d) Some total of all the treatments that counts **Ans. b**

34. The effect of day length on plant development is called
(a) Photosynthesis (b) Photoperiodism
(c) Phototropism (d) Chemotropism **Ans. b**

35. The effect of daily light period on flowering is called
(a) Photoxidation (b) Phototropism
(c) Photoperiodism (d) Photorespiration **Ans. c**

36. In many plants, the changeover from vegetative to reproductive phase takes place in response to
(a) The severity of temperature
(b) The length of the day
(c) The oxygen present in the air
(d) Mainly the food material available in the soil **Ans. b**

37. Garner and Allard are related with
(a) Photosynthesis (b) Photophosphorylation
(c) Phototropism (d) Photoperiodism **Ans. d**

38. Photoperiodism is
(a) Recurrence of day and night
(b) Effect of day length on plant growth
(c) Effect of day length on flowering
(d) Growth curvature in response to light **Ans. c**

39. When the dark period of short day plant is interrupted by a brief exposure of light then the plant
(a) Flowers immediately (b) Turns into a long day plant
(c) Will not flower at all (d) Gives more flowers **Ans. c**

40. If dark period of Sugarcane (a short day plant) is interrupted by a flash of light then it will
(a) Flower immediately (b) Produce abnormal flowers
(c) Not flower (d) Produce profuse flowers
Ans. c

41. The flowering in short day plants does not take place when
(a) The critical period is interrupted by white or red light
(b) An uninterrupted critical dark period
(c) The critical period is interrupted by light
(d) Dark period is interrupted by red light followed by far red light
Ans. a

42. A plant which needs not less than 10 hours of light to flower is
(a) Long day plant (b) Short day plant
(c) Day neutral plant (d) None of these **Ans. a**

43. Which of the following is not a short-day plant
(a) *Saccharum officinarum* (b) *Solanum tuberosum*
(c) *Glycine max* (d) *Brassica campestris* **Ans. b**

44. In short day plants (SDP) flowering is induced by
(a) Long night
(b) Photoperiod less than 12 hours
(c) Photoperiod shorter than initial value and uninterrupted long night
(d) Short photoperiod and interrupted long night **Ans. c**

45. Which of the following are short day plants
(a) Radish (b) Maize
(c) Soya-bean (d) Sugar beet **Ans. b**

46. Chemical agent which has important role in flowering
(a) Fluorocarbon (b) GA_3
(c) Florigen (d) Auxin **Ans. c**

47. A hypothetical chemical involved in the flowering of plants is
(a) Indole acetic acid (b) Florigen
(c) Gibberellin (d) Kinetin **Ans. b**

48. Which of the following plant hormone substitutes for long photoperiod in flowering plants
(a) Auxin (b) Gibberellin
(c) Cytokinin (d) Ethylene **Ans. b**

49. The name 'Florigen' was given by
(a) Miller (b) Cajlakjan
(c) Stanley (d) Lysenko **Ans. b**

50. Which of the following factors influence the flowering in plants
(a) Acidity of soil (b) Water in the soil
(c) Amount of green pigment (d) Photoperiod **Ans. d**

51. Hormone/complex which absorbs light for plant growth is
(a) Phytochrome (b) Cytochrome
(c) Phytohormone (d) Florigen **Ans. a**

52. Which one of the following is a long day plant

(a) Wheat (b) Jowar

(c) Bajra (d) Soybean **Ans. a**

53. *Chrysanthemum* flowers in winter because

(a) They love cool nights of winter

(b) They have warm nights of summer

(c) They are short day plants

(d) They are long day plants **Ans. c**

54. The practice of subjecting seeds to low temperatures for a period of time in order to cause growth and flowering during summer season is called (or induction of flowering by low temperature in plants is)

(a) Wintering (b) Vernalization

(c) Devernalization (d) Thermolysis **Ans. b**

55. Vernalization is

(a) Flowering induction by low temperature treatment of moist seeds

(b) Growth curvature in response to light

(c) Cycles of day and night

(d) Effect of day length on plant growth **Ans. a**

56. Flowering dependent on cold treatment is known as (or cold treatment given to plants for inducing flowering is)

(a) Cryoscopy (b) Cryostat

(c) Vernalization (d) Thermotrophy **Ans. c**

57. The phenomemon of vernalization was described for the first time by

(a) Went (b) Lysenko

(c) Kurosawa (d) Aristotle **Ans. b**

58. Hormone responsible for vernalization is

(a) Absiscin (b) Vernalin

(c) Florigen (d) Colchicine **Ans. b**

59. Which of the following hormones can replace vernalization

(a) Cytokinins (b) Auxins

(c) Ethylene (d) Gibberellins **Ans. d**

60. Enzymes, vitamins and hormones can be classified into a single category of biological chemicals because all of them
(a) Aid in regulating metabolism
(b) Are proteins
(c) Are synthesized within the body
(d) Enhance the oxidative metabolism **Ans. a**

61. DCMU is a herbicide which kills the plant by
(a) Inhibiting photolysis of water as it is a strong inhibitor of pigment system I
(b) Inhibiting carbon dioxide function as it is a strong inhibitor of pigment system II
(c) Checking respiration
(d) Destroying chloroplast **Ans. a**

62. High concentration of hormone will
(a) Kill the plant (b) Cause chlorosis in leaves
(c) Cause stunting in plants (d) Show no effect on the plant **Ans. a**

63. The stored food in germinating cotton seed is acted upon by
(a) Diastase (b) Lipase
(c) Maltase (d) Trypsin **Ans. b**

64. Which of the following is essential for pea seed germination
(a) CO_2 (b) N_2
(c) O_2 (d) H_2 **Ans. c**

65. Lettuce seeds germinate only when
(a) Any colour of light is available
(b) Blue colour light is available
(c) Red colour light is available
(d) Green colour light is available **Ans. c**

66. Germination of the seed is promoted by
(a) Green light (b) Red light
(c) Blue light (d) Infra red light **Ans. b**

67. Weed-killers generally have properties much like that of
(a) Vitamins (b) Enzymes
(c) Hormones (d) Insecticides **Ans. c**

68. Several horticultural techniques are followed for the production of 'Bonsai' plants. One of them is drastic pruning of root system. Which of the following physiological factors is involved in that method

(a) Inadequacy of mineral nutrients
(b) Deficiency of auxins
(c) Impairment of water absorption
(d) Deficiency of cytokinins **Ans. d**

69. Leaf fall starts when the amount of

(a) Auxin increase (b) Auxin decrease
(c) Abscisic acid decreases (d) Gibberellic acid decreases
Ans. b

70. Storage sprouting of potato can be prevented by

(a) Maleic hydrazide (b) Gibberellins
(c) IAA (d) Cytokinin **Ans. a**

71. Among plants 'pheromones' are secreted by the cells of following plants for the given function

(a) All plants for growth and development
(b) Yeast for facilitating mating
(c) All fungi for sexual reproduction
(d) *Rhizopus* for the formation of zygospore **Ans. d**

72. Protoplasmic streaming movements are referred as

(a) Movements of curvature
(b) Movements of locomotion (spontaneous or autonomic)
(c) Thigmonastic
(d) Photonastic **Ans. b**

73. Continuous movement of protoplasm in *Hydrilla* leaves is an example of

(a) Spontaneous movement (b) Induced movement
(c) Paratonic movement (d) Tropic movement **Ans. a**

74. Best material for demonstration of streaming movements of protoplasm within living cells is

(a) Onion peelings (b) Stamina hairs of *Tradescantia*
(c) Sensitive plants (d) Pith cells **Ans. b**

75. Cyclosis is a term referred to
(a) The active movement of *Paramecium*
(b) The active movement of the protoplasm in living cells
(c) The active movement of blood cells
(d) None of these **Ans. b**

76. Movements due to light are shown by
(a) Flowering plants (b) Lower plants
(c) All land plants (d) All the plants **Ans. d**

77. Curvature movement of certain plants in response to unilateral touch is called
(a) Chemotropism (b) Heliotropism
(c) Phototropism (d) Thigmotropism **Ans. d**

78. Plant movement initiated by external stimuli is known as
(a) Paratonic movement (b) Autonomic movement
(c) Seismonasty (d) Nyctinasty **Ans. a**

79. Thigmotropism is the response of the plant
(a) Water (b) Gravity
(c) Contact (d) Light **Ans. c**

80. When tendrils of a stem climb about a solid object, it is known as (or twining of tendrils around a support is a good example of)
(a) Phototropism (b) Thigmotropism
(c) Geotropism (d) Chemotropism **Ans. b**

81. The movements of hairs in *Drosera* is
(a) Thermonastic (b) Thigmonastic
(c) Seismonastic (d) Photonastic **Ans. b**

82. Greater growth on the lower surface of a bifacial leaf is
(a) Nyctinasty (b) Photonasty
(c) Hyponasty (d) Epinasty **Ans. c**

83. Horizontally placed stem exhibit apogeotropism due to
(a) The accumulation of auxin on the upper side
(b) The accumulation of auxin on the lower side
(c) Cell enlargement of the upper side
(d) Cell shrinkage on the lower side **Ans. b**

84. Generally the plant grows towards water and this phenomenon is known as

(a) Hydrotropism (b) Phototropism

(c) Thigmotropism (d) Geotropism **Ans. a**

85. The growth of pollen tubes towards ovules is due to

(a) Geotropism (b) Thigmotropism

(c) Chemotropism (d) Hydrotropism **Ans. c**

86. The wavelength of lights is most effective in phototropism

(a) Blue & violet (b) Blue & red

(c) Red & far-red (d) Green & yellow **Ans. a**

87. Nastic movements differ from tropic movements in being

(a) Controlled by chemicals

(b) Controlled by turgor pressure

(c) Directional with respect to stimulus

(d) Non-directional with respect to stimulus **Ans. d**

88. Opening of a flower and dropping of a bud are examples of

(a) Movement of curvature (b) Epinastic movement

(c) Hyponastic movement (d) Spontaneous movement **Ans. b**

89. Opening and closing of flowers represents a kind of

(a) Autonomic movement (b) Nutation

(c) Tropic movement (d) Nastic movement **Ans. d**

90. Sleep movements of leaves in certain plants are due to

(a) Decrease in intensity of light

(b) Differential growth at the base of leaf

(c) Excess of transpiration

(d) The leaves getting tired of carrying on photosynthesis **Ans. b**

91. Nyctinasty and thigmonasty movements are seen in

(a) *Mimosa* (b) *Drosera*

(c) *Cuscuta* (d) *Utricularia* **Ans. b**

92. Nastic movements is not caused by

(a) Light (b) Temperature

(c) Chemical (d) Sugar **Ans. d**

93. Which of the following movements in plants is not related to changes in auxin levels?

(a) Nyctinastic leaf movements

(b) Movement of roots towards soil

(c) Movement of sunflower tracking the direction of sun

(d) Movement of shoot towards light **Ans. a**

94. Sleeping movements of leguminous plants are

(a) Tropic movements (b) Turgor movements

(c) Growth movements (d) Nastic movements **Ans. d**

95. The type of movement found in the *Mimosa pudica* is

(a) Phototropism (b) Thigmotropism

(c) Geotropism (d) Hydrotropism **Ans. b**

96. Which of the following is an example of turgor movement?

(a) Chemotactic (b) Thermotropic

(c) Seismonasty (d) Haptotropic **Ans. c**

97. The leaves *Mimosa pudica* plant droop down on touch because

(a) The leaf tissues are injured

(b) The turgor pressure of the leaf base changes

(c) The plant has nervous system

(d) The leaves are very tender (delicate) **Ans. b**

98. Some plant movements occur in response to internal stimuli. These include

(a) Contraction movements shown by onion bulb's contracted roots

(b) Ripening of fruits

(c) Negative gravitropism

(d) Shutting of venus's fly trap **Ans. a**

99. Movement in the leaf of the Touch-me-not (*Mimosa*) plant is

(a) Photonasty (b) Seismonasty

(c) Nyctinasty (d) Epinasty **Ans. b**

100. Movement exhibited by telegraphic plant (*Desmodium gyrans*) comes under

(a) Hydrotropic movement

(b) Geotropic movement

(c) Negative geotropic movement
(d) None of the above **Ans. d**

101. The closure of the lid of pitcher in pitcher plant is
(a) A paratonic movement (b) A tropic movement
(c) A turgor movement (d) An autonomous movement
Ans. a

102. Leaves of flowering plants are known to carry on all of the following except
(a) Vegetative propagation (b) Transpiration
(c) Photosynthesis (d) Nerve-impulse conduction
Ans. d

103. Daily movements of leaves is an example of
(a) Lunar rhythm (b) Circardian rhythm
(c) Ultradian rhythm (d) Phototropism **Ans. b**

104. Endogenous rhythmic changes that occur in an organism with a periodicity of twenty four hours is known as
(a) Diurnal rhythm (b) Circardian rhythm
(c) Both of the above (d) None of the above **Ans. c**

105. During germination, stem grows upward and root goes downward because
(a) It depends upon light (b) Of auxin
(c) It does not depend on light (d) Of epinasty and hyponasty
Ans. b

106. Apples are usually wrapped in wax paper to
(a) Prevent change in colour due to sunlight
(b) Prevent aerobic respiration by checking the entry of oxygen
(c) Prevent ethylene formed by fruit to escape and thereby hasten ripening
(d) Improve the appearance of apples **Ans. b**

107. The leaves show
(a) Apogeotropism (b) Plageotropism
(c) Diageotropism (d) Orthotropism **Ans. a**

108. Movements induced by injury are called
(a) Aerotropism (b) Tromatropism
(c) Tromonsty (d) Tromatacty **Ans. b**

109. Time period from perception of stimulus and reaction is called
(a) Restitution time (b) Presentation time
(c) Relaxation time (d) Reaction time **Ans. b**

110. A higher proportion of ethylene is found in
(a) Ripening banana (b) Green banana
(c) Green apple (d) Fresh potato tube **Ans. a**

111. The shifting of chloroplasts to lateral walls of cells when mesophyll of a cell is exposed to bright light, is an example of
(a) Phototropism (b) Photonastism
(c) Phototactism (d) All the above **Ans. c**

112. In rhizome, the growth is
(a) Ageotropic (b) Diageotropic
(c) Plageotropic (d) Geotropic **Ans. b**

113. The effect of unilateral light may be neutralized by fixing plotted plant in the clinostat and
(a) Keeping it horizontally
(b) Keeping horizontally and rotating it
(c) Keeping it vertically
(d) Keeping vertically and rotating it **Ans. d**

114. Which of the following process is never affected by light in a vegetable garden
(a) Seed germination (b) Flowering
(c) Osmosis and fertilization (d) Manufacture of food **Ans. c**

115. Which of the following exerts profound effect on the reproductive growth of a flowering plant
(a) Quality of light (b) Quantity of light
(c) Direction of light (d) Duration of light cycles **Ans. d**

116. Higher concentration of auxin is present in
(a) Root apex (b) Petiole and leaves
(c) Stem apex (d) Internodes **Ans. c**

117. Two seeds are germinated, A in light and B in dark. Which of these would grow much taller in initial stage
(a) A (b) B
(c) Both would grow equally (d) None would grow **Ans. b**

118. For germination of photosensitive seeds, the most effective region of the spectrum which promotes germination is

(a) Red region (b) Far red region

(c) Blue region (d) Green region **Ans. a**

119. Tall variety of pea is treated with gibberellic acid. It would

(a) Become taller (b) Remain unaffected

(c) Become dwarf (d) Behavior cannot be predicted **Ans. b**

120. IAA is applied to the root tip of a seedling in high concentration. The growth of root would

(a) Be inhibited (b) Be increased

(c) Be decreased (d) Remain unaffected **Ans. a**

121. A short day plant is grafted on a long day plant. The composite is kept in short day conditions. Flowering would be induced in the branches of

(a) Short day plant (b) Long day plant

(c) Both (d) None **Ans. c**

122. *Volvox* colonies placed in cold water move towards warmer side on heating the container from one side. It is due to

(a) Chemotaxis (b) Thermotaxis

(c) Thermonasty (d) Thigmotropism **Ans. b**

123. The effect of stimulus subsidies after a time called

(a) Reaction time (b) Relaxation time

(c) Presentation time (d) Conduction time **Ans. b**

124. The minimum quantity of a stimulus required for bringing about a visible responses is called

(a) Threshold value (b) Sensitivity

(c) Perceptibility (d) Summation **Ans. a**

125. A movement in plant stimulated by oxygen is called

(a) Thermotaxis (b) Sensitivity

(c) Nutation (d) Aerotaxis **Ans. d**

126. The correct sequence for cellular growth stages is

(a) Division, differentiation, elongation

(b) Differentiation, division, elongation

(c) Division, elongation, differentiation
(d) Elongation, division, differentiation **Ans. c**

127. Tropic movements are basically due to
(a) Cell elongation (b) Cell division
(c) Sudden water gain (d) Sudden water loss **Ans. a**

128. In *Mimosa*, the turgor changes take place in
(a) Leaflet (b) Rachis
(c) Pulvinus (d) Stem apex **Ans. c**

129. Which of the following is a growth retardant
(a) Adenine (b) Ascorbic acid
(c) Morphactin (d) Zeatin **Ans. c**

130. In plants, growth is
(a) Restricted to certain regions or structures
(b) Irreversible
(c) Change in size
(d) All the above **Ans. d**

131. Plants differ from animals in having
(a) Localized growth
(b) Localized growth and indefinite life span
(c) Indefinite life span
(d) Definite life span **Ans. b**

132. Which one of the following is the motivative force for growth?
(a) Turgor pressure (b) Root pressure
(c) Atmospheric pressure (d) Osmotic pressure **Ans. a**

133. Growth is primarily affected by two climatic factors which are
(a) Light and temperature
(b) Temperature and relative humidity
(c) Light and wind
(d) Rainfall and temperature **Ans. a**

134. Maximum amount of growth in root occurs
(a) In the presence of light (b) At its apex
(c) Behind the apex (d) In the presence of soil **Ans. c**

135. The instrument by which the rate of growth of stem is accurately measured is

(a) Hydrometer (b) Auxanometer

(c) Osmometer (d) Photometer **Ans. b**

136. Auxanometer is an instrument with which we can measure the

(a) Rate of respiration (b) Rate of photosynthesis

(c) Rate of growth (d) Rate of transpiration **Ans. c**

137. Which of the following instruments can be used to record plant growth by seconds

(a) Arc auxanometer (b) Arc indicator

(c) Space marker disc (d) Crescograph **Ans. d**

138. Phytotron is

(a) A device in which plants are grown in soil

(b) A device in which plants are grown in water

(c) A device in which plants are grown under controlled conditions

(d) A device in which plants are grown in rocks **Ans. c**

139. Distribution of growth in a root by marking it at equal intervals with Indian ink was originally studied by

(a) Sachs (b) Strasburger

(c) Went (d) Nitsch **Ans. b**

140. In a growing plant, the very first phase during the process of growth is

(a) Cell division (b) Cell enlargement

(c) Cell differentiation (d) Cell maturation **Ans. a**

141. An apple tree can be made to produce larger fruits than usual by

(a) Applying hormones (b) Thinning of blossoms

(c) Decapitating the stem apex (d) Defoliating **Ans. b**

142. Maximum growth occurs in

(a) Lag phase (b) Senescent phase

(c) Exponential phase (d) Decline phase **Ans. c**

143. The growth involves

(a) Cell division (b) Cell enlargement

(c) Cell maturation (d) All the above **Ans. d**

144. Phytohormones are
(a) Hormones regulating growth from seed to adulthood
(b) Hormones regulating secondary growth
(c) Growth regulators synthesized by plants and influencing physiological processes
(d) Hormones regulating flowering **Ans. c**

145. The classical experiments on growth were performed by
(a) Lamarck and Boysen Jensen (b) Boysen Jensen and Darwin
(c) Darwin and Lamarck (d) De Vries and Paal **Ans. b**

146. *Avena* coleoptile test to find out the amount of growth promoting hormones was discovered by (or Avene coleoptile test was conducted by)
(a) F.W. Went (b) L.J. Oudus
(c) K.V. Thimann (d) F. Skoog **Ans. a**

147. The natural plant hormones were first isolated from
(a) Cotton fruits, spinach leaves and rice plants
(b) Avena coleoptiles, spinach leaves and fungus *Gibberella*
(c) Corn germ oil and human urine
(d) Human urine and rice plants **Ans. c**

148. Indole-3-acetic acid, called as auxin, was first isolated from
(a) Human urine (b) Corn germ oil
(c) Fusarium (d) *Rhizopus* **Ans. a**

149. Bioassay for auxin is
(a) Avena curvature test (b) Green leaf test
(c) Dwarf maize test (d) Cell division test **Ans. a**

150. The biological activity of IAA is tested by
(a) α-amylase set (b) Avena curvature set
(c) Soyabean callus test (d) *Xanthium* leaf disc test **Ans. b**

151. Which one of the following nutrients is concerned with the growth of the plants in view of their role in synthesis of auxin
(a) S (b) Mn
(c) Zn (d) K **Ans. c**

152. Primary precursor of IAA is
(a) Phenylalanine (b) Tyrosin
(c) Tryptophan (d) Leucine **Ans. c**

153. Plant hormone which is translocated to other parts for growth of the plant is known as
(a) Gibberellin (b) Cytokinin
(c) Indole acetic acid (d) None of the above **Ans. c**

154. The movement of auxin is largely
(a) Basipetal (b) Acropetal
(c) Both a and b (d) Centripetal **Ans. a**

155. The principal and naturally occurring auxin in higher plants is
(a) NAA (b) MHI (Maleic hydrazide)
(c) IAA (Indole acetic acid) (d) 2, 4-D **Ans. c**

156. Auxins are
(a) Growth catalysts (b) Destroyers of growth
(c) Growth hormones (d) Inhibitor of growth **Ans. c**

157. Auxins are abundantly produced in
(a) Root
(b) Meristematic region of the root
(c) Shoot
(d) Meristematic region of the shoot **Ans. d**

158. Substances which originate at the tip of stem and control growth elsewhere are
(a) Food materials (b) Auxins
(c) Vitamins (d) Enzymes **Ans. b**

159. A substance which accelerates the growth in stem is
(a) Etiolin (b) Auxin
(c) Vitamin (d) Enzyme **Ans. b**

160. Auxins promote growth
(a) Rarely (b) Always
(c) Mostly (d) Never **Ans. c**

161. Which of the following stimulates the plant for growth
(a) Cytokinin (b) Ethylene
(c) Auxin (d) Gibberellin **Ans. c**

162. Plants bend towards the light because
(a) They need light for photosynthesis
(b) They need light for respiration

(c) Light attracts them
(d) Cells on the shaded side elongated more **Ans. d**

163. If the tip of a seedling is cut off, growth as well as bending cases because it hampers
(a) Perception of light stimulus (b) Transpiration
(c) Respiration (d) Photosynthesis **Ans. a**

164. The plants bend towards the source of light on account of the movement of curvature known as
(a) Geotropism (b) Thigmotropism
(c) Chemotropism (d) Phototropism **Ans. d**

165. Hormone involved in phototropism is
(a) IAA (b) Gibberellin
(c) Kinetin (d) 2, 4-D **Ans. a**

166. Phototropic movements of roots and stems are due to
(a) Epinasty and hyponasty (b) Differential hormonal effect
(c) Action of light (d) Effect of light **Ans. b**

167. A plant bends towards the source of light when exposed to the light on only one side. Which of the following is the best explanation of the phenomenon
(a) They need light for photosynthesis
(b) The apices of their stems are attracted by light
(c) Some auxin accumulates on the shaded side to induce greater cell elongation on that side
(d) Light stimulates the cells on the illuminated side to increase in length **Ans. c**

168. Phototropism in shoot is attributed to
(a) Abscisic acid (b) Cytokinins
(c) Auxins (d) Gibberellins **Ans. c**

169. A plant placed near a window bends outwards because
(a) Its tip is able to get more oxygen
(b) Its tip is able to obtain more light
(c) Its tip is able to receive necessary warmth
(d) The auxin content on the shaded side is higher than on the illuminated side and as a result the cells on the shaded side elongated more than the cells on the illuminated side and tip bends outward **Ans. d**

170. If the stem goes towards sunlight and root goes just opposite to it, then stem movement is called
(a) Phototropic movement
(b) Positive phototropic movement
(c) Negative phototropic movement
(d) None of these **Ans. b**

171. A plant is kept on the window. It bends towards the outside because of
(a) More elongation of cells on the shaded portion
(b) Apical dominance
(c) Osmotic pressure
(d) Photosynthesis **Ans. a**

172. In phototropic movements, curvature takes place
(a) Far behind the tip (b) At the tip
(c) Behind the tip (d) None of these **Ans. c**

173. Phototropic curvature is promoted by
(a) Red light (b) Violet light
(c) Blue light (d) Green light **Ans. c**

174. Bending of growing shoot towards sunlight is termed as
(a) Photonasty (b) Photoperiodism
(c) Heliotropism (d) Hydrotropism **Ans. c**

175. An apparatus commonly used to demonstrate phototropism is
(a) Arc auxanometer (b) Heliotropic chamber
(c) Photometer (d) Clinostat **Ans. b**

176. Roots are positively geotropic
(a) Always (b) Mostly
(c) Rarely (d) Never **Ans. b**

177. The movement of plant organs in response to the force of gravity is called
(a) Hydrotropism (b) Geotropism
(c) Heliotropism (d) Phototropism **Ans. b**

178. Negative geotropism in horizontal stems is caused by
(a) More cells elongation on the upper side
(b) More cells differentiation on the upper side

(c) Accumulation of auxins on lower side

(d) Accumulation of auxins on upper side **Ans. a**

179. Geotropism movements are include as a result of

(a) Bringing down of auxin content on the lower side of the stem

(b) Bringing down of auxin content in the stem apex

(c) Bringing down of auxin content in the root apex

(d) None of the above **Ans. d**

180. Tertiary roots are

(a) Plageotropic (b) Ageotropic

(c) Positive geotropic (d) Negative geotropic **Ans. a**

181. Important factors that modify geotropic responses in plants are

(a) Root and shoot apices, humidity, temperature and light

(b) Root and shoot apices, temperature, light and CO_2 content of air

(c) Phytochrome, protein content of leaves, humidity and temperature

(d) None of the above **Ans. d**

182. Plant organs can detect a change in their orientation with respect to gravity

(a) After at least they have to be kept for a week

(b) After at least they have to be kept for a day

(c) After at least they have to be kept for a minimum of thirty minutes

(d) Within a few seconds **Ans. d**

183. A clinostat is associated with

(a) Turgor movements (b) Measurement of leaf area

(c) Thigmotropism (d) Geotropism **Ans. d**

184. Clinostat is used in studies on

(a) Photosynthesis (b) Respiration

(c) Osmosis (d) Growth movements **Ans. d**

185. Two plants are fixed in two clinostats in a horizontal position. Clinostat A is rotated while clinostat B is stationary. What will happen

(a) In A, roots will move downwards and stem upwards and in B, roots will move upwards and stem downwards

(b) In A, roots and shoots will move upwards

(c) In A, roots and stem will grow horizontally and in B, roots will move downwards and the stem will move upwards

(d) In A, roots and shoots will move downwards and in B, roots and shoots will move upwards **Ans. c**

186. Apical dominance in higher plants is due to

(a) Hormones (b) Enzymes

(c) Carbohydrates (d) Photoperiodism **Ans. a**

187. A good crop of tea leaves from a single plant can be obtained by

(a) Feeding auxin to the plant through roots

(b) Removing the apical buds of the main shoot and the branches

(c) Cutting off the top of the plant and then applying auxin to the cut ends

(d) Supplying auxin from tip of the plant as well as through roots **Ans. b**

188. Apical dominance is due to

(a) Gibberellins in the lateral buds

(b) Cytokinin in the leaf tip

(c) Auxin in the shoot tip

(d) Abscisic acid in the lateral bud **Ans. c**

189. In decapitated plants, axillary buds are activated because

(a) Amount of cytokinin increases in them

(b) Auxin is made available to them

(c) They get more light

(d) More food material is available to them **Ans. a**

190. What is the cause of excurrent habit in Pinus

(a) Presence of gibberellin

(b) Presence of apical dominance

(c) High concentration of cytokinin

(d) High concentration of ABA **Ans. b**

191. Apical dominance is related to (or caused by)

(a) Gibberellin (b) Auxin

(c) Cytokinin (d) Abscisic acid **Ans. b**

192. Which of the following is responsible for apical dominance

(a) GA_3 (b) IAA

(c) ABA (d) Florigen **Ans. b**

193. Auxin inhibits the growth of

(a) Apical buds

(b) Lateral axillary buds

(c) Roots on stem cuttings

(d) Parthenocarpic development of fruits **Ans. b**

194. When the apical buds of a plant are removed then

(a) Plant dies immediately (b) Plant dies after sometime

(c) Lateral bud is formed (d) Length of stem increases

Ans. c

195. If a growing plant is decapitated

(a) Leaves become yellow and have a tendency to fall off

(b) Growth stops

(c) Axillary buds are inactivated

(d) Axillary buds are activated **Ans. d**

196. Leaf fall occurs when the content of

(a) Auxin decreases (b) Auxin increases

(c) Gibberellic acid decreases (d) Abscisic acid decreases **Ans. a**

197. Pruning makes the hedge plant dense because

(a) Injury induces dense growth

(b) Apical dominance is removed

(c) Root sprouts additional branches

(d) Pruning removes shade and allow germination of new seedlings to import a dense growth **Ans. b**

198. Reactivation of cambial cells in spring season is caused by

(a) Auxin (b) Gibberellin

(c) Cytokinin (d) Ethylene **Ans. a**

199. Which one of the following plant functions is not generally governed or controlled by auxin

(a) Apical dominance (b) Phototropism

(c) Photosynthesis (d) Growth **Ans. c**

200. Which of the following hormone is concerned chiefly with root initiation

(a) IBA (b) GA_3

(c) ABA (d) Kinetin **Ans. a**

201. Induction of rooting in stem cutting by auxin treatment would be most beneficial in which of the following

(a) *Cuscuta* (b) *Bougainvillea*

(c) *Marchantia* (d) *Triticum* **Ans. b**

202. Which of the following effects of auxins on plants is the basis for commercial application

(a) Callus formation

(b) Curvature of stem

(c) Induction of root formation in stem cuttings

(d) All of the above **Ans. c**

203. By hormone application, it is possible to obtain seedless fruits. In which of the following it is desirable to obtain seedless fruits

(a) Tomato (b) Orange

(c) Watermelon (d) Pomegranate (Anar) **Ans. a**

204. Seedless tomatoes are produced by

(a) Growing tomato plants in mineral solution

(b) Supplying radioactive compounds

(c) Spraying hormones on flowers

(d) Propagating tomato plants from cuttings **Ans. c**

205. Parthenocarpic fruits can be produced by the application of the following auxin

(a) Indole butyric acid (IBA)

(b) 2, 4-Dichlorophenoxyacetic acid (2, 4 D)

(c) Indole acetic acid (IAA)

(d) All these **Ans. c**

206. Supra-optimal concentration of auxins

(a) Promotes flowering

(b) Kills the plant

(c) Prevents shortening of the internodes

(d) Promotes growth in both stem and root apex **Ans. b**

207. 2,4-Dichlorophenoxy acetic acid is a

(a) Flowering hormone (b) Herbicide

(c) Pesticide (d) Rooting hormone **Ans. b**

208. A high concentration of synthetic auxins is generally used for
- (a) Weed control
- (b) Enhancing root initiation
- (c) Controlling of cell enlargement
- (c) Preventing the growth of the lateral buds **Ans. a**

209. Which of the following is a weed-killer
- (a) NAA (b) 2, 4-D
- (c) ABA (d) GA **Ans. b**

210. Which of the following is not naturally occurring plant hormone
- (a) IAA (b) GA_2
- (c) 2, 4-D (d) Gibberellin **Ans. c**

211. Which of these is correct statement
- (a) Both stems and roots show positive geotropism
- (b) Both stem and roots show negative geotropism
- (c) Only stem show positive geotropism
- (d) Only root show positive geotropism **Ans. d**

212. If a leaf segment of *Bryophyllum* plant is in soil, it produces new plants. This is because its leaf has
- (a) Small embryos
- (b) Buds
- (c) High concentration of auxin and cytokinins
- (d) High concentration of cytokinins **Ans. b**

213. In the absence of light, which one of the following would increase
- (a) Uptake of water (b) Elongation of internodes
- (c) Uptake of mineral salts (d) Extension of lamina **Ans. b**

214. Auxins are known to induce
- (a) Female flowers in male plant
- (b) Male flowers in female plant
- (c) Male sterility in bisexual plants
- (d) Early bolting **Ans. a**

215. Which of the following is not true about auxins
- (a) It inhibits lateral growth (b) It promotes stem elongation
- (c) It promotes cell division (d) It helps hyperelongation

Ans. d

216. Gibberellins was first isolated from

(a) Rice roots (b) Ascomycetous fungus

(c) Coconut milk (d) Oat coleoptiles **Ans. b**

217. If the seedlings are grown in darkness

(a) They are of the same size as those grown in light

(b) They are much healthier than those grown in light

(c) They are similar to those grown in light

(d) They are taller than those grown in light **Ans. d**

218. Gibberellins was isolated from

(a) An alga (b) A fungus

(c) A bacterium (d) A virus **Ans. b**

219. Gibberellins was first extracted from

(a) *Gelidium* (b) *Aspergillus*

(c) *Geberella fuzikuroi* (d) *Gracilaria* **Ans. c**

220. Specific property attributed to gibberellins is

(a) Elongation of genetically dwarf plants

(b) Promotion of abscission in fruits

(c) Shortening of genetically tall plants

(d) Rooting of stem cuttings **Ans. a**

221. Bakanae disease in the rice plants (Paddy) is caused by

(a) Naphthalene acetic acid (NAA)

(b) 2, 4-Dichlorophenoxyacetic acid (2, 4-D)

(c) Gibberellic acid

(d) Indole acetic acid (IAA) **Ans. c**

222. GA was discovered by

(a) Yabuta and Sumiki (b) Utrochet and Dolk

(c) Donoho and Walker (d) Hoshimata and Rappaport **Ans. a**

223. Elongation of a genetically dwarf plant is possible by application of

(a) Vitamin (b) Kinetin

(c) Gibberellins (d) X-rays **Ans. c**

224. Elongation of genetically dwarf plants can be caused by (or can be done with the help of)

(a) Auxin (b) Gibberellin

(c) Florigen (d) None **Ans. b**

225. Internodal elongation is stimulated by

(a) Auxin (b) Cytokine

(c) Gibberellin (d) Phenol **Ans. c**

226. Which hormone is related with internodal elongation

(a) Auxin (b) Cytokinin

(c) Gibberellin (d) ABA **Ans. c**

227. A genetically dwarf plant can be converted into tall plant by the use of

(a) Kinetin (b) GA_3

(c) IAA/X-ray (d) 2, 4-D **Ans. b**

228. Gibberellins differ from auxins since they produce

(a) Cell division (b) Stem elongation

(c) Root initiation (d) Shortening of internodes

Ans. b

229. Dwarfness of plant can be controlled by treating it with

(a) Gibberellic acid (b) Cytokinin

(c) Antigibberillin (d) Auxin **Ans. a**

230. Bioassay for gibberellin is

(a) Green leaf test (b) Cell division test

(c) Dwarf maize test (d) Avena curvature test **Ans. c**

231. The habit of a cabbage plant can be changed drastically by the application of

(a) IAA (b) GA_3

(c) ABA (d) 2, 4-D **Ans. b**

232. Gibberellins are transported through

(a) Xylem (b) Phloem

(c) Xylem and phloem (d) Pith **Ans. a**

233. Gibberellic acid has been successfully employed to induce flowering

(a) In long day plants under short day conditions

(b) In short day plants under long day conditions

(c) For some plants

(d) None of the above **Ans. a**

234. Exogenous application of gibberellins induces male flowers' formation on genetically female plants in

(a) *Cucumis* (b) *Cucurbita*

(c) *Carica* (d) *Coccinia* **Ans. a**

235. Which of the following plant hormones stimulates the synthesis of α-amylase

(a) Auxin (b) Abscisic acid

(c) Ethylene (d) Gibberellin **Ans. d**

236. Cytokinin

(a) Refers to cell movements

(b) Causes dormancy

(c) Is a hormone whose main function is the induction of cell division and delay the senescence

(d) Is the process of cell division **Ans. c**

237. Cytokinins have the ability to

(a) Substitute cold treatment

(b) Cause root formation

(c) Cause morphogenesis

(d) Convert dwarf plants into tall plants **Ans. c**

238. Natural cytokinin was discovered by

(a) Yabuta and Overbeek (b) Miller and Skoog

(c) Letham (d) F.W. Went and Brown **Ans. b**

239. The term cytokinin was proposed by

(a) Went (b) Skoog

(c) Letham (d) Cholodny **Ans. c**

240. Hormone concerned primarily with cell division is

(a) GA_3 (b) IAA

(c) Cytokinin (d) NAA **Ans. c**

241. Cytokinins are known to

(a) Inhibit cytoplasmic movement

(b) Help in retention of chlorophyll

(c) Influence water movement
(d) Promote abscission layer formation **Ans. b**

242. Which of the following is capable of delaying yellowing of leaves
(a) Gibberellins (b) Cytokinins
(c) ABA (d) Auxins **Ans. b**

243. The growth regulators that retards ageing of plant organs is (or which of the following is capable of delaying senescence)
(a) Auxin (b) Gibberellin
(c) Cytokinin (d) Abscisic acid **Ans. c**

244. The following hormone is concerned chiefly with cell division in plants
(a) 2, 4-D (b) IAA
(c) GA (d) Kinetin (Zeatin) **Ans. d**

245. All the cytokinins are
(a) Acidic (b) Aminopurines
(c) Phenol (d) Glucosides **Ans. b**

246. From which of the following was kinetin isolated
(a) Coconut milk (b) Soybean
(c) Ground nut (d) Pineapple **Ans. a**

247. Which of the following is indispensible in cell culture
(a) Gibberellin (b) Kinetin
(c) Ethylene (d) Auxin **Ans. b**

248. For plant tissue culture among the following which one is required
(a) Kinetin (b) Coumarine
(c) Trypsin (d) Caffeine **Ans. a**

249. Which of the following plant material is widely used in the preparation of culture medium
(a) *Pinus longifolia* (b) *Cocos nucifera*
(c) *Borassus flabellifer* (d) *Cycas revolute* **Ans. b**

250. Flowering in some short day plants is promoted by
(a) ABA (b) Giberrellins
(c) Ethylene (d) Auxins **Ans. a**

251. Richmond-Lang effect in shown by
(a) Auxins (b) Gibberellins
(c) Kinetin (d) Sugars **Ans. c**

252. A natural growth regulator (hormone) is
(a) Benzaldehyde (b) 2, 4-D
(c) NAA (d) Ethylene **Ans. d**

253. Ethylene is a
(a) Solid hormone (b) Gaseous enzyme
(c) Gaseous hormone (d) Liquid gas mixture **Ans. c**

254. Ethylene is formed from the amino acid
(a) Glycine (b) Methionine
(c) Tryptophan (d) Leucine **Ans. b**

255. A higher proportion of ethylene is found in
(a) Green banana (b) Ripening banana
(c) Fresh potato tuber (d) Green apple **Ans. b**

256. Hormones
(a) Act as coenzyme
(b) Act as enzyme
(c) Influence synthesis of enzymes
(d) Belong to B-complex group **Ans. c**

257. Hormone that binds to intracellular receptor is
(a) Adrenocorticotropic hormone (b) Thyroxine
(c) Follicle stimulating hormone (d) Glucagon **Ans. b**

258. Hormone that bind to cell surface receptor and require the second messenger camp is
(a) Antidiuretic hormone (b) Cholecystokinin
(c) Calcitriol (d) Gastrin **Ans. a**

259. A hormone secreted from anterior pituitary is
(a) Growth hormone (b) Vasopressin
(c) Oxytocin (d) Epinephrine **Ans. a**

260. A hormone secreted from posterior pituitary is
(a) Vasopressin (b) Thyrotropic hormone
(c) Prolactin (d) Adrenocorticotropic hormone
Ans. a

261. The number of amino acids in human growth hormone is
(a) 91 (b) 151
(c) 191 (d) 291 **Ans. c**

262. Growth hormone causes hyperglycemia. It is a result of
(a) Decreased peripheral utilization of glucose
(b) Decreased hepatic production via gluconeogenesis
(c) Increased glycolysis in muscle
(d) Decrersed lipolysis **Ans. a**

263. Acromegaly results due to excessive release of
(a) Thyroxine (b) Growth hormone
(c) Insulin (d) Glucagon **Ans. b**

264. Growth hormone is released by
(a) Somatostatin
(b) Growth hormone releasing hormone
(c) Prolactin release inhibiting hormone
(d) Luteinizing releasing hormone **Ans. b**

265. The number of amino acids in prolactin is
(a) 134 (b) 146
(c) 172 (d) 199 **Ans. d**

266. Adrenocorticotropic hormone (ACTH) is a single polypeptide containing
(a) 25 amino acid (b) 39 amino acid
(c) 49 amino acid (d) 52 amino acid **Ans. b**

267. Biological activity of ACTH requires
(a) 10-N-terminal amino acid (b) 24-N-terminal amino acid
(c) 24-C-terminal amino acid (d) 15-C-terminal amino acid
Ans. b

268. ACTH stimulates the secretion of
(a) Glucocorticoids (b) Epinephrine
(c) Thyroxine (d) Luteinizing hormone **Ans. a**

269. Excessive secretion of ACTH causes
(a) Cushing's syndrome (b) Addison's disease
(c) Myxoedema (d) Thyrotoxicosis **Ans. a**

270. In Cushing's syndrome-a tumour associated disease of adrenal cortex, there is

(a) Decreased epinephrine production
(b) Excessive cortisol production
(c) Excessive epinephrine production
(d) Decreased cortisol production **Ans. b**

271. ACTH induces rise in

(a) Cyclic AMP (b) Cyclic GMP
(c) Calcium (d) Magnesium **Ans. a**

272. The circulating concentration of ACTH in plasma is

(a) 0.05 m μ/100 ml (b) 0.1–2.0 m μ/100 ml
(c) 2.5–3.5 m μ/100 ml (d) 3.0–5.0 m μ/100 ml **Ans. b**

273. Hyperglycemic effect of glucocorticoids is due to

(a) Inactivation of protein phosphatase
(b) Inactivation of fructose 1,6-biphosphatase
(c) Stimulation of synthesis of pyruvate carboxylase
(d) Stimulation of synthesis of amylase. **Ans. c**

274. The predominant glucocorticoid is

(a) Cortisol (b) Aldosterone
(c) Dehydroephiandrosterone (d) Androstenedione **Ans. a**

275. A specific cortisol binding protein, transcortin is a

(a) Albumin (b) α_1-Globulin
(c) α_2-Globulin (d) β-Globulin **Ans. c**

276. Cortisol is synthesized in

(a) *Zona fasiculata* (b) *Zona glomerulosa*
(c) *Zona reticularis* (d) *Chromaffin* cells **Ans. a**

277. All mammalian steroid hormones are formed from

(a) Purine (b) Pyrimidine
(c) Cholesterol (d) Pyrrole **Ans. c**

278. A very efficient inhibitor of steroid biosynthesis is

(a) Aminoglutethimide
(b) Aminoimidazole
(c) Aminoimidazolesuccinyl carboxamine
(d) Aminopterin **Ans. a**

279. In adrenal gland the cholesterol is stored
(a) Mostly in the free form
(b) Mostly in esterified form
(c) Large amount of free form and less amount of esterified form
(d) Equal amounts of free and esterified form **Ans. b**

280. Aldosterone synthesis occurs in
(a) *Zona reticularis* (b) *Zona fasciculata*
(c) *Zona glomerulosa* (d) *Chromaffian* cells **Ans. c**

281. In the biosynthesis of cortisol, the sequence of enzymes involved is
(a) Hydroxylase–dehydrogenase + isomerase –hydroxylase
(b) Dehydrogenase–hydroxylase–isomerase
(c) Hydroxylase–lyase–dehydrogenase isomerase
(d) Isomerase–lyase–hydroxylase–dehydrogenase **Ans. a**

282. The defect in adrenal cortex responsible for lack of glucocorticoids and mineralcorticoids is
(a) Androstenedione deficiency
(b) 17 α-OH progesterone deficiency
(c) C-21 hydroxylase deficiency
(d) Testosterone deficiency **Ans. c**

283. 3-β-Hydroxysteroid dehydrogenase and $\Delta^{5,4}$ isomerase catalyse the conversion of the weak androgen DHEA to
(a) Androstenedione (b) Testosterone
(c) Progesterone (d) Estrone **Ans. a**

284. In the resting state plasma concentration of cortisol is
(a) 0.6–2.0 μg/100 ml (b) 2.0–4.0 μg/100 ml
(c) 5.0–15.0 μg/100 ml (d) 20.0–25.0 μg/100 ml **Ans. c**

285. The most important effect of aldosterone is to
(a) Increase the rate of tubular reabsorption of sodium
(b) Decrease the rate of tubular reabsorption of potassium
(c) Decrease the reabsorption of chloride
(d) Decrease the renal reabsorption of sodium **Ans. a**

286. One of the potent stimulators of aldosterone secretion is
(a) Increased sodium concentration
(b) Decreased potassium concentration

(c) Increased potassium concentration
(d) Increased ECF volume **Ans. c**

287. In the rennin-angiotensin system the primary hormone is
(a) Angiotensinogen III (b) Angiotensin I
(c) Angiotensin II (d) Angiotensin III **Ans. c**

288. Aldosterone release is stimulated by
(a) α_2-Globulin (b) Renin
(c) Angiotensin II (d) Growth hormone **Ans. c**

289. In the synthesis of Angiotensin I, rennin acts on Angiotensinogen and cleaves the
(a) Leucine - leucine at 10 and 11 position
(b) Valine - tyrosine at 3 and 4 position
(c) Isoleucine - histidine at 5 and 6 position
(d) Proline - histidine at 7 and 8 position **Ans. a**

290. Catecholamine hormones are synthesized in the
(a) *Chromaffin* cells of adrenal medulla
(b) *Zona glomerulosa* of adrenal cortex
(c) *Zona fasciculate* of adrenal cortex
(d) *Zona reticularis* of adrenal cortex **Ans. a**

291. Catecholamine hormones are
(a) 3, 4-Dihydroxy derivatives of phenylethylamine
(b) p-Hydroxy derivatives of phenylacetate
(c) p-Hydroxy derivatives of phenylpyruvate
(d) p-Hydroxy derivatives of phenyllactate **Ans. a**

292. The sequential steps in the conversion of tyrosine to epinephrine are
(a) Ring hydroxylation-decarboxylation-side chain hydroxylation-N-methylation
(b) Side chain hydroxylation-decarboxylation-ring hydroxylation N-methylation
(c) Decarboxylation-ring hydroxylation-side chain hydroxylation-N-methylation
(d) N-methylation-decarboxylation-ring and side chain hydroxylation **Ans. a**

293. The hormone required for uterine muscle contraction for child birth is
(a) Progesterone (b) Estrogen
(c) Oxytocin (d) Vasopressin **Ans. c**

294. The number of amino acids in the hormone oxytocin is
(a) 5 (b) 9
(c) 14 (d) 10 **Ans. b**

295. Vasopressin and oxytocin circulate unbound to proteins and have very short plasma half lives, on the order of
(a) 1–2 minutes (b) 2–4 minutes
(c) 6–8 minutes (d) 10–12 minutes **Ans. b**

296. Melanogenesis is stimulated by
(a) MSH (b) FSH
(c) LSH (d) LMSH **Ans. a**

297. The number of amino acids in antidiuretic hormone is
(a) 9 (b) 18
(c) 22 (d) 26 **Ans. a**

298. ADH
(a) Reabsorbs water from renal tubules
(b) Excretes water from renal tubules
(c) Excretes hypotonic urine
(d) Causes low specific gravity of urine **Ans. a**

299. Increased reabsorption of water from the kidney is the major consequence of the secretion of the hormone
(a) Cortisol (b) Insulin
(c) Vasopressin (d) Aldosterone **Ans. c**

300. An increase in the osmolality of extracellular compartment will
(a) Inhibit ADH secretion
(b) Stimulate ADH secretion
(c) Cause no change in ADH secretion
(d) Stimulate the volume and osmoreceptor and inhibit ADH secretion
Ans. b

301. For Catecholamine biosynthesis the rate limiting enzyme is
(a) DOPA decarboxylase (b) Dopamine β-hydroxylase
(c) Tyrosine hydroxylase (d) Phenylalanine hydroxylase
Ans. c

302. A hormone which cannot cross the blood brain barrier is
(a) Epinephrine (b) Aldosterone
(c) ACTH (d) TSH **Ans. a**

303. The plasma level of epinephrine is less than
(a) 0.1 ng/ml (b) 0.2 ng/ml
(c) 0.4 ng/ml (d) 0.8 ng/ml **Ans. a**

304. Epinephrine is rapidly metabolized by
(a) Monoamine oxidase (b) Deaminase
(b) Transaminase (c) Decarboxylase **Ans. a**

305. Pheochromocytomas are tumours of
(a) Adrenal cortex (b) Adrenal medulla
(c) Pancreas (d) Bone **Ans. b**

306. A characteristic of pheochromocytoma is elevated urinary excretion of
(a) Dopamine (b) Tyrosine
(c) Vinylmandelic acid (d) Phenylalanine **Ans. c**

307. In the synthetic pathway of epinephrine, disulfiram (antabuse) inhibits the enzyme
(a) Tyrosine hydroxylase (b) Dopamine β-hydroxylase
(c) DOPA decarboxylase (d) N-methyl transferase **Ans. b**

308. The biosynthesis of both Catecholamine and serotonin require
(a) Tyrosine hydroxylase
(b) N-methyl transferase
(c) Aromatic amino acid decarboxylase
(d) Tryptophan pyrrolase **Ans. b**

309. Epinephrine stimulates glycogenolysis in
(a) Liver (b) Muscle
(c) Liver and muscle (d) Kidney **Ans. c**

310. A cup of strong coffee would be expected to
(a) Interfere with the synthesis of prostaglandins
(b) Decrease the effect of glucagon
(c) Enhance the effect of epinephrine
(d) Provide the vitamin nicotinic acid **Ans. c**

311. Epinephrine is derived from norepinephrine by
(a) Decarboxylation (b) Hydroxylation
(c) Oxidation (d) N-methylation **Ans. d**

312. 5 HIAA test is negative if patient is taking
(a) Aspirin (b) Colchicine
(c) Phenothiazone (d) Methotrexate **Ans. c**

313. Presence of significant amount of 5-HIAA in urine indicates
(a) Carcinoid in liver
(b) Carcinoid in appendix
(c) Metastasis of carcinoma of liver
(d) Hepatoma **Ans. c**

314. The normal serum level of triiodothyronine (T_3) is
(a) 0.2–0.5 ng/ml (b) 0.7–2.0 ng/ml
(c) 2.0–4.0 ng/ml (d) 5.0–8.0 ng/ml **Ans. b**

315. The normal serum level of thyroxine (T_4) is
(a) 2.0–4.0 μg/100 ml (b) 5.5–13.5 μg/100 ml
(c) 14.0–20.3 μg/100 ml (d) 20.0–25.0 μg/100 ml **Ans. b**

316. Excess secretion of thyroid hormones causes
(a) Hyperthyroidism (b) Myxoedema
(c) Cretinism (d) Cushing syndrome **Ans. a**

317. Insufficient free T_3 and T_4 results in
(a) Grave's disease (b) Mysoedema
(c) Cushing syndrome (d) Gigantism **Ans. b**

318. In primary hypothyroidism the useful estimation is of
(a) T_3 (b) T_4
(c) TBG (d) Autoantibodies **Ans. d**

319. When iodine supplies are sufficient the T_3 and T_4 ratio in thyroglobulin is
(a) 1 : 2 (b) 1 : 4
(c) 1 : 7 (d) 1 : 10 **Ans. c**

320. A substance which competes with iodide uptake mechanism by thyroid gland is
(a) Thiocynate (b) Iodoacetate
(c) Fluoride (d) Fluoroacetate **Ans. a**

321. Thyroperoxidase enzyme contains
(a) Heme (b) Copper
(c) Zinc (d) Magnesium **Ans. a**

322. Thyroproxidase requires hydrogen peroxide as oxidizing agent. The H_2O_2 is produced by
(a) $FADH_2$ dependent enzyme (b) NADH dependent enzyme
(c) NADP dependent enzyme (d) NADPH dependent enzyme
Ans. d

323. Thyroid stimulating hormone is a dimer. The α-subunits of TSH, LH, FSH are identical. Thus the biological specificity must therefore be β-subunit in which the number of amino acids is
(a) 78 (b) 112
(c) 130 (d) 199 **Ans. b**

324. TSH stimulates the synthesis delete
(a) Thyroxine (b) Adrenocorticoids
(c) Epinephrine (d) Insulin **Ans. a**

325. Thyroid hormones are synthesized by the iodination of the amino acid:
(a) Glycine (b) Phenylalanine
(c) Alanine (d) Tyrosine **Ans. d**

326. The tyrosine residues per molecule of thyroglobulin is
(a) 85 (b) 95
(c) 115 (d) 135 **Ans.c**

327. The percentage of inactive precursors (monoiodotyrosine and diiodotyrosine) in thyroglobulin is
(a) 30 (b) 40
(c) 50 (d) 70 **Ans. d**

328. The number of amino acids in parathormone is
(a) 65 (b) 84
(c) 115 (d) 122 **Ans. b**

329. The sequence of amino acid in which the biological value of parathormone is
(a) 1–15 (b) 1–34
(c) 30–50 (d) 50–84 **Ans. b**

330. PTH
(a) Reduces the renal clearance or excretion of calcium
(b) Increases renal phosphate clearance
(c) Increases the renal clearance of calcium
(d) Decreases the renal phosphate clearance **Ans. a**

331. The number of amino acids in the peptide hormone calcitonin is
(a) 16 (b) 24
(c) 32 (d) 40 **Ans. c**

332. Calcitonin causes
(a) Calcinuria and phosphaturia
(b) Decrease in urinary calcium
(c) Decrease in urinary phosphorous
(d) Increase in blood calcium level **Ans. a**

333. The characteristic of hyperparathyroidism is
(a) Low serum calcium
(b) High serum phosphorous
(c) Low serum calcium and high serum phosphorous
(d) High serum calcium and low serum phosphate **Ans. d**

334. Parathyroid hormone
(a) Is released when serum Ca^{++} is too high
(b) Inactivates vitamin D
(c) Is secreted when Ca^{++} is too low
(d) Depends on vitamin K for adequate activity **Ans. c**

335. δ-Cells of islet of langerhans of pancreas produce
(a) Pancreatic polypeptide (b) Pancreatic lipase
(c) Somatostatin (d) Steapsin **Ans. c**

336. β-cells of islet of langerhans of the pancreas secrete
(a) Insulin (b) Glucagon
(c) Somatostatin (d) Pancreatic polypeptide **Ans. a**

337. Target tissue of insulin is
(a) Red blood cells (b) Renal tubular cells
(c) GI tract epithelial cells (d) Liver **Ans. d**

338. Insulin is a dimer. The number of amino acids in the A and B chain respectively is

(a) 19 and 28 (b) 21 and 30
(c) 25 and 35 (d) 29 and 38 **Ans. b**

339. In A chain of the insulin molecule the N-terminal amino acid is

(a) Glycine (b) Valine
(c) Serine (d) Phenylalanine **Ans. a**

340. In the A chain of insulin molecule the C-terminal amino acid is

(a) Asparagine (b) Threonine
(c) Valine (d) Tyrosine **Ans. a**

341. In the B chain of insulin molecule, the N-terminal amino acid is

(a) Proline (b) Threonine
(c) Phenylalanine (d) Lysine **Ans. c**

342. In the B chain of insulin molecule, the C-terminal amino acid

(a) Threonine (b) Tyrosine
(c) Glutamate (d) Valine **Ans. a**

343. In the insulin molecule, the number of interchain disulphide brides is

(a) 1 (b) 2
(c) 3 (d) 4 **Ans. b**

344. In the insulin molecule, the number of intrachain disulphide bridges is

(a) 1 (b) 2
(c) 3 (d) 4 **Ans. a**

345. Insulin exists in polymeric forms, for polymerization it requires

(a) Calcium (b) Magnesium
(c) Manganese (d) Zinc **Ans. d**

346. The number of amino acids in pre-pro insulin is

(a) 51 (b) 86
(c) 109 (d) 132 **Ans. c**

347. Proinsulin has

(a) 74 amino acids (b) 86 amino acids
(c) 105 amino acids (d) 109 amino acids **Ans. b**

348. Daily secretion of insulin in a normal adult man is about

(a) 10 units (b) 20 units

(c) 30 units (d) 50 units **Ans. d**

349. The insulin content of pancreas is about

(a) 50–70 units (b) 100–150 units

(c) 150–180 units (d) 200–250 units **Ans. d**

350. The half life of insulin is

(a) < 3–5 minutes (b) < 8–10 minutes

(c) < 15 minutes (d) < 15 minutes **Ans. a**

351. Insulin stimulates

(a) Hepatic glycogenolysis (b) Hepatic glycogenesis

(c) Lipolysis (d) Gluconeogenesis **Ans. b**

352. Action of insulin on lipid metabolism is

(a) It increases lipolysis and increases triglyceride synthesis

(b) It decreases lipolysis and increases triglyceride synthesis

(c) It decreases lipolysis and decreases triglyceride synthesis

(d) It increases synthesis of triglyceride and increased Ketogenesis

Ans. b

353. Insulin increases the activity of

(a) Pyruvate kinase (b) Phosphorylase

(c) Triacylglycerol kinase (d) Fructose 2, 6-bisphosphatase

Ans. a

354. Insulin decreases the activity of

(a) cAMP dependent protein kinase

(b) HMG CoA-reductas

(c) Phosphodiesterase

(d) Acetyl CoA-carboxylase **Ans. a**

355. The human insulin gene located on the short arm of chromosome

(a) 11 (b) 17

(c) 18 (d) 20 **Ans. a**

356. Normal serum insulin level varies between

(a) 4–25 μU/ml (b) 25–50 μU/ml

(c) 70–90 μU/ml (d) 100–120 μU /ml **Ans. a**

357. Following is a normal overnight fast and a cup of black coffee, a diabetic woman feels slightly nausious and decides to skip breakfast. However she does take her shot of insulin. This may result in

(a) Heightened glycogenolysis (b) Hypoglycemia
(c) Increased lipolysis (d) Glycosuria **Ans. b**

358. Deficiency of insulin results in

(a) Rapid uptake of sugar (b) Low blood glucose level
(c) Decrease urine output (d) Presence of glucose in urine **Ans. d**

359. The primary stimulus for insulin secretion is increased

(a) Blood level of epinephrine (b) Blood level of glucagon
(c) Blood level of glucose (d) Water intake **Ans. c**

360. The α-cells of pancreas islets produce

(a) Insulin (b) Glucagon
(c) Somatostatin (d) Pancreatic polypeptide **Ans. b**

361. The number of amino acids in single chain polypeptide glucagons is

(a) 21 (b) 29
(c) 31 (d) 39 **Ans. b**

362. Glucagon enhances

(a) Hepatic glycogenolysis (b) Muscle glycogenolysis
(c) Hepatic glycogenesis (d) Lipogenesis **Ans. a**

363. Normal serum glucagons level in fasting state varies between

(a) 0–10 pg/ml (b) 20–100 pg/ml
(c) 200–300 pg/ml (d) 400–500 pg/ml **Ans. b**

364. Glucagon

(a) Increases protein synthesis
(b) Inhibits lipolysis in adipocytes
(c) Increases gluconeogenesis in liver
(d) Stimulates muscle glycogenolysis **Ans. c**

365. The serum estradiol level in women during 1–10 days of menstrual cycle is

(a) 0–10 pg/ml (b) 12–20 pg/ml
(c) 24–68 pg/ml (d) 80–100 pg/ml **Ans. c**

366. The serum estradiol level in women during 11–20 days of menstrual cycle is

(a) 5–30 pg/ml (b) 50–300 pg/ml

(c) 500–900 pg/ml (d) 1000 pg/ml **Ans. b**

367. The serum progesterone level in follicular phase is about

(a) 0.2–1.5 ng/100 ml (b) 2.0–2.5 ng/100 ml

(c) 3.5–4.5 ng/100 ml (d) 5.0–6.5 ng/100 ml **Ans. a**

368. Serum progesterone level during pregnancy is

(a) < 12 ng/ml (b) > 12 ng/ml

(c) < 20 ng/ml (d) >24 ng/ml **Ans. d**

369. Androgens are produced by

(a) Cells of sertoli (b) Leydig cells

(c) Rete testis (d) Efferent ductules **Ans. b**

370. The leading cell activity is controlled by

(a) Intestitial cell stimulating hormone

(b) Adernocortex stimulating hormone

(c) Thyroid stimulating hormone

(d) Melanocyte stimulating hormone **Ans. a**

371. Stein-leventhal syndrome is due to overproduction of

(a) Estrogens (b) Androgens

(c) Gastogens (d) Ethinyl estradiol **Ans. b**

372. In the biosynthesis of testosterone the rate limiting step is conversion of

(a) Cholesterol to pregnenolone

(b) Pregnenolone to progesterone

(c) Progesterone to 17 α-hydroxy progesterone

(d) 17 α-Hydroxy progesterone to androstenedione **Ans. a**

373. Urinary 17 ketosteroids

(a) Are not found in women

(b) Reflect the total production of androgenic substances

(c) Indicate the total production of sex hormone

(d) Are highly active androgens **Ans. b**

374. Total number of amino acids in human chorionic gonadotropin is
(a) 53 (b) 92
(c) 145 (d) 237 **Ans. d**

375. A hormone produced by corpus luteum and placenta, concerned with relaxation of pelvis tissue is
(a) HCG (b) Chorionic somatommotropin
(c) Relaxin (d) Progestins **Ans. c**

376. Synthetic progesterone used in oral contraceptive is
(a) Norethindrone (b) Pregnenolone
(c) Androstenodione (d) Stilbestrol **Ans. a**

377. Young women are protected against myocardial infarction because of the activity of
(a) Estrogen (b) Progesterone
(c) Growth hormone (d) Oxytocin **Ans. a**

378. Hormone receptors possess all the following properties except
(a) All of them are proteins
(b) They possess a recognition domain
(c) They bind hormones with a high degree of specificity
(d) Number of receptors in a target cell is constant **Ans. d**

379. The only correct statement about hormone receptors is
(a) Receptors for protein hormones are present in cytosol
(b) Receptors for steroid hormones are membrane bound
(c) Hormone-receptor binding is irreversible
(d) Receptors can undergo down regulation and up regulation **Ans. d**

380. Down regulation is
(a) Increased destruction of a hormone
(b) Feedback inhibition of hormone secretion
(c) Decreased concentration of a hormone in blood
(d) Decrease in number of receptors for a hormone **Ans. d**

381. All the following statements about hormones are true except
(a) All of them require specific carriers in plasma
(b) All of them require specific receptors in target cells

(c) Some of them are subject to feedback regulation
(d) Some of them increase the transcription of certain genes **Ans. a**

382. All the following statements about steroid hormones are true except
(a) They are hydrophobic
(b) They require carriers to transport them in circulation
(c) Their receptors are intracellular
(d) They require cyclic AMP as second messenger **Ans. d**

383. Cyclic AMP acts as the second messenger for all of the following except
(a) Oxytocin (b) TSH
(c) ACTH (d) FSH **Ans. a**

384. Cyclic GMP acts as the second messenger for
(a) Nerve growth factor (b) Atrial natriuretic factor
(c) Epinephrine (d) Norepinephrine **Ans. b**

385. Some hormones produce their intracellular effects by activating
(a) Phospholipase A_1 (b) Phospholipase B
(c) Phospholipase C (d) All of these **Ans. c**

386. G-proteins act as
(a) Hormone carriers (b) Hormone receptors
(b) Second messengers (d) Signal transducers **Ans. d**

387. Signal transducer for glucagons is a
(a) Cyclic nucleotide (b) Phosphoinositide
(c) Stimulatory G-protein (d) Inhibitory G-protein **Ans. c**

388. G-proteins have a nucleotide binding site for
(a) ADP/ATP (b) GDP/GTP
(c) CDP/CTP (d) UDP/UTP **Ans. b**

389. Adenylate cyclase is activated by
(a) GDP-bearing α-Subunit of G-protein
(b) GTP-bearing α-Subunit of G-protein
(c) GDP-bearing γ-Subunit of G-protein
(d) GTP-bearing α-Subunit of G-protein **Ans. b**

390. Tyrosine kinase activity is present in
(a) α-Adrenergic receptors (b) β-Adrenergic receptors
(c) Cholinergic receptors (d) Insulin receptors **Ans. d**

391. Tyrosine kinase activity is present in
(a) Acetylcholine receptor (b) PDGF receptor
(c) ADH receptor (d) All of these **Ans. b**

392. Melatonin is synthesised from
(a) Phenylalanine (b) Tyrosine
(c) Tryptophan (d) None of these **Ans. c**

393. Melanocyte stimulating hormone is secreted by
(a) Pineal gland
(b) Anterior lobe of pituitary gland
(c) Posterior lobe of pituitary gland
(d) Intermediate lobe of pituitary gland **Ans. d**

394. MSH causes
(a) Dispersal of melanin granules in melanocytes
(b) Increase in melanin concentration in melanocytes
(c) Decerease in melanin concentration in melanocytes
(d) Increase in number of melanocytes **Ans. b**

395. Secretion of MSH is regulated by
(a) Pineal gland (b) Melatonin
(c) Hypothalamic hormones (d) ACTH **Ans. c**

396. A hormone synthesized in the hypothalamus is
(a) Hypothalamic Melatonin
(b) Melanocyte stimulating hormone
(c) Vasopressin
(d) Prolactin **Ans. c**

397. Posterior pituitary gland secretes
(a) Catecholamines (b) Oxytocin
(c) Follicle stimulating hormone (d) Serotonin **Ans. b**

398. A nonapeptide among the following is
(a) Antidiuretic hormone
(b) Insulin
(c) ACTH
(d) Thyrotropin releasing hormone **Ans. a**

399. Peripheral vasoconstriction is caused by high concentrations of
(a) Antidiuretic hormone (b) Melatonin
(c) Glucagon (d) Oxytocin **Ans. a**

400. Secretion of Insulin-like Growth Factor-I is promoted by
(a) Insulin (b) Glucagon
(c) Growth hormone (d) Somatomedin C **Ans. c**

401. Growth hormone increases
(a) Protein synthesis (b) Lipogenesis activity
(c) Glycogenolysis (d) All of these **Ans. a**

402. Secretion of growth hormone is inhibited by
(a) Somatomedin C (b) Somatostatin
(c) Feedback inhibition (d) All of these **Ans. b**

403. Secretion of somatotrophin is promoted by
(a) Somatomedin C
(b) Somatostatin C
(c) Growth hormone releasing hormone
(d) Hypoglycaemia **Ans. c**

404. Human growth hormone has
(a) One polypeptide chain and one intra-chain disulphide bond
(b) One polypeptide chain and two intra-chain disulphide bond
(c) Two polypeptide chains joined by one disulphide bond
(d) Two polypeptide chains joined by two disulphide bond **Ans. b**

405. Number of amino acid residues in human growth hormone is
(a) 51 (b) 84
(c) 191 (d) 198 **Ans. c**

406. Number of amino acid residues in prolactin is
(a) 51 (b) 84
(c) 191 (d) 198 **Ans. d**

407. Secretion of prolactin is regulated by
(a) Feedback inhibition
(b) Prolactin releasing hormone
(c) Prolactin release inhibiting hormone
(d) Feedback releasing hormone **Ans. c**

408. All of the following can be formed from pro-opiomelanocortin except

(a) α-and β-MSH (b) β-and γ-Lipotropins

(c) α-and β-Endorphins (d) FSH **Ans. d**

409. All the following statements about proopiomelanocortin are true except

(a) It is made up of 285 amino acids

(b) It is synthesised in pars intermedia and anterior lobe of pituitary gland

(c) It is the precursor of ACTH and melatonin

(d) It is the precursor of corticotropin like intermediate lobe peptide and endorphins **Ans. c**

410. All the following statements about ACTH are true except

(a) It is a tropic hormone

(b) Its target cells are located in adrenal cortex

(c) Its receptors are located in the cell membrane

(d) Its second messenger is inositol triphosphate **Ans. d**

411. Regulation of ACTH secretion occurs through

(a) Corticotropin releasing hormone (CRH) and corticotropin release inhibiting hormone (CRIH) of hypothalamus

(b) Feedback inhibition by cortisol

(c) CRH and feedback inhibition by cortisol

(d) CRIH and feedback inhibition by cortisol **Ans. c**

412. Hormonal activity of ACTH is completely lost on removal of

(a) 5 C-terminal amino acids (b) 10 C-terminal amino acids

(c) 15 C-terminal amino acids (d) None of these **Ans. d**

413. All the following statements about TSH are true except

(a) It is a glycoprotein

(b) It is made up of α-subunits and β–subunits

(c) Receptor recognition involves both the subunits

(d) Its subunit is identical with those of FSH and LH **Ans. d**

414. All the following statements about TSH are true except

(a) It is a tropic hormone

(b) It acts on para-follicular cells of thyroid glands

(c) Its receptors are membrane-bound

(d) Its second messenger is cyclic AMP **Ans. b**

415. All the following statements about thyrotropin releasing hormone are true except

(a) It is secreted by hypothalamus

(b) It is a pentapeptide

(c) It increases the secretion of TSH

(d) Its secretion is inhibited by high level of T_3 and T_4 in blood **Ans. b**

416. All the following statements about FSH are true except

(a) It is a tropic hormone secreted by anterior pituitary

(b) Its secretion is increased by gonadotropin releasing hormone

(c) It acts on Sertoli cells

(d) It increases the synthesis of testosterone **Ans. d**

417. In males, secretion of luteinising hormone is inhibited by

(a) Gonadotropin releasing hormone

(b) FSH

(c) High blood level of testosterone

(d) Inhibin **Ans. c**

418. In structure and function, HCG resembles

(a) FSH (b) LH

(c) GnRH (d) Progesterone **Ans. b**

419. Acromegaly results from overproduction of

(a) ACTH during childhood

(b) TSH during adult life

(c) Growth hormone during childhood

(d) Growth hormone during adult life **Ans. d**

420. Acromegaly results in all the following except

(a) Overgrowth of the bones of face, hands and feet

(b) Increased stature

(c) Enlargements of viscera

(d) Impaired glucose tolerance **Ans. b**

421. Overproduction of growth hormone during childhood causes

(a) Acromegaly (b) Gigantism

(c) Cushing's disease (d) Simmond's disease **Ans. b**

422. Decreased secretion of growth hormone during childhood causes
(a) Simmond's disease (b) Cushing's disease
(c) Dwarfism (d) Cretinism **Ans. c**

423. An amino acid used for the synthesis of thyroid hormone is
(a) Tyrosine (b) Tryptophan
(c) Histidine (d) Proline **Ans. a**

424. Thyroperoxidase iodinates
(a) Free tyrosine in thyroid gland
(b) Tyrosine residues of thyroglobulin
(c) Tyrosine residues of thyroxine binding globulin
(d) Tyrosine residues of thyroxine binding prealbumin **Ans. b**

425. In thyroxine, tyrosine residues are iodinated at positions
(a) 1 and 3 (b) 2 and 4
(c) 3 and 5 (d) 4 and 6 **Ans. c**

426. Thyroid hormones are present in blood
(a) In free form
(b) In association with thyroxine binding globulin (TBG)
(c) In association with thyroxine binding prealbumin (TBPA)
(d) Mainly in association with TBG, partly in free form and sometimes in association with TBPA also **Ans. d**

427. When thyroxine binding globulin and thyroxine binding pre-albumin are saturated with thyroxine, the excess hormone is transported by
(a) Albumin (b) Gamma globulins
(c) Transcortin (d) None of these **Ans. a**

428. Receptors for thyroid hormones are present
(a) On the cell membrane (b) Across the cell membrane
(c) Inside the cells (d) In association with G-proteins **Ans. c**

429. Binding of thyroxine to its receptors
(a) Activates Adenylate cyclase
(b) Activates guanylate cyclase
(c) Activates a stimulatory G-protein
(d) Increases transcription **Ans. d**

430. Secretion of thyroid hormones is regulated by
(a) Hypothalamus (b) Anterior pituitary
(c) Feedback regulation (d) All of these **Ans. d**

431. Clinical features of hyperthyroidism include
(a) Goitre, heat intolerance, weight loss and tachycardia
(b) Goitre, tremors, tachycardia and cold intolerance
(c) Exophthalmos, goiter, tachycardia and loss of appetite
(d) Exophthalmos, goiter, tremors and obesity **Ans. a**

432. All the following may occur in hyperthyroidism except
(a) Goitre (b) Increased appetite
(c) Loss of weight (d) Low BMR **Ans. d**

433. All the following may occur in myxoedema except
(a) Cold intolerance (b) Low BMR
(c) Tachycardia (d) Dry and coarse skin **Ans. c**

434. Mental retardation can occur in
(a) Cretinism (b) Juvenile myxoedema
(c) Myxoedema (d) Juvenile thyrotoxicosis **Ans. a**

435. Parathyroid hormone (PTH) is synthesized in
(a) Chief cells of parathyroid glands
(b) Oxyphil cells of parathyroid glands
(c) Para follicular cells of thyroid glands
(d) Follicular cells of thyroid gland **Ans. a**

436. The number of amino acid residues in PTH
(a) 51 (b) 84
(c) 90 (d) 115 **Ans. b**

437. Amino acid residues which are essential for the biological activity of PTH are
(a) N-terminal 34 amino acids (b) N-terminal 50 amino acids
(c) C-terminal 34 amino acids (d) C-terminal 50 amino acids
Ans. a

438. Half-life of PTH is
(a) A few seconds (b) A few minutes
(c) A few hours (d) A few days **Ans. b**

439. The second messenger for PTH is

(a) Cyclic AMP (b) Cyclic GMP

(c) Diacylglycerol (d) Inositol triphosphate **Ans. a**

440. PTH causes all of the following except

(a) Increased intestinal absorption of calcium

(b) Increased intestinal absorption of phosphate

(c) Increased tubular reabsorption of calcium

(d) Increased tubular reabsorption of phosphate **Ans. d**

441. A high concentration of PTH in blood causes

(a) Increase in plasma calcium and inorganic phosphorous

(b) Decrease in plasma calcium and inorganic phosphorous

(c) Increase in plasma calcium and decrease in plasma inorganic phosphorous

(d) Decrease in plasma calcium and increase in plasma inorganic phosphorous **Ans. c**

442. Tetany can occur

(a) In primary hyperparathyroidism

(b) In secondary hyperparathyroidism

(c) In idiopathic hypoparathyroidism

(d) After accidental removal of parathyroid glands **Ans. d**

443. Crystallisation of insulin occurs in the presence of

(a) Chromium (b) Copper

(c) Zinc (d) Calcium **Ans. c**

444. Daily secretion of insulin is about δ–

(a) 10–20 mg (b) 40–50 mg

(c) 10–20 units (d) 40–50 units **Ans. d**

445. Insulin receptors are decreased in number in

(a) Obesity (b) Starvation

(c) Hyperinsulinism (d) Kwashiorkor **Ans. a**

446. Insulin binding sites are present on the

(a) α-subunits of insulin receptor

(b) β-subunits of insulin receptor

(c) γ-subunits of insulin receptor
(d) α-and β–subunits of insulin receptor **Ans. a**

447. α-Subunits of insulin receptor are present
(a) Outside the cell membrane (b) In the cell membrane
(c) Across the cell membrane (d) In the cytosol **Ans. a**

448. β-Subunits of insulin receptor are present
(a) Outside the cell membrane (b) In the cell membrane
(c) Across the cell membrane (d) In the cytosol **Ans. c**

449. In the insulin receptor, tyrosine kinase domain is present in
(a) α-Subunits (b) β-Subunits
(c) γ-Subunits (d) δ-Subunits **Ans. b**

450. Binding of insulin to its receptor activates
(a) Adenylate cyclase (b) Guanylate cyclase
(c) Phospholipase C (d) Tyrosine kinase **Ans. d**

451. Insulin receptor is made up of
(a) One α-and one β-subunit
(b) Two α-and two β-subunit
(c) Two α- two β-and two γ-subunit
(d) One α, one β -one γ -and one δ-subunit **Ans. b**

452. Insulin is required for the active uptake of glucose by most of the cells except
(a) Muscle cells (b) Renal tubular cells
(c) Adipocytes (d) Liver cells **Ans. d**

453. Insulin decreases
(a) Glycogenesis
(b) Glyolysis
(c) Gluconeogenesis
(d) Tubular reabsorption of glucose **Ans. c**

454. Insulin increases
(a) Glycogenesis (b) Gluconeogenesis
(c) Lipolysis (d) Blood glucose **Ans. a**

455. Insulin increases
(a) Protein synthesis (b) Fatty acid synthesis
(c) Glycogen synthesis (d) All of these **Ans. d**

456. Insulin decreases the synthesis of
(a) Hexokinase (b) Glucokinase
(c) PEP carboxykinase (d) Glycogen synthetase **Ans. c**

457. Diabetes mellitus can occur due to all of the following except
(a) Deficient insulin secretion
(b) Tumour of β″cells
(c) Decrease in number of insulin receptors
(d) Formation of insulin antibodies **Ans. b**

458. Hypoglycaemic coma can occur
(a) In untreated diabetes mellitus
(b) In starvation
(c) After overdose of oral hypoglycaemic drugs
(d) After overdose of insulin **Ans. d**

459. The maineffecting of glucagons is to increase
(a) Glycolysis in muscles (b) Glycogenolysis in muscles
(c) Glycogenolysis in liver (d) Glycogenesis in liver **Ans. c**

460. Tyrosine is required for the synthesis of all of the following except
(a) Melatonin (b) Epinephrine
(c) Norepinephrine (d) Thyroxine **Ans. a**

461. Dopamine is synthesised from
(a) Dihydroxyphenylalanine (b) Epinephrine
(c) Norepinephrine (d) Metanephrine **Ans. a**

462. Epinephrine is synthesised in
(a) Chromaffin cells of adrenal medulla
(b) Sympathetic ganglia
(c) Brain
(d) All of these **Ans. d**

463. Immediate precursor of epinephrine is
(a) Metanephrine (b) Norepinephrine
(c) Dopa (d) Dopamine **Ans. b**

464. The chief metabolite of catecholamines is

(a) Metanephrine (b) Normetanephrine
(c) 3, 4-Dihydroxymandelic acid (d) Vanillylmandelic acid **Ans. d**

465. An enzyme involved in catabolism of

(a) Catecholamines is Dopa decarboxylase
(b) Aromatic amino acid decarboxylase
(c) Monoamine oxidase
(d) Catechol oxidase **Ans. c**

466. Norepinephrine binds mainly to

(a) α-Adrenergic receptors (b) β-Adrenergic receptrors
(c) Muscarinic receptors (d) Nicotinic receptors **Ans. a**

467. A stimulatory G-protein transduces the signals from

(a) α_1-and β_1-adrenergic receptors
(b) α_2-and β_2-adrenergic receptors
(c) α_1-and α_2-adrenergic receptors
(d) β_1-and β_2-adrenergic receptors **Ans. d**

468. Binding of catecholamines to α_2–adrenergic receptors

(a) Increases the intracellular concentration of cAMP
(b) Increases the intracellular concentration of cGMP
(c) Decreases the intracellular concentration of cAMP
(d) Decreases the intracellular concentration of cGMP **Ans. c**

469. Phosphoinositide cascade is activated on binding of catecholamines to

(a) α_1-Adrenergic receptors (b) α_2-Adrenergic receptors
(c) β_1-Adrenergic receptors (d) β_2-Adrenergic receptors **Ans. c**

470. Epinephrine decreases

(a) Glycogenesis (b) Glycogenolysis
(c) Gluconeogenesis (d) Lipolysis **Ans. a**

471. Epinephrine increases the concentration of free fatty acids in plasma by increasing

(a) Extramitochondrial fatty acid synthesis
(b) Mitochondrial fatty acid chain elongation

(c) Microsomal fatty acid chain elongation
(d) Lipolysis in adipose tissue **Ans. d**

472. Epinephrine increases all of the following except
(a) Glycogenolysis in muscles (b) Lipolysis in adipose tissue
(c) Gluconeogenesis in muscles (d) Glucagon secretion **Ans. c**

473. Secretion of catecholamines is increased in
(a) Cushing's syndrome (b) Addison's disease
(c) Phaeochromocytoma (d) Simmond's disease **Ans. c**

474. Zona glomerulosa of adrenal cortex synthesises
(a) Glucocorticoids (b) Mineralocorticoids
(c) Androgens (d) Estrogen and progesterone
Ans. a

475. Cortisol is a
(a) Glucocorticoid (b) Mineralocorticoid
(c) Androgen (d) Estrogen **Ans. b**

476. Steroid hormones are synthesised in all of the following except
(a) Testes (b) Ovaries
(c) Adrenal medulla (d) Adrenal cortex **Ans. a**

477. Steroid hormones are synthesised from
(a) Cholesterol (b) 7-Dehydrocholesterol
(c) Calcitriol (d) 7-Hydroxycholesterol **Ans. a**

478. A common intermediate in the synthesis of cortisol and aldosterone is
(a) Progesterone (b) Testosterone
(c) Estradiol (d) None of these **Ans. a**

479. Glucocorticoids are transported in blood
(a) In association with transcortin chiefly
(b) In association with albumin to some extent
(c) In free form partly
(d) All of these **Ans. d**

480. All the following statements about transcortin are true except
(a) It is synthesised in liver (b) It transports glucocorticoids
(c) It transports aldosterone (d) It transports progesterone
Ans. c

481. The second messenger for glucocorticoids is
(a) Cyclic AMP
(b) Cyclic GMP
(c) Inositol triphosphate
(d) No second messenger is required **Ans. d**

482. Glucocorticoids increase all of the following except
(a) Gluconeogenesis (b) Lipolysis in extremities
(c) Synthesis of elcosanoida (d) Hepatic glycogenesis **Ans. c**

483. Glucocorticoids increase the synthesis of all of the following except
(a) Glucokinase (b) Glucose-6-phosphatase
(c) Fructose-1, 6-biphosphatase (d) Pyruvate carboxylase **Ans. a**

484. Secretion of glucocorticoida is regulated by all the following except
(a) Hypothalamus
(b) Anterior pituitary
(c) Feedback control by blood glucose
(d) Feedback control by glucocorticoids **Ans. c**

485. Excessive secretion of glucocorticoids raises blood glucose by
(a) Decreasing glycogenesis (b) Increasing glycogenolysis
(c) Increasing gluconeogenesis (d) Inhibiting HMP shunt **Ans. d**

486. Mineralcorticoids regulate the metabolism of all of the following except
(a) Sodium (b) Potassium
(c) Calcium (d) Chloride **Ans. c**

487. Mineralocorticoids increase the tubular reabsorption of
(a) Sodium and calcium (b) Sodium and potassium
(c) Sodium and chloride (d) Potassium and chloride **Ans. c**

488. Mineralocorticoids increase the tubular secretion of
(a) Sodium (b) Potassium
(c) Chloride (d) Bicarbonate **Ans. b**

489. In Addison's disease, there is excessive retention of
(a) Potassium (b) Sodium
(c) Chloride (d) Water **Ans. a**

490. In adrenogenital syndrome due to total absence of 21-hydroxylase in adrenal cortex, there is
(a) Deficient secretion of glucocorticoids
(b) Deficient secretion of mineralcorticoids
(c) Excessive secretion of androgens
(d) All of these **Ans. d**

491. Spironolactone is an antagonist of
(a) Cortisol (b) Hydrocortisone
(c) Aldosterone (d) Testosterone **Ans. c**

492. Androgens are synthesised in
(a) Leydig cells in testes (b) Sertoli cells in testes
(c) Seminiferous tubules (d) Prostate gland **Ans. a**

493. Testosterone is transported in blood by
(a) Transcortin
(b) Testosterone binding globulin
(c) Testosterone estrogen binding globulin
(d) Albumin **Ans. c**

494. Secretion of androgens is increased by
(a) LH (b) FSH
(c) ACTH (d) Growth hormone **Ans. a**

495. During late pregnancy, the major source of progesterone is
(a) Adrenal cortex (b) Placenta
(c) Corpus luteum (d) Graafian follicles **Ans. b**

496. Progesterone is transported in blood by
(a) Transcortin
(b) Sex hormone binding globulin
(c) Albumin
(d) Testosterone estrogen binding globulin **Ans. a**

497. Secretion of progesterone
(a) Is more in first half of menstrual cycle than in second half
(b) Is more in second half of menstrual cycle than in first half
(c) Remains constant during menstrual cycle
(d) Decreases during pregnancy **Ans. b**

498. Women become susceptible to osteoporosis after menopause due to decreased

(a) Secretion of Parathormone

(b) Conversion of vitamin D into calcitriol

(c) Secretion of estrogen

(d) Secretion of progesterone **Ans. c**

499. Placenta secretes all of the following except

(a) FSH (b) Progesterone

(b) Estrogen (c) Chorionic gonadotropin **Ans. a**

500. Gastrin is a polypeptide made up of

(a) Five amino acids (b) Twelve amino acids

(c) Seventeen amino acids (d) Twenty amino acids **Ans. c**

501. Biological activity of gastrin is present in the

(a) Four N-terminal amino acids (b) Four C-terminal amino acids

(c) Five N-terminal amino acids (d) Five C-terminal amino acids

Ans. b

502. All the following statements about β-endorphin are true except ì

(a) It is a polypeptide

(b) Its precursor is pro-opio-melanocortin

(c) Its receptors are represent in brain

(d) Its action is blocked by morphine **Ans. d**

503. All the following statements about epidermal growth factor are true except

(a) It is a protein

(b) It possess quaternary structure

(c) Its receptor is made up of a single polypeptide chain

(d) Its receptor posseses tyrosine kinase domain **Ans. b**

504. Met-enkephalin is a

(a) Tripeptide (b) Pentapeptide

(c) Octapeptide (d) Decapeptide **Ans. b**

505. The rate limiting step in catecholamine synthesis is catalysed by

(a) Phenylalanine hydroxylase

(b) Tyrosine hydroxylase

(c) Dopa decarboxylase

(d) Phenylethanolamine N-methyl transferase **Ans. b**

506. Dopa decarboxylase is inhibited by

(a) Epinephrine (b) Norepinephrine

(c) α–Methyldopa (d) None of these **Ans. c**

507. Tyrosine hydroxylase is inhibited by

(a) Catecholamines (b) α–Methyldopa

(c) Phenylalanine (d) Vanillyl mandelic acid **Ans. a**

508. Binding of growth hormone to its receptor results in phosphorylation of

(a) JAK-2 (b) Growth hormone receptor

(c) STATs (d) All of these **Ans. d**

509. Binding of growth hormone to its receptor results in increased transcription of

(a) c-fos gene (b) c-myc gene

(c) p-53 gene (d) None of these **Ans. a**

510. Activation of IRS-1, PI-3 kinase and GRB-2 is brought about by

(a) Glucagon (b) Insulin

(c) Prolactin (d) IGF-2 **Ans. b**

511. The protein IRS-1 is phosphorylated by

(a) Protein kinase A

(b) Protein kinase C

(c) Tyrosine kinase activity of insulin receptor

(d) Tyrosine kinase activity of IGF-1 receptor **Ans. b**

512. Phosphorylated IRS-1 activates GRB-2 which is

(a) G-protein receptor binding protein-2

(b) Growth factor receptor binding protein-2

(c) Growth hormone receptor binding protein-2

(d) Glucocorticoid receptor binding protein-2 **Ans. b**

513. Activated phospholipase C acts on

(a) Phosphatidyl inositol-4, 5-biphosphate

(b) Inositol-1, 4, 5-triphosphate

(c) Protein kinase C
(d) Pl-3 kinase **Ans. a**

514. Phospholipase C is activated by
(a) G_s proteins (b) G_i proteins
(c) G_q proteins (d) G_{12} proteins **Ans. c**

515. Proteoglycans are made up of proteins and
(a) Glucosamine (b) Mannosamine
(c) Sialic acid (d) Mucopolysaccharides **Ans. d**

516. All the following statements about cystic fibrosis are correct except
(a) It is inherited as an autosomal recessive disease
(b) It affects a number of exocrine glands
(c) It causes increased sweating
(d) Sweat chlorides are above 60 mEq/L in this disease **Ans. c**

517. Radioactive iodine uptake by thyroid gland is increased in
(a) Endemic goitre (b) Hyperthyroidism
(c) Myxoedema (d) Creatinism **Ans. b**

518. Normal range of total tri-iodothyronine in serum is
(a) 0.1–0.2 ng/dl (b) 0.1–0.2 μg/dl
(c) 0.8–2.4 ng/dl (d) 0.8–2.4 μg/dl **Ans. b**

519. Administration of TSH increases serum T_3 and T_4 in
(a) Hyperthyroidism of pituitary origin
(b) Hyperthyroidism of thyroid origin
(c) Hypothyroidism of pituitary origin
(d) Hypothyroidism of thyroid origin **Ans. c**

520. High level of T_3 and T_4 and low TSH in serum indicates
(a) Hyperthyroidism of pituitary origin
(b) Hypothyroidism of pituitary origin
(c) Hyperthyroidism of thyroid origin
(d) Hypothyroidism of thyroid origin **Ans. c**

521. BMR is increased in
(a) Endemic goitre (b) Thyrotoxicosis
(c) Myxoedema (d) Cretinism **Ans. b**

522. Which one of the following statements correctly describes eukaryotic DNA?
(a) If uses DNA polymerase with nuclease activities
(b) It is replicated bidirectionally at many points
(c) It contains no repetitive DNA
(d) It is nonlinear **Ans. c**

523. Which one of the following causes frame shift mutation?
(a) Transition
(b) Transversion
(c) Deletion
(d) Substitution of purine to pyrimidine **Ans. c**

524. The second messenger for many hormones is
(a) ATP (b) cyclic AMP
(c) cGMP (d) UTP **Ans. b**

525. The most potent hormone concerned with the retention of sodium in the body is
(a) Cortisone (b) Aldosterone
(c) Corticosterone (d) Cortisol **Ans. b**

526. Aspirin blocks the synthesis of
(a) Prostaglandins only (b) Prostacyclins only
(c) Thromboxanes only (d) All of these **Ans. d**

527. cAMP is so called because it is formed during
(a) TCA cycle (b) Urea cycle
(c) Rhodopsin cycle (d) It has a cyclic structure **Ans. d**

528. In hypophysectonized animals, fasting produces
(a) Severe hyperglycemia (b) Hypoglycemia
(c) No change in blood sugar (d) Mild hyper glycemia **Ans. b**

529. Calcitomica is antagonist to
(a) Serotonin (b) Thyroxine
(c) Tri iodo thyronine (d) Para thyroid hormone **Ans. d**

530. Insulin resistance is encountered in
(a) Addison's disease (b) Hypothyroidism
(c) Hypopituctarism (d) Acromegaly **Ans. d**

531. Richest source of prostaglandins in a human male is
(a) Blood (b) Urine
(c) Semen (d) C.S.F. **Ans. c**

532. One of the following is not used as a second messenger by hormones
(a) mRNA
(b) cAMP
(c) Calcium ions
(d) Myoinisotol 1, 4, 5 triphosphate **Ans. a**

533. This pancreatic hormone increases the blood-sugar level
(a) Insulin (b) Glucagon
(c) Pancreozymin (d) Pancreatic polypeptide **Ans. b**

534. Which one of the following statements is fully correct?
(a) Hormones are needed in the diet
(b) Hormones can be elaborated only by endocrine glands
(c) All the hormones enter the cells and perform their function
(d) Hormones are substance synthesized in the body in small quantities and control and regulate metabolic events **Ans. d**

535. T_3 is
(a) Thyroxine (b) Triodo thyronine
(c) Triodo tyrosine (d) Reverse tri iodo thyronine
Ans. b

536. Which of the following hormone is a peptide of less than ten amino acids?
(a) Insulin (b) Growth hormone
(c) Oxytocin (d) Parathyroid hormone **Ans. c**

537. Which of the following hormone does not activate adenylate cyclase?
(a) Epinephrine (b) Glucagon
(c) Parathyroid hormone (d) Insulin **Ans. d**

538. Pheochromacytoma is a tumor of
(a) adrenal medulla (b) bone
(c) head of Pancreas (d) pituitary **Ans. a**

539. Which one of the following statements is incorrect?
(a) Insulin increases glucose phosphorylation
(b) Insulin increases glycolysis

(c) Insulin augments HMP shunt
(d) Insulin promotes gluconeogenesis **Ans. d**

540. Which of one ring in the structure of the following is aromatic?
(a) Androgens (b) Estrogens
(c) Cholesterol (d) Bile acids **Ans. b**

541. Which of one of the following is not GUT hormone?
(a) Motiline (b) Secretion
(c) Gastrin (d) Calcitonin **Ans. d**

542. Which of the following hormones are synthesized as prehormones
(a) Vasopressin and oxytocin
(b) Growth hormone and insulin
(c) Insulin and parathyroid hormone
(d) Insulin and Glucagon **Ans. c**

543. This hormone has disulphide group:
(a) Glucagon (b) Insulin
(c) T_4 (d) Epinephrine **Ans. b**

544. The blood sugar raising action of the hormone of suprarenal cortex is due to
(a) Glyconeogenesis
(b) Glycogenolysis
(c) Glucagon like activity
(d) due to inhibition of glomerular filtration of Glucose **Ans. a**

545. Hyper insulinism can cause coma since
(a) The chief nutrient for the brain is glucose
(b) The chief nutrient for the heart is glucose
(c) The glucostatic role of the liver is damaged
(d) The kidneys are damaged **Ans. a**

546. Which of the following property of prostaglandins has been utilized by chinicians in hospital for
(a) Inducing fever
(b) Causing inflammation
(c) Effecting smooth muscle contraction
(d) Disaggregation of spermatozoa **Ans. c**

547. A major structural difference between estrogens and androgens is the fact that

(a) The androgens are usually C_{21} steroids
(b) The estrogens are usually digitonin -precipitable
(c) The androgens have an aromatic ring
(d) The estrogens have an aromatic ring **Ans. d**

548. Alloxan can experimentally induce diabetes mellitus due to

(a) Stimulation of α–cells of the islets of langerhans
(b) Necrosis of the α–cells of the islets
(c) Potentiation of insulinase activity
(d) Epinephrine like action **Ans. b**

549. Which of the following alleviates asthma?

(a) PGE_1 only (b) PGE_1 and PGE_2
(c) PGF_2 (d) PGA **Ans. b**

550. Thyroxine is derived from

(a) Tyrosine (b) Tyranine
(c) Taurine (d) Tryptaine **Ans. a**

551. Adrneal cortical response is poor in

(a) Kwashiorkor (b) Marasmus
(c) Fatty liver (d) Atherosclerosis **Ans. a**

552. Prostaglandins are

(a) C_2 unsaturated acids (b) C_{27} saturated alcohols
(c) C_{20} saturated acids (d) C_{27} saturated alcohols **Ans. a**

553. Which one of the following scientists has not worked in the field of prostaglandins?

(a) Voneuler (b) Sultan Karim
(c) Andre robet (d) Kendal **Ans. d**

554. The suffix number in the names of prostaglandins gives the number of

(a) OH groups (b) Double bonds
(c) Acid groups (d) Ketoacids **Ans. b**

555. One of the important functions of prostacyclins is

(a) Inhibition of platelet aggregation
(b) Contraction of uterus

(c) Decrease of gastric secretion
(d) Relieving asthma **Ans. a**

556. Vasopressin is also known as
(a) Antidiabetogenic hormone (b) Antidiuretic hormone
(c) Somatotropic hormone (d) Pitoxin **Ans. a**

557. Which of the following is used for inducing labour?
(a) Prostaglandins (b) Prostacyclins
(c) Vasopressin (d) Thromboxanes **Ans. a**

558. Which of the following does not have disulphide bond?
(a) Oxytocin (b) Vasopressin
(c) Insulin (d) Glucagon **Ans. d**

559. Which is incorrect? Epinephrin promotes the glycogenolysis in
(a) Muscle (b) Liver
(c) Heart (d) None of these **Ans. c**

560. Which of one of the following is released by hypothalamus?
(a) Somatostatin (b) Somatotropic hormone
(c) Somato medin C (d) Luteinising hormone **Ans. a**

561. Which one of the following is not liberated by the adenohypophysis?
(a) Growth hormone (b) TSH
(c) ACTH (d) Gonadotropin **Ans. d**

562. Which of the following hormone is not under the control of ACTH?
(a) Aldosterone (b) Cortisol
(c) Corticosterone (d) Deoxycorticosterone **Ans. a**

563. Which of the following organ prefers fructose to glucose
(a) Liver (b) Testes
(c) Pancreas (d) Heart **Ans. b**

564. Total synthesis of creatine can be done by
(a) Liver (b) Kidneys
(c) Pancreas (d) Heart **Ans. c**

565. Thyrotropin releasing hormone is a
(a) Dipeptide (b) Tripeptide
(c) Octapeptide (d) Decapeptide **Ans. b**

566. Hypthalamo ________ gonadal oxis, fill up the blank with the suitable word.

(a) Adrenal (b) Thyroid
(c) Hypophyseal (d) Pancreatic **Ans. c**

567. The sequence of amino acids in human growth hormone and the synthesis were done by

(a) Sanger (b) Krebs
(c) Chah Holi (d) Molisch **Ans. c**

568. Proopiomelanocortin is the precussor of

(a) ACTH (b) β-tropin
(c) Endorphins (d) All of these **Ans. d**

569. Adrenalin is synthesized from

(a) Adenine (b) Adenosine
(c) Tyrosine (d) Tryptophan **Ans. c**

570. Corticotropin releasing hormone controls the direct release of

(a) Pro-opiomelanocortin (b) α MSH
(c) α MSH (d) Endorphins **Ans. a**

571. The immediate parent of α, β and γ endorphins is

(a) Pro-opiomelanocortin (b) β-lipotropin
(c) ATCH (d) Lipoprotein **Ans. b**

572. Which one of the following is not a symptom of cushing's disease?

(a) Hyperglycemia (b) Hypernatremia
(c) Hirsutism (d) Hyperkalemia **Ans. d**

573. Insulin increases the permeability of glucose across the plasma membrane of muscle cells by

(a) Acting on adenylate cycle
(b) By loosening the integrity of the membrane
(c) Through Ca^{2+} ions
(d) By membrane cruting the hexose carries of intracellular organelles and making them fuse with the plasma membrane **Ans. d**

574. Insulin like growth hormones are produced by

(a) Hypophysis (b) Liver
(c) Pancreas (d) Thyroid **Ans. b**

575. In pheochromocytoma, urine will have
(a) FILGU (b) VMA
(c) 5 HIAA (d) Lysine and Arginine **Ans. b**

576. Aldosteronism will present the chemical pathology of
(a) Addison's (b) Cushing's
(c) Grave's (d) Hartnup's **Ans. b**

577. One of the following does not bind T_3 and T_4
(a) Albumin (b) TBG
(c) TBPA (d) Haptoglobin **Ans. d**

578. Epinephrine causes in muscle
(a) Gluconeogenesis (b) Glycogenesis
(c) Glycolysis (d) Glycogenolysis **Ans. d**

579. Reverse T_3 is
(a) A synthetic compound given counter the effects of T_3
(b) Formed from T4 but has no hormone function
(c) Formed by isomerisation of T_3
(d) Formed from T_4 and has hormone function **Ans. b**

580. This pancreatic hormone promotes hypogenesis
(a) Insulin (b) Glucagon
(c) Stomato station (d) Pancreozymine **Ans. a**

581. It is unique that the following single antidiabetogenic hormone effectively counter acts the several diabetogenic hormones
(a) Glucagon (b) Glucocorticoids
(c) Insulin (d) Growth hormone **Ans. c**

582. Which of the following statements is correct?
(a) Thyroxine inhibits utilization of glucose
(b) Insulin increases utilization of glucose
(c) Glucagon promotes muscle glycogenolysis
(d) Insulin inhibits lipogenesis from carbohydrates **Ans. b**

583. Steroid hormones are synthesized from
(a) Adenine (b) Protein
(c) Vitamin (d) Cholesterol **Ans. d**

584. Hormones act only on specific organs or tissues. These are called

(a) Active sites (b) Reaction centre

(c) Target organ/Tissue (d) Physiological site **Ans. c**

585. Which of the following is noted in cushing's syndrome, a tumor associated disease of the adrenal cortex?

(a) Decreased production of epinephrine

(b) Excessive production of epinephrine

(c) Excessive production of vasopressin

(d) Excessive production of cortisol **Ans. d**

586. A cup of strong coffee would be expected to

(a) Interfere with synthesis of prostaglandins

(b) Decrease the effects of Glucagon

(c) Enhance the effects of epinephrine

(d) Provide the vitamin nicotinic acid **Ans. c**

587. Increased reabsorption of water from the kidney is the major consequence of which of the following hormones?

(a) Cortisol (b) Insulin

(c) Vasopressin (d) Aldosterone **Ans. c**

588. Lack of Glucocorticoids and mineral corticoids might be consequence of which of the following defects in the adrenal cortex?

(a) Androstenadione deficiency

(b) Estrone deficiency

(c) 17 α-OH progesterone deficiency

(d) C- α-Hydroxylase deficiency **Ans. d**

589. ADP ribosylation is the mode of action of

(a) Cholera toxin (b) Acetyl choline

(c) Muscerinic receptors (d) Cyclic AMP **Ans. a**

590. Which one of the following hormones is derived most completely from tyrosine?

(a) Glucagon (b) Thyroxine

(c) Insulin (d) Prostaglandins **Ans. b**

591. Insulin regulates fatty acid synthesis by

(a) Dephosphorylating of acetyl CoA carboxylase

(b) Activating phosphorylase

(c) Inhibiting malonyl CoA formation

(d) Controlling carnitine-Acyl CoA transferase activity **Ans. a**

592. Hormonal stimulation of the formation of the second messenger inositol 1,4,5 triphosphate (IP3) quickly leads to the release of which other intracellular messenger?

(a) cAMP (b) Prostaglandin

(c) Calcinon (d) Leukotriene **Ans. c**

593. Hormone receptors that stimulate cAMP production

(a) are part of a complex of two proteins that transform the external signal into internal cAMP production

(b) are proteins distinct and separate from those that catalyze the production of cAMP

(a) cause release of the catalytic subunit upon binding of the hormone

(d) are not very specific and bind a number of different hormones

Ans. b

594. All the following hormones use cAMP as a second messenger except

(a) Estrogen (b) FSH

(c) Luteinizing (d) Glucagon **Ans. a**

595. All the following hormones promote hyperglycemia except

(a) Epinephrine (b) Norepinephrine

(c) Insulin (d) Glucagon **Ans. c**

596. Glucagon activates the enzyme adenylcyclase which causes the increase of blood sugar level. Hence this hormone is called

(a) Hypoglycemic factor (b) Hyper glycemic factor

(c) Antidiauritic factor (d) Thyrotropin-releasing factor

Ans. b

597. TSH hormone biochemically is a

(a) Protein (b) Fat

(c) Glycoprotein (d) Carbohydrate **Ans. c**

598. A hypochromic microcytic anaemia which increases Fe, store in the bone marrow may be

(a) Folic acid responsive (b) Vitamin B_{12} responsive

(c) Pyridoxine responsive (d) Vitamin C responsive **Ans. d**

599. Which of the following hormones is not involved in carbohydrate metabolism?

(a) ACTH (b) Glucagon
(c) Vasopressin (d) Growth hormone **Ans. c**

600. In the process of transcription, the flow of genetic information is from

(a) DNA to DNA (b) DNA to protein
(c) RNA to protein (d) DNA to RNA **Ans. d**

601. The effect of auxins of coleoptite tips was demonstrated by

(a) Kreb (b) Went
(c) Skoog (d) Bose **Ans. b**

602. The last treatment of long-day plant with far-red (FR) light

(a) Promotes flowering (b) Promotes abscission
(c) Promotes will division (d) Promotes apical dominance
Ans. a

603. Which hormone suppresses the growth of a plant

(a) IAA (b) GA
(c) Abscisic acid (d) Kinetin **Ans. c**

604. Which of the following of the plant growth substances is not a growth promoter ?

(a) IAA (b) GA
(c) Kinetin (d) Abscisic acid **Ans. d**

605. The curvature in *Avena* coleoptile is associated with the growth substance

(a) Abscisic acid (b) GA
(c) IAA (d) Kinetin **Ans. c**

❑❑❑

Chapter-19

Pigments

Fill in the blanks

1. The presumed size of chlorophyll molecules is..................

Ans. Head 15 x 15 Å, Tail - 20 Å

2. The chemical formula for chlorophyll a is.......................

Ans. $C_{55}H_{72}O_5N_4Mg$

3. The correct molecular formula for chlorophyll b is..............

Ans. $C_{55}H_{72}O_4N_3Mg$

4. ....and................elements are needed for chlorophyll formation in plants.

Ans. Iron, Magnesium

5. Chlorophyll absorbs.........as well as.........light only.

Ans. blue, red

6. ....light gives maximum absorption peak by chlorophyll a.

Ans. Blue

7. ....metal is present in chlorophyll.

Ans. Mg

8. Chlorophyll is soluble in.............

Ans. organic solvents

9. Protochlorophyll or...............is the precursor of chlorophyll.

Ans. chlorophylloids

10. The wavelength of green light is...........

Ans. 500-580 mμ

11. In a chlorophyll molecule the magnesium is situated in the..............

Ans. centre of porphyrin

12. Basic structure of all chlorophylls comprises......................

Ans. porphyrin system

13. Chloroplasts fix.............

Ans. CO_2

14. The green wavelength of light is absorbed by.....................

Ans. phycoerythrims

15. To pigment system theory of photosynthesis was proposed by ....

Ans. Emerson

16. ....pigments which pick up nascent oxygen released during photo-oxidation of water and change them into molecular state.

Ans. Carotenoids

17. The porphyrin ring of chlorophyll molecule is a complex ring structure of alternating .... and .... bonds

Ans. single, double

18. Most plants contain chlorophyll a which in amount is........... or ....more than chlorophyll b.

Ans. twice, thrice

19. Chlorophyll a molecule has........... as at its carbon atom 3 of the pyrrole ring II.

Ans. methyl group

20. ....is core metal of chlorophyll molecule.

Ans. Magnesium

21. Cytochrome term was coined by ....

Ans. Keilin (1925)

22. Orange wavelengths (586-647 nm) are least absorbed and also .... effective

Ans. least

23. ....was first to separate pure fractions of chlorophyll a and chlorophyll b.

Ans. CG. Stokes (1864)

24. Protochlorophyllidea is reduced to chlorophyllidea in presence of light and then esterified with phytol group to form.................

Ans. chlorophyll

25. Chlorophyll is soluble in ethyl ether and ....

Ans. acetone

Multiple Choice Questions

1. Chlorophyll b is found in
   (a) Bacteria (b) Prokaryotes
   (c) Eukaryotes (d) Both (b) and (c) **Ans. c**
2. Chlorophylls are
   (a) Tetrapyrroles (b) Chromoproteins
   (c) Sterols (d) Steroids **Ans. a**
3. The phenomenon involved with phytochrome is
   (a) Photoperiodism (b) Leaf fall
   (c) Phototropism (d) Photorespiration **Ans. a**
4. The light inhibits seed germination of
   (a) *Nigella indica* (b) *Lactuca sativa*
   (c) *Lepidium sativum* (d) *Digitalis purpurea* **Ans. a**
5. Photomorphogenetic pigment in plants is
   (a) Chlorophyll (b) Phytochrome
   (c) Carotene (d) Xanthophyll **Ans. b**
6. Hormone responsible for vernalisation is
   (a) Abscissin (b) Vernalin
   (c) Florigen (d) Colchicine **Ans. b**
7. Which of the following factors influence the flowering in plants
   (a) Acidity of soil (b) Water in the soil
   (c) Amount of green pigments (d) Photoperiods **Ans. d**
8. Flowering in short day plants is promoted by
   (a) GA (b) ABA
   (c) IAA (d) Cytokinins **Ans. d**
9. The nature of florigen in short day and long plant is
   (a) Similar (b) Different
   (c) Both (a) and (b) (d) None of these **Ans. a**
10. Henbane (*Hyoscyamus niger*) flowers when phytochrome is in
   (a) P_r form (b) P_{fr} form
   (c) Either of the two (d) None of these **Ans. b**

11. Induction of flowering by low temperature is called
(a) Photoperiodism (b) Cryobiology
(c) Vernalisation (d) Photomorphogenesis **Ans. c**

12. Flowering in long day plants in favoured by
(a) GA (b) IAA
(c) CK (d) ABA **Ans. a**

13. Photoperiodism in plants is related to the formation of
(a) Auxin (b) GA
(c) Florigen (d) Vernalin **Ans. c**

14. Phytochrome is associated with
(a) Absorption of red light with wavelength 660 nm
(b) Absorption of far red light with wavelength 730 nm
(c) Absorption of 660 nm and 730 nm in reversible manner
(d) Absorption of red and blue light for photosynthesis **Ans. c**

15. Flowering in short day plant is induced by
(a) Photoperiod less than tweleve hours
(b) Long night
(c) Short photoperiod and interrupted long night
(d) Photoperiod shorter than initial value and uninterrupted long night **Ans. b**

16. Pigment involved in photoperception in flowering is
(a) Cytochrome (b) Carotene
(c) Phytochrome (d) Phytotron **Ans. c**

17. Phytochrome is most frequently located in
(a) Old leaves (b) Young leaves
(c) Both (a) and (b) (d) Flowers **Ans. a**

18. A plant which requires not less than ten hours of light to flower is a
(a) Long day plant (b) Short day plant
(c) Day neutral plant (d) None of these **Ans. a**

19. The pigment that absorbs both red and far-red light is
(a) Xanthophyll (b) Phytochrome
(c) Cytochrome (d) Chlorophyll **Ans. b**

20. Phytochrome plays important role in
(a) Seed germination and stem elongation
(b) Transpiration only
(c) Flowering only
(d) All of these **Ans. d**

21. Phytomorphogenetic pigment in plant is
(a) Xanthophyll (b) Phytochrome
(c) Chlorophyll (d) Carotene **Ans. b**

22. Phytochrome is involved in
(a) Photoperiodism (b) Transpiration
(c) Photorespiration (d) All of these **Ans. a**

23. Photoperiodism, seed germination, stomatal opening, transpiration and photomorphogenesis are controlled by a pigment known as
(a) Anthocyanin (b) Phytochrome
(c) Cytochrome (d) Carotenoids **Ans. b**

24. Phytotron is a device by which
(a) Mutation is induced in the plants
(b) Electrons are bombarded
(c) Protons are bombarded
(d) Plants are grown under controlled conditions **Ans. d**

25. The yellow pigment in papaya is
(a) Xanthophylls (b) Carcicaxanthin
(c) Athocyanin (d) Carotene **Ans. b**

26. The yellow colour in onion is due to the pigment of
(a) Lycopene (b) Carotene
(c) Quereitin (d) Anthocyanin **Ans. c**

27. Yellow coloured fruits and vegetables are rich source of
(a) Vitamin A (b) Vitamin B
(c) Vitamin K (d) Vitamin D **Ans. a**

28. Pink colour of onion is due to
(a) Anthyocyanin (b) Qucertin
(c) Carotene (d) Chlorophyll **Ans. b**

29. Pigment in Jamun is due to
(a) Anthocyanin (b) Carotene
(c) Chlorophyll b (d) Xanthophylls **Ans. a**

30. Which of the following is responsible for sourness of reddish?
(a) Isothiocynate (b) Cytokinin
(c) Auxin (d) Anthocyanin **Ans. a**

31. Most commonly used oil extraction method from flower is
(a) Distillation (b) Maceration
(c) Solvent extraction (d) None of the above **Ans. c**

32. Which vegetables are rich source of niacin
(a) Spinach (b) Radish
(c) Cow pea (d) Sweet potato **Ans. b**

33. Which vegetable has high calorific value
(a) Fenugreek (b) Potato
(c) Tomato (d) Amaranths **Ans. b**

34. In yellow fruits which pigment is precursor of vitamin A?
(a) Carotenoid (b) Anthocyanin
(c) Flavonoid (d) Xanthonens **Ans. a**

35. The main difference between chl a and b is that chl a
(a) Is linear whereas chl b is branched
(b) Has methyl group whereas chl b has -CHO group
(c) Is more oxidised form
(d) Has porphyrin and chl b has a phytol tail **Ans. b**

36. The presumed size of chlorophyll molecules is
(a) Head 15 × 15 Å Tail - 20Å (b) Head 10×12Å Tail - 15Å
(c) Head 20 × 20Å Tail - 25Å (d) Head 15 × 15Å Tail - 25Å
Ans. d

37. Chlorophyll a is found in all
(a) Oxygen like rating photosynthetic organisms
(b) Autotrophs
(c) Higher plants
(d) Algae **Ans. a**

38. The core metal of chlorophyll is

(a) Fe (b) Mg

(c) Cu (d) Ni **Ans. b**

39. The role of chlorophyll in photosynthesis is

(a) Absorption of water

(b) Absorption of CO_2

(c) Absorption of light and photolysis of water

(d) Absorption of light **Ans. d**

40. Phycobilins pigments are soluble in

(a) Water (b) Alcohol

(c) Ethanol (d) Non soluble **Ans. a**

❑❑❑

Chapter-20

Tools and Techniques

Fill in the blanks

1. The constituents of a mixture containing an insoluble substance suspended in a solvent can be separated by..........

Ans. filtration

2. The liquid/solution which passes through a porous membrane is termed.........

Ans. filtrate

3. Distillation is employed for separating two..........which differ in their..........

Ans. liquids, boiling points

4. Thermally unstable liquids are purified by...............

Ans. distillation under reduced pressure

5. Impure solids may be purified by.........

Ans. sublimation

6. The.........can be used for separating the constituents of a solid mixture only when they differ in their solubilities.

Ans. fractional crystallization

7. The Prussian blue colour obtained during Lassaigne's test for nitrogen is due to the formation of..........

Ans. ferric ferrocyanide

8. In Carius method, the sulphur present in the organic compound is converted into.............

Ans. SO_4^{2-}

9. The...........gives the simplest ratio of the atoms of various elements present in one molecule of the compound.

Ans. empirical formula

10. Sodium extract is boiled with HNO_3 to destroy.........before testing for the presence of chlorine.

Ans. CN^- and/or S^{2-}

11. The histochemical reaction used to detect DNA is............................

Ans. fuelgen reaction

12. Iodine test was developed by....................

Ans. F.V. Raspail

13. The resolving power of light microscope is...........................

Ans. 0-25 μm

14. ....invented phase contrast microscope.

Ans. F. Zernike

15. ....microscopy is useful for detecting nucleic acids.

Ans. Ultraviolet

16. ....microscopy is used for studying mitotically dividing cells.

Ans. Nomarski interference

17. Extremely thin sections for electron microscopic studies are obtained with the help of....................

Ans. ultramicrotome

18. The technique of tracing metabolic pathways using radioactive isotopes is known as................

Ans. autoradiography

19. Centrifugation using a density gradient is called....................

Ans. gradient centrifugation

20. Separation of components of a mixture based on their differential movement is called....................

Ans. chromatography

21. Fuelgen reaction was developed by................. and ....

Ans. Fuelgen, Rossenbeck

22. Microtome was invented by................

Ans. W. His

23. Fine details of surface features of objects can be studied with the help of...............electron microscopy.

Ans. scanning

24. Resolving power of electron microscope is more than................times that of light microscope.

Ans. 100

25. The microscopy involving the focusing of electrons through the specimens is called..............

Ans. transmission electron microscopy

26. Fuelgen reaction is used for detecting.....................

Ans. DNA

27. ....technique is used for determining the arrangement of atoms in the molecular structure of macromolecules.

Ans. X-ray crystallography

28. ....microscopy is useful for studying highly ordered structures like spindle fibres.

Ans. Polarisation

29. The resolution upto...............can be achieved by ultramicroscopy.

Ans. 0-1 μm

30. Cell fractionation includes............... and centrifugation.

Ans. homogenisation

31. Synthetic vaccines belong to............... generation.

Ans. third

32. ....obtained interferon through biotechnology.

Ans. Charles Weismann

33. Vitamin B_2 is obtained on large scale from............

Ans. *Ashbya gosspi*

34. The idea of hybridomas was conceived by............. and ....

Ans. Kohler, Milstein

35. The term 'antibiotic' was coined by...........

Ans. Waksman

36. *Streptomyces griseus* yields more than......... different antibiotics.

Ans. 41

37. The DNA which contains genes from more than one source is called ....

Ans. Recombinant DNA

38. ....organisms are those free-living organisms in the environment which have had a foreign gene inserted into them.

Ans. Transgenic

39. ....are the vehicles used to transfer foreign genetic material into a cell (e.g. plasmid).

Ans. Vectors

40. ....are the known sequences of DNA which are used to find complementary DNA strands.

Ans. DNA probe

41. Limit of resolution of light microscope is...................

Ans. 0.3 μ

42. Lipids in cells can be located by the stain named................

Ans. Sudan black

43. Chemical substances which absorb UV rays and emit visible light are called...........

Ans. fluorescent substances

44. An ultracentrifuge revolves at the rate of.............

Ans. 50,000 revolution per minute

45. Modern compound microscope was devised by..............

Ans. Wilson

46. Basic fuchsin (pararosanillin) is used to stain................

Ans. DNA

47. The isotopes most commonly used in acetoradiography are.............., ...and.................

Ans. C^{14}, H^3 and P^{32}

48. Ion exchange resins are used to separate compounds with low.............. such as..............

Ans. molecular weight, amino acid

49. The process of rupturing the cell membranes and isolating its contents is called............

Ans. homogenization

50. Centrifugation using a density gradient is known as................

Ans. gradient centrifugation

Multiple Choice Questions

1. Impure table salt can be purified by
   (a) Filtration (b) Sublimation
   (c) Crystallization (d) Distillation **Ans. c**
2. A solid which is soluble in a solvent can be obtained from its solution by
   (a) Crystallization (b) Fractional crystallization
   (c) Sublimation (d) Filtration **Ans. a**
3. Fractional crystallization is based on the difference between
   (a) Melting points (b) Solubilities
   (c) Vapour pressure (d) Boiling points **Ans. b**
4. Separation of two substances by fractional crystallization depends upon difference in their
   (a) Densities (b) Volatility
   (c) Solubility (d) Crystalline shape **Ans. c**
5. Chromatography may be used for the purification/separation of the constituents of a mixture consisting of
   (a) Gases (b) Liquids
   (c) Solids (d) All the above **Ans. d**
6. Essential oils can be extracted by
   (a) Sublimation (b) Steam distillation
   (c) Simple distillation (d) Fractional distillation **Ans. b**
7. A mixture of four liquids having their boiling points differing by only a few degrees can be separated by
   (a) Separating funnel (b) Steam distillation
   (c) Fractional distillation (d) Simple distillation **Ans. c**
8. Aniline has high boiling point, high vapour pressure at 100^{0}C and is insoluble in water. Aniline is therefore separated by
   (a) Simple distillation (b) Vacuum distillation
   (c) Steam distillation (d) Alkali treatment **Ans. c**
9. In Dumas method for the estimation of nitrogen, the gas collected in nitrometer is
   (a) N_2 (b) NH_3
   (c) $N + CO_2$ (d) CO_2 **Ans. a**

10. Dumas method is used for the estimation of

(a) Carbon (b) Hydrogen

(c) Nitrogen (d) Sulphur **Ans. c**

11. In Lassaigne's test, when both N and S are present, a blood red colour is obtained. This is due to the formation of

(a) Ferric ferrocyanide (b) Ferric cyanide

(c) Ferric thiocyanate (d) Ferric hydroxide **Ans. c**

12. Liebig's method is used for the estimation of

(a) Nitrogen (b) Carbon and hydrogen

(c) Sulphur (d) Halogens **Ans. b**

13. Carius method is used for the estimation of

(a) Nitrogen (b) Carbon and hydrogen

(c) Halogens (d) None of these **Ans. c**

14. Molecular mass of a volatile compound is generally measured by

(a) Hofmann's method (b) Carius method

(c) Silver salt method (d) Victor Meyer's method **Ans. d**

15. The first microscope was invented by

(a) Leeuwenhoek (b) Hooke

(c) F. Janssen and Z. Janssen (d) Schleiden **Ans. c**

16. Father of microscopy and microbiology is

(a) Theophrastus (b) Janssen

(c) Leeuwenhoek (d) Hooker **Ans. c**

17. In electron microscope, the lenses are

(a) Electric (b) Photoelectric

(c) Electromagnetic (d) Quartz **Ans. c**

18. Which is a pre-requisite for electron microscope

(a) Vacuum (b) Living material

(c) Source of light (d) Reflector **Ans. a**

19. Scanning electron microscope is important for its images which are

(a) Very large and sharp (b) Three-dimensional

(c) Fluorescent (d) Two-dimensional **Ans. b**

20. Which one of the following microscopes can be used for studying unstained and dividing cells

(a) Electron microscope (b) Phase-contrast microscope

(c) Polarizing microscope (d) All the above **Ans. b**

21. Ability to distinguish two closely placed points is

(a) Resolving power (b) Video power

(c) Distinguishing power (d) Magnifying power **Ans. a**

22. Resolving powers of human eye, compound microscope and electron microscope are respectively

(a) 100 µm, 0.25 µm and 10 Å (b) 100 µm, 100 nm and 100 Å

(c) 1 mm, 10 µm and 100 nm (d) 1 mm, 0.5 µm and 1 Å **Ans. a**

23. As compared to light microscope, the resolving power of electron microscope is

(a) 10 times (b) 50 times

(c) 1000 times (d) 5000 times **Ans. c**

24. An Å is equal to

(a) 10^{-5} cm (b) 10^{-6} cm

(c) 10^{-7} cm (d) 10^{-8} cm **Ans. d**

25. The term microscope was coined by

(a) Janssen and Janssen (b) Faber

(c) Robert Hooke (d) Leeuwenhoek **Ans. b**

26. Class room compound microscope was invented by

(a) Kepler and Galileo (b) Janssen and Janssen

(c) Tolles and Abbe (d) Knoll and Ruska **Ans. a**

27. Compound microscope used by Robert Hooke has a magnification of

(a) 110-20 (b) 14-42

(c) 8-16 (d) 16-64 **Ans. b**

28. Dyes that make cellular structures glow under UV-radiations are

(a) Fluorescent (b) Fluorechrome

(c) Autofluorescent (d) Phosphorescent **Ans. b**

29. High magnification of electron microscope is due to

(a) Electron beam (b) Electromagnets

(c) High voltage (d) Vacuum **Ans. b**

30. High resolving power of electron microscope is related to
(a) Electromagnets (b) Long wavelength of electrons
(c) Short wavelength of electrons (d) High voltage **Ans. c**

31. Which one is the most important in the microscopy
(a) Resolving power (b) Magnification
(c) Study of living cells (d) Study of biochemicals **Ans. a**

32. Reflector of classroom microscope is
(a) Convex lens (b) Concave lens
(c) Concavo-convex lens (d) Plano-concave lens **Ans. d**

33. Ocular lens occurs
(a) Nearer the slide
(b) On the path between reflector and object
(c) Nearer the eye
(d) Nearer the source of light **Ans. c**

34. Nose piece of microscope is
(a) Eye piece (b) Capable of revolving
(c) Site for objective (d) Both (b) and (c) **Ans. d**

35. Present-day condenser of the optical microscope was developed by
(a) Abbe (b) Tolles
(c) Gesner (d) Galileo **Ans. a**

36. Scientist to first employ magnifying glass for studying living beings was
(a) Robert Hooke (b) Huxley
(c) Gesner (d) Tolles **Ans. c**

37. Robert B. Tolles is famous in microscopy for invention of
(a) Diaphragm (b) Stage of the microscope
(c) Oil immersion lens (d) Inbuilt illumination **Ans. c**

38. Dark field microscope was invented by
(a) Zsigmondy (b) Zernicke
(c) Merten et al (d) Caspersson **Ans. a**

39. Which of the following microscope is best for studying the process of mitosis
(a) Dark field microscope (b) Phase-contrast microscope
(c) Ultraviolet microscope (d) Electron microscope **Ans. b**

40. Which one requires cooling

(a) Electron microscope (b) Ultraviolet microscope

(c) Dark field microscope (d) Interference microscope **Ans. a**

41. One micrometer (micron, μm) is equal to

(a) 0.1 mm (b) 0.01 mm

(c) 0.001 mm (d) 0.0001 mm **Ans. c**

42. A nanometer (nm) is

(a) 10^{-9} m (b) 10^{-7} cm

(c) 10 Å (d) All the above **Ans. d**

43. Which one is a proper unit of measurement

(a) Micron (b) Angstrom

(c) Nanometer (d) All the above **Ans. c**

44. Types of lenses used in a compound microscope are

(a) 3 (b) 2

(c) 1 (d) 4 **Ans. a**

45. Lenses used in ultraviolet microscope are

(a) Glass lenses (b) Quartz fluorite

(c) Lithium fluoride (d) Both (b) and (c) **Ans. d**

46. In which microscope direct observation is not possible

(a) Compound microscope (b) Dark field microscope

(c) Phase-contrast microscope (d) Ultraviolet microscope **Ans. d**

47. Fluorescent screen or photographic plate is used for observation in

(a) Electron microscope (b) Fluorescent microscope

(c) Ultraviolet microscope (d) All the above **Ans. d**

48. Knoll and Ruska are famous for developing

(a) Ultraviolet microscope (b) Polarizing microscope

(c) Electron microscope (d) X-ray crystallography **Ans. c**

49. Magnification of common oil immersion lens is

(a) 45 X (b) 100 X

(c) 450 X (d) 1000 X **Ans. b**

50. Which one combination of eye piece and objective would have widest field

(a) 5X, 5X (b) 5X, 10X

(c) 10X, 45X (d) 10X, 100X **Ans. a**

51. Organelle not possible to observe without electron microscope is

(a) Chloroplast (b) Ribosome

(c) Mitochondrion (d) Nucleolus **Ans. b**

52. Details of mitochondrial structure could be studied only after the discovery of

(a) Oil immersion lens (b) Fluorescent microscope

(c) Electron microscope (d) All the above **Ans. c**

53. Fluorescent microscope employs

(a) UV rays (b) Polarized light

(c) Beam of electrons (d) Both (a) and (b) **Ans. a**

54. A chemical which emits radiations of visible light after irradiation with UV rays in

(a) Fluorochrome (b) Luminescent

(c) Metachrome (d) Autofluorescent **Ans. d**

55. A fluorochrome if

(a) Quinacrine (b) Fluorescein

(c) Rhodamine and auramine (d) All the above **Ans. d**

56. A section prepared for examination under electron microscope is

(a) Ultrathin (b) Macerated

(c) Thin (d) Of any thickness **Ans. a**

57. Preparation of section for electron microscopy requires

(a) Microtomy (b) Dehydration

(c) Metallic impregnation (d) All the above **Ans. d**

58. Palladium or gold salts are used in staining section for

(a) Superior optical microscope (b) Electron microscope

(c) Ultraviolet microscope (d) Interference microscope

Ans. b

59. Microtomy is

(a) Surgery of tissues

(b) Surgery of endocrine glands

(c) A new technique of surgery where incision is not visible
(d) Fine sectioning by machine **Ans. d**

60. Ultramicrotome cuts sections with the thickness of
(a) 1-5 μm (b) 0.5-2.0 μm
(c) 100-500 nm (d) 20-100 nm **Ans. d**

61. Microtome was invented by
(a) Abbe (b) Tolles
(c) His (d) Malpighi **Ans. c**

62. X-ray crystallography was developed by
(a) Bragg (b) Kirkpatrick
(c) Astbury and Franklin (d) Watson and Crick **Ans. a**

63. X-ray diffraction photomicrographs of DNA used for preparing DNA model by Watson and Crick were obtained by
(a) Bragg (b) Kirkpatrick
(c) Astbury and Franklin (d) White **Ans. c**

64. A scientist wants to study the structure of a protein. What technique would be useful
(a) Electrophoresis
(b) X-ray crystallography
(c) Density gradient centrifugation
(d) Microspectrophotometry **Ans. b**

65. Sudan III is used for identification of
(a) Glycogen (b) Protein
(c) DNA (d) Fat **Ans. d**

66. Chromosomes are distinguished by
(a) Chlor-zinc-iodine (b) Azure B
(c) Pyronin (d) Acetocarmine or Orcein **Ans. d**

67. Schiff's reagent is specific for
(a) Cytoplasm (b) DNA
(c) Pectin (d) Suberin **Ans. b**

68. Schiff's reaction was developed by
(a) Taylor (b) Schiff
(c) Fuelgen and Rossenbeck (d) Schultze **Ans. c**

69. Schultze's reagent can stain differentially
(a) Cellulose (b) Starch
(c) Lignin, cutin and suberin (d) All the above **Ans. d**

70. Schultze's reagent is
(a) Safranin (b) Fast green
(c) Phlorglucinol + HCl (d) Chlor-zinc-iodine **Ans. d**

71. Safranin stains
(a) Lignified walls green
(b) Lignified walls blue
(c) Lignified walls and nuclei red
(d) Lignified walls and nuclei yellow **Ans. c**

72. Fast green stains
(a) Cellulose wall green (b) Cytoplasm green
(c) Mitochondria orange (d) Both (a) and (b) **Ans. d**

73. Which one is a redox dye
(a) Janus green (b) Methylene blue
(c) Neutral red (d) Aniline blue **Ans. a**

74. Janus green dye is specific for
(a) Mitochondria (b) Golgi apparatus
(c) Cell walls (d) Cytoplasm **Ans. a**

75. Metachromasis is
(a) Staining living cells by vital dyes
(b) Staining particular organelles
(c) Staining a structure different by same dye
(d) All the above **Ans. c**

76. A heavy isotope is
(a) ^{14}C (b) ^{3}H
(c) ^{15}N (d) ^{32}P **Ans. c**

77. A radioactive element widely used in study of photosynthesis is
(a) ^{35}S (b) ^{14}C
(c) ^{3}H (d) ^{42}K **Ans. b**

78. Scintillation counter is used for locating
(a) Radioisotope (b) Heavy isotope
(c) DNA (d) mRNA **Ans. a**

79. Nonradioactive heavy isotopes are analysed by
(a) Geiger counter
(b) Mass spectroscope
(c) Density gradient centrifugation
(d) Both (b) and (c) **Ans. d**

80. A radioactive isotope can be traced because it
(a) Emits charged particles (b) Emits light
(c) Gives radio waves (d) Sublimes slowly **Ans. a**

81. Metabolic processes can be studied by means of
(a) Light/compound microscope
(b) Autoradiography
(c) Phase-contrast microscope
(d) Electron microscope **Ans. b**

82. In autoradiography, radioactive isotope is located by
(a) Geiger counter (b) Scintillation counter
(c) Photographic emulsion (d) Illumination screen **Ans. c**

83. For studying RNA synthesis, the radioactive intermediate used is
(a) Uridine (b) Thymidine
(c) Glycine (d) Tryptophan **Ans. a**

84. Taylor (1958) studied chromosome replication by using radioactive
(a) Uridine (b) Thymidine
(c) Histones (d) Phosphate **Ans. b**

85. Homogenization is
(a) Breaking a tissue into cells
(b) Crushing a tissue to separate cells from intercellular substances
(c) Breaking of cells and suspending them in 0.25 M sucrose solution
(d) Separation of different cellular components **Ans. c**

86. In homogenization the suspension is kept at temperature of
(a) 15-20°C (b) 25-35 °C
(c) -4 to 0 °C (d) 0-4 °C **Ans. d**

87. In cell fractionation, the various fractions obtained during differential centrifugation are
(a) Plastidial, mitochondrial and microsomal
(b) Nuclear, mitochondrial and microsomal
(c) Nuclear, plastidial and cytosol
(d) Plastisial, microsomal and cytosol **Ans. b**

88. Which one is a new structure obtained during cell fractionation
(a) Microsomes (b) Microbodies
(c) Endoplasmic reticulum (d) Ribosome subunits **Ans. a**

89. Svedberg unit is for
(a) Molecular weight (b) Sedimentation coefficient
(c) Density (d) Surface tension **Ans. b**

90. Instrument for measuring Svedberg units is
(a) Homogenizer (b) Photometer
(c) Analytical centrifuge (d) Density gradient centrifuge
Ans. c

91. A technique used to study substances present inside cells, their extracts and emissions is
(a) Chromatography (b) Electrophoresis
(c) Specific stains (d) Microspectrophotometry
Ans. d

92. Microspectrophotometry analyses chemicals with the help of
(a) Emission spectra
(b) Absorption spectra
(c) Ability to form gaseous components
(d) Both (a) and (c) **Ans. b**

93. Chromatography is based on the principle of
(a) Passage at different rates (b) Adsorption at different rates
(c) Reactivity of biochemicals (d) Both (a) and (b) **Ans. d**

94. A thin layer of cellulose or alumina powder is used for separating chemicals in
(a) Column chromatography (b) Paper chromatography
(c) Thin layer chromatography (d) Partition chromatography
Ans. c

95. The classroom technique used in separating different photosynthetic pigments from a pigment solution is

(a) Paper chromatography (b) Column chromatography

(c) Partition chromatography (d) Thin layer chromatography

Ans. a

96. A method using paper or film with different electric poles at the two ends and used for separation of charged particles is

(a) Electrolysis (b) Electrophoresis

(c) Electroplating (d) Thin layer chromatography

Ans. b

97. What will you call to the study of plant and animal tissues outside the body in a glass tube

(a) *In vivo* (b) *In vitro*

(c) Innate (d) Ecdysis **Ans. b**

98. Magnification of a microscope having objective lens of 45X and eye piece of 15X would be

(a) 350X (b) 675X

(c) 2275X (d) 1475X **Ans. b**

99. Which type of objective lens will have the least working distance

(a) 5X (b) 10X

(c) 45X (d) 100X **Ans. d**

100. In air the numerical aperture of objective lens would be

(a) 1.6 (b) 1.35

(c) 1.0 (d) Less than one **Ans. d**

101. Stop disc is used in the microscope

(a) Compound light microscope

(b) Differential interference contrast microscope

(c) Dark field microscope

(d) Phase contrast microscope **Ans. c**

102. Stop disc is meant for

(a) Blocking light

(b) Stopping objective in pressing slide

(c) Altering light rays

(d) Splitting light rays **Ans. a**

103. Light is split into two beams in
(a) Fluorescent microscope
(b) Differential interference contrast microscope
(c) Phase-contrast microscope
(d) Dark field microscope **Ans. b**

104. Fluorescent microscope uses
(a) UV radiations (b) Complementary filters
(c) Both (a) and (b) (d) Annular diaphragm **Ans. c**

105. The specimen for scanning electron microscope is first super cooled in liquid propane at
(a) -20°C (b) -60 °C
(c) -120 °C (d) 180 °C **Ans. d**

106. Dehydration of specimen for SEM is carried out in alcohol at
(a) -70 °C (b) -35 °C
(c) -5 °C (d) 15 °C **Ans. a**

107. Microscope used for detecting defects in electrical conductors and computer chips is
(a) Atomic force microscope
(b) Scanning tunneling microscope
(c) X-ray crystallography
(d) Scanning electron microscope **Ans. b**

108. Detailed structure of biological molecules can be known with the help of
(a) Atomic force microscope
(b) Scanning probe microscope
(c) Scanning tunneling microscope
(d) All the above **Ans. a**

109. Magnification of scanning probe microscope is
(a) 1-5 million (b) 25-30 million
(c) 100 million (d) 5-10 lakhs **Ans. c**

110. Isopyknic or equal density centrifugation is achieved in
(a) Buoyant density centrifugation
(b) Density gradient centrifugation

(c) Electrophoresis
(d) Differential centrifugation **Ans. b**

111. Buoyant density centrifugation is carried out at centrifugal force of
(a) 50,000 g for 1-3 hours (b) 50,000 g for 20 hours
(c) 1,00,000 g for 1-3 hours (d) 1,00,000 g for 20 hours **Ans. d**

112. Stationary phase of adsorption/column chromatography is
(a) Charcoal
(b) Silica or alumina
(c) Calcium carbonate or magnesium oxide
(d) All the above **Ans. d**

113. Beads of negative and positive charges used in ion exchange chromatography are of
(a) Cellulose (b) Silica
(c) Alumina (d) All the above **Ans. a**

114. Ion exchange chromatography is used in
(a) Purification of insulin (b) Plasma fractionation
(c) Separation of proteins (d) All the above **Ans. d**

115. Method of chromatography used in determining molecular weight of proteins is
(a) Adsorption chromatography
(b) Gel fractionation chromatography
(c) Affinity chromatography
(d) Thin layer chromatography **Ans. b**

116. Gel used in gel fractionation chromatography is
(a) Agar (b) Carrageenin
(c) Sephadex (d) Acacia gel **Ans. c**

117. Type of chromatography which uses a column of ligands is
(a) Affinity chromatography
(b) Adsorption chromatography
(c) Ion exchange chromatography
(d) Gel fractionation chromatography **Ans. a**

118. Electric field is used in separating chemicals in
(a) Ion exchange chromatography
(b) Electrophoresis

(c) Thin layer chromatography

(d) Column chromatography **Ans. b**

119. The base material used in electrophoresis is

(a) Polyacrylamide (b) Agarose

(c) Carragenin (d) Both (a) and (b) **Ans. d**

120. The technique used in the detection of particular protein is

(a) Buoyant density centrifugation

(b) Density gradient centrifugation

(c) Immune-electrophoresis

(d) Both (b) and (c) **Ans. c**

121. Separation of DNA fractions having $^{15}N/^{15}N$, $^{15}N/^{14}N$ and $^{14}N/^{14}N$ was carried out by Meselson and Stahl through

(a) Buoyant density centrifugation

(b) Density gradient centrifugation

(c) Ion exchange chromatography

(d) Affinity chromatography **Ans. a**

122. Which of the following process can be used to bring two genetically different parents (cells) together for common desirable characters

(a) Hybridization (b) Protoplasm fusion

(c) PCR (d) Sexual reproduction **Ans. b**

123. Protoplasm fusion can be achieved by which of the following agents

(a) Electric current (b) PEG

(c) Sandi virus (d) All the above **Ans. d**

124. The product of protoplasm fusion is a

(a) Uninucleated cell (b) Homokaryon

(c) Heterokaryon (d) Multikaryon **Ans. c**

125. The sequential events during the process of protoplasmic fusion are

(i) Membrane contact

(ii) Nuclear fusion

(iii) Heterokaryon formation

(iv) Membrane fusion at contact area

(v) Hybrid formation

Choose the correct answer

(a) i, ii, iii, iv, v (b) i, iii, ii, v, iv
(c) i, iv, iii, ii, v (d) i, v, iv, iii, ii **Ans. c**

126. The Biolistic PDS - 1000 is a/an

(a) Mutation antibiotics drugs
(b) Gene therapy operation technique
(c) Instrument used to transfer DNA to wide range of cells and tissues
(d) Important cell fusion technique **Ans. c**

127. DIG based technique is used for the detection of certain known DNA in an animal cell, it is preferred over radio labeling techniques because of

(i) Quick results (ii) Safety
(iii) Stability (iv) Easy handling

Choose the correct answer

(a) i, iii, iv only (b) iii, ii only
(c) i, ii, iii, iv (d) i, ii, iii only **Ans. c**

128. Polymerase Chain Reaction (PCR) was invented by

(a) L. Pauling (b) Kary Mullis
(c) M. Gilbert (d) I. S. Johnson **Ans. b**

129. PCR is exclusively used for

(a) DNA identification (b) DNA recombination
(c) DNA amplification (d) DNA repair **Ans. c**

130. The unique feature of the enzyme Taq polymerase used in PCR is

(a) High speed (b) High fidelity
(c) High thermal stability (d) Low thermal stability **Ans. c**

131. Using a PCR to amplify a given segment of DNA, after 25 cycles how many fold it the given segment has been amplified?

(a) 10^3 (b) 10^5
(c) 10^6 (d) 10^9 **Ans. c**

132. DNA polymerase enzyme is concerned with PCR, it's role is

(i) Initiation of DNA synthesis
(ii) Termination of DNA synthesis
(iii) Proof-reading of DNA synthesized by Taq polymerase
(iv) Inhibition of Taq polymerase

Choose correct answer

(a) i and ii only (b) i only

(c) iii only (d) i, ii, iii and iv **Ans. c**

133. Maxum Gilbert and Sanger dideoxy methods are concerned with

(a) DNA amplification (b) DNA damage

(c) DNA repair (d) DNA sequencing **Ans. d**

134. Addition of ^{32}P to one end of one DNA strand is the primary step in Maxam Gilbert method of DNA sequencing, the addition of ^{32}P is brought about by the enzyme

(a) Taq polymerase

(b) Restriction endonuclease

(c) Polynucleotide dehydrogenase

(d) Polynucleotide kinase **Ans. d**

135. Polyacrylamide gel electrophoresis is used for DNA sequencing because, it

(a) Can separate different DNAs

(b) Can resolve DNA molecules differing in length

(c) Can break DNA into small fragments

(d) Can separate purines and pyrimidines **Ans. b**

136. Purines and pyrimidines are cleaved by

(a) Dimethyl sulphate

(b) Hydrazine

(c) Piperazine

(d) Dimethyl sulphates and hydrazine respectively **Ans. d**

137. Maxam Gilberts method differs from Sanger's methods because

(i) It uses ^{32}P

(ii) It does not uses 2, 3 dideoxy analog

(iii) It uses DNA polymerase I

(iv) It can be autoradiogramed

Choose correct answer

(a) i and ii only (b) i and iii only

(c) i, ii and iii (d) i, ii, iii and iv **Ans. a**

138. Amino acids that gives blue colouration with ninhydrin, are

(a) α type (b) β type

(c) γ type (d) δ type

Choose correct answer

(a) i only (b) i and ii only

(c) i, ii and iii only (d) i, ii, iii and iv **Ans. a**

139. During amino acid sequencing terminal amino acid labeling can be done with

(i) Flurodinitro benzene (FDNB) (ii) Dansyl chloride

(iii) Phenyl isothiocyanate (iv) Ninhydrin

Choose correct answer

(a) i, ii, iii and iv (b) i and ii only

(c) i, ii and iii only (d) iii only **Ans. c**

140. Isoelectric focusing technique used to separate proteins works on the principle of

(a) Electrophoretic separation based on relative content of acidic and basic residues

(b) Mass of the protein molecules

(c) Number of amino acids containing sulphur

(d) Coagulation capacity of protein molecules **Ans. a**

141. Isoelectric point is a point at which

(a) The mass of protein is maximum

(b) The net charge of protein is zero

(c) The speed of mobility is maximum

(d) The protein loses its structure **Ans. b**

142. Gel filtration chromatography separates proteins on the basis of

(a) Charge (b) Size

(c) Mass (d) Structure **Ans. b**

143. The solid phases used in gel filtration chromatography are many, which one of the following can not be used as solid phase

(a) Polyacrylamide (b) Sepharose

(c) Agarose (d) Agar **Ans. d**

144. Ion exchange chromatography is used to separate proteins, the basic principle underlying is

(a) Mass difference (d) Charge difference

(c) Structure difference (d) Size difference **Ans. b**

145. Affinity chromatography works on the basic principle; that

(a) A specific protein with high affinity for the supporting system is retained and rest sample flows down

(b) Proteins with acidic charge separates first and basic protein later on

(c) Tertiary structural protein are retained and others flow smoothly through column

(d) All the above are partially true **Ans. a**

146. R_f value is related to

(a) TLC (b) GLC

(c) IEC (d) GFC **Ans. a**

147. The quickest method to detect the presence of an antigen in a frozen sample or viable cell is

(a) TLC (b) GLC

(c) IEC (d) IFI **Ans. d**

148. The technique commonly used to estimate the concentration of plasma proteins is

(a) Double immunodiffusion (b) Simple immunodiffusion

(c) Immunodiffusion (d) ELISA **Ans. b**

149. Purity of particular antigen in sera, culture filtrates or tissues can be best obtained by

(a) Simple immunodiffusion (b) Double immunodiffusion

(c) Both (a) and (b) (d) None of the above **Ans. b**

150. A vector should have which of the following properties

(i) Small size

(ii) Multiple origin of replication

(iii) Multiple restriction sites

(iv) Low replication speed to avoid errors

Choose the correct answer

(a) i, ii, iii (b) i, ii, iv

(c) ii, iii, iv (d) i, iii, iv **Ans. a**

151. A cosmid is a

(a) Circular DNA (b) Plasmid with unique 'Cos' site

(c) Larger plasmid (d) Smaller plasmid **Ans. b**

152. Biosensors are

(a) Devices that can convert biological or biochemical signal or response into a quantifiable electrical signal

(b) Substances that destroy biologically hazardous materials

(c) Simple enzyme like substances

(d) Biological substances that can check pollution **Ans. a**

153. Nicardia erythropolis is a good example of

(a) Tissue based biosensor (b) Cell based biosensor

(c) Enzyme based biosensor (d) All of the above **Ans. b**

154. CsCl gradient centrifugation helps in separation of DNA fragments, the basic principle involved is, that of

(a) DNA fragments can move and accumulate at a position where the density of the two (CsCl and DNA) is same

(b) DNA binds firmly to CsCl

(c) Only smaller fragments of DNA can bind with CsCl

(d) CsCl degrades DNA into desired length fragments **Ans. a**

155. Nucleic acids strongly absorb UV light at

(a) 250 nm (b) 260 nm

(c) 270 nm (d) 300 nm **Ans. b**

156. In autoradiography

(a) β-particles are used to expose silver halide grains of film

(b) α-particles are used to expose silver halide grains of film

(c) Both β and β particles are used

(d) None of the above **Ans. a**

157. Which of the radio isotope of iodine is generally used in autoradiography technique

(a) ^{127}I (b) ^{125}I

(c) ^{126}I (d) ^{114}I **Ans. b**

158. Precipitin is formed when

(a) Antigen and antibody meet

(b) Antigen coagulates in the living system

(c) Antibodies gets coagulated

(d) Antibodies and antigens lose their functions **Ans. a**

159. Fuelgen staining is an identifying feature for

(a) Unicellular organisms (b) Chromosome/DNA

(c) Nucleus (d) Nucleolus **Ans. b**

160. Fuelgen staining is specific for DNA only because

(a) The hydrolysis produces free aldehyde group in deoxyribose which can react with Schiff's reagent

(b) DNA is the true genetic material

(c) It stains a double stranded structure only

(d) DNA is found in nucleus where this stain is most active **Ans. a**

161. Q-banding, the technique of staining chromosome uses which of the following dye

(a) Acridine (b) Ninhydrin

(c) Quinacrine (d) Acetocarmine **Ans. c**

162. R-bands of chromosomes are visualized as green with the help of staining

(a) Acridine (b) Acridine orange

(c) Giemsa (d) Quinacrine **Ans. b**

163. Separation of different amino acids paper chromatography is clearly visualized by spraying to get a purple dot

(a) Acetocarmine (b) Giemsa

(c) Ninhydrin (d) Safranin **Ans. c**

164. Ion exchange chromatography can be used to separate the polypeptide or protein mixture that differ in over all

(a) Charge (b) Mass

(c) Molecular weight (d) Structure **Ans. b**

165. A column packed with sieved particles are used in which of the following technique to separate smaller and larger protein molecules

(a) Affinity chromatography

(b) Gel electrophoresis

(c) Molecular exclusion chromatography

(d) All of the above **Ans. c**

166. The most commonly used technique to detected the presence HIV-the AIDS virus is

(a) FIA (b) RIA

(c) ELISA (d) HPLC **Ans. c**

167. Which of the following is a template dependent enzyme

(a) DNA polymerase (b) RNA polymerase

(c) Oligonucleotidase (d) DNA ligase **Ans. a**

168. Sucrose gradient centrifugation method can be used to estimate the size of

(i) Proteins (ii) RNA molecules

(iii) Ribosomes (iv) Ribosome subunits

Choose the correct answer

(a) i, ii, iii and iv (b) i, ii and iii

(c) ii, iii and iv (d) iii and iv **Ans. a**

169. 60 S, the letter 'S' represents the sedimentation coefficient and is expressed as Svedberg units. Greater the size of ribosome unit

(a) Smaller will be the value of 'S'

(b) Greater will be the value of 'S'

(c) Same will be the value of 'S'

(d) It has no direct correlation with the size **Ans. b**

170. The commonly used 'DNA finger printing technique' in forensic studies is simply a method

(a) Southern blotting (b) Northern blotting

(c) Eastern blotting (d) Western blotting **Ans. a**

171. The method of detecting a very small sample of proteins by immunoassay technique is called as

(a) Southern blotting (b) Northern blotting

(c) Eastern blotting (d) Western blotting **Ans. d**

172. Pregnancy can be detected within very few days of conception by immunoassaying urine for the presence of

(a) FSH (b) LH

(c) HCGH (d) Progesterone **Ans. c**

173. RIA stands for

(a) Regulated immune assessment

(b) Regular immunoassay

(c) Radio immunoassay
(d) Repeated immunoassay **Ans. c**

174. ELISA stands for

(a) Enzyme linked immunosorbent assay
(b) Energy linked immunosorbent assay
(c) Enzyme labelled immunosorbent assessment
(d) Energy linked immunosorbent assessment **Ans. a**

175. NMR is a

(a) Technique (b) Device
(c) Series of device (d) Catalyst **Ans. a**

176. A technique that detects the nuclear spins, reorientation in an applied field and provides us structural information of the materials is

(a) ELISA (b) NMP
(b) FIA (d) RIA **Ans. b**

177. Which of the following enzyme is used in DNA footing technique

(a) DNA polymerase (b) RNA polymerase
(c) DNA ligase (d) Permase **Ans. b**

178. Edman's degradation technique is used for sequencing

(a) Proteins (b) Carbohydrates
(c) Nucleic acids (d) Polysaccharides **Ans. a**

179. In an expression where V = velocity, E_2 = electric force and F = fractional force, the velocity with which the proteins move under electric field is given by

(a) $V = E_z \cdot F$ (b) $V = F/E_z$
(c) $V = E_z/F$ (d) $V = E_z + F$ **Ans. c**

180. PBR 322 is a most commonly use

(a) Plasmid (b) Cosmid
(c) Bacteriophage (d) Phage **Ans. a**

181. Which of the following is not a restriction endonuclease

(a) Eco RI (b) Hind III
(c) DNA ligase (d) Bam HI **Ans. c**

182. Thin layer chromatography is

(a) Partition chromatography

(b) Electrical mobility of ionic species

(c) Adsorption chromatography

(d) None of the above **Ans. c**

183. In gas chromatography, the basis for separation of the components of the volatile material is the difference in

(a) Partition coefficients (b) Conductivity

(c) Molecular weight (d) Molarity **Ans. a**

184. In reverse phase chromatography, the stationary phase is made

(a) Non-polar (b) Polar

(c) Either non-polar or polar (d) None of these **Ans. a**

185. Ion exchange chromatography is based on the

(a) Electrostatic attraction

(b) Electrical mobility of ionic species

(c) Adsorption chromatography

(d) Partition chromatography **Ans. a**

186. The general expression for the appearance of a solute in an effluent is (where V is the elution volume of a substance, V_0 void volume, k_D distribution constant and V_i internal water volume)

(a) $V = V_0 + k_D V_i$ (b) $V = V_0/V_i$

(c) $V = V_0 - k_D V_i$ (d) $V/V_0 = k_D V_i$ **Ans. a**

187. A combination of paper chromatography and electrophoresis involves

(a) Partition chromatography

(b) Electrical mobility of the ionic species

(c) Both (a) and (b)

(d) None of these **Ans. c**

188. HPLC stands for

(a) High pressure liquid chromatography

(b) High performance liquid chromatography

(c) Both (a) and (b)

(d) Highly placed liquid chromatography **Ans. c**

189. The eluent strength is a measure of

(a) Solvent adsorption energy (b) Solvent absorption energy

(c) Solvent diffusivity (d) Solvent mixing index **Ans. a**

190. HPLC methods include

(a) Liquid/liquid (partition) chromatography

(b) Liquid/solid (adsorption) chromatography

(c) Ion exchange and size exclusion chromatography

(d) All the above **Ans. d**

191. Column efficiency is measured in terms of number of plates which is

(a) Inversely related to the square of the peak width

(b) Directly related to the square of the peak width

(c) Inversely related to the cube root of the peak width

(d) Directly related to the square of the peak width **Ans. a**

192. For a typical adsorbent such as silica gel, the most popular pore diameters are

(a) 10 and 50 Å (b) 60 and 100 Å

(c) 100 and 150 Å (d) 150 and 200 Å **Ans. b**

193. An isocratic elution in HPLC is one in which the composition of the solvent

(a) Remains constant (b) Changes continuously

(c) Changes in a series of steps (d) None of these **Ans. a**

194. In reversed phase HPLC, there is a

(a) Non polar solvent/polar column

(b) Polar solvent/non-polar column

(c) Non polar solvent/non-polar column

(d) Any of the above **Ans. b**

195. Which of the following statements is true for a refractive index detector in HPLC

(a) It is more sensitive than a UV detector

(b) It can only be used for isocratic elutions

(c) It does not respond to many solutes

(d) None of above **Ans. c**

196. A gradient elution in HPLC is one in which the composition of the solvent
(a) Remains constant
(b) Is changes continuously or in a series of steps
(c) Both (a) and (b)
(d) None of above **Ans. b**

197. An eluotropic series
(a) Ranks solvents by their relative abilities to displace solutes from a given absorbent
(b) Ranks column packing material by their relative abilities to retain solutes on the column
(c) Is a measure of the solvent adsorption energy
(d) None of the above **Ans. a**

198. Dwell volume is defined as
(a) The volume of solvent contained in liquid chromatographic column
(b) The time required for the gradient to reach the column
(c) The volume of the column between the point at which solvents are mixed and the beginning of the column
(d) None of these **Ans. c**

199. Void volume refers to the
(a) Total volume of eluent in the column the remainder being taken up by the packing material
(b) The volume of solvent contained in a liquid chromatographic column
(c) The time required or the gradient to reach the column
(d) The volume of the column between the point at which solvents are mixed and the beginning of the column **Ans. a**

200. In normal phase HPLC, there is a
(a) Non polar solvent/polar column
(b) Polar solvent/non-polar column
(c) Non polar solvent/non-polar column
(d) Any of the above **Ans. a**

201. Which of the following(s) is/are the advantage of HPLC over traditional LPLC (low-pressure liquid chromatography)?
(a) Greater sensitivity and reusable columns
(b) Ideal for ionic species and large molecules

(c) Sample recovery
(d) All of the above **Ans. d**

202. Which can be used as a mobile phase in HPLC applications
(a) Any compound with solubility in liquid
(b) Any compound with limited solubility in liquid
(c) Any compound with non-solubility in liquid
(d) Any of the above **Ans. a**

203. The first compound microscope was fabricated by
(a) Janssen and Janssen (b) Malpighi
(c) Robert Hooke (d) Leeuwenhoek **Ans. a**

204. The first compound microscope was fabricated in
(a) England (b) Holland
(c) Germany (d) France **Ans. b**

205. The first compound microscope was designed in the year
(a) 1550 (b) 1665
(c) 1690 (d) 1590 **Ans. d**

206. Which of the following objectives has the minimum magnification
(a) LP (b) SLP
(c) HP (d) Oil immersion **Ans. b**

207. The maximum magnification is achieved when the objective used is
(a) Low power (b) Short low power
(c) High power (d) Oil immersion **Ans. d**

208. A bright shining picture in a comparatively darker background of an object is obtained through
(a) Polarising microscope (b) Fluorescent microscope
(c) Electron microscope (d) Interference microscope
Ans. b

209. By the help of SEM we can examine the
(a) Surface (b) Hypodermal zone
(c) Interior of tissue (d) Entire tissue **Ans. a**

210. The resolving power of the electron microscope is
(a) 10Å (b) 5Å
(c) 20Å (d) 100Å **Ans. a**

211. In the freeze fracture technique the material is frozen in
(a) Liquid oxygen (b) Liquid nitrogen
(c) CFC (d) Ozone **Ans. a**

212. The maximum magnification possible with an electron microscope is
(a) 200 times (b) 2000 times
(c) 20000 times (d) 200000 times **Ans. d**

213. Cells organelles can be separated by method of
(a) Autoradiography
(b) Thin section microtomy
(c) Differential and gradient centrifugation
(d) X-ray diffraction **Ans. c**

214. To determine the ultrastructure of a cell organelle, the most likely method to be used would be
(a) Autoradiography (b) Phase contrast microscopy
(c) Electron microscopy (d) Microdissection **Ans. c**

215. Electron microscope has an advantage over light microscope in
(a) Having greater magnification power
(b) That it requires no light
(c) Giving depth focus
(d) Not advantageous at all **Ans. a**

216. An ultracentrifuge rotates at a speed
(a) 10,000 rpm (b) 20,000 rpm
(c) 5,000 rpm (d) >30,000 rpm **Ans. d**

217. For sedimentation of nuclei, the minimum force required is
(a) 100 g (b) 200 g
(c) 250 g (d) 500 g **Ans. d**

218. For sedimenting ribosomes, a minimum force required is
(a) 5,000 g (b) 50,000 g
(c) 1,00,000 g (d) 50,00,000 g **Ans. c**

219. During differential centrifugation the organelle that separates at the end is
(a) Mitochondrion (b) Ribosome
(c) Golgi (d) m-RNA **Ans. b**

220. The tissue like animal is homogenized in the following medium

(a) 0.25 molar sucrose solution (b) Buffer

(c) 1% isopropyl alcohol (d) 0.1%NaCl solution **Ans. a**

221. The identification of a compound on the basis of absorption of particular wavelength of light is done through

(a) Autoradiography (b) X-ray crystallography

(c) Spectrophotometry (d) Chromatography **Ans. c**

222. The R_f value indicates ratio of the distances travelled by

(a) Two solutes (b) Two solvents

(c) Solute to solvent (d) Solvent to solute **Ans. c**

223. The property of specific binding of molecules in a solid chromatographic surface is the basic principle of

(a) Adsorption chromatography (b) Ion-exchange chromatography

(c) Thin layer chromatography (d) Affinity chromatography **Ans. d**

224. The movement of charged particles to a pole in an electric field is the principle involved in

(a) Autoradiography (b) Electrophoresis

(c) Colorimetry (d) X-ray crystallography **Ans. b**

225. A biologically significant radioactive isotope is

(a) ^{14}C (b) ^{11}C

(c) ^{15}O (d) ^{2}H **Ans. d**

226. In autoradiography the following precursor is generally used

(a) 14C (b) 3H

(c) 32P (d) 15N **Ans. b**

227. During X-ray crystallography the diffraction pattern of X-rays is known through

(a) Colorimetry (b) Chromatography

(c) G.M. Counter (d) Photography **Ans. d**

228. Using the technique X-ray crystallography, molecular details of the following chemical become known

(a) m-RNA (b) t-RNA

(c) DNA (d) r-RNA **Ans. b**

229. By using the technique of X-ray crystallography. Kendrew studied the molecule of

(a) Ubiquitin (b) Myoglobin

(c) Tubulin (d) All of these **Ans. b**

230. Which of the following isotopes has very long half life

(a) ^{18}O (b) ^{14}C

(c) ^{15}N (d) $^{36}C1$ **Ans. d**

231. The Southern technique do not involve

(a) The detection of RNA fragments on membranes by specific radioactive antibodies

(b) The detection of DNA fragments on membranes by a radioactive DNA probe

(c) The detection of proteins on membranes using a radioactive DNA probe

(d) Both (a) and (c) **Ans. d**

232. Southern bloting is used for

(a) DNA finger printing (b) RNA transfer

(c) Detection of proteins (d) None of these **Ans. a**

❑❑❑

Chapter-21

Nature of Gene : Its Expression and Regulation

Fill in the blanks

1. ....enzymes unwind DNA helices.

Ans. Helicase

2. ....enzymes break and reseal the strands.

Ans. Topoisomerase

3. New strands of DNA are formed in the........... direction only.

Ans. $5' \rightarrow 3'$

4. The experiment on DNA using ^{15}N isotope proved that its replication is..................

Ans. semiconservative

5. One gene one enzyme hypothesis has been modified as one gene one................hypothesis.

Ans. polypeptide

6. In some viruses, genetic information is carried from RNA to DNA by the process called.............

Ans. reverse transcription

7. The formation of a peptide bond is catalysed by the enzyme................

Ans. peptidyl transferase

8. The unidirectional flow of genetic information (from DNA→RNA→Protein) is referred to as..........in molecular biology.

Ans. central dogma

9. Protein synthesis takes places on the............

Ans. ribosomes

10. t-RNA helps in the synthesis of....................

Ans. proteins

11. The transfer of genetic message from DNA to mRNA is described as................

Ans. transcription

12. The process of RNA synthesis on a DNA template is known as..............................

Ans. transcription

13. In bacteria, regulation of gene expression is affected usually through......................

Ans. operons

14. The eukaryotic gene is split into.............and ....

Ans. exons, introns

15. The exonic portions of the RNA are spliced together to form functional.......................

Ans. mRNA

16. In recombinant DNA technology, the................ enzymes are used to cut DNA into specific fragments containing identifiable genes.

Ans. restriction

17. DNA molecules produced by *in vitro* synthesis in the presence of enzyme.......................are biologically active.

Ans. DNA ligase

18. The short pieces of lagging strand are joined together by the enzyme................after replacing the................with DNA.

Ans. DNA ligase, RNA primer

19. 4A, 4a arrangement of ascospores in *Neurospora ascus* shows .... division segregation.

Ans. first

20. In ciliate protozoa, UAG codes for...........

Ans. glutamine

21. UAA is known as ochre while UAG is called.................

Ans. amber

22. Overlapping of genes has been shown in the bacteriophage..................

Ans. ϕ x 174

23. The smallest mRNA molecule is about.............. nucleotide unit long.

Ans. 300

24. tRNA molecule may have unusual bases like inosine and.....................

Ans. pseudouracil

25. In *E. coli* RNA polymerase is constituted by............subunits.

Ans. five

26. The antibiotic.......................prevents RNA synthesis in prokaryotes and eukaryotes.

Ans. actinomycin

27. The synthesis of DNA on RNA template in tumour virus was demonstrated by Temin and............. independently.

Ans. Baltimore

28. The enzymes permease, β-galactosidase and...... are required for the catabolism of lactose in *E. coli*.

Ans. transacetylase

29. A long mRNA molecule covering all the cistrons is called..............

Ans. polycistronic

30. In 1972................transplanted DNA of virus SB 40 in *E. coli*.

Ans. Paul Berg

31. The shortest segment of chromosome which can undergo mutation is called...............

Ans. Muton

32. ....first time isolated are purified the enzyme DNA polymerase.

Ans. Kornberg

33. ....presented the concept of Archotypal genetic code.

Ans. Jakes

34. The genetic code of which microorganism has been completely determined..............

Ans. phage ϕ x 174

35. ....proposed the wobble hypothesis.

Ans. F.H.C. Crick

36. ....postulated the existence of mRNA.

Ans. Jacob and Monod

37. ....synthesized human leucocyte interferon gene.

Ans. Edge (1981)

38. ....an example of topoisomerases.

Ans. DNA gyrase

39. The ends of a fragmented DNA molecule are sticky due to............

Ans. unpaired bases

40. *In situ* DNA hybridisation is used for locating...................... on ....

Ans. a gene, a chromosome

41. The ratio of DNA in bacterial cell and eukaryotic cell is about ....

Ans. 1:1000

42. During DNA replication the okazaki fragment on the lagging strand are joined together by................

Ans. DNA ligase

43. The number of nucleotides constituting tRNA is usually.................

Ans. 80

44. After completion of transcription, the...............molecule reassociates to form its original structure.

Ans. DNA

45. The hybrid cells of same clone themselves are known as.....................

Ans. Hybridomas

Multiple Choice Questions

1. The scientists involved in discovery of DNA as chemical basis of heredity were
   (a) Hershey and Chase
   (b) Griffith and Avery
   (c) Avery, Mac Leod and Mc Carty
   (d) Watson and Crick **Ans. c**
2. One turn of DNA possesses
   (a) One base pair (b) Two base pairs
   (c) Five base pairs (d) Ten base pairs **Ans. d**
3. Number of codons in the genetic triplet code is
   (a) 4 (b) 16
   (c) 32 (d) 64 **Ans. d**
4. Initiation codons for protein synthesis are
   (a) UUU and GGG (b) AAU and UAA
   (c) AUG and GUA (d) GUG and AUG **Ans. d**
5. Termination codons for protein synthesis are
   (a) AUU, AUG and GUU (b) UGA, UAA and UAG
   (c) UAU, UAG and UUA (d) AAA, UUU and UGA **Ans. b**
6. The two antiparallel strands of DNA are
   (a) Equidistant and run in $5' \rightarrow 3'$ direction
   (b) Equidistant and run in $5' \rightarrow 3'$ and $3' \rightarrow 5'$ directions
   (c) Unequal and run in opposite directions
   (d) Unequal and diverge from each other **Ans. b**
7. The process of multiplication of DNA from DNA is known as
   (a) Replication (b) Duplication
   (c) Transcription (d) Translation **Ans. a**
8. Formation of RNA over the template of DNA is
   (a) Replication (b) Translation
   (c) Transversion (d) Transcription **Ans. d**
9. The area of unwinding and separation of DNA strands during replication is called
   (a) Origin (b) Initiation point
   (c) Primer (d) Replication fork **Ans. a**

10. Topoisomerase is involved in

(a) Producing RNA primer (b) Joining of DNA segments
(c) Producing nick in DNA (d) Separation of DNA strands

Ans. c

11. In DNA replication, the primer is

(a) Small deoxyribonucleotide polymer
(b) Small ribonucleotide polymer
(c) Helix destabilising protein
(d) Enzyme taking part in joining nucleotides to their complementary template bases

Ans. b

12. DNA strand is synthesized in the direction

(a) $5'\to3'$ (b) $3'\to5'$
(c) $1'\to4'$ (d) $6'\to1'$

Ans. a

13. Okazaki segments are

(a) Small segments of RNA
(b) Small peptides
(a) Small DNA segments
(d) Small DNA segments formed over DNA template running in $3'\to5'$ direction

Ans. d

14. Okazaki fragments are joined by

(a) DNA polymerase III (b) DNA ligase
(c) DNA polymerase II (d) DNA polymerase I

Ans. b

15. Okazaki fragments give rise to

(a) Master strand (b) Sense strand
(c) Lagging strand (d) Leading strand

Ans. c

16. Leading strand during DNA replication is formed

(a) Continuously (b) In short segments
(c) First (d) Ahead of replication

Ans. a

17. In proof reading during DNA replication

(a) Wrong nucleotides are inserted
(b) Wrong nucleotides are taken out
(c) Wrong nucleotides are removes and correct ones inserted
(d) Mutations are prevented

Ans. c

18. Nonsense codons take part in
(a) Helping protein synthesis
(b) Termination gene message for polypeptide synthesis
(c) Initiating gene message for polypeptide synthesis
(d) Synthesis of nonprotein amino acids **Ans. b**

19. The two strands of DNA are
(a) Similar and parallel
(b) Similar but antiparallel
(c) Complementary and antiparallel
(d) Complementary and parallel **Ans. c**

20. Transcription involves
(a) Protein synthesis over ribosomes
(b) Removal of worn out organelles by lysosomes
(c) Synthesis of RNA over DNA
(d) Synthesis of DNA over DNA **Ans. c**

21. DNA acts as a template for synthesis of
(a) DNA (b) RNA
(c) Both (a) and (b) (d) Protein **Ans. c**

22. Code transfer for synthesis of polypeptide involves
(a) DNA, tRNA, rRNA and mRNA
(b) mRNA, tRNA, rRNA and DNA
(c) tRNA, DNA, mRNA and rRNA
(d) DNA, mRNA, tRNA and amino acids **Ans. d**

23. In polypeptide synthesis, amino acids are brought over ribosome-mRNA complex by
(a) rRNA (b) tRNA
(c) DNA (d) Nucleotides **Ans. b**

24. tRNA attaches amino acid at its
(a) 3′ end (b) 5′ end
(c) Anticodon (d) Loop **Ans. a**

25. Blender experiment to prove DNA as genetic material was performed by
(a) Hershey and Chase (b) Messelson and Stahl
(c) Watson and Crick (d) Rosalind and Franklin **Ans. a**

26. The strain of Neurospora, which grows on minimal medium is called
(a) Autotroph (b) Prototroph
(c) Auxotroph (d) Heterotroph **Ans. b**

27. For their contribution in biochemical genetics the scientists who received Nobel prize in 1958 were
(a) Beadle and Tatum (b) Beadle and Lederberg
(c) Lederberg and Zinder (d) Zinder and Morgan **Ans. a**

28. Genetic information is carried out by long chain molecule made up of
(a) Amino acids (b) Enzymes
(c) Nucleotides (d) Histone proteins **Ans. c**

29. One gene one enzyme hypothesis was proposed by
(a) Khorana and Nirenberg (b) Beadle and Tatum
(c) Bateson and Punnet (d) Bridges **Ans. b**

30. A mutant strain of *Neurospora* which fails to grow on a minimal medium unless supplemented with a nutrient is called
(a) Auxotroph (b) Autotroph
(c) Heterotroph (d) Prototroph **Ans. a**

31. The terms triplet code and genetic code were coined by
(a) Watson and Crick (b) Nirenberg
(c) Gamow (d) Conrat **Ans. c**

32. Beadle and Tatum produced mutant strain of *Neurospora* by
(a) X-rays (b) UV rays
(c) Beta rays (d) Gamma rays **Ans. a**

33. The technique of detecting and screening the nutritional mutants in *Neurospora crassa* was developed by
(a) Mendel (b) Morgan
(c) Bateson and Punnet (d) Beadle and Tatum **Ans. d**

34. A gene that takes part in the synthesis of polypeptide is
(a) Structural gene (b) Regulator gene
(c) Operator gene (d) Promoter gene **Ans. a**

35. Components of an operon are
(a) Operator, promoter and regulator genes
(b) Regulator, promoter, operator and structural genes

(c) Operator, regulator and structural genes
(d) Regulator, promoter and structural genes **Ans. b**

36. Regulated unit of genetic material is termed as
(a) Operon (b) Regulator gene
(c) Operator gene (d) Okazaki segment **Ans. a**

37. Part of operon producing repressor is known as
(a) Repressor gene (b) Operator gene
(c) Regulator gene (d) Promoter gene **Ans. c**

38. Lactose operon of *Escherichia coli* is
(a) Monocistronic (b) Inducible
(c) Repressible (d) Both b and c **Ans. b**

39. Tryptophan operon of *Escherichia coli* is
(a) Monocistronic (b) Inducible
(c) Unregulated (d) Repressible **Ans. d**

40. Corepressor binds with
(a) Promoter gene (b) Operator gene
(c) Aporepressor (d) Regulator gene **Ans. d**

41. External supply of tryptophan in *Escherichia coli* brings about
(a) Switching off of tryptophan operon
(b) Switching on of tryptophan operon
(c) Switching off of lac operon
(d) Switching on of lac operon **Ans. a**

42. The essential components of eukaryotic cistron are
(a) Introns (b) Exons
(c) Operons (d) Operator and regulator regions
Ans. b

43. Segments of mRNA removed during splicing are
(a) Introns (b) Exons
(c) Promoter regions (d) Integrator regions **Ans. a**

44. Bacterial transformation was discovered by
(a) Avery et al. (b) Watson and Crick
(c) Griffith (d) Hershey and Chase **Ans. c**

45. Cistron is

(a) Functional unit of DNA (b) Functional unit of RNA

(c) Nonfunctional unit of DNA (d) Nonfunctional unit of RNA

Ans. a

46. House keeping genes

(a) Produce antibodies

(b) Constantly operate for cellular activity

(c) Form hormones

(d) Function only at the time of reproduction **Ans. b**

47. The term gene was coined by

(a) Nirenberg (b) Khorana

(c) Kornberg (d) Johannsen **Ans. d**

48. The segment of DNA which participates in crossing over is known as

(a) Muton (b) Recon

(c) Cistron (d) Replicon **Ans. b**

49. Which of the following got Nobel prize for discovering mobile genes, jumping genes or transposons?

(a) Jacob and Monod (b) Khorana

(c) Beadle and Tatum (d) Barbara and Clintock **Ans. d**

50. mRNA coded language is known as

(a) Cryptogam (b) Cryptoanalysis

(c) Cryptogram (d) Codogens **Ans. c**

51. Which one is soluble RNA

(a) tRNA (b) mRNA

(c) rRNA (d) snRNA **Ans. a**

52. Adaptor molecule is

(a) mRNA (b) rRNA

(c) DNA (d) tRNA **Ans. d**

53. Griffith effect is

(a) DNA transcription (b) RNA translation

(c) Bacterial transformation (d) Bacterial transduction **Ans. c**

54. Anticodon is made of
(a) Three adjacent nitrogen bases
(b) Unpaired triplet bases of mRNA
(c) Paired triplet of nitrogen bases on tRNA
(d) Unpaired triplet of nitrogen bases at one end of tRNA **Ans. d**

55. RNA which is present in minimum amount in a cell is
(a) tRNA (b) mRNA
(c) rRNA (d) Both (b) and (c) **Ans. b**

56. Protein cannot be a genetic material because it is not
(a) Large molecule
(b) Ubiquitous
(c) Having diversity and specificity
(d) Able to replicate **Ans. d**

57. Methionine carrying tRNA has an anticodon
(a) AUG (b) UAG
(c) UAC (d) UAA **Ans. c**

58. Bacterium in which transformation was first studied is
(a) *Salmonella typhimurium* (b) *Streptococcus pneumonia*
(c) *Escherichia coli* (d) *Streptococcus aureus* **Ans. b**

59. Experimental material used by Hershey and Chase for proving that DNA controls heredity was
(a) *Diplococcus pneumoniac* (b) *Salmonella typhimurium*
(c) T_2 *phage* (d) TMV **Ans. c**

60. A codon specifies the same amino acid in *Brassica* and Homo because codons are
(a) Nonoverlapping (b) Commaless
(c) Universal (d) Nonambiguous **Ans. c**

61. In eukaryotes mRNA is synthesized with the aid of
(a) RNA polymerase III (b) RNA polymerase II
(c) RNA polymerase I (d) Reverse transcriptase **Ans. b**

62. A bacterium grown over medium having radioactive ^{35}S incorporates radioactivity in
(a) Carbohydrates (b) Proteins
(c) DNA (d) RNA **Ans. b**

63. GUG specifies amino acid valine. However, when functioning as initiation codon it specifies

(a) Methionine (b) Valine

(c) Lysine (d) Isoleucine **Ans. a**

64. New strand formation on a DNA template can be initiated only by

(a) DNA polymerase I (b) DNA polymerase III

(c) RNA primer (d) DNA primer **Ans. c**

65. What is true about ori

(a) One in all organisms

(b) Several in all organisms

(c) One in eukaryotes and several in prokaryotes

(d) One in prokaryotes and several in eukaryotes **Ans. d**

66. DNA has alternate grooves

(a) One major and one minor (b) Two major and one minor

(c) One major and two minor (d) Two major and two minor

Ans. a

67. Number of DNA coils of a nucleosome is

(a) 2 ¾ (b) 2 ¼

(c) 1 ¾ (d) 1 ¼ **Ans. c**

68. Energy for activation of amino acids during protein synthesis comes from

(a) ATP (b) GTP

(c) CTP (d) UTP **Ans. a**

69. AUG initiation codon occurs over

(a) 3′ end of mRNA (b) 5′ end of mRNA

(c) Short arm of tRNA (d) Long arm of tRNA **Ans. b**

70. Central dogma is not directly connected with synthesis of

(a) mRNA (b) Polypeptide

(c) Both a and b (d) Amino acids **Ans. d**

71. In *Streptococcus pneumoniae*

(a) Virulent form is smooth

(b) Virulent form is rough

(c) Nonvirulent form is capsulated

(d) All forms are rough **Ans. a**

72. Codon is triplet of
(a) Template strand of DNA (b) Non-template strand of DNA
(c) mRNA (d) tRNA **Ans. b**

73. Enzyme required during DNA replication is
(a) DNA polymerase (b) DNA ligase
(c) Both a and b (d) Sigma factor **Ans. c**

74. A codon is read in
(a) $3' \rightarrow 5'$ direction (b) $5' \rightarrow 3'$ direction
(c) $3' \rightarrow 6'$ direction (d) $6' \rightarrow 3'$ direction **Ans. b**

75. All nonsense codons have first base
(a) Adenine (b) Cytosine
(c) Uracil (d) Guanine **Ans. c**

76. First triplet codon to be deciphered was
(a) UUU (b) AAA
(c) CTA (d) CCC **Ans. a**

77. Triplet pair deciphered by Khorana was
(a) Serine and Threonine (b) Threonine and Histidine
(c) Cysteine and Valine (d) Phenylalanine and Isoleucine
Ans. c

78. The codon for anticodon 3′ UUUA - 5′ is
(a) 5′ AAAU - 3′ (b) 5′ UAAA - 3′
(c) 3′ UAAD - 5′ (d) 3′ AAAU - 5′ **Ans. a**

79. Central dogma of modern genetics is
(a) RNA → RNA → DNA → Protein
(b) DNA → DNA → RNA → Protein
(c) DNA → RNA → Protein
(d) RNA → DNA → RNA → Protein **Ans. b**

80. What is true of Watson and Crick's model of DNA. It is duplex with
(a) 10 base pairs and 34 Å distance for every turn
(b) 10 base pairs and 3.4 Å distance for each turn of spiral
(c) 20 base pairs and 34 Å for each turn
(d) None of the above **Ans. a**

81. Protein cover of virus is

(a) Capsid (b) Virion

(c) Viroid (d) Mucopeptide **Ans. a**

82. A gene is

(a) Three letter code in DNA

(b) Equivalent to chromosome

(c) Part of chromosome producing a set of enzymes (protein)

(d) Part of chromosome producing only one enzymes (protein) **Ans. d**

83. tRNA recognizes ribosome by

(a) TψC loop (b) DHU loop

(c) Anticodon (d) AA-site **Ans. a**

84. tRNA recognizes amino acyl synthetase enzyme by

(a) Anticodon (b) DHU loop

(c) TψC loop (d) AA-site **Ans. b**

85. Amino acid binding site of tRNA is

(a) 5′ end (b) Anticodon loop

(c) DHU loop (d) -CCA 3′ end **Ans. d**

86. Type of coiling in DNA is

(a) Right handed (b) Left handed

(c) Zigzag (d) Opposite **Ans. a**

87. Hereditary characteristics are passed on from parent to the offspring mainly through

(a) Enzymes (b) Genes

(c) Mutants (d) Centrosomes **Ans. b**

88. The pneumococcus experiment proves that

(a) RNA sometimes controls the production of DNA and proteins

(b) Bacteria under go binary fission

(c) DNA is the genetic material

(d) Bacteria do not reproduce sexually **Ans. c**

89. The strongest evidence that DNA is the genetic material comes from

(a) The fact that chromosomes are made of DNA

(b) Studies on the transformation of bacterial cells

(c) The knowledge that DNA is present in the nucleus

(d) The findings that DNA is present in the cytoplasm **Ans. b**

90. When DNA is transformed from a culture of capsulated bacteria to a culture of non-capsulated bacteria converting the latter into the former type, the process is known as

(a) Transcription (b) Translocation

(c) Transduction (d) Transformation **Ans. d**

91. A gene is made up of

(a) DNA (b) RNA

(c) Either DNA or RNA (d) Amino acids **Ans. c**

92. Genes are chemically

(a) Polynucleotides (b) Histones

(c) Lipoproteins (d) Hydrocarbons **Ans. a**

93. Which is entirely responsible for the transfer of hereditary traits

(a) DNA (b) RNA

(c) Chloroplast (d) Aestivation **Ans. a**

94. A link between generations is provided precisely by

(a) Chromosomes (b) Nucleic acids

(c) Nucleus (d) Cytoplasm **Ans. b**

95. The first undoubted evidence about DNA being the genetic material came from

(a) Transformation of rough-coated strain of *Diplococcus pneumonia* into smooth coated strain

(b) The establishment of DNA as the chief chemical constituent of chromosomes

(c) The establishment of DNA as a self-replicating substance

(d) Transduction of bacteria by action of bacteriophage **Ans. a**

96. Duplication of DNA is called

(a) Replication (b) Transduction

(c) Transcription (d) Translation **Ans. a**

97. The usual method of DNA replication is

(a) Conservative (b) Dispersive

(c) Non-conservative (d) Semi-conservative **Ans. d**

98. A DNA molecule in which both strands have radioactive thymidine is allowed to duplicate in an environment containing non-radioactive thymidine. What will be the exact number of DNA molecules that contain some radioactive thymidine after 3 duplications

(a) One (b) Two

(c) Four (d) Eight **Ans. b**

99. A bacterium containing 100% N^{15} nitrogen bases is allowed to replicate in a medium containing N^{14} bases. After one round of duplication, the result would be

(a) All individuals would be identical to parents

(b) All individuals would be radioactive but the percentage of radioactivity in DNA would be 50%

(c) Only 50% individuals would be radioactive

(d) All individuals would be similar to parents but different among themselves **Ans. b**

100. A bacterium *E. coli* with completely radioactive DNA was allowed to replicate in a non-radioactive medium for two generations. What % of the bacteria should contain radioactive DNA

(a) 100% (b) 50%

(c) 25% (d) 12.5% **Ans. b**

101. Semi-conservative mode of replication of chromosome was demonstrated by (or experimental demonstration of the semi-conservative model of DNA replication was given by)

(a) Messelson and Stahl (b) Watson and crick

(c) Stahl and Urey (d) Bawden and Ririe **Ans. a**

102. If one wants to obtain precise information regarding the exact time and location of synthesis of new DNA, which of the following methods would be most effective for such a study

(a) Electron microscopy

(b) Carbon dating

(c) Isolating and extracting DNA after regular intervals and estimating DNA amount

(d) Using radioactive precursors of nucleic acids **Ans. d**

103. The experimental system used in studies of the discovery of replication of DNA has been (or DNA replication was conducted in)

(a) *Drosophila melanogaster* (b) *Pneumococcus*

(c) *Escherichia coli* (d) None of these **Ans. c**

104. Experiments using ^{15}N (heavy nitrogen) to confirm the semi-conservative replication of DNA were carried out by

(a) Messelson and Stahl (b) Hershey and Chase
(c) Beadle and Tatum (d) Watson and Crick **Ans. a**

105. A DNA molecule of *Escherichia coli* is heavy (full labeled with N^{15}) and is allowed to duplicate in a medium containing N^{14}. After one generation, the two daughter strands molecules

(a) Have same density but do not resemble their parent DNA
(b) Have different densities but resemble their parent DNA
(c) Have different densities and also do not resemble the parent DNA
(d) Both the strands differ in density, but resemble their parent DNA
Ans. a

106. Replication of DNA takes place with the help of

(a) DNA polymerase (b) Lyase
(c) RNAase (d) DNAase **Ans. a**

107. The following is needed during DNA replication

(a) DNA polymerase and DNA ligase
(b) RNA polymerase and translocase
(c) DNA polymerase only
(d) DNA ligase only **Ans. a**

108. DNA polymerase is needed for

(a) Replication of DNA (b) Synthesis of DNA
(c) Elongation of DNA (d) All the above **Ans. a**

109. DNA polymerase is required for the synthesis of

(a) DNA from DNA (b) DNA from RNA
(c) DNA from nucleotides (d) DNA from nucleosides **Ans. c**

110. The enzyme DNA polymerase can add nucleotide to the

(a) 3′ carbon position
(b) 5′ carbon position
(c) Both 3′ and 5′ carbon positions
(d) 4′ carbon position **Ans. a**

111. The enzyme(s) responsible for unwinding of DNA helix during replication is/are

(a) Helicases (b) Topisomerases

(c) DNA polymerase (d) Primase **Ans. a**

112. The enzyme topoisomerase is helpful in

(a) Unwinding of the helix

(b) Breaking of DNA strand

(c) Proof reading of replicating DNA

(d) Synthesis of primer **Ans. b**

113. The expected frequency of introduction of a wrong base during DNA replication is

(a) 1:100 (b) 1:1000

(c) 1:1 million (d) 1:1 billion **Ans. d**

114. Ligase enzyme is used for

(a) Denaturation of DNA (b) Splitting of DNA into small bits

(c) Joining bits of DNA (d) Digestion of lipids **Ans. c**

115. The bits of DNA segments formed are joined with each other by an enzyme

(a) Polymerase (b) Ligase

(c) Lipase (d) Kinase **Ans. b**

116. Okazaki segments are

(a) DNA segments capable of free replication

(b) DNA segments formed during replication

(c) Nucleotide segments formed during transcription

(d) Segments of genes which undergo mutation and recombination

Ans. b

117. DNA polymerase enzyme was discovered by

(a) Kornberg (b) Nirenberg

(c) Watson (d) Crick **Ans. a**

118. Synthesis of DNA takes place by

(a) Transduction (b) Transcription

(c) Transformation (d) Replication **Ans. d**

119. Replication of DNA in eukaryotes commences from

(a) One end of the chromatid extending to the other end

(b) Both ends of the chromatid simultaneously

(c) The centromere to either of the ends of chromatids

(d) Several sites along the DNA of the chromatid simultaneously

Ans. d

120. The modern concept of gene is that it is

(a) A segment of DNA capable of crossing over

(b) A functional unit of DNA

(c) A segment of DNA

(d) A segment of chromosome **Ans. b**

121. Eukaryotes differ from prokaryotes in the mechanism of DNA replication due to

(a) Different enzymes (instead of same enzyme) for synthesis of lagging and leading strands

(b) Discontinuous rather than semi-discontinuous replication

(c) Use of DNA primers rather than RNA primers

(d) Unidirectional rather than bidirectional replication **Ans. a**

122. These are special proteins which help to open up DNA double helix in front of the replication fork. These proteins are

(a) DNA gyrase (b) DNA polymerase I

(c) DNa ligase (d) DNA topoisomerase **Ans. a**

123. The nuclease enzyme which begins its attack from a free end of a polynucleotide is

(a) Exonuclease (b) Kinase

(c) Polymerase (d) Endonuclease **Ans. a**

124. One gene-one polypeptide hypothesis was proposed by

(a) Linus Pauling (b) V. Ingram

(c) Yanofsky (d) Brenner **Ans. c**

125. A functional unit of gene which specifies the synthesis of one polypeptide is known as

(a) Recon (b) Clone

(c) Codon (d) Cistron **Ans. d**

126. The smallest gene affected by mutation is

(a) Exon (b) Recon

(c) Cistron (d) Muton **Ans. d**

127. The smallest unit of DNA which is capable of undergoing crossing over and recombination is called

(a) Recon (b) Muton

(c) Cistron (d) Intron **Ans. a**

128. The terms cistron, recon anc muton were proposed by

(a) W. Ingram (b) Bateson

(c) J. Lederberg (d) S. Benzer **Ans. d**

129. Gene is

(a) A segment of DNA

(b) A segment of DNA and histone

(c) A segment of DNA, RNA and Histone

(d) All of the above **Ans. a**

130. The idea that genes control the production of enzyme was given by

(a) E.L. Tatum (b) T.H. Morgan

(c) A.E. Garrod (d) R.S. Kornberg **Ans. c**

131. One gene one enzyme theory (hypothesis) was proposed by

(a) Beadle and Tatum (b) Avery and McCarty

(c) Jacob and Monad (d) Luria and Delburck **Ans. a**

132. Co-linearity between genes and polypeptides was established by

(a) Linus Pauling (b) V. Ingram

(c) Yanofsky (d) Brenner **Ans. c**

133. Garrod's views on the relationship between genes and enzymes can be best stated as

(a) One mutant gene-one metabolic block concept

(b) One gene-one enzyme hypothesis

(c) One gene-one polypeptide hypothesis

(d) One cistron-one polypeptide hypothesis **Ans. a**

134. Gene controls

(a) Protein synthesis but not heredity

(b) Protein synthesis and heredity

(c) Heredity but not protein synthesis

(d) Biochemical reaction of some enzymes **Ans. b**

135. Beadle and Tatum worked on genetics of *Neurospora* which resulted in the development of a new science called

(a) Genetic engineering (b) Biochemical genetics

(c) Biotechnology (d) Dendrochronology **Ans. b**

136. The chief advantage of the linear arrangement of the ascospores in *Neurospora* is that, in genetic studies, it permits

(a) Easy inference of the orientation of chromatids during meiosis

(b) Ready conservation of mutant phenotype

(c) Accurate counting of spores

(d) Easy collection of ascospores **Ans. b**

137. In *Neurospora,* 8 ascospores are formed. These are 2a, 4A, 2a...it shows that

(a) First generation division (b) Second generation division

(c) No crossing over (d) Some meiosis occurs **Ans. b**

138. RNA and DNA were artificially synthesized in vitro by

(a) Ochoa and Nirenberg (b) Ochoa and Kornberg

(c) Kornberg and Nirenberg (d) Nirenberg and Khorana

Ans. b

139. Who suggested that an intermediate RNA molecule would be needed to read the codons on messenger RNA

(a) M. Nirenberg (b) H.G. Khorana

(c) F. Crick (d) Kornberg **Ans. c**

140. The process of RNA synthesis on the DNA template is

(a) Transduction (b) Translation

(c) Transcription (d) Transformation **Ans. c**

141. Transcription is synthesis of

(a) Protein (b) mRNA

(c) tRNA (d) rRNA **Ans. b**

142. Transcription is the process by which

(a) RNA molecule is synthesized on a DNA strand (or template)

(b) Two daughter strands of DNA are synthesized

(c) Amino acids are joined to form polypeptides
(d) RNA molecule is synthesized within a ribosome **Ans. a**

143. Genetic code determines
(a) Structural pattern of an organism
(b) Sequence of amino acid in protein chain
(c) Variation in offsprings
(d) Constancy of morphological trait **Ans. b**

144. Genetic code was discovered by
(a) Nirenberg and Mathei (b) Novick and Szilard
(c) Kornberg (d) Willkins **Ans. a**

145. Site of protein synthesis
(a) Lysosome (b) Peroxisome
(c) Ribosome (d) Splisosome **Ans. c**

146. Dr. Hargobind Khorana has been awarded Noble Prize for research on
(a) Oral contraceptives (b) Hormones
(c) Genetic code (d) Immunology **Ans. c**

147. There are 64 codons in genetic code dictionary because
(a) There are 64 type of tRNAs found in the cell
(b) There are 44 meaningless and 20 codons for amino acids
(c) There are 64 amino acids to be coded
(d) Genetic code is triplet **Ans. d**

148. Genetic code was deciphered through chemical synthesis of trinucleotides by
(a) Watson & Crick (b) Beadle & Tatum
(c) Briggs & King (d) M.W. Nirenberg **Ans. d**

149. Nirenberg synthesized a mRNA containing 34 poly-adenine (A-A-A-A-A-A......) and found a polypeptide formed of 11 polylysine. It proved that the genetic code for lysine is
(a) Lone adenine (b) A-A doublet
(c) A-A-A triplet (d) Many adenines **Ans. c**

150. Khorana and his colleagues synthesized an RNA molecule with ripening sequence of UG N-bases (UGUGUGUGUGUG). It produced a tetrapeptide with altering sequence of cystien and valine. It proves that codon for cysteine and valine is

(a) UGG and GUU (b) UUG and GGU

(c) UGU and GUG (d) GUG and UGU **Ans. c**

151. Dr. H.G. Khorana deciphered first the triplet codon of

(a) Serine and isoleucine (b) Phenylalanine and methionine

(c) Threonine and histidine (d) Tyrosine and tryptophan

Ans. a

152. 5′ end of the tRNA always ends in the base

(a) Adenine (b) Guanine

(c) Cytosine (d) Thymine **Ans. b**

153. UGA in the yeast mitochondria codes for

(a) Stop signal (b) Tryptophan

(c) Glutamine (d) Aspartic acid **Ans. d**

154. The amount of which of these is least in a cell

(a) mRNA (b) rRNA

(c) tRNA (d) Nothing can be said definitely

Ans. a

155. The genetic code is

(a) Universal

(b) Nearly universal

(c) Similar in the members of a genus

(d) Different for every species **Ans. b**

156. The minimum length of cistron in base pair which synthesizes a polypeptide of 50 amino acids is

(a) 50 bp (b) 100 bp

(c) 150 bp (d) 200 bp **Ans. c**

157. Considering that we have four nucleotides A, G, C, T, the number of base substitutions that can occur in the amino acid codons are

(a) 549 (b) 535

(c) 261 (d) 264 **Ans. a**

158. In the DNA codons are ATG ATG ATG and a cytosine base is inserted at the beginning, which of the following will result

(a) A non-sense mutation (b) CA TGA TGA TG

(c) CAT GAT GAT G (d) C ATG ATG ATG **Ans. c**

159. Genetic code consists of

(a) Adenine and Guanine (b) Guanine and Cytosine

(c) Cytosine and Uracil (d) All **Ans. d**

160. A single anticodon can recognize more than one codon of mRNA. This phenomenon is termed as

(a) Richmond and Lang effect (b) Gene flow hypothesis

(c) Wobble hypothesis (d) Template hypothesis **Ans. c**

161. What becomes established from the Wobble hypothesis?

(a) Process of peptide chain elongation

(b) Economy of the number of tRNA molecules

(c) Process of peptide chain initiation

(d) Process of chain termination **Ans. b**

162. In a codon, Wobbling is generally restricted to

(a) First N base (b) Second N base

(c) Third N base (d) Aromatic amino acids **Ans. c**

163. DNA is the major source of genetic information which is transmitted by transcription into RNA molecules. These RNA molecules are responsible to get this genetic information translated into proteins and thus the central dogma of molecular biology is

(a) RNA → DNA → Proteins (b) DNA → RNA → Proteins

(c) RNA → Proteins (d) RNA → Proteins → DNA

Ans. b

164. Which of the following is called amber

(a) AUG (b) UAA

(c) UAG (d) UGA **Ans. c**

165. Transcription is

(a) Assembly of amino acids by mRNA in the form of polypeptides

(b) Recognition of amino acids by RNA synthesis

(c) Transfer of genetic information from DNA to mRNA

(d) Recognition of base sequence on mRNA **Ans. c**

166. The genetic information if carried by long chain molecules of
(a) Enzymes (b) Amino acids
(c) Nucleotides (d) Chromosomes **Ans. c**

167. DNA generally acts as a template for the synthesis of
(a) Only protein (b) Only DNA
(c) Only RNA (d) Both DNA and RNA **Ans. d**

168. In order to enable a chemical to serve as a genetic code, it is essential that the chemical should be
(a) Able to duplicate itself
(b) Able to form itself into long spiral molecules
(c) A compound of pyrimidines and purines
(d) Easily changed **Ans. a**

169. *E. coli* RNA polymerase consists of different polypeptide chains which are
(a) Two in number (b) Three in number
(c) Four in number (d) Five in number **Ans. d**

170. The core enzyme constituting the *E. coli* RNA polymerase consists of following subunits
(a) $\beta, \beta', \sigma, \alpha$ (b) $\beta, \beta', \alpha, \omega$
(c) $\beta, \beta', \sigma, \omega$ (d) $\beta, \sigma, \alpha, \omega$ **Ans. b**

171. How many different types of RNA polymerases catalyse the synthesis of RNA in eukaryotes
(a) Two (b) Three
(c) Four (d) Five **Ans. b**

172. Formation of RNA from DNA is known as
(a) Translation (b) Translocation
(c) Transformation (d) Transcription **Ans. d**

173. The process by which DNA of the nucleus passes genetic information to mRNA is called
(a) Transcription (b) Transportation
(c) Translocation (d) Translation **Ans. a**

174. Transcription is
(a) Synthesis of DNA on RNA
(b) Synthesis of RNA on DNA

(c) Production of proteins on RNA
(d) Replication of DNA **Ans. b**

175. Enzyme necessary for transcription is (or transcription of DNA is aided by)
(a) DNA polymerase (b) RNA polymerase
(c) Endonuclease (d) RNAase **Ans. b**

176. A DNA strand is directly involved in the synthesis of all the following except
(a) tRNA molecule (b) mRNA molecule
(c) Another DNA strand (d) Protein synthesis **Ans. d**

177. The mRNA is formed
(a) In the nucleus
(b) By free ribosomes
(c) From the ribosomes on endoplasmic reticulum
(d) From DNA in nucleus **Ans. d**

178. In which of the following places, messenger RNA is found in a living cell?
(a) Inside the endoplasmic reticulum
(b) Inside the mitochondria
(c) Inside the nucleus but outside the nucleolus
(d) Inside the nucleolus **Ans. c**

179. Which is the correct sequence of code transfer involved in the formation of a polypeptide?
(a) DNA, tRNA, rRNA, mRNA
(b) rRNA, DNA, mRNA, tRNA
(c) mRNA, tRNA, DNA, amino acids
(d) DNA, mRNA, tRNA, amino acids **Ans. d**

180. The formation of polyribosomes from ribosomes is done in the presence of
(a) Na^+ ions (b) K^+ ions
(c) Ca^{++} ions (d) Mg^{++} ions **Ans. d**

181. In protein synthesis, the codon used as a start signal is
(a) AUG (b) UGA
(c) GUA (d) UAG, UAA **Ans. a**

182. If the DNA strand has the nitrogenous base sequence ATT GCC, the mRNA will have

(a) ATT GCA (b) ATC GCC
(c) UGG ACC (d) UAA CGG **Ans. d**

183. If the sequence of bases in DNA is ATTCGATG, then the sequence of bases in its transcript will be

(a) GUAGCUUA (b) UAAGCUAC
(c) CAUCGAAU (d) AUUCGAUG **Ans. b**

184. From a DNA template with the sequence CTGATAGC, the mRNA sequence formed would be

(a) GUCTUTCG (b) GACUAUCG
(c) UACTATCU (d) GAUTATUG **Ans. b**

185. The function of a non-sense codon is

(a) To release polypeptide chain from tRNA
(b) To form an unspecified amino acid
(c) To terminate the message of gene controlled protein synthesis
(d) To convert a sense DNA into non-sense DNA **Ans. c**

186. Termination of chain growth in protein synthesis is brought about by

(a) UUG, UGC, UCA (b) UCG, GCG, ACC
(c) UAA, UAG, UGA (d) UUG, UAG, UCG **Ans. c**

187. Each codon present on mRNA and anticodon present on tRNA is composed of

(a) One N base only (b) A set of two N base
(c) A set of three N base (d) A set of three out of U, C, A, G **Ans. d**

188. The process by which proteins are synthesized in a cell is called

(a) Transcription (b) Translation
(c) Translocation (d) Transduction **Ans. b**

189. During protein synthesis, amino acids recognize and get attached to tRNA with the help of

(a) Ribosomes; Sigma and Rho factors
(b) mRNA
(c) Amino acyl tRNA synthetase
(d) tRNA **Ans. c**

190. Khorana synthesized a biologically functional tyrosine suppressor tRNA of *E. coli* in 1979 it contained

(a) 77 nucleotide pairs (b) 207 nucleotide pairs

(c) 312 nucleotide pairs (d) 333 nucleotide pairs **Ans. b**

191. An antibiotic which inhibits translation in eukaryotes is

(a) Chloromycetin (b) Penicillin

(c) Puromycin (d) Tetracycline **Ans. c**

192. Initiation of polypeptide chain in protein synthesis is induced by

(a) Methionine (b) Glycine

(c) Leucine (d) Lysine **Ans. a**

193. Who among the following established that RNA is the genetic material

(a) Lederberg (b) Griffith

(c) Nirenberg and Holley (d) Frankel Conrat **Ans. d**

194. Which one of the following chemical characteristics is not common to all living beings

(a) Similar triplet codes for amino acids

(b) Energy is store by high energy-rich phosphate bonds

(c) Types of protein present in the body

(d) Ribosomes act as sites for protein synthesis **Ans. c**

195. The functional unit in the protein synthesis is

(a) Peroxisome (b) Dictyosome

(c) Lysosome (d) Polysome **Ans. d**

196. In protein synthesis, polymerization of amino acids involves three steps. Which one of the following is not involved in protein synthesis

(a) Elongation (b) Transcription

(c) Termination (d) Initiation **Ans. b**

197. Reverse transcriptase is

(a) RNA dependent RNA polymerase

(b) DNA dependent DNA polymerase

(c) RNA dependent DNA polymerase

(d) DNA dependent RNA polymerase **Ans. d**

198. Genes which are inactive for long periods, have the tendency to be bound to

(a) Each other (b) Methyl groups

(c) Actin and myosin (d) Nucleolus **Ans. b**

199. The percentage of DNA in a eukaryotic cell expressed at a given time is about

(a) 1% (b) 20%

(c) 50% (d) 80% **Ans. a**

200. The scientist who first synthesized DNA *in vitro* was

(a) A. Kornberg (b) A. Garrod

(c) J.D. Watson (d) H.G. Khorana **Ans. a**

201. Artificial synthesis of DNA was done by

(a) Nirenberg (b) Kornberg

(c) Khorana (d) Watson and Crick **Ans. b**

202. Nobel prize to A. Kornberg and S. Ochoa was given for

(a) Artificial synthesis of DNA and RNA

(b) Theory of natural selection

(c) One gene one enzyme theory

(d) Mutation theory **Ans. a**

203. In 1980, F. Sanger was awarded Nobel prize second time to be shared by Gilbert and Moseum for their work on

(a) Genetic mapping of chromosomes

(b) Determining amino acid sequence of insulin

(c) Determining the base sequences of DNA of a virus

(d) Determining the structure of DNA **Ans. c**

204. Operon model for enzyme activity (or the concept of operon) was proposed by

(a) Jacob (b) Monod

(c) Boveri (d) Jacob and Monod **Ans. d**

205. The lac operon is an example of

(a) Arabonise operon (b) Inducible operon

(c) Repressible operon (d) Overlapping gene **Ans. b**

206. In *E. coli*, "Lac" operon is induced by

(a) β-galactosidase (b) Lactose

(c) 'I' gene (d) Promoter gene **Ans. b**

207. Wild type *E. coli* cells are growing in normal medium with glucose. They are transferred to a medium containing lactose only as the sugar. Which one of the following changes take place

(a) The lac operon is repressed (b) All operons are induced

(c) *E. coli* cells stop dividing (d) The lac operon is induced

Ans. d

208. In an *E. coli* cell, according to the operon theory an operator gene combines with

(a) Inducer gene to "switch on" structural gene transcription

(b) Regulator gene "switch off" structural gene transcription

(c) Regulator protein to "switch off" structural gene transcription

(d) Regulator protein to "switch on" structural gene transcription

Ans. c

209. According to operon concept, the regulatory gene regulates biochemical reaction in a cell by

(a) Inhibiting transcription (b) Inactivating enzymes

(c) Inactivating substrate (d) Inhibiting migration of mRNA

Ans. a

210. Rapid transcription of lac operon in *E. coli* requires

(a) Presence of lactose

(b) Presence of glucose

(c) Presence of glucose but absence of lactose

(d) Presence of lactose but absence of glucose **Ans. d**

211. A gene which synthesizes a repressor protein is

(a) Regulator gene (b) Promoter gene

(c) Operator gene (d) Structural gene **Ans. a**

212. In an operon, the RNA polymerase binds to

(a) Regulator (b) Promoter gene

(c) Operator gene (d) Constitutive gene **Ans. b**

213. Lac operon is related to

(a) Synthesis of enzyme of lactose anabolism

(b) Synthesis of enzyme of lactose catabolism

(c) Synthesis of lac by lac insect

(d) Degradation of lac in the body of lac insect **Ans. b**

214. The genes are responsible for growth and differentiation in an organism through regulation of

(a) Translocation (b) Transformation

(c) Transduction and translation (d) Translation and transcription

Ans. d

215. The lac operon requires a 'helper' protein which, by binding to the promoter and by facilitating the attachment of RNA polymerase, accelerates the rate of transcription. The protein is

(a) Amino acid activating enzyme

(b) Essential metabolite

(c) Inactive repression protein

(d) Catabolite activator protein **Ans. d**

216. Which of the following is first recombinant DNA

(a) DNA of one bacteria with another bacteria

(b) DNA of a virus and a bacterium

(c) DNA of bacteria and man

(d) DNA of two viruses **Ans. b**

217. Which of the following is employed in recombinant DNA technology

(a) Plastids (b) Plasmids

(c) Ribosomes (d) Histones **Ans. b**

218. Which one of the following unit is unrelated to DNA or gene

(a) Rishon (b) Recon

(c) Cistron (d) Operon **Ans. a**

219. Genetic engineering would not have been possible if one of these were absent

(a) DNA polymerase (b) Reverse transcriptase

(c) DNA ligase (d) RNA synthetase **Ans. c**

220. Which of the following is associated with genetic engineering

(a) Plastid (b) Plasmids

(c) Mutations (d) Hybrid vigour **Ans. b**

221. In genetic engineering, for cutting DNA segment, which of the enzyme is used

(a) Restriction nuclease (b) Ligase

(c) ATPase (d) DNA polymerase **Ans. a**

222. In *Rous sarcoma* virus, the flow of information is

(a) DNA $\rightarrow$ RNA $\rightarrow$ Protein

(b) DNA $\rightarrow$ Protein $\rightarrow$ RNA

(c) RNA $\rightarrow$ DNA $\rightarrow$ RNA $\rightarrow$ Protein

(d) RNA $\rightarrow$ DNA $\rightarrow$ Protein **Ans. c**

223. What is cDNA

(a) Circular DNA

(b) Cloned DNA

(c) DNA produced from reverse transcription of RNA

(d) Cytoplasmic DNA **Ans. c**

224. Teminism i.e. the synthesis of DNA on RNA template was observed in

(a) TMV (b) Rice dwarf virus

(c) Reovirus (d) *Rous sarcoma* virus **Ans. d**

225. Lac operon in *E. coli* consists of three structural genes. Out of these, one codes for transacetylase. The function of transacetylase is

(a) To carry lactose into the cell

(b) To convert lactose into glucose

(c) To convert lactose into galactose

(d) Not known **Ans. d**

226. In split genes, the coding sequence are called

(a) Cistrons (b) Operons

(c) Exons (d) Introns **Ans. c**

227. Restriction enzyme are used in genetic engineering because

(a) They can cut DNA at specific base sequence

(b) They are proteolytic enzymes which can degrade harmful proteins

(c) They are nucleases that cut DNA at variable sites
(d) They can join different DNA fragments **Ans. a**

228. A piece of DNA, cut by a restriction enzyme, forms bonds with other DNA molecules which have
(a) Been fragmented by the same restriction enzyme
(b) Unpaired bases
(c) Plasmid components
(d) Methyl groups attached to them **Ans. a**

229. It is preferable to use yeasts rather than bacteria as recipient cells for recombination of eukaryotic DNA because
(a) Yeast can produce restriction enzymes
(b) Yeast can excise introns from the RNA transcript
(c) Yeast can remove methyl groups
(d) Yeast can reproduce at a faster rate **Ans. b**

230. A bacterium modifies its DNA by adding methyl groups to the DNA. It does so to
(a) Clone its DNA
(b) Be able to transcribe many genes simultaneously
(c) Turn its gene on
(d) Protect its DNA from its own restriction enzymes **Ans. d**

231. The operon model of gene regulation and organization in prokaryotes was proposed by
(a) Jacob and Monod (b) Beadle and Tatum
(c) Meselson and Stahl (d) Wilkins and Franklin **Ans. a**

232. Which site of a tRNA molecule hydrogen binds to an mRNA molecule?
(a) Codon (b) Anticodon
(c) 5′ end of the tRNA molecule (d) 3′ end of the tRNA molecule
Ans. b

233. An environmental agent that triggers transcription from an operon is a
(a) Derepressor (b) Inducer
(c) Regulator (d) Controlling element **Ans. b**

234. Which of the following is used in getting specific part in genetic engineering

(a) Restriction endonuclease (b) Ligase

(c) Protein (d) Lipid **Ans. a**

235. Which has function similar to UGA, UAA?

(a) UAG (b) AGG

(c) AUG (d) GGG **Ans. a**

236. In negative operon

(a) Corepressor binds with repressor

(b) Corepressor does not bind with repressor

(c) Corepressor binds with inducer

(d) cAMP has negative effect on lac operon **Ans. a**

237. Molecular scissor used in genetic engineering is

(a) DNA ligase (b) Helicase

(c) Restriction endonuclease (d) DNA polymerase **Ans. c**

238. The technique involving insertion of a desired gene into the DNA of plasmid vector known as

(a) Gene dressing (b) Gene cloning

(c) Gene drafting (d) Gene splicing **Ans. b**

239. In operon concept, regulator gene functions as

(a) Repressor (b) Inhibitor

(c) Regulator (d) All the above **Ans. a**

240. The information are carried from gene to cytoplasm by

(a) tRNA (b) mRNA

(c) rRNA (d) None **Ans. b**

241. Thermal cycle is used in

(a) Radioactivity (b) Enzyme-catalyzed reaction

(c) Chemical reaction (d) Polymerase chain reaction

Ans. d

242. Producing a "giant mouse" in the laboratory was possible through

(a) Gene mutation (c) Gene manipulation

(c) Gene synthesis (d) Gene duplication **Ans. b**

243. Genetic engineering means

(a) Meiotic division of cells

(b) Nucleotide transfer

(c) Deletion and repair mechanism

(d) None **Ans. b**

244. DNA finger printing has proved useful in forensic science. It involves the use of

(a) cDNA (b) Bacterial DNA

(c) Minisatellites (d) Ribosomal RNA **Ans. c**

245. Introduction of foreign genes for improving genotype is

(a) Genetic engineering (b) Biotechnology

(c) Tissue culture (d) Immunisation **Ans. a**

❑❑❑

Chapter-22

Agricultural Biochemistry

Fill in the blanks

1. Rice bran contains about..........percent fat.

Ans. 17-19%

2. Pulses contain............percent of protein by weight.

Ans. 17-25%

3. Shelled groundnuts contains...........percent oil content.

Ans. 45-50%

4. Soybean contains...........percent protein and...........percent oil content.

Ans. 40, 20

5. The oils and fats are composed of different mixture of...............

Ans. glycerides

6. The amount of energy released by oil and fats is more than protein and carbohydrate by..............

Ans. 2.25 times

7. The essential fatty acid provided by the oils is.............

Ans. linoleic acid

8. Percentage of carbohydrate in wheat grain is.............

Ans. 69%

9. The temperature at which denaturation of protein in cereals takes place............

Ans. 50^0C

10. Porosity percent in wheat is..........

Ans. 50-55

11. Percentage of bran in wheat milling is........

Ans. 12%

12. Oil percentage in parboiled rice bran is........

Ans. 20-28%

13. Rice bran oil contains saturated fatty acids about...........

Ans. 30-42%

14. Composition of unsaturated fatty acid in rice bran oil is..........

Ans. 80-84%

15. Paddy corn wheat & buck wheat seeds are specially rich source of ....

Ans. carbohydrate

16. Useful storage life of dried fruit and dried seed is upto..................

Ans. 360 days or more

17. The temperature at which denaturation and coagulation of cereals protein take place above..................

Ans. 50^0C

18. The endosperm of cereals grain contains the highest amount of carbohydrates in the form of..................

Ans. starch

19. The protein content in pulse is about......... than cereals.

Ans. double

20. The protein efficiency ratio (PER) of pulses varies................

Ans. 0.7-2.1

21. Oil content in raw bran varies between..............

Ans. 12-18%

22. The percentage of fresh juice in fruit squash is added about..................

Ans. 40-45%

23. The invert sugar in fruit jam should not be more than.............

Ans. 40%

24. Jam contains.........percent acid and invert sugar should not be more than.........percent.

Ans. 0.5-0.6, 40

Multiple Choice Questions

1. Rich source of fatty acid is
   (a) Milk (b) Cereals
   (c) Pulse (d) Dry vegetables **Ans. a**
2. Sugar content in fruit jelly is
   (a) 75% (b) 70%
   (c) 80% (d) 65% **Ans. d**
3. How many carbons are constituted in vegetable oils?
   (a) 12 and 14 (b) 16 and 18
   (c) 18 and 22 (d) 25 and 30 **Ans. b**
4. The ratio of sweet orange and khatta orange in jelly marmalade is subjected to
   (a) 2:1 (b) 1:2
   (c) 1:1 (d) 3:1 **Ans. a**
5. Richest source of vitamin C among the fruits
   (a) Lemon (b) Barbados cherry
   (c) Aonla (d) Guava **Ans. a**
6. The___________fruit is also known as butter fruit
   (a) Mango (b) Banana
   (c) Aonla (d) Papaya **Ans. a**
7. Garlic contain amino acid
   (a) Allicin (b) Alliin
   (c) Allinage (d) Diallyl disulphide **Ans. c**
8. Red colour in tomato is due to pigment
   (a) Lycopene (b) Capsaicin
   (c) Isothiocyanate (d) Chlorophyll **Ans. a**
9. Which fruit is richest source of riboflavin?
   (a) Ber (b) Bael
   (c) Orange (d) Mango **Ans. b**
10. The __________is richest source of iron
    (a) Green pea (b) Spinach
    (c) Parsley (d) Celery **Ans. b**

11. Papaya is rich source of vitamin
(a) C (b) B
(c) A (d) K **Ans. c**

12. Sweetest sugar in fruit is
(a) Fructose (b) Glucose
(c) Sucrose (d) Galactose **Ans. a**

13. Which one of the following fruits is a rich source of fat
(a) Cashew nut (b) Walnut
(c) Mango (d) Guava **Ans. b**

14. Rich source of protein is
(a) Banana (b) Cashew nut
(c) Almond (d) Apple **Ans. c**

15. Which of the following is a non-reducing carbohydrate?
(a) Maltose (b) Sucrose
(c) Lactose (d) Glucose **Ans. c**

16. Fat soluble vitamins are
(a) A, D, E and K (b) B and C
(c) A, D and C (d) A and D **Ans. a**

17. The avoid bruising, the onion should not be stored than
(a) 14 m (b) 24 m
(c) 4 m (d) 6 m **Ans. b**

18. Aroma in coffee is created due to
(a) Fermentation (b) Roasting
(c) Drying (d) Cold treatment **Ans. b**

19. In litchi, which protein is edible?
(a) Endocarp (b) Epicarp
(c) Rind (d) Mesocarp **Ans. d**

20. Chilies are good source of
(a) Protein (b) Fat
(c) Carbohydrates (d) Vitamin A and C **Ans. d**

21. Guava is a rich source of
(a) Carbohydrate (b) Vitamin C
(c) Vitamin A (d) β-carotene **Ans. b**

22. Dry fruits like cashew and almond are richest source of
(a) Vitamin B_{12} (b) Carbohydrate
(c) Vitamin D (d) Fat **Ans. d**

23. Ripe papaya fruits contain
(a) Vitamin D (b) Vitamin A
(c) Vitamin C (d) Vitamin B **Ans. c**

24. Papaya contain __________unit of vitamin A
(a) 2500 IU (b) 2020 IU
(c) 3000 IU (d) 3500 IU **Ans. b**

25. Potato is the rich source of
(a) Starch (b) Vitamin
(c) Fat (d) Protein **Ans. a**

26. Potato contain __________percentage of protein
(a) 1.8 (b) 1.4
(c) 2.6 (d) 3.0 **Ans. b**

27. Ripening fruits exist which gas is
(a) NH_3 (b) Ethylene
(c) Methane (d) Nitrogen oxide **Ans. b**

28. Phosphoric acid is added in
(a) Jam (b) Juice
(c) Soft drink (d) Jelly **Ans. c**

29. Enzymatic browning in apple is due to
(a) Oxidase (b) Protease
(c) Catalase (d) Poly phenol oxidase **Ans. d**

30. Monosaccharides is
(a) Triose (b) Hexose
(c) Heptose (d) All of above **Ans. d**

31. Disaccharides is
(a) Sucrose (b) Lactose
(c) Maltose (d) All of above **Ans. d**

32. Starch is
(a) Monosaccharide carbohydrate
(b) Disaccharide carbohydrate

(c) Tri saccharide carbohydrate

(d) Poly saccharide carbohydrate **Ans. d**

33. Amylase % in the starch is?

(a) 10-12% (b) 20-30%

(c) 40-50% (d) 60-65% **Ans. b**

34. Which is fat soluble vitamin

(a) Vitamin A (b) Vitamin D

(c) Vitamin E (d) All of above **Ans. d**

35. Choose the correct answer

(a) Vitamin A - Retinol (b) Vitamin B_1 - Thiamin

(c) Vitamin B_2 - Riboflavin (d) Vitamin C - Biotin **Ans. a**

36. Which is tri saccharide carbohydrate?

(a) Stachyose (b) Raffinose

(c) Docase (d) Glucose **Ans. b**

37. Which amino acid is essential for human being

(a) Isoleucine (b) Leucine

(c) Lysine (d) All of above **Ans. d**

38. Which of the following vitamin is not found in vegetables

(a) Vitamin C (b) Vitamin A

(c) Vitamin E (d) Vitamin D **Ans. d**

39. Which vitamin is water soluble

(a) Vitamin B (b) Vitamin C

(c) Both (d) None of above **Ans. d**

40. Which substances is essential for certain animals species

(a) Choline (b) Myo-inositol

(c) Both (d) None of above **Ans. c**

41. As per ICMR (1981) net calories are recommended for a man per day for moderate work

(a) 2400 (b) 2800

(c) 3900 (d) 4500 **Ans. d**

42. Quantity of protein/day required for a man

(a) 35 gram (b) 45 gram

(c) 55 gram (d) 75 gram **Ans. c**

43. Requirement of ascorbic acid per capita per day is

(a) 20 mg (b) 30 mg
(c) 40 mg (d) 60 mg **Ans. c**

44. 0.45 kg of body fat contains a potential energy of approximately

(a) 2500 kcal (b) 3500 kcal
(c) 4500 kcal (d) 5500 kcal **Ans. b**

45. The metabolism oxidation of carbohydrate, fat and protein molecules release energy in the form of

(a) ADP (b) ATP
(c) Both (d) None of above **Ans. b**

46. The transduction of food energy to mechanical waste take place at a maximal efficiency of about

(a) 15% (b) 25%
(c) 35% (d) 45% **Ans. b**

47. Match the following

	Sources		Energy kcal through human
1.	Protein	I.	7.1 oxdation
2.	Fat	II.	4.1
3.	Carbohydrate	III.	9.3
4.	Ethanol	IV.	4.1

	1	2	3	4		1	2	3	4
(a)	IV	II	III	I	(b)	IV	III	II	I
(c)	III	I	II	IV	(d)	IV	II	III	I

Ans. b

48. Which is essential fatty acid

(a) Linoleic acid (b) Linolenic acid
(c) Both (b) None of these **Ans. c**

49. Match the following

1.	Thiamin	I.	Vitamin B_2
2.	Pyridoxal	II.	Vitamin B_{12}
3.	Cobalamin	III.	Vitamin B_6
4.	Riboflavin	IV.	Vitamin B_1

	1	2	3	4		1	2	3	4
(a)	I	II	III	IV	(b)	IV	III	I	II
(c)	IV	III	II	I	(d)	III	IV	II	I

Ans. c

50. Body building foods are rich in
(a) Sucrose (b) Protein
(c) Minerals (d) Fat **Ans. b**

51. Which of following compounds used for promote the sucrose accumulation
(a) CEPA (b) Glyphosome
(c) Glyphosate (d) All of these **Ans. d**

52. Which is the ideal transport form of carbohydrate
(a) Glucose (b) Cellulose
(c) Sucrose (d) Mannose **Ans. c**

53. Which of the following is alkaloid
(a) Caffeine (b) Nicotine
(c) Strychnine (d) All of these **Ans. d**

54. Monosaccharides in linked with each other by
(a) Glycosidic bonds (b) Glycopyranoside bonds
(c) Phosphopyrine bond (d) None of these **Ans. a**

55. Which are hetero poly saccharide
(a) Cellulose (b) Galactomannans
(c) Chitin (d) None of these **Ans. b**

56. Which are the homo polysaccharide
(a) Cellulose (b) Polymers
(c) Starch (d) All of these **Ans. d**

57. Which are glycosides
(a) Linamarin (b) Dhurrin
(c) Prunasin (d) All of these **Ans. d**

58. Major use of cytokinins is
(a) Cell division
(b) Cell elongation of plant part
(c) Retaral senescence in harvested plant parts
(d) Increase droughtness of plant **Ans. c**

59. Fat and oils are hydrolyzed by lipases to
(a) Fatty acids (b) Glycerol
(c) Both (d) None of these **Ans. c**

60. Citric acid is changed to aconitic acid by the enzyme
(a) Oxidase (b) Lipase
(c) Aconitose (d) Triose **Ans. c**

61. Which of the following element is not essential but it is beneficial to a number of plants
(a) Zinc (b) Mangnese
(c) Sodium (d) Copper **Ans. c**

62. Which of the following is micro nutrient
(a) Copper (b) Magnesium
(c) Sulphur (d) Iron **Ans. a**

63. Abnormal development of terminal buds is deficiency of
(a) Zinc (b) Boron
(c) Molybdenum (d) Mangnese **Ans. b**

64. Abscisic acid causes
(a) Bud dormancy
(b) Senescence
(c) Abscission of leaves and flower
(d) All of these **Ans. d**

65. Which of the following is growth retardant
(a) Daminozide (b) Maleic hydrazide
(c) Both (d) None of these **Ans. c**

66. To determine the total nitrogen present in the feed and multiply this amount by the factor
(a) 4.15 (b) 9.15
(c) 6.25 (d) 6.50 **Ans. c**

67. Which of the following is mercurial fungicide
(a) Bordeaux mixture (b) Ceresan
(c) Kethane (d) None of these **Ans. b**

68. Which is organic compound
(a) Parathion (b) Chlordane
(c) Both (d) None of these **Ans. c**

69. Main carrier of genetic information is
(a) Deoxyribose (b) Deoxyribo nucleic acid
(c) Ribo nucleic acid (d) All of these **Ans. b**

70. Soil solution containing optimal concentration of nutrients called as
(a) Normal solution (b) Eutrophic
(c) Buffer (d) None of these **Ans. b**

71. Simple sugar also known as
(a) Manose (b) Dextrose
(c) Sucrose (d) All of these **Ans. c**

72. $C_6H_{12}O_6 + 6O_2 \rightarrow 6\ CO_2 + 6H_2O$ = 673 cal energy the reaction indicators
(a) Photosynthesis (b) Aerobic respiration
(c) Anaerobic respiration (d) Photo respiration **Ans. b**

73. The ratio between CO_2 produced and O_2 consumed indicates
(a) Photosynthesis ratio (b) Respiration ratio
(c) Energy ratio (d) Respiration quitent **Ans. d**

74. 'Gossipol' is a name of
(a) Chemical (b) Pigment
(c) Alkaloid (d) None of these **Ans. b**

75. Hemicelluloses contain
(a) Araban (b) Xylan
(c) Polyuronides (d) All of these **Ans. d**

76. Protein name proposed by
(a) Mulder (b) Stainley
(c) Robert brown (d) None of these **Ans. a**

77. Nitrogen % in protein is
(a) 7-8% (b) 10-12%
(c) 15-18% (d) 20-22% **Ans. c**

78. Hydrogen % in protein is
(a) 7-8% (b) 10-12%
(c) 15-18% (d) 22-24% **Ans. a**

79. Protein are made of
(a) Carbon (b) Hydrogen
(c) Nitrogen (d) All of these **Ans. d**

80. Nature of nucleo protein is

(a) Acidic (b) Alkaline

(c) Both (d) None of these **Ans. a**

81. Amino acids are linked together by

(a) Peptone (b) Peptide

(c) Peptides (d) Proteases **Ans. b**

82. Mono carboxylic amino acid is

(a) Glycine (b) Alanine

(c) Valine (d) All of these **Ans. d**

83. Which of the following is dicarboxylic amino acid

(a) Glutanic acid (b) Aspartic acid

(c) Both (d) None of these **Ans. c**

84. Which of the following is aromatic amino acid

(a) Phenylalanine (b) Tyrosine

(c) Tryptophan (d) All of these **Ans. d**

85. Which of the following is polar amino acid

(a) Glycine (b) Tyrosine

(c) Glutamine (d) All of these **Ans. d**

86. Which of the following is acidic amino acid

(a) Glutanic acid (b) Aspartic acid

(c) Both (d) None of these **Ans. c**

87. Which amino acid is not essential for adults but it is essential for young infant

(a) Arginine (b) Histidine

(c) Both (d) None of these **Ans. c**

88. Match the following

Product	lysine amino acid content
1. Wheat	I. 0.40
2. Milk	II. 0.38
3. Egg	III. 0.47
4. Soybean	IV. 0.14

	1	2	3	4		1	2	3	4
(a)	IV	I	III	II	(b)	IV	III	I	II
(c)	I	II	III	IV	(d)	IV	III	II	I

Ans. b

89. Lipids are soluble in

(a) Chloroform (b) Ether

(c) Benzene (d) All of these **Ans. d**

90. Simple lipids contain

(a) Carbon (b) Hydrogen

(c) Oxygen (d) All of these **Ans. d**

91. Compound lipids contain

(a) Nitrogen (b) Phorphorus

(c) C,H,O (d) All of these **Ans. d**

92. Ester of the fatty acid and glycerol called

(a) Fats (b) Glycerides

(c) Both (d) None of both **Ans. c**

93. Match the following

	Fatty acids		Carbon atm. No.
1.	Butyric acid	I.	12
2.	Caprylic acid	II.	10
3.	Capric acid	III.	8
4.	Lauric acid	IV.	4

	1	2	3	4		1	2	3	4
(a)	IV	III	I	II	(b)	IV	III	II	I
(c)	I	II	III	IV	(d)	I	III	II	IV

Ans. b

94. Formula of coconut oil is

(a) $CH_3(CH)_4COOH$ (b) $CH_3(CH_2)_2COOH$

(c) $CH_3(CH_2)_{12}COOH$ (d) None of these **Ans. a**

95. Which of the following is non polar amino acid

(a) Proline (b) Lysine

(c) Arginine (d) Trysine **Ans. a**

96. Which of the following is saturated fatty acid

(a) Butyric acid (b) Caproic acid

(c) Capric acid (d) All of these **Ans. d**

97. Which of the following is unsaturated fatty acid

(a) Ethanoides (b) Stearic acid

(c) Palmitic acid (d) Arachidonic acid **Ans. a**

98. Which is monosaturated fatty acid

(a) Linoleic acid (b) Linolic acid

(c) Oleic acid (d) All of these **Ans. c**

99. Which of the following is molecular formula

(a) $C_{18}H_{30}O_2$ (b) $C_{18}H_{34}O_2$

(c) $C_{18}H24O_2$ (d) $C_{16}H_{34}O_2$ **Ans. b**

100. Match the following

Fatty acid		Name	
1.	Mono unsaturated	I.	Linoleic acid
2.	Diunsaturated	II.	Oleic acid
3.	Triunsaturated	III.	Recinalic acid
4.	Hydroxyl fatty acid	IV.	Linolenic acid

	1	2	3	4		1	2	3	4
(a)	II	I	III	IV	(b)	II	I	IV	III
(c)	I	II	III	IV	(d)	II	III	I	IV

Ans. b

101. Glutelins protein is soluble in

(a) Acids (b) Alkalines

(c) Both (d) None of above **Ans. c**

102. Which of the following type of protein are found in potato?

(a) Legumin (b) Tuperin

(c) Prolamines (d) Histone **Ans. b**

103. Which of the following principal centre of nitrate reduction and protein synthesis in plant is

(a) Green leaves (b) Finer roots

(c) Both (d) None of these **Ans. c**

104. Which is the main site of amino acid synthesis in cells

(a) Mitochondria (b) Chloroplasts

(c) Both (d) None of these **Ans. c**

105. Which statement is correct

(a) Hormone is inorganic substances other than nutrients

(b) Site of synthesis of a hormone is same from the site of its action

(c) Vitamins are supposed to be used by the same plant part which produces it without the necessity of being transported

(d) Hormones act as cofactors of enzymes **Ans. c**

106. Where does the National Institute of Nutrition located in India
(a) Mumbai (b) Hyderabad
(c) New Delhi (d) Calcutta **Ans. b**

107. Pulses are rich in
(a) Calories (b) Proteins
(c) Minerals (d) Vitamins **Ans. b**

108. Nitrogen is an essential structural component of
(a) Amino acids (b) Nucleotides
(c) Porphyrins (d) All of these **Ans. d**

109. Match the following

Protein	Solubility
1. Albumin	I. Soluble in alkaline acid
2. Globulin	II. Alcohol soluble
3. Prolamine	III. Salt soluble
4. Glutelin	IV. Water soluble

	1	2	3	4
(a)	I	II	III	IV
(b)	II	III	IV	I
(c)	IV	III	II	I
(d)	IV	II	III	I

Ans. c

110. Protein % of dry weight in rice is
(a) 6-9 (b) 8-10
(c) 10-13 (d) 14-18 **Ans. b**

111. Match the following

Cereal grains	Proteins % of dry weight
1. Wheat	I. 10-15
2. Maize	II. 10-16
3. Barley	III. 7-13
4. Sorghum	IV. 9-13

	1	2	3	4
(a)	I	III	II	IV
(b)	II	III	I	IV
(c)	III	II	I	IV
(d)	IV	I	III	II

Ans. a

112. Ideal protein requirement per kg body weight per day for adults is
(a) 0.5 gm (b) 1.0 gm
(c) 1.5 gm (d) 2.0 gm **Ans. a**

113. Protein efficiency ratio in rice is
(a) 1.53 (b) 2.10
(c) 2.32 (d) 2.77 **Ans. b**

114. Cyanide content mg/100g in pea is
(a) 0.5 (b) 0.8
(c) 2.3 (d) 14.4 **Ans. c**

115. Deficiency of nicotinic acid (Niocin) in the diet leads to deficiency disease is
(a) Beri-beri (b) Pellagra
(c) Ekxanthima (d) None of these **Ans. b**

116. Waxes are
(a) Esters of fatty acids
(b) Long chain aliphatic monohydric alcohols
(c) Have protective coatings
(d) All of these **Ans. d**

117. Oil % in coconut is
(a) 35% (b) 48%
(c) 50% (d) 63% **Ans. d**

118. Auxins were discovered to be chemical substances and hormone by
(a) W.A. Allard (b) F.W. Went
(c) W.B. Watson (d) Mullar **Ans. b**

119. Auxins were discovered in the year of
(a) 1902 (b) 1928
(c) 1936 (d) 1945 **Ans. b**

120. Gibberllins can cause induction of
(a) Seed germination (b) Cell elongation
(c) Control of sex expression (d) All of these **Ans. d**

121. Ethylene play role in
(a) Promoting swelling in the subapical regions
(b) Hastens abscission of leaves
(c) Ripening of fruits
(d) All of these **Ans. d**

122. Cytokinins play a role in plant except one
(a) Regulate seed germination (b) Regulate bud differentiation
(c) Regulate cell enlargement (d) Regulate sex expression
Ans. d

123. Abscisic acid play a role in

(a) Water stress (b) Nucleic acid synthesis
(c) Synthesis of enzymes (d) All of these **Ans. d**

124. Starch are synthesized in

(a) Plastids (b) Mitochondria
(c) Cell wall (d) None of these **Ans. a**

125. The denaturation of proteins by certain agents is due to

(a) Loss in solubility
(b) Loss in enzyme activity
(c) Cleavage of peptide bond
(d) Change in protein conformation only **Ans. d**

126. Isozymes is the name given to

(a) A group of enzymes present at the same location on cell
(b) Different proteins that catalyse the same reaction
(c) Two or more different molecular forms of an enzymes catalyzing the same reaction
(d) Enzymes that have the same optimum pH for their reactions **Ans. c**

127. T_m denotes

(a) Thymine present in m-RNA
(b) The temperature at which DNA is totally melted
(c) Temperature at which DNA is 50% denaturated
(d) Maximum temperature of an enzyme reaction **Ans. c**

128. A peptide isolated from a brain tissue has sequence Glu-His-Trp-Ser-Tyr-Gly-Leu-Arg-Pro-Gly. What is the net charge on this molecule at pH-3.

(a) +2 (b) +3
(c) -2 (d) -3 **Ans. b**

129. If 0.01 M sodium malonate is added to actively respiring isolated mitochondria using pyruvate as a fuel source, respiration soon stops because

(a) Succinate is accumulated due to enzyme inhibition
(b) Conversion to malate does not take place
(c) NADH formation stops due to block in citric acid cycle
(d) Malonte acts as an uncoupler **Ans. a**

130. A child was observed to have signs of night blindness, which could be due to

(a) Very low level (insufficient) of serum vitamin A

(b) Non-consumption of carotenoid containing diet

(c) Failure in the conversion of beta-carotenes of food to vitamin A in the intestine

(d) Inability of liver to store vitamin A **Ans. a**

131. The relatively high amounts potassium required for normal plant growth do not correlate with the observed functions of potassium. The element seems to function mostly as a catalytic agent in enzyme reactions. High amount requirement could be therefore, to

(a) Increase translocation of sugars

(b) Act as an activator for enzyme system by maintaining favourable ionic environment required for optimal activity

(c) To help in stomatal opening

(d) To regulate water in plant cells under stress **Ans. b**

132. During senescence and ripening, an increased concentration of ethylene precursors would be made available by protein degradation. These could be

(a) Methional (b) 2-Keto-4-methyl thiobutyrate

(c) Methionine (d) Methane sulfinic acid **Ans. c**

133. Embryo seed dormancy is mainly found in

(a) Seed of tropical climate (b) Seed of subtropical climate

(c) Seed of temperate climate (d) In all climate regions **Ans. c**

134. The glucose and fructose content in the grains during storage are

(a) Increases (b) Decreases

(c) Constant (d) None of these **Ans. a**

135. Dark reaction of photosynthesis occurs in

(a) Granum (b) Stroma

(c) In both (d) None of these **Ans. b**

136. Which of the following substance controls the cell elongation, apical dominance, root initiation etc

(a) Auxins (b) Gibberellins

(c) Ethylene (d) Abscisic acid **Ans. a**

137. How much portion of plant tissue are made up of carbon, hydrogen and oxygen

(a) 70-75% (b) 80-85%

(c) 95-99% (d) 100% **Ans. c**

138. A nitrogen bacterium that forms a loose association with roots of crop plants is

(a) *Azospirillum* (b) *Azotobacter*

(c) *Clostridium* (d) *Bacilus polymyxa* **Ans. a**

139. Which weed used has been eradicated by bioherbicides

(a) *Cactus* (b) *Eichhomia*

(c) *Parthenium* (d) *Chrysamthemum* **Ans. a**

140. Which element plays an important role in nitrogen fixation ?

(a) Mn (b) Mo

(c) Zn (d) Cu **Ans. b**

141. Mycorrhiza represents

(a) Antagonism (b) Endemism

(c) Symbiosis (d) Parasitism **Ans. c**

142. Nitrogen fixation is

(a) Nitrogen $\rightarrow$ Ammonia (b) Nitrogen $\rightarrow$ Amino Acid

(c) Nitrogen $\rightarrow$ Nitrates (d) Both (a) and (b) **Ans. d**

143. Leghaemaoglobin occurs in

(a) Coralloid Root (b) BGA

(c) Around bacteroids (d) Mycorrhiza **Ans. c**

144. Cochineal insects have been employed in checking over growth of

(a) Eichhornia (b) Cacti

(c) Weeds (d) Insects **Ans. b**

145. The free living soil bacteria which fixes nitrogen

(a) Azotobacter (b) Nitrosomonas

(c) Nitrobacter (d) Pseudomonas **Ans. a**

❑❑❑

Chapter-23

Vitamins

Fill in the blanks

1. Deficiency of vitamin A causes........................

Ans. xeropthalmia

2. Important function of vitamin A is....................

Ans. maintain the integrity of epithelial tissue

3. Deficiency of vitamin D causes..................and ....

Ans. Ricket, Osteomalacia

4. The requirement of vitamin E is increased with greater intake of ....

Ans. polyunsaturated fat

5. Tocopherols prevent the oxidation of.................

Ans. vitamin A

6. Creatinuria is caused due to the deficiency of vitamin..................

Ans. E

7. Deficiency of vitamin B_6 may occur in................

Ans. alcoholics

8. Dietary requirement of vitamin D is....................

Ans. 400 IU

9. Vitamin B_{12} has a complex ring structure (corrin ring) consisting of four...............

Ans. pyrrole rings

10. A deficiency of vitamin B_{12} causes......................

Ans. Pernicious anemia

11. Folate deficiency causes.....................

Ans. Megaloblastic anaemia

12. Niacin contains a.........................group.

Ans. carboxyl

13. Daily requirement of thiamine is..............

Ans. 0.5 mg/1000 calories

14. Chemically lipoic acid is..................... fatty acid.

Ans. sulphur containing

15. Intrinsic factor is chemically a...................

Ans. glycoprotein

16. Deficiency of vitamin C causes.................

Ans. scurvy

17. Vitamin C is required for the synthesis of................. from ....

Ans. bile acid, cholesterol

18. Vitamin B_{12} is stored in....................

Ans. liver

19. One international unit of vitamin A is the activity present in ....

Ans. 0.3 μg of retinol

20. Retinoic acid is involved in the synthesis of..................

Ans. glycoproteins

21. Presently vitamin B_{12} is produced directly during the course of fermentation by bacteria called ....

Ans. Propionibacteria

22. Riboflavin is known is Vitamin.....................

Ans. B_2

23. Some vitamins like B_2, B_{12} are products of ....

Ans. Micro-organisms.

24. Cobalamine is known as Vitamin ....

Ans. B_2

25. Vitamin D deficiency is children causes ....

Ans. Rickets

Multiple Choice Questions

1. Vitamins are
   (a) Accessory food factors
   (b) Generally synthesized in the body
   (c) Produced in endocrine glands
   (d) Proteins in nature **Ans. a**
2. Vitamin D is derived from which of the following precursors by the action of UV light?
   (a) 7-dehydrocholesterol (b) Lanosterol
   (c) Glycocholate (d) Squalene epoxide **Ans. a**
3. Vitamin A or retinal is a
   (a) Steroid
   (b) Polyisoprenoid compound containing a cyclohexenyl ring
   (c) Benzoquinone derivative
   (d) 6-Hydroxychromane **Ans. b**
4. β-Carotene, precursor of vitamin A, is oxidatively cleaved by
   (a) β-Carotene dioxygenase (b) Oxygenase
   (c) Hydroxylase (d) Transferase **Ans. a**
5. Retinal is reduced to retinol in intestinal mucosa by a specific retinaldehyde reductase utilising
   (a) $NADPH + H^+$ (b) FAD
   (c) NAD (d) $NADH + H^+$ **Ans. a**
6. Preformed Vitamin A is supplied by
   (a) Milk, fat and liver (b) All yellow vegetables
   (c) All yellow fruits (d) Leafy green vegetables **Ans. a**
7. Retinol and retinal are interconverted requiring dehydrogenase or reductase in the presence of
   (a) NAD or NADP (b) $NADH + H^+$
   (c) NADPH (d) FAD **Ans. a**
8. Fat soluble vitamins are
   (a) Soluble in alcohol (b) One or more Propene units
   (c) Stored in liver (d) All of these **Ans. d**

9. Most vitamins are
(a) Covalently linked to enzymes
(b) Synthesized by humans
(c) Precursors to coenzyme
(d) Precursors to proteins **Ans. c**

10. The international unit of vitamin A is equivalent to the activity caused by
(a) 0.3 μg of Vitamin A alcohol
(b) 0.344 μg of Vitamin A alcohol
(c) 0.6 μg of Vitamin A alcohol
(d) 1.0 μg of Vitamin A alcohol **Ans. a**

11. Lumirhodopsin is stable only at temperature below
(a) -10°C (b) -20°C
(c) -40°C (d) -50°C **Ans. d**

12. Retinol is transported in blood bound to
(a) Aporetinol binding protein (b) α_2-Globulin
(c) β-Globulin (d) Albumin **Ans. a**

13. The normal serum concentration of vitamin A in mg/100 ml is
(a) 5-10 (b) 15-60
(c) 100-150 (d) 0-5 **Ans. b**

14. Which vitamin is necessary for coenzyme A synthesis
(a) Ascorbic acid (b) Pyridoxine
(c) Biotin (d) Pantothenic acid **Ans. d**

15. One manifestation of vitamin A deficiency is
(a) Painful joints (b) Night blindness
(c) Loss of hair (d) Thickening of long bones
Ans. b

16. Deficiency of Vitamin A causes
(a) Xeropthalmia (b) Hypoprothrombinemia
(c) Megaloblastic anemia (d) Pernicious anemia **Ans. a**

17. An important function of vitamin A is
(a) To act as coenzyme for a few enzymes
(b) To play an integral role in protein synthesis

(c) To prevent hemorrhages
(d) To maintain the integrity of epithelial tissue **Ans. d**

18. Retinal is a component of
(a) Iodopsin (b) Rhodopsin
(c) Cardiolipin (d) Glycoproteins **Ans. b**

19. Retinoic acid participates in the synthesis of
(a) Iodopsin (b) Rhodopsin
(c) Glycoprotein (d) Cardiolipin **Ans. c**

20. On exposure to light rhodopsin forms
(a) All trans-retinal (b) Cis-retinal
(c) Retinol (d) Retinoic acid **Ans. a**

21. Carr-Price reaction is used to detect
(a) Vitamin A (b) Vitamin D
(c) Ascorbic acid (d) Vitamin E **Ans. a**

22. The structure shown below is of

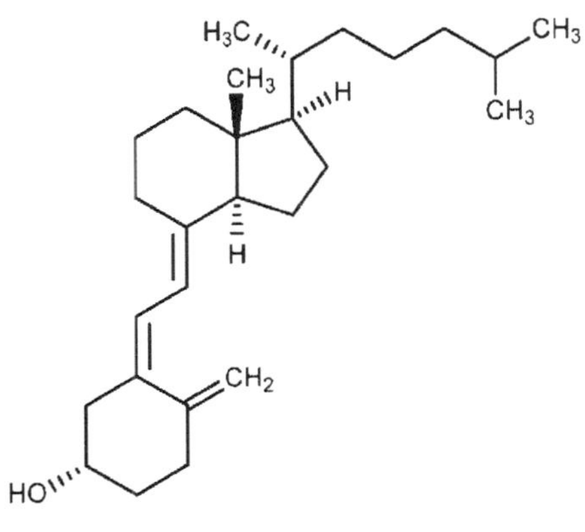

(a) Cholecalciferol (b) 25-Hydroxycholecalciferol
(c) Ergocalciferol (d) 7-Dehydrocholesterol **Ans. a**

23. Vitamin D absorption is increased in
(a) Acid pH of intestine (b) Alkaline pH of intestine
(c) Impaired fat absorption (d) Contents of diet **Ans. a**

24. The most potent Vitamin D metabolite is
(a) 25-Hydroxycholecalciferol
(b) 1, 25-Dihydroxycholecalciferol

(c) 24, 25-Dihydroxycholecalciferol
(d) 7-Dehydrocholesterol **Ans. b**

25. Which of the following coenzymes does not contain adenine
(a) Thiamine pyrophosphate (b) S- adenosylmethionine
(c) Coenzyme A (d) NADH **Ans. a**

26. The normal serum concentration of 25-hydroxycholecalciferol in ng/ml is
(a) 0-8 (b) 60-100
(c) 100-150 (d) 8-55 **Ans. d**

27. The normal serum concentration of 1, 25-dihydroxycholecalciferol in pg/ml is
(a) 26-65 (b) 1-5
(c) 5-20 (d) 80-100 **Ans. a**

28. The normal serum concentration of 24,25-dihydroxycholecalciferol in ng/ml is
(a) 8-20 (b) 25-50
(c) 1-5 (d) 60-100 **Ans. c**

29. A poor source of Vitamin D is
(a) Egg (b) Butter
(c) Milk (d) Liver **Ans. c**

30. Richest source of Vitamin D is
(a) Fish liver oils (b) Margarine
(c) Egg yolk (d) Butter **Ans. a**

31. Deficiency of vitamin D causes
(a) Ricket and osteomalacia (b) Tuberculosis of bone
(c) Hypothyroidism (d) Skin cancer **Ans. a**

32. One international unit (I.U) of vitamin D is defined as the biological activity of
(a) 0.025 μg of cholecalciferol
(b) 0.025 μg of 7-dehydrocholecalciferol
(c) 0.025 μg of ergosterol
(d) 0.025 μg of ergocalciferol **Ans. a**

33. Which of the following answers complete the sentence correctly? During the reduction of FAD

(a) A flavin group is transferred

(b) An equivalent of a hydride ion is transferred

(c) The isoalloxazine ring becomes charged

(d) Two hydrogen atoms are added to the isoalloxazine ring **Ans. d**

34. The β-ring of 7-dehydrocholesterol is cleaved to form cholecalciferol by

(a) Infrared light

(b) Dim light

(c) Ultraviolet irradiation with sunlight

(d) Light of the tube lights **Ans. c**

35. Calcitriol synthesis involves

(a) Both liver and kidney (b) Intestine

(c) Adipose tissue (d) Muscle **Ans. a**

36. Insignificant amount of Vitamin E is present in

(a) Wheat germ oil (b) Sunflower seed oil

(c) Safflower seed oil (d) Fish liver oil **Ans. d**

37. The activity of tocopherols is destroyed by

(a) Commercial cooking (b) Reduction

(c) Conjugation (d) All of these **Ans. a**

38. The requirement of vitamin E is increased with greater intake of

(a) Carbohydrates (b) Proteins

(c) Polyunsaturated fat (d) Saturated fat **Ans. c**

39. Vitamin E reduces the requirement of

(a) Iron (b) Zinc

(c) Selenium (d) Magnesium **Ans. c**

40. The most important natural antioxidant is

(a) Vitamin D (b) Vitamin E

(c) Vitamin B_{12} (d) Vitamin K **Ans. b**

41. Tocopherols prevent the oxidation of

(a) Vitamin A (b) Vitamin D

(c) Vitamin K (d) Vitamin C **Ans. a**

42. Creatinuria is caused due to the deficiency of vitamin
(a) A (b) K
(c) E (d) D **Ans. c**

43. All the following conditions produce a real or functional deficiency of vitamin K except
(a) Prolonged oral, broad spectrum antibiotic therapy
(b) Total lack of red meat in the diet
(c) The total lack of green leafy vegetables in the diet
(d) Being a new born infant **Ans. b**

44. Vitamin K is found in
(a) Green leafy plants (b) Meat
(c) Fish (d) Milk **Ans. a**

45. Function of Vitamin A
(a) Healing epithelial tissues (b) Protein synthesis regulation
(c) Cell growth (d) All of these **Ans. d**

46. Vitamin K_2 was originally isolated from
(a) Soyabean (b) Wheat gram
(c) *Alfa Alfa* (d) Putrid fish meal **Ans. d**

47. Vitamin synthesized by bacterial in the intestine is
(a) A (b) C
(c) D (d) K **Ans. d**

48. Vitamin K is involved in posttranslational modification of the blood clotting factors by acting as cofactor for the enzyme
(a) Carboxylase (b) Decarboxylase
(c) Hydroxylase (d) Oxidase **Ans. a**

49. Vitamin K is a cofactor for
(a) Gamma carboxylation of glutamic acid residue
(b) β-Oxidation of fatty acid
(c) Formation of γ-amino butyrate
(d) Synthesis of tryptophan **Ans. a**

50. Hypervitaminosis K in neonates may cause
(a) Porphyria (b) Jaundice
(c) Pellagra (d) Prolonged bleeding **Ans. b**

51. Dicoumarol is antagonist to

(a) Riboflavin (b) Retinol

(c) Menadione (d) Tocopherol **Ans. c**

52. In the individuals who are given liberal quantities of vitamin C, the serum ascorbic acid level is

(a) 1-1.4 μg/100 ml (b) 2-4 μg/100 ml

(c) 1-10 μg/100 ml (d) 10-20 μg/100 ml **Ans. a**

53. The vitamin which would most likely become deficient in an individual who develop a completely carnivorous life style is

(a) Thiamin (b) Niacin

(c) Vitamin C (d) Cobalamin **Ans. c**

54. In human body highest concentration of ascorbic acid is found in

(a) Liver (b) Adrenal cortex

(c) Adrenal medulla (d) Spleen **Ans. b**

55. The vitamin required for the formation of hydroxyproline (in collagen) is

(a) Vitamin C (b) Vitamin A

(c) Vitamin D (d) Vitamin E **Ans. a**

56. Which of the following is not a dietary antioxidant

(a) Vitamin C (b) Lipoic acid

(c) Vitamin K (d) Beta-carotene **Ans. c**

57. Vitamin required for the conversion of p-hydroxyphenylpyruvate to homogentisate is

(a) Folacin (b) Cobalamin

(c) Ascorbic acid (d) Niacin **Ans. d**

58. Vitamin required in conversion of folic acid to folinic acid is

(a) Biotin (b) Cobalamin

(c) Ascorbic acid (d) Niacin **Ans. c**

59. Ascorbic acid can reduce

(a) 2, 6-Dibromobenzene

(b) 2, 6-Diiodoxypyridine

(c) 2, 6-Dichlorophenol indophenol

(d) 2, 4-Dinitrobenzene **Ans. c**

60. Sterilized milk lacks in

(a) Vitamin A (b) Vitamin D

(c) Vitamin C (d) Thiamin **Ans. c**

61. Scurvy is caused due to the deficiency of

(a) Vitamin A (b) Vitamin D

(c) Vitamin K (d) Vitamin C **Ans. d**

62. Both Wernicke's disease and beriberi can be reversed by administrating

(a) Retinol (b) Thiamin

(c) Pyridoxine (d) Vitamin B_{12} **Ans. b**

63. The Vitamin B_1 deficiency causes

(a) Ricket (b) Nyctalopia

(c) Beriberi (d) Pellagra **Ans. c**

64. Concentration of pyruvic acid and lactic acid in blood is increased due to deficiency of the vitamin

(a) Thiamin (b) Riboflavin

(c) Niacin (d) Pantothenic acid **Ans. a**

65. Vitamin B_1 coenzyme (TPP) is involved in

(a) Oxidative decarboxylation (b) Hydroxylation

(c) Transamination (d) Carboxylation **Ans. a**

66. Increased glucose consumption increases the dietary requirement for

(a) Pyridoxine (b) Niacin

(c) Biotin (d) Thiamin **Ans. d**

67. Thiamin is oxidized to thiochrome in alkaline solution by

(a) Potassium permanganate (b) Potassium ferricyanide

(c) Potassium chlorate (d) Potassium dichromate **Ans. b**

68. Riboflavin is a coenzyme in the reaction catalyzed by the enzyme

(a) Acyl CoA synthetase (b) Acyl CoA dehydrogenase

(c) β-Hydroxy acyl CoA (d) Enoyl CoA dehydrogenase

Ans. b

69. The daily requirement of riboflavin for adult in mg is

(a) 0-1.0 (b) 1.2-1.7

(c) 2.0-3.5 (d) 4.0-8.0 **Ans. b**

70. In new born infants phototherapy may cause hyperbilirubinemia with deficiency of

(a) Thiamin (b) Riboflavin

(c) Ascorbic acid (d) Pantothenic acid **Ans. b**

71. Riboflavin deficiency causes

(a) Cheilosis (b) Loss of weight

(c) Mental deterioration (d) Dermatitis **Ans. a**

72. Magenta tongue is found in the deficiency of the vitamin

(a) Riboflavin (b) Thiamin

(c) Nicotinic acid (d) Pyridoxine **Ans. a**

73. Corneal vascularisation is found in deficiency of the vitamin

(a) B_1 (b) B_2

(c) B_3 (D) B_6 **Ans. b**

74. The pellagra preventive factor is

(a) Riboflavin (b) Pantothenic acid

(c) Niacin (d) Pyridoxine **Ans. c**

75. Pellagra is caused due to the deficiency of

(a) Ascorbic acid (b) Pantothenic acid

(c) Pyridoxine (d) Niacin **Ans. d**

76. Which of the following cofactor is required for conversion of proline residues into hydroxyproline in procollagen

(a) TPP (b) Pantothenate

(c) Niacin (d) Ascorbic Acid **Ans. d**

77. Niacin or nicotinic acid is a monocarboxylic acid derivative of

(a) Pyridine (b) Pyrimidine

(c) Flavin (d) Adenine **Ans. a**

78. Niacin is synthesized in the body from

(a) Tryptophan (b) Tyrosine

(c) Glutamate (d) Aspartate **Ans. a**

79. The proteins present in maize are deficient in

(a) Lysine (b) Threonine

(c) Tryptophan (d) Tyrosine **Ans. c**

80. Niacin is present in maize in the form of
(a) Niatin (b) Nicotin
(c) Niacytin (d) Nicyn **Ans. c**

81. In the body 1 mg of niacin can be produced from
(a) 60 mg of pyridoxine (b) 60 mg of tryptophan
(c) 30 mg of tryptophan (d) 30 mg of pantothenic acid
Ans. b

82. Pellagra occurs in population dependent on
(a) Wheat (b) Rice
(c) Maize (d) Milk **Ans. c**

83. The enzymes with which nicotinamide act as coenzyme are
(a) Dehydrogenases (b) Transaminases
(c) Decarboxylases (d) Carboxylases **Ans. a**

84. Dietary requirement of Vitamin D
(a) 400 I.U. (b) 1000 I.U.
(c) 6000 I.U. (d) 700 I.U. **Ans. a**

85. The Vitamin which does not contain a ring in the structure is
(a) Pantothenic acid (b) Vitamin D
(c) Riboflavin (d) Thiamin **Ans. a**

86. Pantothenic acid is a constituent of the coenzyme involved in
(a) Decarboxylation (b) Dehydrogenation
(c) Acetylation (d) Oxidation **Ans. c**

87. The precursor of CoA is
(a) Riboflavin (b) Pyridoxamine
(c) Thiamin (d) Pantothenate **Ans. d**

88. 'Burning foot syndrome' has been ascribed to the deficiency of
(a) Pantothenic acid (b) Thiamin
(c) Cobalamin (d) Pyridoxine **Ans. a**

89. Pyridoxal phosphate is central to
(a) Deamination (b) Amidation
(c) Carboxylation (d) Transamination **Ans. d**

90. The vitamin required as coenzyme for the action of transaminases is

(a) Niacin (b) Pantothenic acid

(c) Pyridoxal phosphate (d) Riboflavin **Ans. c**

91. Vitamin B_6 deficiency may occur during therapy with

(a) Isoniazid (b) Terramycin

(c) Sulpha drugs (d) Aspirin **Ans. a**

92. Deficiency of vitamin B_6 may occur in

(a) Obese person (b) Thin person

(c) Alcoholics (d) Diabetics **Ans. c**

93. 'Xanthurenic acid index' is a reliable criterion for the deficiency of the vitamin

(a) Pyridoxal (b) Thiamin

(c) Pantothenic acid (d) Cobalamin **Ans. a**

94. Epileptiform convulsion in human infants have been attributed to the deficiency of the vitamin

(a) B_1 (b) B_2

(c) B_6 (d) B_{12} **Ans. c**

95. Which of the following coenzymes acts as an electron sink to promote catalysis

(a) Biotin (b) Lipoamide

(c) CoA (d) Pyridoxal phosphate **Ans. d**

96. Biotin is a coenzyme of the enzyme

(a) Carboxylase (b) Hydroxylase

(c) Decarboxylase (d) Deaminase **Ans. a**

97. The coenzyme required for conversion of pyruvate to oxaloacetate is

(a) FAD (b) NAD

(c) TPP (d) Biotin **Ans. d**

98. In biotin-containing enzymes, the biotin is bound to the enzyme by

(a) An amide linkage to carboxyl group of glutamine

(b) A covalent bond with CO_2

(c) An amide linkage to an amino group of lysine

(d) An amide linkage to α-carboxyl group of protein **Ans. c**

99. A molecule of CO_2 is captured by biotin when it acts as coenzyme for carboxylation reaction. The carboxyl group is covalently attached to

(a) A nitrogen (N_1) of the biotin molecule
(b) Sulphur of thiophene ring
(c) α-Amino group of lysine
(d) α-Amino group of protein **Ans. a**

100. Consumption of raw eggs can cause deficiency of

(a) Biotin (b) Pantothenic acid
(c) Riboflavin (d) Thiamin **Ans. a**

101. The cofactor or its derivative required for the conversion of acetyl CoA to malonyl-CoA is

(a) FAD (b) ACP
(c) NAD^+ (d) Biotin **Ans. d**

102. A cofactor required in oxidative decarboxylation of pyruvate is

(a) Lipoate (b) Pantothenic acid
(c) Biotin (d) Para aminobenzoic acid **Ans. a**

103. The central structure of B_{12} referred to as corrin ring system consists of

(a) Cobalt (b) Manganese
(c) Magnesium (d) Iron **Ans. a**

104. The central heavy metal cobalt of vitamin B_{12} is coordinately bound to

(a) Cyanide group (b) Amino group
(c) Carboxyl group (d) Sulphide group **Ans. a**

105. Vitamin B_{12} has a complex ring structure (corrin ring) consisting of four

(a) Purine rings (b) Pyrimidine rings
(c) Pyrrole rings (d) Pteridine rings **Ans. c**

106. Emperical formula of cobalamin is

(a) $C_{63}H_{88}N_{12}O_{14}P.CO$ (b) $C_{61}H_{82}N_{12}O_{12}P.CO$
(c) $C_{61}H_{88}N_{12}O_{14}P.CO$ (d) $C_{63}H_{88}N_{14}O_{14}P.CO$ **Ans. d**

107. A deficiency of vitamin B_{12} causes

(a) Beri-Beri (b) Scurvy
(c) Perniciuos anemia (d) Ricket **Ans. c**

108. Vitamin B_{12} deficiency can be diagnosed by urinary excretion of
(a) Pyruvate (b) Methylmalonate
(c) Malate (d) Lactate **Ans. b**

109. Subacute combined degeneration of cord is caused due to deficiency of
(a) Niacin (b) Cobalamin
(c) Biotin (d) Thiamin **Ans. b**

110. Vitamin required for metabolism of diols e.g. conversion of ethylene glycol to acetaldehyde is
(a) Thiamin (b) Cobalamin
(c) Pyridoxine (d) Folic acid **Ans. b**

111. Both folic acid and methyl cobalamin (vitamin B_{12}) are required in
(a) Deamination of serine
(b) Deamination of threonine
(c) Conversion of pyridoxal phosphate to pyridoxamine phosphate
(d) Methylation of homocystein to methionine **Ans. d**

112. Folic acid or folate consists of the
(a) Base pteridine, p-amino benzoic acid and asparate
(b) Base purine, p-amino benzoic acid and glutamate
(c) Base pteridine, p-amino benzoic acid and glutamate
(d) Base purine, p-hydroxy benzoic acid and glutamate **Ans. c**

113. Folate as a coenzyme is involved in the transfer and utilization of
(a) Amino group (b) Hydroxyl group
(c) Single carbon moiety (d) Amido group **Ans. c**

114. Folic acid deficiency can be diagnosed by increased urinary excretion of
(a) Methylmalonate (b) Figlu
(c) Cystathionine (d) Creatinine **Ans. b**

115. Sulpha drugs interfere with bacterial synthesis of
(a) Lipoate (b) Vitamin E
(c) Tetrahydrofolate (d) Ascorbic acid **Ans. c**

116. Folate deficiency causes
(a) Microcytic anemia (b) Hemolytic anemia
(c) Iron deficiency anemia (d) Megaloblastic anemia **Ans. d**

117. Thiamin is heat stable in

(a) Acidic medium (b) Alkaline medium

(c) Both (a) and (b) (d) None of these **Ans. a**

118. Thiamin deficiency includes

(a) Mental depression (b) Fatigue

(c) Beriberi (d) All of these **Ans. d**

119. Thiamin diphosphate is required for oxidative decarboxylation of

(a) α-Keto acids (b) α-Amino acids

(c) Fatty acids (d) All of these **Ans. a**

120. The visual pigment retinal binds to the protein rhodopsin by the same type of linkage as that used by another coenzyme. What is this coenzyme

(a) Thiamine pyrophosphate (b) Tetrahydrofolate

(c) Pyridoxal phosphate (d) S- adenosylmethionine **Ans. c**

121. Loss of thiamin can be decreased by using

(a) Unpolished rice (b) Parboiled rice

(c) Whole wheat flour (d) All of these **Ans. d**

122. Daily requirement of thiamin is

(a) 0.1 mg/1,000 Calories (b) 0.5 mg/1,000 Calories

(c) 0.8 mg/1,000 Calories (d) 1.0 mg/1,000 Calories **Ans. b**

123. Thiamin requirement is greater in

(a) Non-vegetarians (b) Alcoholics

(c) Pregnant women (d) Both (b) and (c) **Ans. d**

124. People consuming polished rice as their staple food are prone to

(a) Beriberi (b) Pellagra

(c) Both (a) and (b) (d) None of these **Ans. a**

125. Riboflavin is heat stable in

(a) Acidic medium (b) Alkaline medium

(c) Neutral medium (d) Both (a) and (c) **Ans. d**

126. FAD is a coenzyme for

(a) Succinate dehydrogenase

(b) Glycerol-3-phosphate dehydrogenase

(c) Sphingosine reductase

(d) All of these **Ans. d**

127. Riboflavin deficiency can cause
(a) Peripheral neuritis (b) Diarrhoea
(c) Angular stomatitis (d) None of these **Ans. c**

128. Pellagra preventing factor is
(a) Thiamin (b) Riboflavin
(c) Niacin (d) Pyridoxine **Ans. c**

129. Niacin contains a
(a) Sulphydryl group (b) Carboxyl group
(c) Amide group (d) All of these **Ans. b**

130. NADP is required as a coenzyme in
(a) Glycolysis (b) Citric acid cycle
(c) HMP shunt (d) Gluconeogenesis **Ans. c**

131. NAD is required as a coenzyme for
(a) Malate dehydrogenase
(b) Succinate dehydrogenase
(c Glucose-6-phosphate dehydrogenase
(d) HMG CoA reductae **Ans. a**

132. NAD is required as a coenzyme in
(a) Citric acid cycle (b) HMP shunt
(c) β-Oxidation of fatty acids (d) Both (a) and (c) **Ans. d**

133. Niacin can be synthesised in human beings from
(a) Histidine (b) Phenylalanine
(c) Tyrosine (d) Tryptophan **Ans. d**

134. Daily requirement of niacin is
(a) 5 mg (b) 10 mg
(c) 20 mg (d) 30 mg **Ans. c**

135. Niacin deficiency is common in people whose staple food is
(a) Wheat (b) Polished rice
(c) Maize and / or sorghum (d) None of these **Ans. c**

136. In pellagra, dermatitis usually affects
(a) Exposed parts of body (b) Covered parts of body
(c) Trunk only (d) All parts of the body **Ans. a**

137. Niacin deficiency can occur in
(a) Hartnup disease (b) Phenylketonuria
(c) Alkaptonuria (d) None of these **Ans. a**

138. Pantothenic acid contains an amino acid which is
(a) Aspartic acid (b) Glutamic acid
(c) β-Alanine (d) β-Aminoisobutyric acid
Ans. c

139. Sulphydryl group of coenzyme a is contributed by
(a) β-Alanine (b) β-Aminoisobutyric acid
(c) Methionine (d) Thioethanolamine **Ans. d**

140. Coenzyme A contains a nitrogenous base which is
(a) Adenine (b) Guanine
(c) Choline (d) Ethanolamine **Ans. a**

141. The following is required for the formation of coenyzme A
(a) ATP (b) GTP
(c) CTP (d) None of these **Ans. a**

142. Coenzyme A is required for catabolism of
(a) Leucine (b) Isoleucine
(c) Valine (d) All of these **Ans. d**

143. Deficiency of pantothenic acid in human beings can affect
(a) Nervous system (b) Digestive system
(c) Both (a) and (b) (d) None of these **Ans. c**

144. Pyridoxal phosphate is a coenzyme for
(a) Glutamate oxaloacetate transaminase
(b) Glutamate pyruvate transaminase
(c) Tyrosine transaminase
(d) All of these **Ans. d**

145. Pyridoxal phosphate is required as a coenzyme in
(a) Transamination (b) Transulphuration
(c) Desulphydration (d) All of these **Ans. d**

146. Pyridoxal phosphate is a coenzyme for
(a) Glycogen synthetase (b) Phosphorylase
(c) Both (a) and (b) (d) None of these **Ans. b**

147. Pyridoxine deficiency can be diagnosed by measuring urinary excretion of

(a) Pyruvic acid (b) Oxaloacetic acid

(c) Xanthurenic acid (d) None of these **Ans. c**

148. Pyridoxine deficiency can be diagnosed by measuring the urinary excretion of xanthurenic acid following a test dose of

(a) Glycine (b) Histidine

(c) Tryptophan (d) Pyridoxine **Ans. c**

149. Pyridoxine requirement depends upon the intake of

(a) Carbohydrates (b) Proteins

(c) Fats (d) None of these **Ans. b**

150. Anti-egg white injury factor is

(a) Pyridoxine (b) Biotin

(c) Thiamin (d) Liponic acid **Ans. b**

151. When eggs are cooked

(a) Biotin is destroyed but avidin remains unaffected

(b) Avidin is inactivated but biotin remains unaffected

(c) Both avidin and biotin are inactivated

(d) Both avidin and biotin remain unaffected **Ans. b**

152. Biotin is required as a coenzyme by

(a) Anaerobic dehydrogenases (b) Decarboxylases

(c) Aerobic dehydrogenases (d) Carboxylases **Ans. d**

153. Biotin is a coenzyme for

(a) Pyruvate carboxylase (b) Acetyl CoA carboxylase

(c) Propionyl CoA carboxylase (d) All of these **Ans. d**

154. Lipoic acid is a conenzyme for

(a) Pyruvate dehydrogenase (b) α-Ketoglutarate dehydrogenae

(c) Both (a) and (b) (d) None of these **Ans. c**

155. Chemically, lipoic acid is

(a) Saturated fatty acid (b) Unsaturated fatty acid

(c) Amino acid (d) Sulphur containing fatty acid

Ans. d

156. Folic acid contains
(a) Pteridine (b) p-Amino benzoic acid
(c) Glutamic acid (d) All of these **Ans. d**

157. Conversion of folate into tetrahydrofolate requires
(a) NADH (b) NADPH
(c) $FMNH_2$ (d) $FADH_2$ **Ans. b**

158. Riboflavin deficiency symptoms are
(a) Glossitis (b) stomatis
(c) Vomitting (d) Both (a) and (b) **Ans. d**

159. Vitamin B_{12} forms coenzymes known as
(a) Cobamide (b) Transcobalamin I
(c) Transcobalamin II (d) Both (b) and (c) **Ans. a**

160. Methylcobalamin is required for formation of
(a) Serin from glycine
(b) Glycine from serine
(c) Methionine from homocysteine
(d) All of these **Ans. c**

161. The coenzyme involved in oxidative decarboxylation is
(a) Thiamine pyrophosphate (b) Biotin
(c) NAD+ (d) Pyridoxal phosphate **Ans. a**

162. Absorption of Vitamin B_{12} requires the presence of
(a) Pepsin (b) Hydrochloric acid
(c) Intrinsic factor (d) Both (b) and (c) **Ans. d**

163. Intrinsic factor is chemically a
(a) Protein (b) Glycoprotein
(c) Mucopolysaccaride (d) Peptide **Ans. b**

164. Chemically, Extrinsic Factor of Castle is a
(a) Mucoprotein (b) Glycoprotein
(c) Mucopolysaccharide (d) Cyanocobalamin **Ans. d**

165. Vitamin B_{12} is
(a) Not stored in the body (b) Stored in bone marrow
(c) Stored in liver (d) Stored in RE cells **Ans. c**

166. Vitamin B_{12} is transported in blood by
(a) Albumin (b) Transcortin
(c) Transcobalamin I (d) Transcobalamin II **Ans. d**

167. Vitamin B_{12} is synthesized by
(a) Bacteria only (b) Plants only
(c) Animals only (d) Both (a) and (c) **Ans. a**

168. Deficiency of vitamin B_{12} can occur because of
(a) Decreased intake of vitamin B_{12}
(b) Atrophy of gastric mucosa
(c) Intestinal malabsorption
(d) All of these **Ans. d**

169. Deficiency of vitamin B_{12} can be diagonised by
(a) Carr-Price reaction (b) Ames assay
(c) Watson-Schwartz test (d) Schilling test **Ans. d**

170. Gastyrectomy leads to megaloblastic anaemia within a few
(a) Days (b) Weeks
(c) Months (d) Years **Ans. d**

171. Ascorbic acid is required to synthesize all of the following except
(a) Collagen (b) Bile acids
(c) Bile pigments (d) Epinephrine **Ans. c**

172. Vitamin C enhances the intestinal absorption of
(a) Potassium (b) Iodine
(c) Iron (d) None of these **Ans. c**

173. Vitamin C activity is present in
(a) D-Ascorbic acid (b) D-Dehydroascorbic acid
(c) L-Ascorbic acid (d) Both (a) and (b) **Ans. c**

174. Vitamin C is required for the synthesis of
(a) Bile acids from cholesterol (b) Bile salts from bile acids
(c) Vitamin D from cholesterol (d) All of these **Ans. a**

175. Deficiency of vitamin C causes
(a) Beriberi (b) Pellagra
(c) Pernicious anaemia (d) Scurvy **Ans. d**

176. An early diagnosis of vitamin C deficiency can be made by

(a) Measuring plasma ascorbic acid

(b) Measuring urinary ascorbic acid

(c) Ascorbic acid saturation test

(d) All of these **Ans. c**

177. Daily requirement of vitamin C in adults is about

(a) 100 mg (b) 25 mg

(c) 70 mg (d) 100 mg **Ans. c**

178. The vitamin having the highest daily requirement among the following is

(a) Thiamin (b) Ribovflavin

(c) Pyridoxine (d) Ascorbic acid **Ans. d**

179. Anaemia can occur due to the deficiency of all the following except

(a) Thiamin (b) Pyridoxine

(c) Folic acid (d) Cyanocobalamin **Ans. a**

180. A vitamin which can be synthesized by human beings is

(a) Thiamin (b) Niacin

(c) Folic acid (d) Cyanocobalamin **Ans. b**

181. Laboratory diagnosis of vitamin B_{12} deficiency can be made by measuring the urinary excretion of

(a) Xanthurenic acid (b) Formiminoglutamic acid

(c) Methylmalonic acid (d) Homogentisic acid **Ans. c**

182. The molecule of vitamin A_1 contains

(a) Benzene ring (b) β-Ionone ring

(c) β-Carotene ring (d) None of these **Ans. b**

183. Precursor of Vitamin A is

(a) α-Carotene (b) β-Carotene

(c) γ-Carotene (d) All of these **Ans. d**

184. Two molecules of vitamin A can be formed from 1 molecule of

(a) α-Carotene (b) β-Carotene

(c) γ-Carotene (d) All of these **Ans. b**

185. Conversion of β-carotene into retinal requires the presence of

(a) β-Carotene dioxygenase (b) Bile salts

(c) Molecular oxygen (d) All of these **Ans. d**

186. Conversion of retinal into ritonal requires the presence of
(a) NADH (b) NADPH
(c) $FADH_2$ (d) Lipoic acid **Ans. b**

187. Retinal is converted into retinoic acid in the presence of
(a) Retinal oxidase (b) Retinal carboxylase
(c) Retinene reductase (d) Spontaneously **Ans. d**

188. Vitamin A absorbed in intestine is released into
(a) Portal circulation (b) Lacteals
(c) Both (a) and (b) (d) None of these **Ans. b**

189. Vitamin A is stored in the body in
(a) Liver (b) Adipose tissue
(c) Reticuloendothelial cells (d) All of these **Ans. a**

190. Rhodopsin contains opsin and
(a) 11-cis-retinal (b) 11-trans-retinal
(c) All-cis-retinal (d) All trans-retinal **Ans. a**

191. When light falls on rod cells
(a) All-cis-retinal is converted into all-trans-retinal
(b) 11-cis-retinal is converted into 11-trans-retinal
(c) 11-trans-retinal is converted into all-transretinal
(d) 11-cis-retinal is converted into all-trans-retinal **Ans. d**

192. Conversion of all-trans-retinal into all trans-retinol requires
(a) NAD (b) NADH
(c) NADP (d) NADPH **Ans. d**

193. Retinol isomerase is present in
(a) Retina (b) Liver
(c) Both (a) and (b) (d) None of these **Ans. b**

194. Anti-oxidant activity is present in
(a) β-Carotene (b) Retinol
(c) Retinoic acid (d) All of these **Ans. a**

195. One international Unit of vitamin A is the activity present in
(a) 0.3 μg of β-Carotene (b) 0.3 μg of retinol
(c) 0.6 μg of retinoic acid (d) All of these **Ans. b**

196. Daily requirement of vitamin A in an adult man can be expressed as

(a) 400 IU (b) 1,000 IU

(c) 5,000 IU (d) 10,000 IU **Ans. c**

197. Vitamin B_6 includes

(a) Pyridoxal (b) Pyridoxamine

(c) Pyridoxine (d) All of these **Ans. d**

198. An early effect of vitamin a deficiency is

(a) Xerophthalmia

(b) Keratomalacia

(c) Prolonged dark adaptation time

(d) Follicular hyperkeratosis **Ans. c**

199. Nyctalopia is

(a) Drying of eyes (b) Destruction of cornea

(c) Blindness (d) Inability to see in dimlight

Ans. d

200. Rod cells possess a trans-membrane protein which is

(a) Adenylate cyclase (b) Transducin

(c) Rhodopsin (d) (b) as well as (c) **Ans. c**

201. Provitamins A include

(a) Retinal (b) Retionic acid

(c) Carotenes (d) All of these **Ans. c**

202. Retinoic acid can

(a) Act as a photo receptor

(b) Support growth and differentiation

(c) Act as an anti-oxidant

(d) None of these **Ans. b**

203. Prosthetic group in cone cell phototreceptors is

(a) Iodine (b) Opsin

(c) 11-cis-retinal (d) All-trans-retinal **Ans. c**

204. Retinoic acid is involved in the synthesis of

(a) Rhodopsin (b) Iodopsin

(c) Porphyrinopsin (d) Glycoproteins **Ans. d**

205. Transducin is a
(a) Signal transducer (b) Stimulatory G-protein
(c) Trimer (d) All of these **Ans. d**

206. Vitamin D_3 is
(a) Cholecalciferol (b) Ergosterol
(c) 7-Dehydrocholesterol (d) Ergocaliferol **Ans. c**

207. Ergosterol is found in
(a) Animals (b) Plants
(c) Bacteria (d) All of these **Ans. b**

208. A provitamin D synthesized in human beings is
(a) Ergosterol (b) 7-Dehydrocholesterol
(c) Cholecalciferol (d) 25-Hydroxycholecalciferol
Ans. b

209. 25-Hydroxylation of vitamin D occurs in
(a) Skin (b) Liver
(c) Kidneys (d) Intestinal mucosa **Ans. b**

210. Tubular reabsorption of calcium is increased by
(a) Cholecalciferol (b) 25-Hydroxycholecalciferol
(c) Calcitriol (d) All of these **Ans. c**

211. Parathormone is required for the conversion of
(a) Cholecalciferol into 1-hydroxycholecalciferol
(b) Cholecalciferol into 25-hydroxycholecalciferol
(c) 25-Hydroxycholecalciferol into calcitriol
(d) Cholesterol into 7-dehydrocholesterol **Ans. c**

212. Vitamin D_2 is also said to be
(a) Activated ergosterol (b) Fergocalciferol
(c) Viosterol (d) All of these **Ans. d**

213. The poor sources of vitamin D
(a) Eggs (b) Butter
(c) Milk (d) Liver **Ans. c**

214. The activity of tocopherols is destroyed by
(a) Oxidation (b) Reduction
(c) Conjugation (d) All of these **Ans. a**

215. Some tocopherols are

(a) Terpenoid in structure (b) Dional in structure

(c) Isoprenoid in structure (d) Farnesyl in structure **Ans. a**

216. The methyl groups in the aromatic nucleus of a tocopherols are

(a) 2 (b) 3

(c) 4 (d) 5 **Ans. b**

217. Vitamin E stored in

(a) Mitochondria (b) Microsomes

(c) Both (a) and (b) (d) None of these **Ans. c**

218. Vitamin E protects the polyunsaturated fatty acids from oxidation by molecular oxygen in the formation of

(a) Superoxide (b) Peroxide

(c) Trioxide (d) All of these **Ans. b**

219. The tocopherols prevent the oxidation of

(a) Vitamin A (b) Vitamin D

(c) Vitamin K (d) Vitamin C **Ans. a**

220. Vitamin E protects enzymes from destruction in

(a) Muscles (b) Nerves

(c) Gonads (d) All of these **Ans. d**

221. Vitamin K regulates the synthesis of blood clotting factors

(a) VII (B) IX

(c) X (d) All of these **Ans. d**

222. Ascorbic acid can reduce

(a) 2, 4-dinitro benzene

(b) 2, 6-Dichlorophenol Indophenol

(c) 2, 4-dibromobenzene

(d) 2, 6-dibromo benzene **Ans. b**

223. Sterilized milk is devoid of

(a) Vitamin A (b) Vitamin B_1

(c) Vitamin C (d) Vitamin D **Ans. c**

224. The symptoms of scurvy are

(a) Poor healing of wounds (b) Loosening of teeth

(c) Anaemia (d) All of these **Ans. b**

225. Kwashiorkor results from

(a) Vitamin A deficiency

(b) Vitamin D deficiency

(c) Deficiency of minerals in diet

(d) Protein and caloric deficiency in diet **Ans. d**

226. Which among the following fatty acids is an essential fatty acid for man?

(a) Palmitic acid (b) Oleic acid

(c) Linoleic acid (d) None of these **Ans. c**

227. The number of nutritionally essential amino acids for man is

(a) 6 (b) 8

(c) 10 (d) 12 **Ans. b**

228. Avidin is present in

(a) Cow's milk (b) Raw egg

(c) Green leafy vegetables (d) Carrots **Ans. b**

229. Marasmus is due to malnutrition of

(a) Proteins (b) Proteins and calories

(c) Proteins and vitamins (d) Proteins and minerals **Ans. b**

230. Energy value in kilocalorie per gram of fat in the body is

(a) 1 (b) 4

(c) 9 (d) 18 **Ans. c**

231. The action of Vitamin K in formation of clotting factor is through

(a) Post transcription (b) Post translation

(c) Golgi complex (d) Endoplasmic reticulum

Ans. b

232. Vitamin necessary for CoA synthesis

(a) Pantothenic acid (b) Vitamin C

(c) B_6 (d) B_{12} **Ans. a**

233. Cofactor for transamination is

(a) Thymine (b) Riboflavin

(c) Pyridoxine (d) Niacin **Ans. c**

234. During deficiency of thiamine the concentration of the following compound rises in blood and intracellular fluid

(a) Glycogen (b) Sugar
(c) Amino acids (d) Pyruvic acid **Ans. d**

235. The conversion of carotenoids to Vitamin A takes place in

(a) Intestine (b) Liver
(c) Kidney (d) Skin **Ans. a**

236. Man cannot synthesize vitamin

(a) A (b) B
(c) C (d) D **Ans. c**

237. Vitamin A is required for the formation of a light receptor protein known as

(a) Globulin (b) Lypoprotein
(c) Chomoprotein (d) Rhodospin **Ans. d**

238. Excessive vitamin A in children produces

(a) Irritability (b) Anorexia
(c) Headache (d) All of these **Ans. d**

239. Tocopherols prevent the oxidation of

(a) Vitamin A (b) Vitamin D
(c) Vitamin K (d) Vitamin C **Ans. a**

240. Vitamin K regulates the synthesis of blood clotting factors

(a) VII (b) IX
(c) X (d) All of these **Ans. d**

241. The colour of cyanomethmoglobin is

(a) Pale yellow (b) Pink
(c) Brown (d) Bright red **Ans. d**

242. Transketolase activity is affected in

(a) Bitoin deficiency (b) Pyridoxine deficiency
(c) PABA deficiency (d) Thiamine deficiency **Ans. d**

243. The hydrolysis of glucose-6-PO_4 is catalyzed by a phosphatase that is not found in which of the following?

(a) Liver (b) Kidney
(c) Muscle (d) Small intestine **Ans. c**

244. Vitamin K_2 was originally isolated from

(a) Soyabean (b) Putrid fishmeal

(c) Alfa alfa (d) Oysters **Ans. b**

245. The following form of vitamin A is used in the visual cycle

(a) Retinol (b) Retinoic acid

(c) Retinaldehyde (d) Retinyl acetate **Ans. c**

246. Which among the following is an essential amino acid for man?

(a) Alanine (b) Serine

(c) Valine (d) Glutamic acid **Ans. c**

247. Under what condition to basal metabolic rate goes up?

(a) Cold environment (b) Hot environment

(c) Intake of base forming foods (d) Hypothyroidism **Ans. a**

248. What is the major form of caloric storage in human body?

(a) ATP (b) Glycogen

(c) Creatine phosphate (d) triacylglycerol **Ans. c**

249. The phosphoprotein of milk is

(a) Lactalbumin (b) Lactoglobulin

(c) Vitellin (d) Casien **Ans. d**

250. Dietary deficiency of this vitamin leads to night blindness:

(a) Retinol (b) Niacin

(c) Ascorbic acid (d) Cholecalciferol **Ans. a**

251. A non essential amino acid is not

(a) Absorbed in the intestines (b) Required in the diet

(c) Incorporated into the protein (d) Metabolized by the body **Ans. b**

252. The deficiency of Vitamin B_{12} leads to

(a) Pernicious anaemia (b) Megablastic anaemia

(c) Both (a) and (b) (d) None of these **Ans. c**

253. Which among the following is a nutritionally essential amino acid for man?

(a) Alanine (b) Glycine

(c) Tyrosine (d) Isoleucine **Ans. d**

254. The maximum specific dynamic action of food stuff is exerted by
(a) Carbohydrates (b) Fats
(c) Proteins (d) Vitamins **Ans. c**

255. The essential amino acids
(a) Must be supplied in the diet because the organism has lost the capacity to aminate the corresponding ketoacids
(b) Must be supplied in the diet because the human has an impaired ability to synthesize the carbon chain of the corresponding ketoacids
(c) Are identical in all species studied
(d) Are defined as these amino acids which cannot be synthesized by the organism at a rate adequate to meet metabolic requirements
Ans. b

256. Fibre in the diet is beneficial in
(a) Hyper glycemia (b) Hyper cholseteremia
(c) Colon cancer (d) All of these **Ans. d**

257. Sucrose intolerance leads to
(a) Hyper glycemia (b) Glycosuria
(c) Diarrhoea (d) Hypoglycemia **Ans. c**

258. There can be intolerance with respect to the following sugar
(a) Glucose (b) Lactose
(c) Maltose (d) Xylose **Ans. b**

259. Milk contains very poor amounts of
(a) Calcium (b) Phosphate
(c) Iron (d) Riboflavin **Ans. c**

260. Egg contains very little
(a) Fat (b) Proteins
(c) Carbohydrates (d) Calcium and phosphorus
Ans. c

261. BMR (Basal Metabolic rate) is elevated in
(a) Hyper thyroidism (b) Under nutrition
(c) Starvation (d) Hypothyroidism **Ans. a**

262. Soyabean proteins are rich in
(a) Lysine (b) Alanine
(c) Glcyine (d) Aspartic acid **Ans. b**

263. Corn and gliadin are low in

(a) Lysine (b) Alanine

(c) Glycine (d) Aspartic acid **Ans. d**

264. What is the disease caused by thiamine deficiency?

(a) Nycalopia (b) Scurvy

(c) Rickets (d) Beriberi **Ans. d**

265. Retinol and retinol -binding protein (RBP) bound with this protein

(a) Albumin (b) Prealbumin

(c) α_2-globulin (d) β-globulin **Ans. b**

266. Megaloblastic anemia is caused by the deficiency of

(a) Folic acid (b) Vitamin B_6

(c) Iron (d) Protein **Ans. a**

267. This vitamin acts as anti-oxidant

(a) Vitamin A (b) Vitamin D

(c) Vitamin E (d) Vitamin K **Ans. c**

268. Calcitriol is

(a) 1-OH-cholecalciferol (b) 25-OH-cholecalciferol

(c) 24, 25-diOH cholecalciferol (d) 1, 25-diOH cholecalciferol

Ans. d

269. 1-hydroxylation of 25-OH vitamin D_3 takes place in

(a) Liver (b) Kidneys

(c) Intestine (d) Pancreas **Ans. b**

270. 25-hydroxylation of vitamin D_3 takes place in

(a) Liver (b) Kidneys

(c) Intestine (d) Pancreas **Ans. a**

271. Hydroxylation of 25-hydroxy cholecalciferol is promoted by

(a) Cytochrome - a (b) Parathyroid hormone

(c) Cytochrome-b (d) cAMP **Ans. b**

272. The egg injury factor in raw egg white is

(a) Biotin (b) Avidin

(c) Albumin (d) Calcium salts **Ans. b**

273. The following has cyanide

(a) Vitamin B_{12} (b) Adenyl cobamide

(c) Benzimidazole cobamide (d) Methyl cobamide **Ans. a**

274. The human species can biosynthesize

(a) Vitamin C (b) Vitamin B_{12}

(c) Thiamine (d) Niacin **Ans. d**

275. Retina contains this photosensitive pigment

(a) Rhodopsin (b) Opsin

(c) Retinol (d) Melanin **Ans. a**

276. Anti xerophthalmic vitamin is

(a) Vitamin B_1 (b) Vitamin B_2

(c) Vitamin B_6 (d) Vitamin A **Ans. d**

277. One of the following is not a symptom of addison's disease

(a) Hypoglycemia (b) Hyponatremia

(c) Hypokalemia (d) Hypochoremia **Ans. c**

278. Gammaxane is an antimetabolite of

(a) Thiamine (b) Riboflavin

(c) Pyridoxin (d) Inositol **Ans. d**

279. Pyridoxin deficiency may lead to convulsions as it is needed for the synthesis of

(a) GABA (b) PABA

(c) EFA (d) SAM **Ans. a**

280. Sulpha drugs are antimetabolites of

(a) Vitamin K (b) Pyridoxin

(c) Folic acid (d) Vitamin B_2 **Ans. c**

281. This abnormal metabolite may be responsible for the neurological manifestation of pernicious anemia

(a) Taurine (b) Methyl malonic acid

(c) Xantherunic acid (d) Phenyl pyruvic acid **Ans. a**

282. The vitamin in leafy vegetables

(a) D (b) K

(c) A (d) Both (b) and (c) **Ans. d**

283. Isonicotinic acid hydrazide given in the treatment of tuberculosis may lead to a deficiency of

(a) Vitamin A (b) Pyridoxin
(c) Folate (d) Inositol **Ans. d**

284. Biotin is required for the reaction of CO_2 with

(a) Water
(b) Acetyl CoA
(a) NH_3
(d) Incorporation of carbon 6 in purine **Ans. b**

285. A deficiency of folate leads to

(a) Megaloblastic anemia
(b) Aplastic anemia
(c) Pernicious anemia
(d) Hypochromic microcytic anemia **Ans. b**

286. A deficiency of Iron leads to

(a) Megaloblastic anemia
(b) Aplastic anemia
(c) Pernicious anemia
(d) Hypochromic microcytic anemia **Ans. a**

287. Corninoid coenzymes are coenzymes of

(a) Vitamin B_{12} (b) Vitamin B_6
(c) Vitamin B_2 (d) Vitamin B_1 **Ans. d**

288. Vitamin B_{12} initially binds to the proteins known as

(a) Transcobalamin I (b) R-Proteins
(c) Transcobalamin II (d) Intrinsic factor of castle
Ans. b

289. Extrinsic factor of castle is

(a) Vitamin B_{12} (b) Glycoprotein
(c) R-Proteins (d) Sigma protein **Ans. b**

290. Intrinsic factor of castle is

(a) Vitamin B_{12} (b) Glycoprotein
(c) R-Proteins (d) Sigma protein **Ans. a**

291. Pernicious means

(a) Prolonged (b) Dangerous

(c) Intermittent (d) Idiopathic **Ans. b**

292. Reduction of D-ribonucleotides to Ddeoxyribonucleotides in prokaryotes requires

(a) 5, 6 dimethyl benzimidazole cobamide

(b) Thioredoxin

(c) Tetra hydrobiopterin

(d) Tetra hydrofolate **Ans. b**

293. Biotin is also known as

(a) Anti egg white injury factor (b) Rutin

(c) Both (a) and (b) (d) None of these **Ans. a**

294. Angular stomatosis is due to

(a) Ariboflavinoses (b) Deficiency of Vitamin C

(c) Deficiency of Vitamin B_1 (d) Deficiency of folate **Ans. b**

295. Which of the following vitamins does not act as a precursor for coenzymes

(a) Biotin (b) Thiamin

(c) Folic acid (d) Ascorbic acid **Ans. d**

296. One of the main functions of Vitamin K is cofactor for

(a) Carboxylate for the formation of γ-carboxyglutamate

(b) Methylation of δ-adenosyl methionine

(c) Carboxylation of biotin

(d) One carbon transfer by tetrahydrofolate **Ans. a**

297. Prothrombin time is prolonged by administering

(a) Vitamin K (b) Dicoumarol

(c) Calcium (d) Prothrombin **Ans. a**

298. This vitamin acts as antioxidant

(a) Vitamin A (b) Vitamin D

(c) Vitamin E (d) Vitamin K **Ans. b**

299. This is a photo-labile vitamin

(a) Thiamine (b) Riboflavin

(c) Niacin (d) Cholecalciferol **Ans. c**

300. Convulsive episodes occur when there is a severe deficiency of

(a) Pyridoxine (b) Folic acid

(c) Thiamine (d) Riboflavin **Ans. b**

301. Metastatic classification is seen in hypervitaminosis

(a) A (b) K

(c) D (d) E **Ans. a**

302. The anti vitamin for para aminobenzoic acid is

(a) Aminopterin (b) Dicoumarol

(c) Sulphonamides (d) Thiopanic acid **Ans. c**

303. Several pantothenic acid deficiency in man has been reported to cause

(a) Burning feet syndrome (b) Scurvy

(c) Cataract (d) Xerophthalmia **Ans. c**

304. Cholesterol is a precursor in the biogenesis of

(a) Vitamin A (b) Vitamin D

(c) Vitamin E (d) None of these **Ans. a**

305. This vitamin is a potent antioxidant of vitamin A

(a) Vitamin C (b) Vitamin E

(c) Vitamin K (d) Vitamin D **Ans. b**

306. In retinal rickets, the following hydroxylation of Vitamin D3 does not take place

(a) 25 (b) 1

(c) 24 (d) 7 **Ans. b**

307. The following does not have phosphorous

(a) Riboflavin (b) TPP

(c) NAD^+ (d) COASH **Ans. a**

308. Convulsions and delirium could be caused by a severe deficiency of

(a) Thiamine (b) Glutamate

(c) Niacin (d) Magnesium **Ans. d**

309. Rice polishings contain this vitamin

(a) Riboflavin (b) Niacin

(c) Thiamine (d) Vitamin B_{12} **Ans. d**

310. In beri beri there will be accumulation of _______ in blood
(a) Aceto acetic acid (b) β-OH butyric acid
(c) Pyruvic acid (d) Methyl malonic acid **Ans. c**

311. Symptoms of pellagra are
(a) Dermatitis and diarrhea only
(b) Dermatitis and dementia only
(c) Diarrhea, dermatitis and dementia
(d) Diarrhea and elements only **Ans. c**

312. Pyridoxine deficiency leads to
(a) Megaloblastic anemia
(b) Aplastic anemia
(c) Hypochromic microcytic anemia
(d) Permicious anemia **Ans. c**

313. The significant ocular lesion in arbo flovinosis
(a) Keratomalacia (b) Bitot's spots
(c) Vascularisation of the cornea (d) lachrymal metaplasia **Ans. b**

314. Irradiation of foods raises the content of
(a) Vitamin A (b) Vitamin D
(c) Vitamin E (d) Vitamin K **Ans. c**

315. An anti-vitamin for folic acid is
(a) Amethoptesin (b) Dicoumarol
(c) Pyrithoamine (d) Isoniazid **Ans. a**

316. Thiamin is
(a) Water soluble vitamin (b) Fat soluble vitamin
(c) Purine base (d) Pyrimidine base **Ans. d**

317. The anti-vitamin for para amino benzoic acid is
(a) Aminopterrin (b) Dicoumarol
(c) INH (d) Sulphonamides **Ans. d**

318. The sulphur-containing vitamins among the following B-Vitamin is
(a) Thiamine (b) Riboflavin
(c) Niacin (d) Pyridoxine **Ans. a**

319. Taurinuria may be encountered in

(a) Pernicious anemia (b) Beriberi

(c) Pellegra (d) Folate deficiency **Ans. a**

320. The three vitamins which are specially required for proper nerve functions are acid

(a) Thiamine, niacin and riboflavin

(b) Thiamine, folic acid, choline

(c) Thiamine, riboflavin, patothenic acid

(d) Thiamine, pyridoxin, vitamin B_{12} **Ans. d**

321. This is a rich source for vitamin C

(a) Rice (b) Milk

(c) Egg (d) Lemon **Ans. d**

322. The following vitamin is involved in coenzyme function in transaminations

(a) Nicotinamide (b) Pyridoxine

(c) Thiamine (d) Riboflavin **Ans. b**

323. Methyl malonic aciduria is seen in the deficiency of

(a) Vitamin B_6 (b) Folic acid

(c) Thiamine (d) Vitamin B_{12} **Ans. d**

324. Deficiency of Vitamin C leads to

(a) Rickets (b) Scurvy

(c) Night blindness (d) All of these **Ans. c**

325. If no primer DNA was given, the following scientist could not have synthesized DNA

(a) Ochoa (b) Okazaki

(c) Kornberg (d) Monod **Ans. c**

326. Antisterility vitamin is

(a) Vitamin B_1 (b) Vitamin B_2

(c) Vitamin E (d) Vitamin K **Ans. c**

327. All the following vitamins give rise to cofactors that are phosphorylated in the active form except

(a) Vitamin A (b) Vitamin B_1

(c) Vitamin D (d) Vitamin E **Ans. b**

328. Molecular Iron, Fe, is

(a) Stored in the body in combination with Ferritin
(b) Stored primarily in the spleen
(c) Excreted in the urine as Fe^{2+}
(c) absorbed in the intestine by albumin **Ans. a**

329. Humans most easily tolerate a lack of which of the following nutrients?

(a) Protein (b) Iodine
(c) Carbohydrate (d) Lipid **Ans. c**

330. The deficiency of which one of the following vitamin causes creatinuria?

(a) Vitamin E (b) Vitamin K
(c) Vitamin A (d) Vitamin B_6 **Ans. c**

331. In adults a severe deficiency of vitamin D causes

(a) Night blindness (b) Osteomalacia
(c) Rickets (d) Skin cancer **Ans. b**

332. Which of the following statements regarding Vitamin A is true?

(a) It is not an essential Vitamin
(b) It is related to tocopherol
(c) It is a component of rhodopsin
(d) It is also known as Opsin **Ans. c**

333. Fully activated pyruvate carboxylase depends upon the presence of

(a) Malate and Niacin
(b) Acetyl CoA and biotin
(c) Acetyl CoA and thiamine pyrophosphate
(d) Oxaloacetate and biotin **Ans. b**

334. Pantothenic acid is a constituent of coenzyme involved in

(a) Acetylation (b) Decarboxylation
(c) Dehydrogenation (d) Oxidation **Ans. a**

335. Biotin is involved in which of the following types of reactions?

(a) Hydroxylation (b) Carboxylation
(c) Decarboxylation (d) Deamination **Ans. b**

336. Which of the following vitamins is the precurssor of CoA?

(a) Riboflavin (b) Pantothenate
(c) Thiamine (d) Cobamide **Ans. b**

337. Vitamins that function as dinucleotide derivatives include all the following except

(a) Thiamine (b) Niacin

(c) Nicotinate (d) Vitamin B_2 **Ans. a**

338. Methyl malonic aciduria is seen in a deficiency of

(a) Vitamin B_6 (b) Folic acid

(c) Thiamine (d) Vitamin B_{12} **Ans. d**

339. Retinol and Retinol binding protein are bound with this protein

(a) Albumin (b) Prealbumin

(c) α-globulin (d) β-globulin **Ans. b**

340. Megaloblastic anemia is caused by the deficiency of

(a) Folic acid (b) Vitamin B_6

(c) Iron (d) Protein **Ans. a**

341. This vitamin acts as anti oxidant.

(a) Vitamin A (b) Vitamin D

(c) Vitamin E (d) Vitamin K **Ans. d**

342. Calcitriol is

(a) 1-hydroxy cholecalciferol

(b) 25-hydroxy cholecalciferol

(c) 24, 25-dihydroxy cholecalciferol

(d) 1, 25-dihydroxy cholecalciferol **Ans. d**

343. 1-hydroxylation of 25-hydroxy Vitamin D3 takes place in

(a) Liver (b) Kidneys

(c) Intestine (d) Pancreas **Ans. b**

344. 25-hydroxylation of Vitamin D_3 takes place in

(a) Liver (b) Kidneys

(c) Intestines (d) Pancreas **Ans. a**

345. Hydroxylation of 25-hydroxy cholecalciferol is promoted by

(a) Cytochrome A (b) Panthyroid hormone

(c) Cytochrome b (d) cAMP **Ans. b**

346. The egg injury factor in raw egg white is

(a) Biotin (b) Avidin

(c) Albumin (d) Calcium salts **Ans. b**

347. The following has cyanide
(a) Vitamin B_{12} (b) Adenyl cobamide
(c) Benzimidazole cobamide (d) Methyl cobamide **Ans. a**

348. The human species can biosynthesize
(a) Vitamin C (b) Vitamin B_{12}
(c) Thiamin (d) Niacin **Ans. d**

349. Retina contains this photo sensitive pigment
(a) Rhodopsin (b) Opsin
(c) Retinol (d) Malanin **Ans. a**

350. Antixerophthalmic vitamin is
(a) Vitamin B_1 (b) Vitamin B_2
(c) Vitamin B_6 (d) Vitamin A **Ans. d**

351. One of the following is not symptom of Addison's disease
(a) Hypoglycemia (b) Hyponatremia
(c) Hypokalemia (d) Hypochloremia **Ans. c**

352. Gammaxine is an antimetabolite of
(a) Thiamine (b) Riboflavin
(c) Pyridoxin (d) Inositol **Ans. d**

353. Pyridoxine deficiency may lead to convulsions as it is needed for the synthesis of
(a) GABA (b) PABA
(c) EFA (d) SAM **Ans. a**

354. Sulpha drugs are antimetabolites of
(a) PABA (b) Pyridoxin
(c) Vitamin B_2 (d) Pantothenic acid **Ans. a**

355. This abnormal metabolite may be responsible for the neurological manifestation of pernicious anemia
(a) Taurine (b) Methyl malonic acid
(c) Xanthurenic acid (d) Phenyl pyruvic acid **Ans. b**

356. Choline is not required for the formation of
(a) Lecithins (b) Acetyl choline
(c) Sphingomyelin (d) Cholic acid **Ans. d**

357. Isonicotinic acid hydrazide given in the treatment of tuberculosis may lead to a deficiency of

(a) Vitamin A (b) Pyridoxin

(c) Folate (d) Inositol **Ans. b**

358. Steroidal prohormone is

(a) Vitamin A (b) Vitamin C

(c) Vitamin D (d) None of these **Ans. c**

359. A deficiency of folate leads to

(a) Megaloblastic anemia

(b) Aplastic anemia

(c) Pernicious anemia

(d) Hypochromic microcytic anemia **Ans. a**

360. Deficiency of Iron leads to

(a) Megaloblastic anemia

(b) Aplastic anemia

(c) Pernicious anemia

(d) Hypochromic microcytic anemia **Ans. d**

361. Corrinoid coenzymes are coenzymes of

(a) Vitamin B_6 (b) Vitamin B_{12}

(c) Vitamin B_2 (d) Vitamin B_1 **Ans. b**

362. Vitamin B_{12} initially binds to the proteins known as

(a) Transcobalamin I (b) R-proteins

(c) Transcobalamin II (d) Intrinsic factor of castle **Ans. b**

363. Extrinsic factor of castle is

(A) Vitamin B_{12} (b) Glycoprotein

(c) R-proteins (d) Sigma protein **Ans. a**

364. Intrinsic factor of castle is

(a) Vitamin B_{12} (b) Glycoprotein

(c) R-proteins (d) Sigma protein **Ans. b**

365. Pernicious means

(a) Prolonged (b) Dangerous

(c) Intermittent (d) Idiopathic **Ans. b**

366. Reduction of D-ribonucleotides to D-deoxyribonucleotides in prokaryotes requires

(a) 5, 6 dimethyl benzimindazole cobamide

(b) Thiredoxin

(c) Tetra hydrobiopterin

(d) Tetra hydrofolate **Ans. a**

367. Antirachitic vitamin is

(a) Vitamin A (b) Vitamin D

(c) Vitamin E (d) Vitamin K **Ans. b**

368. Angular stomatitis is due to

(a) Ariboflavinosis (b) Deficiency of Vitamin C

(c Deficiency of Vitamin B_1 (d) Deficiency of folate **Ans. a**

369. Increased carbohydrate consumption increases the dietary requirement for

(a) Thiamine (b) Riboflavine

(c) Pyridoxine (d) Folic acid **Ans. a**

370. One of the main functions of Vitamin K is the cofactor for

(a) Carboxylase for the formation of γ-carboxy glutamate

(b) Methylation by S-adenosyl methionine

(c) Carboxylation by biotin

(d) One carbon transfer by tetra hydrofolate **Ans. a**

371. Prothrombin time is prolonged by administering

(a) Vitamin K (b) Dicoumarol

(c) Calcium (d) Prothrombin **Ans. b**

372. This is photo labile vitamin

(a) Thiamine (b) Riboflavin

(c) Niacin (d) Cholecalciferol **Ans. b**

373. Convulsive episodes occur when there is a severe deficiency of

(a) Pyridoxine (b) Folic acid

(c) Thiamine (d) Riboflavin **Ans. a**

374. Metastatic calcification is seen in hypervitaminosis

(a) A (b) K

(c) D (d) E **Ans. c**

375. The anti-vitamin for para amino benzoic acid is
(a) Aminopterin (b) Dicoumasol
(c) Sulphanomides (d) Thiopamic acid **Ans. c**

376. Severe patothemic acid deficiency in man has been reported to cause
(a) Burning feet syndrome (b) Scurvy
(c) Cataract (d) Xeropththalmia **Ans. a**

377. In renal rickets, the following hydroxylation of Vitamin D_3 does not take place
(a) 25 (b) 1
(c) 24 (d) 7 **Ans. b**

378. Which of the following does not have phosphorous?
(a) Riboflavin (b) TPP
(c) NAD^+ (d) CaASH **Ans. a**

379. Rice-polishings contain whcih of the following Vitamin?
(a) Riboflavin (b) Niacin
(c) Thiamine (d) Vitamin B_{12} **Ans. c**

380. In beri beri there will be accumulation of ________ in blood.
(a) Aceto acetic acid (b) hydroxy butyric acid
(c) Pyruvic acid (d) Methyl malonic acid **Ans. c**

381. Symptoms of pellagra are
(a) Dermatitis and diarrhea only
(b) Dermatitis and Dermentia only
(c) Diarrhea and dermentia only
(d) Diarrhea, Dermatitis and dementia **Ans. d**

382. Pyridoxine deficiency leads to
(a) Megaloblastic anemia
(b) Aplastic anemia
(c) Hypochromic microcytic anemia
(d) Pernicious anemia **Ans. c**

383. The significant ocular lesion in a riboflavinosis is
(a) Keratomalacia (b) Bitot's spots
(c) Vascularisation of the cornea (d) Lachrynal metaplasia **Ans. c**

384. An anti-vitamin for folic acid is

(a) Aminopterin (b) Dicoumarol

(c) Pyrithiamine (d) Isoniazid **Ans. a**

385. Thiamine is

(a) Water-soluble vitamin (b) Fat soluble vitamin

(c) Purine base (d) Pyrimidine base **Ans. d**

386. The anti-vitamin for para amino benzoic acid is

(a) Aminopterin (b) Dicoumarol

(c) INH (d) Sulphanomides **Ans. d**

387. The sulphur containing vitamins among the following B Vitamin is

(a) Thiamine (b) Riboflavin

(c) Niacin (d) Pyridoxine **Ans. a**

388. Taurinuria may be encountered in

(a) Pernicious anemia (b) Beriberi

(c) Pellegra (d) Folate deficiency **Ans. a**

389. The three vitamins which are specially required for proper nerve functions are

(a) Thiamine, Niacin and Riboflavin

(b) Thiamin, Folic acid, Choline

(c) Thiamine, Riboflavin, Pantothenic acid

(d) Thiamine, Pyridoxin, Vitamin B_{12} **Ans. d**

390. A biochemical indication of vitamin B_{12} deficiency can be obtained by measuring the urinary excretion of

(a) Pyruvic acid (b) Malic acid

(c) Methyl malonic acid (d) Urocanic acid **Ans. c**

391. This is a rich source for Vitamin C

(a) Rice (b) Milk

(c) Egg (d) Lemon **Ans. d**

392. Which of the following vitamin is involved in coenzyme function in transaminations?

(a) Nicotinamide (b) Pyridoxine

(c) Thiamine (d) Riboflavin **Ans. b**

393. Methyl malonic aciduria is seen in a deficiency of

(a) Vitamin B_6 (b) Folic acid

(c) Thiamine (d) Vitamin B_{12} **Ans. d**

394. In pernicious anemia, Urine contains high amounts of

(a) Methyl malonic acid (b) FIGLU

(c) VMA (d) 5 HIAA **Ans. a**

395. Anti sterility Vitamin is

(a) Vitamin B_1 (b) Vitamin B_2

(c) Vitamin E (d) Vitamin K **Ans. c**

396. Deficiency of thiamine causes

(a) Beri beri (b) Scurvy

(c) Night blindness (d) Rickets **Ans. a**

397. Deficiency of Vitamin D leads to

(a) Rickets (b) Osteomalacia

(c) Xeropthalmia (d) Both (a) and (b) **Ans. d**

398. The vitamin that is useful in cancer is

(a) A (b) B complex

(c) C (d) E **Ans. a**

399. Vitamin A over dosage causes injury to

(a) Mitochondria (b) Microtubules

(c) Lysosomes (d) E.R **Ans. c**

400. Which is a pro vitamin or vitamin that has antioxidant properties?

(a) Beta carotene (b) Vitamin E

(c) Vitamin C (d) Vitamin D **Ans. b**

401. The vitamin required for carboxylation reaction is

(a) Vitamin B_2 (b) Vitamin B_6

(c) Biotin (d) Vitamin B_{12} **Ans. c**

402. Biological activity of tocopherols has been attributed in part to their action as

(a) Antioxidant

(b) Anticoagulents

(c) Provitamin

(d) Carriers in electron transport system **Ans. b**

403. Biotin is essential for
(a) Translation (b) Carboxylation
(c) Hydroxylation (d) Transamination **Ans. b**

404. Which of the following vitamin act as a respiratory catalyst?
(a) B_2 (b) Pyridoxine
(c) B_{12} (d) C **Ans. a**

405. Metal in Vitamin B_{12} is
(a) Copper (b) Cobalt
(c) Iron (d) Zinc **Ans. b**

406. Increased protein intake is accompanied by an increased dietary requirement for
(a) Thiamine (b) Riboflavine
(c) Folic acid (d) Nicotininic acid **Ans. a**

407. Whole wheat is an excellent source of
(a) Vitamin D (b) Vitamin C
(c) Vitamin A (d) Thiamine **Ans. d**

408. Vitamin used in the treatment of homocystinuria is
(a) B_1 (b) B_5
(c) B_{12} (d) B_6 **Ans. d**

409. Which of the following is not a component of coenzyme A?
(a) Pantothenic acid (b) Adenylic acid
(c) Acetic acid (d) Sulfhydryl group **Ans. c**

410. The most active form of Vitamin D is
(a) 25-Hydroxycholecalciferol (b) 1, 25-dihydroxycholecalciferol
(c) 25-dihydroxyergocalciferol (d) None of these **Ans. a**

411. The important part in the structure of flavoprotein is
(a) Vitamin B_6 (b) Vitamin B_2
(c) Vitamin B_1 (d) Vitamin A **Ans. b**

412. Vitamin essential for transamination is
(a) B_1 (b) B_2
(c) B_6 (d) B_{12} **Ans. c**

413. Biotin deficiency is characterized by the following except

(a) Muscular pain (b) Anaemia

(c) Nausea (d) Dermatitis **Ans. b**

414. Calcitriol inhibits the conversion of

(a) Cholesterol into 7-dehydrocholesterol

(b) Cholecalciferol into 1-hydroxycholecalciferol

(c) Cholecalciferol into 25-hydroxycholecalciferol

(d) 25-Hydroxycholecalciferol into 1,25- dihydroxycholecalciferol

Ans. d

415. Bowlegs and knock-knees can occur in

(a) Rickets (b) Osteomalacia

(c) Both (a) and (b) (d) Hypervitaminosis D **Ans. a**

416. Calcification of soft tissues can occur in

(a) Osteomalacia (b) Rickets

(c) Hypervitaminosis D (d) None of these **Ans. c**

417. Levels of serum calcium and inorganic phosphorus are increased in

(a) Hypervitaminosis D (b) Hypoparathyroidism

(c) Hypovitaminosis D (d) None of these **Ans. a**

418. Requirement of vitamin E increases with the increasing intake of

(a) Calories (b) Proteins

(c) PUFA (d) Cholesterol **Ans. c**

419. In human beings, vitamin E prevents

(a) Sterility (b) Hepatic necrosis

(c) Muscular dystrophy (d) None of these **Ans. d**

420. Vitamin E protects

(a) Polyunsaturated fatty acids against aperoxidation

(b) Vitamin A and carotenes against oxidation

(c) Lung tissue against atmospheric pollutants

(d) All of these **Ans. d**

421. Intestinal bacteria can synthesise

(a) Phyllogquinone (b) Farnoquinone

(c) Both (a) and (b) (d) Menadione **Ans. b**

422. A water soluble form of vitamin K is
(a) Phylloquinone (b) Farnoquinone
(c) Menadione (d) None of these **Ans. c**

423. Prothrombin time is prolonged in
(a) Vitamin K deficiency (b) Liver damage
(c) Both (a) and (b) (d) None of these **Ans. c**

424. A synthetic form of vitamin K is
(a) Menadione (b) Farnoquinone
(c) Phylloquinone (d) None of these **Ans. a**

425. Retinal is reduced to retinol by retinene reductase in presence of the coenzyme
(a) NAD^+ (b) $NADP^+$
(c) $NADH + H^+$ (d) $NADPH + H^+$ **Ans. c**

426. Retinal exists as an ester with higher fatty acids in the
(a) Liver (b) Kidney
(c) Lung (d) All of these **Ans. d**

427. Retinol is transported to the blood as retinol attached to
(a) α_1-globulin (b) α_2-globulin
(c) β-globulin (d) γ-globulin **Ans. a**

428. Carotenes are transported with the
(a) Minerals (b) Proteins
(c) Lipids (d) Lipoproteins **Ans. d**

429. The drugs that form complexes with pyridoxal are
(a) Isoniazid (b) Penicillamine
(c) Rifampicin (d) Both (a) and (b) **Ans. d**

430. In the blood the vitamin esters are attached to
(a) α_1-lipoproteins (b) α_2-lipoproteins
(c) β-lipoproteins (d) γ-lipoproteins **Ans. c**

431. The percentage of Vitamin A in the form of esters is stored in the liver
(a) 80 (b) 85
(c) 90 (d) 95 **Ans. d**

432. The performed Vitamin A is supplied by foods such as

(a) Butter (b) Eggs

(c) Fish liver oil (d) All of these **Ans. d**

433. The non-protein part of rhodopsin is

(a) Retinal (b) Retinol

(c) Carotene (d) Repsin **Ans. a**

434. Lumirhodopsin is stable only at a temperature below

(a) -35°C (b) -40°C

(c) -45°C (d) -50°C **Ans. d**

435. The normal concentration of vitamin A in blood in I.V/dl

(a) 20-55 (b) 24-60

(c) 30-65 (d) 35-70 **Ans. b**

436. Continued intake of excessive amounts of vitamin A especially in children produces

(a) Irritability (b) Anorexia

(c) Headache (d) All of these **Ans. d**

437. Match the metal ions, vitamins and related compounds given in Column A with their specific functions in microbial metabolism listed in Column B. Choose the mismatch

Column A	Column B
a. p- aminobenzoic acid	A. Precursor of tetrahydrofolic acid, a coenzyme involved in transfer of one-carbon units
b. Pantothenic acid	B. Precursor of coenzyme A and of the prosthetic group of acyl carrier proteins
c. Vitamin B_{12}	C. Involved in rearrangement reaction (e.g., glutamate mutase)
d. Cu^{2+}	D. Present in carbonic anhydrase

Ans. d

438. The first vitamin produced during a fermentation process using a wild bacterium is

(a) Vitamin C (b) Thiamine

(c) Vitamin A (d) Vitamin K **Ans. a**

439. Vitamin B_1 was first isolated by

(a) Mc Collum (b) Funk

(c) Mellanby (d) Albert Van Szent Gyorgy

Ans. b

440. Which of the following vitamin is produced by microflora of the digestive system

(a) Vitamin A (b) Vitamin C

(c) Cobalamine (d) Vitamin D **Ans. c**

❑❑❑

About the authors

Dr. R.S. Sengar the author of the book is Associate Professor and Officer - In-Charge, Dept. of Ag. Biotechnology, College of Agriculture in Sardar Vallabhbhai Patel University of Agriculture & Technology, Meerut. Dr. Sengar has 20 years teaching, research and extension experience to his credit. He remained associated with the teaching for a period of about 10 years at G.B. Pant University of Agri. & Tech. Pantnagar, U.S. Nagar (Uttrakhand). He has supervised 5 Ph.D. and 4 M.Sc. students and published more than 48 research papers in India and foreign journals of repute. Dr. Sengar contributed 105 abstracts/ papers to different conferences/ symposiums/ workshop organized at National and International forums. He has published 780 popular articles in Hindi and English languages in important journals and magazines. He has also published 6 books for scientific community and farmers. Dr. Sengar has delivered more than 200 talks and live programms at All India Redio (A.I.R.) and T.V. at Delhi, Lucknow, Rampur and Bareilly centers. Dr. Sengar has keenly worked as one of the active member of advisory committee of Doordarshan, New Delhi. Dr. Sengar is life member of several professional societies. He is a member of the editorial boards and review committees of few journals. Dr. Sengar is on the penal of examiners of several colleges/ institution and universities. He is one of the members of RDC of Uttrakhand Technical University, Dehradun (Uttrakhand). He has been the Principal and Co-Investigator of 7 National Projects financed by various government departments. Dr. Sengar is recipient of the 'Best writer' award from Vishwa Agro Marketing and Communication Kota Rajasthan, 'Kunwar Saxena Bahadur SRDA' award from society for Recent Development in Agriculture, Meerut, 'Man of the Year' & 'Research Board of Advisors' awards from American Biographical Institute, Inc. U.S.A., 'Aryabhat' 2010 award from Vigyan Bharti, New Delhi and 'Fellow of Society of Plant Research', Bareilly. Dr. Sengar has also received "Gold medal 2011" from Hitech Horticulture Society, Meerut, "Dr. J.C. Edward Medal 2012" from Bioved Research Society, Allahabad. "Distinguished Scientist Award 2013" from The Academy of Environmental Biology, Lucknow.

Dr. Reshu Chaudhary, M.Sc. Biochemistry and Ph. D. from Chaudhary Charan Singh University, Meerut, is presently working as Senior Research Fellow in Tissue Culture Laboratory, College of Biotechnology, Sardar Vallabhbhai Patel University of Agriculture & Technology, Meerut. She has a Research experience of nearly 3 years. She is a good researcher and published more than 20 scientific research papers/articles in different National & International Journals of repute. Dr. Reshu Chaudhary is recipient of "Young Scientist Award- 2012" from Bioved Research Society, Allahabad, and "Scientist of the Year 2013" from The Academy of Environmental Biology, Lucknow. Her areas of interest are Biochemistry, Biotechnology and Tissue Culture.

MIX
Papier | Fördert gute Waldnutzung
FSC® C083411

Zeitfracht Medien GmbH
Ferdinand-Jühlke-Straße 7
99095 Erfurt, Deutschland
produktsicherheit@kolibri360.de